T0231437

Cropping Systems:
Trends and Advances

Anil Shrestha, PhD
Editor

Cropping Systems: Trends and Advances has been co-published simultaneously as *Journal of Crop Production*, Volume 8, Numbers 1/2 (#15/16) and Volume 9, Numbers 1/2 (#17/18) 2003.

CRC Press
Taylor & Francis Group
Boca Raton London New York

CRC Press is an imprint of the
Taylor & Francis Group, an **informa** business

CRC Press
Taylor & Francis Group
6000 Broken Sound Parkway NW, Suite 300
Boca Raton, FL 33487-2742

© 2003 by Taylor & Francis Group, LLC
CRC Press is an imprint of Taylor & Francis Group, an Informa business

No claim to original U.S. Government works

ISBN-13: 9781560221067 (hbk)
ISBN-13: 9781560221074 (pbk)

Visit the Taylor & Francis Web site at
http://www.taylorandfrancis.com

and the CRC Press Web site at
http://www.crcpress.com

Library or Congress Cataloging-in-Publication Data
Cropping systems : trends and advances/ Anil Shrestha. editor.

p.cm.
Includes bibliographical references (p.   ).
ISBN 9781560221067 (hbk) ISBN 9781560221074 (pbk) I. Cropping systems. 2.
Cropping systems-Research. I. Shrestha. Anil. II. Journal of crop production

S602.5.C78 2003
631.5'8-ilc21

2003010337

Implications of Elevated CO_2-Induced Changes
   in Agroecosystem Productivity   217
     S. A. Prior
     H. A. Torbert
     G. B. Runion
     H. H. Rogers

Weed Biology, Cropping Systems, and Weed Management   245
     Douglas D. Buhler

Cropping Systems and Integrated Pest Management:
   Examples from Selected Crops   271
     K. R. Barker
     C. Sorenson

Conceptual Model for Sustainable Cropping Systems
   in the Southeast: Cotton System   307
     Harry H. Schomberg
     Joe Lewis
     Glynn Tillman
     Dawn Olson
     Patricia Timper
     Don Wauchope
     Sharad Phatak
     Marion Jay

Cropping Systems and Water Quality Concerns   329
     D. Brook Harker
     Brian McConkey
     Helen H. McDuffie

The Role of Precision Agriculture in Cropping Systems   361
     Bradley Koch
     Rajiv Khosla

Trends in Decision Support Systems for Cropping Systems
   Analysis: Examples from Nebraska   383
     R. M. Caldwell

A Framework for Economic Analysis of Cropping Systems:
   Profitability, Risk Management, and Resource Allocation   409
     Carl R. Dillon

Cropping Systems:
Trends and Advances

CONTENTS

Emerging Trends in Cropping Systems Research   1
Anil Shrestha
David R. Clements

Advances in the Design of Resource-Efficient Cropping Systems   15
Charles A. Francis

Cropping Systems and Soil Quality   33
R. Lal

The Role of Cover Crops in North American Cropping Systems   53
Marianne Sarrantonio
Eric Gallandt

Advances in Tillage Research in North American Cropping
Systems   75
D. C. Reicosky
R. R. Allmaras

The Importance of Root Dynamics in Cropping Systems
Research   127
M. J. Goss
C. A. Watson

Key Indicators for Assessing Nitrogen Use Efficiency
in Cereal-Based Agroecosystems   157
D. R. Huggins
W. L. Pan

Forage Legumes for Sustainable Cropping Systems   187
Craig C. Sheaffer
Philippe Seguin

Conceptual Framework for Evaluating Sustainable Agriculture   433
Murari Suvedi
Christoffel den Biggelaar
Shawn Morford

From Chemical Ecology to Agronomy: Cropping Systems
in the Humid Northeast   455
D. L. Smith
C. Costa
B. Ma
C. Madakadze
B. Prithiviraj
F. Zhang
X. Zhou

Current and Potential Role of Transgenic Crops
in U.S. Agriculture   501
C. S. Silvers
L. P. Gianessi
J. E. Carpenter
S. Sankula

Problems and Perspectives of Yam-Based Cropping
Systems in Africa   531
Indira J. Ekanayake
Robert Asiedu

Crop Technology Introduction in Semiarid West Africa:
Performance and Future Strategy   559
John H. Sanders
Barry I. Shapiro

Brazilian Agriculture: The Transition to Sustainability   593
Robert M. Boddey
Deise F. Xavier
Bruno J. R. Alves
Segundo Urquiaga

Cropping Systems in Eastern Europe: Past, Present, and Future   623
Imre Molnar

Socioeconomic and Agricultural Factors Associated
with Mixed Cropping Systems in Small Farms
of Southwestern Guatemala   649
    Francisco J. Morales
    Edin Palma
    Carlos Paiz
    Edgardo Carrillo
    Iván Esquivel
    Vianey Guillespie
    Abelardo Viana

The Future of Cereal Yields and Prices: Implications
for Research and Policy   661
    Mark W. Rosegrant
    Michael S. Paisner
    Siet Meijer

Index   691

ABOUT THE EDITOR

Anil Shrestha, PhD, is currently a Weed Ecologist with the University of California's Statewide IPM Program. He is based at the University's Kearney Agricultural Center in Parlier, California. Dr. Shrestha was born in Nepal and he received his BSc from Narendra Dev University of Agriculture and Technology in Faizabad, India. After completing his undergraduate degree, he worked several years with the Department of Agriculture and with the FAO Fertilizer Program in Nepal as Assistant Agronomist and Regional Supervisor, respectively. He completed his MS in crop and soil sciences as a Fulbright Fellow at Cornell University and his PhD as a C. S. Mott Fellow of Sustainable Agriculture in the Department of Crop and Soil Sciences at Michigan State University. Dr. Shrestha worked briefly as a research associate at the Center for Evaluative Studies of the Department of Agricultural and Extension Education at Michigan State. Later, he worked for five years as a post-doctoral fellow and as a research associate in the Department of Plant Agriculture's weed science program at the University of Guelph in Canada. Dr. Shrestha is committed to extension and research efforts in vegetation management, cropping systems, and agroecology, and has authored more than 25 refereed papers and several extension articles in these fields.

Emerging Trends
in Cropping Systems Research

Anil Shrestha

David R. Clements

SUMMARY. The demand for producers to farm in an environmentally sound, ethical, economically viable, and socially acceptable manner has driven cropping systems research beyond basic production agronomy. Cropping systems research today encompasses a broad range of topics ranging from agronomic to ecological, environmental, social, and economic dimensions of systems and should also effectively integrate these factors. New trends are emerging among technical aspects and new research tools are being developed. The biological basis of sustainability is also being revisited. Humanity may be seen as following a learning circle that began with the inception of agriculture as a slightly modified form of harvesting from nature, to input-intensive agriculture, and now back again to Natural Systems Agriculture where an attempt is made to learn from and even mimic nature and to accentuate natural processes. This contribution briefly outlines the trends and paradigm shifts that have taken place globally in agricultural cropping systems research. *[Article copies available for a fee from The Haworth Document Delivery Service: 1-800-HAWORTH. E-mail address: <docdelivery@haworthpress.com> Website: <http://www.HaworthPress.com> © 2003 by The Haworth Press, Inc. All rights reserved.]*

Anil Shrestha is IPM Weed Ecologist, University of California, Kearney Agricultural Center, 9240 South Riverbend Avenue, Parlier, CA 93648 USA (E-mail: anil@uckac.edu).

David R. Clements is Associate Professor, Department of Biology, Trinity Western University, Langley, BC, Canada V2Y 1Y1 (E-mail: clements@twu.ca).

[Haworth co-indexing entry note]: "Emerging Trends in Cropping Systems Research." Shrestha, Anil, and David R. Clements. Co-published simultaneously in *Journal of Crop Production* (Food Products Press, an imprint of The Haworth Press, Inc.) Vol. 8, No. 1/2 (#15/16), 2003, pp. 1-13; and: *Cropping Systems: Trends and Advances* (ed: Anil Shrestha) Food Products Press, an imprint of The Haworth Press, Inc., 2003, pp. 1-13. Single or multiple copies of this article are available for a fee from The Haworth Document Delivery Service [1-800-HAWORTH, 9:00 a.m. - 5:00 p.m. (EST). E-mail address: docdelivery@haworthpress.com].

KEYWORDS. Agroecology, holistic model, interdisciplinary research, systems approach, sustainable agriculture

INTRODUCTION

Agriculture systems today may be described as goal-oriented manipulations of ecosystems for human gains (Ohlander, Lagerberg, and Gertsson, 1999). In this context, many would agree that the "yield and profit maximization" approach or the "conventional" approach to agricultural cropping systems in North America has led to serious environmental, ecological, economic, and social problems. Farm and rural populations have steadily declined leading to an overall economic and social demise of many rural communities (Fisk, Hesterman, and Thorburn, 1998). This process has been further aggravated by rapid urbanization and lack of economic incentives to farm.

Human society today is dominated by rapid technological innovations and political transformations, summed up by terms like "globalization" or "information age." The information disseminated by the media on human and environmental health is making people more conscious of the quality of the food they eat and the environmental consequences of the systems under which these foods were produced. A cropping system adopted on a farm today has more obvious and detectable social, ecological, economic, and environmental implications than ever before. Because of the demand for producers to farm in an environmentally sound, ethical, and socially acceptable manner, these opinions of society and legislation have driven cropping systems research beyond basic production agronomy. Many researchers now feel the need for an approach that takes into account these concerns and the construction of interfaces that link activities across a range of disciplines associated with agricultural cropping systems (Park and Seaton, 1996). In this process, it is necessary to understand the biological principles underlying the entire agroecosystem in order to design profitable, resource efficient and environmentally sound cropping systems (Bushnell, Francis, and King, 1991).

Thus, cropping systems research today encompasses a broad range of topics ranging from agronomic to ecological, environmental, social, and economic aspects with an accompanying challenge to adequately integrate these factors. New trends are emerging that link technical aspects of research and new methods and tools are being developed, while at the same time the biological basis of sustainability is being revisited. It is often difficult to see how these various trends can be better studied through a systems approach. As we discuss below, a systems approach is necessary to meet the challenges of a rapidly growing global population that relies on a diminishing population of farmers who have to sustain a profitable operation on an increasingly vulnerable land base. The objective of this contribution is to outline the trends and paradigm shifts that have taken place in agricultural cropping systems research in North America. Other articles in this volume will further expand and help illustrate these trends.

CROPPING SYSTEMS FOR MAXIMIZING CROP YIELDS

Cropping systems have been traditionally defined as the cropping patterns on a farm and their interactions with farm resources, other farm enterprises, and available technology that determine their make up (Andrews and Kassam, 1976). Okigbo (1980) defined cropping system as the pattern of growing crops in combination and sequences in time and space, in addition to the practices and technologies with which crops are produced. In other words, the definitions of cropping systems have been limited to the systematic arrangement of crops as influenced by local factors for crop production. The aims of cropping systems research have traditionally been to design patterns in time and space that would maximize crop production. Cooke (1979) emphasized the need for research and development to achieve yields nearer to known potential crop yields. Similarly, Olsen (1982) stressed the need to understand biological processes in cropping systems to overcome factors limiting crop productivity. Up to the early 1980s the goal of cropping systems design was almost exclusively to maximize crop yields on a farm to maximize profits, with the ultimate goal to feed the growing population worldwide.

The goal of breaking barriers to crop productivity and maximizing crop yields was successfully attained, more or less, in many parts of the world. There was a sevenfold increase in gross agricultural production between 1880 and 1980 in the US alone (Johnson, 1984). However, many researchers believed that this cropping systems design paradigm was a dangerously reductionist approach (Bawden, 1991; Prihar et al., 2000) because the focus was on maximizing crop production with overspecialization on factors limiting crop yields. In developing countries, this approach led to intensive cropping systems that caused degradation of the production base, i.e., soil, environment, and water (Prihar et al., 2000). In developed countries such as the US and those in the European Union, crop production increased to such levels that governments in these countries are now seeking ways to limit burgeoning overall production and the resulting environmental cost (Becker, 2000). In this process of maximizing crop yields, it was recognized that yields were influenced by factors beyond the cropping pattern and that an interdisciplinary approach was required in designing cropping systems and evaluating their impacts.

DEVELOPMENT
OF AN INTERDISCIPLINARY RESEARCH APPROACH

Axinn and Axinn (1984) suggest that, the high degree of specialization among agricultural scientists has resulted in professional researchers who have been trained not to see all of the components, but to focus on only one, and those aspects of the ecosystem which interact with that one. Academic criticism of these extremely specialized, reductionist, crop-yield oriented cropping systems were enjoined by well documented biophysical, environmental, economic, and sociocultural consequences (Bawden, 1991; Bonnen, 1983; Clancy, 1986). For example, Sen (1988) suggested that cropping systems research required an interdisciplinary approach because crop production is affected by several physical factors that are not easily changed. Fageria (1992) broadly categorized these factors as climatic, soil, plant, and socioeconomic effects. Edwards (1987) further described the complexity of the interactions among the various factors in crop production systems and emphasized the need to identify the relative importance of these interactions. Many of these interactions can only be explained by collaborative efforts among agricultural scientists of various disciplines. Hence, knowledge of the involvement of various factors in maximizing crop productivity led to an emphasis on interdisciplinary research (Olsen, 1982) and a philosophy of a systems approach to agriculture was developed (Bawden, 1991; Conway, 1990; Spedding, 1979).

The Philosophy of a Systems Approach

A system has been defined as a group of interacting components, operating together for a common purpose, capable of reacting as a whole to external stimuli. It is further believed that the system is unaffected directly by its own outputs and has a specified boundary (Spedding, 1988). Bawden (1991) suggested that a systems approach is a new paradigm in agriculture that embraces both "production enhancement" and "impact assessment" factors. Rambo and Sajise (1985) emphasized five system properties to assess agroecosystem performance: productivity, stability, sustainability, equitability, and autonomy. Hence, the philosophy of a systems approach to agriculture, the scope and boundary, and the properties of agroecosystem analysis were developed in the mid- to late-1980s.

Gradually, with increased globalization, free trade, and environmental awareness, the typical farm, and its associated cropping systems, is no longer viewed as a system developed to meet local food production needs. Rather, cropping systems design should look beyond sustainability to focus also on conservation, and improvement of environmental quality and human health (Prihar et al., 2000). Concerns of environmental conservation in some instances have led restoration of agricultural lands to their "natural" habitats.

Large areas in the US have been taken out of agricultural production under the Conservation Reserve Program and in some cases, restoration to their "natural" habitat is being attempted (Clements and Swanton, 2001). Other large-scale trends have improved the ecological health of land that is still farmed. For example, the adoption of conservation tillage over a substantial percentage of North American farmland in a relatively brief period has produced a landscape with less soil erosion, more proactive weed management and more natural organisms. These landscapes have greater biological sustainability while employing less energy and improving economic returns (Blumberg and Crossley, 1983; Clements et al., 1995; House and Parmalee, 1985; Swanton, Clements and Derksen, 1993; Weersink et al., 1992).

The current and future need may be for a balance between food security and the environment (Cassman and Harwood, 1995). However, such a balance means consideration of factors beyond the farm. Envisioning the need to encompass various extrinsic and intrinsic factors affecting a farm, Cavigelli et al. (1998) have proposed a "holistic model" that takes into account a multitude of factors that needs to be considered for farm change and cropping systems design (Figure 1).

A Holistic Model

The holistic model for farm change (Figure 1) envisaged by Cavigelli et al. (1998) takes into account social, political, physical, biological, and economic factors and suggests that the economic return to farm owners would be driven by balancing these factors, cost of production, and market for the products. This model addresses concerns far beyond the simplistic view frequently inherent within cropping systems research. Conway (1990) stressed that agriculture is one of the most complex human activities, one that required multiple decisions related to crops and livestock, labor, inputs, sales, consumption, and each decision involved consequences on the other activities in the short and long term.

The activities within the farm are also influenced by external factors, for example, it may be suggested that the evolution of large, highly mechanized farms in industrialized countries may be seen as a consequence of government policies, especially subsidy programs. Cavigelli et al. (1998) recognize these external factors in the social, political, and regulatory environment, and show how they impact factors such as farm size, labor, capital, debt, and land (Figure 1). Similarly, Spedding (1979) recognized a two-way interaction between agriculture and the environment. In their holistic model, Cavigelli et al. (1998) described this interaction as the physical and biological environment affecting the farm. Furthermore, the model stresses the integration of ecology in agricultural system design and outlines various management practices and outcomes

FIGURE 1. A holisitic model for farm change.

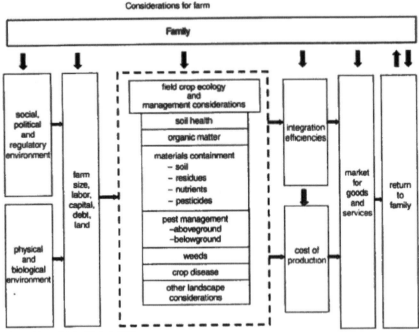

Adapted from Cavigelli et al. 1998

such as soil health, organic matter content, materials containment, and pest management (Figure 1). The model is also sensitive to changes in markets, opportunities for providing high value goods and services, landscape aesthetics, environmental protection, and economic efficiency. Thus, the model goes beyond the traditional approach of maximizing crop yields and considers biology along with economics, engineering, and human well-being. Within the farm, the authors prioritized managing the habitats of biological organisms in cropping systems, i.e., managing plants, organic inputs, and their residues.

It can be implied that this complex model for farm change, although not in its exact form, was considered earlier. Francis (1994) stated that, "the boundaries among biological, physical, and social sciences are blurring as we attempt to solve complex challenges" (p. 151). He concluded that long-term resource availability, environmental impacts of farming, and social equity were issues beyond immediate biological and economic considerations of crop production. Thus, from the early-1990s cropping systems research has encompassed a multitude of factors ranging from biological and economic to social factors related with agricultural production and environmental conservation.

These visions of cropping systems have motivated the development of new research directions. Throughout this volume, we present examples of the trends and advances in cropping systems research.

RESEARCH TRENDS

The potential for incorporating a multitude of factors in cropping systems research has been demonstrated in several studies. For example, Colvin, Erbach, and Kemper (1990) included socioeconomic and Koo et al. (2000) included environmental and economic factors in cropping systems research. In the process, Colvin, Erbach, and Kemper (1990) were able to identify the socioeconomic constraints for farmers converting from monocropping to crop rotations. Similarly, Koo et al. (2000) provided scenarios of economic and environmental consequences of five cropping systems in northeast Kansas. Several other examples exist of systems research by interdisciplinary teams (Bryant et al., 1992; Temple et al., 1994; Young et al., 1994). Within any agricultural discipline, Francis (1994) identified that expansion of the "classical component focus" may serve as an application of a systems approach. He provided several examples of a "classical component research" approach, such as effect of cropping sequence and nitrogen fertilizer rates on crop yield and economic return, search for pest and disease tolerant crop cultivars with greater profitability than traditional cultivars. Such an approach involved wider interpretation of field data. He identified several research/education directions for cropping systems in the immediate future. These directions included factors such as soil quality, tillage systems, rhizosphere dynamics, nutrient cycling, crop adaptation to environmental stresses, pest dynamics, crop/animal interactions, crop simulation, genetic engineering, and economics. He also stressed the need to measure environmental impact, resource use, risk, and quality of life while designing cropping systems.

Similarly, within a farm, the holistic model described earlier identified landscape ecology as an important research component consisting of several sub-components for the design of resource-efficient, sustainable cropping systems (Figure 1). This model stresses the importance of integrating ecology in cropping systems design and recognizes the relationship between soil quality and crop health as the heart of field crop ecology (Cavigelli et al., 1998). Altieri and Francis (1992) and Gliessman (1998) also highlight the importance of incorporating ecological principles in cropping systems design. Francis (2003) recommends the emerging discipline of agroecology as an innovative and integrative approach to design of future cropping systems and highlights "Natural Systems Agriculture" as a new paradigm to save the soil while producing the crop and animal output needed for a growing global population.

"Natural Systems Agriculture" is an agricultural system where nature is mimicked rather than subdued and ignored (Piper, 1999). Examples of this system of agriculture include growing of perennials in mixtures because perennial roots hold the soil and provide a diversity of species to thwart insect or pathogen attacks (Jackson and Jackson, 1999). Similarly, the potential for restoring unproductive cropland to natural habitats has been demonstrated (Gilbert et al., 2000).

Within a farm, agronomic and mechanical practices enhancing soil quality and growth and development of crops under adverse and favorable environmental conditions are being assessed. Efforts are being made to enhance our knowledge of nutrient cycles, crop root dynamics, crop rotations, pest management, and effect of cropping systems on water quality. Research and education efforts underway in these areas will be highlighted throughout the remainder of this volume. Some examples are also provided of regional problems associated with the cropping systems in the US, Canada, and other countries, and research efforts directed to address these local issues are discussed.

As well as advances in agronomy and systems thinking, innovative technical tools are helping to revolutionize approaches to design of cropping systems. Examples of such tools include precision farming through the use of global positioning systems (GPS), decision support systems (DSS), and biotechnology. Similarly, interpretation of experimental data in cropping systems research is being enhanced by improved biometric and econometric techniques. Used properly, these technical innovations serve to enhance the systems approach by linking the various components of cropping systems.

WHERE ARE WE ON THE LEARNING CYCLE?

The paradigm shifts in agriculture and cropping systems research effort over the years can be viewed as a learning cycle (Figure 2). Agriculture began with human intervention of natural systems consisting of plant communities. Plants that served human and livestock needs were domesticated and grown as crops. Natural systems that had traditionally supplied food for human needs became agroecosystems, constituting at first only slightly modified versions of the natural ecosystems. As human and livestock populations increased, the objective became to maximize crop yields to feed the growing population. This process was further influenced by the growing number of non-farming sectors in society. In response to meet the growing food and economic needs, agricultural scientists responded by developing high input, highly mechanized cropping systems. However, these systems produced unintended consequences in terms of environmental health issues such as off-site or non-target pesticide effects, reduced groundwater quality or soil degradation. Both the non-farming

FIGURE 2. The learning cycle in the development of agriculture cropping systems.

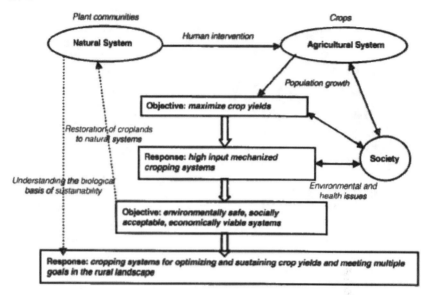

population and farmers themselves who were concerned about environmental ethics have raised concerns over environmental safety of cropping systems. The objective of cropping systems research thereon became identification of environmentally safe, socially acceptable, and economically viable systems.

Cropping systems researchers responded to this newly formulated objective by determining ways to optimize and sustain crop yields by taking a systems approach to agriculture. Also, in this process, unproductive and degraded croplands are being restored as 'natural' habitats, as researchers grapple with the biological and ecological basis of sustainability, and how best to mimic natural systems. In other words, we have almost completed a cycle in our quest for feeding the world population and have become aware of the consequences associated with our current goals and cropping systems.

The importance of cropping system design has never been so important. We are presently engaged in the process of initiating the next cycle of research that must lead to strategies for feeding the growing world population amidst deteriorating environment and market uncertainties. Current trends in cropping systems research are based on a systems approach or an expansion of the classical component focus by wider interpretation of field data. The importance of ecological principles in cropping systems design has been recognized. Research tools such as GPS, DSS, and biometrics are being developed and utilized, and the profitability and risk associated with cropping systems are being assessed. However, various countries are at different stages in this learning cycle, while various regions within a country are faced with their own unique local problems. Countries that are behind others in this learning cycle would do well to learn from the experiences of the forerunners and avoid making similar mistakes.

The current trend in agricultural research points to the emergence of a systems approach to cropping systems design. Traditional boundaries between various disciplines are blurring and interdisciplinary research groups are being established. Environmental, social, and economic effects of the production system are being considered in the choice of enterprises and the design of cropping systems. Anticipated effects of global climate change on crop production are being investigated. In this whole process, the biological principles within agroecosystems are being investigated and research is being supplemented by the development and use of various forms of information technology. Serious challenges prevent the wider acceptance of these emerging trends, however. It is difficult to integrate various branches of knowledge effectively, in order to realize fully the complexity that systems thinking demands. At the farm level, this means employing dynamic systems more responsive to change and to the incorporation of economic, biological and social knowledge than present conventional systems. There is limited research funding from commercial sources for supporting systems research that does not result in recommending more inputs or experimental technologies. Cropping systems designs cannot be patented or sold (C. A. Francis, personal communications). Finally, there is the challenge of acceptance and adoption of these new ways of thinking by agricultural researchers and practitioners. In the 20th century, agricultural technology advanced through a serious of "silver bullets" such as the development of pesticides, mechanization, fertilizer technology or high-yielding hybrid crop varieties. A holistic, systems approach is not a "silver bullet," but it may be our best hope in the 21st century. Considerably more guidance is necessary to enable this more amorphous approach to hit the target. There is much to learn and we trust that the other contributions in this volume will provide some useful insights as we continue to travel the circle of learning.

REFERENCES

Altieri, M. A. and C. A. Francis. (1992). Incorporating agroecology into the conventional agricultural curriculum. *American Journal of Alternative Agriculture* 7:89-93.

Andrews, D. J. and A. H. Kassam. (1976). The importance of multiple cropping to increasing world food supplies. In *Multiple Cropping*, ed. M. Stelly, ASA Special Publication 27, Madison, WI, pp. 1-10.

Axinn, G. H. and N. W. Axinn. (1984). Energy and food relationships in developing countries: A perspective from the social sciences. In *Food and Energy Relationships*, ed. C. W. Hall and D. Pimentel, Orlando, FL: Academic Press., Inc., pp. 121-146.

Bawden, R. J. (1991). Systems thinking and practice in agriculture. *Journal of Dairy Science* 74:2362-2373.

Becker, J. (2000). Can sustainable agriculture/habitat management pay off? *Journal of Sustainable Agriculture* 17:113-128.

Blumberg, A. Y. and D. A. Crossley, Jr. (1983). Comparison of soil surface arthropod populations in conventional tillage, no-tillage and old field systems. *Agri-Ecosystems* 8: 247-253.

Bonnen, J. T. (1983). Historical sources of U. S. agricultural productivity: Implications for R & D policy and social science research. *American Journal of Agricultural Economics* 65:958.

Bryant, K. J., S. M. Masud, R. D. Lacewell, and J. W. Keeling. (1992). Profit, cost, and soil erosion implications of dryland crop production systems: Texas High Plains. *Journal of Production Agriculture* 5:307-312.

Bushnell, J., C. Francis, and J. King. (1991). Design of resource efficient, environmentally sound cropping systems. *Journal of Sustainable Agriculture* 1(4):49-65.

Cassman, K. G. and R. R. Harwood. (1995). The nature of agricultural systems: Food security and environmental balance. *Food Policy* 20:439-454.

Cavigelli, M. A., S. R. Deming, L. K. Probyn, and R. R. Harwood (eds.). (1998). *Michigan Field Crop Ecology: Managing Biological Processes for Productivity and Environmental Quality*. East Lansing, MI: Michigan State University Extension Bulletin E-2646.

Clancy, K. L. (1986). Human nutrition, agriculture and human value. *Agriculture and Human Values* 1:10.

Clements, D. R., C. J. Swanton, S. F. Weise, R. Brown, D. Hume, and P. Stonehouse. (1995). Energy analysis of tillage and herbicide inputs in alternative weed management systems. *Agriculture, Ecosystems and Environment* 52:119-128.

Clements, D. R. and C. J. Swanton. (2001). Agriculture: Healthy, sick or left in the waiting room? In *Malthus and the Third Millennium*, ed. W. Chesworth, M. R. Moss, and V. G. Thomas. Guelph, ON, Faculty of Environmental Sciences, University of Guelph, Guelph, ON. pp. 125-140.

Colvin, T. S., D. C. Erbach, and W. D. Kemper. (1990). Socioeconomic aspects of machinery requirements for rotational agriculture. In *Sustainable Agricultural Systems*, ed. C. A. Edwards, R. Lal, P. Madden, R. H. Miller, and G. House, Soil and Water Conservation Society, Ankeny, IA. pp. 533-545.

Conway, G. R. (1990). Agroecosystems. In *Systems Theory Applied to Agriculture and the Food Chain*, ed. J. G. W. Jones, P. R. Street, and C. R. W. Spedding, Elsevier Applied Science, New York, NY.

Cooke, G. W. (1979). Some priorities for British soil science. *Journal of Soil Science* 30:187-273.

Edwards, C. A. (1987). The concept of integrated systems in lower input sustainable agriculture. *American Journal of Alternative Agriculture* 2(4):148-152.

Fageria, N. K. (1992). *Maximizing Crop Yields*, Marcel Dekker Inc., New York, NY.

Fisk, J. W., O. B. Hesterman, and T. L. Thorburn. (1998). Integrated farming systems: A sustainable agriculture learning community in the USA. In *Facilitating Sustainable Agriculture*, ed. N. G. Roling and M. A. E. Wagemakers, Cambridge, UK: Cambridge University Press, pp. 217-231.

Francis, C. A. (1994). Practical applications of agricultural systems research in temperate countries. *Journal of Production Agriculture* 7:151-157.

Francis, C. A. (2003). Advances in the design of resource-efficient cropping systems. *Journal of Crop Production* 8(1/2):15-32.

Gilbert, J. C., D. J. G. Gowing, P. R. G. Higginbottom, and R. J. Godwin. (2000). The habitat creation model: A decision support system to assess the viability of converting arable land into semi natural habitat. *Computers and Electronics in Agriculture* 28:67-85.

Gliessman, S. R. (1998). *Agroecology: Ecological Processes in Sustainable Agriculture*. Chelsea, MI: Ann Arbor Press.

House, G. J. and R. W. Parmelee. (1985). Comparison of soil arthropods and earthworms from conventional and no-tillage agroecosystems. *Soil Tillage Research* 5: 351-360.

Jackson, W. and L. L. Jackson. (1999). Developing high seed yielding perennial polycultures as a mimic of mid-grass prairie. In *Agriculture as a Mimic of Natural Ecosystems*, ed. E. C. Lefroy, R. J. Hobbs, M. H. O'Connor, and J. S. Pate, Dordrecht, NL: Kluwer Academic Publishers. pp. 1-38.

Johnson, G. L. (1984). Academia needs a new covenant for serving agriculture. *Mississippi Agriculture and Forestry Experiment Station Special Publication*, Mississippi State.

Koo, S., J. R. Williams, B. W. Schurle, and M. R. Langemeier. (2000). Environmental and economic tradeoffs of alternative cropping systems. *Journal of Sustainable Agriculture* 15:35-38.

Marten, G. R. (1988). Productivity, stability, sustainability, equitability, and autonomy as properties for agroecosystem assessment. *Agricultural Systems* 26:291-316.

Ohlander, L., C. Lagerberg, and U. Gertsson. (1999). Visions for ecologically sound agricultural systems. *Journal of Sustainable Agriculture* 14(1):73-79.

Okigbo, B. N. (1980). The importance of mixed stands in tropical agriculture. In *Opportunities for Increasing Crop Yields*, ed. R. G. Hurd, P. V. Biscoe, and C. Dennis, London, UK: Pitman Publishing. pp. 233-244.

Olsen, S. R. (1982). Removing barriers to crop productivity. *Agronomy Journal* 74:1-4.

Park, J. and R. A. F. Seaton. (1996). Integrative research and sustainable agriculture. *AgricultureSystems* 50:81-100.

Piper, J. (1999). Natural systems agriculture. In *Biodiversity in Agroecosystems*, ed. W. W. Collins and C. O. Qualset, Boca Raton, FL: CRC Press. pp.167-195.

Prihar, S. S., P. R. Gajri, D. K. Benbi, and V. K. Arora. (2000). *Intensive Cropping: Efficient Use of Water, Nutrients, and Tillage*. Binghamton, NY: Food Products Press.

Rambo, A. T. and P. E. Sajise. (1985). Developing a regional network for interdisciplinary research on rural ecology: The Southeast Asian Universities Agroecosystem Network (SUAN) experience. *The Environmental Professional* 7:289-298.

Sen, D. N. (1988). Key factors affecting weed-crop balance in agroecosystems. In *Weed Management in Agroecosystems*, ed. M. A. Altieri and M. Liebman, Boca Raton, FL: CRC Press, pp. 157-182.

Spedding, C. R. W. (1979). *An Introduction to Agricultural Systems.* Applied Science Publishers, London.

Swanton, C. J., D. R. Clements, and D. A. Derksen. (1993). Weed succession under conservation tillage: A hierarchical framework for research and management. *Weed Technology* 7:286-297.

Temple, S. R., D. B. Friedman, O. Somasco, H. Ferris, K. Scow, and K. Klonsky. (1994). An interdisciplinary, experiment station-based participatory comparison of alternative crop management systems for California's Sacramento Valley. *American Journal of Alternative Agriculture* 9(1/2):64-71.

Weersink, A., M. Walker, C. Swanton, and J. E. Shaw. (1992). Costs of conventional and conservation tillage systems. *Journal of Soil and Water Conservation* 47: 328-334.

Young, F. L., A. G. Ogg Jr., and R. I. Papendick. (1994). Case studies of integrated-whole farm system designs: Field-scale replicated IPM trials. *American Journal of Alternative Agriculture* 9(1/2):52-56.

Advances in the Design
of Resource-Efficient Cropping Systems

Charles A. Francis

SUMMARY. Cropping systems in the Midwest USA are evolving as farmers seek labor-efficient designs to capture and use available light and precipitation. Heavy reliance on fossil fuels and other energy-intensive production inputs coupled with uncertain markets for commodities have encouraged the search for greater energy efficiency as well as alternative crops, markets, and production systems. Agroecology has emerged as an innovative and integrative approach to evaluating systems more suited to the often harsh and unpredictable environment, using native prairie structure and function as one guide to design of future systems. A more specific approach, Natural Systems Agriculture, is being explored as a new paradigm to saving soil while producing the crop and animal output needed for a growing global population. Whole-farm and landscape-level design and planning will become more important as society recognizes and values multifunctional rural landscapes. *[Article copies available for a fee from The Haworth Document Delivery Service: 1-800-HAWORTH. E-mail address: <docdelivery@haworthpress.com> Website: <http://www. HaworthPress.com> © 2003 by The Haworth Press, Inc. All rights reserved.]*

KEYWORDS. Crop rotations, intercropping, woody buffers, crop/animal integration, nutrient cycling, Natural Systems Agriculture, perennial polyculture, holistic resource management, rural landscapes, corn-soybean systems

---

Charles A. Francis is Professor, Department of Agronomy, 255 Keim Hall, University of Nebraska, Lincoln, NE 68583-0910.

Contribution of the Agricultural Research Division, University of Nebraska, Lincoln, NE, USA. Journal Series No. 13373.

[Haworth co-indexing entry note]: "Advances in the Design of Resource-Efficient Cropping Systems." Francis, Charles A. Co-published simultaneously in *Journal of Crop Production* (Food Products Press, an imprint of The Haworth Press, Inc.) Vol. 8, No. 1/2 (#15/16), 2003, pp. 15-32; and: *Cropping Systems: Trends and Advances* (ed: Anil Shrestha) Food Products Press, an imprint of The Haworth Press, Inc., 2003, pp. 15-32. Single or multiple copies of this article are available for a fee from The Haworth Document Delivery Service [1-800-HAWORTH, 9:00 a.m. - 5:00 p.m. (EST). E-mail address: docdelivery@haworthpress.com].

10.1300/J144v08n01_02

INTRODUCTION

Cropping systems in the Midwest of the U.S. have developed through a gradual shift from highly diverse, integrated, crop/animal farms of modest size to highly specialized, industrial-model, crop farms with much larger area under control of one manager. The Homestead Act of 1862 provided that "any person who is the head of a family, or who has arrived at the age of twenty-one years, and is a citizen of the United States, or who shall have filed his declaration of intention to become such . . . be entitled to enter one quarter-section or a less quantity of unappropriated public lands . . . subject to pre-emption at one dollar and twenty-five cents, or less, per acre . . . " (U.S. Government, 1862). This act provided land for many who had smaller, worn-out farms in the East and for landless families looking for a new start. For a century, settlers on these lands developed diverse systems for family subsistence, and later for sale of food grains and other products locally, and eventually sale to more distant markets. Wheat (*Triticum aestivum* L.), corn (*Zea mays* L.), oats (*Avena sativa* L.), and forages predominated in an agriculture that depended on animal draft power until the introduction of small tractors in the first half of the last century. The Homestead Act that forced people to live on their own land plus the continuing need to care for livestock contributed to a dispersed pattern of homesteads in contrast to the concentrated rural communities in some parts of northern Europe.

Structural changes in agriculture and modifications in cropping patterns have been implemented more quickly since 1950. In four decades leading to 1990, the number of U.S. farms declined from over 5 million to less than 2 million, a 64% reduction, while at the same time the farm population declined from over 15 million to under 2 million people, a reduction of 88% (Lyson et al., 1999). What allowed the reduction in people and consolidation of farmland was rapid mechanization and specialization in a small number of commodities, with most production exported from the farm. Production today continues to skew toward large farms, with over half the cropland harvested on less than 10% of the farms, and these each control 1,000 acres or more (Bureau of the Census, 1994). Farmers in the U.S. on average currently own about half the land they farm. Demographics and ownership patterns influence current cropping practices, systems, landscape and conservation decisions, as well as long-term investment in farm infrastructure.

This pattern of farm ownership, production, and intensification in a few commodities, especially in the Midwest, has resulted in a cropping system pattern that is summarized in the first column in Table 1 (adapted from Stinner and Blair, 1990). There is low crop diversity, with many farms specialized in only corn and soybean (*Glycine max* L. Merr.); in some irrigated areas of Nebraska, continuous corn monoculture with irrigation has been the most profit-

TABLE 1. Comparison between current conventional cropping and future, re-source-efficient cropping and crop/animal systems (adapted from Stinner and Blair, 1990).

System Characteristic	Current	Future
Fossil fuel energy	High	Moderate-low
Fertilizers	Inorganic	Organic
Tillage	Moderate	Low-none
Pest protection	Inorganic chemical	System design
Crop diversity	Low	High (+ animals)
Types of crops	Annual	Annual/perennial
Animals in systems	None	Multiple species
Temporal diversity	Low-moderate	High
Spatial diversity	Low	Moderate-high
Nutrient cycling	Open/pulsed	Closed
Labor use	Low	Moderate
Complexity of management	Low	Moderate-high
Marketing	Distant, commodities	Local, diverse crops
Family, community involvement	Low	High

able pattern for several decades, in large part due to government commodity support programs. There is a high use of fossil fuels, especially for tillage and irrigation, and of chemical pesticides and inorganic fertilizers. Annual crops dominate the system, due to high levels of production per acre, available markets, rapid return on investment, federal commodity programs, and some flexibility to meet market changes. There is low spatial and temporal diversity, an open nutrient cycle (fertilizer applications and erosion plus leaching losses), and major flow-through of materials (import of purchased inputs) and energy into and away from the farm (export of raw grain commodities). Although farms are operated with a high investment of technology, complexity of management is simplified on the farm by use of few crops, purchase of needed inputs, and reliance on consultants in choice of crop cultivars, chemical and fertilizer inputs, and irrigation scheduling. Marketing can be complex, since most crops are exported from the farm for distant use. Family and community involvement in the farming operation is low, aside from the principal operator.

This paper briefly reviews the advances in crop production technologies that are designed to make current systems more efficient, and concentrates on future options for cropping systems and rural landscapes. Changes in technologies must be considered in the context of the evolving ecological, economic, and social milieu of rural families and communities and how these changes impact future cropping and system decisions. The corn-soybean rotation system is used for convenient comparison since this is the dominant conventional cropping pattern in the region. Agricultural systems in the U.S. are highly impacted by federal farm policy and support payments, as well as export markets for the major commodities. Thus, any discussion without consideration of these factors would be shallow and unrealistic. The complexity of decisions facing today's crop producer is a daunting challenge, and one that must be discussed in the broader context of the future of the rural landscape and community.

IMPROVED CROP PRODUCTION EFFICIENCY

In this review, the conventional methods of improving crop production efficiency are described. These include fine-tuning varieties and cultural practices, making efficient use of expensive inputs, and applying global positioning systems (GPS) technology to place inputs where they are most effective and profitable. Beyond improving current systems, futuristic researchers are exploring how ecological principles can be used to design farming systems through the emerging field of agroecology, the combining of principles and processes from agriculture and ecology. Evaluation of system productivity and efficiency should increasingly be viewed in the context of environmental and social impacts and how sustainable these systems will be for the long term.

This means that farming systems must be viewed as components of a multifunctional agroecological landscape that is closely linked to local rural communities. To be sustainable for the indefinite future, food production systems will need to evolve toward alternatives that are more dependent only on contemporary solar energy, and Natural Systems Agriculture provides one model to meet these goals. Several dimensions of resource efficient cropping systems are discussed as alternative strategies needed to meet global food challenges.

Improving Current Systems

Researchers are contributing to improved production efficiency of specialized crop production systems by increasing genetic yield potential, seeking precise rates and timing of fertilizer use for specific sites, exploring alternatives to chemical pesticide applications, and devising schedules for judicious use of irrigation water. Many technical reports on current research are found in *Agronomy Journal, Journal of Soil and Water Conservation*, and other specialized journals in horticulture, fruit and vine crops, and pastures and forages. More popular articles on reduced inputs and environmentally sound production practices are seen in *Successful Farming, Organic Gardening*, and other monthly mainstream publications. For more than a decade, two additional publications have provided focus on new options in agriculture and food systems: *American Journal of Alternative Agriculture* and *Journal of Sustainable Agriculture*. Information from these sources is finding its way into the current teaching of agriculture and cropping systems in universities (e.g., Mason, 1999) and into Extension materials in the sustainable agriculture arena (e.g., Carter, Olson, and Francis, 1998).

Farmers contribute to increased production efficiency by translating these research results into applications for specific fields and farms. The trend over several decades has been to homogenize the production environment with widespread recommendations of new hybrids and varieties, new fertilizer formulations, and broad spectrum insecticides and herbicides that lead to greater efficiency in terms of labor use. Over the past decade, we have focused more on the unique nature and characteristics of each field and even areas within fields. Farmers have been attracted to new GPS systems that can precisely locate equipment position in the field and provide information on crop yields that can be translated into yield maps each year. These in turn can be used to determine the more precise placement of inputs, especially fertilizers, to maximize yields or return to fertilizer investment based on yield potential and crop response in specific areas of each field. The GPS technology at present is ahead of our ability to determine exactly what those fertilizer rates should be. There are parallel applications of the technology to fine-tuning weed management with chemical and other alternative treatments.

Choosing the most efficient crops and especially the cultivars within those crops is an important management decision. Most farmers feel locked into current crops because of experience, equipment on hand, familiarity with accessible markets, and commodity support programs. Most confine their decisions to choosing the best among new hybrids or varieties of corn, soybean, wheat, or other crops that come onto the market each year. These decisions are informed by the commercial seed producers who do excellent marketing through company employees as well as farmer-dealers who often purchase at some discount as well as promote a particular company's products to relatives and neighbors. Also available in each state are uniform variety and hybrid test results that compare cultivars submitted for trials in multiple locations, with costs partially paid by the companies (e.g., Nelson et al., 2000). Crop cultivar decisions have been based primarily on productivity, with some attention to maturity, standing ability, and tolerance to major insect and pathogen problems. More recently farmers have included among their criteria nitrogen (N) use efficiency, response to irrigation, and adaptation to limited tillage systems.

Two issues surrounding cultivar choice have emerged in the last decade. One is the availability of crop cultivars, particularly corn, with special traits that enhance their value in the marketplace. Examples are high oil, high lysine, and white or blue endosperm cultivars for specific products. There will be more of these cultivars available to farmers, and they will help meet demand for special products in the future.

This issue is linked to the current debate about the availability of hybrids and varieties produced through transgenic methods, most often known by the misleading label "GMO" (genetically modified organism) cultivars. This is a misnomer because all of our major crops have undergone genetic modification, first at the hands of women in prehistoric times who chose the best individual plants for seed (Plucknett and Smith, 1986), up to the present-day plant breeders with sophisticated technologies and selection techniques. Proponents of new biotechnologies promote the potentials of cultivars that will provide their own pest protection, ease management, and bring new traits into specialty cultivars with premium prices. While admitting these possible advantages, many scientists and consumers are concerned about the potential non-intended side effects of interspecific and especially interfamily gene transfer, such as gene escape to wild populations, loss of genetic resistance to insects or pathogens, or unexpected food allergies. With respect to crop yields, there is some evidence that transgenic cultivars lag behind sister materials selected for yield and not the transgenic traits (yield lag) and may actually have yields depressed by the introduced genes (yield drag), as shown in soybean (Elmore et al., 2001). There is still much to learn about the potentials and problems of these new technologies.

Reduced tillage to conserve scarce water resources and lower energy costs in crop production was born of necessity in the Great Plains. Early research at University of Nebraska and commercial development and testing in cooperation with Fleischer Manufacturing in Columbus (Nebraska) led to the first widely used no-till Buffalo© planters and cultivators. The planting equipment was designed to move stubble a few cm away from the planting furrow, with either a disk opener or shoe then used to place seed at appropriate depth into moist soil, and following press wheels to assure the necessary seed-soil contact. This planter has been modified and perfected over more than four decades, and similar models are now available from other major equipment manufacturers.

To replace or complement herbicide applications, no-till and high residue cultivators were developed for planting systems where no land preparation is used prior to planting. Cultivators are used to catch those weeds that emerge in the no-till system. In practice, there are far fewer weeds in the no-till system since there is no prior tillage to bring up weed seed into conditions favorable for germination. The no-till cultivators are designed with shanks further apart, back to front, to allow residue to flow around and between them and not get pushed cross-ways to take out a row of planted crop. Shields are used with early cultivation to protect small seedlings. Later cultivation can hill up crops and cover small weeds that come up in the planter row. Similar to planters, these cultivators for use in high residue are now made by several major manufacturers. Experienced agronomists and farmers in the Great Plains estimate that for each field pass eliminated (disking, planting, cultivation) at least one cm of moisture is conserved. In dryland farming with only 30-50 cm of total rain and snow moisture, it is essential to minimize field operations that cause loss of this scarce resource (Fenster, 1980).

Fertilizer is the second most expensive input in irrigated corn production, yet excess applications of relatively inexpensive N over several decades have led to both extra farmer expense and major non-point source contaminants of surface and ground water in many agricultural regions of the U.S. Both to save costs and to correct this environmental problem, farmers are seeking ways to fine-tune fertilizer applications, especially N. This has been achieved by careful soil testing, nutrient budgeting to consider all available sources, late-spring soil N testing, and leaf chlorophyll and stalk N testing. Soil testing using a grid system can provide samples for a useful estimate of available nutrients as well as reveal spatial variation if each sample is analyzed separately. Nutrient budgeting takes into account yield goals and the expected crop needs for nutrients. For N, this includes available nutrients in the root zone, that which will arrive with rain and snow, N applied with irrigation water, and that from previous crops. By factoring in all these sources along with yield goal, it is possible to anticipate the N needs of the crop and assure that over application will not occur. This saves both money and the environment. Current high energy prices are causing more farmers to carefully budget their N applications.

Innovations on the Horizon

Geographic Information Systems (GIS) are being applied through global positioning to precisely locate equipment and unique features of individual fields. The goals are to make more efficient use of expensive inputs–fertilizers, pesticides, irrigation water–in the areas of fields where they will produce the greatest return; to reduce inputs where yield potential does not warrant their use; and to prevent excessive applications that cost the farmer more than is expected in returns and have potential to cause some negative environmental impact. The equipment to draw yield maps of fields and to apply variable rates of inputs is available, but there is yet much research and testing needed to make best use of these ideas.

Concurrent with this interest in site-specific application of production inputs is a continuing growth in farm size, field equipment, and needed investment in farm infrastructure. To farm more hectares and maintain efficiency of labor use, farmers in the Midwest continually seek ways to simplify and homogenize their production practices to be able to cover expanded areas with as small a number of field operations as possible. This means planting the most productive and specifically adapted cultivars over the largest area that is feasible, consistent with the need to provide some diversity in maturity and other genetic traits to avoid potential disasters from insects, pathogens, and adverse weather. It also means preventive treatments with pesticides and application of fertilizers to avoid problems that might occur, and that would incur larger labor costs later in the season to remedy the problems. This is one cost to larger farm size, since operators who manage fewer acres can visit each field more often, fine-tune input use for smaller areas, and respond more quickly to adverse biotic or weather conditions.

Reduced tillage systems described above are now standard practice in the western part of the U.S. Corn Belt, with farmers driven by the necessity to conserve limited soil moisture in as many ways as possible. These are still relatively new practices farther east in the region. With more erratic weather patterns, and recognition of the vital role of soil moisture and the effects of deficit in the plant even before any symptoms occur, these no-till, till-plant, and reduced-till systems are gaining more favor even in the higher rainfall areas of this region. Availability of a wider range of equipment options also promotes a change to less tillage, as well as the recognition of increased carbon (C) sequestration and reduced surface soil erosion when these planting strategies are used.

Fresh water is potentially a limiting resource on the global scale. Agriculture is entirely dependent on rainfall, moisture from winter snows, and irrigation from surface sources and underground aquifers. It is difficult to compete for scarce water sources with industry, commercial activities, or residences, since all these human needs add more value and people are willing to pay more for water than we can afford in farming. Thus, it is essential that we make the most efficient possible use of available precipitation as well as irrigation water. The reduced tillage strategies already described are one approach to water use efficiency. Another is to choose crops that are most effective in converting water to dry matter and useful production for food, feed, and fiber. Within crop species, there are genetic differences in efficiency of water use and ability to survive short periods of water deficit. There is active selection in plant breeding programs to find those cultivars most suited for stress conditions. Some genetic work on barley (*Hordeum vulgare* L.), tomato (*Lycopersicon esculentum* Mill.), and other species that can grow with partial use of sea water is suggesting new directions for exploitation of the tremendous ocean water resource. Irrigation scheduling to meet crop needs and not exceed the immediate requirements of growing crops can help reduce producer costs for fossil fuels and labor. Unique methods of irrigation such as center pivots with low pressure and drop nozzles, pulse applications, skip-row watering (every other furrow in surface systems), and drip application either on the surface or underground are all techniques that have been developed in arid regions of the world and have potential for the Midwest. There are different costs associated with these improved methods, and the most efficient for a given situation must be determined. Improved methods are more efficient in water use and conversion to crop dry matter than the conventional flood or furrow irrigation methods we have traditionally used when this resource was less limiting.

Finally, the conventional farmer is seeking diversity in crop cultivars and their non-traditional uses to find new markets and higher value from well-known crops. Corn with high oil, elevated levels of specific amino acids, and special characteristics such as white endosperm for food products make these attractive alternatives to traditional feed grain corn hybrids. Ethanol production from corn grain is another new product that uses a substantial amount of production, and elaboration of sweeteners from corn is another alternative use. Use of crop residues for production of fiberboard, for fuel, and for other products can boost income per hectare, although there are trade-offs for the removal of this C source from the cropping system and the eventual depletion of organic matter as a result. These are examples from corn, our most widely-grown cereal crop, and there are specific new cultivars and new uses from a number of other conventional and traditional crop species such as wheat, grain sorghum [*Sorghum bicolor* (L.) Moench], and soybean.

DESIGN USING AGROECOLOGICAL PRINCIPLES

Innovative farming alternatives for the future in the Midwest are emerging from studies of agroecology, a blending of concepts and understanding of mechanisms from both agriculture and ecology (Altieri and Francis, 1992; Gliessman, 1998). Narrowly defined, agroecology is the application of ideas from the study of ecology to the design and management of agricultural production systems (Altieri, 1987). Natural systems are relatively stable in terms of species mix and productivity, are closely adapted to specific soils and other uniqueness of sites, and are self-sustaining in nutrients. A better knowledge base on how these systems operate, how system function is related to structure, and how they are resilient in the face of unpredictable weather events may lead to design of more productive and sustainable agroecosystems, those we use to produce outputs useful for the human enterprise.

One key characteristic of natural ecosystems is diversity. Prairie systems comprise a complex mix of grasses, forbs, and legumes (Weaver, 1968). The limited spatial diversity in today's conventional corn production in this region comes from weeds that are controlled to the greatest extent possible, different hybrids in adjacent fields, grassy species in the field margins, and the occasional cover crop planted after corn is harvested. Likewise the temporal diversity is limited to rotation with soybeans in a two-year pattern, or to longer-term rotations on some mixed farms where livestock enterprises are still included. Many fields under irrigation in the western part of the region have been in continuous corn because of mechanization and the ability to farm large areas with specialized equipment, the opportunities for nearby markets for feed grains, ethanol, and sweeteners, and the relative profitability of this crop compared to others.

Studies in ecology suggest that more diverse patterns, both in space and time, would be both more stable and more resilient to changes in weather. Rotations of dissimilar species, especially cereals with legumes or summer with winter annuals, have been shown over many sites and combinations to produce higher yields of all crops in the rotation compared to continuous culture of any of them. The increases due to rotation vary, but appear to average about 10% (Francis and Clegg, 1990). Crop rotations that create temporal diversity in midwest production systems can control some insect pests, help the farmer to manage weeds, and reduce total nutrient needs if one or more crops in the sequence are legumes (Cruse and Dinnes, 1995; Francis and Clegg, 1990). Even the rotation of different corn hybrids has been shown to produce yields that differ from continuous culture of the same hybrid (Hicks and Peterson, 1981). Significant increases in yields were found in three of five hybrids tested, compared to their continuous plantings, although these yield boosts were less than the same hybrids in rotation with soybean.

Another approach is use of spatial diversity both within fields and within the farm. Within-field diversity has been a characteristic of cropping systems since the invention of agriculture some ten thousand years ago (Plucknett and Smith, 1986). The use of multiple cropping systems is now more prevalent on small farms where production focuses on subsistence and on a diverse array of products for local markets (Francis, 1986). A range of multiple cropping strategies such as strip cropping, relay cropping, and intensive mixed intercropping has been used in both tropical and temperate climates, and these appear to have great potential for future sustainable production systems. Diversity within the field promotes nutrient cycling, efficient resource use by crops of different growth cycles and rooting patterns, and protection against some damaging insects and pathogens. Within the farm, enterprise diversity and smaller fields permit greater biological interactions among crops, providing habitat for predators that can help in control of insects that reduce crop yields. More crops permit greater imagination in creating beneficial rotations whose advantages were described above. How much diversity is needed in crop and crop/animal systems was explored by the authors of a book from a symposium (Olson, Francis, and Kaffka, 1995). Much is known about the intercropping and rotational practices described here, but there is still much to be learned about diversity in soil biota, impacts of soil organisms on crops, and the complex biogeochemical reactions and interactions that provide a foundation for all agroecosystems (Kennedy, 1995; Neher, 1995).

More complex systems of rotation are used in other regions of the world that have similar soils and climate to the midwest. The four-year grass/legume pasture followed by four-year sequence of annual crops is a common practice in Argentina. This management pattern controls most insects and some weeds, as well as providing most of the nutrients needed during the cropping part of the cycle. Such rotations are less common in the midwest where the majority of cattle are fed in confinement and these enterprises are on different farms from where crops are produced, or may be distant from the cropping fields. Longer-term rotations that involve crops and livestock do provide a useful opportunity to create a more diverse set of enterprises, use natural resources efficiently, and reduce the use of chemical fertilizer and pesticide inputs.

Mixed farming, the combination of crops with animal enterprises, provides a number of other structural advantages and potentials to better use plant resources on the farm. When animals forage for dropped corn ears and soybean seeds, as well as consume some of the residue left after harvest, they reduce the amount of volunteer plants in the following season. By ingesting the residue in the field and depositing urine and feces there, the livestock speed nutrient cycling and minimize extraction from that site. Other advantages of having livestock integral to the farm include using discard or low-quality grain that would otherwise not be marketable; providing a destination for hay from waterways, roadsides, or stream filter strips; and consuming damaged crops, for example, after a severe hail storm or during a drought that makes grain production unfeasible. Economically, the livestock provide a different income source, one that potentially can be flexible and available through most of the year. Major commodity crops are harvested in July (wheat) or October (corn, sorghum, soybean) and proceeds are used to pay off operating loans. There is very uneven cash flow through the year. Grazing multiple livestock species takes advantage of different grazing patterns, diverse nutrient needs of the species, and unique markets available for a range of products. These grazing practices have been reported to increase total animal yield by up to 90% (Cook, 1985). A conference on multiple species grazing, including wildlife, was sponsored by Winrock and the proceedings edited by Baker and Jones (1985).

Intensive rotational grazing is another innovation that is providing greater total forage for livestock and gain as well as improving the pasture resource (Murphy, 1990). Moving animals frequently to new areas based on amount of remaining forage improves pasture growth, places animals on areas with new growth, and maximizes total forage production through the season. There are still peak production times in late spring and early summer as well as in early fall, and some provision must be made to harvest and stockpile this excess for other times of the year when forage and crop residues are short. These strategies are more realistic than frequent changes in animal numbers. Again, the complementarities of crop production with forages and livestock are considerable, and more closely mimic the natural plant/animal balance in a natural ecosystem compared to farms with only crops or only feedlots.

Incorporation of perennial forages and woody trees and shrubs introduces another level of diversity and permanence into agricultural production systems. The seminal book *Permaculture* (Mollison, 1990) introduced the time-honored multiple species systems based on woody perennials in the tropics and subtropics to a broad audience in temperate regions. Mollison (1990) advocated and described the transformation of some areas of annual cropping into annual/perennial mixtures that would gradually evolve into perennial plantings for food, feed, and fiber. These permanent systems could continue to include some annual component crops, have animal interactions that improve nutrient cycling, feature animal use of forages and crop products that are not useful to humans, and replacement or renewal of the perennial elements as they become non-productive over time. Some of the principles of permaculture have been incorporated into what is now more widely called agroforestry in the U.S., and numerous systems using these principles have been described in two recent books (Buck, Lassoie, and Fernandes, 1999; Garrett, Rietveld, and Fisher, 2000). Expanding the earlier reports that focused on tropical regions and especially the forest zone and savanna, these new references include extensive descriptions of examples from temperate regions where monocultures and high technology have replaced much of the diversity from agriculture.

One highly creative application of agroecology to design of farming systems is the development of Natural Systems Agriculture. First described in the landmark book *New Roots for Agriculture* (Jackson, 1980), this concept uses the natural prairie as a model for perennial cropping and crop/animal systems that would supply their own fertility, protect against unwanted insects, pathogens, and weeds, and run entirely on contemporary energy. This suggests that no subsidies of fossil fuels would be used. The perennialization of several major crops [wheat, rye (*Secale cereale* L.), barley, sorghum, sunflower (*Helianthus* spp.)] is under way at The Land Institute near Salina, Kansas, with more species being considered. To stop with perennial monocultures would miss the larger point, which is the integration of several species in such a way as to mimic the functional groups featured in native prairie. For eventual comparison with perennial grain polycultures, the productivity of conventional annual grains is being assessed in the Sunshine Farm at The Land Institute, again run on contemporary energy and internal fertility and pest control. Current results (Marty Bender, personal communication, 2000) indicate that about 25% of the tilled land is needed for fuel production, either for horses or for biodiesel, and another 25% of the land for N fixation in order to operate the remaining 50% of the land for extractive crop production. This should be considered a maximum level of land for energy and N production, since very limited research has been conducted on the system and much greater efficiencies are likely to be achieved with further work. More of the background information for this type of system is found in the book *Farming in Nature's Image: An Ecological Approach to Agriculture* (Soule and Piper, 1992).

FUTURE RECOGNITION
OF MULTIFUNCTIONAL RURAL LANDSCAPES

There is growing appreciation among the general public that rural landscapes provide much more than food for people and feed for livestock. According to Brandt, Tress, and Tress (2000) in the introduction to a recent European conference on multifunctional landscapes,

> From a human perspective, many of the earth's landscapes are increasingly being used simultaneously for several purposes. During the postwar period, intensified land use has been furthered primarily by spatial segregation of functions. Growing land pressure and environmental problems have made this strategy problematic and a paradigm for multifunctionality is emerging. Thus, there will be high demands on the landscapes of the future, which will have to serve simultaneously the following functions: ecological (as an area for living), economic (as an area for production), socio-cultural (as an area for recreation and identification), historical (as an area for settlement and identity), and aesthetic (as an area for experiences). (p. 15)

These several functions of rural landscapes have always been important, but it is the recognition of their value to society and a willingness to pay for multiple functions that is changing. The recent recognition that C capture by growing plants can be drastically changed by different tillage management systems and the quantitative measure of C sequestration can lead to C credits or payments to farmers for using these practices (Lavendel, 2000). He cites the case of Illinois farmer Jim Kinsella who has been using no-till planting for corn and soybean over 26 years, and has been able to capture and store about 25 tons of C per hectare in the top 30 cm of soil over that time. Increasing soil C can improve soil quality and increase its capacity to filter and degrade pollutants, trap sediments, and reduce emissions of gases as organic matter is mineralized to inorganic components. There is current discussion of farmers receiving payments for "carbon credits" from other enterprises such as power plants that must either reduce their own emissions or buy a comparable number of credits from others.

Growing plants absorb carbon dioxide and emit oxygen, and this is another major contribution of the managed agricultural landscape. To the degree that this area is covered with growing vegetation for the longest possible part of the year, more photosynthesis can take place and more C fixed in the field, and this activity can be recognized and paid for. The practical windows to increase C capture in a field planted to corn are the weeks prior to crop planting in the spring and the weeks during which the crop is senescing and after harvest in the fall. Grass cover crops during these periods can capture soluble nutrients and transform them into organic form, reducing potential for loss by erosion with soil or by infiltration through the soil profile. Leguminous cover crops have the additional advantage of some N fixation, plus the nutrient capture already mentioned. Keeping soil in place and water capture and absorption by the soil are both promoted by crop residues and cover crops, and this function is a valuable one to keep soil in fields and away from road ditches, culverts, streams, lakes, and other public-maintained sites in the landscape. Air quality can also be improved by windbreaks or other vegetation that slow the wind, absorb soil particles, and reduce pesticide drift impacts on nearby fields or living areas.

Multiple human activities for recreation include hunting, bird watching, horseback riding, hiking, camping, skiing, and snowmobiling, activities that could be especially attractive if there is a varied landscape on the farm with streams and forested areas. Some enterprising farmers have developed a clientele for bed and breakfast or farm weekends to participate in agricultural activities, with perhaps the ultimate a cattle round-up and chuckwagon meal on one ranch in western Nebraska. Farmers producing fruits and vegetables have used farm visits as an opportunity for sale of products, either from a roadside stand or a pick-your-own type of operation. When customers visit the farm, they may also purchase other crafts or products that are elaborated from home-grown natural resources. When a mixed farm produces livestock, fruits, and vegetables, as well as organizing a weekend resident package, the food can have tremendous value added if prepared and served to guests. The more cultural and historic places and learning dimensions that are built into the farm stay, the more value can be added to the location and its resources. These are ways to broaden and diversify the income stream to the farm family, and the only real limitation is the creative imagination needed to think up new possibilities for marketing and attracting customers.

CONCLUSIONS: FUTURE AGRICULTURAL SYSTEMS

From this discussion emerges a vision of improving the efficiency of current farming systems, changing systems to make more creative use of resources, and adding value to the rural landscape. There are ways to make current crop production more resource-efficient, and ways to redesign systems so that they are both more environmentally sound as well as profitable. As society recognizes the multiple and crucial functions of rural areas, people will be more willing to pay for ecosystem services that today are taken for granted.

The characteristics of future cropping and crop/animal systems are shown in the second column of Table 1, modified from Stinner and Blair (1990). By converting to reduced tillage planting and dedicating some fields to perennial forages, there will be reduced fossil fuel consumption by farms in the future. If more cover crops and diverse grain legumes are included in rotations, the needs for purchased inorganic fertilizers can be reduced due to N from the legumes as well as the "nitrogen-sparing" effects of planting fewer high-N consuming cereal grains in continuous systems. Both tillage and application of fossil-fuel dense chemical pesticides can be reduced by creative design of rotations, multiple crop systems, and cultural practices appropriate to these systems. Diversity can be markedly increased by including more crop species in rotations and by introducing ruminant livestock into the system. Using some perennial crops and woody species can also increase diversity and reduce planting costs that recur each year with entirely annual crops. The resulting temporal and spatial diversity can help close nutrient cycles from the current open cycles in annual row crop culture. Labor use is likely to increase due to more enterprises, but this can be spread through more of the year to take advantage of a human resource that is currently under utilized on many cash grain farms. Complexity of management increases with the number of crops and animal species, each requiring a set of efficient practices and a marketing scheme, yet there is more opportunity to create local markets for diverse products in ways that are not possible with large plantings of single commodity crops. Finally, there is potential for more involvement of the family in the farming operation, if some of the enterprises can be managed by children as 4-H or FFA projects, there are livestock that require daily care, and there are value-added activities that can provide a valuable use of human labor on the farm to add value to natural products grown there.

Such an idealistic farm of the future suggests multiple changes that would have to be introduced into the commodity-oriented, homogeneous farm of today. The most difficult change is away from the "mental monoculture" that dominates our present conventional thinking and planning in agriculture. It is comfortable to stay with a known routine, even though the financial and environmental results may not appear to be positive for the long term. Often people stay with the systems that are known because of lack of knowledge about alternatives, limited financial or human capital to invest in new directions, or hesitation with more risky ventures that are less known in an area. In this decision making environment, the role of education is critical–both in schools and in the continuing education available for adult learners. Creating an innovative series of educational opportunities that helps people explore the full range of human potentials can open possibilities that can benefit the individual, family, and community. Perhaps the largest change needed is for people in the rural areas to see this full range of potential enterprises and opportunities for economic returns, and to appreciate the possibilities of working together in communities to help design and achieve a positive future. Although many of the current trends toward larger farms and fewer people in rural areas may be discouraging, we should remember the words of Nobel Laureate biologist René de Bos, "Trend is not destiny!"

REFERENCES

Altieri, M.A. (1987). *Agroecology: The Scientific Basis of Alternative Agriculture.* Boulder, CO: Westview Press.

Altieri, M.A. and C.A. Francis. (1992). Incorporating agroecology into the conventional agricultural curriculum. *American Journal of Alternative Agriculture* 7:89-93.

Baker, F.H. and R.J. Jones. (1985). In *Proceedings of a Conference on Multi-Species Grazing,* ed. F.H. Baker and R.J. Jones, Morrilton, AK: Winrock International.

Brandt, J., B. Tress, and G. Tress. (2001). Introduction to the conference theme. In *Multifunctional Landscapes: Interdisciplinary Approaches to Landscape Research and Management,* ed. J. Brandt, B. Tress and G. Tress. Conference, Univ. Roskilde, Roskilde, Denmark, Oct. 18-21.

Buck, L.E., J.P. Lassoie, and E.C.M. Fernandes. (1999). *Agroforestry in Sustainable Agriculture Systems.* Boca Raton, FL: Lewis Publishers.

Bureau of the Census. (1994). Census of Agriculture, 1992. Vol. 1, Geographic Area Series: Pt. 51. U.S. Summary, Final Report. Government Printing Office, Washington, DC.

Carter, H., R. Olson, and C. Francis. (1998). *Linking People, Purpose, and Place: An Ecological Approach to Agriculture.* Extension and Educational Materials for Sustainable Agriculture, Vol. 7. North Central Region and Cooperative Extension Division., Center for Sustainable Agricultural Systems. Lincoln, NE: University of Nebraska. 266 p.

Cook, C.W. (1985). Biological efficiency from rangelands through management strategies. In *Proceedings of Biological Efficiency from Rangelands Through Management Strategies,* ed. F.H. Baker and R.J. Jones. Morrilton, AK: Winrock International.

Cruse, R.M. and D.L. Dinnes. (1995). Spatial and temporal diversity in production fields. In *Exploring the Role of Diversity in Sustainable Agriculture,* ed. R.K. Olson, C.A. Francis, and S. Kaffka. Madison, WI: American Society of Agronomy, pp. 73-94.

Elmore, R.W., F.W. Roeth, L.A. Nelson, C.A. Shapiro, R.N. Klein, S.Z. Knezevic, and A. Martin. (2001). Glyphosate-resistant soybean cultivar yields compared with sister lines. *Agronomy Journal* 73:408-412.

Fenster, C.R. (1980). Protect soils with vegetative residues. NebGuide G80-513-A, Cooperative Extension, Lincoln, NE: University of Nebraska.

Francis, C.A. (1986). *Multiple Cropping Systems*. New York, NY: Macmillan Publishing Company.

Francis, C.A. and M.D. Clegg. (1990). Crop rotations in sustainable agricultural systems. In *Sustainable Agricultural Systems*, ed. C.A. Edwards, R. Lal, P. Madden, R.H. Miller, and G. House. Ankeney, IA: Soil and Water Conservation Society. pp. 107-122.

Garrett, H.E., W.J. Rietveld, and R.F. Fisher. (2000). *North American Agroforestry: An Integrated Science and Practice*. Madison, WI: American Society of Agronomy.

Gliessman, S.R. (1998). *Agroecology: Ecological Processes in Sustainable Agriculture*. Chelsea, MI: Ann Arbor Press.

Hicks, D.R. and R.H. Peterson. (1981). Effect of corn variety and soybean rotation on corn yield. In *Proceedings 36th Annual Corn and Sorghum Research Conference*, Washington, DC. pp. 84-93.

Jackson, W. (1980). *New Roots for Agriculture*. Lincoln, NE: University of Nebraska Press.

Kennedy, A.C. (1995). Soil microbial diversity in agricultural systems. In *Exploring the Role of Diversity in Sustainable Agriculture*, ed. R.K. Olson, C.A. Francis, and S. Kaffka. Madison, WI: American Society of Agronomy. pp. 35-54.

Lavendel, B. (2000). Carbon in our soil. *Conservation Voices*, August/September. pp. 12-15.

Lyson, T.A., C.C. Geisler, and C. Schlough. (1999). Preserving community agriculture in a global economy. In *Under the Blade: The Conversion of Agricultural Landscapes*, ed. R.K. Olson and T.A. Lyson. Boulder, CO: Westview Press. pp. 181-216.

Mason, S.C. (1999). *Study Guides for Resource-Efficient Crop Management*. Dubuque, IA: Kendall/Hunt Publishing Company.

Mollison, B. (1990). *Permaculture*. Washington, DC: Island Press.

Murphy, B. (1990). Pasture management. In *Sustainable Agriculture in Temperate Zones*, ed. C.A. Francis, C.B. Flora, and L.D. King. New York, NY: John Wiley & Sons. pp. 231-262.

Neher, D.A. (1995). Biological diversity in soils of agricultural and natural ecosystems. In *Exploring the Role of Diversity in Sustainable Agriculture*, ed. R.K. Olson, C.A. Francis, and S. Kaffka. Madison, WI: American Society of Agronomy. pp. 55-72.

Nelson, L.A., R.N. Klein, R.W. Elmore, D.D. Baltensperger, C. Shapiro, S. Knezevic, and J. Krall. (2000). *Nebraska Corn Hybrid Tests*. Extension Circular 00-105, Cooperative Extension. Lincoln, NE: University of Nebraska.

Olson, R.K., C.A. Francis, and S. Kaffka. (1995). *Exploring the Role of Diversity in Sustainable Agriculture*. Madison, WI: American Society of Agronomy.

Plucknett, D.L. and N.J.H. Smith. (1986). Historical perspectives on multiple cropping. In *Multiple Cropping Systems*, ed. C.A. Francis. New York, NY: Macmillan Publishing Company. pp. 20-39.

Soule, J.D. and J.K. Piper. (1992). *Farming in Nature's Image: An Ecological Approach to Agriculture*. Washington, DC: Island Press.

Stinner, B.R. and J.M. Blair. (1990). Ecological and agronomic characteristics of innovative cropping systems. In *Sustainable Agricultural Systems*, ed. C.A. Edwards, R.

Lal, P. Madden, R.H. Miller, and G. House. Ankeney, IA: Soil and Water Conservation Society. pp. 123-140.

U.S. Government. (1862). The Homestead Act: May 20, 1862. U.S. Statutes at Large, Vol. XII, p. 392 ff. (from Archives of the West, 1868-1874, www.pbs.org/weta/thewest/wpages/wpgs650/homestd.htm).

Weaver, J.E. (1968). *Prairie Plants and Their Environment: A Fifty-Year Study in the Midwest*. Lincoln, NE: University of Nebraska Press.

Cropping Systems and Soil Quality

R. Lal

SUMMARY. Cropping system refers to temporal and spatial arrangements of crops, and management of soil, water and vegetation in order to optimize the biomass/agronomic production per unit area, per unit time and per unit input. Soil quality refers to its intrinsic attributes that govern biomass productivity and environment moderating capacity. It is the ability of soil to perform specific functions of interest to humans. Three components of soil quality (e.g., physical, chemical and biological) are determined by inherent soil characteristics, some of which can be altered by management. Soil quality and soil resilience are inter-related but dissimilar attributes. Resilient soils, which have the ability to restore their quality following a perturbation, have high soil quality and vice versa. Decline in soil quality sets-in-motion degradative processes, which are also of three types, namely physical (e.g., compaction, erosion), chemical (e.g., acidification, salinization) and biological (e.g., depletion of soil organic matter content). Soil degradation, a biophysical process but driven by socioeconomic and political causes, adversely affects biomass productivity and environment quality. Determinants of soil quality are influenced by cropping systems and related components. Dramatic increases in crop yields during the 20th century are attributed to genetic improvements in crops, fertilizer use, and improved cropping systems. Dependence on fertilizers and other input, however, need to be reduced by adopting cropping systems to enhance biological nitrogen fixation and use efficiency of water and nutrients through conservation tillage,

R. Lal is Professor, School of Natural Resources, The Ohio State University, Columbus, OH 43210.

[Haworth co-indexing entry note]: "Cropping Systems and Soil Quality." Lal, R. Co-published simultaneously in *Journal of Crop Production* (Food Products Press, an imprint of The Haworth Press, Inc.) Vol. 8, No. 1/2 (#15/16), 2003, pp. 33-52; and: *Cropping Systems: Trends and Advances* (ed: Anil Shrestha) Food Products Press, an imprint of The Haworth Press, Inc., 2003, pp. 33-52. Single or multiple copies of this article are available for a fee from The Haworth Document Delivery Service [1-800-HAWORTH, 9:00 a.m. - 5:00 p.m. (EST). E-mail address: docdelivery@haworthpress.com].

cover crops, and improved methods of soil structure and nutrient management. *[Article copies available for a fee from The Haworth Document Delivery Service: 1-800-HAWORTH. E-mail address: <docdelivery@haworthpress. com> Website: <http://www.HaworthPress.com> © 2003 by The Haworth Press, Inc. All rights reserved.]*

KEYWORDS. Soil degradation, agronomic productivity, conservation tillage, cover crops, integrated nutrient management

INTRODUCTION

While the supporters of the Malthusian concept have been proven wrong during the 19th and 20th centuries, meeting the challenge of food security to all inhabitants of the Earth during the 21st century will require a careful assessment of world's soil resources and intensification of strategies for sustainable management of this crucial resource. Soil is non-renewable over the human time scale of decades to centuries, is fragile and easily degraded when misused and mismanaged, is unequally distributed among geographical regions of the world because of limitation of climate and slope gradient, and is fixed in location and cannot be transported. Whereas the world population has increased 20 times from 0.3 billion in 1 AD to 6 billion in 2000, soil resources available for food production have decreased because of degradation and conversion to non-agricultural uses. Consequently, the per capita arable land area is declining. Global average per capita arable land area was 0.23 ha in 1995 and will be 0.14 ha in 2050 (Engelman and LeRoy, 1995; Lal, 2000).

The soil has traditionally been used to produce feed, fiber, and fuel. In addition to being a medium for plant growth, additional and diverse demands on soil in the 21st century include: as a repository of industrial wastes, as a filter/biomembrane and a reservoir of natural waters, a raw material for industry, foundation for civil structures, and an archive for historical and geological/astronomical events. These diverse functions, indispensable to human existence, are determined by inherent soil characteristic termed "soil quality." The latter implies ability of the soil to perform functions of interest to humans (Papendick and Parr, 1992; Doran and Parkin, 1994; Lal, 1993; 1997).

BASIC CONCEPTS OF SOIL QUALITY

The concept of soil quality in one form or another has been used for some 2500 years. In ancient Rome, Columella ranked classes of soils for each of the different kinds of terrain (Carter et al., 1997). In Greece, Theophrastus

(327-287 BC) recommended abundant manuring of "thin soils" (Tisdale and Nelson, 1966). Virgil emphasized the importance of "blackish, loose and crumbling" soil. In Moorish Spain, Ibn-Al-Awam wrote a book "Kitab al-Felha" during the 12th century. In this book, dealing with the agricultural issues, the author describes attributes of soils of "good" and "inferior" quality. Farmers and revenue officers in the Indian subcontinent for millennia have used a concept of "relative value" of soil. The soil's relative value, according to this concept, is expressed on a scale similar to the division of the monetary currency such as into 10¢, 25¢, 50¢ or a dollar referring to soil quality of 1/10, 1/4, 1/2 and equal to that of a "normal" soil of good productivity (Lal, 2001).

While the concept of soil quality is not new, it has historically been a vague, qualitative and subjective measure of the soil's ability to produce biomass. An important challenge of modern times is making it a precise, quantitative, and objective tool of evaluating soil's ability to perform diverse functions that are changing rapidly with advances in the modern civilization. The goal is to assess the value of soil for a specific function (e.g., biomass production, grain yield, water quality). Jenny (1980) used the inter-relationship between soil properties and soil type to define "the state of the soil system." Richter (1987) and Lal (1987) proposed an "ecological" approach to maintaining/sustaining soil resource.

The term soil quality received considerable scientific attention during the 1990s (Lal, 1993, 1997, 1998, 1999; Doran and Parkin, 1994; Doran, Sarrantonio, and Lieberg, 1996; Carter et al., 1997; Bezdicek et al., 1996; Karlen et al., 1997). Whereas there have been several attempts at defining soil quality (Table 1), it has also been a debatable concept (Sojka and Upchurch, 1999). In general, the term refers to the capacity of the soil to perform specific functions, and the relevant attributes of soil quality depend on the specific function. Some of these functions, and specific attributes related to soil quality, are listed in Table 2. In ecological terms, soil quality interacts closely with air quality and water quality (Figure 1). In fact, air quality and water quality depends on soil quality, which together determine an ecosystem's health. In terms of agronomic productivity, important attributes of soil quality are the ability to retain and supply water and nutrients, provide a deep and favorable medium for root growth, and facilitate exchange of gases between water and soil (Lal, 1997).

For agronomic functions, there are three distinct components of soil quality: soil physical quality, soil chemical quality and soil biological quality (Figure 2). Each component comprises critical soil attributes, which are in dynamic equilibrium with the environment within a management system. Important attributes affecting soil physical quality are texture, structure, porosity, and pore size distribution, available water holding capacity, infiltration capacity, internal drainage, effective rooting depth, and soil temperature (Lowery et al., 1996; Arshad, Lowery, and Grossman, 1996; Topp et al., 1997). Some of these

TABLE 1. Some definitions of soil quality.

Definition	Reference
(i) Sustained capability of a soil to accept, store and recycle water, nutrients and energy	Anderson & Gregorich (1984)
(ii) Inherent attributes of soil inferred from soil characteristics	SSSA (1987)
(iii) The ability of soil to support crop growth	Powers & Myers (1989)
(iv) The capacity of soil to function in a productive and sustained manner	NCR-59 (1991)
(v) The capacity of soil to function within ecosystem boundaries	Larson & Pierce (1991)
(vi) The capacity of soil to produce safe and nutritious crops	Parr et al. (1992)
(vii) Inherent attributes of soil and characteristics and processes that determine the soil's capacity to produce economic goods and services and regulate the environment	Lal (1993)
(viii) The capacity of soil to function within ecosystem boundaries to sustain biological productivity, maintain environmental quality, and promote plant and animal health	Doran & Parkin (1994)
(ix) Soil's capacity or fitness to support growth without resulting in soil degradation or otherwise harming the environment	Gregorich & Acton (1995)
(x) Soil quality is the capacity of a specific kind of soil to function, within natural or managed ecosystem boundaries, to sustain plant and animal productivity, maintain or enhance water and air quality, and support human health and habitation	SSSA (1995)
(xi) Productivity and environment moderation capacity	Lal (1997)

TABLE 2. Soil quality attributes related to specific soil functions.

Function	Soil quality attributes
1. Crop yield and biomass production	Soil structure, soil organic carbon (SOC) content, texture, least limiting water range or available water capacity, porosity and pore size distribution, rooting depth, cation/anion exchange capacity (CEC/AEC), pH, electrical conductivity, nutrient retention and supply capacity
2. Water quality	SOC and microbial biomass carbon (MBC), clay content, soil structure, erodibility, infiltration rate, CEC/AEC, pH
3. Greenhouse gas emissions	SOC and MBC contents, activity and species diversity of soil micro and macro fauna and flora, structure, aeration, anaerobiosis, soil temperature, and moisture regimes
4. Structural foundation	Uniformity coefficient, soil strength, bearing capacity, permeability, deformation

FIGURE 1. Inter-dependence between soil quality, water quality, and air quality.

properties (e.g., texture) cannot be modified. Others (e.g., structure, water retention, infiltration capacity) can be managed. Important attributes of soil chemical quality are pH, cation exchange capacity (CEC)/anion exchange capacity (AEC), nutrient retention and availability, toxicity of some elements (e.g., Al, Mn) and high concentration of soluble salts as indicated by electrical conductivity of saturated paste (Heil and Sposito, 1997; Doran and Jones, 1996). Most attributes affecting soil chemical quality can be altered by soil management techniques. Important attributes of soil biological quality are soil organic carbon (SOC) and microbial biomass carbon (MBC) contents, soil respiration, activity and species diversity of soil fauna and flora including that of earthworms and termites (Rice, Moorman, and Beare, 1996; Parkin, Doran, and Franco-Vizcaino, 1996; Gregorich et al., 1997). There are some inherent characteristics (Figure 2) that involve attributes of all three components, and it is their interaction with environment and management that defines quality of a soil for a specific function. Some argue that SOC content is as vital for maintaining soil quality as blood is for the human body (Martius, Tiessesn, and Vlek, 1999). Albrecht (1938) stated that "soil organic matter (SOC) is one of our most important natural resources; its unwise exploitation has been devastating; and it must be given its proper rank in any conservation policy as one of the major factors affecting the level of crop production in the future." Indeed, SOC con-

FIGURE 2. Components of soil quality.

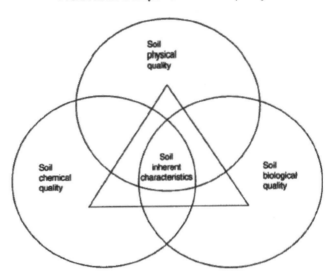

tent and its dynamics may be a key indicator of all three components of soil quality.

DETERMINANTS OF SOIL QUALITY

There are several determinants of soil quality. Important among these are SOC content, soil structure and aggregation, water and nutrient retention capacity and the ability of the soil to restore its life support processes following a drastic perturbation. Soil resilience, an important attribute of soil quality, depends on a wide range of endogenous and exogenous factors (Lal, 1997). Two principal soil functions, biomass productivity, and environment moderating capacity, depend on these attributes of soil quality.

Soil quality (S_q), as a management objective, has to be determined quantitatively and objectively. For agronomic functions, principal attributes, and processes affecting soil quality can be determined quantitatively as per Equation 1 (Lal, 1993),

$$S_q = f(P_i, S_c, R_d, e_d, N_c, B_d)_t \qquad \text{(Eq. 1)}$$

where P_i is productivity determined in appropriate units, S_c is an index of structural characteristics including porosity and pore size distribution, R_d is rooting depth, e_d is charge density (e.g., CEC, AEC), N_c is nutrient reserve, B_d is a measure of soil biodiversity including SOC and MBC contents, and t is time. Specific parameters involved (such as those affecting crop yield) in developing such a function may be specific to soil and cropping system (Equation 2) (Lal, 1997),

$$Y = f(SOC, S_c, Rd, e_d.N_c, B_d) \qquad (Eq. 2)$$

where Y is crop yield. Critical limits of a specific parameter, beyond which crop yield or biomass production declines sharply, may also be specific to soil and cropping systems.

There exists a close relationship between soil quality and soil resilience (S_r) (Lal, 1997). Resilient soils have high soil quality and vice versa (Equation 3).

However, indicators of soil resilience

$$S_r \alpha S_q \qquad (Eq. 3)$$

may be different than that of the soil quality, and denote the magnitude and ease of change in key attributes (Equation 4) (Lal, 1997),

$$S_r = f(SOC', MBC', S_c', R_d', e_d', N_c', B_d')t \qquad (Eq. 4)$$

where the symbol prime (e.g., SOC') denotes rate of change in the specific variable because of the management input. Soil quality and resilience also differ in terms of the critical limits and threshold value of key soil properties. Both soil quality and soil resilience can be managed. Therefore, soil quality enhancement will remain a major goal of soil and crop management during the 21st century (Lal, 1998, 1999).

Agronomic yield is an indirect but an integrative indicator of soil quality. High yields are obtained on soils of high intrinsic quality. There are large differences in yield of crops on a continental basis (Table 3). The average yield of corn (*Zea mays* L.) ranges from 1556 kg ha^{-1} in Africa to 6492 kg ha^{-1} in North-Central America, with yields in North America being 4.2 times those in Africa. Similarly, the ratio of average yield to the maximum yield on a continental basis is 1:1.74 for wheat (*Triticum aestivum* L.), 1:2.43 for rice (*Oryza sativa* L.), 1:2.34 for millet, 1:5.50 for sorghum (*Sorghum bicolor* L. Moench.), and 1:2.71 for soybean (*Glycine max* L. Merr). The least average yield of all crops in the world is reported for Africa, and the low yield there is also in accord with the large area affected by water erosion and the prevalence of severe soil chemical degradation (e.g., nutrient depletion). It is apparent to realize, therefore, that improvements in crop yields and agronomic productivity in Africa necessitate reversal of soil degradative trends and enhancement of soil quality on a continental scale.

Impacts of soil quality on changes in crop yield in the U.S. during the 20th century are shown in Table 4. During the 20th century crop yield increased by a factor of 5.6 for corn, 3.2 for wheat, 2.2 for soybean, 3.9 for rice, and 3.8 for peanuts (*Arachis hypogaea* L.) (Table 4). This impressive increase in crop yields is attributed to a combination of both biological (genetic improvement) and managerial factors. The managerial factors, with strong impact on soil quality, include mechanical system of soil management (e.g., mulch or conservation tillage), chemical input (e.g., fertilizer, pesticides) and crop management (e.g., crop rotations and systems) (Paarlberg and Paarlberg, 2000). In addition, water management (including both drainage and irrigation) also played an important role in enhancing agronomic productivity.

TABLE 3. Average crop yield of different crops in 1998 (FAO, 1998).

Region	Wheat	Rice	Corn	Millet	Sorghum	Soybean
	--------------------------------- kg ha^{-1} ---------------------------------					
World	2624	3747	4395	777	1428	2240
Africa	1836	2183	1556	642	886	965
Asia	2520	3824	3886	979	1162	1353
NC America	2740	5313	6492	1501	3625	2619
S. America	2090	3189	2853	1211	3542	2472
Europe	3203	5188	5025	606	4870	1984
Oceania	2112	9755	6379	1167	1874	1788

TABLE 4. Average yield of crops in the U.S. during the 20th century (Paarlberg and Paarlberg, 2000; USDA, 2000).

Year	Corn	Wheat	Soybean	Rice	Peanuts
	---------------------------------- kg ha^{-1} ----------------------------------				
1900	1696	969	--	1680	--
1910	1633	949	--	1881	862
1920	1696	942	--	2072	793
1930	1507	895	1084	2442	793
1940	2135	1151	1272	2330	788
1950	2763	1326	1440	3147	1093
1960	4459	1777	1676	4536	1702
1970	5589	2113	1878	5096	2666
1980	6657	2409	2039	5757	2778
1990	7599	2530	2456	6451	2677
2000	8402	2873	2422	6625	2987
Multiples of increased yield	5.6	3.2	2.2	3.9	3.8

SOIL DEGRADATION

Soil degradation (S_d) is reverse or decline of soil quality (Equation 5) (Lal, 1997).

$$S_d = -dS_q/dt \qquad \text{(Eq. 5)}$$

It refers to the decline in soils inherent capacity to produce economic goods and provide ecological services. Similar to soil quality, there are also three types of soil degradation: physical, chemical, and biological degradation (Figure 3; Lal, Hall, and Miller, 1989). Soil physical degradation includes decline in soil structure with attendant ramifications of crusting, compaction, anaerobiosis, accelerated soil erosion, and desertification. Important chemical processes include acidification, leaching, nutrient/elemental imbalance, salinization, reduction in CEC/AEC, and depletion of soil fertility. Soil biological degradation encompasses reduction in SOC and MBC and decline in soil biodiversity.

Soil degradation is a biophysical process but it is driven by socioeconomic and political causes. Important socioeconomic causes of soil degradation are land tenure, access to market, poverty, institutional support and human health. Political stability, policies and regulations are important political causes affecting human health (Lal, 1997).

The data in Table 5 show estimates of the strong and extreme forms of soil

FIGURE 3. Different types of soil degradation processes with adverse impact on soil quality.

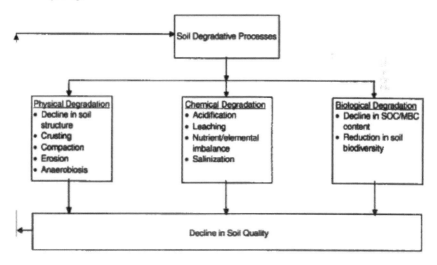

TABLE 5. Global extent of strong and extreme forms of soil degradation (adapted from Oldeman, 1994).

Region	Water erosion	Wind erosion	Physical degradation*	Chemical degradation*
			10^6 ha	
Africa	102	9	19	62
Asia	73	15	12	74
S. America	12	--	8	70
C. America	23	1	5	7
N. America	--	1	1	--
Europe	12	1	36	26
Oceania	--	--	2	1
World	223	26	83	240

*Total degradation

degradation estimated to affect land area of 223 million ha by water erosion and 26 million ha by wind erosion. In addition, land area affected by all levels of severity (e.g., slight, moderate, severe, and strong) is estimated at 83 million ha for physical degradation (e.g., compaction, crusting, waterlogging) and 240 million ha for chemical degradation (e.g., nutrient depletion, salinization, etc.) (Oldeman, 1994). Degradative trends can be reversed through soil quality enhancement using soil's resilience through adoption of appropriate cropping systems and the attendant management. There is a close relationship between soil degradation, soil quality, and soil resilience (Lal, 1997).

SOIL QUALITY AND GLOBAL HUNGER

Soil quality has a direct bearing on productivity, food security, and human health. All other factors remaining the same, agronomic/biomass productivity decreases with increase in extent and severity of soil degradation. Despite impressive gains in agricultural production during the second half of the 20th century, the global problem of hunger and malnourishment persists, especially in sub-Saharan Africa and South Asia. The number of undernourished people in all developing countries was 918 million (35% of the total population) in 1970, 906 million (28%) in 1980, 841 million (20%) in 1990 and is estimated to be 680 million (12%) in 2010 (Table 6) (Kracht, 1998; Kracht and Schulz, 1998; Blanckenburg, 1998). Although food security depends on numerous factors (e.g., access to food as determined by income, availability of food, etc.), the importance of agronomic productivity as influenced by soil quality cannot be overemphasized.

Soil resources are finite, unequally distributed among ecological regions and prone to degradation. The global arable land area is about 1379 million ha,

of which 268 million ha or about 20% is irrigated (Table 7). Because of the
rapid increase in population, especially of those in developing countries, the
per capita arable land area is declining even without considering the risks of
soil degradation. The per capita arable land area in 1995 was 0.23 ha in the
world, 0.23 ha in Africa, 0.20 ha in Latin America, and 0.12 ha in Asia. By
2050, even if there is no soil degradation and conversion to non-agricultural
uses, the per capita arable land area will decrease to 0.14 ha in the world, 0.08
ha in Africa, 0.11 ha in Latin America, and 0.07 ha in Asia (Lal, 2000). Meeting
basic necessities of life (e.g., food, feed, fuel, and fiber) from a small per capita
arable land area necessitates adoption of those improved cropping systems
which enhance/maintain soil quality to ensure high and sustained productivity.

TABLE 6. Trends in number and percentage of people undernourished (adapted
from Foster and Leathers, 1999; Kracht, 1998).

Region	1970 10^6	1970 % of population	1980 10^6	1980 % of population	1990 10^6	1990 % of population	2010 10^6	2010 % of population
Sub-Saharan Africa	103	38	148	41	215	43	264	30
Near East and North Africa	48	27	27	12	37	12	53	10
East and Southeast Asia	476	41	379	27	269	16	123	6
South Asia	238	33	303	34	255	22	200	12
Latin America and Caribbean	53	19	48	14	64	15	40	7
All developing countries	918	35	906	28	841	20	680	12

TABLE 7. Global trends in agricultural land use (FAO, 1998; Foster and Leathers,
1999; Postel, 1999).

Year	Arable land (10^6 ha)	Irrigated land (10^6 ha)
1960	1266	142
1965	1280	162
1970	1302	182
1975	1313	201
1980	1332	220
1985	1348	231
1990	1357	242
1995	1376	260
1997	1379	268

CROPPING SYSTEMS AND SOIL QUALITY

There are two principal strategies of enhancing soil quality (Figure 4). One, restoring degraded soils by conversion of highly erodible lands to a non-agricultural (forestry, natural vegetation) uses. For example, the Conservation Reserve Program (CRP), adopted on some 16 million ha of highly erodible land, has reduced soil erosion and improved soil quality (Follett et al., 2001). Two, adopting improved cropping systems along with recommended management practices for erosion control, soil structure improvements, SOC enhancement, and strengthening the nutrient cycling mechanism (Figure 4).

Cropping systems play an important role in improving soil quality. In a narrow sense, "cropping system" implies a temporal sequence in which different crops are grown on the same land. In this context, the term cropping system is synonymous with "crop rotation." In a broad sense, however, cropping system implies both temporal sequences of crop and the management (soil and crop) practices adopted to grow them (Figure 5). Important among these are tillage methods and crop residue management, drainage and irrigation, integrated nutrient management, erosion control, integrated pest management, and crop rotations and cover crops. The ideal combination of these components is specific to soil and ecoregion, and has to be researched and fine-tuned under site-specific conditions.

Three components outlined in Figure 4 require special mention. These are conservation tillage, soil fertility and nutrient management, and growing cover crops.

FIGURE 4. Technological options for soil fertility enhancement.

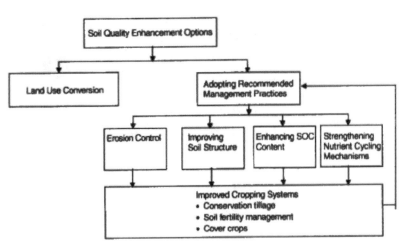

FIGURE 5. Components of cropping systems that affect soil quality and agronomic productivity (INM = Integrated nutrient management, IPM = Integrated pest management).

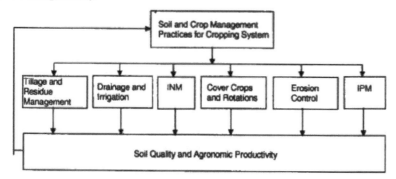

Conservation Tillage

Soil quality can be enhanced by decreasing the intensity and frequency of mechanical disturbance to soil by tillage operations, and by increasing the amount of crop residues and biomass left on the soil surface. Conservation tillage implies any tillage method that reduces soil erosion and maintains at least 30% of the ground cover by returning crop residue on the soil surface (Uri, 1999; Michalson, Papendick, and Carlson, 1999). A continuous use of conservation tillage reduces soil erosion, conserves water, and increases SOC content in the surface layer (Lal et al., 1999). The proportion of cropland area under conservation tillage in the U.S. has increased from 3.8% in 1970 to 36.7% in 2000, and the area has remained practically constant since 1995 (Table 8). Despite numerous advantages (e.g., saving in energy input, C sequestration, etc.), crop yields can be low under conservation tillage due to weed infestation, poor drainage, and low soil temperatures during spring. Alleviating these constraints may increase the area and also lead toward a continuous rather than rotational use of conservation tillage.

Soil Fertility Management

Soil fertility management is a critical component of any cropping system designed to enhance and sustain productivity. Increase in crop yields in the second half of the 20th century (Table 4) necessitated increasing use of the chemical fertilizers. Total elemental use of chemical fertilizers in the U.S., of about 20 million tons of NPK (Table 9), is the highest national fertilizer use in the world. Excessive and indiscriminate use of fertilizers and other chemicals have adverse impacts on water quality and flux of trace gases from soil to the

TABLE 8. Percent of planted area in conservation tillage in the USA (modified from Uri, 1999; CTIC, 2000).

Year	% of planted area
1970	3.8
1975	6.3
1980	12.5
1985	23.8
1990	26.1
1992	31.4
1994	35.0
1995	35.5
1996	35.8
1997	37.3
1998	37.2
2000	36.7

TABLE 9. Use of chemical fertilizers in the USA (adapted from Paarlberg and Paarlberg, 2000; IFDC, 2000).

Year	N	P	K	Total NPK
			1,000 tons	
1930	378	794	354	1,526
1940	419	912	435	1,766
1950	1,237	2,110	1,381	4,728
1960	3,024	2,617	2,179	7,820
1970	7,459	4,576	4,040	16,075
1980	11,502	4,642	5,511	21,655
1990	10,048	3,841	4,537	18,426
1995	11,111	4,090	4,738	19,939
1997	11,162	4,195	4,847	20,204

atmosphere. Thus, there is a strong need to reduce dependence on chemical fertilizers while maintaining or enhancing productivity through improving soil quality. The amount of fertilizer use can be reduced by: (i) improving efficiency through improved formulations, mode and time of application, etc., (ii) decreasing losses due to erosion, leaching and volatilization, and (iii) strengthening nutrient recycling mechanisms. Identifying alternatives to chemical fertilizers is another important option. Judicious use of biosolids (Rechcigl and MacKinnon, 1997; Sopper, 1993) can complement chemical fertilizers. Organic fertilizers derived from animal products contain 1 to 5% N, 0.4 to 2% P and 1 to 3% K (Barker, 1997; Huntley, Barker, and Stratton, 1997). The amount of potentially compostable organic materials produced in the U.S. is about 1 billion tons per year. Properly used, biosolids can be a great resource in enhancing soil quality and decreasing the use of chemical fertilizers. Biological nitrogen fixation, through use of legumes in the rotation cycle, is another important strategy of reducing dependence on chemical fertilizers.

Cover Crops

Cover crops are legumes, cereals or an appropriate mixture grown specifically to protect the soil against erosion, ameliorate soil structure, enhance soil fertility and suppress pests (Lal et al., 1991). Cover crops are grown to fill gaps in either time or space when grain crops would leave the ground bare. In the U.S., most cover crops are grown during the cold season. Cover crops planted during the fall include rye (*Secale cereale* L.), clover (*Trifolium* spp.), or vetch (*Coronilla* and *Vicia* spp.). Incorporation of cover crops in the rotation cycle reduces incidence of pests, improves soil structure, controls erosion and enhances soil quality. Effectiveness of conservation tillage is improved by incorporation of cover crop in the rotation cycle. An important environmental benefit of growing cover crops is soil carbon sequestration to reduce net emissions of greenhouse gases into the atmosphere.

Productivity of a rotation cycle can be measured by several indices including total biomass, agronomic yield, economic yield, resource use efficiency, energy flux, thermodynamics of input and output (Lal, 1994). With frequent use of cover crops (planted fallows) in the rotation cycle, the land use efficiency of a cropping system can be assessed by one or the other of the following indices:

Land use factor (L): The factor L is defined as the ratio of cropping period C plus fallow (cover crop) period to the cropping period (Okigbo, 1978) (Equation 6).

$$L = \frac{C+F}{C}$$

(Eq. 6)

The value of actor L decreases with more frequent use of the cover crops in the rotation cycle. Another index of productivity assessment for computing efficiency of mixed crops is the Land Equivalent Ratio (LER). The mixture may consist of cereals with legumes, cereals with root crops or with perennials. The LER is calculated by Equation 7 (Willey and Osiru, 1972).

$$LER = \sum_{i=1}^{n} \left[\frac{Y_i}{Y_m} \right]$$

(Eq. 7)

Where Y_i and Y_m are yields of component crops in the intercrop and monoculture system, respectively, and n is the number of crops involved. Since crops involved have different maturity period, the Area Time Equivalent Ratio

(ATER) is used to assess productivity of a cropping system (Hiebsch and McCollum, 1987) (Equation 8).

$$\text{ATER} = \frac{1}{t}\left[\overset{n}{\underset{i=1}{\bullet}} \left[\frac{dYi}{Ym} \right] \right] \qquad \text{(Eq. 8)}$$

Where d is the growth period of crop in days and t is the time in days for which the field remains occupied, which is the growth period of the longest duration crop. If t = d, ATER = LER.

CONCLUSION

In the old Roman Empire, all roads led to Rome. In a mission to enhance agricultural production and improve environment, all roads lead back to soil quality and its management. Soil quality must be managed, improved, and restored. Decline in soil quality undermines soil resilience and accentuates soil degradation. Important indicators of soil quality may be SOC content, soil structure, nutrient and water retention capacity. These indicators are specific to soil type and functions. Crop yield and biomass productivity depends on soil quality. An important objective of soil management is to enhance soil quality.

Cropping systems play an important role in maintaining and enhancing soil quality. Choice of appropriate cropping systems is critical to maintaining/enhancing soil quality and agronomic sustainability. Global food security depends on the premise of reversing soil degradative trends and improving soil quality. In addition to crop rotation or sequence of cropping, three principal components of cropping systems in relation to soil quality are: (i) conservation tillage and residue management to conserve soil and water, (ii) soil fertility management through use of chemical fertilizers, biosolids, and biological nitrogen fixation, and (iii) growing cover crops to provide ground cover. Improving soil quality through judicious management can reduce dependence on fertilizers and other chemicals. In addition to enhancing productivity, choice of appropriate cropping systems improves soil quality, sequesters C and improves environment, and sustains agronomic productivity.

REFERENCES

Albrecht, W. (1938). *Loss of Soil Organic Matter and Its Restoration*. Soils and Men. Yearbook of Agriculture. Washington, DC: USDA, pp. 347-360.

Anderson, D.W. and E.G. Gregorich. (1984). Effect of soil erosion on soil quality and productivity. In *Soil Erosion and Degradation*, Proceedings of the Second Annual

Western Provincial Conference on Rationalization of Water and Soil Research and Management, Saskatchewan, Canada, pp. 2-10.

Arshad, M.A., B. Lowery, and B. Grossman. (1996). Physical tests for monitoring soil quality. In *Methods for Assessing Soil Quality*, ed., J.W. Doran and A.J. Jones, Special Publication No. 49, Madison, WI: Soil Science Society of America, pp. 123-141.

Barker, A.V. (1997). Composition and uses of compost. In *Agricultural Uses of By-Products and Wastes*, ed., J.E. Rechcigl and H.C. MacKinnon, Washington, DC: American Chemical Society, pp. 140-162.

Bezdicek, D.F., R.I. Papendick, and R. Lal. (1996). Importance of soil quality to health and sustainable land management. In *Methods for Assessing Soil Quality*, ed., J.W. Doran and A.J. Jones, Special Publication No. 49, Madison, WI: Soil Science Society of America, pp. 1-7.

Blanckenburg, P.V. (1998). The feeding capacity of the planet earth: Development, potentials and restrictions. In *Food Security and Nutrition*, ed., U. Kracht and M. Schulz, New York, NY: St. Martin's Press, pp. 91-106.

Carter, M.R., E.G. Gregorich, D.W. Anderson, J.W. Doran, H.H. Janzen, and F.J. Pierce. (1997). Concepts of soil quality and their significance. In *Soil Quality for Crop Production and Ecosystem Health*, ed., E.G. Gregorich and M.R. Carter, Amsterdam: Elsevier, pp. 1-19.

CTIC (2000). National Crop Residue Management Survey. Conservation Tillage Information Center, National Association of Conservation Districts Lafayette, IN.

Doran, J.W. and T.B. Parkin. (1994). Defining and assessing soil quality. In *Defining Soil Quality for a Sustainable Environment*, ed., J.W. Doran, D.C. Coleman, D.F. Bezedick, and B.A. Stewart, Special Publication No. 35, Madison, WI: Soil Science Society of America, pp. 3-21.

Doran, J.W., M. Sarrantonio, and M.A. Lieberg. (1996). Soil health and sustainability. *Advances in Agronomy* 56:1-54.

Doran, J.W. and A. Jones (1996). *Methods for Assessing Soil Quality*. Special Publication No. 49, Madison, WI: Soil Science Society of America.

Engelman, R. and P. LeRoy (1995). *Conserving Land: Population and Sustainable Food Production*. Washington, DC: Population Action International.

FAO (1998). *Yearbook on Agriculture*, Rome, Italy: Food and Agriculture Organization.

Follett, R.F., E. Samson-Liebig, J.M. Kimble, E.G. Pruessner, and S. Waltman. (2001). Carbon sequestration under CRP in the historic grassland soils of the USA. In *Soil Carbon Sequestration and the Greenhouse Effect*, ed., R. Lal, Special Publication No. 57, Madison, WI: Soil Science Society of America, pp. 27-39.

Foster, P. and H.D. Leathers. (1999). *The World Food Problem: Tackling the Causes of Undernutrition in the Third World*. Boulder, CO: Lynne Reimer Publishers.

Gregorich, E.G. and D.F. Acton. (1995). Understanding soil's health. In *The Health of Our Soils: Towards Sustainable Agriculture in Canada*, ed., D.F. Acton and L.J. Gregorich, Ottawa, Canada: Center for Land and Biological Resources Research, Research Branch, Agriculture and Agri-Food Canada, pp. 5-10.

Gregorich, E.G., M.R. Carter, J.W. Doran, C.E. Pankhurst, and L.M. Dwyer. (1997). Biological attributes of soil quality. In *Soil Quality for Crop Production and Eco-*

system Health, ed., E.G. Gregorich and M.R. Carter, Amsterdam: Elsevier, pp. 81-113.

Heil, D. and G. Sposito. (1997). Chemical attributes and processes affecting soil quality. In *Soil Quality for Crop Production and Ecosystem Health*, ed., E.G. Gregorich and M.R. Carter, Amsterdam: Elsevier, pp. 59-79.

Hiebsch, C.K. and R.E. McCollum. (1987). Area x Time Equivalency Ratio: A method for evaluating the productivity of intercrops. *Agronomy Journal* 79:15-22.

Huntley, E.E., A.V. Barker, and M.L. Stratton. (1997). Composition and uses of organic fertilizers. In *Agricultural Uses of By-Products and Wastes*, ed., J.E. Rechcigl and H.C. MacKinnon, Washington, DC: American Chemical Society, pp. 120-139.

IFDC (2000). Fertilizer use statistics. Muscle Shoals, AL, International Fertilizer Development Center.

Jenny, H. (1980). *The Soil Resource*. New York, NY: Springer-Verlag.

Karlen, D.L., M.J. Mausbach, J.W. Doan, R.G. Cline, R.F. Harris, and G.E. Schuman. (1997). Soil quality: A concept, definition and framework for evaluation. *Soil Science Society of America Journal* 61:4-10.

Kracht, U. (1998). Hunger, malnutrition and poverty: trends and prospects towards the 21st century. In *Food Security and Nutrition*, ed., U. Kracht and M. Schulz, New York, NY: St. Martin's Press, pp. 55-74.

Kracht, U. and M. Schulz. (1998). *Food Security and Nutrition*. New York, NY: St. Martin's Press.

Lal, R. (1987). *Tropical Ecology and Physical Edaphology*. Chichester, UK: John Wiley and Sons.

Lal, R. (1993). Tillage effects on soil degradation, soil resilience, soil quality and sustainability. *Soil and Tillage Research* 27:1-8.

Lal, R. (1994). Methods and Guidelines for Assessing Sustainable Use of Soil and Water Resources in the Tropics. SMSS Tech. Monograph 21, The Ohio State University, Columbus, OH, 73 pp.

Lal, R. (1997). Degradation and resilience of soils. *Philosophical Transactions of Royal Society (Biology) London*, 352:997-1010.

Lal, R. (1998). *Soil Quality and Agricultural Sustainability*. Chelsea, MI: Ann Arbor Press.

Lal, R. (1999). *Soil Quality and Soil Erosion*. Boca Raton, FL: CRC Press.

Lal, R. (2000). Soil management in the developing countries. *Soil Science* 165: 57-72.

Lal, R. (2001). Managing world soils for food security and environment quality. *Advances in Agronomy* 74: 155-192.

Lal, R., G.F. Hall, and F.P. Miller. (1989). Soil degradation. I. Basic processes. *Land Degradation and Rehabilitation* 1:51-69.

Lal, R., E. Regnier, D.J. Eckert, W.M. Edwards, and R. Hammond. (1991). Expectations of cover crops for sustainable agriculture. In *Cover Crops for Lean Water*, ed. W. Hagrove, Ankeny, IA: Soil Water Conservation Society, pp. 1-10.

Lal, R., R.F. Follett, J. Kimble, and C.V. Cole. (1999). Managing U.S. cropland to sequester carbon in soil. *Journal of Soil Water Conservation* 54:374-381.

Larson, W.E. and F.J. Pierce. (1991). Conservation and enhancement of soil quality. In *Evaluation for Sustainable Land Management in the Developing World*. Bangkok, Thailand: IBSRAM.

Lowery, B., M.A. Arshad, R. Lal, and W.J. Hickey. (1996). Soil water parameters and soil quality. In *Methods for Assessing Soil Quality*, ed., J.W. Doran and A.J. Jones, Special Publication No. 49, Madison, WI: Soil Science Society of America, pp. 143-155.

Martius, C., H. Tiessen, and P. Vlek. (1999). The challenge of management soil organic matter in the tropics. ZEF news #2 (Sept. 1999), Bonn, Germany: University of Bonn, pp. 3-4.

Michalson, E.L., R.I. Papendick, and J.E. Carlson. (1999). *Conservation Farming in the United States: The Methods and Accomplishments of the STEEP Program.* Boca Raton, FL: CRC Press.

NRC-59. (1995). Soil quality. Cited by J.W. Doran and T.B. Parkin "Defining and Assessing Soil Quality." Special Publication No. 35, Madison, WI: Soil Science Society of America, pp. 3-21.

Okigbo, B.N. (1978). Cropping systems and related research in Africa. AAAS Occasional Publication Series OT-1, Addis Ababa, Ethiopia, 81 pp.

Oldeman, L.R. (1994). The global extent of soil degradation. In *Soil Resilience and Sustainable Land Use*, ed., I. Szabolcs and D.J. Greenland, Wallingford, UK: CAB International, pp. 99-118.

Paarlberg, D. and P. Paarlberg. (2000). *The Agricultural Revolution of the 20th Century.* Ames, IA: Iowa State University Press.

Papendick, R.I. and J.F. Parr. (1992). Soil quality–the key to a sustainable agriculture. *American Journal of Alternative Agriculture* 7:2-3.

Parkin, T.B., J.W. Doran, and E. Franco-Vizcaino. (1996). Field and laboratory tests of soil respiration. In *Methods for Assessing Soil Quality*, ed., J.W. Doran and A.J. Jones, Special Publication No. 49, Madison, WI: Soil Science Society of America, pp. 231-245.

Parr, J.F., R.I. Papendick, S.B. Hornick, and R.E. Meyer. (1992). Soil quality: Attributes and relationship to alternative and sustainable agriculture. *American Journal of Alternative Agriculture* 7: 5-11.

Postel, S. (1999). *Pillar of Sand: Can the Irrigation Miracle Last?* New York, NY: W.W. Norton.

Powers, J.F. and R.J.K. Myers. (1989). The maintenance or improvements of farming systems in North America and Australia. In *Soil Quality in Semi-Arid Agriculture*, ed., J.W.B. Stewart, Proceedings of an International Conference Sponsored by the Canadian International Development Agency, Saskatoon, Saskatchewan, Canada, 11-16 June 1989, University of Saskatchewan, Saskatoon, Canada, pp. 273-292.

Rechcigl, J.E. and H.C. MacKinnon. (1997). *Agricultural Uses of By-Products and Wastes.* Washington, DC: American Chemical Society.

Rice, C.W., T.B. Moorman, and M. Beare. (1996). Role of microbial biomass carbon and nitrogen in soil quality. In *Methods for Assessing Soil Quality*, ed., J.W. Doran and A.J. Jones, Special Publication No. 49, Madison, WI: Soil Science Society of America, pp. 203-215.

Richter, J. (1987). *The Soil as a Reactor: Modelling Processes in the Soil.* Cremlingen, West Germany: Catena-Verlag.

Sojka, R.E. and D.R. Upchurch. (1999). Reservations regarding the soil quality concept. *Soil Science Society of America Journal* 63:1039-1054.

Sopper, W.E. (1993). *Municipal Sludge Use in Reclamation.* Boca Raton, FL: Lewis Publishers.

SSSA (1987). *Glossary of Soil Science Terms,* Madison, WI: Soil Science Society of America.

SSSA (1995). Statement on Soil Quality. Agronomy News, June 1995, Madison, WI: American Society of Agronomy/Crop Science Society of America/Soil Science Society of America.

Tisdale, S.L. and W.L. Nelson. (1966). *Soil Fertility and Fertilizers.* Third Edition, New York, NY: Macmillan.

Topp, G.C., W.D. Reynolds, F.J. Cook, J.M. Kirby, and M.R. Carter. (1997). Physical attributes of soil quality. In *Soil Quality for Crop Production and Ecosystem Health,* ed., E.G. Gregorich and M.R. Carter, Amsterdam: Elsevier, pp. 21-58.

Uri, N.D. (1999). *Conservation Tillage in U.S. Agriculture: Environmental, Economic and Policy Issues.* Binghamton, NY: Food Products Press.

USDA (2000). *National Statistics.* Washington, DC: United States Department of Agriculture.

Willey, R.W. and D.S.O. Osiru. (1972). Studies on mixtures of maize and beans with particular reference to plant population. *Journal of Agricultural Science* (Cambridge) 79:519-529.

The Role of Cover Crops
in North American Cropping Systems

Marianne Sarrantonio
Eric Gallandt

SUMMARY. The benefits of cover crops in cropping systems have long been recognized. Legumes have historically been used to provide biologically fixed nitrogen to cash crops, and it has been shown that soil erosion can be slowed significantly with even minimal amounts of soil cover during vulnerable times of year. The role of cover crops in North American farming systems is expanding to include management of weeds, disease and pests, and overall enhancement of soil quality through organic matter enrichment, improved nutrient cycling and reduction of soil compaction. While the predominant temporal niche for cover crops in North America remains the winter, other opportunities in diverse cropping systems exist for cover crop inclusion, such as summer fallow, living mulches or full-year fallow crops. To date, the use of cover crops is constrained by economic, biological, and farm operational factors, but farmer education, continued research, and government policy changes can aid in overcoming existing barriers to adoption. *[Article copies available for a fee from The Haworth Document Delivery Service: 1-800-HAWORTH. E-mail address: <docdelivery@haworthpress.com> Website: <http://www. HaworthPress.com> © 2003 by The Haworth Press, Inc. All rights reserved.]*

Marianne Sarrantonio and Eric Gallandt are Assistant Professors, Department of Plant, Soil, and Environmental Sciences, University of Maine, Orono, ME.

Address correspondence to: Marianne Sarrantonio, Assistant Professor, Department of Plant, Soil, and Environmental Sciences, University of Maine, 102 Deering Hall, Orono, ME 04473 (E-mail: mariann2@maine.edu).

[Haworth co-indexing entry note]: "The Role of Cover Crops in North American Cropping Systems." Sarrantonio, Marianne, and Eric Gallandt. Co-published simultaneously in *Journal of Crop Production* (Food Products Press, an imprint of The Haworth Press, Inc.) Vol. 8, No. 1/2 (#15/16), 2003, pp. 53-74; and: *Cropping Systems: Trends and Advances* (ed: Anil Shrestha) Food Products Press, an imprint of The Haworth Press, Inc., 2003, pp. 53-74. Single or multiple copies of this article are available for a fee from The Haworth Document Delivery Service [1-800-HAWORTH. 9:00 a.m. - 5:00 p.m. (EST). E-mail address: docdelivery@ haworthpress.com].

KEYWORDS. Erosion control, nitrogen, soil quality, weed management

INTRODUCTION

A cover crop is defined as a crop grown primarily to cover the soil in order to protect it from soil erosion and nutrient losses between periods of regular crop production, or between trees and vines in orchards and vineyards (Pieters and McKee, 1938; Brady and Weil, 1999). The term is also commonly used to include legumes grown as "green manures" to enhance the short-term nitrogen (N) fertility of the soil, crops used to catch or re-cycle excess soluble nutrients in the soil, those used to improve overall soil quality through organic matter (OM) addition or by the action of deep roots, competitive crops used to suppress weeds, and even crops used to enhance the activity of biological control agents, such as predatory insects, in a cropping system. The use of cover crops has a long and rich history in agriculture. As early as the 5th century BC there were reports in China of crops whose benefits to the soil were as "good as silk-worm excrement," and Virgil advised Roman farmers to sow their grain "where grew the bean, the slender vetch, or the fragile stalks of the bitter lupine" (Pieters, 1927). Earliest reports of green manuring in England occurred in the days of Jethro Tull (the original one–18th century) when clovers and other legumes began to replace the fallow period in grain-turnip rotations (Russell, 1913). Winter annual cover crops were regularly part of many North American cropping systems through the early 20th century (Pieters, 1927). The use of crops strictly for soil improvement declined after World War II in most industrialized countries due to the ready availability of inorganic N fertilizers and herbicides, as well as the push for intensification of cropping to feed a rapidly expanding world population. The practice of cover cropping has reemerged in recent decades in response to growing knowledge of ecological principles, increasing costs of agricultural inputs, and the global recognition of serious and widespread soil degradation. The objectives of this paper are to provide an overview of the potential role of cover crops in cropping systems, to describe some of the most common and successful systems utilizing cover crops in North America, and to identify some constraints which have inhibited their expanded use.

EROSION CONTROL

Cover crops contribute both directly and indirectly to reducing soil erosion rates. The presence of additional crop residue, whether dead or alive, serves to decrease raindrop impact and soil detachment, to physically stabilize topsoil with roots, and to create a tortuous path for surface water, slowing its momentum and ability to carry soil. Indirectly, the soil organic matter (SOM) and enhanced microbial activity associated with cover crops may, over time, increase soil aggregation and water infiltration rates (McVay, Radcliffe, and Hargrove, 1989; Roberson, Sarig, and Firestone, 1991; Friedman, 1993; Dapaah and Vyn, 1998), thus allowing water to move into, rather than across the soil. A recent survey of vegetable growers utilizing cover crops in western New York reported that the most often cited reason for using cover crops was for erosion control (Stivers-Young and Tucker, 1999). According to the revised Universal Soil Loss Equation (RUSLE) used to predict rates of erosion, soil residue cover (SRC) of as little as 10% can reduce erosion rates by about 30%; at 50% SRC soil erosion reduction can be greater than 80% (Moldenhauer and Langdale, 1995). While there are several other ways to reduce erosion, Nyakatawa, Reddy, and Lemunyon (2001) point out that manipulating the cropping management factor in the soil loss equation is the most cost effective and easiest method of reducing overall soil loss. The addition of a cover crop to a cropping system not only enhances the total residue cover, but also generally extends cover into the more vulnerable months of the year, when the soil might otherwise have little or no protection.

Cereal rye (*Secale cereale* L.), also known as winter or grain rye, has proven to be the most useful species throughout the colder areas of North America as a winter annual cover for erosion protection. The plant's ability to germinate and grow quickly in cool weather, along with its extensive and deep fibrous root system gives it an unrivaled capacity to provide soil cover after the summer cropping season is over. Growers who are too occupied in early fall harvesting the summer's crop can still generally establish rye into October as far north as USDA Hardiness Zone 5 (Sarrantonio, 1994). Although rye will germinate and grow in temperatures as low as 4°C, the cropping system still benefits when the rye is given a chance to produce more roots and ground cover by an earlier planting. In a study conducted in the Chesapeake Bay, fall ground cover development and root depth were reduced by about 50% when planting date of rye was delayed one month, from September 1 until October 1 (Brinsfield and Staver, 1991).

Some vegetable and field crop growers take advantage of residue produced by winter annuals as mulch for no-tillage summer crops, thus prolonging erosion control through the summer. Because the residues keep the soil from warming quickly, these conservation practices are more common in southern states of the US, particularly in corn (*Zea mays* L.) production (Frye, Smith, and Williams, 1985; Corak et al., 1987; Ott and Hargrove, 1989; Ranells and Wagger, 1991; Hanson et al., 1993), but also in sorghum (*Sorghum bicolor* L.) (Hargrove, 1986; Boquet and Dabney, 1991), soybean (*Glycine max* L.) (Eckert, 1988; Zhu et al., 1989), and cotton (*Gossypium hirsutum* L.) (Grisso,

Johnson, and Dumas, 1985; Giesler, Paxton, and Millhollon, 1993). With the exception of the soybean system, legumes are often the selected cover crops for no-tillage production, providing the added benefit of supplying N to the cash crop.

While in most cases the no-tillage cover crop is chemically or mechanically killed prior to seeding, researchers at the Rodale Institute developed a system for direct-seeding field corn into a living cover of hairy vetch (*Vicia villosa* Roth) in the spring using sharp coulters and extra weight on a standard no-tillage planter. The vetch could be killed by mowing 7 to 10 days later when the corn was emerging but still had its growing point underground. The cover crop thus had extra time to produce biomass, fix N, and control weeds.

The use of cover crops has been also expanded to no-tillage vegetable production. Researchers in Maryland found that a hairy vetch cover crop compared favorably with black plastic for tomato (*Lycopersicon esculentum* L.) production, in terms of both yield and economics (Kelly et al., 1995). Researchers at Virginia Polytechnic Institute have developed a no-tillage vegetable transplanter that can slice through mowed cereal rye and other cover crop residues to deliver tomato or other vegetable transplants into the soil, an important tool toward future development of no-tillage vegetable systems (Morse, 1999).

SOIL QUALITY ENHANCEMENT

Soil quality is a concept that arose in the early 1990s in an effort to define and measure soil ecological functional ability in a holistic way. A commonly used working definition of soil quality is "The capacity of the soil to function in a productive and sustained manner while maintaining or improving the resource base, environment, and plant, animal and human health" (Doran and Parkin, 1994). The commonly used indicators of soil quality encompass soil biological, chemical and physical measurables, including the following: soil texture, bulk density, water infiltration rates, water holding capacity, pH, electrical conductivity, soil organic carbon content, nutrient status, soil ecology, including biomass and activity of soil microbes and larger organisms. In an agricultural soil, nearly all of these indicators are affected by the crops grown and by the quantity and quality of plant residues returned to that soil on a yearly basis.

The inclusion of cover crops in a cropping system may provide multiple benefits toward improving overall soil quality (Doran, Sarrantonio, and Liebig, 1996). A grain cropping systems with a significant legume cover crop component at the Rodale Institute Farming System Trial exhibited higher OM content and microbial biomass (Wander et al., 1994), greater water stable aggregation

(Friedman, 1993), and reduced nitrate leaching (Harris et al., 1994) as compared to a similar system that relied on chemical N fertilizers. The legume-based system accumulated organic N over time, whereas the fertilizer-based system without cover crops lost organic N over the same period (Drinkwater, Wagoner and Sarrantonio, 1998). In a long-term cotton rotation study in Alabama, rotations that included crimson clover (*Trifolium incarnatum* L.) had higher SOM, soil microbial biomass carbon (C) and crop yield than the rotations without a cover crop after 99 years (Entry, Mitchell, and Backman, 1996). The presence of a grass cover in a prune (*Prunus domestica* L.) orchard in California led to significant improvement in soil physical characteristics, including aggregate stability and saturated hydraulic conductivity, a measure of the soil's capacity to drain water (Roberson, Sarig, and Firestone, 1991). Water-stable aggregation also increased when cover crops of various types preceded corn in Georgia (McVay, Radcliffe, and Hargrove, 1989) and Ontario (Dapaah and Vyn, 1998), relative to corn with no cover crops.

NUTRIENT MANAGEMENT

Although legumes have long been known to contribute to soil N when used as green manures, the practice gained considerable popularity in recent decades among organic growers and others wishing to reduce reliance on chemical fertilizers. In temperate regions, winter annual legumes have been most commonly used as N sources, since they can often be placed into cropping system with little interference with cash crop production. Various members of the vetch (*Vicia*) and pea (*Pisum*) genera have been popular, as have some annual clovers, such as crimson, subterranean (*T. subterraneum* L.), and berseem (*T. alexandrinum* L.), but perennials such as red and white clover (*T. pratense* L. and *T. repens* L.) and biennial sweetclovers (*Melilotus* spp.) are also utilized, particularly in cropping systems where the goal is to simultaneously improve soil health and short-term N management.

The N-fixing capabilities and biomass production of these legumes varies by climate and soil condition, ranging between 80 and 250 kg N ha^{-1} for hairy vetch (*V. villosa* Roth) and 70-150 kg N ha^{-1} for most annual clovers (Reeves, 1994; Sarrantonio, 1994). Most of the N contained in these plants will be found in the foliage at any given time, with about 15-25% found in the roots (Scott et al., 1987; Reeves, Wood, and Touchton, 1993). Because no legume is able to fix 100% of its own N needs (Alexander, 1977), a convenient rule of thumb dictates that the N in the aboveground portion of the plant is roughly equivalent to the N fixed by the plant, assuming conditions are optimal for N-fixation to occur (Sarrantonio, 1994).

The availability of plant-usable forms of N following a green manure is largely microbially-mediated, and therefore dependent, in the short term, on factors which affect microbial (particularly bacterial) activity (Alexander, 1977). These factors include moisture, temperature, microbial access to the substrate, and pH, which in turn are affected by weather, soil type, tillage, and residue size and composition, among other things. We are still unable to predict with any great certainty the quantity and timing of N availability following green manure crops, a fact which often serves as a deterrent to growers' replacing fertilizer N with biologically-fixed N. Huntington, Grove, and Frye (1985) reported that the majority of inorganic N released from a winter hairy vetch crop in Kentucky occurred too late in the season for corn to effectively utilize it. Several other researchers similarly found a slow or inadequate rate of N availability following winter annual legumes in corn (Karlen and Doran, 1991) and sorghum production (Groffman, Hendrix, and Crossley, 1987). Other research, however, indicates that inorganic N can appear quickly and in large flushes following winter annual legumes--as much as 140 kg ha^{-1} of nitrate N can be made available within one week after a green manure crop is killed (Dabney et al., 1987; Sarrantonio and Scott, 1988; Lathwell, 1990; Raynes and Lennartsson, 1995; Sarrantonio, 1995). The absence of plant roots to take up this soluble flush of N in the ensuing weeks makes it vulnerable to loss through leaching. McCracken et al. (1994) recorded a loss of nearly 50 kg ha^{-1} of leached N from a vetch-corn system over three years. Grain rotations based on long-term legume use, however, leached only about half as much nitrate as those based on inorganic fertilizers over a five-year period, with similar yields, indicating that efficiency of N use from leguminous cover crops improves significantly with continued use (Harris et al., 1994; Drinkwater, Wagoner, and Sarrantonio, 1998).

Cover crops can also be used effectively to catch excess soluble nutrients in the soil profile, preventing their loss from the soil ecosystem. In a Chesapeake Bay study, cereal rye planted in early September was able to absorb nearly 70 kg ha^{-1} of soil N by early December (Brinsfield and Staver, 1991). Brassicas and other cool season grasses were only about half as effective at fall N uptake in the same study, and winter annual legumes took up very little fall N. McCracken (1989) reported than N leaching was reduced nearly 100% when cereal rye was planted after no-tillage corn. Both cereal rye and annual ryegrass (*Lolium multiflorum* Lam.) were far superior to legume covers at recovering corn fertilizer N in Maryland (Meisinger, Shipley, and Decker, 1990).

WEED MANAGEMENT

Cover cropping can stress weed populations at multiple points: reducing or preventing propagule production, reducing seedling establishment, and minimizing the competitive ability of individual weeds that do successfully establish (Gallandt, Liebman, and Huggins, 1999). These stresses are a result of disturbances associated with cover crop management, the competitive ability of the cover crop, and effects of the residues following termination of cover crop growth.

Relative to monocropping, crop rotations generally experience less weed pressure, presumably because of varying instead of predictable patterns of soil disturbance, crop competition, cultivation, and herbicide use (Liebman and Dyck, 1993). By inserting cover crops into selected niches within cropping systems, the number and diversity of these stresses may be expanded. Establishment of a fall cover crop, for example, may require preparation of a seedbed, thereby terminating the growth and seed production of late-season weeds. Warm-season cover crops permit repeated soil disturbance prior to planting, thereby encouraging, for certain weed species, high levels of germination and thus depletion of seed bank reserves.

The choice of cover crop species should consider emergence rate, time to canopy closure, and the opportunity for mowing to prevent weed seed production. Cereals are well-known for their rapid establishment and ability to compete with weeds (van Heemst, 1985), and winter types (for example, cereal rye) are perhaps the most widely used type of cover crop throughout temperate cropping regions. Oat (*Avena sativa* L.), another competitive cereal, may be planted in open "windows" throughout the temperate part of the year. Buckwheat (*Fagopyrum esculentum* Moench) grows rapidly in warm weather and can be seeded as a competitive summer cover crop, following early season crops or preceding fall-planted crops (Sarrantonio, 1994).

The ability of a cover crop species to suppress weed growth is, not surprisingly, proportional to the amount of cover crop canopy produced (Liebman and Davis, 2000). Consequently, cover crop species with complementary spatial or temporal growth patterns may be grown together–intercropped–to increase total biomass production (Vandermeer, 1990), leaving fewer resources to support weed growth (Liebman and Dyck, 1993). By extension, this concept has been investigated in efforts to improve the competitive ability of a growing cover crop with mixtures such as hairy vetch and crimson clover grown with small grains (Creamer, Bennet, and Stinner, 1997) or Austrian winter peas (*Pisum sativum* spp. *arvense* L.), oats and hairy vetch grown together (Jannink, Liebman, and Merrick, 1996).

Retained on the soil surface, the residues of non-cash crop species such as rye and hairy vetch physically and chemically suppress weeds, leading to reduced densities (Mohler and Teasdale, 1993). Teasdale (1996), however, concluded that the cover crop residues in such no-tillage systems, while reducing soil erosion and improving soil quality, alone do not provide acceptable levels of weed control.

As some cover crop residues decay, they release phytotoxins that can cause mortality of non-emerged seedlings, an effect known as allelopathy. Blackshaw et al. (2001), in Alberta, Canada, found that weed suppression from yellow sweetclover (*Melilotus officinalis* (L.) Desr.) was similar whether the cover crop residues were left on the soil surface, incorporated, or even harvested for hay. Emergence of several small-seeded broadleaf weed species following incorporation of red clover was reduced to a similar extent when the green manure residue was flail mowed and incorporated by rototilling or left intact and incorporated with an Imantas Spader (Gallandt, unpublished results). Early in their work with cereal mulch systems for weed control in vegetable crops, Putnam and DeFrank (1983) found that residue-mediated effects were related to seed size. Growth of smaller-seeded crops and weeds was reduced by cereal residue whereas larger-seeded species grew normally.

Incorporated rye residues release benzoxazolinones which tend to be particularly phytotoxic to small-seeded species (Putnam, 1994). Brassica species contain glucosinolate compounds phytotoxic to certain weed species. Incorporated residues of rapeseed (*Brassica napus* L.), for example, reduced weed densities 73-85% in a subsequent potato (*Solanum tuberosum* L.) crop (Boydston and Hang, 1995). Many leguminous species grown primarily for their N contribution to subsequent crops may also help to reduce emergence of certain weed species. Incorporated crimson clover and hairy vetch reduced establishment of morning-glory (*Ipomoea lacunosa* L.), and further decreased the biomass of weeds that did establish (White, Worsham, and Blum, 1989). Likewise, Lehman and Blum (1997) found in laboratory studies that crimson clover and subterranean clover reduced emergence of both morning-glory and redroot pigweed (*Amaranthus retroflexus* L.).

Reductions in weed seedling densities following incorporation of cover crop residues may also involve soil borne pathogens. Sampling a field experiment conducted in Maine, Conklin (2000) used a soil bioassay developed by Dabney et al. (1996) to assess the phytotoxicity of contrasting soil management practices. In soils bioassayed during the first few weeks following incorporation of compost and red clover green manure, wild mustard had a higher incidence of *Pythium* and, in one of two years, reduced growth. Bioassays using the larger-seeded sweet corn crop showed low levels of *Pythium* infection and no differences in root elongation between the compost/green manure and the conventionally fertilized soil management systems.

INSECT AND PATHOGEN MANAGEMENT

The use of cover crops in rotation to manage specific crop disease and insect pests has been hampered by the complexity of the agroecosystem interactions involved. Cover crops can impact disease and insect damage by changing soil chemical and physical properties, by releasing exudates and decomposition products that directly affect pathogens, by serving as hosts for competitors, parasites and predators, by changing above and belowground environmental factors, such as moisture levels and air movement, or by affecting the overall health of succeeding or concurrent crops. In addition, cover crops can confuse insect and vertebrate visual or olfactory clues or create mechanical barriers to movement (Pickett and Bugg, 1998). Unfortunately, these same strategies can work against crop managers if the cover crop attracts additional pests, or acts as an alternate host for pathogens and insect pests in the field.

Few generalizations can be made to date about the effect of cover crops on diseases of commodity crops. Sorghum-Sudan grass (*Sorghum bicolor* (L.) Moench × *S. bicolor* var. *sudanense*) and other covers in an apple (*Malus domestica* Borkh.) orchard resulted in lower incidence of fire blight (*Erwinia amylovora*), but increased incidence of some other diseases, and led to an overall decrease in total and marketable yield, attributed largely to competition for water (Brown and Glenn, 1999). In a California study, both Lana woolypod vetch (*Vicia dasycarpa*) and purple vetch (*V. benghalensis*) served as hosts for *Sclerotinia minor*, but fava beans (*V. faba*), a member of the same genus, did not (Koike et al., 1996). Elmer and LaMondia (1999) reported that the ability of cereal cover crops to suppress pathogens of strawberry (*Fragaria chiloensis* Duchesne) was dependent on the type of fertilizer used.

Studies on cover crop effects on pest occurrence in rotations that include a mechanistic approach will be the most useful for future management recommendations. In a field trial that resulted in lower incidence of the fungus *Alternaria brassicae* in no-tillage cabbage (*Brassica oleracea* var. *capitata*) production with a legume cover crop versus in a similar system with chemical fertilizer and no cover, researchers speculated that the disease inhibition was due primarily to reduced contact of the cabbage leaves with the soil (Hoyt and Walgenbach, 1995). A significant decrease in black root rot (*Thielaviopsis basicola*) in cotton following hairy vetch incorporation was attributed to the fumigant action of ammonia release from decomposing vetch residues in the soil (Candole and Rothrock, 1997). Meanwhile, researchers in Ohio demonstrated that a living cover of Sudan grass (*S. bicolor* var. *sudanense*) significantly reduced splash dispersal of *Colletotrichum acutatum* conidia relative to bare ground under simulated rainfall (Ntahimpera et al., 1998).

Several studies indicate strong bio-active properties associated with sorghum-Sudan grass that lead to changes in soil microbial populations. *Verticillium* of potato was reduce by 24-29% following sorghum-Sudan grass (Stark, 1995). It also significantly reduced damage due to the nematode *Meloidogyne hapla* when grown preceding lettuce (*Lactuca sativa* L.) probably by serving as a non-host in the rotation (Viaene and Abawi, 1998). Elmer and LaMondia

(1999) found that sorghum-Sudan grass and 'Saia' oats (*Avena strigosa*) helped reduce lesion nematodes (*Pratylenchis penetrans*) in strawberry production.

Much of the research pertaining to insect-pest control with cover crops has focused on attracting beneficial organisms to a cover crop that will then feed on or parasitize insect-pests. Such systems are more successful when a living cover crop is maintained for all or most of the growing season as a living mulch ground cover or as a "refuge strip," a narrow band of cover crop grown between rows of the commodity crop. Because of these constraints, much research in this area has been conducted in perennial crops, where maintenance of living cover crops presents less of a challenge. Various studies demonstrated success in attracting beneficials, such as carabid beetles (Carmona and Landis, 1999), coccinellids (Rice et al., 1998) and arthropods (Smith et al., 1996) to perennial flowers or orchard crops. However, movement into the crop canopy and subsequent effect on the insect pests was generally not great. One exception was a system in which a legume cover crop used to breed *Eusesius tularensis*, a predatory mite, was cut and placed into the branches of young citrus trees (Grafton-Cardwell, Ouyang and Bugg, 1999), leading to enhanced predator mite population in the citrus canopy.

In annual crops, legume living mulch between rows of broccoli (*Brassica oleracea* var. *botrytis*) was shown to inhibit aphids in the broccoli, but also inhibited their parasitoid *Diaeretiella rapae* (Costello and Altieri, 1995), highlighting the complexity of managing these systems effectively. Parajulee and Slosser (1999) found that relay strips of hairy vetch, wheat (*Triticum aestivum* L.) and canola (*Brassica rapa* L.) enhanced predator number in adjacent cotton rows. Killed rye mulch was effective in attracting parasitoids and armyworm (*Pseudaletia unipuncta*) in a subsequent no-tillage corn crop (Laub and Luna, 1992), and suppressing Colorado potato beetle (*Leptinotarsa decemlineata* Say) in no-tillage tomato production (Hunt, 1998).

In some cases, the cover crop can create new pest and disease problems in the cropping system. Many growers have complained that cover crops increase wireworms in the succeeding crops. Wireworms of four different species were higher in Florida potato crop when sorgum-sudangrass was used as summer fallow than when the field was mechanically fallowed, leading to substantial economic losses in the cover crop system (Jansson and Lecrone, 1991). Use of crimson clover preceding sorghum led to increased damage to the grain due to *Rhizoctonia solani* Kuhn (Dabney et al., 1996). The effect was greatest when sorghum planting occurred immediately after killing of the clover. Millet used for weed control in California strawberries attracted European corn borer (*Ostrinia nubilalis* Hubner), a pest previously unreported in that crop (Maas et al., 1998).

COVER CROP ESTABLISHMENT

The inclusion of a cover crop into an established cropping system can present numerous logistical challenges to a grower. Some scenarios for cropping systems that incorporate cover crops can be found in Figure 1. Typically, growers prefer for the cover crop to occupy an underutilized temporal or spatial niche in the cropping system, thus avoiding the need to significantly alter their crop rotation. In most cropping systems in temperate regions the only lengthy underutilized time niche occurs from mid-fall until the spring planting season. Where mild fall weather is prolonged, cover crops can be sown after harvest of summer annual crops. Additionally, harvest of silage corn and winter and spring grain crops typically leave adequate time for establishment and growth of a wide range of winter annual cover crop species (Samson, Foulds, and Patriquin, 1991; Stute and Posner, 1995).

In most northern regions of North America, the existing cropping system leaves only a narrow window for establishment of a cover crop before harsh weather begins. In such systems, overseeding (also known as undersowing) the cover crop into the cash crop before harvest can allow the cover crop sufficient time to germinate and establish a viable root system. Red clover and an-

FIGURE 1. Five management options for the inclusion of cover crops in annual cropping systems

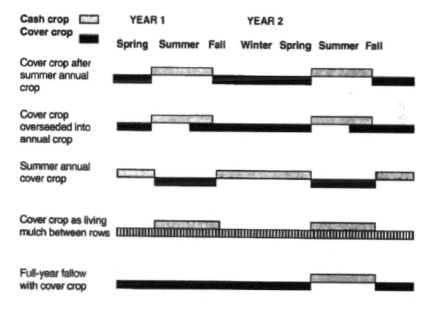

nual ryegrass have been successfully seeded into field corn in New York by broadcasting cover crop seed at final cultivation (Mt. Pleasant, 1982; Scott et al., 1987). Corn can also be overseeded at mid-silk or later (Corak et al., 1987; Scott et al., 1987), although specialized equipment, such as a high-clearance tractor is generally necessary. Janke et al. (1987) reported successful over-seeding of both grass and legume covers into soybeans at the leaf-yellowing stage in late summer. Some vegetable row crops offer fewer mechanical constraints to overseeding because of canopy architecture. Broccoli was over-seeded with a variety of legumes 4-5 weeks after transplanting (Foulds, Stewart, and Samson, 1991) and researchers in New Mexico reported over-seeding chile (*Capsicum annuum* L.) in mid-season with various covers (Guldan et al., 1996).

A traditional method of establishing hayfields in colder areas can be used to establish cover crops into winter grains in the spring. This method of spreading seed on frozen ground over the dormant winter grain is known as 'frost-seeding' and relies on the thawing and freezing action of the soil in the early spring to pull the seed into crevices that develop in the soil (Samson, Foulds, and Patriquin, 1991; Hesterman et al., 1992). The cover crop, typically a perennial or biennial legume, is already well-established by the time the grain is removed in the summer and can grow until the following spring or beyond. Concurrent or companion seeding a legume with a spring grain can produce the same effects, with the exception that the legume has more potential to become a competitor for resources with the grain.

In more southern areas of North America, a more common period for cover crops may be the summer, when it is too hot or dry to produce economic crops. The use of sorghum-Sudan grass as a summer fallow preceding potatoes in Florida has already been mentioned (Jansson and Lecrone, 1991). Researchers in North Carolina have evaluated a broad range of summer legumes and grains as potential summer cover crops for weed control and possible N management before fall vegetables (Creamer and Baldwin, 2000). In addition, in dryland areas, cover crops may replace bare summer fallow in a wheat cropping sequence (Moyer et al., 2000).

Other models for cover crop establishment include full-year fallow, in which a field or a portion of it is removed from cash crop production for a period of time, generally one to two years, and planted to a biennial or perennial cover. While few growers feel able to remove fields from crops for such long periods, those who practice this system feel that the short-term income loss is eventually balanced by long-term improvement in soil and reduction of pest and weed problems (Nordell and Nordell, 1998). The cover crop may serve as forage, although the intent is not to use the field for intensive hay production (Coleson, personal communication).

Living mulches are cover crops that are maintained as an understory in a cash crop, generally for erosion protection, and to a lesser extent, for weed management and attraction of beneficial organisms. Living mulches can easily become competitive with the growing crop (Box et al., 1980; Grubinger and Minotti, 1990) and have been managed most successfully in perennial cultures, such as orchards and vineyards, where rows are wide and the cover crop can be suppressed periodically by mowing or other means (Bugg et al., 1991; Bugg et al., 1996).

Selection of the cover crop to fit the cropping system must also be based on the often challenging climatic conditions under which it must be established and grow. A cover crop might be needed, for example, not only for its cold tolerance, but ability to germinate and establish when seeded on the soil surface under a canopy that restricts light. In cropping systems in which covers are established as an overseeding, the young cover crop may need to withstand significant foot or tractor traffic in the harvesting process. As such, the available options of species tolerant of a wide range of inhospitable conditions while producing sufficient biomass and remaining affordable is still relatively small (Sarrantonio, 1994).

CONSTRAINTS TO COVER CROPPING

While the benefits of cover crops have long been recognized, their adoption by North American farmers has been slow. Grower reluctance to incorporate cover crops into their cropping systems is based on complex economic, biological and operational issues (Stivers-Young and Tucker, 1999; Mallory, Posner, and Baldock, 1998).

Most economic analyses of cover crop systems are based on simple models that compare establishment costs to income changes from crop yield improvements, or to calculations of N fertilizer replacement values (N-FRV) for legume cover crops. The establishment costs include labor, machinery, fuel, seed, and any inoculants or fertilizer used specifically for the cover crop; seed costs generally constitutes the major portion of establishment costs (Allison and Ott, 1987; Mallory, Posner and Baldock, 1998). When N-FRV is used as a sole criterium, legume cover crops rarely appear economically beneficial, due largely to high cost of seed and relatively low prevailing chemical N prices (Frye, Smith, and Williams, 1985; Hargrove, 1986; Ott, 1987; Mallory, Posner, and Baldock, 1998). If commodity income were used for the analysis, net economic benefit of the cover crop would be apparent only when the value of any increased yield to the following crop is greater than these establishment costs. As such, the perceived value of the cover crop can shift from year to year as commodity prices or seeding costs fluctuate.

Conservation tillage systems can become more profitable by using cover crops that reseed between rows of annual crops. Crimson clover and subterranean clover have been used successfully in this way (Evers, Smith, and Beale, 1985; Oyer and Touchton, 1990; Boquet and Dabney, 1991; Ranells and Wagger, 1991) but the logistics of maintaining a living cover long enough to reseed may be a considerable constraint in many annual cropping systems.

In studies of no-tillage systems, legume residues, especially vetches, often boost corn or sorghum yield over and above any N fertilizer value they contribute, and combined with some supplemental fertilizer, often lead to higher net profit than optimum rates of fertilizer alone (Shurley, 1987; Hanson et al., 1993; Roberts et al., 1998). Some recent studies have taken into account the risk factor involved in cover crop use, which is related to the yearly variation in profit due to climatic interactions, cover crop failure, opportunity costs due to delayed planting of commodity crops, or reduction of commodity quality or yield from diseases, insects or resource competition associated with the cover crop use. Risk-averse farmers are willing to accept lower average yields in return for greater financial stability. Surprisingly, most of these studies, which focused on no-tillage planting of corn into legume residue, determined that the cover crop was still economically beneficial, even for risk-averse growers; risk was generally lower with some level of supplemental N fertilizer (Ott and Hargrove, 1989; Hanson et al., 1993; Roberts et al., 1998; Lu, Watkins, and Teasdale, 1999). Kelly et al. (1995) reported that tomatoes transplanted into hairy vetch mulch was economically superior to black plastic and no-cover treatments, even for risk-averse growers.

The commonality in these economic analyses is the implication that the value of cover crops to the farming system is solely a function of yearly variations in commodity and N fertilizer prices. Economic analyses that attempt to look at systems-level and long-term benefits are still lacking. Pest management with cover crops may save growers the cost of pesticide applications, but few studies will venture into this area yet. Natural resource accounting methods, which include long term on- and off-site changes in environmental resources into economic analyses, attempt to place monetary value on a ton of soil saved, on SOM accrued, or on more intangible qualities of the whole farm to cover crop systems (Faeth, 1993; King, Bohlen, and Crosson, 1995) and may eventually be able to assign a more comprehensive value to cover crops. Government policies that provide economic incentives for cover cropping would help overcome the primary short-term economic barriers to widespread adoption of these systems (Faeth et al., 1991).

Several of the biological constraints associated with cover crops have already been discussed, such as the possibility of introducing new pest species, or competing with crops for water or other resources (Corak, Smith, and Frye, 1991; Mallory, Posner, and Baldock, 1998; Brown and Glenn, 1999). For many growers, the timing of planting and harvesting of economic crops severely limits opportunities to include cover crops in the rotation (Giesler, Paxton, and Millhollon, 1993). Others reject the opportunity costs of reducing harvest of other crops in their rotation, such as the loss of small grain straw harvest when red clover is frost seeded into the grain (Mallory, Posner, and Baldock, 1998). Some cover crops, such as hairy vetch or annual ryegrass, which have favorable life cycles and tolerances are rejected by growers who fear they may become weeds on their farms (Sarrantonio, 1994) due to their seeding habits.

The existence of substantial cover crop residue in the field prior to cash crop establishment can lead to mechanical difficulties in mowing, plowing, and cultivation or no-tillage planting (Grisso, Johnson, and Dumas, 1985; Campbell, Karlen, and Sojka, 1984; Eckert, 1988). Because of the difficulty of mowing vetch, which is viney and tends to wrap around mower parts or mat on the ground, many growers prefer to grow vetch along with rye, which lends support and makes mowing easier. Allowing time for cover crop desiccation before field operations is key to reducing mechanical problems that can result in poor stands or equipment failure (Reeves, 1994). The establishment of the cover crop as an overseeding into an established crop canopy may present a sizeable mechanical barrier to growers with large acreage who do not own high-clearance tractors or other specialized equipment.

CONCLUSIONS

Cover crops have long been included in North American cropping systems to provide protection from soil erosion and improve N cycling abilities. Challenges facing the research and extension communities are to explore and promote additional benefits of cover cropping for pest, disease and weed management, as well as overall soil and environmental quality. Economic analyses which rely strictly on N-replacement values for green manures or improvement in yield of the following crop should be replaced by whole resource accounting, which will include less easily measured benefits to the whole farm ecosystems and surrounding watersheds. Research is needed to identify a broader range of cover crops to fit widely varying cropping systems and climatic conditions, and to pinpoint ecological inefficiencies in current cover crop management. Government policies in the future should provide economic incentives for growers willing to overcome managerial and biological constraints to include cover crops on their farms.

REFERENCES

Alexander, M. (1977). *Introduction to Soil Microbiology*. New York, NY: Wiley and Son.
Allison, J.R. and S.L. Ott. (1987). Economics of using legumes as a nitrogen source in conservation tillage systems. In *The Role of Legumes in Conservation Tillage Sys-*

tems, ed. J.F. Power, Ankeny, IA: Soil Conservation Society of America, pp. 145-150.

Blackshaw, R.E., J.R. Moyer, R.C. Doram, and A.L. Boswell. (2001). Yellow sweet-clover, green manure, and its residues effectively suppress weeds during fallow. *Weed Science* 49:406-413.

Boquet, D.J. and S.M. Dabney. (1991). Reseeding, biomass, and nitrogen content of selected winter legumes in grain sorghum culture. *Agronomy Journal* 83:144-148.

Box, J.E., S.R. Wilkerson, R.N. Dawson, and J. Kozachyn. (1980). Soil water effects on no-till corn production in strip and completely killed mulches. *Agronomy Journal* 72:797-802.

Boydston, R.A. and A. Hang. (1995). Rapeseed (*Brassica napus*) green manure crop suppresses weeds in potato (*Solanum tuberosum*). *Weed Technology* 9:669-675.

Brady, N.C. and R.R. Weil. (1999). *The Nature and Properties of Soil*, 12th Ed. Upper Saddle River, NJ: Prentice Hall.

Brinsfield, R. and K. Staver. (1991). Role of Cover Crops in Reduction of Cropland Nonpoint Source Pollution. *Final report to USDA/SCS, Cooperative Agreement #25087.*

Brown, M.W. and D.M. Glenn. (1999). Ground cover and selective insecticides as pest management tools in apple orchards. *Journal of Economic Entomology* 92:899-905.

Bugg, R.L., G. McGourty, M. Sarrantonio, W.T. Lanini, and R. Bartolucci. (1996). Comparison of 32 cover crops in an organic vineyard on the north coast of California. *Biological Agriculture and Horticulture* 13:63-81.

Bugg, R.L., M. Sarrantonio, J.D. Dutcher, and S.C. Phatak. (1991). Understory cover crops in pecan orchards: Possible management systems. *Journal of Alternative Agriculture* 6:50-60.

Campbell, R.B., D.L. Karlen, and R.E. Sojka. (1984). Conservation tillage for maize production in the U.S. Southeastern Coastal Plain. *Soil Tillage Research* 4:511-529.

Candole, B.L. and C.S. Rothrock. (1997). Characterization of the suppressiveness of hairy vetch-amended soils to *Thielaviopsis basicola*. *Phytopathology* 87:197-202.

Carmona, D.M. and D.A. Landis. (1999). Influence of refuge habitats and cover crops on seasonal-density of ground beetles (Coveoptera: Carabidae) in field crops. *Environmental Entomology* 28:1145-1153.

Conklin, A.E. (2000). Effects of red clover (*Trifolium pratense*) green manure and compost soil amendments on the growth and health of wild mustard (*Brassica kaber*) seedlings. *M.S. Thesis*, Orono, ME: University of Maine.

Corak, S.J., W.W. Frye, and M.S. Smith. (1991). Legume mulch and nitrogen fertilizer effects on soil water and corn production. *Soil Science Society of America Journal* 55:1395-1400.

Corak, S.J., W.W. Frye, M.S. Smith, J.H. Grove, and C.T. MacKown. (1987). Fertilizer nitrogen recovery by no-till corn as influenced by a legume cover crop. In *The Role of Legumes in Conservation Tillage Systems*, ed. J.F. Power, Ankeny, IA: Soil Conservation Society of America, pp. 43-44.

Costello, M.J. and M.A. Altieri. (1995). Abundance, growth rate and parasitism of *Brevicoryne brassicae* and *Myzus perisicae* (Homoptera: Aphididae) on broccoli grown in living mulches. *Agriculture, Ecosystems and Environment* 52:187-196.

Creamer, N.G. and K.R. Baldwin. (2000). An evaluation of summer cover crops for use in vegetable production systems in North Carolina. *HortScience* 35:600-603.

Creamer, N.G., M.A. Bennett, and B.A. Stinner. (1997). Evaluation of cover crop mixtures for use in vegetable production systems. *Horticulture Science* 32:866-870.

Dabney, S.M., G.A. Breitenbeck, B.J. Hoff, J.L. Griffin, and M.R. Milam. (1987). Management of subterranean clover as a source of nitrogen for a subsequent rice crop. In *The Role of Legumes in Conservation Tillage Systems*, ed., J.F. Power, Ankeny, Iowa: Soil Conservation Society of America, pp. 54-55.

Dabney, S.M., J.D. Schreiber, C.S. Rothrock, and J.R. Johnson. (1996). Cover crops affect sorghum seedling growth. *Agronomy Journal* 88:961-970.

Dapaah, H.K. and T.J. Vyn. (1998). Nitrogen fertilization and cover crops effects on soil structural stability and corn performance. *Communications in Soil Science and Plant Analysis* 29:2557-2569.

Doran, J.W. and T.B. Parkin. (1994). Defining and assessing soil quality. In *Defining Soil Quality for a Sustainable Environment*, ed., D.W. Doran, D.C. Coleman, D.F. Bezdicek, and B.A. Stewart. Special Publication No. 35, Madison, WI: Soil Science Society of America, pp. 3-21.

Doran, J.W., M. Sarrantonio, and M.A. Liebig. (1996). Soil health and sustainability. *Advances in Agronomy* 56:1-54.

Drinkwater, L.E., P. Wagoner, and M. Sarrantonio. (1998). Legume-based cropping systems have reduced carbon and nitrogen losses. *Nature* 396:262-265.

Eckert, D.J. (1988). Rye cover crops for no-tillage corn and soybean production. *Journal of Production Agriculture* 1:207-210.

Elmer, W.H. and J.A. LaMondia. (1999). Influence of ammonium sulfate and rotation crops on strawberry black root rot. *Plant Disease* 83:119-123.

Entry, J.A., C.C. Mitchell, and C.B. Backman. (1996). Influence of management practices on soil organic matter, microbial biomass and cotton yield in Alabama's "Old Rotation." *Biology and Fertility of Soils* 23:353-358.

Evers, G.W., G.R. Smith, and P.E. Beale. (1988). Subterranean clover reseeding. *Agronomy Journal* 80:855-859.

Faeth, P. (1993). An economic framework for evaluating agricultural policy and the sustainability of production systems. *Agriculture, Ecosystems and Environment* 46:161-173.

Faeth, P., R. Repetto, K. Kroll, Q. Dai, and G. Helmers. (1991). Paying the Farm Bill: U.S. Agricultural Policy and the Transition to Sustainable Agriculture. World Resources Institute, Washington, DC.

Foulds, C.M., K.A. Stewart, and R.A. Samson. (1991). On-farm evaluation of legume interseedings in broccoli. In *Cover Crops for Clean Water*, Proceedings of a Conference, West Tennessee Experiment Station, April 9-11, 1991, Jackson, TN, ed., W.L. Hargrove, Ankeny, IA: Soil and Water Conservation Society, pp. 179-180.

Friedman, D.B. (1993). Carbon, nitrogen, and aggregation dynamics in low-input and reduced tillage cropping systems. *M.S. Thesis*, Ithaca, NY: Cornell University.

Frye, W.W., W.G. Smith, and R.J. Williams. (1985). Economics of winter cover crops as a source of nitrogen for no-till corn. *Journal of Soil and Water Conservation* 40:246-249.

Gallandt, E.R., M. Liebman, and D.R. Huggins. (1999). Improving soil quality: Implications for weed management. *Journal of Crop Production* 2:95-121.

Giesler, G.G., K.W. Paxton, and E.P. Millhollon. (1993). A GSD estimation of the relative worth of cover crops in cotton production systems. *Journal of Agricultural and Resource Economics* 18:47-56.

Grafton-Cardwell, E.E., Y. Ouyang, and R.L. Bugg. (1999). Leguminous cover crops to enhance population development of *Euseius tularensis* (Acari: Phytoseiidae) in citrus. *Biological Control: Theory and Applications in Pest Management* 16:73-80.

Grisso, R., C.E. Johnson, and W.T. Dumas. (1985). Influence of four cover conditions on cotton production. *Transactions of American Society of Agricultural Engineers* 28:435-439.

Groffman, P.M., P.F. Hendrix, and D.A. Crossley Jr. (1987). Nitrogen dynamics in conventional and no-tillage agroecosystems with inorganic fertilized and legume nitrogen inputs. *Plant and Soil* 97:315-332.

Grubinger, V.P. and P.L. Minotti. (1990). Managing white clover living mulch for sweet corn production with partial rototilling. *American Journal of Alternative Agriculture* 5:4-5.

Guldan, S.J., C.A. Martin, J. Cueto-Wong, and R.L. Steiner. (1996). Interseeding legumes into chile: Legume productivity and effect on chile yield. *HortScience* 31:1126-1128.

Hanson, J.C., Lichtenberg, A.M. Decker, and A.J. Clark. (1993). Agricultural economics: Profitability of no-tillage corn following a hairy vetch cover crop. *Journal of Production Agriculture* 6:432-437.

Hargrove, W.L. (1986). Winter legumes as a nitrogen source for no-till grain sorghum. *Agronomy Journal* 78:70-74.

Harris, G.H., O.B. Hesterman, E.A. Paul, S.E. Peters, and R.R. Janke. (1994). Fate and behavior of legume and fertilizer ^{15}N in a long-term cropping system experiment. *Agronomy Journal* 86:910-915.

Hesterman, O.B., T.S. Griffin, P.T. Williams, G.H. Harris, and D.R. Christenson. (1992). Forage legume-small grain intercrops: Nitrogen production and response of subsequent corn. *Journal of Production Agriculture* 5:340-348.

Hoyt, G.D. and J.F. Walgenbach. (1995). Pest evaluation in sustainable cabbage production systems. *HortScience* 30:1046-1048.

Hunt, D.W.A. (1998). Reduced tillage practices for managing the Colorado potato beetle in processing tomato production. *HortScience* 33:279-282.

Huntington, T.G., J.H. Grove, and W.W. Frye. (1985). Release and recovery of nitrogen from winter annual cover crops in no-till corn production. *Communications in Soil Science and Plant Analysis* 16:193-211.

Janke, R.R., R. Hofstetter, B. Volak, and J.K. Radke. (1987). Legume interseeding research at the Rodale Research Center. In *The Role of Legumes in Conservation Tillage Systems*, ed., J.F. Power, Ankeny, IA: Soil Conservation Society of America, pp. 90-91.

Jannink, J.L., M. Liebman, and L.C. Merrick. (1996). Biomass production and nitrogen accumulation in pea, oat, and vetch green manure mixtures. *Agronomy Journal* 88:231-240.

Jansson, R.K. and S.H. Lecrone. (1991). Effects of summer cover crop management on wireworm (Coleoptera:Elateridae) abundance and damage to potato. *Journal of Economic Entomology* 84:581-586.

Karlen D.L. and J.W. Doran. (1991). Cover crop management effects on soybean and corn growth and nitrogen dynamics in anon-farm study. *American Journal of Alternative Agriculture* 6:71-82.

Kelly, T.C., Y.C. Lu, A.A. Abdul-Baki, and J.R. Teasdale. (1995). Economics of a hairy vetch mulch system for producing fresh-market tomatoes in the mid-Atlantic region. *Journal of the American Society for Horticultural Science* 120:854-860.

King, D.M., C. Bohlen, and P.R. Crosson. (1995). Natural resource accounting and sustainable watershed management. Solomons, MD: University of Maryland, Center for Environmental and Estuarine Chesapeake Biological Laboratory.

Koike, S.T., R.F. Smith, L.E. Jackson, L.J. Wyland, J.I. Inman, and W.E. Chaney. (1996). Phacelia, Lana woollypod vetch, and Austrian winter pea: Three new cover crop hosts of *Sclerotinia minor* in California. *Plant Disease* 80:1409-1412.

Lathwell, D.J. (1990). Legume green manures: Principles for management based on recent research. *Tropsoils Bulletin No. 90-01*, Raleigh, NC: Soil Management Collaborative Research Support Program.

Laub, C.A. and J.M. Luna. (1992). Winter cover crop suppression practices and natural enemies of armyworm (Lepidoptera: Noctuidae) in no-till corn. *Environmental Entomology* 21:41-49.

Lehman, M.E. and U. Blum. (1997). Cover crop debris effects on weed emergence as modified by environmental factors. *Allelopathy Journal* 4:69-88.

Liebman, M. and A.S. Davis. (2000). Integration of soil, crop, and weed management in low-external-input farming systems. *Weed Research* 40:27-47.

Liebman, M. and E. Dyck. (1993). Crop rotation and intercropping strategies for weed management. *Ecological Applications* 3:92-122.

Lu, Y.C., B. Watkins, and J. Teasdale. (1999). Economic analysis of sustainable agricultural cropping systems for mid-Atlantic states. *Journal of Sustainable agriculture* 15:77-93.

Maas, J.L., J.M. Enns, S.C. Hokanson, and R.L. Hellmich. (1998). Injury to strawberry crowns caused by European corn borer larva. *HortScience* 33:866-867.

Mallory, E.B., J.L. Posner, and J.O. Baldock. (1998). Performance, economics, and adoption of cover crops in Wisconsin cash grain rotations: On-farm trials. *American Journal of Alternative Agriculture* 13:2-11.

McCracken, D. (1989). Control of nitrate leaching with winter annual cover crops. *University of Kentucky Agronomy and Soil Science News and Views* 10(2):1-2.

McCracken, D.V., M.S. Smith, J.H. Grove, C.T. Mackown, and R.L. Blevins. (1994). Nitrate leaching as influenced by cover cropping and nitrogen source. *Soil Science Society of America Journal* 58:1013-1020.

McVay, K.A., D.E. Radcliffe, and W.L. Hargrove. (1989). Winter legume effects on soil properties and nitrogen fertilizer requirements. *Soil Science Society of America Journal* 53:1856-1862.

Meisinger, J.J., P.R. Shipley, and A.M. Decker. (1990). Using winter cover crops to recycle nitrogen and reduce leaching. In *Conservation Tillage for Agriculture in the*

1990's, ed. J.P. Mueller and M.G. Wagger, Raleigh, NC: North Carolina State University, pp. 3-6.

Mohler, C.L. and J.R. Teasdale. (1993). Response of weed emergence to rate of *Vicia villosa* Roth and *Secale cereale* L. residue. *Weed Research*. 33:487-499.

Moldenhauer, W.C. and G.W. Langdale. (1995). Crop residue management to reduce soil erosion and improve soil quality. *USDA/ARS Report No. 39*, January 1995, pp. 1-47.

Morse, R.D. (1999). No-till vegetable production--its time is now. *HortTechnology* 9:373-379.

Moyer, J.R., R.E. Blackshaw, E.G. Smith, and S.M. McGinn. (2000). Cereal cover crops for weed suppression in a summer fallow-wheat cropping sequence. *Canadian Journal of Plant Science* 80:441-449.

Mt. Pleasant, J. (1982). Corn polyculture systems in New York. *M.S. Thesis*, Ithaca, NY: Cornell University.

Nordell, A. and E. Nordell. (1998). Cultivating questions concerning the bio-extensive market garden. *Small Farmer's Journal* 22:29-33.

Ntahimpera, N., M.A. Ellis, L.L. Wilson, and L.V. Madden. (1998). Effects of a cover crop on splash dispersal of *Colletotrichum acutatum* conidia. *Phytopathology* 88:536-543.

Nyakatawa, E.Z., K.C. Reddy, and J.L. Lemunyon. (2001). Predicting soil erosion in conservation tillage cotton production using the revised universal soil loss equation (RUSLE). *Soil and Tillage Research* 57: 213-224.

Ott, S.L. (1987). An economic and energy analysis of crimson clover as a nitrogen fertilizer substitute in grain sorghum production. In *The Role of Legumes in Conservation Tillage Systems*, ed., J.F. Power, Ankeny, IA: Soil Conservation Society of America, pp. 150-151.

Ott, S.L. and W.L. Hargrove. (1989). Profits and risks of using crimson clover and hairy vetch cover crops in no-till corn production. *American Journal of Alternative Agriculture* 4:65-70.

Oyer, L.J. and J.T. Touchton. (1990). Utilizing legume cropping systems to reduce nitrogen fertilizer requirements for conservation-tilled corn. *Agronomy Journal* 82: 1123-1127.

Parajulee, M.N. and J.E. Slosser. (1999). Evaluation of potential relay strip crops for predator enhancement in Texas cotton. *International Journal of Pest Management* 45:275-286.

Pickett, C.H. and R.L. Bugg. (1998). Enhancing biological control--habitat management to promote natural enemies of agricultural pests, In *Enhancing Biological Control*, ed., C.H. Pickett and R.L. Bugg, Berkeley, CA: University of California Press, pp. 1-24.

Pieters, A.J. (1927). *Green Manuring*. New York, NY: John Wiley and Sons, Inc.

Pieters, A.J. and R. McKee. (1938). The use of cover and green-manure crops. In *Soils and Men*, USDA Yearbook of Agriculture, Washington, DC: US Government Printing Office, pp. 431-444.

Putnam, A.R. (1994). Phytotoxicity of plant residues. In *Managing Agricultural Residue*, ed., P.W. Unger, Boca Raton, FL: Lewis Publishers, pp. 285-314.

Putnam, A.R. and J. DeFrank. (1983). Use of phytotoxic plant residues for selective weed control. *Crop Protection* 2:173-181.

Ranells, N.N. and M.G. Wagger. (1991). Strip management of crimson clover as a reseeding cover crop in no-till corn. In *Cover Crops for Clean Water*, Proceedings of a Conference, West Tennessee Experiment Station, April 9-11, 1991, Jackson, TN, ed., W.L. Hargrove, Ankeny, IA: Soil and Water Conservation Society, pp. 174-175.

Raynes, F.W. and E.K.M. Lennartsson. (1995). The nitrogen dynamics of winter green manures. In *Soil Management in Sustainable Agriculture*, ed., H.F. Cook and H.C. Lee, Wye, UK: Wye College Press, pp. 308-311.

Reeves, D.W. (1994). Cover crops and rotations. In *Crop Residue Management*, ed., J.L. Hatfield and B.A. Stewart, Boca Raton: Lewis Publishers, pp. 125-172.

Reeves, D.W., C.W. Wood, and J.T. Touchton. (1993). Timing nitrogen applications for corn in a winter legume conservation-tillage system. *Agronomy Journal* 85:98-106.

Rice, N.R., M.W. Smith, R.D. Eikenbary, D. Arnold, W.L. Tedders, B. Wood, B.S. Landgraf, G.G. Taylor, and G.E. Barlow. (1998). Assessment of legume and nonlegume ground covers on Coleoptera:Coccinellidae density for low-input pecan management. *American Journal of Alternative Agriculture* 13:111-123.

Roberson, E.B., S. Sarig, and M.K. Firestone. (1991). Cover crop management of polysaccharide-mediated aggregation in an orchard soil. *Soil Science Society of America Journal* 55:734-738.

Roberts, R.K., J.A. Larson, D.D. Tyler, B.N. Duck, and K.D. Dillivan. (1998). Economic analysis of the effects of winter cover crops on no-tillage corn yield response to applied nitrogen. *Journal of Soil and Water Conservation* 53:280-284.

Russell, E.J. (1913). *The Fertility of the Soil*. London, UK: Cambridge University Press.

Samson, R.A., C.M. Foulds, and D.G. Patriquin. (1991). Effects of cover crops on cycling of nitrogen and phosphorus in a winter wheat-corn sequence. In *Cover Crops for Clean Water*, ed., W.L. Hargrove, Ankeny, IA: Soil and Water Conservation Society, pp. 106-107.

Sarrantonio, M. (1995). Microbial activity and nitrogen dynamics following and winter annual green manure. In *Soil Management in Sustainable Agriculture*, ed., H.F. Cook and H.C. Lee, Wye, UK: Wye College Press, pp. 300-307.

Sarrantonio, M. (1994). *Northeast Cover Crop Handbook*. Emmaus, PA: Rodale Institute.

Sarrantonio, M. and T.W. Scott. (1988). Tillage effects on the availability of nitrogen to corn following a winter green manure crop. *Soil Science Society of America Journal* 2:1661-1668.

Scott, T.W., J. Mt. Pleasant, R.F. Burt, and D.J. Otis. (1987). Contributions of ground cover, dry matter, and nitrogen from intercrops and cover crops in a corn polyculture system. *Agronomy Journal* 79:792-798.

Shurley, D.W. (1987). Economics of legume cover crops in corn production. In *The Role of Legumes in Conservation Tillage Systems*, ed., J.F. Power, Ankeny, IA: Soil Conservation Society of America, pp. 152-153.

Smith, M.W., D.C. Arnold, R.D. Eikenbary, N.R. Rice, A. Shiferaw, B.S. Cheary, and B.L. Carroll. (1996). Influence of ground cover on beneficial arthropods in pecan. *Biological Control: Theory and Applications in Pest Management* 6:164-176.

Stark, J.C. (1995). Development of sustainable production systems for the Pacific NW. *Project Report No. LW91-29.* Logan, UT: Western Region SARE.

Stivers-Young, L.J. and F.A. Tucker. (1999). Cover-cropping practices of vegetable producers in western New York. *HortTechnology* 9:459-465.

Stute, J.K. and J.L. Posner. (1993). Legume cover crops options for grain rotations in Wisconsin. *Agronomy Journal* 85:1128-1132.

Teasdale, J.R. (1996). Contribution of cover crops to weed management in sustainable systems. *Journal of Production Agriculture* 9: 475-479.

van Heemst, H.D.J. (1985). The influence of weed competition on crop yield. *Agricultural Systems* 18:81-93.

Vandermeer, J.H. (1990). Intercropping. In *Agroecology,* ed., C.R. Carroll, J.H. Vandermeer, and P. Rosset, New York, NY: McGraw-Hill Publishing Co., pp. 481-516.

Viaene, N.M. and G.S. Abawi. (1998). Management of *Meloidogyne hapla* on lettuce in organic soil with sudangrass as a cover crop. *Plant Disease* 82:945-952.

Wander, M.M., S.J. Traina, B.R. Stinner, and S.E. Peters. (1994). The effects of organic and conventional management on biologically-active soil organic matter pools. *Soil Science Society of America Journal* 58:1130-1139.

White, R.H., A.D. Worsham, and U. Blum. (1989). Allelopathic potential of legume debris and aqueous extracts. *Weed Science* 37:674-679.

Zhu, J.C., C.J. Gantzer, S.H. Anderson, E.E. Alberts, and P.R. Beuselinck. (1989). Runoff, soil, and dissolved nutrient losses from no-till soybean with winter cover crops. *Soil Science Society of America Journal* 53:1210-1214.

Advances in Tillage Research in North American Cropping Systems

D. C. Reicosky

R. R. Allmaras

SUMMARY. Numerous innovations in tillage systems have significantly altered agricultural production in North America. Mechanical, biological, and chemical innovations reduced labor requirements, increased yields and crop residues, and reduced pest impacts. Regional trends in tillage systems and equipment are the result of evolving design driven by soil, plant and climate factors that affect erosion, water conservation, and offsite nutrient control. Within the past three decades, technological advances led to an increased interest in conservation tillage systems to replace intensive conventional tillage practices. For agriculture to be sustainable, it requires improved soil tillage and residue management systems. New technology consisting of precision agricultural techniques

D. C. Reicosky is Soil Scientist, USDA-Agricultural Research Service, North Central Soil Conservation Research Laboratory, 803 Iowa Avenue, Morris, MN 56267 USA (E-mail: reicosky@morris.ars.usda.gov).

R. R. Allmaras is Soil Scientist (retired), USDA-ARS Soil and Water Science Department, 152 Borlaug Hall, 1991 Upper Buford Circle, University of Minnesota, St. Paul, MN 55108 USA (E-mail: allmaras@soils.umn.edu).

The authors would like to acknowledge the helpful comments and suggestions from several colleagues and especially those who provided recent reprints for this review: Ken Potter, John Morrison, Henry Janzen, Martin Carter, Harold van Es, Con Campbell, Jeff Mitchell, Wayne Reeves, Warren Busscher, Roger Veseth, Denis Angers, J. R. Salinas-Garcia, Mike Lindstrom and many others who contributed to this chapter.

[Haworth co-indexing entry note]: "Advances in Tillage Research in North American Cropping Systems." Reicosky, D. C., and R. R. Allmaras. Co-published simultaneously in *Journal of Crop Production* (Food Products Press, an imprint of The Haworth Press, Inc.) Vol. 8, No. 1/2 (#15/16), 2003, pp. 75-125; and: *Cropping Systems: Trends and Advances* (ed: Anil Shrestha) Food Products Press, an imprint of The Haworth Press, Inc., 2003, pp. 75-125. Single or multiple copies of this article are available for a fee from The Haworth Document Delivery Service [1-800-HAWORTH, 9:00 a.m. - 5:00 p.m. (EST). E-mail address: docdelivery@haworthpress.com].

and yield maps has already begun to change tillage systems. Agriculture's impact on global increase of carbon dioxide (CO_2) requires more sequestration and maintenance of high soil carbon (C) levels for enhanced soil quality. The best soil management systems involve less soil disturbance and more focus on residue management within a geographical location as driven by economic and environmental considerations. *[Article copies available for a fee from The Haworth Document Delivery Service: 1-800-HAWORTH. E-mail address: <docdelivery@haworthpress.com> Website: <http://www.HaworthPress.com>]*

KEYWORDS. Physiographic regions, climate regions, management land resource areas, conventional tillage, conservation tillage, mechanical, chemical, and biological innovations and controlled traffic concepts, tillage systems

INTRODUCTION

Tillage or soil preparation has been an integral part of traditional agricultural production. Tillage is the mechanical manipulation of the soil and plant residue to prepare a seedbed where crop seeds are planted to produce grain for our consumption. Tillage fragments soil, enhances the release of soil nutrients for crop growth, kills weeds, and modifies the circulation of water and air within the soil. Intensive tillage can adversely affect soil structure and cause excessive break down of aggregates leading to potential soil movement via tillage/water erosion. Intensive tillage accelerates soil carbon (C) loss and greenhouse gas emissions, that impact environmental quality. New knowledge is required to minimize agricultural tillage impacts on the environment. This review emphasizes more precise definitions of "conventional" and "conservation" systems in various regions of North America. Emphasis will be placed on conservation tillage impacts on soil C management and environmental quality issues considering the general trend from intensive tillage to more conservation-based tillage practices over the last thirty years.

Tillage Terminology–Conventional versus Conservation

When most of the primary conventional tillage (CT) used the moldboard plow, the use of "conventional" terminology was not a problem. However, as technology changed and shifted toward "conservation" tillage, accurate terminology has become increasingly important. The term "conventional" is not universal in its meaning and "conservation" is an umbrella term that includes many other types of tillage [including no-till (NT)] and residue management systems that require specific definition of the tillage tools, soil interaction, and residue response to the tillage operation. Conservation tillage encompasses combinations of cultural practices that result in the protection of soil resources while crops are grown (Allmaras and Dowdy, 1985). The definition of conventional versus conservation tillage also depends on the geographic location and specific soil conditions. As of 1994, Agriculture and Agri-Food Canada defined conservation tillage as "tillage methods that leave most of the crop residue (trash) on the surface of the soil (including minimum tillage methods)" and conventional tillage as "the tillage methods that incorporate most of the crop residue into the soil." This is in contrast to the commonly accepted US definition of 30% cover at seeding. Presently, reduced soil disturbance involving a combination of tillage and herbicides for weed control and land preparation for the next crop are being widely used in the Great Plains and could be appropriately called the conventional system for that region. In the Texas High Plains, Unger and Skidmore (1994) suggest that stubble mulching is now the conventional form of tillage replacing the moldboard plow and disk harrow. In South Carolina, Bauer and colleagues (personal communication, 2000) suggest that no soil surface disturbance has become conventional in the Coastal Plains replacing the disk harrow as the primary tool. In the Midwest region of the US, our observations suggest that fewer and fewer farmers use the moldboard plow and are now using deep chisel plows or combination tillage tools composed of residue cutting disks, chisel plows, "subsoil" shanks and covering disks. As new technology develops and new tillage and planting equipment become available, we should discontinue the use of the vague and nondescript terms of "conventional" and "conservation" terminology and provide explicit descriptions of equipment for tillage, residue management and planting. Terminology for conservation tillage practices became confusing when research agencies and manufacturing companies reporting and promoting "minimum-tillage," "reduced-tillage," "ridge-tillage (RT)," "mulch-tillage," "zero-tillage," and "NT" without defining attributes of soil and residue mixing. An accurate description requires listing all operations in the system and should be the prime consideration in discussing various conservation tillage systems. Because of the number and diversity of available conservation tillage systems across North America, following these guidelines will provide a clearer understanding of soil disturbance or mixing and residue management when the specific tillage tools used in the system are described.

Merits and Demerits of Conservation Tillage

The merits of conservation tillage systems are recognized throughout North America. Increased interest in conservation tillage arises from economic and environmental advantages of these systems over CT practices. Concern over tillage/water/wind erosion and increased pressure to farm land unsuitable for conventional tillage practices has led to the development of "reduced" tillage and residue management systems commonly referred to as conservation tillage systems. Both depth and frequency of tillage are reduced. The emphasis is placed on conserving crop residue and leaving 30% cover on the soil surface to protect the soil from raindrop or wind impacts and minimize soil erosion (CTIC, 1995). Mulch-till, no-till, and ridge-till are conservation tillage-planting systems because they fulfill the 30% surface residue cover requirement and they maintain crop residue in the upper 10 cm of the soil (Allmaras et al., 1998). The 30% cover is a function of tillage, but there are more functions that cannot be determined unless the tillage tools are known. A common problem related to 30% cover criterion as the only method to determine tillage is that less than 30% cover usually occurs with the moldboard tillage, but non-moldboard tillage after sparse crop residue can also produce less than 30% cover. Yet the crop residue environment below the surface is markedly different.

Some advantages of conservation tillage are improved timing of planting and harvesting and increased potential for double cropping and conservation of soil water through decreased evaporation and increased infiltration, and reduced fuel, labor, and machinery requirements. One disadvantage is that residue on the soil surface delays soil temperature increases which in temperate and cold climate regions impedes the germination and early crop growth and allows increased potential for insect and disease damage to crops. With conservation tillage systems, there is a need for more precise management of soil fertility and weed control to achieve maximum yields. One limitation for social acceptance of conservation tillage is an old aesthetics about the "trashy residue" on the soil surface relative to that of clean, tilled fields based on tradition, but this aesthetic is changing.

Definition of Tillage Systems

Tillage systems can be identified according to their overall objectives such as conventional or conservation, or systems can be described according to the primary tillage implement, e.g., moldboard plow, chisel plow, or disk harrow (Reeder, 2000). Defining tillage systems by objective gives a sense of why a particular system is being used and/or the desired outcome. Terms such as NT or RT refer to the system's basic strategy to meet an objective. Conservation tillage should include NT, zero till, direct seeding (drilling), slot plant, strip till, RT, and mulch till. Conventional till should include clean tillage, and may or may not include minimum till or reduced till concepts.

Conventional tillage is often thought of as two major operations: (1) primary tillage and (2) secondary tillage. Allmaras et al. (1998) suggest two categories to distinguish the use of the moldboard plow versus other tillage tools for primary tillage. Primary tillage displaces and shatters the soil as it reduces soil strength and tends to bury and mix plant residues and fertilizers within the tilled zone. Primary tillage is more aggressive, deeper and leaves a rough surface relative to secondary tillage operations. Primary tillage tools are the moldboard plow, chisel plow and various types of combination disc-chisel-subsoil tools designed to disturb the soil to greater depths. However, depending on the features, options and operation of that implement, results of primary tillage will be different as soil conditions change. Secondary tillage varies widely with type and number of operations, penetration is nearly always shallower than with primary tillage tools. Secondary tillage provides additional soil breakup, levels and firms the soil, closes some air pockets and kills some weeds. Tillage equipment often associated with secondary tillage includes disk harrows, field cultivators, spring-tooth harrows, levelers, packers, and other types of finishing equipment. Interpreting tillage system results may be confusing or meaningless when little information about secondary tillage operations is available.

As a result of the potential terminology confusion in different parts of North America, a list of tillage system alternatives can be used to describe the type of tillage operation (after Morrison, 2000). Reeder (2000) and American Society of Agricultural Engineers (ASAE) Standards provided tillage implement descriptions and definitions of agricultural tillage implements (ASAE, 1998a; 1998b) and for soil engaging components of conservation tillage planters, drills and seeders (ASAE, 1999). Generalized definitions of tillage alternatives (based mainly on surface residue coverage) are provided by the Conservation Technology Information Center (CTIC) as follows (CTIC, 1995):

Conventional till (CT): Tillage types that leave less than 15% residue uncovered after planting, or less than 560 kg ha^{-1} of small grain residue equivalent throughout the critical wind erosion period. This generally involves moldboard plowing and sometimes a large number of tillage passes. Sparse crop residue can produce this cover with any tillage system.

Reduced till/Minimum till: Tillage types that leave 15-30% residue coverage after planting or 560-1120 kg ha^{-1} of small grain residue equivalent throughout the critical wind erosion period. This system fails to specify the tillage implements involved.

Ridge till (RT): The soil is left undisturbed from harvest to planting except for nutrient injection. Planting is completed in a seedbed prepared on ridges with sweeps, disk openers, coulters, or row cleaners. Residue is left on the surface between ridges. Weed control is accomplished with herbicides and/or cultivation. This full width tillage usually produces surface residue cover more than 30%, but maybe less when crop residue is sparse. Ridges are rebuilt during cultivation. Specified amount of surface residue is usually from 30 to 50%.

Mulch till: The soil is disturbed prior to planting using tillage tools such as chisels, field cultivators, disks (harrows), sweeps, or blades. Weed control is accomplished with herbicides and/or cultivation. Surface residue cover is usually greater than 30%, and the tillage is full-width.

No-till (NT)/Zero-till: The soil is left undisturbed from harvest to planting except for nutrient injection. Planting or drilling is accomplished in a narrow seedbed or slot created by coulter, row cleaners, disk openers, in-row chisels, or roto-tillers. Weed control is accomplished primarily with herbicides. Cultivation may be used for emergency weed control.

Strip till/Zone till: A modification of NT, sometimes similar to RT. Row width disturbance of less than 25% is necessary to fulfill the surface residue coverage. This variant of no till provides for traffic control in row crops.

The Economic Research Service had developed a tillage-system survey that identified the tillage tools, residue cover and depth of incorporated crop residue (Allmaras et al., 1998, 2000). It also identified surface cover based upon crop residue production.

Tillage Implements

Moldboard Plow

The moldboard plow was the most widespread and important implement of primary tillage in North America until the late 1960s. In humid climates, the soil inversion action is highly effective in burying and killing annual and perennial weeds as well as volunteer crops. The soil loosening action of the moldboard plow is excellent. When used during favorable soil conditions, moldboard plowing results in a clean surface that facilitates secondary tillage and precision seeding, but also leaves the surface susceptible to wind and water erosion. On light, sandy soils, there is a capacity to deeply plow under organic residues, manure and semi-liquid manure mixed with straw in order to amend the fertility status. As tractors and plows got larger, there was more of an effort to plow deeper to eliminate plow pans and subsoil compaction from previous tillage. The moldboard plow promotes pan formation that can limit root development. As horsepower (HP) of tractors and plow share size increased, plowing depth progressively increased to a depth of about 30 cm leaving the topsoil looser and the upper subsoil more dense than before. The direct impact of the moldboard plow is evidenced in the abrupt decrease in porosity at the plow pan (Logsdon et al., 1990). Moreover, loosening decreases the bearing strength of the tilled or topsoil layer. Together, increased load of common agricultural equipment, diminished surface bearing strength and trafficability caused increases in the depth of soil compaction under heavy machines particularly when the soils were wet (Logsdon et al., 1992; Voorhees, 1992).

Chisel Plow

A modified chisel plow in western US wheat (*Triticum aestivum* L.) lands replaced the moldboard and disc plows (Meyer and Mannering, 1961). Chisel plows bury about 40% of the crop residues and leaves the soil surface in a rough condition for infiltration of rainfall and protection from wind erosion. Point type and sweep type tools are used on chisel spring-shanks, which are typically staggered and spaced 30 to 40 cm apart for flow of residue between the shanks. Depth control wheels regulate operational depth. The chisel plow can be considered a reduced-tillage or minimum-tillage implement. The chisel plow is the dominant primary tillage implement in America in 2000. Johnson (1988) field demonstrated that the tool blade on chisel type implements had a marked influence on surface residue cover and surface roughness.

Blade Plow or Sweep Plow

A blade plow or sweep plow is used primarily in the western US Great Plains to cut roots of weeds following small grains. The implement has a large diameter coulter ahead of the shovel to cut residue; most residue remains on the surface. Several V-shaped sweeps (0.76 to 1.83 m wide) are mounted on standards attached to a tool bar or frame. Operating depth ranges from 6 to 13 cm controlled by both width of the sweeps and depth control wheels. Secondary tillage may consist of several rod weeding or harrowing operations for additional weed control and soil firming before planting. This system is called stubble-mulching in some regions.

Field Cultivator

The field cultivator was introduced as a light-weight secondary tillage and seedbed preparation implement with staggered, spaced spring-shanks (Nelson, 1997). Point type and sweep type tools are used on the shanks. Operation depth is regulated by depth control wheels. This implement continues to be the dominant secondary tillage implement. It also facilitates application of fertilizer and pesticides during secondary tillage.

Tandem Disk Harrow

The tandem disk harrow, established as a major tillage implement, was designed with angled gangs of concave disk blades to cut residues and soil and to bury 40 to 60% of the residues. The disk harrow is known to form a subsurface compacted soil layer, but it continues to be used as a major implement to cut crop residues, uproot crop stubble, roughen the soil surface, and to accomplish either primary or secondary tillage. The disk harrow is often considered a reduced-/minimum-tillage implement. It often replaces the chisel system in the more arid climates of the northern Great Plains (Allmaras et al., 1994).

Coulter Carts

Coulter carts consist of a multiple tubular steel mainframes for versatility mounted on a chassis for transport. The cart carries multiple coulters, residue movers and other tillage tools to enhance seeding under extreme residue and NT field conditions. Generally, seed drills and planters are attached to the coulter carts for a one-pass operation. Many of these carts are used in unique reduced tillage systems where zone till is used instead of pure NT or direct seeding. Under certain soil conditions, zone till will give better structure and seed-soil contact and help warm up the soil in cooler areas.

Trends in Tillage Equipment

Over the past 60 years, the design of tillage systems and associated equipment has evolved to achieve once over field operations. New equipment design, new herbicides, more fertilizer application, genetically modified (GM) crops, short stature wheat, higher plant density, larger harvest equipment, and innovative farmers all interact to develop new tillage implements. The general trend is toward more surface residue management (Unger, 1999) and less soil manipulation mainly related to economic issues and partly related to the energy requirements and environmental issues. As plant populations and biomass production are increasing (Allmaras et al., 1998), there is more crop residue to manage. The combination of new computer-aided technology that incorporates precision agricultural techniques and yield maps will further change tillage equipment design in North America.

Allmaras et al. (1998) summarized changes in technology and resource conservation from 1940 to 1990. Advances in tillage and planting equipment have directly impacted management of crop residue for soil erosion control and other benefits to soil quality. Crop residues may provide excellent erosion control, both when on the surface or buried near the surface, but also present significant challenges for machinery system design. Adopted technology for soil and water conservation relates to crop residue return, conservation tillage and soil water response to tillage/residue management, crop rotation, machinery, weed management, crop improvement, fertilizer and cropping nutrition management, and disease management technology.

The early popularity of moldboard plows grew as crop production increased until the peak plow production in the US occurred in the 1950s and '60s when 75,000 to 140,000 units were shipped annually (USDA, 1965; USDA, 1977). Gradually, the moldboard plow was replaced by chisel plows, sweeps, combination tools, and disks for primary tillage. In the late 1980s to 1990, fewer than

3,000 moldboard plows were shipped annually in the US. The Department of Commerce reports that the number of moldboard plows shipped by manufacturers dropped from 46,300 in 1977 to 1,400 in 1991 (USDC, 1992) (see Figure 1). Some of the impetus for change came from the new farm bills and stewardship incentives that encourage conservation farming. The primary reasons given by the farmers for this transition away from the plow are efficiency, equipment width, and speed which the multiple combination tillage tools can be pulled through the soil. Other reasons for going away from the moldboard plow range from no more "dead furrows," no headlands, higher skilled operators, leaving residue on the surface for decreased erosion to overall economics. The moldboard plow may have special uses depending on soil type and wetness, but combination tillage tools have become more prevalent over much of the US.

Maintaining crop residue on the surface for erosion control has required a significant modification of seeding equipment; the most obvious change is increasing plant biomass associated with increased yields (Allmaras et al., 1998). Straw choppers and chaff spreaders on combines must uniformly spread the residue at harvest to improve subsequent residue management

FIGURE 1. Number of moldboard plows shipped in the U.S. from 1977 to 1991.

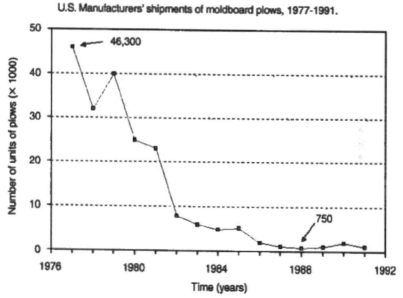

U.S. Manufacturers' shipments of moldboard plows, 1977-1991.

Source: Economic Research Service–U.S. Dept. of Commerce, 1982. [AERI/production inputs]

(Douglas et al., 1989). Seeding equipment requires more downward force in NT than other forms of reduced tillage because of soil strength and more soil engaging tools per seeded row. Devices have been added to seeders to move crop residues away from the openers and seed furrow. These include coulters to cut residue ahead of openers, scuffer wheels or angled disks and residue managers to move residue to the side (Morrison et al., 1988; ASAE, 1998a, 1998b). Coulter carts with triple disk openers, large diameter disks, offset double disks, and angled single disks have been incorporated to improve cutting through surface residue and soil (Tessier et al., 1991). This type equipment is mounted to precede the air seeders and enables rapid seeding of large areas.

As farms get larger and are managed by fewer people, the need for larger equipment is evident to keep all operations timely. Increased HP is required to pull the larger implement and to power the hydraulics needed to achieve once over operations. The average power of tractors sold in the US increased dramatically from 20 kW in 1950 to over 50 kW in 1982 (Allmaras et al., 1998). At present, various types of tracked or wheeled tractors are available that have average power well over 100 kW. The largest track tractors have a power rating of 306 kW (410 HP) and the largest 4WD wheeled tractors have a power rating of 317 kW (425 HP). These larger power units pull larger tillage, planting, transporting, and harvesting machines often at faster speeds. Numerous technological advances in the last 20 years have helped deliver engine power to the appropriate mechanical operation. These technological advances in tractor design also have impacted the adoption of conservation practices.

Axle weight, contact pressure and soil conditions are major factors controlling the degree of soil compaction (Voorhees, 1992). Axle weight of tractors has increased as much as 500% with tractor power in the last 60 years, but contact pressure of wheel tractors has not changed greatly. As a result, subsoil compaction has increased. This increased axle weight is often problematic in spring when soil is wet and more conducive to compaction. To minimize soil compaction, the use of triple wheels on all-wheel drive tractors or crawler tractors to pull the tillage equipment is now common. Many of these trends are driven by timeliness of operations for economic reasons; that include earlier seeding and ability to cover more land in less time. The weight of loaded harvest equipment and soil water content at harvest are big factors to minimize soil compaction.

Conservation tillage systems that leave more crop residue on the soil surface in wet, heavy soils may be more susceptible to compaction unless controlled traffic techniques are used (Reeder, 2000). Controlled traffic is a soil compaction management concept that separates traffic zones from cropping zones within a field. Compaction is managed, but not eliminated. Controlled traffic improves traction, flotation, and timeliness of planting, spraying, and harvesting while minimizing potential yield losses from compaction. Controlled traffic eliminates overlaps and skips during application pesticides and fertilizers and during seeding that typically waste 10 to 15% of chemicals, seed, fuel, and other operational expenses. The first requirement of this concept is to make all equipment cover the same width, or multiples of that width with common wheel and traffic patterns. Adapting controlled traffic is no simple change, but eliminating waste from overlaps, and reduced yields from any gaps in application, may quickly pay for the extra investment in controlled traffic (Reeder, 2000).

Today, most weed control is accomplished using herbicides instead of mechanical cultivation. A major change in tillage for weed control has occurred with the development of Genetically Modified Organism (GMO) that includes crops resistance to various herbicides. Alternative management practices that limit negative herbicide and environmental impacts provide more environmentally sound options. Similar GMOs with incorporated insect resistance also affect tillage decisions. The increase in flexibility and economic returns are a result of the decreasing cost of using this new technology to replace cultivation for weed and insect management.

Crop rotations are an integral of tillage systems. When cover crops and double cropping are used in climates where the cold season permits plant growth, the rotation cycle length includes cover crops, double crops, and full season crops. Crop rotation is highly regionalized whereas the adoption of tillage systems is not significantly regionalized (Allmaras et al., 1998). More of the non-moldboard systems are being used successfully in all crop rotations. Crop rotation is perhaps the most notable technology where practice and research recommendation differ. The additional benefit of cover crops is in the control of pests while continuing with conservation tillage innovations and reducing chemical use (Karlen et al., 1994a; Reeves, 1994).

Tillage rotation is practiced somewhat in corn-soybean rotations (Hill, 2001). Uninterrupted NT ranged from 2.5 (Indiana) to 1.4 (Minnesota) years in soybean-corn rotations of the Corn Belt; 77% of farmers have tried NT in Indiana. The difference between Indiana and Minnesota is likely a thermal effect.

The amount of crop residue on the surface after tillage is influenced by the tillage tool. This information about surface cover is widely published (e.g., Unger, 1999). The residue burial pattern of depth distribution and concentration differs between tillage tools such at each tillage tool has a specific pattern (Staricka et al., 1991, 1992; Allmaras et al., 1988, 1996). These specific patterns persist when primary tillage is followed by secondary tillage (Allmaras et al., 1996). The moldboard plow causes about 75% inversion for each passage (e.g., each year in continuous moldboard systems) so that the older residue is placed near the surface while fresh residue is buried below about 15 cm. Other tillage tools including chisel and disk harrow buried the new residue with the old residue near the surface. These residue positions when using the same primary tillage tool year after year helped to explain erosion control based on soil surface roughness, surface residue and residue incorporated in the 0 to 15-cm depth.

REGIONAL OVERVIEW OF TILLAGE SYSTEMS IN NORTH AMERICA

Tillage and cropping systems are interdependent. The type of tillage depends on crops and crop rotations adopted in a given geographic region. The crops grown are selected based on climate variables, including water and temperature limits suitable for crop growth. The type of soil and tillage tool interaction depends on clay content and type of clay as it affects the cation exchange capacity, water holding capacity, soil organic carbon (C) and fluxes of water, air, and solutes. The spatial variation of the climate regions and soil types and their direct interaction in crop production require that we discuss advances in tillage research in North American cropping systems based on loosely defined geographic regions.

Any discussion of tillage systems must include purposes and tools used. A "tillage system" is a sequence of operations that manipulate the soil and residue to produce a crop. Reeder (2000) compiled an up-to-date discussion of tillage systems and related issues in the US. Morrison (2000) presented a brief review of the history of NT farming in America and discussed the recent developments in conservation tillage. Allmaras et al. (1991, 1994) discussed tillage management regions within the US based on soil types and management land resource areas. Allmaras et al. (1998) reviewed research technology inputs that have transformed American agriculture in the last 60 years to better understand the contributions of science and technology to soil and water conservation. In Canada, Carter (1994a) provided an excellent review on conservation tillage systems. In Mexico, Claverán et al. (1997) and Tapia-Vargas et al. (2001) summarized recent tillage system and erosion research. Excellent overviews of tillage and residue management are also presented by Unger and Skidmore (1994), Hatfield and Stewart (1994), Paul et al. (1997), Lal et al. (1998), Magdoff and van Es (2000), and Allmaras et al. (2000). All of these reports were heavily drawn upon in this review.

Allmaras, Unger, and Wilkins (1985) delineated nine tillage management regions in the US. The boundaries of these tillage management regions did not follow state boundaries, but instead geographical classification used common climate, topography, soils, land-use practices to solve conservation problems. Within a tillage management region, the technological problems and potential for development of conservation tillage should apply throughout the region. Several reviews (Allmaras et al., 1991; Allmaras et al., 1998) have shown conservation tillage to be adapted sooner in those tillage management regions where more summer than winter crops are adapted. In 1998, reduced-tillage was used on 27% of America's farmland, while NT, strip-tillage, RT, and mulch-tillage totaled 37% of the land (CTIC, 2000). Moldboard plowing and an excessive number of operations ("conventional" tillage) were used on 36% of the land. Using statistics that identify the specific tillage tools, present day moldboard tillage is about 7% (Allmaras et al., 2000).

Factors influencing interaction of tillage tools and soil properties are the same factors that affect soil formation and variability across North America. The general forces of weather as described by Jenny (1941) are an active expression of the five major factors: climate, living organisms, parent material, topography and time, which largely control the kind of soil that develops and how that soil might interact in a tillage operation. These five factors for soil formation are critical in determining what types of tillage tools are required for achieving expected outcomes. The corresponding factors for determining tillage type required for a given soil are summarized by Reeder (2000). Selecting the most appropriate system for a particular soil and cropping sequence requires matching the operations to the elements that include crop sequence, topography, soil type, drainage, and weather conditions.

A specific agroecosystem depends on many factors that can be characterized using physical, chemical, biological, and socioeconomic principles. At present, there is not a unified approach to describe agricultural cropping or tillage systems that addresses the above factors. However, in most inventories of agricultural and tillage systems, climate and soil type are major factors. They play an important role in describing various tillage systems and any changes that have taken place in the last 20 years. Soil conservation problems and tillage management systems can be grouped geographically. Soil water and temperature must be managed in all regions. As a result of these complications, we will attempt to define tillage and crop production systems within geographic regions with similar soils and climate. Based on common climate, topography, soil, land use, and cultural practice for a given geographic area, the North American continent was divided along state or province boundaries. This report summarizes the findings for each of these regions to offer a glimpse at the regional characteristics and changes in tillage equipment associated with water and temperature regimes and natural variability across the continent. In many cases, research has been done on a regional level to solve problems within specific physiographic areas. Six geographic regions in the US, three in Canada and one in Mexico are defined (Figure 2) to discuss tillage systems.

Northeastern United States
(ME, VT, NH, RI, CT, MA, NY, NJ, MD, PA, DE, WV, VA)

A general description of agriculture in the Northeast US was provided by Bennett (1977) and Blevins and Moldenhauer (1995). The soil properties and

FIGURE 2. Geographic regions of North America identified for advances in tillage research.

interaction with high intensity rainfall make the area susceptible to soil erosion in the Appalachian Mountains and the Coastal Plain. The soil degradation and loss that took place with CT has now been reduced with forms of conservation tillage. The Appalachian Mountains' rugged terrain makes the land less attractive for agricultural development because much cropland consists of small fields and narrow valleys. Livestock-based agriculture predominates on the steep hillsides. The Northeast generally receives adequate rainfall for crop production. This precipitation is not always uniformly distributed through the growing season. The growing season is short to intermediate in length, but is generally long enough for growth of major crops that include corn (*Zea mays* L.), wheat, soybean (*Glycine max* L. Merr.), and horticultural crops.

Row crops in the Northeast US traditionally have been produced under clean cultivation with moldboard plowing and disking for seedbed preparation with several tillage operations after planting for weed control. However, on the sloping lands, wind and water erosion became a serious problem with continued clean cultivation. In the early 1960s, when the advantages of NT or reduced till became evident, researchers questioned the need for extensive secondary tillage as reliable herbicides and NT planting equipment were developed. Conservation tillage practices reduced wind and water erosion significantly compared to clean tillage. Double cropping with soybeans after barley (*Hordeum vulgare* L.) or wheat has become popular in the Piedmont area. Plowing and disking the soil in double cropping systems cause excess soil moisture losses due to additional exposure to sun and wind. Conservation tillage and NT significantly reduced the time between the small grain harvest and soybean planting resulting in a longer growing season for maximum grain production. Planting directly into small grain stubble, conserved soil moisture, reduced soil erosion and caused the soybean plants to set pods higher. The higher pod set aids in harvesting, particularly with wet soil conditions (Blevins and Moldenhauer, 1995).

Cool spring temperatures are major constraints to corn production in Northeast US and a NT system intensifies the problem. Cox et al. (1990) evaluated the influence of a fall moldboard plowed (CT), NT and RT systems on subsequent growth, development and yield of continuous corn. They found yields from NT averaged 10% lower and yields from RT average 5% lower than the CT treatments with drainage. In the undrained experiment, where some flooding occurred in the three-year experiment, grain yields were significantly higher under RT than under CT and NT. They suggested ridge configuration reduced duration of surface flooding from two days to one day, which resulted in increased plant survival, faster growth and development and higher grain yields that led to the conclusion that RT appeared to be a well adopted conservation tillage system.

Karunatilake, Van Es, and Schindlebeck (2000) evaluated the performance of reduced tillage systems after rotation from a perennial crop (alfalfa, *Medicago sativa* L.) on a clay loam soil in northern New York. They compared corn production with plow till, zone till, NT, and RT. They found corn yield was higher under plow till in one year but was similar to NT in the remaining years of the study. They concluded that reduced tillage systems can perform equally or better compared to fall moldboard plow tillage on this clay loam soil if adequate consideration is given to maintaining soil structure.

Tillage during winter is seldom possible due to frozen or excessively wet soil conditions. "Frost tillage" developed by Van Es and Schindlebeck (1995) was a primary tillage practice performed when a frozen layer exists at the surface and the underlying soil is tillable. The frozen layer needs to be sufficiently thick to support the field equipment (generally 5 to 10 cm thick), but still thin enough to be easily shattered by a tillage tool. Van Es, De Gaetano, and Wilks

(1998) determined and mapped the number of days frost tillage can be performed at various recurrence periods. Frost tillage resulted in a rough soil surface, even after thawing, thereby facilitating water infiltration. Soil drying was improved in certain years and residue cover was greater with frost tillage compared to spring tillage. Yields were similar in both treatments, suggesting that frost tillage may be an alternative management option to move more of the fieldwork into the winter months.

Southeastern United States (NC, SC, TN, GA, AL, LA, MS, AR, FL)

The Southeastern US historically has had severe soil erosion and subsequent surface water quality problems. Many soils have root restrictive genetic horizons and hardpans or are underlain by shallow limestone, sand or heavy clay. From a physiographic perspective, the Coastal Plains are flatter than sloping land of the Piedmont and Appalachian Plateau and do not possess a severe soil erosion hazard as described by Campbell, Reicosky, and Doty (1974), Tyler et al. (1994), and Langdale and Moldenhauer (1995). These unique soils have many restrictions that have been exacerbated through moldboard plowing and from compaction with heavy farm equipment. Generally, the coarse textured A_p horizon holds small amounts of soil C (Hendrix, Franzluebbers and McCracken, 1998), plant available water and the hardpan restricts root growth into the subsoil, thereby, limiting water and nutrient uptake of cotton (*Gossypium hirsutum* L.), corn, and soybean.

Conservation tillage is relatively new in the Southeast with many of the aspects not fully understood because soils, climate and farms differ widely across the area. The design of one tillage system to fit all conditions is unrealistic. Annual deep tillage requiring more energy, usually in-row subsoiling, is often necessary in southeastern Coastal Plains hardpan soils to maintain a suitable rooting environment (Busscher et al., 1995). They found that residual "slit" tillage (shallow subsoiling to 15 cm) did not out yield NT the following year. Most of the slits did not persist with only 10% identifiable one year after tillage. Lack of slit persistence in the soil was due to soil collapse and infilling by sand particles. Even though the slits did not persist, slit tillage may be better than deeper subsoiling (46 cm) if performed annually because it conserved energy and maintained yields. Chiseling and subsoiling to increase the rooting depth is rapidly gaining popularity in the Southeast particularly in the Coastal Plains' soils. Yield responses are weather dependent with increases most likely in dry years when water stress occurs at critical stages of plant development. Equipment locally referred to as the "ripper hipper" received some acceptance. This in-row subsoiler loosens hard pans and allows deeper root penetration for water and nutrient extraction. The bedding component of this tool improves water and soil temperature relations for seed germination in the early spring. Results with the "super seeder," which subsoils under the row with a minimum disruption of surface residue, were also weather dependent (Karlen et al., 1991). It maintained residue on the surface to protect the soil from raindrop impact and minimize soil erosion.

Because of a longer growing season, double cropping soybean after wheat harvest is a viable alternative to monocropped soybean production in the Southeast US. The amount of soil fracture by in-row subsoiling may not be sufficient for soybean grown in narrow rows. Frederick et al. (1998) evaluated the yield response of double crop soybean to surface tillage (disk harrow twice to 18 cm) and deep tillage (four-shank Para Till[1] to 41 cm) when grown in 19- and 76-cm row widths. Soybean seed yields were normally higher for narrow row widths. Seed-yield increases due to deep tillage were greatest when plots were deep tilled before planting both crops and when no surface tillage and narrow rows were used. More recently, Busscher, Frederick, and Bauer (2000) found that spring-only deep tillage maintained lower conc indices in the fol- lowing double-cropped growing season than fall-only deep tillage. Compared to non-disked treatments, disked treatments produced equal or higher mean- profile cone indices (a measure of penetration resistance). They concluded that if producers deep till only once a year, spring tillage appears to be better for the Goldsboro loamy sand.

Research on rainfed crop production in Coastal Plain soils indicate some form of deep tillage is needed for roots to efficiently extract subsoil moisture. For cotton production using conservation tillage, researchers have reported equipment and cultural practice effects (Burmeister, Patterson, and Reeves, 1995; Naderman, 1993), on soil strength (Busscher and Bauer, 1995), irriga- tion and tillage (Camp, Bauer, and Busscher, 1999; McConnell et al., 1995). With deep tillage, Frederick and Bauer (1996) reported 25% greater winter wheat yield with no surface tillage in a dry year and no effect in another year with adequate rainfall. They concluded that the probability of yield increases from deep tillage should be greater with no surface tillage than with disking.

Central United States (OH, IN, MI, IL, WI, MN, IA, MO)

The Central US is referred to as the Corn Belt. The soils in the western por- tion of the Central US are predominantly Mollisols with Alfisols on the eastern edge. Characteristics of tillage interactions with soils and cropping systems are provided by Griffith, Mannering, and Moldenhauer (1977), Allmaras et al. (1994), Lal et al. (1994a), Moldenhauer and Black (1994), Moldenhauer and Mielke (1995), Amemiya (1997), and Reeder (2000). Soils in the northern three-fourths of the region are derived from glacial till while many of the soils in the western part of the region are derived from loess deposits. Corn and soy- bean are the principle crops in each of the states and account for 75% of total cropland in the Central US, which is prone to accelerated erosion with risks of both off and on-site damage. Nearly 40% of the land in Minnesota was mold-board plowed before soybean planting in the early 1990s (Allmaras et al., 1994). The remainder of planted land was moldboard plowed or NT or RT depending on the soil type. This marked change in tillage management occurred because moldboard plowing was the primary tillage for corn and soybeans as late as 1975. Permanent pasture and small grains have increased in the southern sections with a limited number of horticulture crops grown in the northern states.

Tillage systems used in the Central US vary widely in terms of equipment type, tillage depth, and amount of soil disturbance. Griffith, Mannering, and Moldenhauer (1977) stated that tillage systems during the early 1970s used moldboard plowing in the fall and spring with various forms of secondary tillage using field cultivators or disk harrows. Other forms of spring plowing followed by wheel track planting were employed in some areas, but no longer used. Lal et al. (1994a) discussed the gradual transition from intensive tillage to conservation tillage. They reported that about 45% of the land area in the Central US was in some form of conservation tillage that included NT, RT, strip till, mulch till, and "reduced" till. The principle advantage of using conservation tillage was soil erosion control. Conservation tillage was adapted primarily because reduced time and labor in seedbed preparation. Types of conservation tillage adapted depended on crops, rotations, specific soil type, climate, and drainage conditions. This was illustrated by Iqbal et al. (1995) who studied in-row soil disturbance effects from the use of NT, a single coulter or triple coulters in NT. The triple coulter unit produced the zone of lowest bulk density and penetration resistance, but corn plant emergence and growth rate was the slowest, apparently as a result of soil compaction at the base of the coulters. The results showed the planting method was not compatible with a strip tillage technique and existing soil properties. Conservation tillage methods are generally easier to adapt for crops on soils that are well drained (Allmaras et al., 1991).

The RT system was often used in soils with slow internal drainage (Fausey, 1991; Eckert, 1990). In Ohio, RT was considered appropriate for poorly drained, heavy textured soils where crop growth was limited by cool soil and anaerobic conditions (Erickson, 1982). Data of Fausey and Lal (1989a, 1989b) showed that mean daily maximum soil temperatures measured in April were highest atop the undrained ridges. Ridge tillage and raised beds further away from the drain had higher maximum temperatures than moldboard plow tilled or NT methods of seedbed preparation. An overview by Lal (1990) states that RT reduced labor costs, enhanced soil fertility, improved water management, improved water and wind erosion control, as well as facilitated multiple cropping, enhanced rooting depth and improved pest management. Given all these benefits, why was RT used on less than 4% of the total cultivated land in the Central US (ERS 1994)?

Hatfield et al. (1998) discussed RT research that evaluated trace chemical movement from site of application within the soil to groundwater aquifers. The environmental impacts were generally positive, but depended on soil and climatic factors. Ridge till changed soil temperature and soil water patterns compared to NT and full width moldboard plow/chisel plow and/or disking for the primary tillage. These changes led to improved environment for crop emergence and early growth because of warmer soil temperatures in cool climate and better water relations in moderately or poorly drained soils. Hatfield et al. (1998) concluded that crop production would likely be enhanced by RT systems. Similar benefits were noted on a sandy soil in Minnesota (Lamb et al., 1998).

In humid climates with corn/soybean cropping systems, the moldboard plow system has been the traditional tillage system. Karlen et al. (1994b) found in 12 years of tillage system comparisons with continuous corn, the NT and chisel systems accumulated C in the 30 cm soil layer relative to the moldboard plow system. Carbon return from the primary production was nearly the same for all three tillage systems. Similar results were found in other long-term tillage comparisons of continuous corn and a corn/soybean sequence in Ohio (Lal, Mahboubi, and Fausey, 1994; Dick and Durkalski, 1997; Dick et al., 1998; Huggins et al., 1998). Wander, Bidart, and Aref (1998) demonstrated that tillage methods impacted the depth distribution of soil organic matter (OM) differently in three Illinois soils. Generally, NT increased soil C and particulate OM 25 and 70% compared to CT (moldboard plow after corn and chisel plow after soybean) in the surface 5 cm at the expense of soil C in the 5 to 17.5-cm depth. The results demonstrate the importance of soil type variation and tillage tool interactions in C accumulation.

Deep soil compaction sometimes requires subsoiling clay or clay loam soils to modify soil properties and enhance crop growth (Wu et al., 1995). Evans et al. (1996) found that subsoiling in Minnesota had very little effect on plant growth and no effect on grain yield over three cropping seasons. Subsoiling to 41 cm had significant effects on bulk density and volumetric water content the year after tillage, but in subsequent years, these effects were not significant with random field traffic. Volumetric soil moisture content generally increased in relation to soil bulk density increases. Subsoil tillage impacted crop residue accumulation, but did not affect soil bulk density, volumetric water content or grain yield. Results indicated that subsoiling does not necessarily improve yields or soil moisture availability particularly if pre-existing compaction does not limit root development.

Great Plains of the United States
(ND, SD, KS, NE, OK, TX, CO, WY, MT)

The Great Plains has become an important agricultural region in the US because of N fertilizers, weed control, and less summer fallow. This region extends from the Canadian border to southernmost extremes of Texas. The area is often characterized by relatively low rainfall occurring most of which occurs during the summer months. The moisture environment ranges from moist subhumid in the east to semiarid in the west. There is a wide range in temperature from very cold in the north to moderate temperatures in the south. Soils and cropping system characteristics for this area have been reviewed by Fenster (1977), Unger, Wiese and Allen (1977), Unger and Skidmore (1994), Moldenhauer and Black (1994), Stewart and Moldenhauer (1994), and Allmaras et al. (1994). The soils are mostly deep Mollisols with smaller areas of Alfisols in the southeastern portion. Irridisols occur more frequently in the western parts where the cropland is often a small percentage of the total land area. Corn and soybean are the major crops in the western Corn Belt and eastern Great Plains while wheat is the major crop in the drier areas. Allmaras et al. (1994) reported primary tillage with a chisel, disk or a sweep plow was used on most land planted to wheat, corn, and soybean. The sweep plow has been a primary tillage tool since the 1940s in the wheat fallow rotations of the semiarid regions. In the more humid regions, various forms of conservation tillage are more widely used which improves water conservation and reduces soil erosion and degradation. The main limitations of crop production in these semiarid areas are water supply and soil water storage capacity.

Winter wheat and grain sorghum (*Sorghum* ×*alum* [L.] Moench) are the major crops of this region. Cotton is another major crop in the southern portions of the Great Plains, which is often irrigated if ground water is available. Other important crops include corn, sugar beets (*Beta vulgaris* L. subsp. *vulgaris*), various vegetable crops in the southern regions and alfalfa, peanuts (*Arachis hypogaea* L.) and some soybean and a limited amount of oats (*Avena sativa* L.). Crop selections in a rotation depend on water availability and the potential for irrigation.

As in other regions, tillage in the southern Great Plains once considered "conventional" changed with advances in technology. "Fallow" treatments were designed to conserve water by plowing and frequent shallow tillage during the season without a crop. With subsequent wind and water erosion on fallow areas, straw mulching became the "conventional" system in that area (Jones and Johnson, 1983; Unger and Skidmore, 1994). A combination of limited tillage and herbicides for weed control is now widely used and could be appropriately called the conventional system for growing dryland wheat and grain sorghum in the region. With the new straw mulch tillage, large sweeps or blades undercut the soil surface at 5 to 10 cm while retaining most of the crop residues on the soil surface. Straw mulch tillage is effective for controlling wind erosion, provided adequate crop residues are available. Chisel plows are widely used for dryland wheat production, but tillage systems involving the chisel plow seldom retain enough residue on the surface in parts of the region to be classified as conservation tillage (Unger and Skidmore, 1994). For irrigated wheat, the land is usually disked, chiseled and perhaps disked again and furrowed to control irrigation water flow. Moldboard plowing was used by some producers, but has been on a decline (Unger and Skidmore, 1994).

Field experiments in Oklahoma on continuous winter wheat over a 10-year period, compared six different tillage systems: the moldboard plow, chisel plow, disk harrow, sweep, and NT (Epplin, Al-Sakkat, and Peeper, 1994). Wheat yields from moldboard plow tillage systems were consistently greater and showed less variability than yields from three intermediate tillage systems and NT. The lowest yield was from NT. Yield was inversely related to the amount of crop residue cover on the field prior to planting. The yield declines were attributed to rootborne and soilborne pathogens, secondary toxins, and increased weed competition associated with higher crop residue cover. Raun et al. (1998) showed substantial N fertilizer effects on soil C and N in continuous wheat related to biomass and grain yield.

Cropping systems involving fallow are sometimes used in drier regions and can influence soil C storage (Paustian, Elliott, and Carter, 1998). Fallowing has decreased as the role of carbon (C) became more important in maintaining soil physical properties. Salinas-Garcia et al. (1997) evaluated long-term and seasonal changes in soil organic C, soil microbial biomass, soil microbial N, and mineralizable C and N in continuous corn under several different tillage systems in Texas. The tillage treatments examined were CT (shredding and disking stalks after harvest, followed by lifting out the crown stubble in old plant rows and re-bedding with row middles and beds cultivated during fall and winter to control weeds); moldboard plow (shredding and disking stalks after harvest, followed by moldboard plow to 30 cm and field cultivation, then bedding, with row middles and beds cultivated during fall and winter to control weeds); chisel plow (the same operations as moldboard plow but using a chisel plow to 30 cm instead of the moldboard); and "minimum" tillage (shredding and disking stalks after harvest, followed by root and plant stubble lifting and forming low-profile beds with herbicides for weed control); and NT (shredding stalks and spraying herbicides as needed for fall and winter weed control). They found that seasonal distribution of soil microbial biomass C and mineralizable C were consistently greater in non-moldboard plowed tillage systems, averaging 22 and 34% greater than moldboard plow treatments at planting. The greater amount of crop residues remaining with "minimum" tillage and NT may have provided available substrate for maintenance of the larger soil microbial biomass pool and the higher C and N mineralization in the

0 to 20 cm depth during the growing season. Reduced tillage systems that promote surface residue accumulation provided opportunities for increasing C sequestration and mineralizable nutrients within the soil microbial biomass (Franzluebbers, Hons, and Zuberer, 1998).

Potter et al. (1997, 1998) compared NT and stubble mulch management on four dryland cropping systems and found that fallow limits C accumulation. They found NT treatments resulted in significant differences in soil organic carbon (SOC) distribution in the soil profile compared to stubble mulch tillage. The SOC differences were largest in continuous cropping systems. No-till management with continuous crops sequestered more C compared to double mulch treatments. Similarly, Dao (1998) found that with continuous winter wheat, greater SOC storage occurred in NT compared to moldboard plow tillage. Carbon storage in the top 20 cm with NT was increased as available wheat residue increased, but soil C storage in the moldboard system was never sensitive to the amount of wheat residue returned. Peterson et al. (1998) noted smaller SOC losses in NT versus non-moldboard plow tillage in the Great Plains for wheat or sorghum/fallow, continuous wheat or sorghum, and some wheat/fallow/sorghum cropping systems. Differences in the net primary production (C input) can also modify the relative organic C storage in these non-moldboard tillage systems (Wienhold and Halvorson, 1998). Reeder, Schuman, and Bowman (1998) described soil C and N changes in Conservation Reserve Program lands in the Great Plains without tillage for at least 10 years, suggesting some C input from native grasses.

Strip-tillage/zone-tillage was promoted and rapidly adopted in the 1990s for soils which were not friable and benefited from localized loosening and drying in the seed zone prior to seeding. Halvorson and Hartman (1984) reported that as much as 20% of the area of sugar beet production, used shallow powered row-zone tillers (7 to 10 cm deep) with no yield differences compared to conventional and NT systems. Sugar beet strip-tillage systems were also compared on a very fine sandy loam by Smith et al. (1995) who found that both minimum tillage and powered tiller row-zone systems reduced total energy requirements by 60% over conventional moldboard plow systems. Placing fertilizers in the soil prior to seeding is one of several alternatives for strip-tillage (Morrison, 1999). Tillage (10 to 20-cm wide strips) is accomplished with various tools such as knives, sweeps, or rotary tillers. Residue rakes, rolling coulter blades, residue wheels, or other tools may precede the tillage tool to clear a path through residues and control depth. The tilled strips are somewhat similar in condition to the seedbed produced by CT, so that conventional row-crop seeders can be used without the need for excessive ballast weight or downforce springs. Residue rakes are usually attached to the seeders to clear any loose residues from the path of the tillage tool. Compared to NT, strip-tillage usually requires one more field operation to facilitate conservation tillage on difficult soils. Strip-tillage technology promises to be readily accepted by farmers and useful for many years. As equipment designs are improved and more experience is gained more and more tillage operations become less intensive and more effective for controlling both wind and water erosion in the Great Plains. The continued improvements in conservation tillage in a stepwise matter will lead to even better ways to minimize soil erosion and enhance soil physical properties.

There is uncertainty about conservation tillage in furrow irrigated fields with ample surface residue because the residue "dams" irrigation furrows and prevents uniform water distribution (Carter, Berg, and Sanders, 1991). Research was conducted to compare the agronomic and economic performance of conservation tillage during the establishment years with conventional tillage in furrow-irrigated cropping systems of corn, soybean, winter barley (*Hordeum vulgare* L.), and dry bean (*Phaseolus* spp.) in western Colorado (Ashraf et al., 1999). They found conservation tillage can be used successfully in furrow irrigated cropping system to control soil erosion. Surface residues can be managed without adversely affecting crop yields. They concluded that successful adoption of conservation tillage under furrow irrigation would require growers to adopt new production management practices and possibly purchase new equipment to operate in high residue conditions.

Pacific Northwest United States (ID, WA, OR)

The Pacific Northwest US supports significant areas of both irrigated and rainfed agriculture (Papendick and Miller, 1977; Sojka and Carter, 1994; Papendick and Moldenhauer, 1995). Much of the region's cropland is dry farmed, but significant areas in the Columbia Basin and the Snake River Plains are irrigated. The topography varies from nearly level valleys to steep, sloping uplands interior. Much of the steeper land is located in higher precipitation zones susceptible to soil erosion. Up to 80% of cropland in dry land farming commonly has slopes from 8 to 30% with some exceeding 50%. As a result of the topography and soil properties, water erosion and wind erosion are major concerns. Winter wheat is the major non-irrigated crop and other cereal grains and pea (*Pisum sativum* L.) are also important in areas of high precipitation. Major irrigated crops include potato (*Solanum tuberosum* L.), sugar beets, alfalfa, beans, corn, small grains, and various tree fruits. The climate ranges from humid to subhumid and semiarid on various parts of the mountain areas. The surrounding mountain ranges modify local climate so much that seasonal and annual precipitation in areas of a few kilometers apart may differ by as much as 50% (Papendick and Miller, 1977). Some soils are developed from loess and some are derived from glacial drift. Nearly all of the upland soils contain varying amounts of volcanic ash in the surface layers. Many of the coastal area soils are mainly alluvial. Most of the soils are permeable, well drained. Some are not sufficiently deep to store precipitation for a wide range of crops. Soils in the low precipitation zones typically are fine, sandy loams or silt loams.

Most wheat/fallow or sorghum/fallow rotations in semiarid environments showed soil organic C decline related to tillage systems (Rasmussen et al., 1998). The C storage ranked by tillage method showed NT > non-moldboard tillage > moldboard tillage (Smith and Elliot, 1990; Allmaras et al., 1998; Rasmussen, Albrecht, and Smiley, 1998). Primary tillage with a disk or a sweep plow provided a soil organic C storage superior to a moldboard based system. In an adjacent field trial with wheat/pea, Rasmussen, Albrecht, and Smiley (1998) reported aggrading soil organic C with the non-moldboard system but continued decline with a moldboard system that lasted for more than 30 years. Only the treatment with regular applications of animal manure maintained a stable C content throughout the study.

Tillage for soil conservation on irrigated row crops may not always be NT or even residue maintenance on the soil surface. Subsoiling in furrow or sprinkler irrigation, and small damming basins under sprinkler irrigation are examples of tillage practices that may replace otherwise conventional tillage systems to improve infiltration, reduce runoff, and prevent erosion. Only in the past 8-10 years have NT systems been introduced to irrigated land in the Pacific Northwest (Sojka and Carter, 1994). No-till systems for irrigated land were developed and evaluated by Carter and Berg (1991) and Carter, Berg, and Sanders (1991). They showed cereal or corn can be grown following alfalfa, corn following cereal or corn and cereal following corn without tillage using the same furrows for irrigating the subsequent NT crop. Both erosion and sediment losses were greatly reduced and in many cases completely eliminated. Crop yields were nearly identical without tillage as with traditional tillage. No-till conserved soil by reducing erosion and sedimentation and increased net income as a result of reduced tillage cost. The work of Carter and Berg (1991) and Carter, Berg, and Sanders (1991) demonstrated that conservation tillage can be successful on furrow-irrigated land and is currently the best approach for soil and water conservation in the Pacific Northwest.

Tillage erosion, movement of soil downslope by a mechanical implement, may not be as spectacular or destructive as water erosion from intensive rainfall events. Though recognized as a problem on the steep slopes in the Pacific Northwest, effects of tillage erosion are more subtle and accumulate over long periods of time. Tillage erosion is very severe on the steep slopes especially with moldboard plowing (Papendick and Miller, 1977). On irregular terrain, it is more practical to contour than to farm up and down the hill or across the slopes. Repeated plowing downhill moves large amounts of soil down slope. Plowing accelerates the removal of soils from the hilltops and results in translocation down slope and degradation. Today's faster and more powerful tractors compound this problem by moving the soil even faster and farther with conservation tillage tools.

Arid West (Irrigated) (UT, NV, CA, NM, AZ)

The irrigated valleys of California and Arizona produce numerous high value row crops including many fruits, vegetables, and cotton. Cropping practices are characterized by frequent fertilizer and irrigation inputs and intensive tillage for bed formation. Aggressive tillage practices and associated loss of soil OM have led to recent concerns about soil degradation in the region (Mitchell et al., 1999). Much of the intensive tillage is used to loosen the soil so that it can be reshaped into 1.53 m center-to-center beds for planting various vegetable crops requiring irrigation. Other types of tillage methods are used with emphasis on RT to enable furrow cultivation for the high value irrigated cash crops. An important alternative for potentially reducing such degradation and overall production costs is the use of reduced tillage. In recent years, several conservation tillage systems have been developed including NT, strip till, and RT that are commonly used in the irrigated valleys (Mitchell et al., 1999).

California law requires complete burial of cotton residue to minimize carryover of the boll worm (*Heliothis armigera*). Typically, four to seven CT passes over the field are required to get complete incorporation of the cotton residue to meet legal requirements. Recently, the Rome-Pegasus[1] plow (Lyle Carter, 2000, personal communication) has been invented as a one-pass cotton tillage tool. After the cotton has been picked, the unit opens the furrow, cuts the tap roots, trenches the standing cotton stalks followed by closing discs for a one-pass cotton residue burial and tillage application. The equipment benefits are cost savings in fuel, operator time, and other farm energy costs. The tillage tool greatly reduces field compaction when compared to current conventional farm practices and also reduces dust emissions and topsoil loss. The incorporation of the cotton residue enhances the soil profile and field trials have shown that boll-worm control is equal to or better than that of conventional moldboard plow tillage. The one-pass operation provides a unique form of residue management. It complies with legal requirements and also leaves the field with ridges for the following crop. The one-pass operation also cleans out the irrigation furrows so that the next crop can be easily irrigated and is equivalent to a RT operation.

In New Mexico, Christensen et al. (1994) observed more soil C storage in NT than in sweep-blade tillage systems applied to a sorghum/fallow/wheat system. Both treatments were converted from long-term moldboard systems without change in cropping systems. They noted a 25% increase in stored C

---

1. Names are necessary to report factually on available data; however, the USDA neither guarantees nor warrants the standard of the product, and the use of the name by USDA implies no approval of the product to the exclusion of others that may also be suitable.

within five years that indicated a short-term advantage for NT and sweep-blade systems over the moldboard tillage system.

The impact on soil OM and nutrient cycling are increasing in importance with a current renewed focus on agricultural sustainability. In many parts of the world, rice (*Oryza sativa* L.) straw is burned for disease and pest control and for labor and energy savings. However, air quality concerns have dramatically reduced or banned rice straw burning in California. A search for alternate residue management techniques is essential. Eagle et al. (2000) found that straw management affected grain yield in zero N fertilizer plots. They used a heavy roller to roll, crush and flatten the straw into the soil surface prior to winter flooding. The straw effect was mainly due to the greater yields in the straw retained treatments (incorporated and rolled) compared to the straw removed treatment (burned and baled). This unique form of residue management increased the soil N supply and led to a reduction in N use efficiency in the N fertilized plots suggesting that N fertilizer can be reduced when the rice straw is retained.

Eastern Canadian Provinces (NB, NF, NS, PE)

In eastern Canada, soil is traditionally prepared for seeding field crops after fall moldboard plowing followed by a spring secondary tillage. Ketcheson (1977) and Vyn, Janovicek, and Carter (1994) provide a general description of the soils and climate for the eastern Canadian provinces. Much of the area has a continental climate that is modified by the Great Lakes. Glacial deposition is the parent material for the majority of the soils in eastern Canada. Prompted by concerns about soil erosion and high capital and labor investments with intensive tillage, grain and oil seed producers are increasing the adoption of less intensive tillage systems. The main limitation of many of the soils is imperfect or poor internal drainage that may limit crop production. Many fine textured soils in annual crop production areas are systematically tile drained. Some agricultural soils in Prince Edward Island have major limitations of low fertility and poor soil structure mainly evident as dense subsoil. The combination of poor subsoil structure with humid climate conditions causes excessive soil moisture especially in the spring and autumn. Thus, many of these soils are susceptible to soil compaction. Much of eastern Canada's climate supports a wide range of crops including corn, soybeans, potatoes, cereal crops, hay, and pasture.

Carter (1992) determined the physical condition of the soil profile under reduced tillage conditions for winter wheat in eastern Canada. He assessed the response of winter wheat to different tillage treatments in regard to plant survival and grain production. Three tillage treatments were employed: direct drilling (NT), shallow tillage (two passes with a rotary harrow at 10 cm deep), and moldboard plowing to 25 cm followed by a furrow press to reconsolidate the loosened soil. The latter tillage treatment is a one-pass system that eliminates the need for secondary tillage. All tillage treatments retained straw and stubble from the previous crop that was spread during the harvest operation. Direct drilling increased plant survival but not grain yield in the first year compared to other tillage systems. Leaf diseases significantly reduced crop performance under shallow tillage and direct drilling in the second year. None of the tillage systems had adverse effects on the soil strength and field capacity over the 35 cm soil profile depth. Auxiliary measurements indicated relatively large macro pore-volumes under moldboard plowing followed by a furrow press were less efficient in conducting air than macropores under direct drilling that were most prevalent at the lower soil depth.

Fall moldboard tillage is the conventional primary tillage used with potatoes in eastern Canada. Carter, Sanderson, and McLeod (1998) compared spring moldboard plowing and fall chisel plowing in the potato phase of a 3-year rotation. They evaluated the degree of soil loosening, soil macrostructure, soil density, strength and permeability and crop yield and quality. Although moldboard plowing provided an additional 2 to 10 cm of loose soil at the lower depth compared to chisel plow, there were no differences in soil permeability in sandy loam soils. The total potato yield and marketable yield were not influenced by differences in primary tillage over the 3-year period. Use of spring primary tillage and replacement of moldboard plow with a chisel plow, within the potato phase of the 3-year rotation, caused little change in soil physical quality compared to the CT systems. The use of the chisel plow appears to be a suitable alternative to enhance conservation tillage techniques for sandy loam soils.

With the current interest in C cycling processes and the need to assess soil C stocks, Carter et al. (1997) determined potential storage of soil C and N for a wide range of soils under different agricultural management systems. Information was obtained from agricultural soils under intensive tillage over the last 25 years. Some soils in eastern Canada possessed a relatively high potential for OM storage when the appropriate tillage methods were used. Angers et al. (1997) concluded for a range of soils under continuous corn and small grain cereal production in Eastern Canadian conditions, reduced tillage systems did not increase the storage of soil OM in the entire profile, at least in a 5 to 10 year period. Where crop production and residue inputs are not affected by tillage, they found no differences between tillage treatments in total organic C and N storage down to 60 cm. In the surface 0 to 10 cm, C and N contents were higher under NT than under moldboard plow whereas in the deeper levels (20 to 40 cm), the reverse trend was observed. Placement of the residues was a major factor influencing C and N distribution in the soil profile, especially when erosion and deposition occurred (Gregorich et al., 1998).

Vyn, Janovicek, and Carter (1994) reviewed tillage requirements for predominant annual crops in eastern Canada. They focused on tillage systems for corn, small grains, soybeans and evaluated tillage-induced change in soil properties and their effect on soil erosion and crop performance. The autumn mold-board plow tillage system was used as the conventional treatment and compared to other forms of conservation tillage and NT. They showed that reduced grain yields in NT compared to autumn plowed tillage systems could not be attrib-uted to differences in rates of emergence or corn plant population. They con-cluded that inferior crop performance in the NT system has been attributed to either increased pest and disease problems or poor seedbed conditions. No-till soils have been characterized as having higher bulk densities, higher soil pene-tration resistance, lower macroporosities, and seedbeds with a greater propor-tion of coarse aggregates. These inferior soil conditions could inhibit root growth and water availability over the growing season. They further noted that corn performance was unaffected by tillage systems used for the preceding soybean crop.

Little is known about micronutrient availability under different tillage sys-tems. Carter and Gupta (1997) studied the effects of several minimum tillage methods on micro- and secondary nutrient content of barley and soybean. They identified incipient deficiencies and detrimental tillage-induced changes in plant nutrient accumulation in a fine sandy loam Podzol. They evaluated various forms of spring and fall moldboard plow, paraplow (slant-legged soil loosener) followed by rotary harrow, disk harrow, chisel plow, and direct drill-ing. Reduced tillage had various effects on micronutrient concentration in plant parts, which were probably related to pH changes in the surface soil. Overall, grain micronutrient concentrations in barley and soybean were in the sufficiency range for optimal yield.

Vyn, Janovicek, and Carter (1994) also discussed the development and evaluation of strip tillage systems on corn yields. They evaluated various com-binations of in-row soil loosening, in-row surface residue removal and RT planting systems. They found that on sandy loam soils, strip tillage yielded similarly to the autumn moldboard plowing treatment and were generally greater than the NT systems. On silt loam soils, strip tillage yields were inter-mediate to an autumn moldboard plowing system. On a clay loam soil, strip tillage used as a secondary tillage operation after autumn offset disking re-sulted in corn yields between those with moldboard plow and NT systems. The ability of strip tillage systems to significantly improve corn performance rela-tive to NT alone indicated that suitable soil conditions can be attained by loos-ening a relatively small volume of soil (Janovicek, Vyn, and Voroney, 1997). Similarly, Raimbault, Vyn, and Tollenaar (1991) used powered rotary tiller units to produce shallow 12-cm wide tilled strips into herbicide-killed rye (*Secale cereale* L.) cover crop.

The success of reducing tillage is dependent upon the crop, soil type, and the preceding year's crop residue left on the surface. The diversity of crops and soil types associated with many farming operations in eastern Canada make the selection of a single tillage system very difficult. In these soil types and the cool climates, Vyn, Janovicek, and Carter (1994) concluded that with proper management, few situations exist where an autumn moldboard plow tillage system was necessary to ensure economic crop yields. While the particular tillage system may be specific for a crop and slightly different for soybean versus small grain versus corn, use of less intensive tillage and conservation tillage techniques has merit. Due to frequent and often intense rainfall events combined with the relatively complex topography in eastern Canada, conservation tillage techniques that leave residue on the surface are required to decrease severe water erosion.

Soil compaction, in the surface 0 to 20 cm, is common on light textured sandy soils due to vehicular traffic, animal traffic, and natural soil bulk density that is marginal for crop growth. Carter and Kunelius (1998) evaluated non-inversion tillage using a para plow (a slant legged cultivator that loosens the top 10 to 20 cm) on permanent pasture productivity. Penetrometer resistance profiles showed that the loosened soil condition persisted for three to four years. However, the non-inversion soil loosening caused a negative pasture yield response attributed to root injury following the tillage operation.

Sijtsma et al. (1998) evaluated tillage costs in eastern Canada. They assumed similar crop productivity and input costs in two rotations and found that fuel usage for seedbed preparation and crop establishment was lower with several reduced tillage practices (10.0 to 23.7 l ha^{-1}) than conventional moldboard plowing (27.6 l ha^{-1}). The conventional moldboard plowing combined with secondary tillage was the most costly system in both rotations. Replacement of the moldboard plow with various combinations of alternative tillage systems provided annual tillage cost savings of 44 to 60% for the three-year potato rotation and 10 to 40% for the barley-soybean rotation. They concluded that the adoption of various reduced tillage practices would be more economical than a conventional moldboard plow system.

Central Canadian Provinces (ON, QC)

Central Canada contains a large portion of Canada's agricultural land, where soil is traditionally prepared for seeding field crops using fall moldboard plowing followed by a spring secondary tillage. The agroecosystems in the central Canadian provinces have been described by Ketcheson (1977) and Carter et al. (1997). Major crops are corn and soybean with cereals, pasture, and some forages. Approximately 65% of agricultural land in eastern Québec is under cereal and forage production. Most of this land is still being cropped with conventional methods, including primary tillage in fall and secondary tillage in spring, just prior to seeding. Although some progress has been achieved, particularly in the southwestern part of the province, adoption of conservation tillage practices is still lagging compared with other areas of North America.

Constraints to the adoption of conservation tillage in Québec potentially include cool, wet springs, short growing seasons, and variable precipitation patterns. Légère et al. (1997) considered the interactive effects of rotation, tillage, and weed management intensity simultaneously in the cereal cropping system. Their objective was to determine effects of conservation tillage practices and weed management intensity on populations and dry weights of crop and weeds at midseason and on final grain yields in two rotations of barley and red clover (*Trifolium pratense* L.). Tillage treatments were moldboard plow in the fall (15 to 18 cm), followed by spring secondary tillage; chisel plow in the fall (12 to 15 cm), followed by spring secondary tillage; and NT or direct seeding. Spring secondary tillage consisted of two passes of a rigid-tooth finishing harrow. Yields produced under NT were comparable to those in moldboard plow treatments and did not require a major increase in herbicide use. Their findings demonstrate implementation of conservation tillage for spring barley production is feasible in eastern Québec. However, the benefits of conservation tillage practices can only be fully realized if the proper attention is given to crop establishment and weed management.

The adoption of NT systems on clay soils in North America has been hampered by reports of delayed emergence and growth and reduced grain yields, especially following winter wheat in a rotation. Opoku, Vyn, and Swanton (1997) evaluated several conservation tillage and wheat residue management systems that provided favorable seedbed conditions for emergence, growth, and yield of corn following winter wheat on clay soils in Ontario. They found NT yield potential was affected by the amount of wheat residue present. Modifying the spring NT planting system by adopting fall zone till or fall tandem disk produced corn grain yields no different from fall moldboard plow or fall chisel plow treatments. Similar results were observed for soybean by Vyn, Opoku, and Swanton (1998). No-till soybean growth was delayed and yields were reduced with increasing wheat residue left after planting. They recommended fall zone till and fall tandem disk systems as the best conservation tillage alternatives to fall moldboard plowing.

Tillage operations such as plowing are known to increase loss of soil OM (Janzen et al., 1997). Conversely, reduced tillage frequency and increased surface residue increase soil OM. Angers, N'dayegamiye, and Côté (1993) conducted a silage corn production study to determine the influence of reduced tillage practices on soil OM in particle-size fractions and microbial biomass. The three tillage treatments were "minimum" tillage which consisted of two passes of a field cultivator (spring tines) in the spring; RT in which ridges were reformed each spring before planting; and moldboard plow in the fall followed by harrowing in the spring. Moldboard plowing was performed at 18-cm and spring cultivation at 7-cm depths. The ridges were 15 cm high after formation. Silage yields were essentially the same and were not influenced by tillage. Total organic C did not differ among the tillage treatments at any depth sampled. The study demonstrated that, even in cropping systems involving almost no return of aboveground residue to the soil (typical of silage corn production) a reduced tillage intensity can maintain or increase the more labile fractions of soil OM.

Crop rotation and tillage practices can alter OM accumulation and mineralization by changing the structure of the soil through soil disturbance and mixing. Angers, Samson, and Légère (1993) studied changes in water-stable soil aggregation under different tillage and rotations during a 4-year study. They evaluated two rotations of barley and three tillage treatments. The three tillage methods consisted of fall moldboard plowing (15 to 18 cm) with spring secondary tillage, fall chisel plowing (12 to 15 cm) followed by spring secondary tillage and NT. The mean weight diameter of water-stable aggregates did not vary significantly with time under the NT treatment, but decreased significantly under the moldboard plow and chisel plow treatments. The effect of water content on mean weight diameter was less apparent under NT, suggesting that these aggregates were less susceptible to slaking. In a companion study, Angers et al. (1993) found ratios of microbial biomass C and carbohydrate C to total organic C suggested that there was a significant enrichment of the OM in labile forms as tillage intensity was reduced. The ratio of both mild-acid and hot-water soluble carbohydrates to total organic C was greater under NT than under moldboard plowed soil after only three cropping seasons suggesting an enrichment of labile carbohydrates in OM under reduced tillage. Four years of conservation tillage resulted in greater OM in the topsoil layer compared to more intensive tillage systems.

Western Canadian Provinces (MB, AB, SK, BC)

Agroecosystems in western Canadian prairie provinces have been described by Johnson (1977), Larney et al. (1994), and Janzen et al. (1998). Much of this agricultural land is semiarid and lies east of the Canadian Rockies with a semiarid climate. Soil erosion remains the dominant threat to long-term sustainability of farming and has had an impact on long-term soil productivity via its effect on soil quality. "Plowless summer fallow" whereby crop residues are kept on the soil surface as protection against evaporation and wind erosion has been practiced in western Canada since the area was brought into cultivation. This technique is also called "trash cover farming" or stubble mulching. It is considered CT but has undergone major improvements.

The main crop in the Canadian prairies is spring wheat with barley, canola (*Brassica napus* L.), and flax (*Linum usitatissium* L.) occupying significant portions of the area. Winter wheat is also grown. Mixed farming is prevalent in many provinces and rotations include various proportions of forage legumes and grasses for pasture and hay. A limited amount of irrigation is utilized in southern Alberta where the two main crops are soft spring wheat and alfalfa. Crops grown in selected areas of the Red River Valley in Manitoba include sugar beets, potatoes, corn, field peas (*Pisum sativum* L. subsp. *sativum*), beans, and other specialty crops. Thus, the tillage system used depends on the crop and soil type.

Janzen et al. (1998) reviewed C storage in long-term research sites and found that significant gains in soil C storage occurred with practices such as limited summer fallow, increased use of forage grasses, improved fertility management, and reduced tillage intensity. Cultivation of grassland and forest soils on the Canadian prairie has caused about 20 to 30% loss of soil organic C, apparently related to use of intensive tillage equipment, the disk harrow, chisel plow, and the V-blade plow. Larney et al. (1997) detected a soil organic C increase under NT when compared to intensively tilled systems, but no benefits were apparent relative to the widely used stubble-mulch system.

Hao et al. (2000) studied effects of conventional and reduced till systems on soil physical properties and crop residue conservation for two crop sequences. For wheat and annual legumes, CT consisted of chisel plowing and double disking in fall and light-duty cultivation and harrow packing in spring. Reduced tillage consisted only of light cultivation and harrow packing in spring. For sugar beets, CT consisted of moldboard plowing, double disking, light cultivation, harrow packing, and ridging in fall and spring while reduced tillage consisted of chisel plowing, harrow packing, and ridging in both fall and spring ridging. They found no significant differences between minimum tillage and CT in soil bulk density and penetrometer cone index data. Reduced tillage over CT improved residue cover sufficient to reduce to wind erosion. Crop sequence is crucial to successful implementation of reduced tillage systems for irrigated cropping.

Data from four long-term studies in Saskatchewan were examined for evidence that N fertility was a constraint under NT compared with tilled systems (McConkey et al., 2002). Changes may be needed because of greater water conservation, slower N mineralization, and greater denitrification losses under NT compared to CT. Tillage methods studied were CT systems that involved late fall cultivation and a single pre-seeding cultivation, generally less than 15 cm. In the early years of the study, a disk harrow was used as primary tillage. The reduced tillage system involved only one cultivation prior to seeding. On the CT fallow treatment, one to five tillage operations were performed during the summer with a heavy-duty cultivator and/or rod weeder to control weeds. Herbicides followed by one or two operations with a wide V-blade cultivator or a heavy-duty cultivator provided weed control in reduced tillage fallow. A single pre-seeding cultivation was made on the fallow in the reduced and CT tillage systems. For NT systems, all weed control was accomplished with herbicides and seeding with narrow openers to minimize soil disturbance. They concluded that current fertilizer N recommendations, which were formulated for tilled systems, may be inadequate for maximum wheat production with acceptable grain protein under NT.

Using similar tillage methods, Campbell et al. (2001) found most soil quality responses could be associated with treatment effects on crop residue production and soil inputs. Most of the labile soil quality attributes, e.g., microbial biomass C, light fraction C, light fraction N, and N mineralization were increased by fertilizers, cropping frequency, and by including legumes and rotations with green manure or hay crops with limited impact of tillage method. The greater amount of crop residues and less soil disturbance resulting from changing crop management probably contributed to the observed increase in water stable aggregates after 10 years of NT. As a result of intensive tillage, soils have been degraded and considerable erosion has taken place. Increased adoption of conservation tillage system is seen by many as one of the few options to ensure long-term sustainability and economic viability of the farms across the Canadian Prairie provinces (Dumanski et al., 1986).

Mexico (All States)

Tillage systems and agricultural research in Mexico has largely been driven by crop production as a means to achieve business profitability and a better standard of living for farmers. Claverán et al. (1997) and Tiscareño-López et al. (1999) summarized Mexico's agricultural productivity and much of the recent tillage systems research. Nearly 20 million ha are dedicated to food and fiber production. Agriculture is the main economic livelihood for 25% of the population living in rural areas. Eighty percent of the cropland is cultivated under rain-fed conditions in a gradient of rainfall north-south that ranges from 200 to 2000 mm during the growing season. The dominant soil types are Andisols and some Vertisols. The soils in Mexico are easily eroded under dry or wet conditions due to its lack of structure. At the same time, 85% of the country is classified as arid or semiarid, where potential evapotranspiration exceeds total annual precipitation, so drought imposes a high risk of crop failure. Cropland is also limited by topography such that with intensive tillage, erosion is a major problem. During the 1970s and 1980s, state agencies encouraged expansion of arable land wherever necessary and possible. As a result of such political decisions, temperate and tropical forests have been reduced by 30 and 75%, respectively, since 1960. Today, 65 to 85% of land has been identified as undergoing a degradation process due to soil erosion, nutrient losses, agrochemical pollution, and lake eutrophication. The need to identify appropriate technology to optimize tillage systems and crop production has been realized along with the necessity of implementing soil and water conservation practices to protect the land.

After decades of conventional agriculture based on traditional intensive tillage practices, the steep sloped lands are becoming less productive because soil erosion and nutrient losses have been greatly accelerated. Tiscareño-López et al. (1999) reported on soil conservation methods used with steep slope agriculture conducted at the Lake Patzcuaro Watershed in Central Mexico. Twelve soil and residue management treatments were implemented to evaluate maize with CT (disking stalks after harvest, followed by disk plowing and disk harrowing, then bedding), reduced tillage (shredding stalks and disk harrowing, then bedding), and NT (shredding stalks only) under varying percentages of soil crop residue coverage. In CT, the primary tool was the disk plow. Application of conservation tillage as NT and/or reduced tillage represented a feasible technology to reduce water erosion by 80% and reduced nutrient losses by 60%. They noted that mechanical soil movement from traditional intensive cultivation of the uplands accelerated sedimentation of the lake, with an annual surface area reduction of 70 ha. Conservation tillage technology increased productivity by 24% compared to CT and motivated 1500 farmers to form a regional program to promote NT and crop residue management following a participatory research approach.

Some aspects of crop residue management associated with conservation tillage may be difficult to adopt, especially when farmers need crop residues to feed animals. Any recommendation to leave crop residues in the field requires appropriate shredders or residue cutters for easy residue management and conviction that residue cover is vital for soil protection. However, the largest factor in acceptance of conservation tillage was social limitations, including a strong tradition for farmers to use CT methods. Valdivia and Villarreal (1998) developed the "The Farmer Researcher" program to increase maize productivity in several regions of Mexico with a potential for high productivity.

Salinas-Garcia et al. (2000b, 2001) characterized soil microbial activity, N mineralization, and nutrient distribution in Vertisols and Andisols under rainfed corn production from three tillage experiments located in Michoacán, Mexico. They evaluated CT (disking stalks after harvest, followed by disk plowing and disk harrowing, then bedding), minimum tillage (shredding stalks and disk harrowing, then bedding), and NT (shredding stalks only) under varying percentages of soil crop residue coverage. Reduced tillage and NT treatments significantly increased crop residue accumulation on the soil surface. For all treatments, soil organic C, microbial biomass C and N, N mineralization, total N, and extractable P were higher in 0 to 5-cm depth and decreased with depth, apparently related to residue incorporation. Over the 0 to 20-cm depth of disk plow tillage, the same parameters were generally lower, but more evenly distributed. Salinas-Garcia et al. (2000b, 2001) concluded that soil OM enhancement with conservation tillage was probably the most important beneficial change in those soils.

Conservation tillage in semiarid lands increased the water available to plants, reduced soil erosion and reduced N loss in runoff. Salinas-Garcia (1997) showed the best conservation tillage response when 33 to 66% residue cover is left on the soil. Soil erosion on irrigated semiarid croplands was reduced by 40 and 49% under chisel plow tillage and NT, respectively, in comparison to disk harrow tillage. Conservation tillage production costs were 15 to 30% lower than those of CT. Additional information on economics of conservation tillage in Mexico was presented by Erenstein (1997) and Islas Gutiérrez (1997). The effect of different tillage methods on corn insects (Carrillo Sánchez, Aguilera, and Lourdes Garcia, 1997) and on the water balance (Scopel and Chavez Guerra, 1997) shows numerous benefits of conservation tillage that lead to sustainable production and economic stability.

In Mexico, it is common to grow two cash crops per year on the same piece of land. This decreases the time available for residue degradation and large quantities of residue can build up, especially in irrigated areas. A common practice is to first burn the residue and then till the soil with the moldboard or disk plow. Smoke from burning residue increases air pollution and raises other environmental concerns suggesting a need for alternative residue management systems. Salinas-Garcia et al. (2000a) found that high quantities of surface residue did not allow the coulters and seed openers to penetrate the soil consistently. The residue was pushed into the soil and folded up on either side of the seed to provide a microenvironment that prevented seed germination. This phenomena has been defined as the "taco effect" since the seed is contained inside a residue shape resembling a taco shell. In the Midwest US, this phenomenon is referred to as "hair pinning" of crop residue. This occurred despite cleaning wheels, coulters, and disk furrow openers on the planter. They suggested need for new management or machinery system techniques to leave only the quantity needed on the soil surface for specific objectives such as erosion and evaporation control.

SOIL EROSION

While tillage is often considered necessary in many agricultural production systems, the associated soil erosion and degradation can lead to increased soil variability and yield decline. Three types of soil erosion are defined: tillage, water, and wind erosion where erosion implies soil movement by some external force. Traditional approaches to characterizing wind and water erosion are highly dependent upon climate (rainfall) and soil type (Lal, 1999). While intensive tillage loosens the soil for greater response to erosive forces of wind and water, the tillage process itself also moves the topsoil downslope.

Tillage erosion or tillage-induced translocation, the net movement of soil downslope through action of mechanical implements and gravity acting on loosened soil, has been observed for many years. Papendick, McCool, and Krauss (1983) reported original topsoil on most hilltops had been removed by tillage erosion in the Pacific Northwest. The moldboard plow was identified as the primary cause, but all tillage implements will contribute to this problem (Govers et al., 1994; Lobb and Kachanoski, 1999). Tillage erosion has become an important factor in soil management considerations and is often confused with water erosion.

Soil translocation from moldboard plow tillage operations has been identified as a cause of soil movement from specific landscape positions that can be greater than currently accepted soil loss tolerance levels (Lindstrom, Nelson, and Schumacher, 1992; Govers et al., 1994; Lobb, Kachanoski, and Miller, 1995; Poesen et al., 1997). Soil is not directly lost from the fields by tillage translocation, rather it is moved away from the convex slopes and deposited on concave slope positions. Lindstrom, Nelson, and Schumacher (1992) showed that soil movement on a convex slope in southwestern Minnesota could result in a sustained soil loss level of approximately 30 t ha^{-1} yr^{-1} from annual moldboard plowing. Lobb, Kachanoski, and Miller (1995) estimated soil loss in southwestern Ontario from a shoulder position to be 54 t ha^{-1} yr^{-1} from a tillage sequence of moldboard plowing, tandem disk and a C-tine cultivator. In this case, tillage erosion, as estimated through resident Cesium137, accounted for at least 70% of the total soil loss. Tillage speed increases nonlinearly the rate of tillage erosion.

The relationship between soil productivity and erosion is complex. Soils are not the sole factors controlling crop yields. The degree to which crop yield losses are related to soils is a function of several interacting factors including soil physical, chemical and biological properties, landscape position, crop grown, management practices, and weather conditions before and during the growing season. Schumacher et al. (1999), used modeling procedures to show that tillage erosion caused soil loss from the shoulder position while soil loss from water erosion occurred primarily in the mid to lower backslope position. The decline in overall soil productivity was greater when both processes were combined compared to either process acting alone. Water erosion contributed to nearly all the decline in soil productivity in the backslope position when both tillage and water erosion processes were combined. While there are many other reasons for intensive tillage, tillage sets up the soil to be loose, open and very susceptible to high intensity rainfall and subsequent erosion. The net effect of soil translocation from the combined effects of tillage and water erosion was an increase in spatial variability of crop yield and a likely decline in over all soil productivity (Schumacher et al., 1999).

TILLAGE MANAGEMENT AND ENVIRONMENTAL CONCERNS

Concern for environmental quality and greenhouse gas emissions (carbon dioxide (CO_2), methane, nitrous oxide) requires new knowledge to minimize agriculture's impact. The link between global warming and atmospheric CO_2 has heightened interest in soil C storage in agricultural production systems. Agricultural soils play an important role in C sequestration or storage and thus can help mitigate global warming (Lal et al., 1998). The moldboard plow has been the symbol of US agriculture over the last 150 years. Intensive tillage in the US has mineralized or oxidized between 30 and 50% of the native soil C or soil OM since the pioneers brought the soils into cultivation (Schlesinger, 1985). Tillage processes and mechanisms, e.g., tillage-induced CO_2 losses, lead to C loss and are directly linked to soil productivity, soil properties, and environmental issues (Paustian, Collins, and Paul, 1997). Soil C dynamics can have an indirect effect on climate change through net absorption or release of CO_2 from soil to the atmosphere in the natural C cycle. Carbon comes into the system through photosynthesis and is returned to the atmosphere as CO_2 through microbial respiration and anthropogenic intervention. Good soil C management is vital because of its role in maintaining soil fertility, physical properties, and biological activity required for food production and environmental quality. Good soil C management is also needed to partially offset greenhouse gas emissions from manufacture and use of acid fertilizers, liming, and fossil fuels as well as to minimize the release of more potent nitrous oxide and methane. Minimizing agriculture's impact on the global increase of CO_2 requires that we sequester and maintain high soil C levels through decreased tillage intensities and improved residue management.

Sustainable agriculture requires good crop production and rotations using tillage systems and soil management focused on lower inputs and energy use. Moldboard plow tillage is unique in that it inverts the soil sufficiently to adversely affect soil physical, chemical and biological properties, and processes. Less intensive conservation tillage may reduce problems of soil erosion, OM losses, and structural degradation (Carter, 1994b). Conservation tillage, particularly no-till or direct seeding, has potential to reduce negative effects of plow tillage and to allow better C sequestration. Recent technology advances have led to combined tillage and planting operations (NT) as well as herbicide applications that can replace mechanical cultivation (Allmaras, Unger, and Wilkins 1985; Carter, 1994b; and Reeder, 2000).

Tillage affects soil microbial activity, OM decomposition, and soil C loss in agricultural systems. Much of the C is lost as CO_2, which is the end product of microbial feeding on soil OM. Reicosky and Lindstrom (1993, 1995) showed major short-term gaseous loss of C immediately after moldboard tillage that partially explains long-term C loss from tilled soils. Gas exchange was measured using a large, portable chamber to determine CO_2 loss from various types of tillage. Moldboard plow immediately before CO_2 measurement was the deepest and most intensive tillage, and produced more CO_2 loss than other tillages immediately preceding the CO_2 measurement. No-till or no soil disturbance before CO_2 measurement lost the least amount of CO_2. Repeated moldboard plowing allows rapid CO_2 loss and oxygen entry because it loosens the soil and inverts both soil and residue. Much of the CO_2 released initially is from entrapment and storage, and the remainder from accelerated microbial respiration when oxygen entry is accelerated. Long-term plowing accelerates microbial decomposition in the 15 to 30 cm layer and the lack of shallow residue causes aggregate breakdown to ultimately decrease surface soil C content. Ellert and Janzen (1999) and Rochette and Angers (1999) found similar results for different soils and less intensive tillage methods. This interaction of soil and residue mixing enhances aerobic microbial decomposition of incorporated residue to decrease soil organic C (Reicosky et al., 1995). Allmaras et al. (2000) reviewed many field trials that show the moldboard plow tillage system stores less soil organic C than all other tillage systems. They also found that farmers have now reduced their use the moldboard plow to about 7% of the land prepared for corn, wheat, and soybean.

Reicosky (1997) reported that average short-term C loss from four conservation tillage tools was 31% of the CO_2 from the moldboard plow. The moldboard plow lost 13.8 times more CO_2 as the soil not tilled while conservation tillage tools averaged about 4.3 times more CO_2 loss. The smaller CO_2 loss from conservation tillage tools was significant and suggests progress in equipment development for enhanced soil C management. Conservation tillage reduces the extent, frequency, and magnitude of mechanical disturbance caused by the moldboard plow and reduces the large air-filled soil pores to slow the rate of gas exchange and C oxidation.

Strip tillage tools are designed to minimize soil disturbance (Morrison, 1999). Different strip tillage tools and moldboard plow were compared to quantify short-term tillage-induced CO_2 loss relative to tillage intensity (Reicosky, 1998). Less intensive strip tillage reduced soil CO_2 losses. No-till had the lowest CO_2 loss and moldboard plow had the highest immediately after tillage. Forms of strip tillage had an initial soil CO_2 loss related to tillage intensity intermediate between the extremes of plowing and NT. The cumulative CO_2 losses for 24 hours were directly related to the soil volume disturbed by the tillage tool. Reducing the volume of soil disturbed by tillage should enhance soil and air quality by increasing the soil C content. This suggests that soil and environmental benefits of strip tillage be considered in soil management decisions. The CO_2 released immediately after moldboard plowing suggests little C sequestration. Conservation tillage methods that leave most of the crop residue on the surface with limited soil contact yield better C sequestration to enhance environmental quality.

Tillage, through its effect on traffic and soil OM, affects water quality and water use efficiency (Betz et al., 1998). Water is often the most limiting factor for crop production in North America. The soil's capacity to hold water is a function of the amounts of sand, silt, and clay and the OM content. Water holding capacity of soils having the same sand, silt, and clay contents showed that for each percent increase in OM, the water holding capacity of the soil increases by 3.7% on a volume basis (Hudson, 1994). The inverse relationship between intensive tillage and soil OM content leads to the question of tillage impact on water use efficiency. While the role of tillage on infiltration and runoff is generally understood, we are now getting a better understanding of the impact of tillage and soil C on plant available water holding capacity. The contribution of soil OM to water holding capacity and water quality cannot be underestimated. Both of these are key factors in agricultural watersheds for maintaining environmental quality.

CONCLUSIONS

Traditional agricultural production has involved at least five separate operations: (1) tillage, (2) planting, (3) cultivating, (4) harvesting, and (5) processing, transporting, and storage before final consumption. Tillage is on this list because it has historically been an integral part of the production process. Over the last 30 years, new technology is redefining these operations where tillage and planting are combined in conservation tillage and where mechanical cultivation is being replaced by herbicides. Modern, large farm equipment can perform these operations easily and quickly with one pass. Historically, the moldboard plow was an essential tool for the early pioneers in settling the prairies of central and western US and Canada. The moldboard plow allowed the farmer to create a soil environment in which grain crops could thrive and meet the needs of the increasing population. At the same time, the moldboard plow degraded soil from increased water, wind, and tillage erosion as well as biological oxidation of soil OM. In the drier areas, other types of chisel plows and large sweeps are primary tillage tools based on the crops and available water. Tillage is intimately related to the cropping rotations that are limited by soil and water resources within the physiographic region. New tillage systems with emphasis on crop residue management and soil conservation will encompass new technology and continue to evolve around the best systems within a given geographic location as driven by economic and environmental considerations. As new agricultural tillage and planting practices are developed across North America, their impacts on the environment and energy use will need to be evaluated critically to ensure their compatibility and sustainability with society's needs.

REFERENCES

ASAE EP291.2 (1998a). Terminology and definitions for soil tillage and soil-tool relationships. In *ASAE Standards 2000*, St. Joseph, MI: ASAE, pp. 109-112.

ASAE S414.1. (1998b). Terminology and definitions for agricultural tillage implements. In *ASAE Standards 2000*, St. Joseph, MI: ASAE, pp. 251-262.

ASAE S477. (1999). Terminology for soil-engaging components for conservation-tillage planters, drills and seeders. In *ASAE Standards 2000*, St. Joseph, MI: ASAE, pp. 326-331.

Allmaras, R.R., S.M. Copeland, P.J. Copeland, and M. Oussible. (1996). Spatial relations between oat residue and ceramic spheres when incorporated sequentially by tillage. *Soil Science Society of America Journal* 60:1209-1216.

Allmaras, R.R. and R.H. Dowdy. (1985). Conservation tillage systems and their adoption in the United States. *Soil Tillage Research* 5:197-222.

Allmaras, R.R., G.W. Langdale, P.W. Unger, and R.H. Dowdy. (1991). Adoption of conservation tillage and associated planting systems. In *Soil Management for Sustainability*, ed. R. Lal and F. Pierce, Ankeny, IA: Soil and Water Conservation Society, pp. 53-83.

Allmaras, R.R., J.L. Pikul, Jr., J.M. Kraft, and D.E. Wilkins. (1988). A method for measuring incorporated crop residue and associated soil properties. *Soil Science Society of America Journal* 52:1128-1133.

Allmaras, R.R., J.F. Power, D.L. Tanaka, and S.M. Copeland. (1994). Conservation tillage systems in the northernmost Central United States. In *Conservation Tillage in Temperate Agroecosystems*, ed. M.R. Carter, Boca Raton, FL: Lewis Publishers, pp. 255-284.

Allmaras, R.R., H.H. Schomberg, C.L. Douglas, Jr., and T.H. Dao. (2000). Soil organic carbon sequestration potential of adopting conservation tillage in U.S. croplands. *Journal of Soil and Water Conservation* 55:365-373.

Allmaras, R.R., P.W. Unger, and D.W. Wilkins. (1985). Conservation tillage systems and soil productivity. In *Soil Erosion and Crop Productivity*, ed. R.F. Follett and B.A. Stewart, Madison, WI: Agronomy Society of America, pp. 357-411.

Allmaras, R.R., D.W. Wilkins, O.C. Burnside, and D.J. Mulla. (1998). Agricultural technology and adoption of conservation practices. In *Advances in Soil and Water Conservation*, ed. F.J. Pierce and W.W. Frye, Chelsea, MI: Sleeping Bear Press, Ann Arbor Press, pp. 99-158.

Amemiya, M. (1977). Conservation tillage in the western Corn Belt. *Journal of Soil and Water Conservation* 32:29-36.

Angers, D.A., N. Bissonnette, A. Légère, and N. Samson. (1993). Microbial and bio-chemical changes induced by rotation and tillage in a soil under barley production. *Canadian Journal of Soil Science* 73:39-50.

Angers, D.A., M.A. Bolinder, M.R. Carter, E.G. Gregorich, C.F. Drury, B.C. Liang, R.P. Voroney, R.R. Simard, R.G. Donald, R.P. Beyaert, and J. Martel. (1997). Impact of tillage practices on organic carbon and nitrogen storage in cool, humid soils of Eastern Canada. *Soil Tillage Research* 41:191-201.

Angers, D.A., A. N'dayegamiye, and D. Côté. (1993). Tillage-induced differences in organic matter on particle-size fractions and microbial biomass. *Soil Science Society of America Journal* 57:512-516.

Angers, D.A., N. Samson, and A. Légère. (1993). Early changes in water-stable aggregation induced by rotation and tillage in a soil under barley production. *Canadian Journal of Soil Science* 73:51-59.

Ashraf, M., C.H. Pearson, D.G. Westfall, and R. Sharp. (1999). Effect of conservation tillage on crop yields, soil erosion, and soil properties under furrow irrigation in western Colorado. *American Journal of Alternative Agriculture* 14(2):85-92.

Bennett, O.L. (1977). Conservation tillage in the Northeast. *Journal of Soil and Water Conservation* 32:9-12.

Betz, C.L., R.R. Allmaras, S.M. Copeland, and G.W. Randall. (1998). Least limiting water range: Traffic and long-term tillage influences in a Webster soil. *Soil Science Society of America Journal* 62:1384-1393.

Blevins, R.L. and W.C. Moldenhauer. (1995). Crop residue management to reduce erosion and improve soil quality: Appalachia and Northeast. Conservation Research Report 41.

Burmeister, C.H., M.G. Patterson, and D.W. Reeves. (1995). Challenges of no till cotton production on silty clay soils in Alabama. In *Conservation Tillage Systems for Cotton Special Report 169*, ed. M.R. McClelland, T.D. Valco, and R.E. Frans, Fayetteville, AR: Arkansas Agriculture Experiment Station, University of Arkansas, pp. 5-7.

Busscher, W.J. and P.J. Bauer. (1995). Soil strength of conventional- and conservation-tillage cotton grown with a cover crop. In *Conservation Tillage Systems for Cotton Special Report 169*, ed. M.R. McClelland, T.D. Valco, and R.E. Frans, Fayetteville, AR: Arkansas Agriculture Experiment Station, University of Arkansas, pp. 18-20.

Busscher, W.J., J.H. Edwards, M.J. Vepraskas, and D.L. Karlen. (1995). Residual effects of silt tillage and subsoiling in a hardpan soil. *Soil Tillage Research* 35:115-123.

Busscher, W.J., J.R. Frederick, and P.J. Bauer. (2000). Timing effects of deep tillage on penetration resistance and wheat and soybean yield. *Soil Science Society of America Journal* 64:999-1003.

CTIC. (1995). *Survey Guide: National Crop Residue Management Guide*. West Lafayette, IN: Conservation Technology Information Center.

CTIC. (2000). CTIC News Release, September 27, 2000. Web site <http://www.ctic.purdue.edu>.

Camp, C.R., P.J. Bauer, and W.J. Busscher. (1999). Evaluation of no-tillage crop production with subsurface drip irrigation on soils with compacted layers. *Transactions of the American Society of Agricultural Engineers* 42(4):911-917.

Campbell, R.B., D.C. Reicosky, and C.W. Doty. (1974). Physical properties and tillage of Paleudults in the southeastern Coastal Plains. *Journal of Soil and Water Conservation* 29:220-224.

Campbell, C.A., F. Selles, G.P. Lafond, V.O. Biederbeck, and R.P. Zentner. (2001). Tillage-fertilizer changes: Effect on some soil quality attributes under long-term crop rotations in a thin Black Chernozem. *Canadian Journal of Soil Science* 81:157-165.

Carrillo Sánchez, J.L., M. Aguilera, and M. Lourdes García. (1997). Efecto de diferentes métodos de labranza y cobertura vegetal sobre la incidencia de insects asociados al maíz en la región centro de México. In *Avances de la Investigación en Llabranza de Conservación.* Libro Técnico No. 1, ed. A.R. Claverán, G.J. Velázquez, V.J.A. Muños, M.L. Tiscareño, G.J.R. Salinas and R.M.B. Najera, Michoacán, Mexico. pp. 151-165 (in Spanish).

Carter, D.L. and R.D. Berg. (1991). Crop sequences and conservation tillage to control irrigation, furrow erosion and increase farmer income. *Journal of Soil and Water Conservation* 46:139-142.

Carter, D.L., R.D. Berg, and B.J. Sanders. (1991). Producing no till cereal or corn following alfalfa on furrow-irrigated land. *Journal of Production Agriculture* 4:174-179.

Carter, M.R. (1992). Characterizing the soil physical condition in reduced tillage systems for winter wheat on a fine sandy loam using small cores. *Canadian Journal of Soil Science* 72:395-402.

Carter, M.R. (1994a). *Conservation Tillage in Temperate Agroecosystems.* Boca Raton, FL: Lewis Publishers/CRC Press Inc., 400 pp.

Carter, M.R. (1994b). Strategies to overcome impediments to adoption of conservation tillage. In *Conservation Tillage in Temperate Agroecosystems,* ed. M.R. Carter, Boca Raton, FL: Lewis Publishers, pp. 1-19.

Carter, M.R., D. Angers, E.G. Gregorich, and M.A. Bolinder. (1997). Organic carbon and nitrogen stocks and storage profiles in cool, humid soils of Eastern Canada. *Canadian Journal of Soil Science* 77:205-210.

Carter, M.R. and U.C. Gupta. (1997). Micronutrient concentrations in barley and soybeans under minimum tillage on Podzolic soils in a cool climate. *Acta Agriculturæ Scandinavica, Section B, Soil and Plant Science* 47:7-13.

Carter, M.R. and H.T. Kunelius. (1998). Influence of non-inversion loosening on permanent pasture productivity. *Canadian Journal of Soil Science* 78:237-239.

Carter, M.R., J.B. Sanderson, and J.A. McLeod. (1998). Influence of time of tillage on soil physical attributes in potato rotations in Prince Edward Island. *Soil Tillage Research* 49:127-137.

Christensen, N.B., W.C. Lindeman, E. Salazar-Sosa, and L.R. Gill. (1994). Nitrogen and carbon dynamics in no till and stubble mulch tillage systems. *Agronomy Journal* 86:298-303.

Claverán, A.R., G.J. Velázquez, V.J.A. Muños, M.L. Tiscareño, G.J.R. Salinas, and R.M.B. Najera. (1997). Avances de la investigación en labranza de conservación. Libro Técnico No. 1. Michoacán, Mexico (in Spanish).

Cox, W.J., R.W. Zobel, H.M. Van Es, and D.J. Otis. (1990). Growth development and yield of maize under three tillage systems in the northeastern U.S.A. *Soil Tillage Research* 18:295-310.

Dao, T.H. (1998). Tillage and crop residue affects on carbon dioxide evolution and carbon storage in a Paleustoll. *Soil Science Society of America Journal* 62:250-256.

Dick, W.A., R.L. Blevins, W.W. Frye, S.E. Peters, D.R. Christenson, F.J. Pierce, and M.L. Vitosh. (1998). Impacts of agricultural management practices on C sequestration in forest-derived soils of the eastern Corn Belt. *Soil Tillage Research* 47(3,4): 235-244.

Dick, W.A. and J.T. Durkalski. (1997). No-till production agriculture and carbon sequestration in a typic Fragiudalf soil of northeastern Ohio. In *Management of Carbon Sequestration in Soil*, ed. R. Lal, J.M. Kimball, R.F. Follett, and B.A. Stewart, Boca Raton, FL: CRC Press, pp. 59-71.

Douglas, C.L. Jr., P.E. Rasmussen, and R.R. Allmaras. (1989). Cutting height, yield level, and equipment modification effects on residue distribution by combines. Trans. ASAE. 32:1258-1262.

Dumanski, J., D. Coote, G. Luciuk, and C. Lok. (1986). Soil conservation in Canada. *Journal of Soil and Water Conservation* 41:204-210.

Eagle, A.J., J.A. Bird, W.R. Horwath, B.A. Lindquist, S.M. Brouder, J.E. Hill, and C. van Kessel. (2000). Rice yield and nitrogen utilization efficiency under alternative straw management practices. *Agronomy Journal* 92:1096-1103.

Eckert, D.J. (1990). Ridge planting for row crops on a poorly drained soil. I. Rotation and drainage effects. *Soil Tillage Research* 18:181-188.

Economic Research Service (ERS). (1994). Agricultural resources and environmental indicators. Agriculture Handbook 705. Washington, DC: Economic Research Service, U.S. Department of Agriculture, 205 p.

Ellert, B.H. and H.H. Janzen. (1999). Short-term influence of tillage on CO_2 fluxes from a semi-arid soil on the Canadian Prairies. *Soil Tillage Research* 50:21-32.

Epplin, F.M., G.A. Al-Sakkat and T.F. Peeper. (1994). Impacts of alternate tillage methods for continuous wheat on grain yield and economics: Implications for conservation compliance. *Journal of Soil and Water Conservation* 49:394-399.

Erenstein, O. (1997). La econmía de la labranza de conservación en Mexico. In *Avances de la investigación en labranza de conservación*, Libro Técnico No. 1, ed. A.R. Claverán, G.J. Velázquez, V.J.A. Muños, M.L. Tiscareño, G.J.R. Salinas, and R.M.B. Najera, Michoacán, Mexico. pp. 225-243 (in Spanish).

Erickson, A.E. (1982). Tillage effects on soil aeration. In *Predicting Tillage Affects on Soil Physical Properties and Processes*, ed. P.W. Unger, D.M. Van Doren, Special Publication No. 44, Madison, WI: Agronomy Society of America. pp. 91-104.

Evans, S.D., M.J. Lindstrom, W.B. Voorhees, J.F. Moncrief, and G.A. Nelson. (1996). Effect of subsoiling and subsequent tillage on soil bulk density, soil moisture and corn yield. *Soil Tillage Research* 38:35-46.

Fausey, N.R. (1991). Experience with ridge till on slowly permeable soils in Ohio. *Soil Tillage Research* 18:195-206.

Fausey, N.R. and R. Lal. (1989a). Drainage-tillage effects on Crosby-Kokomo soil association in Ohio. I. Effects on stand and corn yield. *Soil Technology* 2:359-370.

Fausey, N.R. and R. Lal. (1989b). Drainage-tillage effects on Crosby-Kokomo soil association in Ohio. II. Soil temperature regime and infiltrability. *Soil Technology* 2:371-383.

Fenster, C.R. (1977). Conservation tillage in the northern Plains. *Journal of Soil and Water Conservation* 32:37-42.

Franzluebbers, A.J., F.M. Hons, and D.A. Zuberer. (1998). In situ and potential CO_2 evolution from a Fluventic Ustocherpt in southcentral Texas as affected by tillage and cropping intensity. *Soil Tillage Research* 47(3,4):303-308.

Frederick, J.R. and P.J. Bauer. (1996). Winter wheat responses to surface and deep tillage on the southeastern Coastal Plain. *Agronomy Journal* 88:829-833.

Frederick, J.R., P.J. Bauer, W.J. Busscher, and G.S. McCutcheon. (1998). Tillage management for double cropped soybean grown in narrow and wide width culture. *Crop Science* 38:755-762.

Govers, G., K. Vandaele, P.J.J. Desmet, J. Poesen, and K. Bunte. (1994). The role of tillage in soil redistribution on hillslopes. *European Journal of Soil Science* 45: 469-478.

Gregorich, E.G., K.J. Greer, D.W. Anderson, and B.V. Liang. (1998). Carbon distribution and losses: Erosion and deposition effects. *Soil Tillage Research* 47(3,4): 291-302.

Griffith, D.R., J.V. Mannering, and W.C. Moldenhauer. (1977). Conservation tillage in the eastern Corn Belt. *Journal of Soil and Water Conservation* 32:20-28.

Halvorson, A.D. and G.P. Hartman. (1984). Reduced seedbed tillage effects on irrigated sugar beet yield and quality. *Agronomy Journal* 76:603-606.

Hatfield, J.L., R.R. Allmaras, G.W. Rehm, and B. Lowery. (1998). Ridge tillage for corn and soybean production: Environmental quality impacts. *Soil and Tillage Research* 48:145-154.

Hatfield, J.L. and B.A. Stewart. (1994). *Crop Residue Management*. Boca Raton, FL: CRC Press.

Hao, X.Y., C. Chang, F.J. Larney, J. Nitschelm, and P. Regitnig. (2000). Effect of minimum tillage and crop sequence on physical properties of irrigated soil in southern Alberta. *Soil Tillage Research* 57(1-2):53-60.

Hendrix, P.F., A.J. Franzluebbers, and D.V. McCracken. (1998). Management effects on C accumulation and loss in soils of the southern Appalachian Piedmont of Georgia. *Soil and Tillage Research* 47(3,4):245-251.

Hill, P.R. 2001. Use of continuous no till and rotational tillage systems in the central and northern Corn Belt. *Journal of Soil and Water Conservation* 56:286-290.

Hudson, B. (1994). Organic matter and available water holding capacity. *Journal of Soil and Water Conservation* 49:189-194.

Huggins, D.R., G.A. Buyanovsky, G.H. Wagner, J.R. Brown, R.G. Darmody, T.R. Peck, G.W. Lesoing, M.B. Vanotti, and L.G. Bundy. (1998). Soil organic C in the tallgrass prairie-derived region of the Corn Belt: Effects of long-term crop management. *Soil Tillage Research* 47(3,4):219-234.

Islas Gutiérrez, J. (1997). Análisis económico de diferentes métodos de labranza de conservación para maíz de temporal. In *Avances de la investigación en labranza de conservación*, Libro Técnico No. 1., ed. A.R. Claverán, G.J. Velázquez, V.J.A. Muños, M.L. Tiscareño, G.J.R. Salinas, and R.M.B. Najera, Michoacán, Mexico. pp. 167-180 (in Spanish).

Iqbal, M., S.J. Marley, D.C. Erbach, and T.C. Kaspar. (1995). Effects of coulter treatments on seed furrow smearing and early crop response. ASAE Paper No. 95-1322, St. Joseph, MI: ASAE.

Janovicek, K.J., T.J. Vyn, and R.P. Voroney. (1997). No-till corn response to crop rotation and in-row residue placement. *Agronomy Journal* 89:588-596.

Janzen, H.H., C.A. Campbell, E.G. Gregorich, and B.H. Ellert. (1997). Soil carbon dynamics in Canadian agroecyosystems. In *Soil Processes and the Carbon Cycle*, ed. R. Lal, J.M. Kimble, R.F. Follett, and B.A. Stewart, Boca Raton, FL: CRC Press, pp. 57-80.

Janzen, H.H., C.A. Campbell, R.C. Iazurralde, B.H. Ellert, N. Juma, W.B. McGill, and R.P. Zentner. (1998). Management effects on soil C storage on the Canadian prairies. *Soil Tillage Research* 47(3,4):181-195.

Jenny, H. (1941). *Factors of Soil Formation*. New York, NY: McGraw Hill, 541 p.

Johnson, R.R. (1988). Soil engaging-tool effects on surface residue and roughness with chisel-type implements. *Soil Science Society of American Journal* 52: 237-243.

Johnson, W.E. (1977). Conservation tillage in western Canada. *Journal of Soil and Water Conservation* 32:61-65.

Jones, O.R. and W.C. Johnson. (1983). Cropping practices: Southern Great Plains. In *Dryland Agriculture*. Agronomy Monograph 23, ed. H.E. Dregne and W.O. Willis, Madison, WI: Agronomy Society of America, pp. 365-385.

Karlen, D.L., W.J. Busscher, S.A. Hale, R.B. Dodd, E.E. Strickland, and T.H. Garner. (1991). Drought conditions energy requirements and subsoiling effectiveness for selected the tillage implements. *Transactions of the American Society of Agricultural Engineers* 34:1967-1972.

Karlen, D.L., G.E. Varvel, D.E. Bullock, and R.M. Cruse. (1994a). Crop rotations for the 21st century. *Advances in Agronomy* 53:1-45.

Karlen, D.L., N.C. Wollenhaupt, D.C. Erbach, E.C. Barry, J.B. Swan, N.S. Nash, and J.L. Jordahl. (1994b). Long-term tillage affects on soil quality. *Soil Tillage Research* 32:313-227.

Karunatilake, U., H.M. Van Es, and R.R. Schildelbeck. (2000). Soil and maize response to plow and no-tillage after alfalfa-to-maize conversion on a clay loam soil in New York. *Soil Tillage Research* 55:31-42.

Ketcheson, J. (1977). Conservation tillage in eastern Canada. *Journal of Soil Water Conservation* 32:57-60.

Lal, R. (1990). Ridge tillage. *Soil Tillage Research* 18:107-111.

Lal, R. (1999). *Soil Quality and Soil Erosion*. Boca Raton, FL: CRC Press, 329 pp.

Lal, R., J. Kimball. R.F. Follett, and C.V. Cole. (1998). *The Potential of U.S. Cropland to Sequester Carbon and Mitigate the Greenhouse Effect*, Ann Arbor, MI: Sleeping Bear Press, 128 pp.

Lal, R., T.J. Logan, M.J. Shipitalo, D.J. Eckert, and W.A. Dick. (1994). Conservation tillage in the Corn Belt of the United States. In *Conservation Tillage in Temperate Agroecosystems*, ed. M.R. Carter, Boca Raton, FL: Lewis Publishers, pp. 73-114.

Lal, R., A.A. Mahboubi, and N.R. Fausey. (1994). Long-term tillage and rotation effects on properties of a central Ohio soil. *Soil Science Society of America Journal* 58:517-522.

Lamb, J.A., R.H. Dowdy, J.L. Anderson, and R.R. Allmaras. (1998). Water quality in an irrigated sandy soil: Ridge tillage in rotated corn and soybean compared with full-width tillage in continuous corn. *Soil Tillage Research* 48:167-177.

Langdale, G.W. and W.C. Moldenhauer. (1995). Crop residue management to reduce erosion and improve soil quality: Southeast. *Conservation Research Report Number 39*.

Larney, F.J., E. Bremer, H.H. Janzen, A.M. Johnson, and C.W. Lindwall. (1997). Changes in total, mineralizeable and light-fraction soil organic matter with cropping and tillage intensities in semiarid southern Alberta, Canada. *Soil Tillage Research* 42:229-240.

Larney, F.J., C.W. Lindwall, R.C. Izaurralde, and A.P. Moulin. (1994). Tillage systems for soil and water conservation on the Canadian Prairie. In *Conservation Tillage in Temperate Agroecosystems*, ed. M.R. Carter, Boca Raton, FL: Lewis Publishers, pp. 305-328.

Légère, A., N. Samson, R. Rioux, D.A. Angers, and R.R. Simard. (1997). Response of spring barley to crop rotation, conservation tillage, and weed management intensity. *Agronomy Journal* 89:628-638.

Lindstrom, M.J., W.W. Nelson, and T.E. Schumacher. (1992). Quantifying tillage erosion rates due to moldboard plowing. *Soil Tillage Research* 24:243-255.

Lobb, D.A. and R.G. Kachanoski. (1999). Modelling tillage translocation using steppe, near plateau, and exponential functions. *Soil Tillage Research* 51:261-277.

Lobb, D.A., R.J. Kachanoski, and M.H. Miller. (1995). Tillage translocation and tillage erosion on shoulder slope landscape positions measured using 137Cesium as a tracer. *Canadian Journal Soil Science* 75:211-218.

Logsdon, S.D., R.R. Allmaras, W.W. Nelson, and W.B. Voorhees. (1992). Persistence of subsoil compaction from heavy axle loads. *Soil Tillage Research* 23:95-110.

Logsdon, S.D., R.R. Allmaras, L. Wu, J.B. Swan, and G.W. Randall. (1990). Macroporosity and its relation to saturated hydraulic conductivity under different tillage practices. *Soil Science Society of America Journal* 54:1096-1101.

Magdoff, F. and H. van Es. (2000). Building soils for better crops. In *Sustainable Agriculture Network Handbook Series Book 4*. Sustainable Agriculture Publishing, Hills Building, Burlington, VT: University of Vermont, 230 pp.

McConkey, B.G., D. Curtin, C.A. Campbell, S.A. Brant, and F. Selles. (2002). Crop and soil nitrogen status of tilled and no-tillage systems in semiarid regions of Saskatchewan. *Canadian Journal of Soil Science* 82:489-498.

McConnell, J.S., W.H. Baker, C.S. Rothrock, and B.S. Frizzell. (1995). Cotton yield response to irrigation, reduced tillage and cover crops. In *Conservation-Tillage Systems for Cotton Special Report 169*, ed. M.R. McClelland, T.D. Valco, and R.E. Frans, Fayetteville, AR: Arkansas Agriculture Experiment Station, University of Arkansas, pp. 78-81.

Meyer, L.D. and J.V. Mannering. (1961). Minimum tillage for corn. *Agricultural Engineering* 41:72-75.

Mitchell, J., T. Hartz, S. Pettygrove, D. Munk, D. May, F. Menezes, J. Diener, and T. O'Neil. (1999). Organic matter recycling varies with crops grown. *California Agriculture* 53(4):37-40.

Moldenhauer, W.C. and A.L. Black. (1994). Crop residue management to reduce erosion and improve soil quality: Northern Great Plains. *Conservation Research Report Number 38*.

Moldenhauer, W.C. and L.N. Mielke. (1995). Crop residue management to reduce erosion and improve soil quality: North Central. *Conservation Research Report Number 42*.

Morrison, J.E., Jr. (1999). Row-zone alternative to no-till row crop production, St. Joseph, MI: ASAE, ASAE Paper No. 99-1089.

Morrison, J.E., Jr. (2000). Development and future of conservation tillage in America. In *Proceedings of the China International Conference on Dryland Water Saving Farming*, Beijing, P.R. China: Ministry of Agriculture. pp. 15.

Morrison, J.D., Jr., R.R. Allen, D.E. Wilkins, G.M. Powell, R.D. Grisso, D.C. Erbach, L.P. Herndon, D.L. Murray, G.E. Formanek, D.L. Pfost, M.M. Herron, and D.J. Baumert. (1988). Conservation planter, drill and air-type seeder selection guideline. *Applied Engineering Agriculture* 4:300-309.

Naderman, G. (1993). Equipment considerations for reduced tillage cotton production in the Southeast. In *Conservation Tillage Systems for Cotton Special Report 160*, ed. M.R. McClelland, T.D. Valco, and R.E. Frans, Fayetteville, AR: Arkansas Agriculture Experiment Station, University of Arkansas, pp. 13-17.

Nelson, P.J. (1997). To hold the land: soil erosion, agricultural scientists, and the development of conservation tillage techniques. *Agricultural History* 71:71-90.

Opoku, G., T.J. Vyn, and C.J. Swanton. (1997). Modified no-till systems for corn following wheat on clay soils. *Agronomy Journal* 89:549-556.

Papendick, R.I., D.K. McCool, and H.A. Krauss. (1983). Soil conservation: Pacific Northwest. In *Dryland Agriculture*, ed. H.E. Dregne and W.O. Willis, Agronomy 23. Madison, WI: American Society of Agronomy.

Papendick, R.I. and D.E. Miller. (1977). Conservation tillage in the Pacific Northwest. *Journal of Soil and Water Conservation* 32:49-56.

Papendick, R.I. and W.C. Moldenhauer. (1995). Crop residue management to reduce erosion and improve soil quality: Northwest. *Conservation Research Report Number 40*.

Paul, E.A., K. Paustian, E.T. Elliot, and C.V. Cole. (1997). *Soil Organic Matter and Temperate Ecosystems: Long-Term Experiments in North America*. Boca Raton, FL: CRC Press, 414 pp.

Paustian, K., H.P. Collins, and E.A. Paul. (1997). Management controls on soil carbon. In *Soil Organic Matter in Temperate Ecosystems: Long-Term Experiments in North America*, ed. E.A. Paul, K. Paustian, E.T. Elliot, C.V. Cole, Boca Raton, FL: CRC Press, pp. 15-49.

Paustian, K., E.T. Elliott, and M.R. Carter. (1998). Tillage and crop management impacts on soil C storage: Use of long-term experimental data. *Soil Tillage Research* 47(3,4):181-351.

Peterson, G.A., A.D. Halvorson, J.L. Havlin, O.R. Jones, D.J. Lyons, and D.L. Tanaka. (1998). Reduced tillage and increasing cropping intensity in the Great Plains conserves soil C. *Soil Tillage Research* 47(3,4):207-218.

Poesen, J., B. Wesenael, G. Govers, J. Martinez-Fernadez, B. Desmet, K. Vandaele, T. Quine, and G. Degraer. (1997). Patterns of rock fragment covered generated by tillage erosion. *Geomorphology* 18:193-197.

Potter, K.N., O.R. Jones, H.A. Torbert, and P.W. Unger. (1997). Crop rotations and tillage effects on organic carbon sequestration in the semiarid Southern Great Plains. *Soil Science* 162:140-147.

Potter, K.N., H.A. Torbert, O.R. Jones, J.E. Matocha, J.E. Morrison, Jr., and P.W. Unger. (1998). Distribution and amount of soil organic C in long-term management systems in Texas. *Soil Tillage Research* 47(3,4):309-321.

Raimbault, B.A., T.J. Vyn, and M. Tollenaar. (1991). Corn response to rye cover crop, tillage methods, and planter options. *Agronomy Journal* 83: 287-290.

Rasmussen, P.E., S.L. Albrecht, and R.W. Smiley. (1998). Soil C and N changes under tillage and cropping systems in semi-arid Pacific Northwest agriculture. *Soil Tillage Research* 47(3,4):197-205.

Rasmussen, P.E., K.W.T. Goulding, J.R. Brown, P.R. Grace, H.H. Janzen, and M. Körchens. (1998). Long-term agroecosystem experiments: Assessing agricultural sustainability and global change. *Science* 282:893-896.

Raun, W.R., G.V. Johnson, S.B. Phillips, and R.L. Westerman. (1998). Effect of long-term N fertilization on soil organic C and total N in continuous wheat under conventional tillage in Oklahoma. *Soil Tillage Research* 47(3,4):323-330.

Reeder, R. (2000). Conservation tillage systems and management. MWPS-45. Second Ed. 2000. *Crop Residue Management with No-Till, Ridge-Till, Mulch-Till and Strip-Till.* MidWest Plan Service, Iowa State University, Ames, IA. 270 pp.

Reeder, J.D., G.E. Schuman, and R.A. Bowman. (1998). Soil C and N changes on Conservation Reserve Program lands in the Central Great Plains. *Soil Tillage Research* 47(3,4):339-349.

Reeves, D.W. (1994). Cover crops and rotations. In *Crop Residue Management*, ed. J.L. Hatfield and B.A. Stewart, Boca Raton, FL: Lewis Publisher, Inc., pp. 125-172.

Reicosky, D.C. (1997). Tillage-induced CO_2 emissions from soil. *Nutrient Cycling in Agroecosystems* 49:273-285.

Reicosky, D.C. (1998). Strip tillage methods: Impact on soil and air quality. In *Proceedings of the Australian Society of Soil Science Inc., National Soils Conference, Brisbane, Australia,* Brisbane, Australia: Australian Society of Soil Science Inc., pp. 56-60.

Reicosky, D.C., W.D. Kemper, G.W. Langdale, C.L. Douglas, Jr., and P.E. Rasmussen. (1995). Soil organic matter changes resulting from tillage and biomass production. *Journal of Soil and Water Conservation* 50:253-261.

Reicosky, D.C. and M.J. Lindstrom. (1993). Fall tillage methods: Effect on short-term carbon dioxide flux from soil. *Agronomy Journal* 85:1237-1243.

Reicosky, D.C. and M.J. Lindstrom. (1995). Impact of fall tillage and short-term carbon dioxide flux. In *Soil and Global Change*, ed. R. Lal, J. Kimble, E. Levine, and B.A. Stewart, Chelsea, MI: Lewis Publishers, pp. 177-187.

Rochette, P. and D.A. Angers. (1999). Soil surface carbon dioxide fluxes induced by spring, summer, and fall moldboard plowing in a sandy loam. *Soil Science Society of America Journal* 63:621-628.

Salinas-Garcia, J.R. (1997). Adelantos de investigación el labranza de conservación en Mexico. In *Memorias de la IV Reunión Bienal de la Red Latinamericana de*

Labranza Conservacionista, ed. R. Claveran A., y F.O. Rulfo V., Morelia, Michoacan, Mexico: CENAPROS. INIFAP. SAGAR, pp. 231-240 (in Spanish).

Salinas-Garcia, J.R., A.D. Báez-González, M. Tiscareño-López and E. Rosales-Robles. (2001). Residue removal and tillage interaction effects on soil properties under rain-fed corn production in Central Mexico. *Soil Tillage Research* 59:67-79.

Salinas-Garcia, J.R., J.M. Cabrera-Sixto, J.E. Morrision, Jr., W.A. LePori, and A.R. Morales-Martinez. (2000a). Tillage system criteria for high surface residue conditions. In *Proceedings of the 4th International Conference on Soil Dynamics (ICSD-IV)*. Adelaide, South Australia. 8 pp.

Salinas-Garcia, J.R., F.M. Hons, and J.E. Matocha. (1997). Long-term effects of tillage and fertilization on soil organic matter dynamics. *Soil Science Society of America Journal* 61:152-159.

Salinas-Garcia, J.R., J. de J. Velazquez-Garcia, M. Gallardo-Valdez, and F. Caballero-Hernandez. (2000b). Tillage effects on microbial biomass and nutrient distribution in soils under rain-fed corn production in Michoacan, Mexico. In *Proceedings of the 15th Conference ISTRO*, Fort Worth, TX: ISTRO, pp. 12.

Schlesinger, W.H. (1985). Changes in soil carbon storage and associated properties with disturbance and recovery. In *The Changing Carbon Cycle: A Global Analysis*, ed. J.R. Trabalha, D.E. Reichle, New York, NY: Springer-Verlag, pp. 194-220.

Schumacher, T.E., M.J. Lindstrom, J.A. Schumacher, and G.D. Lemme. (1999). Modelling spatial variation and productivity due to tillage and water erosion. *Soil Tillage Research* 51:331-339.

Scopel, E. and E. Chavez Guerra. (1997). Efectos de labranza de conservación sobre el balance hídrico del cultivo de maíz de temporal. In *Avances de la Investigación en Labranza de Conservación*. Libro Técnico No. 1., ed. A.R. Claverán, G.J. Velázquez, V.J.A. Muños, M.L. Tiscareño, G.J.R. Salinas, and R.M.B. Najera, Michoacán, Mexico. pp. 91-106 (in Spanish).

Sijtsma, C.H., A.J. Campbell, N.B. McLaughlin, and M.R. Carter. (1998). Comparative tillage cost for crop rotations utilizing minimum tillage on a farm scale. *Soil Tillage Research* 49:223-231.

Smith, J.L. and L.F. Elliott. (1990). Tillage and residue management effects on soil organic matter dynamics in semiarid regions. *Advances in Soil Science* 13:69-88.

Smith, J.A., C.D. Yonts, D.A. Biere, and M.D. Rath. (1995). Field operation energy use for a corn-dry edible bean-sugar beet rotation. *Applied Engineering Agriculture* 11(2):219-224.

Sojka, R.E. and D.L. Carter. (1994). Constraints on conservation tillage under dryland and irrigated agriculture in the United States Pacific Northwest. In *Conservation Tillage in Temperate Agroecosystems*, ed. M.R. Carter, Boca Raton, FL: Lewis Publishers. pp. 285-304.

Staricka, J.A., R.R. Allmaras, and W.W. Nelson. (1991). Spatial variation of crop residue incorporated by tillage. *Soil Science Society of America Journal* 55:1668-1674.

Staricka, J.A., R.R. Allmaras, W.W. Nelson, and W.E. Larson. (1992). Soil aggregate longevity as determined by the incorporation of ceramic spheres. *Soil Science Society of America Journal* 56:1591-1597.

Stewart, B.A. and W.C. Moldenhauer. (1994). Crop residue management to reduce erosion and improve soil quality: Southern Great Plains. *Conservation Research Report Number 37.*

Tapia-Vargas, M., M. Tiscareño-López, J.J. Stone, and J.L. Velazquez-Valle. (2001). Tillage system effects on runoff and sediment yield in hillslope agriculture. *Field Crops Research* 69:173-182.

Tessier, S., G.M. Hyde, R.I. Papendick, and K.E. Saxton. (1991). No-till seeders effects on seed zone properties and wheat emergence. *Transactions of the American Society of Agricultural Engineers* 34:733-739.

Tiscareño-López, M., A.D. Baez-González, M. Velázquez-Valle, K.N. Potter, J.J. Stone, M. Tapia-Vargas, and R. Claverán-Alonso. (1999). Agricultural research for watershed restoration in central Mexico. *Journal of Soil and Water Conservation* 54:686-692.

Tyler, D.D., M.G. Wagger, D.V. McCracken, and W.L. Hargrove. (1994). Role of conservation tillage in sustainable agriculture in the Southern United States. In *Conservation Tillage in Temperate Agroecosystems*, ed., M.R. Carter, Boca Raton, FL: Lewis Publishers, pp. 209-229.

Unger, P. (1999). *Managing Agricultural Residues.* Boca Raton, FL: CRC Press.

Unger, P.W. and E.L. Skidmore. (1994). Conservation tillage in the southern United States Great Plains. In *Conservation Tillage in Temperate Agroecosystems*, ed., M.R. Carter, Boca Raton, FL: Lewis Publishers, pp. 329-356.

Unger, P.W., A.F. Wiese, and R.R. Allen. (1977). Conservation tillage in the southern Plains. *Journal of Soil and Water Conservation* 32:43-48.

USDA. (1965). *Agricultural Statistics (1965).* Washington, DC: USDA.

USDA. (1977). *Agricultural Statistics (1977).* Washington, DC: USDA.

U.S. Department of Commerce (USDC), Bureau of the Census. (1992). Current industrial reports. MA35A.

Valdivia, R. and E. Villarreal. (1998). Technical diagnostic methodology for maize crop. In *XVII Congress of Plant Genetics*, Acapulco, Mexico.

Van Es, H.M., A.T. DeGaetano, and D.S. Wilks. (1998). Space-time upscaling of plot-based research information: frost tillage. *Nutrient Cycling in Agroecosystems* 50:85-90.

Van Es, H.M. and R.R. Schindelbeck. (1995). Frost tillage for soil management in the Northeastern USA. *Journal of Minnesota Academy of Science* 59(2):37-39.

Voorhees, W.B. (1992). Wheel-induced soil physical limitations to root growth. *Advances in Soil Science* 19:73-95.

Vyn, T.J., K. Janovicek, and M.R. Carter. (1994). Tillage requirements for annual crop production in Eastern Canada. In *Conservation Tillage in Temperate Agroecosystems*, ed., M.R. Carter, Boca Raton, FL: Lewis Publishers, pp. 47-71.

Vyn, T.J., G. Opoku, and C.J. Swanton. (1998). Residue management and minimum tillage systems for soybean following wheat. *Agronomy Journal* 90:131-138.

Wander, M.M., M.G. Bidart, and S. Aref. (1998). Tillage impacts on the depth distribution of total and particulate organic matter in three Illinois soils. *Soil Science America Journal* 62:1704-1711.

Wienhold, B.J. and A.D. Halvorson. (1998). Cropping system influences on several soil quality attributes in the Northern Great Plains. *Journal of Soil and Water Conservation* 53:254-258.

Wu, L., J.B. Swan, R.R. Allmaras, and S.D. Logdson. (1995). Tillage and traffic influences on water and solute transport in corn-soybean systems. *Soil Science Society of America Journal* 59:185-191.

The Importance of Root Dynamics in Cropping Systems Research

M. J. Goss

C. A. Watson

SUMMARY. This paper examines the nature and importance of the dynamics of crop root growth, particularly root turnover, and the application to different cropping systems. Methods now available to investigate root dynamics are summarized, and information being obtained is presented. Effects of physical, chemical, and biological factors on root dynamics are discussed. Growth of new roots and death of older roots can change the initial distribution in soil, allowing roots to exploit zones that have a more favorable nutrient or water supply. In herbaceous crops, the lifespan of roots appears to range between 16 and 36 per cent of the annual growth cycle. However, there is a paucity of data with which results can be compared. Localized enrichment of the water and nutrient supply enhances root turnover, and plants growing in soil well supplied with nutrients tend to have shorter-lived roots than those from nutrient limiting conditions. Both drought and excess water can induce premature root

M. J. Goss is Professor, Department of Land Resource Science, University of Guelph, Guelph, Ontario, Canada.

C. A. Watson is Head, Land Management Department, Environment Division, Scottish Agricultural College, Craibstone Estate, Aberdeen AB21 9YA, Scotland, UK.

Address correspondence to: M. J. Goss, Professor, Department of Land Resource Science, University of Guelph, Guelph, Ontario, Canada N1G 2W1 (E-mail: mgoss@lrs.uoguelph.ca).

[Haworth co-indexing entry note]: "The Importance of Root Dynamics in Cropping Systems Research." Goss, M. J., and C. A. Watson. Co-published simultaneously in *Journal of Crop Production* (Food Products Press, an imprint of The Haworth Press, Inc.) Vol. 8, No. 1/2 (#15/16), 2003, pp. 127-155; and: *Cropping Systems: Trends and Advances* (ed: Anil Shrestha) Food Products Press, an imprint of The Haworth Press, Inc., 2003, pp. 127-155. Single or multiple copies of this article are available for a fee from The Haworth Document Delivery Service [1-800-HAWORTH, 9:00 a.m. - 5:00 p.m. (EST). E-mail address: docdelivery@haworthpress.com].

death, as can the resupply of water after drought. Turnover of roots contributes to carbon deposition in soil through their death and decay, as well as from the release of exudates from those roots during their lifetime. Improved understanding of root turnover is important for the development of more sustainable cropping systems. In particular, it could be used to improve the exploitation of N released from green manure as well as capturing N that has been leached below the rooting zone of staple crops. It is stressed that root turnover has more importance for plants with longer life cycles than in short season annual crops. *[Article copies available for a fee from The Haworth Document Delivery Service: 1-800-HAWORTH. E-mail address: <docdelivery@haworthpress.com> Website: <http://www.HaworthPress.com> © 2003 by The Haworth Press, Inc. All rights reserved.]*

KEYWORDS. Root development, root architecture, root turnover, cropping systems

INTRODUCTION

Roots have been described variously as the 'invisible' or 'hidden' part of a plant (Weaver, 1926; Waisel, Eshel, and Kafkafi, 1996). Just as the vegetative parts of the shoot, stems, and leaves, assimilate carbon dioxide (CO_2) from the air, so roots capture and provide essential nutrient ions and water from the soil. Roots also synthesize growth controlling substances or their precursors (Marschner, 1995).

The promeristem, which gives rise to the main roots, is present in the embryo (Esau, 1965). In arabidopsis (*Arabidopsis thaliana* L.), it is derived from cells generated by the first division of the zygote (Scheres et al., 1994). One tap root or a few fibrous root axes may be formed from primordia present in the seed. Each major root will commonly form lateral branches from cells of the pericycle (Esau, 1965). Initial growth of roots follows a basic pattern: unvacuolated cells in the root apex undergo transverse or tangential divisions to give rise to files of cells. Those cells closest to the root tip form cells of the root cap, while the more distal cells give rise to the root proper (Laux and Jürgens, 1997). Cells of the root cap are readily sloughed off. They secrete significant amounts of mucilage, and influence plant growth by detecting gravity as well as sensing soil physical conditions (Juniper and Pask, 1973).

Cells of the root proper expand both longitudinally and radially following the formation of a central vacuole. Selective transport of mineral ions and organic molecules into the vacuole draws water into the vacuole by osmosis. Controlled softening of cell walls allows cells to expand as water enters and moves into the vacuoles (Taylor and Brar, 1991; Cosgrove, 1997). Nutrient ions move radially across the root from the soil through the cell walls (apoplastic pathway), or from cell to cell in the symplast. At the endodermis, the innermost layer of cells of the root cortex, suberization of cell walls greatly reduces the ease with which ions and water can move through the apoplastic pathway, forcing them to enter the symplastic transport stream (Robards et al., 1973; Sanderson, 1983). From the endodermis, ions and water move into the stele, for longitudinal transport in the xylem and phloem. Ions and water will enter both the xylem, for basipetal transport to the shoot, and the phloem, for transport to the root apex. The living cells of the phloem, sieve tubes and their companion cells, can limit the entry and movement of ions. Mature xylem vessels are part of the apoplastic pathway, but exchange of ions onto cell wall components and uptake by xylem parenchyma can modify the mass of ions that are transported to the shoot (Marschner, 1995). These features not only help buffer fluctuations in the supply of nutrients from the soil, but also contribute to ability of roots to respond to localized enrichment of some nutrients [e.g., phosphorus (P) and nitrogen (N)].

Most of the water taken up by plants is used to cool the above ground parts as it evaporates from the leaves. The rest supports internal transport of nutrients and assimilate, and provides structural strength to soft tissues through the maintenance of turgor. Limitations to the supply of water will commonly cause stomatal closure and reduced assimilation of carbon (C) (Marschner, 1995). The functions of the root system can impact shoots over different time frames, which may be relatively short with respect to the uptake of water but longer for nutrient absorption.

Prior to emergence, C resources are allocated to both roots and shoots, often with the proportion allocated to the root being at its maximum value (Demotes-Mainard and Pellerin, 1982). After emergence, allocation of assimilates to roots often appears to follow a source-sink relationship and depends on the magnitude of competing sinks in the shoot (Marschner, 1995). For example, if the supply of nutrients or water from the roots limits the potential for shoot expansion, there may be more resources available to support root growth over the short term. However, if new sinks develop in the shoot, such as fruits, less assimilate will be available to the root system (Liedgens et al., 2000). These variations in resource allocation can lead to significant changes in mass, distribution, and function of roots within the soil profile over the life of crops (Klepper et al., 1973). This paper examines the nature and importance of the dynamics of root growth, and the significance of root turnover during the life of plant species. We also consider the need to investigate root growth under different cropping systems.

ASSESSMENT OF ROOT DYNAMICS

Comparison of available data on root dynamics in different species is complicated by differences in both methodological approaches and terminology. Mackie-Dawson and Atkinson (1991) categorized methodological approaches into three broad groups: (i) soil sampling methods (i.e., root system excavation), (ii) observation methods, and (iii) indirect methods. These methods have been reviewed extensively elsewhere (e.g., Smit et al., 2000; Harper, Jones, and Sackville-Hamilton, 1991; Mackie-Dawson and Atkinson, 1991). It is worth noting, however, that even where one technique has been used, observations using the minirhizotron method for example, a variety of experimental protocols have been applied. This may help to explain some of the differences in estimates of root longevity. Intervals between observations using the minirhizotron vary from days (e.g., Black et al., 1998; Watson et al., 2000), 2-3 weeks (e.g., Cheng, Coleman, and Box, 1990; Dickmann, Nguyen, and Pregitzer, 1996) to 1 month (e.g., Hendrick and Pregitzer, 1996; Majdi and Kangas, 1997).

Rytter and Rytter (1998) investigated the impact of sampling intensity on fine root development and disappearance over a 2-month period and concluded that in basket willow (*Salix viminalis* L.) bi-weekly observations were adequate. Given the seasonal nature of root mortality (Hendrick and Pregitzer 1993a; Ruess, Hendrick, and Bryant, 1998) and species differences, this warrants further investigation over a longer time period and in other species. Furthermore the definition of fine root diameter in trees differs between studies, e.g., < 1 mm (Thomas et al., 1999), < 2 mm diameter (Hendrick and Pregitzer, 1997), < 3 mm (Aber et al., 1985). It is not only root size that is important but also root order (Eissenstat et al., 2000). Roots without branches generally have a shorter lifespan than roots with branches (Fusseder, 1987).

Differentiating between live and dead roots is a major difficulty in assessing root lifespan or longevity (discussed by Eissenstat and Yanai, 1997; and Comas, Eissenstat, and Lakso, 2000). In practice, determining when a root is no longer functional is very difficult. Roots not only function as organs for uptake of water and nutrients, but also transport materials from one part of the root system to another and from root to shoot. Loss of root cortex may indicate that the root is not contributing significantly to nutrient uptake, but its ability to act as a conduit for nutrients may remain unimpaired. A variety of 'vital' stains have been advocated, but none have proved to have universal application or unequivocal reliability. Observation of individual roots has resulted in decision rules based on color changes (e.g., Smit and Zuin, 1996) and reductions in root diameter. The grazing activity of soil fauna may also give some indication of changes in root viability. Different approaches for determining root viability are described by Hooker, Hendrick, and Atkinson (2000).

Root dynamics have been described using changes in a number of parameters over time; these include length, number, mass, and architecture. These parameters are sometimes, but not always, related to an estimate of the size of the total root system. Table 1 illustrates the range of terms which have been used to describe root turnover. Tingey, Phillips, and Johnson (2000) suggest that the absolute rate of root turnover is the appropriate measure when C and nutrient fluxes are the important aspects of investigation. However, relative turnover would be the more appropriate measure for studies of root demography, when turnover is compared with the root mass of the standing crop. Using two different approaches to calculating turnover, Aber et al. (1985) found the results differed by an order of magnitude on a single site. A number of methods have been used to calculate root production and root mortality, these include the Max-Min method (Persson, 1978), the balancing-transfer method (McClaugherty, Aber, and Melillo, 1982; Fairley and Alexander, 1985) and the compartment flow method (Santantonio and Grace, 1987). Several authors, including Kurz and Kimmins (1987), Publicover and Vogt (1993), and Lehmann and Zech (1998), have compared methods on selected datasets and reported on advantages and errors associated with each method. The ratio of the rate of root death to rate of production gives an index of root-turnover, which ranges from 0 to infinity. Values < 1 indicate that the root system is expanding, and values > 1 indicate that the root system is declining. A value of 1 means that the production rate is keeping up with the death rate.

Minirhizotrons allow population statistics such as birth rate, death rate and median lifespan of roots to be calculated (Majdi, 1996). From this data, the days to 50% mortality (median lifespan) can be derived (Eissenstat and Yanai, 1997). The term 'half-life' has also been applied by some workers (Fitter et al., 1997; Thomas et al., 1999). Survivorship or survival curves, where the initial number of roots within a cohort (that is new roots observed on a particular date), and the number of original members of the cohort surviving at subsequent observation dates is plotted over time, have been used frequently (Hendrick and Pregitzer, 1993a; Pregitzer, Hendrick, and Fogel, 1993; Kosola, Eissenstat, and Graham, 1995; Fitter et al. 1997; Majdi and Kangas, 1997; Thomas et al., 1999; Watson et al., 2000). This is sometimes expressed as the log of the fraction of cohorts surviving at different times (Goins and Russelle, 1996; Fitter et al., 1997). Recently more complex data analysis techniques have been used to describe root longevity. Black et al. (1998) applied the concept of survival analysis (Lee, 1992) to roots for the first time. This is a statistical technique which uses the age at death of all observed roots rather than simply using the number of roots that survive for longer than some arbitrary time (Hendrick and Pregitzer, 1992, 1993a; Hooker et al., 1995). Wells and Eissenstat (2001) recently applied a proportional hazards model permitting the analysis of multiple covariates which influence lifespan.

TABLE 1. Definitions of root 'turnover' used by researchers

Terminology	Definition	Units	Author
Turnover time	Mean fine root age in growth phase + mean fine root age in standing phase + mean fine root age in decay phase	days	Rytter and Rytter (1998)
Turnover rate	365/turnover time (see above)	yr^{-1}	Rytter and Rytter (1998)
Turnover index	(Specific root growth rate + specific root death rate)/2	% day^{-1}	Cheng, Coleman, and Box (1990)
Turnover index	Ratio of fine root mortality to initial fine root length	-	Hendrick and Pregitzer (1992)
Index of turnover	Differences between cumulative births and deaths	-	Fitter et al. (1997)
Root turnover	Ratio of annual growth to total root mass	-	Dahlman and Kucera (1965)
Root turnover	Root production/standing crop rootmass	% yr^{-1}	Aber et al. (1985)
Root turnover	Sum of root length production/average length of roots observed in each year	%	Hayes and Seastedt (1987)
Root turnover	(Sum of length of fine root mortality) × (correction factor for converting length to biomass)	$gm^{-2}yr^{-1}$	Steele et al. (1997)
Root turnover	Inverse of mean root lifespan		Nadelhoffer (2000)

ROOT DYNAMICS AND ROOT ARCHITECTURE

Growth and development of a root system involves the elongation of primordia present in the seed and the formation of new axes and branches. The root system can thereby explore deeper soil layers over time, and the extent and intensity of exploration in each layer will also tend to increase (Figure 1). Elongation of new or existing roots allows the exploitation of relatively immobile nutrients outside the zone of depletion that typically develops in the soil around the original root system (Bhat and Nye, 1973). Increased root length and root-length-density also contribute to improved anchorage of the plant (Goss, 1991). Root growth into deeper horizons or increase in the extent of exploration of a layer can increase the likelihood of different sources of water and nutrients being intercepted (Huck et al., 1986).

As the crop grows, soil factors begin to impact root systems. During dry periods, availability of water declines in the surface layers of the soil. As a result, conditions for growth become more favorable in deeper soil layers. Under such conditions root distribution may change from declining linearly or exponentially with depth, to become uniform or even increase with depth (Klepper et al., 1973). Part of the change comes from the growth of new roots at depth, but there can be significant death of roots in the surface layers. This turnover of roots can be viewed as a means of optimizing the use of the C allocated from

FIGURE 1. Root growth and distribution of maize as affected by stage of development. Data for roots sampled in the row (adapted from Stypa et al., 1987).

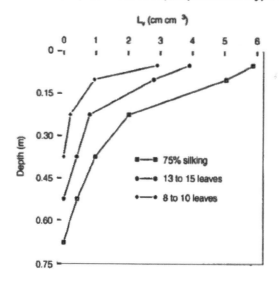

the shoot for the acquisition of water and nutrients. In the short term, the plant response to sub-optimum soil conditions can result in a greater allocation of resources to the root system. Shoot-to-root dry weight ratio will therefore tend to decline, and the weight of roots can be enhanced (Huck et al., 1986). If new resources are not intercepted, total dry matter accumulation will slow until the overall growth of the plant is reduced.

Turnover of roots in response to changes in the availability of water and nutrients as well as to adjustments in the allocation of photo-assimilates contributes to the deposition of C in the soil during the growing season. As roots grow and perform their various functions they release C-rich exudates, and then as they die and decay the C in structural components are released into the soil through microbial activity.

Wells and Eissenstat (2001) found that the risk of a fine root dying was inversely related to its diameter. They also observed that fine roots of smaller diameter were often found in locations where there was an intense root proliferation, while fine roots of larger diameter were often found to be more isolated.

THE SIGNIFICANCE OF ROOT TURNOVER IN DIFFERENT PLANT SPECIES

The importance of root turnover is likely to vary between species and between plants exhibiting different life-strategies. The shorter the life cycle, the less the opportunity for the root system to experience changes in soil conditions so that root turnover would confer little advantage to the plant. Consequently most studies of root turnover have concentrated on investigating activity in perennial pasture, forage, and tree species. But vines such as tomato (*Lycopersicon esculentum* Mill.) and kiwifruit (*Actinidia deliciosa* (A. Chev.) C.F. Lang et A.R. Ferguson) have also received significant attention (e.g., Reid et al., 1996; Reid, Sorensen, and Petrie, 1993). Among field-grown vegetable crops Brussels sprouts (*Brassica oleracea* L.), which have a biennial life cycle and form a tap root, and leeks (*Allium porrum* L.), which also have a biennial life cycle but form a fibrous root system, have been studied (Smit and Zuin, 1996). In annual plants, the survival of roots becomes an important issue, and this has been investigated in cereals (Swinnen, Van Veen, and Merckx, 1995). Even within herbaceous crops a wide range of values have been reported for the longevity of roots (Table 2). For short season crops, such as pea (*Pisum sativum* L.) and spring oats (*Avena sativa* L.), the agespan of roots is about 20 per cent of the plant lifespan. For a perennial crop, such as alfalfa (*Medicago sativa* L.), the lifespan of roots appears to range from about 16 to 36 per cent of the annual growth cycle.

TABLE 2. Longevity of roots of different crop species (adapted from Pritchard and Rogers, 2000)

Plant		Modal root life span (d)	Reference
Spring oats (*Avena sativa*)		32	Atkinson and Watson (2000)
Pea (*Pisum sativum*)		35	
White mustard (*Sinapis alba*)		39	
Sorghum (*Sorghum bicolor*)	Early season roots Late season roots	42-49 23-27	Cheng, Coleman, and Box (1990)
Groundnut (*Arachis hypogaea*)		24-31	Krauss and Deacon (1994)
Faba bean (*Vicia faba*)		69-90	Rengasamy and Reid (1993)
Winter wheat (*Triticum aestivum*)	Early season roots Late season roots	> 135 46	Gibbs and Reid (1992)
Leek (*Allium porrum*)		130	Smit and Zuin (1996)
Sugar beet (*Beta vulgaris*)		60-130	Van Noordwijk et al. (1994)
Brussels sprouts (*Brassica oleraceae*)		70	Smit and Zuin (1996)
Sugarcane (*Saccharum officinarum*)		14-90	Ball-Coelho et al. (1992)
Tomato (*Lycopersicon esculentum*)		56-70	Reid et al. (1996)
Alfalfa (*Medicago sativa*)	Early season roots Mid-season roots Late season roots	58-131 47- 92 7-131	Goins and Russelle (1996)
White clover (*Trifolium repens*)		> 43	Atkinson and Watson (2000)
Red clover (*Trifolium pratense*)		> 43	
Perennial ryegrass (*Lolium perenne*)	Early season roots Late season roots	144 59	Gibbs and Reid (1992)

Consistent with the general concepts outlined above, Ruess, Hendrick, and Bryant (1998) found that the shorter the growing season, the longer the survival of roots. However, Swinnen, Van Veen, and Merckx (1995) found that the mortality of roots produced at tillering by spring barley (*Hordeum vulgare* L.), a short season crop, was greater than those produced by winter wheat (*Triticum aestivum* L.), a long season crop. Nevertheless, if the mortality was calculated relative to the life cycle of the crop rather than in absolute terms, these results would be consistent with the conclusions of Ruess, Hendrick, and Bryant (1998). Within a plant stand, those roots produced early in the season tend to survive the longest (Ruess, Hendrick, and Bryant, 1998). In winter wheat and spring barley, for example, almost twice as many roots produced at tillering survived to crop maturity than did those produced at flowering (Swinnen, Van Veen, and Merckx, 1995).

Total root length may also be important for root turnover. The total length of roots produced by Brussels sprouts was double that of leeks grown under the same conditions, but the longevity of roots in the former was half that of leeks (Smit and Zuin, 1996).

These observations suggest that there may be very significant differences between net root production indicated by 'snap shots' obtained by periodic sampling programs, and gross production that takes account of root turnover. Gibbs and Reid (1992) compared net and gross root production in winter wheat and perennial ryegrass (*Lolium perenne* L.) over a 240 day period. In that experiment net root length production underestimated gross production by 36% in the case of the cereal crop and by 45% in the ryegrass. Such results are clearly important when the amount of C entering the soil is being assessed.

Consequences of Root Turnover for Nutrient Cycling

In deciduous trees, there is a seasonal withdrawal of N from senescent leaves, which is stored over the winter period and remobilized for spring growth (Millard and Neilsen, 1989). Some species, for example sycamore (*Acer pseudoplatanus* L.), are inefficient at withdrawing N from leaves, and much is lost in litter fall (Millard and Proe, 1991). Whether there is a similar, and species dependent, withdrawal of nutrients from roots is unclear. Both Vogt et al. (1995) and Meier, Grier, and Cole (1985) measured significant internal translocation of nutrients from senescent roots, while Nambiar (1987) suggested that this flux was negligible in radiata pine (*Pinus radiata* L.). Understanding and quantifying this potential flux of C and nutrients to the soil is important in terms of modeling belowground processes in cropping systems.

FACTORS AFFECTING ROOT DYNAMICS

Root dynamics are influenced by aboveground and belowground environmental factors but are also under inherent genetic control. This is summarized in Figure 2.

Inherent Factors

General differences in root architecture between crop plants have been studied systematically, beginning with the work of Weaver (1926). Developmental stage has a marked effect on root dynamics in most crops. Root systems of cereal crops tend to expand until flowering, when there is a steady decline in total root length (Liedgens et al., 2000). In groundnut (*Arachis hypogaea* L.), Krauss and Deacon (1994) found that root death began 3-4 weeks after plants were sown. Death of lateral roots was not related to the onset of flowering or pod-filling, but peaks of root death in a specific root region occurred 3-5 weeks after peak production in that region. This is in marked contrast to cucumber

FIGURE 2. Factors that influence root dynamics.

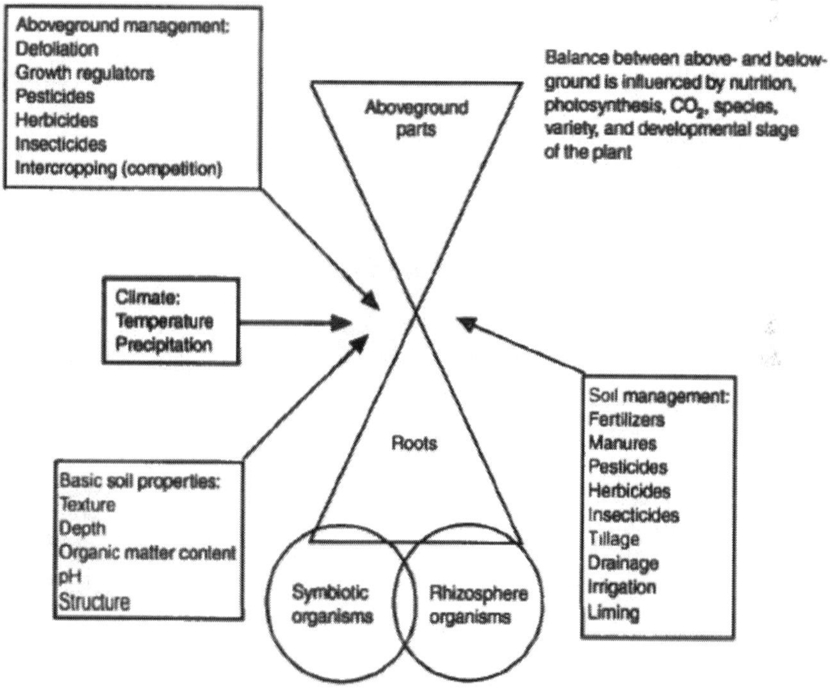

(*Cucumis sativus* L.), where Van der Vlugt (1990) observed root death coinciding with the onset of fruiting. Berntson and Bazzaz (1996, 1997) suggested that in yellow birch (*Betula papyrifera* L.) and red maple (*Acer rubrum* L.) seedlings there was evidence for a correlation between the birth and death rates in roots and shoots. Root:shoot ratio can be depressed by heavy fruit development in fruit trees; turnover of fine roots being enhanced at this time (Buwalda and Lenz, 1992). The increase in sinks for C, which developing seeds and fruits represent, clearly provide competition with roots for photosynthate.

Soil Factors

The soil can impose physical, chemical, and biological constraints on root growth and development. These constraints modify the architecture of the root system. Mechanical impedance, the resistance of soil to the penetration by roots, can be important in determining the volume of soil explored and the intensity of exploration. Although roots have some ability to grow through rigid pores smaller than their own diameter (Goss, 1977; Scholefield and Hall, 1985), this appears to be limited to pores larger than the diameter of the root meristem (Goss, 1977). Main roots and lateral branches of cereals exhibit similar sensitivity to mechanical constraint (Goss, 1977). As soils dry their strength tends to increase, so that in the field there can be variation in the mechanical impedance within a soil layer dependant on the water content (Douglas et al., 1986). Compacted soils tend to offer greater mechanical constraint than loose soils, and transport oxygen to the root and CO_2 away from it are also reduced relative to that in loose soil. Soil aeration also influences the size of a root system and the volume of soil explored (Goss, Barraclough, and Powell, 1989). Similarly, waterlogging can limit root growth through reduced aeration and because of the release of toxic materials in the soil (Marschner, 1995).

Both the concentration and location of nutrients within the profile can modify root growth and distribution, leading to proliferation of roots in zones of localized enrichment of nitrate (NO_3^-), ammonium (NH_4^+), and phosphate ions (P), but not potassium (K) (Drew, 1975). In contrast, aluminium ions that can accumulate in acid soil layers, inhibit root elongation and indirectly restrict cell division in the root meristem leading to restricted root exploration (Marschner, 1995).

Soil fauna, such as nematodes, bacteria, and parasitic fungi can invade roots and impair growth. There are even some families of higher plants that are parasitic on the roots of crop plants (e.g., *Striga asiatica* L. on cereals, and *Orobanche crenata* L. on beans), thereby diverting assimilates and nutrients away from the host plant.

Competition for resources is likely to be a key aspect among soil factors that determine root dynamics. This is important for both competition within or between plant species such as can occur in agroforestry (Schroth, 1995), and competition between crop and weed species (Arnone and Kestenholz, 1997). Allelopathy and the release of phytotoxins may also mediate interspecific competition. Phenolic acids are one group of allelochemicals, and are found in root exudates (Vaughan and Ord, 1991). Particular phenolic acids have been shown to inhibit the extension of main roots and the development of branch roots. Nutrient uptake can also be influenced by phenolics possibly through the alteration of membrane permeability or the electrical potential across the membrane (Glass and Dunlop, 1974). In situations such as agroforestry it may be advantageous to manipulate below ground competition through management such as pruning (see section on defoliation).

Heterogeneity of soil resources may result from inherent variability or may be induced by agricultural practices. Plants can exploit spatially heterogeneous resources by increasing root growth or physiological activity in resource enriched areas (Robinson and van Vuuren, 1998; Huang and Eissenstat, 2000; Schroth, Rodrigues, and D'Angelo, 2000). In field conditions an increased spacing of groundnut plants was found to increase the rate of root production (Rao et al., 1989; Simmonds and Azam-Ali, 1989). This can result from an increase in the total availability of water and nutrient resources, or to the heterogeneity of resources as observed by McKay and Coutts (1989). Greater heterogeneity may favor rapid root exploitation and turnover (Smucker, 1993). Hodge et al. (1999) found that root death rate was accelerated in nutrient rich patches compared to bulk soil, suggesting a shorter lifespan in these microsites. However, the data on lifespan of fine roots in patches is inconsistent. Pregitzer, Hendrick, and Fogel (1993) and Fahey and Hughes (1994) worked in mixed hardwoods under field conditions. They found that roots growing in patches, which were enhanced with water alone or water and nutrients in combination, lived longer than those in unamended soil. In contrast, localized water and nutrient additions resulted in shorter lifespan of old-field herbaceous species in a pot experiment (Gross, Peters, and Pregitzer, 1993).

Resource heterogeneity may also affect mycorrhizal activity (Bending and Read, 1995; Hooker and Black, 1995) allowing small-scale and niche-exploitation of resources by fungal hyphae. This may then affect the longevity of roots. Thus Hooker et al. (1995) have shown that poplar (*Populus generosa interamericana* L.) roots colonized by arbuscular mycorrhizal fungi had a shorter lifespan than non-colonized roots. In contrast, Hodge, Robinson, and Fitter (2000) found lifespan of plantain (*Plantago lanceolata* L.) roots to be unaffected by inoculation with arbuscular mycorrhizal fungi. Arbuscular mycorrhiza can also increase root lifespan under drought conditions (Espelata, Eissenstat, and Graham, 1998). Any changes in longevity resulting from colonization by arbuscular mycorrhizal fungi may be linked to changes in root morphology resulting from development of the mycorrhiza (Berta et al., 1995;

Hooker, Munro, and Atkinson, 1992). However, there can also be changes in the content of growth control substances such as ABA (Danneberg et al., 1992) and cytokinins (Drüge and Schönbeck, 1992), and any improvement in P nutrition may also affect longevity. Nonetheless, colonization with arbuscular mycorrhizal fungi can enhance root proliferation in nutrient rich patches (Hodge, Robinson, and Fitter, 2000). Majdi, Damm, and Nylund (2001) reported that even among mycorrhizal roots of Norway spruce (*Picea abies* L.), those in soils with added N had reduced longevity compared with those in unfertilized soil.

Temporal changes in resource availability, as exemplified by fertilizer application, also appears to be an important factor in root turnover. In Norway spruce, fertilizer application reduced the length of fine root, and the mortality of these roots was greater than that for roots in unfertilized plots (Majdi and Kangas, 1997). In winter wheat there was no effect of soil management on root longevity, but in spring barley, the treatment with the larger fertilizer input resulted in greater root mortality (Swinnen, Van Veen, and Merckx, 1995). In forest trees, the turnover of fine roots increased quadratically with N availability. However, while the effect was very evident in stands fertilized with nitrate-N, there was no significant effect in stands fertilized with ammoniacal-N (Aber et al., 1985).

Water supply also plays a role in root survival. Goss et al. (1984) showed that when the soil water content was replenished in mid-July after a period without rain the uptake of water from surface layers was much less than that found only one month earlier in the season, while uptake at depth was much greater. This suggested that there had been significant root death in the topsoil, consistent with adverse effects of drought on the longevity of lateral roots (Smucker, 1993). Rewetting soil can result in an enhanced death rate of roots (Rengasamy and Reid, 1993). In alfalfa, Goins and Russelle (1996) found that the greatest mortality in fine roots took place in the top 0.1 m of the soil profile, presumably because of the greater drought stress experienced near the soil surface. Reid et al. (1996) considered changes in the demographics of tomato root systems under developing water deficits. They reported greater root death in non-irrigated soil compared with irrigated soil. The gross number of roots produced in the non-irrigated land was also greater than in the irrigated soil, although this was largely due to the numbers produced in the 0.45-0.7 m layer (Figure 3). However, as a fraction of the roots present, the turnover of roots was greater in the top 0.2 m of soil. As the demography of roots was weighted towards a large proportion of young roots (Figure 4), mortality was most common among the youngest roots. However, temperature effects are also important for root development (see below), and could have contributed to such observations. The main period for tree root growth is in the spring, but a second peak often occurs in autumn before leaf fall under temperate conditions

FIGURE 3. Variation in rooting density of tomato in irrigated and non-irrigated soil 120 days after planting. Data from Reid et al. (1996).

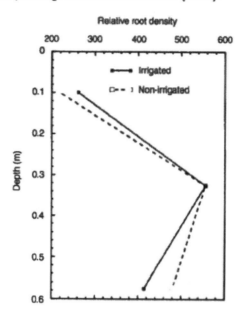

FIGURE 4. Cumulative frequency of root longevity in tomato. Calculated from Reid et al. (1996).

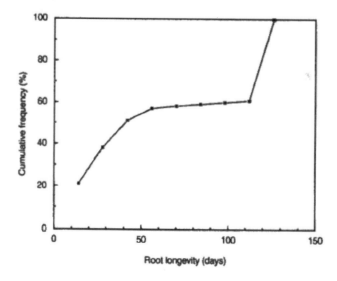

(Atkinson, 1980), and an improved water balance at this time may be an important contributory factor. Excess water can also can also impair root development and the ability of plants to grow into soil once waterlogged conditions have ceased (Thomas and Hartman, 1998).

Soil temperature affects root systems through root initiation and branching as well as the orientation and direction of growth (reviewed by Kaspar and Bland, 1992). Root growth patterns differ between species but are at least partly determined by temperature.

Under controlled conditions, root production in annual plants generally increases with temperature (e.g., Mozafar, Schreiber, and Oertli, 1993) although extremely high temperatures can cause both elongation and initiation of roots to decrease (Paradales, Yamauchi, and Kono, 1991). Root branching in corn (*Zea mays* L.) is also affected by temperature with more branching at higher temperatures (Feil et al., 1991). Temperature has also been shown to affect the angle of root growth of corn roots (Onderdonk and Ketcheson, 1973). At 17EC the roots grew closer to the horizontal, but at 10 and 30EC root growth tended towards the vertical. The influence of temperature on the trajectory of corn roots does not continue indefinitely, but mainly affects the first 0.1 m of growth (Tardieu and Pellerin, 1991). Temperature therefore affects root distribution within the soil profile, and influences the potential for more localized root proliferation at later stages in the plant growth cycle.

For perennials, Hendrick, and Pregitzer (1997) found that fine-root lifespan in temperate forests was strongly influenced by temperature, and Watson et al. (2000) suggest temperature as a major factor in explaining differences between grass and white clover (*Trifolium repens* L.) root longevity at contrasting sites. The response of tree roots to temperature has recently been reviewed by Pregitzer et al. (2000). In forests, the start of root growth in spring generally coincides with increases in soil temperature (Lyr and Hoffman, 1967; Kuhns et al., 1985). Under temperate conditions, most root growth in grasses occurs prior to the peak of aboveground growth (Troughton, 1957; Campbell et al., 1994). An optimum in belowground activity prior to aboveground growth is also generally accepted for trees (Lyr and Hoffman, 1967). The length of lag phase between these events is likely to be linked to site temperature (Steele et al., 1997).

Under controlled environment conditions, longevity has been directly related to temperature; 70% of grass roots survived for more than 35 days at 15EC compared with only 16% at 27EC (Forbes, Black, and Hooker, 1997). Van Noordwijk et al. (1994) related lower sugar beet (*Beta vulgaris* L.) root mortality deeper in the profile to cooler temperatures at depth.

Soil texture may also be a factor affecting root longevity. Given the effects of below ground conditions described in the above sections, soil type might be expected to influence root dynamics. However, largely owing to the difficulties of making soil type comparisons under field conditions there is little published information available. Rytter and Rytter (1998) found no difference in root turnover time between basket willows grown under the same climatic conditions in sand or clay. Effects of soil texture could have been masked in this study because nutrients and water supply were carefully controlled.

Soil structure is likely to be more important than texture alone. Structure not only determines the likelihood of roots experiencing mechanical impedance and poor aeration, but will have a marked influence on the availability of water and nutrients (Goss, 1977, 1991; Passioura, 1991).

Root pathogens and root herbivory are likely to impact root dynamics, but there is relatively little information on the magnitude of their effects (Brown and Gange, 1991). Impacts of root pathogens on root growth have been studied (e.g., Kotcon, Rouse, and Mitchell, 1984; Xiao and Subbarao, 2000) but the associated root dynamics have not. Kosola, Eissenstat, and Graham (1995) observed that fine root longevity of citrus trees was reduced when pathogen levels were higher. However, this was an observational field study and did not address interactions between specific pathogens and root dynamics. Norman, Atkinson, and Hooker (1996) found that strawberry (*Fragaria chiloensis* L.) root system necrosis resulting from infection with *Phytophthora fragariae* was positively correlated with the proportion of the root system made up of higher order roots. Given the known interactions between branching patterns and other biotic and abiotic factors, concurrent changes in root longevity are also likely. In addition to the direct impact of pathogens on root lifespan, it is possible that decaying tissue resulting from root turnover could allow build-up of pathogens in the rhizosphere (Krauss and Deacon, 1994). The use of fungicide and a combination of fungicide and insecticide, but not insecticide alone, have been shown to markedly increase root longevity in sugar-maple (*Acer saccharum* L.) (Wells, 1999, cited by Eissenstat et al., 2000). This suggests that roots will survive longer in the absence of pathogen pressure. This may help to explain some of the long lifespans quoted in the literature for laboratory experiments. However, specialized disease organisms may not increase root turnover. In potato (*Solanum tuberosum* L.), infection with *Globodera pallida* (potato cyst nematode) reduced the rate of root death, although average root length was less than that of uninfected controls (Smit and Vamerali, 1998).

Aboveground Factors

Defoliation by cutting or grazing affects root turnover in forages. Root growth has been shown to be severely reduced by defoliation in a number of species including alfalfa (Luo, Meyerhoff and Loomis, 1995), perennial ryegrass (Jarvis and Macduff, 1989; Mackie-Dawson, 1999), and stargrass (*Cynodon nlemfuensis* L.) (Alcordo, Mislevy, and Rechcigl, 1991). These latter authors found that the reduction in root growth in stargrass was related to the severity of defoliation. Winter clipping of cereal shoots delayed root extension and reduced root length density relative to unclipped plants (Bonachela, 1996), effects were most evident prior to anthesis. However, in poplar, there was no evidence of the anticipated dieback in roots following coppicing (Dickmann, Nguyen, and Pregitzer, 1996). Grazing reduced the production of fine roots in soil under northern boreal forest (Ruess, Hendrick, and Bryant, 1998).

The impact of defoliation on belowground components of legumes varies between species. After the first cut in alfalfa, Goins and Russelle (1996) reported an average loss of more than 20% of fine roots in the top 0.2 m of soil, indicating the close relationship between the supply of C material from the shoot and associated root growth. Butler, Greenwood, and Soper (1959) found that losses of roots and nodules in white clover caused by repeated defoliation was rapidly balanced by new growth. In red clover (*Trifolium pratense* L.) there was a less dramatic loss of roots and nodules than in white clover but much less compensatory regrowth. Loss of nodules following defoliation is also evident in leguminous trees (Fownes and Anderson, 1991).

Most of these results are consistent with a decrease in production of photosynthate caused by defoliation leading to a shift in resource allocation between above and below ground parts to maintain a homeostasis. This initially results in a decline in relative growth of roots until photosynthetic activity in the shoot is fully restored. These results are consistent with those reported by Brouwer (1963), and with the concepts of source-sink relationships. However, more C transport to the roots does not necessarily result in slower root turnover. In some grass species there is evidence that the half-life of roots is shorter under elevated CO_2 than under current ambient levels (Fitter et al., 1996).

Temporal variability in root dynamics has also been identified in some plants. However, there is not enough data available to generalize over seasonal effects. Many workers have shown differences in longevity between root cohorts produced in different seasons. Some studies have shown greater survival of fine roots initiated in spring (Hendrick and Pregitzer, 1993b; Ruess, Hendrick, and Bryant, 1998), while others have shown autumn initiated roots to survive longer (Kosola, Eissenstat, and Graham, 1995). Thomas et al. (1999) found that radiata pine roots first observed after mid-summer had a shorter lifespan than those first observed before mid-summer.

Few workers have observed roots over a period of more than two years. In situations where plant physiology and species dynamics are constantly changing, such as during establishment of a mixed grass sward or system involving woody perennials, this may be important. In perennial crops observation of root dynamics has often been limited to the growing season (e.g., Katterer and

Andrén, 1999); however, ignoring overwinter activity could result in underestimating annual fine root processes (Ruess, Hendrick, and Bryant, 1998).

ROOT DYNAMICS AND SUSTAINABLE CROPPING SYSTEMS

Improvement in the efficiency of nutrient utilization is important for developing more sustainable cropping systems. Not only is the goal to reduce input costs, but also to lessen the risk of environmental contamination. Improved knowledge of root systems is important if these goals are to be attained. For more mobile nutrients, such as NO_3^-, crops can be selected in a sequence that exploits differences in the depth and density of rooting between species or between varieties (Cameron and Haynes, 1986; Meek et al., 1994). That way nutrients that have moved below the rooting zone of one crop can be intercepted by the deeper roots of a subsequent crop (Cameron and Haynes, 1986; Huang, Rickert, and Kephart, 1996). Alternatively, a suitable companion crop or 'living mulch' could be selected that would grow slowly while the staple crop matures, but will then flourish as the main crop senesces (Leary and DeFrank, 2000). In this way the period when nutrients are being removed from the soil can be extended (Cameron and Haynes, 1986), as can the period during which water is extracted. Both aspects will tend to have beneficial effects as nutrient uptake will reduce the source of contaminants from the soil, and water uptake will reduce the amount of deep percolation. To exploit such opportunities we need to identify root systems that can readily explore the soil profile, and are plastic enough to exploit localized zones of nutrient enrichment.

For relatively immobile nutrients such as P, preventing soil from eroding is likely to be important. Root systems that provide good anchorage for the plant will also tend to stabilize the surface soil (Goss, 1991). Where the interest is in improved soil structure and C sequestration, roots with a rapid production rate and short life cycle would be of considerable value (Pritchard and Rogers, 2000).

Preventing the carry over of disease from season to season or crop to crop is another important goal for cropping system development. It is likely that there is a link between above ground defoliation and damage by root pathogens and root dynamics. Huisman (1982) concluded that there was little information on root growth dynamics and disease epidemiology and that root pathologists needed to integrate assessments of root growth dynamics into their research. This is still an area where there has been little activity.

We know that shoot and root systems of the same variety can behave differently as a result of soil conditions. Site specific management of the land aims to develop tools that allow producers to adjust inputs (e.g., fertilizers, pesticides) within the boundaries of a given field. We need to address the importance of within-field variability in relation to different cropping practices and farming systems. Even where agronomic experiments include a control given no fertilizer N, this is unlikely to replicate soil conditions under organic or integrated production systems. As biological, chemical, and physical processes appear to contribute to the variability in fertility within a field, the possibilities for gaining a better understanding through root studies seem likely.

We also need to be able to interpret the impact of varieties on nutrient cycling studies. Varieties resulting from breeding programs designed for high input systems are not likely to develop good mycorrhizal relationships. Older varieties developed during a period of limited inputs are more likely to be colonized because the optimum fertilizer input for crop growth is less. For example, Foulkes, Sylvester-Bradley, and Scott (1998) showed that as a result of breeding successes, 56 kg ha^{-1} more N was required to provide the maximum economic rate of fertilizer application and 42 kg ha^{-1} was removed in the grain by winter wheat in 1988 than in 1969. Siddique, Belford, and Tennant (1990) showed older wheat cultivars invested more dry matter in roots than newer varieties at anthesis. This was related to the fact that the modern cultivars reached anthesis approximately 30 days earlier than the older cultivars. Breeding programs have also had some unexpected impacts on root growth and development. Miralles, Slafer, and Lynch (1997), working on a comparison between standard height wheat cultivars and near-isogenic semidwarf and dwarf lines developed from them, found that the effects of Rht alleles on root length and weight were opposite to those observed on aboveground tissue.

Multi-cropping, growing more than one crop per year in a field, and intercropping, growing more than one crop in strips in the same field, are areas where roots have received too little attention. Examples of these systems include legume-non-legume sequences for better nutrient use, and agroforestry where trees provide fodder or fruit but are also intended to improve nutrient cycling. For all these aspects of cropping systems, there is interest in developing decision support systems, and associated crop models. Many models have been formulated that establish a basic distribution of roots in the profile, but have failed to include differences in the lifespan of root members, especially in relation to soil factors, both biotic and physical.

CONCLUSIONS

Considerable knowledge has been developed in the areas of root architecture, and development. Understanding of the dynamics of root systems is improving rapidly, but much of this has been focused on trees rather than herbaceous species. One reason for that is the greater propensity for root turnover in plants with a longer life cycle. Knowledge of root dynamics offers the possibility of improving the choice of cropping sequences and plant performance to meet the needs of a sustainable agriculture. There are difficulties associated with both the definition and measurement of root turnover, and agreement on both aspects is needed if progress is to be made in comparing the effectiveness of different cropping systems.

REFERENCES

Aber, J.D., J.M. Melillo, K.J. Nadelhoffer, C.A. McClaugherty, and J. Pastor. (1985). Fine root turnover in forest ecosystems in relation to quantity and form of nitrogen availability–a comparison of two methods. *Oecologia* 66:317-321.

Alcordo, I.S., P. Mislevy, and J.E. Rechcigl. (1991). Effect of defoliation on root development of star-grass under greenhouse conditions. *Communications in Soil Science and Plant Analysis* 22:493-504.

Arnone, J.A. and C. Kestenholz. (1997). Root competition and elevated CO_2: Effects on seedling growth in *Linum usitatissimum* populations and *Linum Silene cretica* mixtures. *Functional Ecology* 11:209-214.

Atkinson, D. (1980). The distribution and effectiveness of the roots of tree crops. *Horticultural Review* 2:424-490.

Atkinson, D. and C.A. Watson. (2000). The beneficial rhizosphere: A dynamic entity. *Applied Soil Ecology* 15:99-104.

Ball-Coelho, B., E.V.S.B. Sampaio, H. Tiessen, and J.W.B. Stewart. (1992). Root dynamics in plant and ratoon crops of sugar cane. *Plant and Soil* 142:297-305.

Bending, G.D. and D.J. Read. (1995). The structure and function of the vegetative mycelium of ectomycorrhizal plants. 5. Foraging behavior and translocation of nutrients from exploited litter. *New Phytologist* 130:401-409.

Berntson, G.M. and F.A. Bazzaz. (1996). The allometry of root production and loss in seedlings of *Acer rubrum* (Aceraceae) and *Betula papyrifera* (Betulaceae): Implications for root dynamics in elevated CO_2. *American Journal of Botany* 83:608-616.

Berntson, G.M. and F.A. Bazzaz. (1997). Nitrogen cycling in microcosms of yellow birch exposed to elevated CO_2: Simultaneous positive and negative below-ground feedbacks. *Global Change Biology* 3:247-258.

Berta, G., A. Trotta, A. Fusconi, J.E. Hooker, M. Munro, D. Atkinson, M. Giovannetti, S. Morini, P. Fortuna, B. Tisserant, V. GianinazziPearson, and S. Gianinazzi. (1995). Arbuscular mycorrhizal induced changes to plant-growth and root-system morphology in prunus-cerasifera. *Tree Physiology* 15:281-293.

Bhat, K.K.S. and P.H. Nye. (1973). Diffusion of phosphate to plant roots in soil: I. Quantitative autoradiography of the depletion zone. *Plant and Soil* 123:169-174.

Black, K.E., C.G. Harbron, M. Franklin, D. Atkinson, and J.E. Hooker. (1998). Differences in root longevity of some tree species. *Tree Physiology* 18:259-264.

Bonachela, S. (1996). Root growth of triticale and barley grown for grain or for forage-plus-grain in a Mediterranean climate. *Plant and Soil* 183:239-251.

Brouwer, R. (1963). Some aspects of the equilibrium between overground and underground plant parts. In *Jaarb.* I.B.S. Wageningen, The Netherlands, pp. 31-39.

Brown, V.K. and A.C. Gange. (1991). Effects of herbivory on vegetation dynamics. In *Plant Root Growth–An Ecological Perspective*, ed. D. Atkinson, Oxford, England: Blackwell Scientific Publications, pp. 453-470.

Butler, G.W., R.M. Greenwood, and K. Soper. (1959). Effects of shading and defoliation on the turnover of root and nodule tissue of plants of *Trifolium repens*, *Trifolium pratense* and *Lotus uliginosus*. *New Zealand Journal of Agricultural Research* 2:415-426.

Buwalda, J.G. and F. Lenz. (1992). Effects of cropping, nutrition and water supply on accumulation and distribution of biomass and nutrients for apple trees on 'M9' root systems. *Physiologia Plantarum* 84:21-28.

Cameron, K.C. and R.J. Haynes. (1986). Retention and movement of nitrogen in soils. In *Mineral Nitrogen in the Plant-Soil System*, ed. R.J. Haynes, Orlando, FL: Academic Press, Inc., pp. 166-241.

Campbell, C.D., L.A. Mackie-Dawson, E.J. Reid, S.M. Pratt, E.I. Duff, and S.T. Buckland (1994). Manual recording of minirhizotron data and its application to study the effect of herbicide and nitrogen-fertilizer on tree and pasture root-growth in a silvopastoral system. *Agroforestry Systems* 26:75-87.

Cheng, W., D.C. Coleman, and J.E. Box. (1990). Root dynamics, production and distribution in agroecosystems on the georgia piedmont using minirhizotrons. *Journal of Applied Ecology* 27:592-604.

Comas, L.H., D.M. Eissenstat, and A.N. Lakso. (2000). Assessing root death and root system dynamics in a study of grape canopy pruning. *New Phytologist* 147:171-178.

Cosgrove, D.J. (1997). Relaxation in a high-stress environment: The molecular bases of extensible cell walls and cell enlargement. *The Plant Cell* 9:1031-1041.

Dahlmann, R.C. and C.L. Kucera. (1965). Root productivity and turnover in native prairie. *Ecology* 46:84-89.

Danneberg, G., C. Latus, W. Zimmer, B. Hundeshagen, H.-J. Schneider-Poetsch, and H. Bothe. (1992). Influence of vesicular-arbuscular mycorrhiza on phytohormone balance in maize (*Zea mays* L.). *Journal of Plant Physiology* 141:33-39.

Demotes-Mainard, S. and S. Pellerin. (1992). Effect of mutual shading on the emergence of nodal roots and the root/shoot ratio of maize. *Plant and Soil* 147:87-93.

Dickmann, D.I., P.V. Nguyen, and K.S. Pregitzer. (1996). Effects of irrigation and coppicing on above-ground growth, physiology, and fine-root dynamics of two field-grown hybrid poplar clones. *Forest Ecology and Management* 80:163-174.

Douglas, J.T., M.G. Jarvis, K.R. Howse, and M.J. Goss. (1986). Structure of a silty soil in relation to management. *Journal of Soil Science* 37:137-151.

Drew, M.C. (1975). Comparison of the effects of a localized supply of phosphate, nitrate, ammonium and potassium on the growth of the seminal root system, and the shoot in barley. *New Phytologist* 75:479-490.

Drüge, U. and F. Schönbeck. (1992). Effect of vesicular-arbuscular mycorrhizal infection on transpiration, photosynthesis and growth of flax (*Linum usitatissimum* L.) in relation to cytokinin levels. *Journal of Plant Physiology* 141:40-48.

Eissenstat, D.M. and R.D. Yanai. (1997). The ecology of root lifespan. *Advances in Ecological Research* 27:1-60.

Eissenstat, D.M., C.E. Wells, R.D. Yanai, and J.L. Whitbeck. (2000). Building roots in a changing environment: Implications for root longevity. *New Phytologist* 147:33-42.

Espeleta, J.F., D.M. Eissenstat, and J.H. Graham. (1998). Citrus root responses to localized drying soil: A new approach to studying mycorrhizal effects on the roots of
mature trees. *Plant and Soil* 206:1-10.

Esau, K. (1965). *Plant Anatomy,* Second Edition. New York, NY: John Wiley and
Sons.

Fahey, T.J. and J.W. Hughes. (1994). Fine-root dynamics in a northern hardwood forest ecosystem at Hubbard Brook experimental forest, NH. *Journal of Ecology*
82:533-548.

Fairley, R.I. and I.J. Alexander. (1985). Methods of calculating fine root production in
forests. In *Ecological Interactions in Soil,* ed. A.H. Fitter, D. Atkinson, D.J. Read,
and M.B. Usher, Oxford, England: Blackwell Scientific Publications, pp. 37-42.

Feil, B., R. Thirapon, G. Geisler, and P. Stamp. (1991). The impact of temperature on
seedling root traits of European and Tropical maize (*Zea mays* L.) cultivars. *Journal
of Agronomy and Crop Science* 166:81-89.

Fitter, A.H., G.K. Self, J. Wolfenden, M.M.I. van Vuuren, T.K. Brown, L. Williamson,
J.D. Graves, and D. Robinson. (1996). Root production and mortality under elevated atmospheric carbon dioxide. *Plant and Soil* 187:299-306.

Fitter, A.H., J.D. Graves, J. Wolfenden, G.K. Self, T.K. Brown, D. Bogie, and T.A.
Mansfield. (1997). Root production and turnover and carbon budgets of two contrasting grasslands under ambient and elevated atmospheric carbon dioxide concentrations. *New Phytologist* 137:247-255.

Forbes, P.J., K.E. Black, and J.E. Hooker. (1997). Temperature-induced alteration to
root longevity in *Lolium perenne. Plant and Soil* 190:87-90.

Foulkes, M.J., R. Sylvester-Bradley, and R.K. Scott. (1998). Evidence for differences
between winter wheat cultivars in acquisition of soil mineral nitrogen and uptake
and utilization of applied fertilizer nitrogen. *Journal of Agricultural Science* 130:29-44.

Fownes, J.H. and D.G. Anderson. (1991). Changes in nodule and root biomass of
Sesbania sesban and *Leucana leucocephala* following coppicing. *Plant and Soil*
138:9-16.

Fusseder, A. (1987). The longevity and activity of the primary root of maize. *Plant and
Soil* 101:257-265.

Gibbs, R.J. and J.B. Reid. (1992). Comparison between net and gross root production
by winter wheat and by perennial ryegrass. *New Zealand Journal of Crop and Horticultural Science* 20:483-487.

Glass, A.D.M. and J. Dunlop. (1974). Influence of phenolic acids on ion uptake–IV.
Depolarisation of membrane potentials. *Plant Physiology* 54:855-858.

Goins, D.G. and M.P. Russelle. (1996). Fine root demography in alfalfa (*Medicago
sativa* L.). *Plant and Soil* 185:281-291.

Goss, M.J. (1977). Effects of mechanical impedance on root growth in barley (*Hordeum
vulgare* L.) I. Effects on the elongation and branching of seminal root axes. *Journal
of Experimental Botany* 28:96-111.

Goss, M.J. (1991). Consequences of the effects of roots on soil. In *Plant Root Growth–
An Ecological Perspective,* ed. D. Atkinson, Oxford, England: Blackwell Scientific
Publications, pp. 171-186.

Goss, M.J., K.R. Howse, J.M. Vaughan-Williams, M.A. Ward, and W. Jenkins.
(1984). Water use by winter wheat as affected by soil management. *Journal of Agricultural Science* 103:189-199.

Goss, M.J., P.B. Barraclough, and B.A. Powell. (1989). The extent to which physical factors in the rooting zone limit crop growth. *Aspects of Applied Biology* 22: 173-181.

Gross, K.L., A. Peters, and K.S. Pregitzer. (1993). Fine-root growth demographic responses to nutrient patches in 4 old-field plant-species. *Oecologia* 95:61-64.

Harper, J.L., M. Jones, and N.R. Sackville-Hamilton. (1991). The evolution of roots and the problems of analysing their behaviour. In *Plant Root Growth–An Ecological Perspective*, ed. D. Atkinson, Oxford, England: Blackwell Scientific Publications, pp. 3-22.

Hayes, D.C. and T.R. Seastedt. (1987). Root dynamics of tallgrass prairie in wet and dry years. *Canadian Journal of Botany* 65:787-791.

Hendrick, R.L. and K.S. Pregitzer. (1992). The demography of fine roots in a Northern hardwood forest. *Ecology* 73:1094-1104.

Hendrick, R.L. and K.S. Pregitzer. (1993a). Patterns of fine root mortality in two sugar maple forests. *Nature* 361:59-61.

Hendrick, R.L. and K.S. Pregitzer. (1993b). The dynamics of fine-root length, biomass, and nitrogen-content in two northern hardwood ecosystems. *Canadian Journal of Forest Research* 23:2507-2520.

Hendrick, R.L. and K.S. Pregitzer. (1996). Temporal and depth-related patterns of fine-root dynamics in northern hardwood forests. *Journal of Ecology* 84:167-176.

Hendrick, R.L. and K.S. Pregitzer. (1997). The relationship between fine root demography and the soil environment in northern hardwood forests. *Ecoscience* 4:99-105.

Hodge, A., A.H. Fitter, D. Robinson, B.S. Griffiths, and A.H. Fitter. (1999). Plant, soil fauna and microbial responses to N-rich organic patches of contrasting temporal availability. *Soil Biology and Biochemistry* 31:1517-1530.

Hodge, A., D. Robinson, and A.H. Fitter. (2000). An arbuscular mycorrhizal inoculum enhances root proliferation in, but not nitrogen capture from, nutrient-rich patches in soil. *New Phytologist* 145:575-584.

Hooker, J.E. and K.E. Black. (1995). Arbuscular mycorrhizal fungi as components of sustainable soil-plant systems. *Critical Reviews in Plant Science* 15:201-212.

Hooker, J.E., K.E. Black, R.L. Perry, and D. Atkinson. (1995). Arbuscular mycorrhizal induced alteration to root longevity of poplar. *Plant and Soil* 172:327-329.

Hooker, J.E., R. Hendrick, and D. Atkinson. (2000). The measurement and analysis of fine root longevity. In *Root Methods*, ed. A.I. Smit, A.G. Bengough, C. Engels, M. Van Noordwijk, S. Pellerin, and S.C. Van der Geijn, Berlin, Germany: Springer-Verlag, pp. 273-304.

Hooker, J.E., M. Munro, and D. Atkinson. (1992). Vesicular-arbuscular mycorrhizal fungi induced alteration in poplar root-system morphology. *Plant and Soil* 145: 207-214.

Huang, B. and D.M. Eissenstat. (2000). Root plasticity in exploiting water and nutrient heterogeneity. In *Plant-Environment Interactions*, ed. R.E. Wilkinson, New York, NY: Marcel Dekker Inc., pp. 111-132.

Huang, Y., D.H. Rickert, and K.D. Kephart. (1996). Recovery of deep-point injected soil nitrogen-15 by switchgrass, alfalfa, ineffective alfalfa, and corn. *Journal of Environmental Quality* 25:1394-1400.

Huck, M.G., C.M. Peterson, G. Hoogenboom, and C.D. Busch. (1986). Distribution of dry matter between shoots and roots of irrigated and nonirrigated determinate soybeans. *Agronomy Journal* 78:807-813.

Huisman, O.C. (1982). Interrelations of root growth dynamics to epidemiology of root-invading fungi. *Annual Review of Phytopathology* 20:303-327.

Jarvis, S.C. and J.H. Macduff. (1989). Nitrate nutrition of grasses from steady-state supplies in flowing solution culture following nitrate deprivation and or defoliation. I. Recovery of uptake and growth and their interactions. *Journal of Experimental Botany* 40:965-975.

Juniper, B. E. and G. Pask. (1973). Directional secretion by the golgi bodies in maize root cells. *Planta* (Berlin) 109:225-231.

Kaspar, T.C. and W.L. Bland. (1992). Soil temperature and root growth. *Soil Science* 154:290-298.

Katterer, T. and O. Andrén. (1999). Growth dynamics of reed canarygrass (*Phalaris arundinacea* L.) and its allocation of biomass and nitrogen below ground in a field receiving daily irrigation and fertilisation. *Nutrient Cycling in Agroecosystems* 54:21-29.

Klepper, B., H.M. Taylor, M.G. Huck, and E.L. Fiscus. (1973). Water relations and growth of cotton in drying soil. *Agronomy Journal* 65:307-310.

Kosola, K.R., D.M. Eissenstat, and J.H. Graham. (1995). Root demography of mature citrus trees–the influence of *Phytophthora nicotianae*. *Plant and Soil* 171:283-288.

Kotcon, J.B., D.I. Rouse, and J.E. Mitchell. (1984). Dynamics of root growth in potato fields affected by the early dying syndrome. *Phytopathology* 74:462-467.

Krauss, U. and J.W. Deacon. (1994). Root turnover of groundnut (*Arachis hypogaea* L.) in soil tubes. *Plant and Soil* 166:259-270.

Kuhns, M.R., H.E. Garrett, R.O. Teskey, and T.M. Hinckley. (1985). Root-growth of blackwalnut trees related to soil-temperature, soil-water potential, and leaf water potential. *Forest Science* 31:617-629.

Kurz, W.A. and J.P. Kimmins. (1987). Analysis of some sources of error in methods used to determine fine root production in forest ecosystems:a simulation approach. *Canadian Journal of Forest Research* 17:909-912.

Laux, T. and G. Jurgens. (1997). Embryogenesis: A new start in life. *The Plant Cell* 9:989-1000.

Leary, J. and J. DeFrank. (2000). Living mulches for organic farming systems. *Horticulture Technology* 10:692-697.

Lee, E.T. (1992). *Statistical Methods for Survival Data Analysis*. New York, NY: John Wiley.

Lehmann, J. and W. Zech. (1998). Fine root turnover of irrigated hedgerow intercropping in northern Kenya. *Plant and Soil* 198:19-31.

Liedgens, M., A. Soldati, P. Stamp, and W. Richner. (2000). Root development of maize (*Zea mays* L.) as observed with minirhizotrons in lysimeters. *Crop Science* 40:1665-1672.

Luo, Y., P.A. Meyerhoff, and R.S. Loomis. (1995). Seasonal patterns and vertical distributions of fine roots of alfalfa (*Medicago sativa* L.). *Field Crops Research* 40:119-127.

Lyr, H. and M. Hoffmann. (1967). Growth rates and periodicity of tree roots. *International Review of Forestry Research* 2:181-236.

Mackie-Dawson, L.A. and D. Atkinson. (1991). Methodology for the study of roots in field experiments and the interpretation of results. In *Plant Root Growth–An Ecological Perspective*, ed. D. Atkinson, Oxford, England: Blackwell Scientific Publications, pp. 25-47.

Mackie-Dawson, L.A. (1999). Nitrogen uptake and root morphological responses of defoliated *Lolium perenne* (L.) to a heterogeneous nitrogen supply. *Plant and Soil* 209:111-118.

Majdi, H. (1996). Root sampling methods–applications and limitations of the minirhizotron technique. *Plant and Soil* 185:255-258.

Majdi, H., E. Damm, and J.-E. Nylund. (2001). Longevity of mycorrhizal roots depends on branching order and nutrient availability. *New Phytologist* 150:195-202.

Majdi, H. and P. Kangas. (1997). Demography of fine roots in response to nutrient applications in a Norway spruce stand in southwestern Sweden. *Ecoscience* 4:199-205.

Marschner, H. (1995). *Mineral Nutrition of Higher Plants*, Second Edition. San Diego, CA: Academic Press Inc.

McClaugherty, C.A., J.D. Aber, and J.M. Melillo. (1982). The role of fine roots in the organic-matter and nitrogen budgets of two forested ecosystems. *Ecology* 63:1481-1490.

McKay, H. and M.P. Coutts. (1989). Limitations placed on forestry production by the root system. *Aspects of Applied Biology* 22:245-254.

Meek, B.D., D.L. Carter, D.T. Westermann, and R.E. Peckenpaugh. (1994). Root-zone mineral nitrogen changes as affected by crop sequence and tillage. *Soil Science Society of America Journal* 58:1464-1469.

Meier, C.E., C.C. Grier, and D.W. Cole. (1985). Below- and aboveground N and P use by *Abies amabilis* stands. *Ecology* 66:1928-1942.

Millard, P. and G.G. Neilsen. (1989). The influence of nitrogen supply on the uptake and remobilization of stored N for the seasonal growth of apple trees. *Annals of Botany* 63:301-309.

Millard, P. and M.F. Proe. (1991). Leaf demography and the seasonal internal cycling of nitrogen in sycamore (*Acer pseudoplatanus* L.) seedlings in relation to nitrogen supply. *New Phytologist* 117:587-596.

Miralles, D.J., G.A. Slafer, and V. Lynch. (1997). Rooting pattern in near-isogenic lines of spring wheat for dwarfism. *Plant and Soil* 197:79-86.

Mozafar, A., P. Schreiber, and J.J. Oertli. (1993). Photoperiod and root zone temperature–interacting effects on growth and mineral nutrients of maize. *Plant and Soil* 153:71-78.

Nadelhoffer, K.J. (2000). The potential effects of nitrogen deposition on fine-root production in forest ecosystems. *New Phytologist* 147:131-139.

Nambiar, E.K.S. (1987). Do nutrients retranslocate from fine roots? *Canadian Journal of Forest Research* 17:913-918.

Norman, J.R., D. Atkinson, and J.E. Hooker. (1996). Arbuscular mycorrhizal fungal-induced alteration to root architecture in strawberry and induced resistance to the root pathogen *Phytophthora fragariae*. *Plant and Soil* 185:191-198.

Onderdonk, J. J. and J.W. Ketcheson. (1973). Effect of soil temperature on direction of corn root growth. *Plant and Soil* 39:177-186.

Paradales, J., A.Yamauchi, and Y. Kono. (1991). Growth and development of sorghum roots after exposure to different periods of a hot root-zone temperature. *Environmental and Experimental Botany* 31:397-403.

Passioura, J.B. (1991). Soil structure and plant growth. *Australian Journal of Soil Research* 29:717-728.

Persson, H. (1978). Root dynamics in a young Scots pine stand in Central Sweden. *Oikos* 30:508-519.

Pregitzer, K.S., R.L. Hendrick, and R. Fogel. (1993). The demography of fine roots in response to patches of water and nitrogen. *New Phytologist* 125:575-580.

Pregitzer, K.S., J.A. King, A.J. Burton, and S.E. Brown. (2000). Responses of tree fine roots to temperature. *New Phytologist* 147:105-115.

Pritchard, S.G. and H.H. Rogers. (2000). Spatial and temporal deployment of crop roots in CO_2-enriched environments. *New Phytologist* 147:55-71.

Publicover, D.A. and K.A. Vogt. (1993). A comparison of methods for estimating forest fine-root production with respect to sources of error. *Canadian Journal of Forest Research* 23:1179-1186.

Rao, R.C.N., L.P. Simmonds, S.N. Azam-Ali, and J.H. Williams. (1989). Population, growth and water use of ground-nut maintained on stored water. I. Root and shoot growth. *Experimental Agriculture* 25:51-61.

Reid, J.B., I. Sorensen, and R.A. Petrie. (1993). Root demography of kiwifruit (*Actinidia delicosa*). *Plant, Cell and Environment* 16:949-957.

Reid, J.B., D. Winfield, I. Sorensen, and A.J. Kale. (1996). Water deficit, root demography, and the causes of internal blackening in field-grown tomatoes (*Lycopersicon esculentum* Mill). *Annals of Applied Biology* 129:137-149.

Rengasamy, J.I. and J.B. Reid. (1993). Root system modification of faba beans (*Vicia faba* L.), and its effects on crop performance. II. Role of water stress. *Field Crops Research* 33:197-215.

Robards, A.W., S.M. Jackson, D.T. Clarkson, and J. Sanderson. (1973). The structure of barley roots in relation to the transport of ions into the stele. *Protoplasma* 77:291-311.

Robinson, D. and M.M.I. van Vuuren. (1998). Responses of wild plants to nutrient patches in relation to growth rate and life form. In *Inherent Variation in Plant Growth: Physiological Mechanisms and Ecological Consequences*, ed. H. Lambers, H. Poorter, and M.M.I. van Vuuren, The Netherlands: Backhuys, Leiden, pp. 237-257.

Ruess, R.W., R.L. Hendrick, and J.P. Bryant. (1998). Regulation of fine root dynamics by mammalian browsers in early successional Alaskan taiga forests. *Ecology* 79:2706-2720.

Rytter, R.-M. and L. Rytter. (1998). Growth, decay, and turnover rates of fine roots of basket willows. *Canadian Journal of Forest Research* 28:893-902.

Sanderson, J. (1983). Water uptake by different regions of barley root: Pathway of radial flow in relation to development of the endodermis. *Journal of Experimental Botany* 34:240-253.

Santantonio, D. and J.C. Grace. (1987). Estimating fine-root production and turnover from biomass and decomposition data. A compartment flow model. *Canadian Journal of Forest Research* 17:900-908.

Scheres, B., H. Wolkenfelt, V. Willemsen, M. Terlouw, E. Lawson, C. Dean, and P. Weisbeek. (1994). Embryonic origin of the Arabidopsis primary root and root meristem initials. *Development* 120:2475-2750.

Scholefield, D. and D.M. Hall. (1985). Constricted growth of grass roots through rigid pores. *Plant and Soil* 85:153-162.

Schroth, G. (1995). Tree root characteristics as criteria for species selection and systems-design in agroforestry. *Agroforestry Systems* 30:125-143.

Schroth, G., M.R.L. Rodrigues, and S.A. D'Angelo. (2000). Spatial patterns of nitrogen mineralization, fertilizer distribution and roots explain nitrate leaching from mature Amazonian oil palm plantation. *Soil Use and Management* 16:222-229.

Siddique, K.H.M., R.K. Belford, and D. Tennant. (1990). Root-shoot ratios of old and modern, tall and semidwarf wheats in a mediterranean environment. *Plant and Soil* 121:89-98.

Simmonds, L.P. and S.N. Azam-Ali. (1989). Population, growth and water use of groundnut maintained on stored water. IV. The influence of population on water-supply and demand. *Experimental Agriculture* 25:87-98.

Smit, A.L., A.G. Bengough, C. Engels, M. Van Noordwijk, S. Pellerin, and S.C. Van der Geijn. (2000). *Root Methods*. Berlin, Germany: Springer-Verlag.

Smit, A.L. and A. Zuin. (1996). Root growth dynamics of Brussels sprouts (*Brassica oleracea* var. *gemmifera*) and leeks (*Allium porrum* L.) as reflected by root length, root colour and UV fluorescence. *Plant and Soil* 185:271-280.

Smit, A.L. and T. Vamerali. (1998). The influence of potato cyst nematodes (*Globodera pallida*) and drought on rooting dynamics of potato (*Solanum tuberosum* L.). *European Journal of Agronomy* 9:137-146.

Smucker, A.J.M. (1993). Soil environment modifications of root dynamics and measurement. *Annual Review of Phytopathology* 31:191-216.

Steele, S.J., S.T. Gower, J.G. Vogel, and J.M. Norman. (1997). Root mass, net primary production and turnover in aspen, Jack pine and Black spruce forests in Saskatchewan and Manitoba, Canada. *Tree Physiology* 17:577-587.

Stypa, M., A. Nunez-Barrios, D.A. Barry, M.H. Miller, and W.A. Mitchell. (1987). Effects of subsoil bulk density, nutrient availability and soil moisture on corn root growth in the field. *Canadian Journal of Soil Science* 67:293-308.

Swinnen, J., J.A. Van Veen, and R. Merckx. (1995). Root decay and turnover of rhizodeposits in field-grown winter wheat and spring barley estimated by ^{14}C Pulse-labelling. *Soil Biology and Biochemistry* 27:211-217.

Taylor, H.M. and G.S. Brar. (1991). Effect of soil compaction on root development. *Soil and Tillage Research* 19:111-119.

Tardieu, F. and S. Pellerin. (1991). Influence of soil temperature during root appearance on the trajectory of roots of field grown maize. *Plant and Soil* 131:207-214.

Thomas, F.M. and G. Hartmann. (1998). Tree rooting patterns and soil water relations of healthy and damaged stands of mature oak (*Quercus robur* L. and *Quercus petraea* [Matt] Liebl). *Plant and Soil* 203:145-158.

Thomas, S.M., D. Whitehead, J.B. Reid, F.J. Cook, J.A. Adams, and A.C. Leckie. (1999). Growth, loss, and vertical distribution of *Pinus radiata* fine roots growing at ambient and elevated CO_2 concentration. *Global Change Biology* 5:107-121.

Tingey, D.T., D.L. Phillips, and M.G. Johnson. (2000). Elevated CO_2 and conifer roots: Effects on growth, life span and turnover. *New Phytologist* 147:87-103.

Troughton, A. (1957). *The Underground Organs of Herbage Grasses*. England: Lamport Gilbert and Co.

Van der Vlugt, J.L.F. (1990). VI. Plant growth and chemical analysis. *Norwegian Journal of Agricultural Sciences* 4:77-90.

Van Noordwijk, M., G. Brouwer, H. Koning, F.W. Meijboom, and W. Grzebisz. (1994). Production and decay of structural root material of winter-wheat and sugar-beet in conventional and integrated cropping systems. *Agriculture Ecosystems and Environment* 51:99-113.

Vaughan, D. and B.G. Ord. (1991). Extraction of potential allelochemicals and their effects on root morphology and nutrient contents. In *Plant Root Growth–An Ecological Perspective*, ed. D. Atkinson, Oxford: Blackwell Scientific Publications, pp. 399-421.

Vogt, K.A., D.J. Vogt, H. Asbjornsen, and R.A. Dahlgren. (1995). Roots, nutrients and their relationship to spatial patterns. *Plant and Soil* 169:13-123.

Waisel, Y., A. Eshel, and U. Kafkafi. (1996). *Plant Roots: The Hidden Half*. New York, NY: Marcel Decker.

Watson, C.A., J.M. Ross, U. Bagnaresi, G.F. Minotta, F. Roffi, D. Atkinson, K.E. Black, and J.E. Hooker. (2000). Environment-induced modifications to root longevity in *Lolium perenne* and *Trifolium repens*. *Annals of Botany* 85:397-401.

Weaver, J.E. (1926). *Root Development of Field Crops*. New York, NY: McGraw-Hill.

Wells, C.E. and D. Eissenstat. (2001). Marked differences in survivorship among apple roots of different diameters. *Ecology* 82:882-892.

Wells, C.E. (1999). *Advances in the Root Demography of Woody Species*. PhD Thesis. The Pennsylvania State University, USA.

Xiao, C.L. and K.V. Subbarao. (2000). Effects of irrigation and *Verticillium dahliae* on cauliflower root and shoot growth dynamics. *Phytopathology* 90:995-1004.

Key Indicators
for Assessing Nitrogen Use Efficiency
in Cereal-Based Agroecosystems

D. R. Huggins
W. L. Pan

SUMMARY. Improving nitrogen use efficiency (NUE) is an important objective of agroecosystem management. We define and demonstrate key indicators of NUE that enable a broader assessment of N management strategies. Nitrogen efficiency components and indexes were defined to assess soil and crop physiological processes, and agronomic and environmental factors related to N use. Measurements of grain yield, grain N, aboveground plant N, applied N, post-harvest root-zone soil N, and N losses via subsurface drains were used to assess N retention efficiency, available N uptake efficiency, N utilization efficiency, N harvest index, N yield efficiency, N reliance index, grain N accumulation efficiency, N balance index, N fertilizer utilization efficiency, and N loss index. Nitrogen use indicators were evaluated for two field studies: (1) hard red

D. R. Huggins is affiliated with the USDA-ARS, Land Management and Water Conservation Research Unit, Washington State University, Pullman, WA.

W. L. Pan is affiliated with the Department of Crop and Soil Sciences, Washington State University, Pullman, WA.

Address correspondence to: D. R. Huggins, USDA-ARS, Land Management and Water Conservation Research Unit, 215 Johnson Hall, Washington State University, Pullman, WA 99164-6421 (E-mail: dhuggins@wsu.edu).

[Haworth co-indexing entry note]: "Key Indicators for Assessing Nitrogen Use Efficiency in Cereal-Based Agroecosystems." Huggins, D. R., and W. L. Pan. Co-published simultaneously in *Journal of Crop Production* (Food Products Press, an imprint of The Haworth Press, Inc.) Vol. 8, No. 1/2 (#15/16), 2003, pp. 157-185; and: *Cropping Systems: Trends and Advances* (ed: Anil Shrestha) Food Products Press, an imprint of The Haworth Press, Inc., 2003, pp. 157-185. Single or multiple copies of this article are available for a fee from The Haworth Document Delivery Service [1-800-HAWORTH, 9:00 a.m. - 5:00 p.m. (EST). E-mail address: docdelivery@haworthpress.com].

spring wheat with four N levels and two tillage treatments: no-tillage
(NT) and conventional tillage (CT); and (2) corn in crop sequences of
continuous corn (C-C), corn-soybean (C-S), two years of corn following
alfalfa (ALF-C-C), and two years of corn following perennial grass
(CRP-C-C). Tillage, crop rotation, and applied N had large and variable
effects on different indicators of N use. N efficiency components and in-
dexes were useful for monitoring cropping system N use, assessing N
management strategies, and identifying key areas for improvements in
NUE. *[Article copies available for a fee from The Haworth Document Delivery
Service: 1-800-HAWORTH. E-mail address: <docdelivery@haworthpress.com>
Website: <http://www.HaworthPress.com> © 2003 by The Haworth Press, Inc.
All rights reserved.]*

KEYWORDS. Nitrogen use efficiency, agroecosystem, indicators, nitrogen uptake

INTRODUCTION

Regulating the supply and fate of N in agroecosystems is an important management objective with major economic and environmental consequences.
From a crop production standpoint, N is generally the most limiting plant nutrient and N availability is routinely supplemented through applications of fertilizer. Worldwide, however, removal of N in harvested grain that is derived
from fertilizer sources is estimated to be only 33% of applied N fertilizer
(Raun and Johnson, 1999). Low fertilizer N recovery indicates the occurrence
of many competing pathways in the cycling and flow of N within a cropping
system, the potential for N to move beyond cropping system boundaries and
degrade both water (Burkhart and James, 1999; Huggins, Randall, and Russelle,
2001) and air quality (Mummey, Smith, and Bluhm, 1998; Mosier et al., 1996).
In addition to N fertilizer management, virtually every cultural practice can
be a major regulator of processes that influence the amount and timing of N
availability, crop demands for N, and competing pathways that lead to inefficient N use and undesirable N losses. Major effects on N use have frequently
been reported for crop rotation (Baldock et al., 1981; Hesterman et al., 1987;
Badaruddin and Meyer, 1994; Randall et al., 1997), tillage regime (Legg, Stanford, and Bennett, 1979; Karlen, Hunt, and Matheny, 1996; Rao and Dao,
1996), genotype (Moll, Kamprath, and Jackson, 1982; Tillman, Pan, and
Ullrich, 1991; Eghball and Maranville, 1991; Gauer et al., 1992), water management (Russelle et al., 1981; Wienhold, Trooien, and Reichman, 1995;
Oberle and Keeney, 1990), non-N nutrition (Moss et al., 1981), and various N
management strategies (Fiez, Miller, and Pan, 1994; Mengel, Nelson, and

Huber, 1982; Sowers et al., 1994; Wuest and Cassman, 1992; Mahler, Koehler, and Lutcher, 1994). Furthermore, interaction of cultural practices with ecologic (e.g., climate, soil, topography organisms) and socioeconomic (e.g., commodity prices, agrichemical costs, government programs, land tenure) factors creates tremendous spatial and temporal variability of N cycling and flows in soil and crops as well as in crop and N input economies (Bock, 1984; Pan et al., 1997; Pierce and Nowak, 1999). As a consequence, N management strategies for dryland or rainfed cereals [e.g., corn (*Zea mays* L.), wheat (*Triticum aestivum* L.), and barley (*Hordeum vulgare* L.)] have developed to accommodate the uncertainties associated with an unpredictable N management regime. These circumstances have contributed to: (1) N management strategies and recommendations developed at regional scales and averaged over field-scale temporal and spatial variation; (2) insurance applications of N fertilizer to reduce the risk of N deficiencies over time and space; (3) sub-optimal N use efficiencies due to N supplies that are over or under crop requirements; (4) unbalanced cropping system N budgets with declining reserves of organic N sources, increased reliance on external N sources and undesired N leaching and gaseous losses; and (5) economic losses and environmental degradation. The effectiveness of current N management strategies is becoming increasingly questioned as air, water, and soil quality are degraded at watershed and global scales (Tilman et al., 2001) and as producers seek greater efficiencies in N use to reduce external farm inputs and costs.

Progress towards designing sustainable agroecosystems and specifically, improvements in N management, will likely be realized from a broader assessment of N use and the development of decision support systems (DSS) that promote integrative N management strategies (Figure 1, modified from Johnson and Huggins, 1999). Major elements of a DSS for N management are: (1) development and use of diagnostic tools and models that evaluate N flows and cycles among various N pools (sources and sinks) over the duration of the cropping system; (2) design of N efficient crop management strategies including rotation, tillage and fertilization based on knowledge of N cycling and flow processes within environmental and economic constraints; (3) adoption of N application technologies that enable sufficient N supplies, forms, placement and timing for meeting N management goals; and (4) identification and use of key criteria or indicators for assessing how well an N management strategy has met economic and environmental goals. Our objective is to focus on this last important element of a DSS and demonstrate a framework of key indicators that can be used to monitor and assess N management strategies from a broader perspective of N use.

An assessment of agroecosystem N use should include indicators that evaluate: (1) agronomic practices and their influence on major soil and plant physiological processes that effect N use; (2) economic factors and the optimization

FIGURE 1. Elements of decision support system for integrated N management (adapted from Johnson and Huggins, 1999).

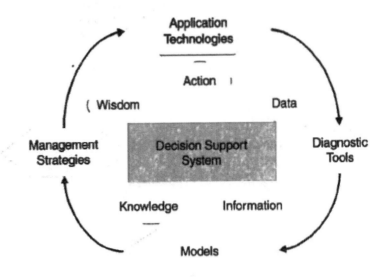

of agronomic inputs to achieve crop performance goals; and (3) environmental considerations including the sustainability of the agricultural resource base and the potential for resource degradation. Tradeoffs among agronomic and economic performance and environmental quality and sustainability are often unavoidable. However, a more thorough evaluation of N use through key indicators should increase understanding of agroecosystem complexity, provide a basis for guiding future technological developments and cultural practices for improved N management, and serve as a warning system for identifying unsustainable N management practices.

The application of indicators for assessing agroecosystem N use is illustrated using published data from: (1) the annual dryland-cropping region of the Pacific Northwest where wheat and grain legumes [e.g., peas (*Pisum sativum* L.) or lentils (*Lens colinaris* Medik.)] are commonly grown; and (2) the rainfed northern Corn Belt where corn-soybean [*Glycine max* (L.) Merr.] rotations are prevalent. While these examples are useful, the indicators presented do not provide an exhaustive evaluation of N use and can be modified or expanded as needed to assess many different agroecosystems. The first study (Huggins and Pan, 1993) located near Pullman, WA, evaluated the response of hard red spring wheat (HRSW) to applied N under conventional moldboard plow tillage (CT) and no-tillage (NT) following a winter wheat crop. In this example, N efficiency components are used as key indicators to assess the combined effects of disturbance regime (tillage) and increasing amounts of applied N fertilizer on N use. The second study (Randall et al., 1997; Huggins, Randall, and Russelle, 2001) located at the University of Minnesota Southwest Research and Outreach Center, Lamberton, MN, evaluated corn under conventional tillage in four different cropping systems: continuous corn (C-C); a two year corn-soybean rotation (C-S); and two years of corn following either six years of alfalfa (*Medicago sativa* L.) (ALF-C-C) or following six years of primarily smooth brome grass (*Bromus inermis* Leyss.) representing a Conservation Reserve Program planting (CRP-C-C). In this study, N fertilization of corn was based on best management practices and indicators of NUE are used to assess crop rotation effects on N use over time.

EVALUATION OF MAJOR N SOURCES AND SINKS

An overall goal of agroecosystem N management is to have available N supplies sufficient to meet crop growth, yield, and grain quality requirements. To achieve this goal, an understanding of N flows and cycles among various N sources and sinks is required. Figure 2 depicts key crop physiological and soil processes that regulate N flows and cycling over a two-year cereal-grain legume rotation typical of the annual cropping region of the Pacific Northwest and the northern Corn Belt. Sources of available N are derived from: (1) internal processes such as N mineralization of labile organic pools and N_2-fixation by legumes or free-living bacteria; and (2) external sources such as applied N fertilizers and animal manures. Nitrogen sinks consist of many competing N pathways including: (1) N flows to crop (N uptake) and non-crop sinks (soil N immobilization, N uptake by weeds); and (2) N losses beyond cropping system boundaries via leaching, runoff, denitrification, volatilization (soil and crop losses), soil erosion, and removal via crop harvest (Figure 2).

In general, two approaches have been used to account for N sources and sinks in agroecosystems: (1) an N mass balance approach (Bock, 1984; Meisinger and Randall, 1991) that attempts to account for all N inputs and outputs; and (2) labeling an input with ^{15}N (Hauck and Bremner, 1976; Legg and Meisinger, 1982) and assessing a balance for the labeled input based on its fate within the agroecosystem. In the former approach, unaccounted for losses are typically attributed to leaching, denitrification or ammonia volatilization as quantitative data for these processes are often lacking. In the latter approach, studies with ^{15}N-labeled inputs assess the fate of the labeled N, but do not evaluate N sources and sinks for the entire agroecosystem. Moll, Kamprath, and Jackson (1982) used an N mass balance approach to assess factors that contribute to NUE. Their approach was expanded by Huggins and Pan (1993) to de-

FIGURE 2. Schematic of N flow and cycling in two year, cereal-legume cropping system emphasizing key crop physiological and soil processes that contribute to N use efficiency (G_w/N_s) and N harvest efficiency (N_g/N_s) components: (1) N retention efficiency (N_{av}/N_s): b, c, and f; (2) available N uptake efficiency (N_t/N_{av}): a, b, c, d, and f; (3) N utilization efficiency (G_w/N_t): a, d, and e; (4) N harvest index (N_g/N_t): a, d, and e; (5) N reliance index (N_t/N_s): b, f; and (6) N fertilizer utilization efficiency (G_w/N_t): e, f.

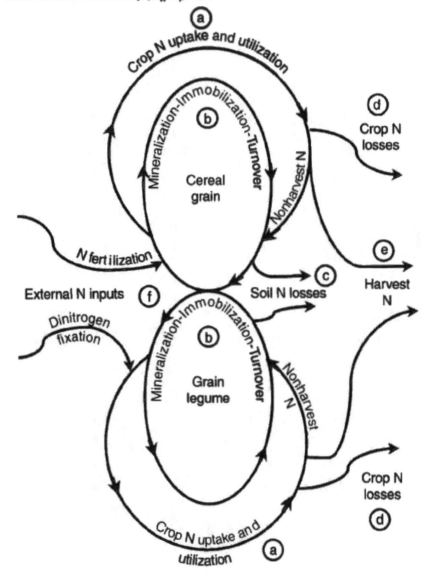

rive NUE components of major soil and plant physiological processes based on commonly obtained soil and plant data. The NUE components defined by Huggins and Pan (1993) can serve as key indicators that evaluate specific N-related soil and plant physiological processes, but need to be augmented to assess overall agroecosystem performance with respect to economic and environmental factors. Proposed additional NUE indicators for assessing economic and environmental components of NUE as well as previously defined NUE terminology (Moll, Kamprath, and Jackson, 1982; Huggins and Pan, 1993) are summarized in Table 1.

Nitrogen Use Efficiency Components Related to Soil Processes

Overall NUE is defined as grain production (G_w) per unit of N supply (N_s) where N_s is the sum of all sources of potentially available N such as N fertilizer (N_f), residual inorganic soil N prior to crop N uptake (N_r), soil mineralized N (N_m), N fixed in clay minerals (N_x), and depositional N (N_d) from atmospheric, irrigation and run-on N. To assess retention of soil N supply, available soil N (N_{av}) is defined as N supply minus soil N losses due to immobilized N (N_{im}), N leached (N_l), N eroded (N_{er}), gaseous N losses (N_{gl}), and N chemically fixed (N_{cf}). These definitions lead to partitioning NUE into the product of two factors: (1) N retention efficiency (N_{av}/N_s), the fraction of N supply that is

TABLE 1. Summary of N use efficiency and index terminology.

G_w = grain yield
N_{av} = available N
N_g = grain N
N_s = N supply
N_t = aboveground crop N at physiological maturity
N_f = applied N fertilizer
N_{lo} = N loss via leaching, denitrification, and volatilization
N_{td} = N loss via artificial subsurface drains
N_{av}/N_s = N retention efficiency
N_t/N_{av} = available N uptake efficiency
N_t/N_s = N uptake efficiency = $(N_t/N_{av})(N_{av}/N_s)$
G_w/N_{av} = available N use efficiency = $(N_t/N_{av})(G_w/N_t)$
G_w/N_t = N utilization efficiency
N_g/N_t = N harvest index
N_f/N_s = N reliance index
G_w/N_f = N fertilizer utilization efficiency
N_{lo}/N_t = N loss index
N_g/N_f = N balance index = $(N_{lo}/N_f)(N_g/N_{lo})$
G_w/N_s = N use efficiency = $(N_{av}/N_s)(G_w/N_{av})$ = $(N_t/N_s)(G_w/N_t)$
N_g/N_s = grain N accumulation efficiency = $(N_g/N_{av})(N_{av}/N_s)$ = $(N_g/N_t)(N_t/N_s)$

available after accounting for N losses; and (2) available N use efficiency (G_w/N_{av}), a measure of the capability of a crop to produce grain from available N:

$$G_w/N_s = N_{av}/N_s \times G_w/N_{av} \qquad [1]$$

Nitrogen Retention Efficiency

Nitrogen retention efficiency (N_{av}/N_s) is an index of the amount of plant available N in proportion to the total available N supplied from all sources. Nitrogen retention efficiency is diminished through N flows and cycling to N pools that are not available to the crop (e.g., leaching, gaseous losses). Maximum N retention efficiency occurs when $N_{av} = N_s$, however, N losses commonly range from 1 to 35% (Kumar and Goh, 2000) and low N retention efficiency can be a major contributor to poor overall NUE.

Nitrogen retention efficiency usually decreases as levels of applied N increase due to the greater potential for N losses (Legg and Meisinger, 1982). In the HRSW study, estimates of N retention efficiency declined from 100% in control plots (zero applied N) to 73% under CT and 67% under NT at the greatest N rate (Table 2). Soil residual levels (0 to 1.5 m) of post-harvest nitrate-N were low and similar across all N levels (15 to 27 kg N ha^{-1}) indicating that losses of N from available pools likely occurred during the growing season. In eastern Washington, denitrification and leaching potentials decrease following spring planting, as precipitation is low and evapotranspiration high. The HRSW followed winter wheat and N immobilization has been reported to be a major competing pathway for N fertilizer in continuous cereal rotations (Fredrickson, Koehler, and Cheng, 1982). Therefore, biological N immobilization is likely a major controller of plant N availability in the HRSW study, as has been found in other studies (Ladd and Amoto, 1986; Haynes, 1997).

Diminished N availability in cereal production can result in grain protein levels below targeted goals, as occurred in the HRSW study. To reduce the risk of not achieving grain protein goals, producers typically compensate by increasing N application rates, further reducing N retention efficiencies. Nitrogen retention efficiencies of 65% or less are probable under this scenario and indicate the potential for large improvements. Immobilization of N could be decreased with fall N applications that allow N movement to occur through surface zones during wet winter months when immobilization potential is low and result in increased N retention efficiency. However, the advantage of fall applied N may be offset by greater N leaching or denitrification losses. Alternatively, changing crop sequences so that HRSW follows a grain legume instead of winter wheat could reduce N immobilization and increase N retention efficiency. Overall, the analysis of N retention efficiency identifies the fate of

TABLE 2. Effect of N rate (N_f), conventional tillage (CT), and no-tillage (NT) on nitrogen use efficiency components of hard red spring wheat (adapted from Huggins and Pan, 1993).

N_f	N_s		N_{av}/N_s		N_u/N_{av}		G_w/N_t		N_u/N_t		G_w/N_s	
	CT	NT	CT	NT	CT	NT	CT	NT	CT	NT	CT	NT
——— kg ha^{-1} ———							——— kg kg^{-1} ———					
0	76.4	61.2	1.00	1.00	0.67	0.63	48.5	38.9	0.83	0.80	32.3	24.7
56	132.4	117.2	0.79	0.76	0.81	0.77	48.5	44.5	0.82	0.80	30.7	25.8
112	188.4	173.2	0.74	0.67	0.86	0.87	42.2	42.2	0.82	0.82	26.8	24.2
168	244.4	229.2	0.73	0.67	0.88	0.83	38.3	34.6	0.83	0.82	24.2	19.2
Analysis of Variance												
Tillage (T)	†		NS		NS		*		*		*	
N rate (NR)	**		**		**		**		NS		**	
T × NR	NS		NS		NS		**		NS		**	

†, *, ** Significant at the 0.1, 0.05, and 0.01 probability levels, respectively. NS = not significant.

immobilized N as an important area of future NUE research in this cereal-based cropping system.

Conservation tillage can reduce N retention efficiency as compared to tillage-based systems. No-tillage can result in greater N losses via denitrification (Rice and Smith, 1982; Linn and Doran, 1984), leaching (Tyler and Thomas, 1977), and lower net N mineralization (Carter and Rennie, 1982; Doran, 1987) as compared to CT. In the HRSW study, differences in N loss and N retention efficiency between NT and CT primarily occurred prior to spring planting (15 kg ha^{-1} less N_s under NT than CT) when soils were wet and the potential for denitrification and leaching was greatest. However, little influence of tillage on N retention efficiency was expressed with increasing levels of spring applied N (Table 2) supporting the conclusion that N immobilization is the major competing pathway for N use under both NT and CT.

Crop rotation can have a large effect on N retention efficiency. In the Minnesota study, estimates of N retention efficiency during the corn segment of different rotations ranged from 40 to over 95% (Tables 3 and 4). High N retention efficiency tended to be sustained with increasing N supply, but large decreases occurred when residual soil nitrate levels increased following dry years and became subject to losses via subsurface drains during wet years (Randall et al., 1997; Huggins, Randall, and Russelle, 2001). During 1988, growing season precipitation was 64% of normal and low N retention efficiencies were due to large estimated amounts of N immobilization (Table 3). Soil sampling for inorganic N in the fall of 1988 and subsequent spring indicated low levels of residual N, and rates of applied N were not reduced to compensate for the large amounts of in-season N mineralization that occurred during 1989. Consequently, post-harvest residual soil nitrate-N levels were high in the fall of 1989; however, N retention efficiencies were also high due to minimal N losses from the soil profile. Applied N was reduced in the 1990 season to compensate for high levels of residual N, however, net N mineralization remained high and large residual levels of post-harvest soil nitrate-N persisted. Greater precipitation during 1991, 1992, and 1993 resulted in increasing subsurface drain flows and losses of residual soil N via artificial drains (Randall et al., 1997). Nitrogen losses reduced N retention efficiencies to 65% during 1993 when precipitation was 166% above normal (Table 3). In this case, low N retention efficiencies were due to net immobilization during a dry year (1988) and losses of residual soil nitrate-N via subsurface drains during wet years. In addition, the unpredicted in-season release of mineralized N following a dry year contributed to the rapid buildup of residual soil nitrate-N and subsequent losses during wet years.

Improvements in N retention efficiency will depend on management strategies that limit residual buildup of soil nitrate-N in these corn-based cropping systems. Diagnostics to improve prediction of in-season N mineralization fol-

TABLE 3. Effect of continuous corn (C-C) and corn-soybean (C-S) rotation on nitrogen use efficiency components of corn (adapted from Randall et al., 1997).

Year	N_f		N_s		N_{av}/N_s		N_v/N_{av}		G_w/N_f		G_w/N_s	
	C-C	C-S	C-C	C-S	C-C	C-S	C-C	C-S	C-C	C-S	C-C	C-S
	kg ha^{-1}								kg kg^{-1}			
1988	179	179	245	235	0.44	0.40	0.62	0.57	41.2	44.6	11.1	10.3
1989	146	146	344	345	0.99	0.99	0.33	0.30	52.0	56.0	17.1	16.5
1990	35	35	279	280	0.98	0.99	0.39	0.43	59.9	54.8	22.8	23.3
1991	137	122	293	319	0.76	0.75	0.61	0.63	57.1	59.9	26.5	28.4
1992	174	134	311	217	0.83	0.85	0.61	0.66	52.7	58.1	26.7	32.7
1993	157	106	263	239	0.66	0.63	0.62	0.70	48.2	52.7	19.7	23.4
Average	138	120	289	273	0.78	0.77	0.53	0.55	51.9	54.4	20.7	22.4

TABLE 4. Effect of crop sequence on nitrogen use efficiency components of corn (adapted from Huggins, Randall, and Russelle, 2001).

Year	Sequence†	N_f	N_s	N_{av}/N_s	N_t/N_{av}	G_w/N_t	N_g/N_t	G_w/N_s
		--- kg ha^{-1} ---		------------------------ kg kg^{-1} ------------------------				
1994	C-C	165	211	0.93	0.61	75.0	0.75	42.6
	S-C	112	174	0.93	0.72	78.3	0.75	52.7
	ALF-C	17	137	0.98	0.80	84.6	0.77	65.7
	CRP-C	177	188	0.94	0.67	71.4	0.75	51.6
	LSD$_{0.05}$		76	0.02*	0.16*	11.1*	0.07	17.3*
1995	C-C	160	248	0.91	0.48	52.4	0.63	23.1
	S-C	108	260	0.90	0.58	52.1	0.78	27.1
	ALF-C-C	77	203	0.93	0.54	57.7	0.57	28.8
	CRP-C-C	149	312	0.95	0.55	44.0	0.53	22.9
	LSD$_{0.05}$		45*	0.03*	0.06*	5.8*	0.05*	3.3*

† C = corn; S = soybean; ALF = alfalfa; CRP = Conservation Reserve Program planting.
* Significant differences occurred according to Fischer's least significant difference (LSD) test (0.05 level of probability).

lowing dry years could have avoided buildup of residual soil nitrate, subsequent N losses via artificial drains and increased N retention efficiency. Including perennial crops in rotation could also improve overall NUE as residual soil nitrate was low and N retention efficiency for subsequent corn crops was greater than 95% for corn following six years of either alfalfa or CRP (Table 4).

Nitrogen Use Efficiency Components Derived from Plant Physiological Processes

Plant physiological components of NUE are derived from partitioning available N use efficiency (G_w/N_{av}) into: (1) available N uptake efficiency (N_t/N_{av}); and (2) N utilization efficiency (G_w/N_t):

$$G_w/N_{av} = N_t/N_{av} \times G_w/N_t \qquad [2]$$

Efficiency components are also defined for grain N (N_g) accumulation efficiency (N_g/N_s) as the product of available grain N accumulation efficiency (N_g/N_{av}) and N retention efficiency (N_{av}/N_s):

$$N_g/N_s = N_g/N_{av} \times N_{av}/N_s \qquad [3]$$

In turn, available grain N accumulation efficiency is divided into components of N harvest index (N_g/N_t) and available N uptake efficiency:

$$N_g/N_{av} = N_g/N_t \times N_t/N_{av} \qquad [4]$$

Available N Uptake Efficiency

Reported values for first year crop recovery of applied N range from 20 to 87% (Kumar and Goh, 2000). Variation in crop N recovery is due to interactions of climate, soil and crop management practices (e.g., fertilizer source and application time, placement, and rate; tillage and crop rotation; crop type and genotype grown) that control N losses and plant uptake. Therefore, fertilizer N recovery is a function of: (1) N retention efficiency determined from processes affecting N losses and the size of available N pools; and (2) available N uptake efficiency (N_t/N_{av}) derived from plant physiological processes that affect the acquisition of available N.

In the HRSW study, greater amounts of applied N decreased N retention efficiency but increased available N uptake efficiency (Table 2). Available N uptake efficiency averaged 86% across tillage treatment at the greatest level of applied N compared to 65% in control plots. In the same region, Sowers et al. (1994) reported available N uptake efficiencies for soft white winter wheat of 78% indicating efficient crop physiological recovery and accumulation of available N. High available N uptake efficiency occurs when the amount and timing of available N supplies are well synchronized with crop demand. Nitrogen applications that exceed the assimilative capacity of the crop can result in less N recovery. However, it has been demonstrated that hard red winter wheat can be an efficient buffer when excess N is applied by accumulating plant N and preventing residual soil N buildup (Raun and Johnson, 1995; Johnson and Raun, 1995). In the HRSW study, overall N uptake efficiency ($N_t/N_s = N_{av}/N_s \times N_t/N_{av}$) slowly declined with increasing N rate from 67 to 64% in CT and 63 to 56% in NT as increasing N uptake efficiencies did not fully compensate for decreasing N retention efficiency (Table 2). In wheat-based dryland agroecosystems, variations in overall N uptake efficiency that occurred with different landscape positions were attributed to different magnitudes of N loss that affect N retention efficiency (Fiez, Pan, and Miller, 1995).

Annual variability in environmental conditions affecting yield and a wide range in N supply had large impacts on available N uptake efficiency of corn (Tables 3 and 4). Available N uptake efficiency was greater than 70% when available N was less than 160 kg ha^{-1} but declined to 30% as available N reached 344 kg ha^{-1}. These data show the physiological limitations of N assimilation when N supplies are greater than corn requirements. Low N fertilizer recoveries of 9% were reported for corn by Liang and MacKenzie (1994) and were associated with high rates of applied N under low yielding environments. These data show that corn, in contrast to wheat, has a low capacity to assimilate excess N supplies (poor buffering capacity) leading to potentially high post-harvest residual soil inorganic N.

Available N uptake was the most efficient for first year corn following either alfalfa (80%) or CRP (72%) under a high yielding environment with relatively low residual levels of soil inorganic N (Table 4). These values are greater than N fertilizer recoveries of 40 to 62% reported for corn in an ^{15}N study (Francis, Schepers, and Vigil, 1993) and at the higher end of the typical range of reported N fertilizer recoveries (Kumar and Goh, 2000). Caution must be used, however, when comparing estimates of N fertilizer uptake made by different methods. Crop recoveries of N calculated by difference methods are usually greater than those estimated using [15]N (Westerman and Kurtz, 1974; Varvel and Peterson, 1990; Torbert et al., 1992) due, in part, to mineralization-immobilization turnover reactions (Jansson and Persson; 1982). Available N uptake efficiencies declined to 55% for second year corn following either alfalfa or CRP due to greater levels of available soil N and lower corn yields.

Although best management practices were followed in the corn study, low available N uptake efficiencies occurred as available N supplies were not well matched with crop N demands. This situation occurred as corn grain yields and available N supplies are difficult to predict under climatic conditions of the northern Corn Belt (Huggins and Alderfer, 1995; Huggins, Randall, and Russelle, 2001). Management practices to increase available N uptake efficiency should focus on strategies that reduce available N supplies including modification of the prevalent corn-soybean rotation to include crops that result in more efficient use of N and water (e.g., perennial and cover crops).

Nitrogen Utilization Efficiency

Differences in N utilization efficiency (G_w/N_t) among cropping systems are expressed when grain yields are either lower or higher at equivalent levels of aboveground N. Maximizing N utilization efficiency is desirable for many cereal crops, however, dilution effects can lead to reduced grain protein levels. Comparisons of yield at equal levels of aboveground N removes most yield effects of cropping systems due to differences in N supply, availability, or uptake. Beneficial "rotation effects" of legumes on yields of subsequent crops have been reported for wheat (Hargrove, Touchton, and Johnson, 1983) and corn (Moschler et al., 1967; Baldock et al., 1981; Ebelhar, Frye, and Blevins, 1984). In these studies, rotation effects represent examples of greater N utilization efficiency where yield increases were largely unexplained but attributed to non-N factors such as differences in water, temperature, soil structure, pests, and phytotoxic or allellopathic effects. Differences in N utilization efficiency can also arise from cultural practices such as tillage that influence yield-determining biotic and abiotic interactions. In the HRSW study, N utilization efficiency decreased with increasing amounts of applied N despite greater grain yields (Table 2). Differences in grain yield between CT and NT, however, were not eliminated with increased levels of applied N and resulted in greater N utilization efficiency under CT (38.3 kg G_w per kg N_t) as compared to NT (34.6 kg G_w per kg N_t). Identification of causal factors that contributed to differences in N utilization efficiency remained elusive in this study, as complex interactions of yield-affecting variables were not evaluated.

Nitrogen utilization efficiency of corn ranged from 41 to 84.6 kg G_w per kg N_t across all cropping systems (Tables 3 and 4). The annual fluctuations in N utilization efficiency show the large influence that climatic conditions can have on overall utilization of N. No significant differences in N utilization efficiency occurred between C-C and C-S rotations, which averaged 55.1, and 57.2 kg G_w per kg N_t, respectively, over eight years. Favorable water and temperature conditions during the 1994 growing season resulted in the greatest N utilization efficiencies for corn averaging 77.3 kg G_w per kg N_t across all crop rotations (Table 4). Despite large efficiencies of N utilization in 1994, corn following alfalfa had the greatest N utilization efficiency (84.6 kg G_w per kg N_t). Management effects on N utilization efficiency can be significant and cropping system strategies should promote N utilization efficiency by integrating crop rotation and other practices that limit pests and soil quality deficiencies while enhancing yield.

Nitrogen Harvest Index

Nitrogen harvest index (NHI, N_g/N_t) is the proportion of aboveground N that is partitioned to the grain. In eastern Washington, NHI has been reported to range from 70 to 77% for soft white wheat (Sowers, 1992; Rasmussen and Rhodes, 1991) and from 70 to 83% for hard red wheat (Huggins, 1991; Koenig, 1993). In other hard red wheat producing regions of the United States, reported NHI has ranged from 32 to 85% (McNeal et al., 1971; Bauer, 1980; Loffler, Rauch, and Busch, 1985; Wuest and Cassman, 1992; Rao and Dao, 1996). In the HRSW study, NHI was not significantly affected by N rate and averaged 83% under CT and 81% under NT (Table 2). Although significant differences in NHI occurred between CT and NT at equal N rates, the relationship between N_g and N_t was the same for both tillage treatments. This occurred as NHI was affected by differences in N supply between CT and NT rather than physiological efficiencies related to mobilization of N to the grain. Efficient remobilization of N to the grain and resultant high NHI indicates that further efficiency gains in NHI may be difficult to achieve under eastern Washington environments. The wide range in NHI reported for other regions, however, indicates that increasing the annual stability of NHI in wheat may be an important management objective particularly if grain quality is a goal.

Reported values for NHI of corn range from 47 to 85% (Olson and Sander, 1988; Oberle and Keeney, 1990; Maskina et al., 1993; Overman, Wilson, and Kamprath, 1994; and Kessavalou and Walters, 1997). Oberle and Keeney (1990) reported increased NHI with greater applied N for corn. In the Minnesota study, NHI of corn varied annually ranging from 52 to 77% (Table 4) and displayed a trend of decreasing NHI with greater N_t. Rotation affects on corn NHI were not significant in this study, and the greatest NHI occurred in 1994 when growing conditions favored yield.

Complicating the interpretation of NHI as well as available N uptake efficiency and N utilization efficiency are gaseous losses of plant N that can occur during or after anthesis (Harper et al., 1987; Parton et al., 1988; Francis, Schepers, and Vigil, 1993). Gaseous losses of plant N would decrease values of N_t that are measured at physiological maturity and this value would not represent maximum N accumulation. Westerman, Raun, and Johnson (2000) discuss the implications of post-anthesis N loss on N balance methods and present modifications of NUE components for grain crops.

Nitrogen Use Efficiency

Nitrogen use efficiency (G_w/N_s) of HRSW ranged from 19 to 32 kg G_w per kg N_s and was significantly influenced by both N level and tillage (Table 2). Nitrogen efficiency components that contributed to the inverse relationship between applied N and NUE were identified as decreasing N retention efficiency and N utilization efficiency (Table 2). Limiting competing sinks for available N or changing crop rotation to improve N utilization efficiency are major N management strategies that could result in improved overall NUE. The greater NUE of CT compared to NT is primarily a function of greater N utilization efficiency which could be improved in NT by diversifying crop rotation, thereby reducing non-N factors that are limiting yield.

The NUE for corn ranged from 10 to 66 kg G_w per kg N_s (Tables 3 and 4), a wide range in NUE considering N inputs were based on best management practices. The C-C rotation displayed the lowest overall NUE averaging 23.5 kg G_w per kg N_s as compared to 27.8 kg G_w per kg N_s for the C-S rotation over eight years. Large annual variability occurred in contributions of N efficiency components to overall NUE, and evidence of previous years predisposing subsequent N efficiency components occurred. For example, the lowest NUE occurred in 1988, a drought year, due to poor N retention (primarily from N immobilization) and utilization efficiency (Table 4). The following year, the major contributor to low NUE was poor available N uptake efficiency as N immobilized during 1988 was mineralized and contributed to an over supply of available N. The greatest NUE occurred across all rotations in 1994 when favorable growing conditions resulted in high corn yields (Table 4). Nevertheless, in 1994, NUE for first year corn after alfalfa (65.7 kg G_w per kg N_s) was significantly greater than the C-C NUE of 42.6 kg G_w per kg N_s. In this case, N retention efficiency and available N uptake efficiency were greater for ALF-C than C-C. These data show the large and variable effects that cropping system, N management and climate interactions can have on NUE.

Grain N Accumulation Efficiency or N Harvest Efficiency

Grain N accumulation efficiency (N_g/N_s) evaluates crop physiological and soil processes that contribute to the overall efficiency at which N accumulates in the grain. Grain N accumulation efficiency is equivalent to N harvest efficiency if only grain is removed from the field, however, N harvest efficiency would also include N in any crop biomass that is removed. In both the HRSW and corn studies, grain N accumulation efficiency is less than one across all treatments (Tables 5, 6, and 7) and is a warning signal of the potential for N surpluses to pollute surface- and groundwaters, if they are not recycled and efficiently used by subsequent crops. In the Minnesota study (1994-1995), grain N accumulation efficiencies averaged 36% for C-C and 39% for C-S rotations and low efficiencies have contributed to substantial losses of nitrate-N to

TABLE 5. Nitrogen rate (N_f), conventional tillage (CT), and no-tillage (NT) effects on nitrogen use efficiency components of hard red spring wheat (adapted from Huggins and Pan, 1993).

N_f	N_f/N_s CT	N_f/N_s NT	G_w/N_f CT	G_w/N_f NT	N_g/N_s CT	N_g/N_s NT	N_g/N_f CT	N_g/N_f NT
kg ha^{-1}				kg kg^{-1}				
0	0.00	0.00	ud	ud	0.55	0.50	ud	ud
56	0.42	0.48	72.5	53.9	0.52	0.47	1.23	0.98
112	0.59	0.65	45.0	37.4	0.52	0.48	0.88	0.73
168	0.69	0.73	35.1	26.2	0.53	0.45	0.77	0.62

Analysis of Variance

Tillage (T)	†		**		*		*	
N rate (NR)	**		**		†		**	
T × NR	NS		*		NS		NS	

ud = undefined.
†, *, ** Significant at the 0.1, 0.05, and 0.01 probability levels, respectively, NS = not significant.

TABLE 6. Continuous corn (C-C) and corn-soybean (C-S) rotation effects on nitrogen use efficiency components of corn (adapted from Randall et al., 1997).

	N_f		N_f/N_a		G_w/N_f		N_{td}/N_f	
Year	C-C	C-S	C-C	C-S	C-C	C-S	C-C	C-S
	--- kg ha^{-1} ---					kg kg^{-1}			
1988	179	179	0.73	0.76	15.3	13.5	0.00	0.00
1989	146	146	0.42	0.42	40.3	39.1	0.00	0.00
1990	35	35	0.13	0.13	182.0	186.3	0.17	0.11
1991	137	122	0.47	0.38	56.8	74.1	0.51	0.65
1992	174	134	0.56	0.62	47.6	52.9	0.30	0.24
1993	157	106	0.60	0.44	33.0	52.7	0.58	0.83
Average	138	120	0.49	0.46	62.5	69.8	0.26	0.28

TABLE 7. Crop sequence effects on nitrogen use efficiency components of corn (adapted from Huggins, Randall, and Russelle, 2001).

Year	Sequence†	N_f	N_f/N_a	G_w/N_f	N_g/N_a	N_g/N_f	N_{td}/N_f
		kg ha^{-1}			kg kg^{-1}		
1994	C-C	165	0.79	52.9	0.43	0.53	0.09
	S-C	112	0.65	81.5	0.50	0.80	0.12
	ALF-C	17	0.12	537.5	0.60	4.87	0.18
	CRP-C	177	0.96	53.0	0.54	0.56	0.01
	LSD$_{0.05}$		0.32*	14.0*	0.15*	0.15*	0.09*
1995	C-C	160	0.65	35.7	0.28	0.43	0.14
	S-C	108	0.42	65.6	0.28	0.67	0.23
	ALF-C-C	77	0.38	75.4	0.28	0.74	0.18
	CRP-C-C	149	0.48	47.6	0.28	0.58	0.11
	LSD$_{0.05}$		0.08*	8.5*	0.04	0.13	0.13

† C = corn; S = soybean; ALF = alfalfa; CRP = Conservation Reserve Program planting.
* Significant differences occurred according to Fischer's least significant difference (LSD) test (0.05 level of probability).

subsurface field drains in these cropping systems (Randall et al., 1997; Huggins, Randall, and Russelle, 2001). The greatest grain N accumulation efficiencies occurred for first year corn following alfalfa (60%) and CRP (52%), however, grain N accumulation efficiencies for second year corn decreased to 28% and was accompanied by increases in residual soil nitrate-N.

Analysis of grain N accumulation efficiency components: N retention efficiency (N_{av}/N_s), available N uptake efficiency (N_f/N_{av}) and N harvest index

(N_g/N_t), provides a means to identify where possible improvements in N management can occur. Grain N accumulation efficiency of corn averaged 40% across all crop sequences (1994-1995). Average values for each N efficiency component were: 93% for N retention efficiency, 62% for available N uptake efficiency, and 69% for N harvest index. Contributions of N efficiency components to the difference between N_g and N_s are N retention efficiency, 11%; available N uptake efficiency, 59%; and N harvest index, 30%. Low efficiencies due to N harvest index are not likely to contribute to N pollution and crop residue N can cycle to supply internal N sources in the future. Therefore, low available N uptake efficiency is a major factor leading to N pollution. As previously discussed, corn does not assimilate available N as supplies increase beyond crop requirements. In the corn study, low available N uptake efficiency is due to an over supply of available N. Nitrogen retention efficiency can remain high if soil nitrate-N is retained within the root zone; however, it is subject to losses via leaching or denitrification if not assimilated by the crop. During earlier years of this experiment, low N retention efficiencies occurred during wet seasons when considerable losses of soil nitrate-N occurred via subsurface drains. Producers can compensate for high residual levels of soil nitrate-N by adjusting N rates according to soil test results; however, soil tests for inorganic N fail to evaluate in-season net N mineralization. In this study, low N retention efficiencies occurred due to wide annual fluctuations in net N mineralization that resulted in over-applications of N, high soil residual levels of nitrate-N, and when combined with low available N uptake efficiencies, substantial N losses. Management strategies to increase grain N accumulation efficiency include improving predictive capabilities for quantifying available N supplies from N mineralization-immobilization turnover or improving available N uptake efficiencies. The latter could be achieved by including crops with greater N uptake efficiencies in rotation with corn or by the use of companion or cover crops that increase overall N uptake efficiencies.

Grain N accumulation efficiency of HRSW decreased with increasing levels of applied N and was greater for CT (53%) than NT (48%) (Table 5). At the greatest N rate (168 kg N ha^{-1}), 85% of the difference in N efficiency components between CT and NT was due to N retention efficiency (73% for CT and 67% for NT). This analysis identifies N losses to competing N pathways as a major area requiring improvement if grain N accumulation efficiency is to increase under NT and CT and is particularly relevant to developing N management strategies for increasing grain protein in HRSW. The identification of competing N pathways takes on added significance as the average N harvest efficiency is 50% and NHI is greater than 80% indicating substantial flows of N to non-crop sinks.

ECONOMIC AND ENVIRONMENTAL QUALITY FACTORS OF N USE

Economic goals of N management have focused on maximizing farm income from the use of N fertilizer. Important factors include crop and fertilizer prices, the cost of fertilizer application, and the relationship between crop yield (and or quality) and N application rate. Bock (1984) defined N use efficiency in terms of fertilizer use where yield efficiency (G_w/N_f) is equal to the product of recovery efficiency (N_f/N_f) and physiological efficiency (G_w/N_t). This definition of NUE emphasizes crop recovery of N fertilizer inputs and yield response to applied N for maximizing N fertilizer use efficiency. In addition to economic objectives, environmental impact and resource sustainability are important N management considerations. Major environmental quality goals of N management include: (1) sustaining or regenerating N resources within the agroecosystem; and (2) limiting N losses that can adversely impact air and water quality. Environmental quality and economic goals of N management are closely interrelated and dependent on NUE as: (1) resources of indigenous or internally derived N can offset requirements for applied N fertilizer; and (2) N not recovered by crops can be lost from the agroecosystem and adversely affect environmental quality and economic returns.

The previous definition of NUE (G_w/N_s) can be expanded to integrate Bock's (1984) definition of NUE (G_w/N_f) and N efficiencies related to environmental quality by partitioning G_w/N_s into two factors: (1) N reliance index (N_f/N_s) a measure of the crops dependence on N fertilizer inputs; and (2) N fertilizer utilization efficiency (G_w/N_f) a measure of crop yield response to N fertilizer inputs:

$$G_w/N_s = N_f/N_s \times G_w/N_f \qquad [5]$$

An analogous approach can be used to partition grain N accumulation efficiency (N_g/N_s) into two factors: (1) N reliance index (N_f/N_s); and (2) N balance efficiency (N_g/N_f) a measure of the balance between N outputs and N inputs:

$$N_g/N_s = N_f/N_s \times N_g/N_f \qquad [6]$$

Nitrogen balance efficiency (N_g/N_f) can be further divided into: (1) an N loss index that evaluates N losses (N_{lo}) from the system via leaching, denitrification and volatilization as a proportion of N fertilizer (N_{lo}/N_f); and (2) the ratio of harvested N, to N losses (N_g/N_{lo}):

$$N_g/N_f = N_{lo}/N_f \times N_g/N_{lo} \qquad [7]$$

Nitrogen Fertilizer Utilization Efficiency

The response of grain yield to applied N is the most common evaluation of N use. Often, yield response curves to applied N are curvilinear and N fertilizer utilization efficiency will vary with the shape of the curve and the portion used for calculations. Prior knowledge of the relationship between yield and applied N enables calculations of returns to N fertilizer costs and determination of optimum economic N rates (Bock, 1984). Evaluation of N fertilizer utilization efficiency based on ratios of yield and applied N (G_w/N_f) enables simple comparisons among treatments and cropping systems on efficiencies of applied N use. In many cropping systems, overall utilization efficiencies decrease as applications of N increase. For example, in the HRSW study, N fertilizer utilization efficiencies decreased as applied N increased averaging 35.1 kg G_w per kg N_f for CT and 26.2 kg G_w per kg N_f for NT at the greatest N rate (Table 5). The decrease in N fertilizer utilization efficiency that occurs as levels of applied N increase is due, in part, to the yield response to indigenous levels of available soil N when no N fertilizer is used. Consequently, as contributions of indigenous soil N decrease, the level of applied N required to obtain yield goals increases and overall utilization efficiency decreases. Increased dependence on N fertilizer and consequent decreases in utilization efficiency can increase the economic sensitivity of the cropping system to changes in grain and fertilizer N prices. In the HRSW study, N fertilizer utilization efficiency was significantly lower for NT than CT at equivalent N rates indicating lower economic returns to N fertilizer inputs.

In the Minnesota study, N fertilizer utilization efficiencies for corn during the first three years ranged from 13.5 to 186.3 kg G_w per kg N_f for the C-S rotation with similar values occurring for C-C (Table 6). Fertilizer utilization efficiencies were very low for 1988 as corn grain yields for C-C and C-S rotations were 30% of estimated yield goals. In 1990, soil testing revealed the buildup of large residual levels of inorganic N likely originating from unrecovered N applied in 1988. Substantial reductions in N fertilizer rates based on soil test values resulted in utilization efficiencies that were 10 times greater in 1990 than 1988. While these data underscore the importance of soil sampling, they also show the instability in utilization efficiency that can occur as corn yields vary due to unpredictable weather conditions (Huggins and Alderfer, 1995). Nitrogen fertilizer utilization efficiency averaged 22% more for the C-S rotation than for C-C over 8 years (Tables 6 and 7). The greater utilization efficiency for C-S was a function of less applied N and greater yields achieved with C-S as compared to C-C during 1991 through 1995. Nitrogen fertilizer utilization efficiencies for corn following alfalfa were 10 times greater than C-C as applied N was 10% of levels applied to C-C. This analysis indicates that considerable improvements in economic return to N fertilizer inputs can be realized by designing crop rotations that include grain and/or forage legumes.

Nitrogen Reliance Index

In many annual, cash crop rotations, reliance on internal N sources such as N_2-fixation, net soil N mineralization and wet and dry N deposition is not feasible as these N supplies are insufficient to meet crop requirements. Nevertheless, increasing N contributions from internal supplies remains attractive and can lead to decreased expenditures on external N fertilizer sources. A greater N reliance index (N_f/N_s) indicates increased dependency on external N supplies to alleviate crop N deficiencies. An economic analysis of this factor would evaluate the costs of internal and external sources of N to determine target proportions of each. Nitrogen cycling within a cropping system could, however, make this accounting process difficult as a portion of external N inputs often cycle and become future internal N contributions. Nevertheless, the proportion of total N supply derived from N fertilizer inputs is likely to be characteristic for a given cropping system. Furthermore, depletion of internal N contributions over time would indicate soil resource degradation and a greater necessity to rely on external N inputs to achieve crop performance goals. This, in turn, would increase the importance of evaluating the N reliance index under scenarios of increasing N fertilizer costs.

In the HRSW study, the N reliance index increased with higher N rates and averaged 69% at the greatest N level under CT and 73% under NT (Table 5). The small difference between the two tillage systems was due to less estimated N_s in NT than CT. These values were similar to the N reliance index of C-C (72%) in southwest Minnesota (Tables 4 and 5). Interestingly, from 1991-1995, the N reliance index was substantially lower for the C-S (50%) compared to C-C (61%), an example of positive legume rotation effects. Not surprisingly, first year corn after alfalfa had the lowest N reliance index averaging 12% while first year corn following CRP (perennial grasses) had the greatest N reliance index (95%). Unanticipated were decreases in N reliance index for second year corn after CRP (48%) resulting from large internally-derived N from mineralization of labile organic N constituents (Huggins, Randall, and Russelle, 2001). Consequently, over two years the average N reliance index for corn after CRP was similar to continuous corn. The N reliance index increased from 12 to 38% for second year corn after alfalfa signifying substantial but decreasing internal contributions of N. Continued cropping to corn would likely result in similar values as C-C over time. The N reliance index has likely increased over time in many agroecosystems as levels of soil organic matter have declined (Huggins et al., 1998), reserves of readily mineralizable forms of organic N decreased, and crop yields increased. Therefore, managing agroecosystems to increase soil organic matter levels as well as inclusion of legumes in rotation are important strategies for decreasing reliance on N fertilizer inputs.

Nitrogen Balance Index

The sustainability of agroecosystems with severe imbalances in N inputs and outputs is questionable as nutrient surpluses can contribute to pollution while deficits lead to impoverishment of the system and decreased productivity. Major N inputs of agroecosystems consist of synthetic fertilizers, organic manures, biological N_2-fixation, deposition of sediment and run-on, and atmospheric deposition. Major N outputs are harvested products, removed crop residues, leaching, gaseous losses and transport in run-off or sediment. The N balance index defined as the ratio of N removal via grain harvest to N fertilizer inputs accounts for major N inputs and outputs directly related to N management. Nitrogen balance indexes greater than one indicate the use of internal N sources and possible soil N resource degradation, whereas indexes less than one indicate the potential for N losses from the agroecosystem. In the HRSW study, values of N balance indexes close to one occur for the first level of applied N and then decrease below one as levels of applied N increase (Table 5). In this case, N requirements for obtaining HRSW yield and grain protein goals result in N balance indexes that are less than one. The positive effect of legumes on N balance index is shown for corn in the Minnesota study (Table 7). Nitrogen balance indexes averaged 0.48 for C-C, increased by 50% for the C-S rotation and were greatest for first year corn following alfalfa where the N balance index was 4.87. Interestingly, the N balance index for second year corn following alfalfa decreased below one and was similar to the C-S rotation. Consequences of an N balance index that is less than one are dependent on the proportion of N that is still cycling within the agroecosystem versus N that has been lost beyond agroecosystem boundaries. In the HRSW study, losses of N are difficult to separate from immobilized N, however, the previous analyses have indicated that a large proportion of unaccounted for N is likely immobilized and retained within the agroecosystem in organic forms. If, however, N balance indexes are less than one for every crop in rotation, applications of N are likely compensating for N losses from the agroecosystem. Therefore, the N balance index should be evaluated for the cropping system as a whole to determine whether or not the crop rotation has an N balance index greater or less than one.

Nitrogen Loss Index

The N loss index evaluates N losses beyond agroecosystem boundaries as a fraction of applied N fertilizer. In the corn study, ratios of N losses via artificial drains to applied N (N_{td}/N_f) ranged from 0 to 83% and are a partial measure of the N loss index (Tables 6 and 7). The N_{td}/N_f index was 22% for C-C and 26% for C-S averaged across eight years. These major N losses expressed as a percentage of applied N indicate that N management strategies need to be seriously reconsidered. Future efforts to design sustainable agroecosystems should explore cropping systems that use water and N more efficiently.

CONCLUSIONS

An important agricultural goal is to design cropping systems that use N efficiently. Many environmental, biological, and economic factors contribute to the overall N use efficiency of an agroecosystem. Consequently, various N efficiency components and indexes that evaluate different aspects of cropping system N use should be developed and used to identify where improvements in NUE can be realized. We applied N efficiency components and indexes to evaluate crop physiological and soil processes, and agronomic and environmental sustainability factors, and demonstrated their use in wheat and corn production systems. In the wheat example, major improvements in NUE could be realized by increasing N retention efficiency and N utilization efficiency. In contrast, improvements in regulating N supplies and increasing available N uptake efficiencies would greatly improve NUE in the corn study. Both examples showed large N reliance indexes for continuous cereals, an indication of the major dependency these cropping systems have on external sources of N fertilizer. The inclusion of legumes in crop sequences substantially reduced the N reliance index for corn and had positive impacts on many N efficiency components. Grain N accumulation efficiencies and N balance ratios of less than one occurred in both wheat and corn studies, indicating large diversions of available N to non-crop pathways. Nitrogen losses via subsurface drains were a significant proportion of N fertilizer applied to corn, as shown by the N loss index. Overall, evaluation of N efficiency components and indexes proved useful for: (1) understanding how major cropping system components such as rotation and tillage influence the efficiency of N use; and (2) identifying N management strategies that can improve NUE.

REFERENCES

Badaruddin, M. and D.W. Meyer. (1994). Grain legume effects on soil nitrogen, grain yield, and nitrogen nutrition of wheat. *Crop Science* 34:1304-1309.

Baldock, J.O., R.J. Higgs, W.H. Paulson, J.A. Jackobs, and W.D. Shrader. (1981). Legume and mineral fertilizer effects on crop yields of several crop sequences in the Upper Mississippi Valley. *Agronomy Journal* 73:885-890.

Bauer, A. (1980). Responses of tall and semidwarf hard red spring wheats to fertilizer nitrogen rates and water supply in North Dakota 1969-1974. North Dakota State University, ND: North Dakota Agricultural Experiment Station.

Bock, B.R. (1984). Efficient use of nitrogen in cropping systems. In *Nitrogen in Crop Production*, ed. R.D. Hauck, Madison, WI: Agronomy Society of America, Crop Science Society of America, and Soil Science Society of America, pp. 273-294.

Burkhart, M.R. and D.E. James. (1999). Agricultural-nitrogen contributions to hypoxia in the Gulf of Mexico. *Journal of Environmental Quality* 28:850-859.

Carter, M.R. and D.A. Rennie. (1982). Changes in soil quality under zero tillage farming systems: distribution of microbial biomass and mineralizable C and N potentials. *Canadian Journal of Soil Science* 62:587-597.

Doran, J.W. (1987). Microbial biomass and mineralizable nitrogen distributions in no-tillage and plowed soils. *Biology and Fertility of Soils* 5:68-75.

Ebelhar, S.A., W.W. Frye, and R.L. Blevins. (1984). Nitrogen for legume cover crops for no tillage corn. *Agronomy Journal* 76:51-55.

Eghball, B. and J.W. Maranville. (1991). Interactive effects of water and nitrogen stresses on nitrogen utilization efficiency, leaf water status and yield of corn genotypes. *Communications in Soil Science and Plant Analysis* 22:1367-1382.

Fiez, T.E., B.C. Miller, and W.L. Pan. (1994). Winter wheat yield and grain protein across varied landscape positions. *Agronomy Journal* 86:1026-1032.

Fiez, T. E., W. L. Pan, and B. C. Miller. (1995). Nitrogen efficiency analysis of winter wheat among landscape positions. *Soil Science Society America Journal* 59:1666-1671.

Francis, D.D., J.S. Schepers, and M.F. Vigil. (1993). Post-anthesis nitrogen loss from corn. *Agronomy Journal* 85:659-663.

Fredrickson, J.K., F.E. Koehler, and H.H. Cheng. (1982). Availability of ^{15}N-labeled nitrogen in fertilizer and in wheat straw to wheat in tilled and no-till soil. *Soil Science Society America Journal* 46:1218-1222.

Gauer, L.E., C.A. Grant, D.T. Gehl, and L.D. Bailey. (1992). Effects of nitrogen fertilization on grain protein content, nitrogen uptake, and nitrogen use efficiency of six spring wheat (*Triticum aestivum* L.) cultivars, in relation to estimated moisture supply. *Canadian Journal of Plant Science* 72:235-241.

Hargrove, W.L., J.T. Touchton, and J.W. Johnson. (1983). Previous crop influence on fertilizer nitrogen requirements for double-cropped wheat. *Agronomy Journal* 75:855-859.

Harper, L.A., R.R. Sharpe, G.W. Langdale, and J.E. Giddens. (1987). Nitrogen cycling in a wheat crop: Soil, plant and aerial nitrogen transport. *Agronomy Journal* 79:965-973.

Haynes, R.J. (1997). Fate and recovery of ^{15}N derived from grass/clover residues when incorporated into a soil and cropped with spring or winter wheat for two succeeding seasons. *Biology and Fertility of Soils* 25:130-135.

Hesterman, O.B., M.P. Russelle, C.C. Sheaffer, and G.H. Heichel. (1987). Nitrogen utilization from fertilizer and legume residues in legume-corn rotations. *Agronomy Journal* 79:726-731.

Hauck, R.D. and J.M. Bremner. (1976). Use of tracers for soil and fertilizer nitrogen research. In *Advances in Agronomy*, ed. N.C. Brady, New York, NY: Academic Press, Agronomy Society of America, Volume 28, pp. 219-266.

Huggins, D.R. (1991). *Redesigning No-Tillage Cropping Systems: Alternatives for Increasing Productivity and Nitrogen Use Efficiency.* Ph.D. Dissertation, Pullman, WA: Washington State University. 235 p.

Huggins, D.R. and R.D. Alderfer. (1995). Yield variability within a long-term corn management study: implications for precision farming. In *Site-Specific Management for Agricultural Systems,* ed. P.C. Robert, R.H. Rust and W.E. Larson, Miscellaneous Publications, Madison, WI: Agronomy Society of America, Crop Science Society of America, and Soil Science Society of America, pp. 417-426.

Huggins, D.R., G.A. Buyanovsky, G.H. Wagner, J.R. Brown, R.G., Darmody, T.R. Peck, G.W. Lesoing, M.B. Vanotti, and L.G. Bundy. (1998). Soil organic carbon in the tall-grass prairie-derived region of the Corn Belt: Effects of long-term management. *Soil & Tillage Research* 47:219-234.

Huggins, D.R. and W.L. Pan. (1993). Nitrogen efficiency component analysis: an evaluation of cropping system differences in productivity. *Agronomy Journal* 85:898-905.

Huggins, D.R., G.W. Randall, and M.P. Russelle. (2001). Surface drain losses of water and nitrate following conversion of perennials to row crops. *Agronomy Journal* 93:477-486.

Jansson, S.L. and J. Persson. (1982). Mineralization and immobilization of soil nitrogen. In *Nitrogen in Agricultural Soils,* ed. F.J. Stevenson, Agronomy Monograph 22. Madison, WI: Agronomy Society of America and Soil Science Society of America, pp. 229-252.

Johnson, G.A. and D.R. Huggins. (1999). Knowledge-based decision support strategies: linking spatial and temporal components within site-specific weed management. *Journal of Crop Production* 2:225-238.

Johnson, G.V. and W.R. Raun. (1995). Nitrate leaching in continuous winter wheat: use of a soil-plant buffering concept to account for fertilizer nitrogen. *Journal of Production Agriculture* 8:486-491.

Karlen, D.L., P.G. Hunt, and T.A. Matheny. (1996). Fertilizer [15]nitrogen recovery by corn, wheat, and cotton grown with and without pre-plant tillage on Norfolk loamy sand. *Crop Science* 36:975-981.

Kessavalou, A. and D.T. Walters. (1997). Winter rye as a cover crop following soybean under conservation tillage. *Agronomy Journal* 89:68-74.

Koenig, R.T. (1993). *Morphological Development, Productivity and Nutrient Use by Spring Wheat in Response to an Enhanced Ammonium Supply.* Ph.D. Dissertation, Pullman, WA: Washington State University. 194 p.

Kumar, K. and K.M. Goh. (2000). Crop residues and management practices: effects on soil quality, soil nitrogen dynamics, crop yield, and nitrogen recovery. In *Advances in Agronomy,* ed. D.L. Sparks, San Diego, CA: Academic Press, pp. 197-319.

Ladd, J.N. and M. Amato. (1986). The fate of nitrogen from legume and fertilizer sources in soils successively cropped into winter wheat under field conditions. *Soil Biology and Biochemistry* 18:417-425.

Legg, J.O. and J.J. Meisinger. (1982). Soil Nitrogen Budgets. In *Nitrogen in Agricultural Soils,* ed. F.J. Stevenson, Madison, WI: American Society of Agronomy. pp. 503-566.

Legg, J.O., G. Stanford, and O.L. Bennett. (1979). Utilization of labeled-N fertilizer by silage corn under conventional and no-till culture. *Agronomy Journal* 71:1009-1015.

Liang, B.C. and A.F. MacKenzie. (1994). Corn yield, nitrogen uptake and nitrogen use efficiency as influenced by nitrogen fertilization. *Canadian Journal of Soil Science* 74: 235-240.

Linn, D.M. and J.W. Doran. (1984). Effect of water-filled pore space on carbon dioxide and nitrous oxide production in tilled and nontilled soils. *Soil Science Society America Journal* 48:1267-1272.

Loffler, C.M., T.L. Rauch, and R.H. Busch. (1985). Grain and plant protein relationships in hard red spring wheat. *Crop Science* 25: 521-524.

Mahler, R.L., F.E. Koehler, and L.K. Lutcher. (1994). Nitrogen source, timing of application, and placement: effects on winter wheat production. *Agronomy Journal* 86:637-642.

Maskina, M.S., J.F. Power, J.W. Doran, and W.W. Wilhelm. (1993). Residual effects of no-till crop residues on corn yield and nitrogen uptake. *Soil Science Society America Journal* 57:1555-1560.

McNeal, F.H., M.A. Berg, P.L. Brown, and C.F. McGuire. (1971). Productivity and quality response of five spring wheat genotypes, *Triticum aestivum* L., to nitrogen fertilizer. *Agronomy Journal* 63:908-910.

Meisinger, J.J. and G.W. Randall. (1991). Estimating nitrogen budgets for soil-crop systems. In *Managing Nitrogen for Groundwater Quality and Farm Profitability*, ed. R.F. Follett, D.R. Keeney, and R.M. Cruse, Madison, WI: Soil Science Society of America, pp. 85-124.

Mengel, D.B., D.W. Nelson, and D.M. Huber. (1982). Placement of nitrogen fertilizers for no-till and conventional till corn. *Agronomy Journal* 74:515-518.

Moschler, W.W., G.M. Shear, D.L. Hallock, R.D. Sears, and G.D. Jones. (1967). Winter cover crops for sod-planted corn: their selection and management. *Agronomy Journal* 59:547-551.

Moll, R.H., E.J. Kamprath, and W.A. Jackson. (1982). Analysis and interpretation of factors which contribute to efficiency of nitrogen utilization. *Agronomy Journal* 74:562-564.

Mosier, A.R., J.M. Duxbury, J.R. Freney, O. Heinemeyer, and K. Minami. (1996). Nitrous oxide emissions from agricultural fields: Assessment, measurement and mitigation. *Plant Soil* 181:95-108.

Moss, H.J., C.W. Wrigley, F. MacRitchie, and P.J. Randall. (1981). Sulfur and nitrogen fertilizer effects on wheat: II. Influence on grain quality. *Australian Journal of Agricultural Research* 32:213-226.

Mummey, D.L., J.L. Smith, and G. Bluhm. (1998). Assessment of alternative soil management practices on N_2O emissions from US agriculture. *Agriculture, Ecosystems and Environment* 70:79-87.

Oberle, S.L. and D.R. Keeney. (1990). Factors influencing corn fertilizer N requirements in the northern U.S. corn belt. *Journal of Production Agriculture* 3:527-534.

Olson, R.A. and D.H. Sander. (1988). Corn production. In *Corn and Corn Improvement. ASA Monograph 18*, eds. G.F. Sprague and J.W. Dudley, Madison, WI: Agronomy Society of America, pp. 639-686.

Overman, A.R., D.M. Wilson, and E.J. Kamprath. (1994). Estimation of yield and nitrogen removal by corn. *Agronomy Journal* 86:1012-1016.

Pan, W.L., D.R. Huggins, G.L. Malzer, C.L. Douglas, Jr., and J.L. Smith. (1997). Field heterogeneity in soil-plant nitrogen relationships: implications for site-specific management. In *The State of Site-Specific Management for Agriculture*, ed. F.J. Pierce and E.J. Sadler, Miscellaneous Publications, Madison, WI: Agronomy Society of America, Crop Science Society of America, and Soil Science Society of America, pp. 81-89.

Parton, W.J., J.A. Morgan, J.M. Altenhofen, and L.A. Harper. (1988). Ammonia volatilization from spring wheat plants. *Agronomy Journal* 80:419-425.

Pierce, F.J. and P. Nowak. (1999). Aspects of precision agriculture. In *Advances in Agronomy*, Volume 67, New York, NY: Academic Press. pp.1-85.

Randall, G.W., D.R. Huggins, M.P. Russelle, D.J. Fuchs, W.W. Nelson, and J.L. Anderson. (1997). Nitrate losses through subsurface tile drainage in Conservation Reserve Program, alfalfa, and row crop systems. *Journal of Environmental Quality* 24:360-366.

Rao, S.C. and T.H. Dao. (1996). Nitrogen placement and tillage effects on dry matter and nitrogen accumulation and redistribution in winter wheat. *Agronomy Journal* 88:365-371.

Rasmussen, P.E. and C.R. Rhodes. (1991). Tillage, soil depth, and precipitation effects on wheat response to nitrogen. *Soil Science Society of America Journal* 55:1221-1224.

Raun, W.R. and G.V. Johnson. (1995). Soil-plant buffering of inorganic nitrogen in continuous winter wheat. *Agronomy Journal* 87:827-834.

Raun, W.R. and G.V. Johnson. (1999). Improving nitrogen use efficiency for cereal production. *Agronomy Journal* 91:357-363.

Rice, C.W. and M.S. Smith. (1982). Denitrification in no-till and plowed soils. *Soil Science Society America Journal* 46:1168-1173.

Russelle, M.P., E.J. Deibert, R.D. Hauck, M. Stevanovic, and R.A. Olson. (1981). Effects of water and nitrogen management on yield and ^{15}N-depleted fertilizer use efficiency of irrigated corn. *Soil Science Society America Journal* 45:553-558.

Sowers, K.E. (1992). *Spring Nitrogen Applications with Point Injection or Topdress to Optimize Nitrogen Use Efficiency, Yield and Protein in Soft White Winter Wheat.* M.S. Thesis. Pullman, WA: Washington State University, 114 p.

Sowers, K.E., W.L. Pan, B.C. Miller, and J.L. Smith. (1994). Nitrogen use efficiency of split nitrogen applications in soft white winter wheat. *Agronomy Journal* 86: 942-948.

Tillman, B.A., W.L. Pan, and S.E. Ullrich. (1991). Nitrogen use by Northern European and Pacific Northwest U.S. barley genotypes under no-till management. *Agronomy Journal* 83:194-200.

Tilman, D., J. Fargione, B. Wolff, C. D'Antonio, A. Dobson, R. Howarth, D. Schindler, W.H. Schlesinger, D. Simberloff, and D. Swackhamer. (2001). Forecasting agriculturally driven global environmental change. *Science* 292:281-284.

Torbert, H.A., R.L. Mulvaney, R.M. VandenHeuvel, and R.G. Hoeft. (1992). Soil type and moisture regime effects on fertilizer efficiency calculation methods in a nitrogen-15 tracer study. *Agronomy Journal* 84:66-70.

Tyler, D.D. and G.W. Thomas. (1977). Lysimeter measurements of nitrate and chloride losses from soil under conventional and no-tillage corn. *Journal of Environmental Quality* 6:63-66.

Varvel, G.E. and T.A. Peterson. (1990). Nitrogen fertilizer recovery by corn in mono-culture and rotation systems. *Agronomy Journal* 82:935-938.

Westerman, R.L. and L.T. Kurtz. (1974). Isotopic and nonisotopic estimations of fertilizer nitrogen uptake by sudangrass in field experiments. *Soil Science Society of America Proceedings* 38:107-109.

Westerman, A.L., W.R. Raun, and G.V. Johnson. (2000). Nutrient and water use efficiency. In *Handbook of Soil Science*, ed. M.E. Sumner, Boca Raton, FL: CRC Press, pp. D-175-189.

Wienhold, B.J., T.P. Trooien, and G.A. Reichman. (1995). Yield and nitrogen use efficiency of irrigated corn in the northern great plains. *Agronomy Journal* 87:842-846.

Wuest, S.B. and K.G. Cassman. (1992). Fertilizer-nitrogen use efficiency of irrigated wheat: I. Uptake efficiency of preplant versus late-season application. *Agronomy Journal* 84:682-688.

Forage Legumes
for Sustainable Cropping Systems

Craig C. Sheaffer
Philippe Seguin

SUMMARY. Perennial and annual forage legumes are important components of sustainable cropping systems. Forage legumes are a primary source of forage to supply protein and fiber for livestock rations. They can be grazed, or stored as hay or silage. They contribute biologically fixed N and sustain the soil by reducing erosion and increasing soil organic matter levels. Diversifying cropping systems by including legumes can also reduce weed, insect, and disease incidence. Potential new uses of legumes include phytoremediation of N contaminated sites and capturing N lost from cropping systems. Legumes also have potential use as a feedstock for renewable energy production. Legumes have traditionally been used in rotation with grain crops but more recently have been shown promise as winter cover crops, intercrops with grain crops, and as living mulches. In this review, we discuss traditional and

Craig C. Sheaffer is Professor, Department of Agronomy and Plant Genetics, University of Minnesota, St. Paul, MN, USA.

Philippe Seguin is Assistant Professor, Department of Plant Science, McGill University, Macdonald Campus, Ste. Anne-de-Bellevue, QC, Canada.

Address correspondence to: Craig C. Sheaffer, Professor, Department of Agronomy and Plant Genetics, University of Minnesota, 411 Borlaug Hall, 1991 Buford Circle, St. Paul, MN 55108 USA.

[Haworth co-indexing entry note]: "Forage Legumes for Sustainable Cropping Systems." Sheaffer, Craig C., and Philippe Seguin. Co-published simultaneously in *Journal of Crop Production* (Food Products Press, an imprint of The Haworth Press, Inc.) Vol. 8, No. 1/2 (#15/16), 2003, pp. 187-216; and: *Cropping Systems: Trends and Advances* (ed: Anil Shrestha) Food Products Press, an imprint of The Haworth Press, Inc., 2003, pp. 187-216. Single or multiple copies of this article are available for a fee from The Haworth Document Delivery Service [1-800-HAWORTH, 9:00 a.m. - 5:00 p.m. (EST). E-mail address: docdelivery@haworthpress.com].

http://www.haworthpress.com/store/product.asp?sku=J144
10.1300/J144v08n01_08

new roles of forage legumes in sustainable cropping systems with examples primarily chosen from northern USA and Canada. *[Article copies available for a fee from The Haworth Document Delivery Service: 1-800-HAWORTH. E-mail address: <docdelivery@haworthpress.com> Website: <http://www. HaworthPress.com> © 2003 by The Haworth Press, Inc. All rights reserved.]*

KEYWORDS. Biological N fixation, soil erosion, intercropping, living mulches, rotations, cover crops, forages

INTRODUCTION

Legumes are essential component of sustainable agro-ecosystems. They contribute high quality forage for livestock rations while providing ecosystem services like organic N, soil stabilization, and nutrient recycling. Legumes were recognized as an essential part of crop rotations in Roman times (Rogers, 1976). In 16th century Europe, the value of legumes to cropping systems was not only recognized, but specific management strategies were elaborated to maximize the benefits derived from their use (De Serres, 1600). In North America, in the early to mid-1900s, small diversified farms that included livestock often relied on forage legumes like red clover (*Trifolium pratense* L.) and sweet clover (*Melilotus* spp.) for livestock feeding and green manuring. However, the industrialization of agriculture that has emphasized specialization and grain production has resulted in cropping systems that lack species diversity. Productivity from these specialized systems depends on extensive control of plant genetics, soil fertility, and pests (weeds, insects, and pathogens) (Tilman, 1999). An example is the alternate year, soybean (*Glycine max* L. Merr.) and corn (*Zea mays* L.) rotation that is widely practiced on Upper Midwestern landscapes.

While superior to monocultures, the corn-soybean rotation is still not sustainable from economic and environmental perspectives. Excessive production has resulted in prices below break-even levels for producers and marginal levels of profit are achieved only through payments from the US government. From an ecological/environmental standpoint, the corn-soybean rotation is "leaky" for nitrates and pesticides that can damage soil and water quality (Randall et al., 1997) and is extremely susceptible to soil erosion because of lack of effective ground cover (Randall, 2001). Because water leaving the landscape in the Midwest feeds into the Mississippi River and its tributaries, this rotation has been recognized as a contributor to the seasonal "dead zone" in the Gulf of Mexico (Feber, 2001). With increasing concerns about the environmental and economic sustainability of agricultural production systems, there has been a renewed interest in diversifying agricultural production systems through the use of forage legumes alone and in mixture with forage grasses. Our objective is to review the potential roles of forage legumes in sustainable cropping systems.

FORAGE LEGUME USE IN AGRO-ECOSYSTEMS

Forage legumes are members of the Fabaceae (or Leguminosae) family that also includes important food crops like edible beans (*Phaseolus* spp. L.), pea (*Pisum sativum* L.), and soybean. Seeds of these grain legumes contain high concentrations of protein and/or oil. Forage legumes that have smaller seeds but protein rich forage are valuable components of livestock rations. Forage legumes are commonly fed after storage as hay or silage or utilized directly by grazing.

For hay and silage production, alfalfa *(Medicago sativa* L.) and red clover that have an upright growth habit adapted to mechanical harvesting are frequently used (Table 1). Of the two, alfalfa represents about 80% of the forage legume seedings with about 11 million ha producing 83 million tons of forage in the US alone (Martin, 1996). Its success is attributable to its wide adaptation, exceptional persistence, and consistently superior productivity when compared to other legumes. Many of these traits have been improved by intensive plant breeding by public institutions and commercial companies. However, on some soils and for some uses where alfalfa is not adapted, red clover is harvested for hay and silage. Red clover is especially adapted to soils with lower soil pH levels than are suited for alfalfa but it has less persistence; it rarely persists more than three years (Leath, 1988; Taylor and Quesenberry, 1996).

While all legumes have potential to be grazed, those best adapted to grazing include white clover (*Trifolium repens* L.), birdsfoot trefoil (*Lotus corniculatus* L.), Kura clover (*Trifolium ambiguum* L.), and cicer milk vetch (*Astragalus cicer* L.). White clover is the most widely grown pasture legume because of its tolerance to a wide range of soil conditions and its extensive ability to regenerate both vegetatively and from seeds (Duke, 1981; Frame, Charlton, and Laidlaw, 1998). Grazing tolerant alfalfa varieties have recently been developed (Smith et al., 2000).

There are several other legumes that have important roles but that are used to a limited extent. For example, because of their growth habit, some legumes like common vetch (*Vicia sativa* L.), hairy vetch (*Vicia villosa* Roth), and crown vetch (*Cornilla varia* L.) are adapted for use as grown covers in agricultural systems or along roadsides. Sweetclover is still among the best legumes to use as a green manure crop. Although legume growth characteristics do make them adaptable to specific uses, selection of a legume within any region

TABLE 1. Characteristics of some commonly used forage legumes used in northern USA and Canada. (Adapted from Sheaffer and Barnes, 1987; and Sheaffer et al., 1993.)

Species	Scientific name	Uses[b]	Life cycle[c]	Growth habit	Max. yield Mg ha^{-1}	Characteristics or tolerance to stresses[a]						
						Winter Hardiness	Drought	Wet	Soil acidity	Frequent harvest	Seedling vigor	Bloat
Alfalfa	*Medicago sativa* L.	H,S,G	P	Bunch	6-22	E	G	P	P	G	F	Yes
Alsike clover	*Trifolium hybridum* L.	H,G	P	Bunch	4-8	P	P	G	G	P	F	Yes
Birdsfoot trefoil	*Lotus corniculatus* L.	G,H,C	P	Bunch	3-11	F	F	G	G	G	P	No
Common vetch	*Vicia sativa* L.	G,H	A	Bunch	2-6	-	P	F	G	F	E	No
Crown vetch	*Coronilla varia* L.	G,H,C	P	Rhizome	1-10	P	G	P	E	P	P	No
Cicer milk vetch	*Astragalus cicer* L.	G,H	P	Rhizome	4-8	E	G	F	G	P	F	No
Kura clover	*Trifolium ambiguum* M.B.	G,C	P	Rhizome	4-8	E	E	G	G	E	P	Yes
Red clover	*Trifolium pratense* L.	H,S,G	P	Bunch	4-13	F	P	F	F	P	G	Yes
Sainfoin	*Onobrychis viciaefolia* Scop.	G,H	P	Bunch	4-12	E	G	P	G	P	F	No
Sweet clover	*Melilotus* spp.	M,H,C	B,A	Bunch	4-9	E	G	P	P	P	F	Yes
White clover	*Trifolium repens* L.	G	P	Stolon	3-6	P	P	G	F	E	F	Yes

[a]E = excellent, G = good, F = fair, P = poor.
[b]C = soil conservation, G = grazing, H = hay, M = green manure, S = silage.
[c]A = annual, B = biennial, P = perennial.

is greatly influenced by soil characteristics and adaptation to climatic conditions (Sheaffer and Barnes, 1987; Sheaffer et al., 1993; Frame, Charlton, and Laidlaw, 1998).

BENEFITS OF LEGUMES TO AGRO-ECOSYSTEMS

There are numerous unique benefits of including forage legumes in cropping systems. These benefits include contribution of nitrogen (N), soil and water conservation, production of high quality forage for livestock feeding, and bioremediation. In addition, temporal or spatial diversification of cropping systems by inclusion of legumes can provide weed, insect and disease control; and can enhance habitat for wildlife.

Dinitrogen Fixation

A distinctive feature of most members of the Fabaceae family is the capability to biologically fix atmospheric N_2. Biological nitrogen fixation (BNF) results from the establishment of a symbiosis between legumes and soil bacteria, collectively known as rhizobia. These include members of the genus *Rhizobium*, *Bradyrhizobium*, *Sinorhizobium*, and *Azorhizobium*. These microorganisms invade legume roots and form nodules, which are structures hosting the rhizobia and where N_2 fixation takes place (Graham, 1998). This invasion of the legume roots is the result of an exchange of molecular signals between legumes and rhizobia. In the nodules, rhizobia reduce N_2 to NH_3 that is converted by plant enzymes to form amino acids that are available to the host plant (Vance, 1998). In return, legume plants provide fixed C compounds (i.e., photosynthates) to the rhizobia.

The quantity of N_2 fixed by legumes varies between 15 and 390 kg N ha yr^{-1} depending on the species and prevailing biotic and abiotic conditions (Table 2). Biotic and abiotic factors affecting N_2 fixation, include stresses such as pests, inadequate soil fertility, soil acidity, salinity, drought, and defoliation. These processes/conditions affect the molecular communication process between rhizobia and legumes and/or reduce the photosynthetic capacity of the plant thus affecting N_2 fixation (Graham, 1992; Zhang and Smith, 2002). High soil N levels will also usually decrease N_2 fixation rates (Streeter, 1988). Usually only a portion of legume N is derived from BNF, the remainder is derived from soil sources (Sheaffer, Barnes, and Heichel, 1989). Peterson and Russelle (1991) assumed that on average for the US Corn Belt, 50% of the N in the herbage of alfalfa was derived from fixation in the seeding year and 80% in succeeding years.

Although legumes have the potential to fix and contribute high quantities of BNF-N, only part of the fixed N is usually returned to the soil for use by a sub-

TABLE 2. Estimates of the range of N_2 biologically fixed by common forage legumes. (Compiled from Ledgard and Steele, 1992; Seguin et al., 2000, 2001; Sheaffer and Barnes, 1987; Vance, 1998; and Zhu et al., 1998.)

Species	kg N ha^{-1} yr^{-1}	%Ndfa[a]
Alfalfa	51-386	46-92
Annual medics (*Medicago* spp.)	100-200	79-86
Birdsfoot trefoil	49-162	30-85
Kura clover	17-158	25-57
Red clover	15-373	35-87
Common vetch	110-184	75
White clover	45-291	62-93

[a] %Ndfa: Percentage of N derived from the atmosphere.

sequent crop. This is because a significant portion of the BNF-N is removed during harvest of herbage or grain. For example, Heichel and Barnes (1984) compared BNF-N incorporated by alfalfa subject to a one- versus a three-cut harvest system in the seeding year. The one-cut system, in which herbage and roots were incorporated in the fall, resulted in the addition of 37% more BNF-N compared to the three-cut system, in which only roots were incorporated. Likewise, they reported that growth of soybean actually resulted in a soil N deficit because of greater N removal in the grain than accrued through fixation. The use of incorporated legume N by subsequent crops has been reported to range from only 17 to 43% (Hesterman et al., 1987; Harris and Hesterman, 1990). Factors such as timing of plowdown to synchronize legume N availability with nutrient need of the following crop (Stute and Posner, 1995) and tillage method influence N use efficiency (Peterson and Russelle, 1991; Huggins, Randall, and Russelle, 2001).

Because of the contribution of BNF-N, agricultural extension services of the Corn Belt recommend that N fertilizer rates for corn grown following alfalfa be reduced (Peterson and Russelle, 1991). The N credit and consequently N fertilizer rate reduction is influenced by several factors including corn yield potential, alfalfa stand age, stand density, and soil type (Voss and Schrader, 1984; Hesterman, 1988).

Legumes also contribute N to associated grasses growing in mixture. Biologically fixed N can be transferred to associated grasses via: (i) release of N from root exudates, (ii) decomposition of roots and nodules, (iii) N leaching from leaves and/or leaf decomposition, (iii) transfer via mycorrhizal fungi, and/or (iv) animal excreta (Ledgard, 1991; Hamel, Furlan, and Smith, 1992; Dubach and Russelle, 1994; Paynel, Murray, and Cliquet, 2001). The amount of N transferred varies between 5 and 80 kg ha^{-1}, representing 20 to 70% of the grasses' N-requirements (Table 3). This proportion will greatly vary de-

TABLE 3. Examples of amounts of N_2 fixed by forage legumes in mixture with grasses and transfer of fixed-N to accompanying grass.

Mixture		N_2 fixed (kg ha^{-1})	Amount of N_2 transferred (kg ha^{-1})	% of legume N_2 fixation	% of grass N coming from transfer	Reference
Transfer belowground and via herbage residues						
Alfalfa-reed canary grass		54	7	13	68	Brophy, Heichel, and Russelle,1987
Alfalfa-reed canary grass		75	30	40	30	Heichel and Henjum, 1991
Alfalfa-timothy		107	10	10	24	Ta and Faris, 1987
Birdsfoot trefoil-reed canary grass		31	5	10	17	Brophy, Heichel, and Russelle,1987
Birdsfoot trefoil-orchard grass		105	21	20	26	Farnham and George, 1994
Red clover-Italian ryegrass		130	30	23	39	Boller and Nosberger, 1987
Total transfer						
White	*via plants*	300	78	26	27	Ledgard, 1991
clover-	*via animals*		+60	+22	+23	
ryegrass	*Total*		138	48	50	

pending on the legume species, environment, and N status of pastures (Legard and Steele, 1992). Ledgard (1991) reported that the proportion of grass N derived from legume transfer would be doubled in cases where manure is returned. This is attributable to the fact that ruminant animals excrete 85% of their ingested N. When legumes are grown in association with grasses, the amounts of N_2 fixed often fluctuate and usually decrease over time. The reason for such fluctuations is the very dynamic and complex nature of legumes-grass mixtures (Haynes, 1980). In the short term, the presence of associated grasses will promote N_2 fixation as they take up soil N. This uptake of N has a positive effect on N_2 fixation as mineral N usually inhibits N_2 fixation. However, this uptake of N stimulates the growth of grasses and thus increases their competitiveness relative to that of legumes. As a result, legume productivity, persistence, and N_2 fixation will be reduced. This reduction in legume content finally results in reduced N inputs into the system, and ultimately reduced grass growth.

Feeding Value

Forage legumes are important components of ruminant livestock rations because they contain high levels of fiber that can be converted to energy by ruminant animals (Buxton and Mertens, 1995). Although fiber is essential to rumen function, it also greatly affects forage intake. Intake of forage is limited by total fiber content (i.e., content of hemicellulose, cellulose, and lignin), and digestibility. Legumes consistently have high forage protein concentration due to their unique capacity to biologically fix N. Legumes thus represent a reliable and inexpensive protein source for animal nutrition. Generally, relative to grasses, legumes are also a superior source of other essential minerals including Ca, Mg, K, Zn, Co, and Cu (Spears, 1994).

Compared to grasses, legumes often have greater rates of digestion and greater voluntary intake because of lower cell wall concentrations. For example, Wilman, Mtengeti, and Moseley (1996) reported that forage intake was 60% lower with grasses than legumes because cell wall content was 45% lower in legumes such as red clover, alfalfa, and sainfoin (*Onobrichis viciifolia* Scop.), compared to grasses such as timothy (*Phleum pratense* L.), ryegrass (*Lolium perenne* L.), and tall fescue (*Festuca arundinacea* Schreb). However, total digestibility of grass and legume forage may sometimes be similar because of the greater lignification of legume cell walls. The differential feeding value of legumes and grasses is in part attributable to differences in their physical structure and anatomy. For example, the average cell wall concentration of cool-season grasses was 500 and 700 g kg^{-1}, in leaves and stems, respectively, compared to 250 and 650 g kg^{-1}, in leaflets and stems of alfalfa (Sheaffer et al., 2000). In addition, the fiber content of legume leaves changes less with increasing maturity than that of grass leaves (Mowat et al., 1965).

Energy Feedstock

Legumes also have potential as a feedstock for electrical generation. For example, alfalfa has been studied as a biomass for generation of electricity (DeLong et al., 1995). In the process, leaves and stems of alfalfa hay were to be separated to produce a leaf pellet for high quality livestock feed and a stem pellet for gasification and production of electricity. While the project to produce electricity from alfalfa was terminated under the economic and energy supply conditions of the late 1990s (Sheaffer et al., 2000), the concept may still be viable as alternative energy and N sources that are less energy costly are sought.

Soil and Water Conservation

Perennial and annual legumes used in cropping systems reduce soil erosion by decreasing surface runoff and by increasing infiltration of precipitation (Hargrove and Frye, 1987; Thomas et al., 1992; Zemenchik et al., 1996). Living plants and their residues on the soil surface protect the soil from impact of raindrops, reduce soil surface pore blockage, and reduce the velocity of running water (Hargrove and Frye, 1987; Thomas et al., 1992). Also, residues from the roots and shoots increase soil organic carbon (C), water stable aggregation, and macroporosity thus increasing the rate of water infiltration and soil water retention (McVay, Radcliffe, and Hargrove, 1989; Angler, 1992; Rasse, Smucker, and Santos, 2000). The value of legumes is illustrated in the "C factor" of the Universal Soil Loss Equation that is used to predict soil loss from different cropping systems. The C factor is greatly influenced by level of mulch cover (Wischmeier and Smith, 1978). Systems including legumes like alfalfa in rotation or as covers typically have C values < 0.10 whereas corn-soybean rotations including tillage may be as high as 1.0. More specifically, in Pennsylvania, living mulch research showed that when corn was no-till seeded into crownvetch and birdsfoot trefoil, soil and herbicide losses were reduced about 98 and 96%, respectively, compared to a conventionally tilled system (Hall, Hartwig, and Hoffman, 1994). Likewise, in Nova Scotia, interseeding red clover into corn reduced soil loss from 46 to 78% compared to a corn monoculture (Wall, Pringle, and Sheard, 1991).

The positive impact of legumes is influenced by the duration of legume growth during a season and over years (Perfect et al., 1990). Although winter annual cover crops like crimson clover (*Trifolium incarnatum* L.) and hairy vetch reduce surface runoff and enhance soil organic matter (OM) (Frye et al., 1988), erosion and runoff from annual legumes and seeding year stands of perennial legumes is likely to be more than for older legume stands because older stands provide increased ground cover and long-term improvement in soil physical properties. For example, Angler (1992) reported that water stable aggregation increased each year as alfalfa stands aged but did not increase after stands were 4 years old.

Legumes that improve soil OM and provide soil coverage may enhance water available for crop growth by reducing evaporation, improving water infiltration, and increasing soil water holding content (Frye et al., 1988). In contrast, legumes may also deplete soil water on wet soils enabling earlier planting. In some cases, use of legumes in rotations can be undesirable because of depletion of soil water reserves (Voss and Schrader, 1984; Frye et al., 1988; Huggins, Randall, and Russelle, 2001). For example, Decker et al. (1994) reported that a heavy winter growth of crimson clover produced a very dry soil that restricted seed placement and limited corn growth. Although soil moisture level and crop growth were better under hairy vetch, it was still less than where no winter cover crop was used.

Bioremediation

Deep rooted legumes like alfalfa can intercept and remove NO_3-N from deep in the soil and prevent its leaching into the groundwater (Kelner, Vessey, and Entz, 1997; Peterson and Russelle, 1991). Although alfalfa obtains some

N from BNF even at high soil N levels, substitution of soil derived-N for BNF-N makes alfalfa effective at scavenging residual fertilizer N following annual crops (Mathers, Steward, and Blair, 1975; Lamb et al., 1995). Randall et al. (1997) reported that annual NO_3-N losses were less for a perennial crop of alfalfa (6 kg ha^{-1}) than for a corn-soybean rotation (180 kg ha^{-1}). These reduced NO_3-N losses were associated with less drainage volume and NO_3-N concentration in the drainage water from alfalfa. However, research by Huggins, Randall, and Russelle (2001) showed that when alfalfa was converted to corn the benefits of alfalfa in reducing concentrations and losses of NO_3-N through subsurface drainage ceased following the second year of corn. Russelle et al. (2001) demonstrated that alfalfa was an effective crop for remediation of a site that was contaminated by a railroad spill of anhydrous ammonia. Over a 3-year period, a non-N fixing (ineffective) alfalfa removed 972 kg N ha^{-1} compared to 287 kg N ha^{-1} removed by annual cereals. Removal of inorganic N by alfalfa also caused subsoil NO_3-N under the spill site to decline markedly.

Weed Control

Inclusion of forage legumes in cropping systems can be an effective approach to weed control (Liebman and Dyck, 1993; Entz, Bullied, and Katepa-Mupondwa, 1995). Mechanisms of weed control vary depending on the cropping system. Generally in crop rotations, crop diversification prevents the development of populations of specific weeds adapted to cultural practices and resource competitiveness associated with a monoculture of a single crop or row crops (Liebman and Dyck, 1993). Also, use of crop specific herbicide and tillage practices results in the proliferation of specific weed species associated with continuous cropping. Legumes like alfalfa that are maintained in stands for several years and that are periodically harvested are especially effective in weed control since weeds can be removed before seed production, and because development of new weed populations from seed is restricted by competition with existing plants. Recently, Ominski, Entz, and Kentel (1998) and Clay and Aguilar (1998) described the reductions in populations of specific weeds through the use of alfalfa in rotation. While populations of certain weeds in subsequent crops may continue to decrease with increasing numbers of years of legumes in rotation (Nag, 1967); populations of other weeds adapted to the legume system may actually increase. For example, in the northern USA and Canada, dandelion (*Taraxacum officinala* Weber) and quack grass (*Elytrigia repens* (L.) Nevski) populations are frequently associated with declines in alfalfa populations with time (Sheaffer and Wyse, 1982; Triplett, Van Keuren, and Walker, 1977).

Intercropping of small seeded legumes with another crop can also lead to greater weed suppression than suppress due to either crop grown alone. For example, legumes used in "smother crop" (DeHaan et al., 1994) or "living mulch" (Zemenchik et al., 2000) systems are grown between rows of an annual grain crop and suppress weeds by competing for water, light, and nutrients. Also, the legume smother crop may provide a habitat conducive to weed seed degradation by insects and microorganisms (Buhler, Liebman, and Obrycki, 2000).

Herbage residues from legume cover crops can also suppress weed emergence and growth (Fisk et al., 2001). For example, desiccated residual of a hairy vetch cover crop reduced velvetleaf (*Abutilon theophrasti* Medik) and green foxtail (*Setaria viridis* (L.) Beauv.) establishment, and severely inhibited common lambsquarter (*Chenopodium album* L.) establishment (Teasdale, 1993). Light extinction, physical impedance, and alteration of the red to far-red ratio of the light and diurnal soil temperature variation all played a role in weed suppression by the hairy vetch residue (Teasdale and Daughtry, 1993; Teasdale and Mohler, 2000). Most research has shown a positive relationship between suppression of weed populations and cover crop mulch biomass (Ross et al., 2001; Teasdale and Mohler, 2000).

Chemical constituents of legume herbage have suppressed germination and growth of some weed species in laboratory tests (Bradow and Connick, 1990; Miller, 1996). For example, White, Worsham, and Blum (1989) reported that water extracts of hairy vetch and crimson clover herbage were allelopathic to morning glory (*Ipomoea lacunosa* L.) and wild mustard (*Brassica kaber* [D.C.] L.C. Wheeler). Although allelopathy by legumes may play a role in weed suppression (Weston, 1996), field trials have failed to validate that it is a major factor in weed control (Teasdale and Mohler, 2000).

Differences in temporal availability of N between legume and synthetic N sources could also influence early season weed growth and competitiveness (Liebman and Dyck, 1993). Since N from decomposing legume herbage is usually released more slowly than N from synthetic N fertilizers, some weeds would be deprived of an early competitive advantage. Dyck, Liebman, and Erich (1995) confirmed this hypothesis using crimson clover as the first crop in a double cropping sequence with sweet corn. However, the effectiveness of this weed control strategy is greatly influenced by cultural practices that influence the timing of legume residue decomposition such as method and time of incorporation (Dyck and Liebman, 1995).

UTILIZATION OF LEGUMES IN CROPPING SYSTEMS

History

There is a long history of legume use in cropping systems. In addition to providing forage for livestock, legumes were promoted to improve land that

TABLE 4. Total seeding year dry matter and N yields of four legumes subject to two harvest regimes and the effect of seeding year treatments on yield of the following Sudan grass crop. (Adapted from Groya and Sheaffer, 1985.)

| Legume | Harvest treatments[a] | Legume DM yield[b] | Legume N Yield | | Sudan grass DM yield |
			Total	Incorporated	
		Mg ha^{-1}	——— kg ha^{-1} ———		Mg ha^{-1}
Non-dormant	H_3	6.8	303	82	6.6
alfalfa	H_0	4.1	183	183	8.0
Dormant alfalfa	H_3	6.5	330	124	6.4
	H_0	5.3	207	207	8.3
Red clover	H_3	5.4	360	162	7.4
	H_0	6.2	224	224	7.3
Sweet clover	H_3	6.2	305	65	7.2
	H_0	6.1	348	348	8.3

[a]H_3, Harvest with herbage removed except fall regrowth; fall herbage regrowth, crowns and roots incorporated; H_0, No harvests, accumulated herbage, crowns and roots incorporated.
[b]DM yield represents potentially harvested forage from July-September in H_3, and in October for H_0.

yield of a subsequent corn crop by 11% compared to continuous corn. Dury et al. (1999) also showed that red clover intercropped with wheat increased yield of a subsequent no-tilled corn crop by 13%. They concluded that the low C:N ratio of red clover herbage resulted in a decrease in surface residue, increased soil temperature, and enhanced plant emergence compared to no-tillage without red clover. Likewise, Vyn et al. (1999) reported that red clover seeded after wheat and barley produced soil NO_3-N levels sufficient for growth of a subsequent corn crop.

Non-dormant alfalfas are another strategy for integrating legumes into northern US systems. The nondormant alfalfa variety 'Nitro' was specifically developed for use as an annual alfalfa in cropping systems to provide both forage and N for subsequent crops (Sheaffer, Barnes, and Heichel, 1989). Following one or more harvest by early September, 'Nitro' provided greater fall growth and N production than dormant alfalfas that normally overwinter in northern US and Canada (Hesterman et al., 1986). When 'Nitro' was seeded in spring and harvested three times by early September, October herbage regrowth, roots, and crowns contained 105 kg ha^{-1} of BNF-N compared to 94 and 63 kg ha^{-1} for other nondormant and dormant varieties.

True annuals legumes that die following one-growing season have also been evaluated. Annual medics (e.g., burr, *Medicago polymorpha* L.; snail, *M. scutellata* (L.) Mill.; barrel, *M. truncatula* Gaertn.), and berseem clover (*Trifolium alexandrium* L.) are generally prostrate winter annuals in Mediterranean climates. They have been recently evaluated as summer annual forages in the midwest US. In contrast to alfalfa and red clover most annual medics flower profusely in response to long daylength in June and have little regrowth following harvest when planted in the spring (Shrestha et al., 1998; Zhu, Sheaffer, and Barnes, 1996). In Minnesota and Michigan, spring seeded annual medic monocultures had summer forage yields ranging from 2.5 to 4.5 Mg ha^{-1} (Shrestha et al., 1998; Zhu, Sheaffer, and Barnes, 1996). Zhu et al. (1998) reported that spring seeded annual medic monoculture produced from 100 to 200 kg ha^{-1} of N at a July harvest. In a barley (*Hordeum vulgare* L.) grain production system, interseeded barrel medic produced 116 kg N ha^{-1} and 3.5 Mg ha^{-1} of biomass in the fall for ground cover (Moynihan, Simmons, and Sheaffer, 1996).

Berseem clover is an erect, non-bloating legume that like medics is used as a winter annual in Mediterranean climates. In the US it has been used as a spring seeded emergency forage and a green manure crop. In Michigan, a spring seeded berseem clover monoculture had similar forage yield and quality as Nitro alfalfa (~4.0 Mg ha^{-1}) at two summer harvests (Shrestha et al., 1998). The yield of berseem clover also exceeded that of barrel, burr, and snail medic. Sheaffer, Simmons, and Schmitt (2001) reported that when spring seeded with oat, berseem clover had fall DM and N production of 4.4 Mg ha^{-1} and 125 kg ha^{-1} that were each about 200% greater than for barrel medics. The legumes had similar effect on corn grain yield increasing yields from 9% on high fertility soil to 82% on low fertility soil when no N was applied. Ghaffarzadeh (1997), in Iowa, reported that berseem clover regrowth following oat harvest produced 2.7 Mg ha^{-1} of forage and 53 kg ha^{-1} of N. Shrestha et al. (1999) compared berseem clover and barrel medic as unharvested green manure crops for winter canola (*Brassica napus* L.) production. Ninety days following planting, berseem clover and barrel medic produced 104 and 85 kg ha^{-1} of N; however, yields of a subsequent winter canola crop were not consistently improved by incorporation of the legumes. In Montana, when berseem cover was managed for production of N for fall plowdown with a single forage harvest in July, it produced up to 200 kg N ha^{-1} while alfalfa produced only 87 kg N ha^{-1} (Westcott et al., 1995).

Many other legumes have been evaluated as annuals. For example, Stute and Posner (1993) in Wisconsin, reported that a diversity of legumes grown with spring-seeded oat provided 0.5 to 2.9 Mg ha^{-1} of herbage for fall incorporation. Hairy vetch produced the greatest fall yields, but medium red clover or Ladino clover were better suited to the system because hairy vetch climbed the oat canopy and caused lodging and grain yield loss.

Although use of forage legumes in annual systems has potential to produce both forage and N for subsequent crops they are not widely utilized except under emergency circumstances because of seed costs and perceived low economic value. In addition, when seeded in small grain systems where the grain is valued, interseeded legumes may reduce grain yields or interfere with harvesting (Stute and Posner, 1993; Moynihan, Simmons, and Sheaffer, 1996).

Intercropping with Corn and Soybean

Intercropping of legumes with corn and soybean is an attractive approach to adding diversity to the midwestern US landscape because of the large area planted to these crops and the general lack of soil cover early in the growing season and during late fall and winter.

Early research on intercropping legumes with corn focused on using corn as a companion crop for alfalfa establishment. Corn provided greater biomass and economic return than small grains and was less prone to lodging; however, alfalfa stands were less uniform than when established with small grains (Tesar, 1957). Also, for successful use of a corn companion crop system, it was recommended to seed corn in 152 cm rows to reduce competition for light and moisture (Tesar, 1957). Triplett (1962) evaluated an alfalfa intercrop as a compromise between traditional rotations and a continuous grain monoculture. He concluded that use of an alfalfa intercrop in continuous corn was impractical because corn yields were lower than in a corn-wheat-alfalfa rotation or when N was applied to continuously cropped corn. When N was applied to intercropped corn, yields were similar to those with no intercrop.

Later, using herbicides for weed control, Scott et al. (1987) evaluated several perennial and annual legumes as well as seeding dates and concluded that interseeding red clover alone or mixed with annual ryegrass when the corn was 15 to 30 cm high provided good fall cover without affecting corn yields (Table 6). Successful herbicide-aided interseedings have also been reported with a diversity of legumes when seeding occurred at the last corn cultivation and moisture was adequate; however, soil moisture deficits and competition with the corn were sometimes noted as causes of legume stand failure (Abdin et al., 1998; Exner and Cruse, 1993; Stute and Posner, 1993).

Reports of intercropping legumes into soybean are less numerous than for corn because soybeans are considered to be less N dependent and are more competitive than corn. Hively and Cox (2001) successfully established Dutch white clover, medium red clover, and alfalfa in soybeans at the last spring cultivation without use of herbicides. All legumes provided greater than 30% ground cover in the fall and did not interfere with soybean harvest or soybean yield. The following spring, the legumes had an average biomass and N yield

TABLE 5. The effect of intercropping of annual legumes with small grains on grain yield, total biomass yield and fall legume yield.

Reference	Location	Legume	Companion Crop	Grain Yield	Total Biomass Yield	Fall Legume Yield	
				kg ha⁻¹	Mg ha⁻¹	DM Mg ha⁻¹	N kg ha⁻¹
Ghaffarzadeh, 1997	IA	Berseem clover	Oat-grain	2186	6.2	2.9	53
Sheaffer, Barnes, and Heichel, 1989	MN	Nitro alfalfa	Herbicide	---	5.8	1.3	85
			None	---	7.8	1.1	---
			Oat-forage	---	5.5	1.1	---
			Oat-grain	2509	1.0	1.1	---
			Winter wheat	---	5.6	0.3	---
Hesterman et al., 1992	MI	Nitro alfalfa	Winter wheat-frost seeded	3004	---	---	48
		Medium red clover	Winter wheat-frost seeded	---	---	---	75
		Nitro alfalfa	Oat	3265	---	---	46
		Medium red clover	Oat	---	---	---	29
Moynihan, Simmons, and Sheaffer, 1996	MN	Barrel medic	Barley	750-3100	---	3.0	75-152

TABLE 6. Examples of annual legumes intercropped with corn.

Reference	Location	Legume Species	Legume Seeding rate	Legume seeding time	Corn Yield (kg ha⁻¹) With legume	Corn Alone No N	Corn Alone N fertilized
Scott et al., 1987	NY	Red clover	---	Corn 15-30 cm	5211[a]	4133	56935
		Alfalfa	---	Corn 15-30 cm	5628		
		Sweet clover	---	Corn 15-30 cm	5518		
		Hairy vetch	---	Corn 15-30 cm	5564		
Jeranyama, Hesterman, and Sheaffer, 1998	MI	Burr medic	6.5 kg ha⁻¹	At corn planting	7828[b]	4500	10000
		Burr medic	6.5 kg ha⁻¹	14 DAP[c]	8652[b]	4500	10000
		Burr medic	6.5 kg ha⁻¹	28 DAP	9064	4500	10000
Buhler, Kohler, and Foster, 1998	IA	Berseem clover	200 seeds m⁻²	At corn planting	8460	8500[a]	---
		Barrel medic	200 seeds m⁻²	At corn planting	7480	8500	
DeHaan, Sheaffer, and Barnes 1997	MN	Burr medics	85 seeds m⁻²	At corn planting	10800	11000[a]	
		Burr medics	260 seeds m⁻²	At corn planting	9000		
		Burr medics	775 seeds m⁻²	At corn planting	8750		
		Burr medics	85 seeds m⁻²	At corn planting	9750	8400	
		Burr medics	260 seeds m⁻²	At corn planting	8400		
		Burr medics	775 seeds m 2	At corn planting	7500		
Abdin et al., 1998	QC	Hairy vetch	30 kg ha⁻¹	10 DAP (11 cm) or 20 DAP (30 cm)	8700[a]	8700[a]	---
		Sub clover	12 kg ha⁻¹		8900		---
		Sweet clover	7 kg ha⁻¹		7700		---
		Black medic	15 kg ha⁻¹		8500		---
		Crimson clover	22 kg ha⁻¹		7000		---
		Alfalfa (Nitro)	12 kg ha⁻¹		7900		---
		Berseem clover	20 kg ha⁻¹		7800		---

[a]Herbicide applied to corn and legumes for weed control.
[b]Weeds not controlled, exception corn alone.
[c]DAP, days after planting.

of 0.7 Mg ha^{-1} and 25 kg ha^{-1}, respectively. Furthermore, corn yield following legumes was increased about 25% compared to a no-cover treatment.

Intercropped legumes can act as smother crops for weeds thereby reducing herbicide use (DeHaan, Sheaffer, and Barnes, 1997). Smother crops are specialized cover crops that suppress weeds without reducing main crop yields (DeHaan et al., 1994). Ideally, smother crops should provide short-term competition for the first 4 weeks of the growing season. DeHaan, Sheaffer, and Barnes (1997) evaluated short-lived (60-90 days), spring-seeded annual medics as smother crops in corn. When planted simultaneously with corn without herbicides, they reduced weed biomass by an average of 55%. Buhler, Kohler, and Foster (1998) also planted medics and berseem clover simultaneously with corn in a 25 cm band centred over the crop row. Sava snail medic and berseem clover reduced weed biomass by about 60%. In soybean, weed suppression was only 15 to 41% with little difference among legume species. Legumes were more effective in weed suppression in corn than in soybean because of better establishment.

When used as smother crops with corn or soybeans, legumes can successfully control weeds but often compete with the grain crop for resources (DeHaan, Sheaffer, and Barnes, 1997), this competition often reduces corn yields, but the extent of competition can be reduced by varying the legume seeding rate and crop row spacing. As noted above, delaying the legume seeding date until the last cultivation of a row crop has been shown to be an effective approach to establish legumes with little effect on row crop yield. Likewise, Jeranyama, Hesterman, and Sheaffer (1998) found that delaying planting of annual medics until 28 days after corn planting could minimize the competition with a corn crop. They also reported a corn grain yield increase of 17% relative to non-N fertilized corn without medics.

Winter Annual Cover Crops

Legumes can be incorporated as cover crops into cropping systems following harvest of summer annuals like corn and soybean or can also be seeded by surface broadcasting into a summer crop just before leaf drop (Frye et al., 1988). The following spring the cover crop can be either plowed or chemically killed to form mulch for conservation tillage. Plowing the legume enhances N recovery by the subsequent crop, but use of dead residue as surface mulch for conservation enhances soil and water conservation. For high residue situations, spring seeding of the subsequent crop is usually delayed by 1 to 2 weeks and special planting equipment is often required. For example, Decker et al. (1994) reported that vertical cutting coulters mound to the tractor toolbar were required to cut through the tangled growth of hairy vetch.

For success of a winter annual cover system, soil moisture must be adequate for legume seed germination and seedling growth and adequate heat units must be available prior to the start of winter. The potential of winter cover crops is maximized in the southern US because of the mild fall and winter conditions that promote legume growth and warm springs that allow for rapid germination and growth or subsequent grain crops. In more Northern latitudes, successful use of winter annual cover crops is less consistent due to a lack of heat units in the fall and severe winters that often kill or injure most fall planted legumes. In addition, because mulches lower soil temperatures in the spring, growth of the succeeding primary crop is often delayed.

Trials have been conducted in diverse regions of the US to evaluate alternative cover crops. Frye et al. (1988) in a comprehensive review of the use of annual legume cover crops indicated that crimson clover and hairy vetch appeared to be the best-adapted legumes for the southeastern US, whereas, in northern areas hairy vetch, which has greater winterhardiness than crimson clover, is more productive and more widely used. In Maryland, Decker et al. (1994) reported that at a warmer coastal plain location crimson clover consistently had greater DM yield than hairy vetch (6.5 versus 5.1 Mg ha^{-1}, respectively), but yields were similar for the two legumes at the colder Piedmont location (average of 3.3 Mg ha^{-1}). However, hairy vetch consistently had greater herbage N content compared to crimson clover (average of 142 and 107 kg ha^{-1}, respectively) due to a greater herbage N concentration. Because of lower temperatures in fall and winter, the performance of annual cover crops is reduced in the North. In New York, Scott et al. (1987) found that hairy vetch seeded in corn at mid-silk and after silage harvest had much smaller spring DM (920 and 93 kg ha^{-1}, respectively) and N yields (32 and 8 kg ha^{-1}) than in Maryland.

Perennial Living Mulch Systems

Living mulch systems involve planting grain crops into a living but suppressed legume sod that can recover following harvest of the grain crop. Legume living mulches can provide ground cover for erosion control and suppress weeds during crop growth (Worsham and White, 1987). Fall and winter ground cover provided by regenerated legumes is especially valuable for reduction of erosion normally associated with row crops and for valuable livestock feed. While living mulches can involve either annual or perennial legumes, the use of perennial legumes is most exciting because it provides for a permanent ground cover without the necessity of annual reseeding.

Competition for resources between a grain crop and a perennial legume living mulch is more severe than with annual mulches because perennial legumes have well-developed crowns and permanent root systems. As a result, crop yields in living mulch systems are often less than in conventional systems where the legume is killed followed by tillage or herbicides (Table 7). Schultz, Erickson, and Bronson (1987) reported that with more than 50% of the ground cover of alfalfa, red clover, birdsfoot trefoil, and crown vetch retained, corn silage yield was reduced by 25% compared to a no-legume control. Martin, Greyson, and Gordon (1999) reported that corn growing in white clover-grass living mulch had delayed corn emergence and development and increased frost sensitivity compared to conventional tillage/herbicide control. Also, in Wisconsin, with a Kura clover-corn system, there was a high risk for loss of corn population and yield in cool springs because Kura clover, a C_3 species, was able to recover from suppression and obtain a competitive advantage over corn, a C_4 species (Zemenchik et al., 2000). Eberlein, Sheaffer, and Oliveira

TABLE 7. Perennial legume living mulches used for corn production.

Reference	Living mulch	Legume treatment	Corn yield kg ha^{-1}	Legume yield kg ha^{-1}
Eberlein, Sheaffer, and Oliveira, 1992	Alfalfa (non-irrigated)	Killed	7025	---[a]
		Partial suppression (broadcast)	5582	---
		Partial suppression (band)	3768	---
	Alfalfa (irrigated)	Killed	7338	---
		Partial suppression (broadcast)	7338	---
		Partial suppression (band)	7087	---
Zemenchik et al., 2000	Kura clover	Killed	10600	170[a,b]
		Band-killed	9200	290
		Suppressed	7900	400
Fischer and Burill, 1993	White clover	Killed	31000[f]	0
		Suppressed (76 cm corn)[e]	23000	1811[d]
		Suppressed (38 cm corn)	42000	711
Martin, Greyson, and Gordon, 1999	White clover-grass	Killed	10972	0
		Till-30 cm band	8376	2197[b,c]
		Herbicide-30 cm band	6208	2699
		Till/herbicide	7425	1819
Mayer and Hartwig, 1986	Corn stalks	Disked-no till	4955	---
	Crown vetch	No till	6774	---
	Birdsfoot trefoil	No till	6147	---
	Red clover	No till	6711	---

[a]Data not available.
[b]Between row legume biomass yield.
[c]Fall sampling.
[d]64 Days after planting.
[e]76 and 38 cm corn row spacings were used.
[f]Sweet corn yield.

(1992) reported that corn grain yields were similar for a killed and living mulch of alfalfa when soil moisture was adequate; however, with no irrigation, corn grain yields were reduced by 50% by a partial suppression treatment.

The challenge with living mulch systems is to suppress the living plants so as to avoid competition during germination and early season growth of the interseeded crop (e.g., corn) while allowing for sufficient sod recovery to provide soil surface cover beneath the crop canopy. However, this balance is difficult to achieve. Therefore, living mulch systems require careful season-long monitoring. Most successful living mulch systems depend on specialized no-till and spraying equipment and strategic use of herbicides and/or tillage because perennial legumes often have greater resistance to herbicides than annuals (Hartwig, 1987; Fischer and Burrill, 1993). Nitrogen fertilization is usually beneficial to yield of grain crops because suppressed legumes compete for soil N (Cardina, Hartwig, and Lukezic, 1986; Mayer and Hartwig, 1986; Martin, Greyson, and Gordon, 1999). Schultz, Erickson, and Bronson (1987) recommended that an effective perennial living mulch system should involve use of residual soil-active herbicides, aggressive control of annual and perennial weeds, 30-cm wide legume kill bands, and application of a starter N fertilizer since suppressed legumes appear to compete with corn for N. In the successful Kura clover living mulch system used in Wisconsin, glyphosate was applied in 61-cm strips centred on corn with 15-cm of untreated Kura clover between rows (Zemenchik et al., 2000).

CONCLUSIONS

There are many benefits of using legumes in cropping systems. Legumes can be managed to provide forage with variable levels of nutrients and fiber for livestock and wildlife. Alone or mixed with grasses, legumes are sources of biologically fixed N and sustain the soil by reducing erosion and increasing OM levels. Additional, non-N rotation effects include reduced weed, insects, and disease incidence. Potential new uses of legumes include phytoremediation of N contaminated sites and capturing N lost from cropping systems. Legumes also have potential use as a feedstock for renewable energy production.

Legumes have traditionally been used in rotation with grain crops but more recently have been shown promise as winter cover crops, intercrops with grain crops, and as living mulches. Agroecosystems that provide food, fiber, and habitat for humans must be sustainable to insure future productivity. From an agronomic standpoint, a sustainable system must maintain and possibly improve environmental quality and provide a suitable economic return so as to achieve a viable enterprise. In addition, there are other important issues like quality of life and rural community viability that are greatly influenced by economic and environmental quality issues. Unfortunately, many current cropping systems fail to meet economic and environmental criteria for sustainability. The widely practiced corn-soybean rotation has been adopted by producers for economic reasons. However, economic returns for that system are artificially enhanced by commodity price supports and the system has been indicted as a major contributor to environmental degradation.

Traditional use of legumes in rotation has declined and there is a need for development of effective approaches for integration of legumes into current cropping systems. Examples of potential strategies for inclusion of legumes include use as intercrops with grain crops, as smother crops, and as living mulches. Unfortunately, decisions on selection of cropping systems are often made based solely on economic conditions that in the US are greatly influenced by government programs. Modification of US government agricultural support programs would play an important role in modifying cropping systems so as to provide a long-term sustainable agriculture.

REFERENCES

Abdin, O., B.E. Coulman, D. Cloutier, M.A. Faris, X. Zhou, and D.L. Smith. (1998). Yield and yield components of corn interseeded with cover crops. *Agronomy Journal* 90:63-68.

Angler, D.A. (1992). Changes in soil aggregation and organic carbon under corn and alfalfa. *Soil Science Society of America Journal* 56:1244-1249.

Baldock, J.O., R.L. Higgs, W.H. Paulson, J.A. Jackobs, and W.D. Shrader. (1981). Legume and mineral N effects on crop yields in several crop sequences in the Upper Mississippi Valley. *Agronomy Journal* 73:885-890.

Baldock, J.O. and R.B. Musgrave. (1980). Manure and mineral fertilizer effects in continuous and rotational crop sequences in Central New York. *Agronomy Journal* 72:511-518.

Blackshaw, R.E., J.R. Moyer, R.C. Doram, A.L. Boswall, and E.G. Smith. (2001). Suitability of undersown sweetclover as a fallow replacement in semiarid cropping systems. *Agronomy Journal* 93:863-868.

Boller, B.C. and J. Nosberger. (1987). Symbiotically fixed nitrogen from field-grown white and red clover mixed with ryegrasses at low levels of ^{15}N-fertilization. *Plant and Soil* 104:219-226.

Bradow, J.M. and W.J. Connick. (1990). Volatile seed germination inhibitors from plant residues. *Journal Chemical Ecology* 16:645-666.

Brophy, L.S., G.H. Heichel, and M.P. Russelle. (1987). Nitrogen transfer from forage legume to grass in a systematic planting design. *Crop Science* 27:753-758.

Buhler, D.D., K.A. Kohler, and M.S. Foster. (1998). Spring-seeded smother plants for weed control in corn and soybean. *Journal of Soil and Water Conservation* 53:272-275.

Buhler, D.D., M. Liebman, and J.J. Obrycki. (2000). Theoretical and practical challenges to an IPM approach to weed management. *Weed Science* 48:274-280.

Buxton, D.R. and D.R. Mertens. (1995). Quality-related characteristics of forages. In *Forages, the Science of Grassland Agriculture*, Vol II, ed., R.F. Barnes, D.A. Miller, and C.J. Nelson, Ames, IA: Iowa State University Press, pp. 83-96.

Cardina, J., N.L. Hartwig, and F.L. Lukezic. (1986). Herbicidal effects on crownvetch rhizobia and nodule activity. *Weed Science* 34:338-343.

Clay, S.A. and I. Aguilar. (1998). Weed seedbanks and corn grown following continuous corn or alfalfa. *Agronomy Journal* 90:813-818.

Cochrane, W.W. (1993). *The Development of American Agriculture*, 2nd ed., Minneapolis, MN: University of Minnesota Press.

Crookston, R.K. (1984). The rotation effect: what causes it to boost yields? *Crops and Soils* 36:12-14.

Decker A.M., A.J. Clark, J.J. Meisinger, F.R. Mulford, and M.S. McIntosh. (1994). Legume cover crop contributions to no-tillage corn production. *Agronomy Journal* 86:126-135.

DeHaan, R.L., C.C. Sheaffer, and D.K. Barnes. (1997). Effect of annual medic smother plants on weed control and yield in corn. *Agronomy Journal* 89:813-821.

DeHaan, R.L., D.L. Wyse, N.J. Ehlke, B.D. Maxwell, and D.H. Putman. (1994). Simulation of spring-seeded smother plants for weed control in corn (*Zea mays*). *Weed Science* 42:35-43.

DeLong, M.M., D.R. Swanberg, E.A. Oelke, C. Hanson, M. Onischak, M.R. Schmid, and B.C. Wiant. (1995). Sustainable biomass energy production and rural economic development using alfalfa as a feedstock. In *Second Biomass Conference of the Americas*, ed., D.L. Klass, Golden, CO: National Renewable Energy Laboratory, pp. 1582-1591.

De Serres, O. (1600). *Le Théâtre d'Agriculture*. Paris, France: Mettayer (in French).

Dubach, M. and M.P. Russelle. (1994). Forage legume roots and nodules and their role in nitrogen transfer. *Agronomy Journal* 86:259-266.

Duke, J.A. (1981). *Handbook of Legumes of World Economic Importance*. New York, NY: Plenum.

Dury, C.F., C.S. Tan, T.W. Wlacky, T. Oloya, A.S. Hamill, and S.E. Weaver. (1999). Red clover and tillage influence on soil temperature, water content, and corn emergence. *Agronomy Journal* 91:101-108.

Dyck, E. and M. Liebman. (1995). Crop-weed interference as influenced by a leguminous or synthetic fertilizer nitrogen source: II. Rotation experiments with crimson clover, field corn, and lambsquarters. *Agriculture, Ecosystems and the Environment* 56:109-120.

Dyck, E., M. Liebman, and M.S. Erich. (1995). Crop-weed interference as influenced by a leguminous or synthetic fertilizer nitrogen source: I. Doublecropping experiments with crimson clover, sweet clover, sweet corn, and lambsquarters. *Agriculture, Ecosystems and the Environment* 56: 93-108.

Eberlein, C.V., C.C. Sheaffer, and V.F. Oliveira. (1992). Corn growth and yield in an alfalfa living mulch system. *Journal of Production Agriculture* 5:332-339.

Entz, M.H., W.J. Bullied, and F. Katepa-Mupondwa. (1995). Rotational benefits of forage crops in Canadian Prairie cropping systems. *Journal of Production Agriculture* 8:521-529.

Exner, D.N. and R.M. Cruse. (1993). Interseeded forage legume potential as winter ground cover, nitrogen source, and competitor. *Journal of Production Agriculture* 6:226-231.

Farnham, D.E. and J.R. George. (1994). Dinitrogen fixation and nitrogen transfer in birdsfoot trefoil-orchardgrass communities. *Agronomy Journal* 86:690-694.

Feber, D. (2001). Keeping the stygian waters at bay. *Science* 291:968-973.

Fischer, A. and L. Burrill. (1993). Managing interference in a sweet corn-white clover living mulch system. *American Journal of Alternative Agriculture* 8:51-56.

Fisk, J.W., O.B. Hesterman, A. Shrestha, J.J. Kells, R.R. Harwood, J.M. Squire, and C.C. Sheaffer. (2001). Weed suppression by annual legume cover crops in no-tillage corn. *Agronomy Journal* 93:319-325.

Frame, J., J.F.L. Charlton, and A.S. Laidlaw. (1998). *Temperate Forage Legumes.* Oxon, UK:CAB International.

Fribourg, H.A. and I.J. Johnson. (1955). Dry matter and nitrogen yields of legume tops and roots in the fall of the seeding year. *Agronomy Journal* 47:73-77.

Frye, W.W., R.L. Blevins, M.S. Smith, S.J. Corak, and J.J. Varco. (1988). Role of annual legume cover crops in efficient use of water and nitrogen. In *Cropping Strategies for Efficient Use of Water and Nitrogen,* ed., W.L. Hargrove, Madison, WI: American Society of Agronomy, pp. 129-154.

Ghaffarzadeh, M. (1997). Economic and biological benefits of intercropping berseem clover with oat in corn-soybean-oat rotations. *Journal of Production Agriculture* 10:314-319.

Graham, P.H. (1992). Stress tolerance in *Rhizobium* and *Bradyrhizobium,* and nodulation under adverse soil conditions. *Canadian Journal of Microbiology* 38:475-484.

Graham, P.H. (1998). Biological dinitrogen fixation: symbiotic. In *Principles and Applications of Soil Microbiology.* ed., D.M. Sylvia, P. Hartel, J. Fuhrmann, and D. Zuberer, Upper Saddle River, NJ: Prentice Hall, pp. 322-345.

Groya, F.L. and C.C. Sheaffer. (1985). Nitrogen from forage legumes: harvest and tillage effects. *Agronomy Journal* 77:105-109.

Hall, J.K., N.L. Hartwig, and L.D. Hoffman. (1994). Cyanazine losses in runoff from no-tillage corn in "living" and dead mulches vs. unmulched conventional tillage. *Journal of Environmental Quality* 13:105-110.

Hamel, C., V. Furlan, and D.L. Smith. (1992). Mycorrhizal effects on interspecific plant competition and nitrogen transfer in legume-grass mixtures. *Crop Science* 32:991-996.

Hargrove, W.L. and W.W. Frye. (1987). The need for legume cover crops in conservation tillage production. In *The Role of Legumes in Conservation Tillage Systems,* ed., J.F. Power, Ankney, IA: Soil Conservation Society of America, pp. 1-4.

Harris, G.H. and O.B. Hesterman. (1990). Quantifying the nitrogen contribution from alfalfa to soil and two succeeding crops using nitrogen-15. *Agronomy Journal* 82:129-134.

Hartwig, N.L. (1987). Cropping practices using crownvetch in conservation tillage. In *The Role of Legumes in Conservation Tillage System,* ed., J.F. Power, Ankney, IA: Soil Conservation Society of America, pp. 109-110.

Haynes, R.J. (1980). Competitive aspects of the grass-legume association. *Advances in Agronomy* 33:227-261.

Heichel, G.H. and D.K. Barnes. (1984). Opportunities for meeting crop nitrogen needs from symbiotic nitrogen fixation. In *Organic Farming: Current Technology and Its Role in a Sustainable Agriculture*, ed., D.F. Bezdicek et al., Madison, WI: American Society of Agronomy. pp. 49-59.

Heichel, G.H. and K.I. Henjum. (1991). Dinitrogen fixation, nitrogen transfer, and productivity of forage legume-grass communities. *Crop Science* 31:202-208.

Hesterman, O.B. (1988). Exploiting forage legumes for nitrogen contribution in cropping systems. In *Cropping Strategies for Efficient Use of Water and Nitrogen*, ed., W.L. Hargrove, Madison, WI: American Society of Agronomy, pp. 155-156.

Hesterman, O.B., T.S. Griffin, P.T. Williams, G.H. Harris, and D.R. Christenson. (1992). Forage legume-small grain intercrops: nitrogen production and response of subsequent corn. *Journal of Production Agriculture* 5:340-348.

Hesterman, O.B., M.P. Russelle, C.C. Sheaffer, and G.H. Heichel. (1987). Nitrogen utilization from fertilizer and legume residues in legume-corn rotations. *Agronomy Journal* 79:726-731.

Hesterman, O.B., C.C. Sheaffer, D.K. Barnes, W.E. Lueschen, and J.H. Ford. (1986). Alfalfa dry matter and nitrogen production, and fertilizer nitrogen response in legume-corn rotations. *Agronomy Journal* 78:19-23.

Hively, W.D. and W.J. Cox. (2001). Interseeding cover crops into soybean and subsequent corn yields. *Agronomy Journal* 93:308-313.

Huggins, D.R., G.W. Randall, and M.P. Russelle. (2001). Subsurface drain losses of water and nitrate following conversion of perennials to row crops. *Agronomy Journal* 93:477-486.

Jeranyama, P., O.B. Hesterman, and C.C. Sheaffer. (1998). Medic planting date effect on dry matter and nitrogen accumulation when clear-seeded or intercropped with corn. *Agronomy Journal* 90:616-622.

Kelner, D.J., J.K. Vessey, and M.H. Entz. (1997). The nitrogen dynamics of 1-, 2-, and 3-year stands of alfalfa in a cropping system. *Agriculture, Ecosystems and the Environment* 64:1-10.

Lamb, J.F.S., D.K. Barnes, M.P. Russelle, C.P. Vance, G.H. Heichel, and K.I. Henjum. (1995). Ineffectively and effectively nodulated alfalfas demonstrate biological nitrogen fixation continues with high nitrogen fertilization. *Crop Science* 35:153-157.

Leath, K.T. (1988). Diseases and forage stand persistence in the United States. In *Persistence of Forage Legume*, ed., G.C. Marten, A.G. Matches, R.F. Barnes, R.W. Brougham, R.J. Clements, and G.W. Sheath, Madison, WI: American Society of Agronomy, Crop Science Society of America, Soil Science Society of America, pp. 465-479.

Ledgard, S.F. (1991). Transfer of fixed nitrogen from white clover to associated grasses using ^{15}N methods in swards grazed by dairy cows. *Plant and Soil* 131:215-223.

Ledgard, S.F. and K.W. Steele. (1992). Biological nitrogen fixation in mixed legume/grass pastures. *Plant and Soil* 141:137-153.

Liebman, M. and E. Dyck. (1993). Crop rotation and intercropping strategies for weed management. *Ecological Applications* 3:92-122.

Martin, N.P. (1996). Forages. In *Quality of U.S. Agricultural Products*. Task Force Report No. 126. Ames, IA: Council for Agricultural Science and Technology, pp. 184-194.

Martin, R.C., P.R. Greyson, and R. Gordon. (1999). Competition between corn and a living mulch. *Canadian Journal of Plant Science* 79:579-596.

Mathers, A.C., B.A. Steward, and B. Blair. (1975). Nitrate-nitrogen removal from soil profiles by alfalfa. *Journal of Environmental Quality* 4:403-405.

Mayer, J.B. and N.L. Hartwig. (1986). Corn yield in crownvetch relative to dead mulches. *Proceedings of the Annual Meeting of the Northeast Weed Science Society* 40:34-35.

McVay, K.A., D.E. Radcliffe, and W.L. Hargrove. (1989). Winter legume effects on soil properties and nitrogen fertilizer requirements. *Soil Science Society of America Journal* 53:1856-1862.

Miller, A.C. (1942). Jefferson as an agriculturalist. *Agricultural History* 16:65-78.

Miller, D.A. (1996). Allelopathy in forage crop systems. *Agronomy Journal* 88:854-859.

Mowat, D.N., R.S. Fulkerson, W.E. Tossell, and J.E. Winch. (1965). The *in vitro* digestibility and protein content of leaf and stem portions of forages. *Canadian Journal of Plant Science* 45:321-331.

Moynihan, J.M., S.R. Simmons, and C.C. Sheaffer. (1996). Intercropping annual medic with conventional height and semidwarf barley grown for grain. *Agronomy Journal* 88:823-828.

Nag, K.C. (1967). Effect of various cropping systems in a crop rotation study on crop performance and weed content. *Dissertation Abstracts* 28:746B.

Ominski, P.E., M.H. Entz, and N. Kenkel. (1998). Weed suppression by *Medicago sativa* in subsequent cereal crops: a comparative survey. *Weed Science* 47:282-290.

Paynel, F., P.J. Murray, and J.B. Cliquet. (2001). Root exudates: a pathway for short-term N transfer from clover and ryegrass. *Plant and Soil* 229:235-243.

Perfect, E., B.D. Kay, W.K.P. van Loom, R.W. Sheard, and T. Pojasok. (1990). Rates of change in soil structural stability under forages and corn. *Soil Science Society of America Journal* 54:179-186.

Peterson, T.A. and M.P. Russelle. (1991). Alfalfa and the nitrogen cycle in the Corn Belt. *Journal of Soil and Water Conservation* 46:229-235.

Randall, G.W. (2001). Our corn-soybean system fails the sustainability test on all fronts. *The Land Stewardship Letter* 19:2-3.

Randall, G.W., D.R. Huggins, M.P. Russelle, D.J. Fuchs, W.W. Nelson, and J.L. Anderson. (1997). Nitrate losses through subsurface tile drainage in CRP, alfalfa, and row crops systems. *Journal of Environmental Quality* 26:1240-1247.

Rasse, D.P., A.J.M. Smucker, and D. Santos. (2000). Alfalfa root and shoot mulching effects on soil hydraulic properties and aggregation. *Soil Science Society of America Journal* 64:725-731.

Rogers, H.H. (1976). Forages legumes (with particular reference to lucerne and red clover). In *Report of the Plant Breeding Institute for 1975*. Cambridge, UK: Plant Breeding Institute. pp. 22-57.

Ross, S.M., J.R. King, R.C. Izaurralde, and J.T. O'Donavan. (2001). Weed suppression by seven clover species. *Agronomy Journal* 93:820-827.

Russelle, M.P., J.F.S. Lamb, B.R. Montgomery, D.W. Elsenheimer, B.S. Miller, and C.P. Vance. (2001). Alfalfa rapidly remediates excess inorganic nitrogen at a fertilizer spill site. *Journal of Environmental Quality* 30:30-36.

Schultz, M.A., A.E. Erickson, and J.A. Bronson. (1987). Intercropping corn and forage legumes in Michigan. In *The Role of Legumes in Conservation Tillage Systems*, ed., J.F. Power, Ankeny, IA: The Soil Conservation Society of America, pp. 91-92.

Scott, T.W., J. Mt. Plesant, R.R. Burt, and D.J. Otis. (1987). Contributions of ground cover, dry matter, and nitrogen from intercrops and cover crops in a corn polyculture system. *Agronomy Journal* 79:792-798.

Seguin, P., M.P. Russelle, C.C. Sheaffer, N.J. Ehlke, and P.H. Graham. (2000). Dinitrogen fixation in Kura clover and birdsfoot trefoil. *Agronomy Journal* 92:1216-1220.

Seguin, P., C.C. Sheaffer, N.J. Ehlke, P.H. Graham, and M.P. Russelle. (2001). Nitrogen fertilization and rhizobial inoculation effects on Kura clover growth. *Agronomy Journal* 93:1262-1268.

Sheaffer, C.C. and D.K. Barnes. (1987). Forage crops. In *CRC Handbook of Plant Science in Agriculture*. Volume II, Boca Raton, FL: CRC Press, pp. 217-249.

Sheaffer, C.C. and D.L. Wyse. (1982). Common dandelion (*Taraxacum officinale*) control in alfalfa (*Medicago sativa*). *Weed Science* 30:216-220.

Sheaffer, C.C., D.K. Barnes, and G.H. Heichel. (1989). *Annual Alfalfa in Crop Rotation*. Bulletin 588-1989. St. Paul, MN: Minnesota Agriculture Experiment Station.

Sheaffer C.C., N.P. Martin, J.F.S. Lamb, G.J. Cuomo, J.G. Jewett, and S.R. Quiry. (2000). Leaf and stem properties of alfalfa entries. *Agronomy Journal* 92:733-739.

Sheaffer, C.C., R.D. Mathison, N.P. Martin, D.L. Rabas, H.J. Ford, and D.R. Swanson. (1993). *Forage Legumes: Clovers, Birdsfoot Trefoil, Cicer Milkvetch, Crownvetch, Sainfoin and Alfalfa*. Station Bulletin 597-1993. St. Paul, MN: Minnesota Agricultural Experiment Station.

Sheaffer, C.C., S.R. Simmons, and M.A. Schmitt. (2001). Annual medic and berseem clover dry matter and nitrogen production in rotation with corn. *Agronomy Journal* 93:1080-1086.

Shrestha, A., O.B. Hesterman, L.O. Copeland, J.M. Squire, J.W. Fisk, and C.C. Sheaffer. (1999). Annual legumes as green manure and forage crops in winter canola (*Brassica napus* L.) rotations. *Canadian Journal of Plant Science* 79:19-25.

Shrestha, A., O.B. Hesterman, J.M. Squire, J.W. Fisk, and C.C. Sheaffer. (1998). Annual medics and berseem clover as emergency forages. *Agronomy Journal* 90:197-201.

Simpson, B.B. and M.C. Ogorzaly. (1995). *Economic Botany Plants in Our World*, 2nd ed., New York, NY: McGraw-Hill Inc.

Singer, J.W. and W.J. Cox. (1998). Agronomics of corn production under different crop rotations. *Journal of Production Agriculture* 11:462-468.

Smith, S.R., Jr., J.H. Bouton, A. Singh, and W.P. McCaughey. (2000). Development and evaluation of grazing-tolerant alfalfa cultivars: A review. *Canadian Journal of Plant Science* 80:503-512.

Spears, J.W. (1994). Minerals in forages. In *Forage Quality, Evaluation, and Utilization*, ed., G.C. Fahey, Jr., Madison, WI: American Society of Agronomy, pp. 281-317.

Storer, F.H. (1910). *Agriculture in Some of Its Relations with Chemistry*, Vol III, 7th edition. New York, NY: Charles Scribner's Sons.

Streeter, J. (1988). Inhibition of legume nodule formation and N_2 fixation by nitrate. *CRC Critical Reviews in Plant Science* 7:1-23.

Stute, J.K. and J.L. Posner. (1993). Legume cover crop options for grain rotations in Wisconsin. *Agronomy Journal* 85:1128-1132.

Stute, J.K. and J.L. Posner. (1995). Synchrony between legume nitrogen release and corn demand in the upper Midwest. *Agronomy Journal* 87:1063-1095.

Ta, T.C. and M.A. Faris. (1987). Effects of alfalfa proportions and clipping frequencies on timothy-alfalfa mixtures. II. Nitrogen fixation and transfer. *Agronomy Journal* 79:820-824.

Taylor, N.L. and K.H. Quesenberry. (1996). *Red Clover Science*. Dordrecht, The Netherlands: Kluwer Academic Publishers.

Teasdale, J.R. (1993). Interaction of light, soil moisture, and temperature with weed suppression by hairy vetch residue. *Weed Science* 41:46-51.

Teasdale, J.R. and C.S.T. Daughtry. (1993). Weed suppression by live and desiccated hairy vetch (*Vica villosa*). *Weed Science* 41:207-212.

Teasdale, J.R. and C.L. Mohler. (2000). The quantitative relationship between weed emergence and physical properties of mulches. *Weed Science* 48:385-392.

Tesar, M.B. (1957). Establishment of alfalfa in wide-row corn. *Agronomy Journal* 49:63-68.

Thomas, M.L., R. Lal, T. Logan, and N.R. Fausey. (1992). Land use and management effects on nonpoint loading from Miamian soil. *Soil Science Society of America Journal* 56:1871-1875.

Tilman, D. (1999). Global environmental impacts of agricultural expansion: the need for sustainable and efficient practices. *Proceeding of the National Academy of Science* 96:5995-6000.

Triplett, G.B. (1962). Intercrops in corn and soybean cropping systems. *Agronomy Journal* 54:106-109.

Triplett, G.B., R.W. Van Keuren, and J.D. Walker. (1977). Influence of 2,4-D, pronamide and simazine on dry matter production and botanical composition of an alfalfa-grass sward. *Crop Science* 17:61-65.

Undersander, D., N. Martin, D. Cosgrove, K. Kelling, M. Schmitt, J. Wedberg, R. Becker, C. Grau, J. Doll, and M.E. Rice. (2000). *Alfalfa Management Guide*. Madison, WI: American Society of Agronomy, Crop Science Society of America, Soil Science Society of America.

Vance, C.P. (1998). Legume symbiotic nitrogen fixation: agronomic aspects. In *The Rhizobiaceae: Molecular Biology of Model Plant-Associated Bacteria*, eds., H.P. Spaink, A. Kondorosi, and P.J.J. Hooykaas, Dordrecht, The Netherlands: Kluwer Academic Publishers, pp. 509-530.

Vanotti, M.B. and L.G. Bundy. (1995). Soybean effects on soil nitrogen availability in crop rotations. *Agronomy Journal* 87:676-680.

Voss, R.D. and W.D. Schrader. (1984). Rotation effects and legume sources of nitrogen for corn. In *Organic Farming: Current Technology and Its Role in a Sustainable Agriculture*, ed., D.F. Bezdicek, J.F. Power, D.R. Keeney, and M.J. Wright,

Special Publication No. 46, Madison, WI: American Society of Agronomy, Crop Science Society of America, Soil Science Society of America, pp. 61-68.

Vyn, T.J., K.J. Janovicek, M.H. Miller, and E.G. Beauchamp. (1999). Soil nitrate accumulation and corn response to preceding small-grain fertilization and cover crops. *Agronomy Journal* 91:17-24.

Wall, G.J., E.A. Pringle, and R.W. Sheard. (1991). Intercropping red clover with silage corn for soil erosion control. *Canadian Journal of Soil Science* 71:137-145.

Westcott, M.P., L.E. Welty, M.L. Knox, and L.S. Prestbye. (1995). Managing alfalfa and berseem clover for forage and plowdown nitrogen in barley rotations. *Agronomy Journal* 87:1176-1181.

Weston, L.A. (1996). Utilization of allelopathy for weed management in agroecosystems. *Agronomy Journal* 88:860-866.

White, R.H., A.D. Worsham, and U. Blum. (1989). Allelopathic potential of legume debris and aqueous extracts. *Weed Science* 37:674-679.

Wilman, D., E.J. Mtengeti, and G. Moseley. (1996). Physical structure of twelve forage species in relation to rate of intake by sheep. *Journal of Agricultural Science, Cambridge* 126:277-285.

Wischmeier, W.H. and D.D. Smith. (1978). *Predicting Rainfall Erosion Losses: A Guide to Conservation Planning*. Agricultural Handbook No. 537. USDA-SEA. Washington, DC: US Government Printing Office.

Worsham, A.D. and R.H. White. (1987). Legume effects of weed control in conservation tillage. In *The Role of Legumes in Conservation Tillage Systems*, ed., J.F. Power, Ankeny, IA: The Soil Conservation Society of America. pp. 113-118.

Zemenchik, R.A., K.A. Albrecht, C.M. Boerboom, and J.G. Lauer. (2000). Corn production with kura clover as a living mulch. *Agronomy Journal* 92:698-705.

Zemenchik, R.A., N.C. Wollenhaupt, K.S. Albrecht, and A.H. Bosworth. (1996). Runoff, erosion, and forage production form established alfalfa and smooth bromegrass. *Agronomy Journal* 88:461-466.

Zhang, F. and D.L. Smith. (2002). Interorganismal signaling in suboptimum environments: The legume-rhizobia symbiosis. *Advances in Agronomy* 76:125-161.

Zhu, Y., C.C. Sheaffer, and D.K. Barnes. (1996). Forage yield and quality of six annual *Medicago* species in the north central USA. *Agronomy Journal* 88:955-960.

Zhu, Y., C.C. Sheaffer, M.P. Russelle, and C.P. Vance. (1998). Dry matter accumulation and dinitrogen fixation of annual *Medicago* species. *Agronomy Journal* 90: 103-108.

Implications
of Elevated CO_2-Induced Changes
in Agroecosystem Productivity

S. A. Prior

H. A. Torbert

G. B. Runion

H. H. Rogers

SUMMARY. Since CO_2 is a primary input for crop growth, there is interest in how increasing atmospheric CO_2 will affect crop productivity and alter cropping system management. Effects of elevated CO_2 on grain and residue production will be influenced by crop selection. This field study evaluated soybean [C_3; *Glycine max* (L.) Merr.] and grain sorghum [C_4; *Sorghum bicolor* (L.) Moench.] cropping systems managed under conservation tillage practices and two atmospheric CO_2 concentrations

S. A. Prior is Plant Physiologist, H. A. Torbert is Soil Scientist, G. B. Runion is Plant Pathologist, and H. H. Rogers is Plant Physiologist, USDA-Agricultural Research Service, National Soil Dynamics Laboratory, Auburn, AL.

Address correspondence to: S. A. Prior, USDA-Agricultural Research Service, National Soil Dynamics Laboratory, 411 South Donahue Drive, Auburn, AL 36832 USA (E-mail: sprior@acesag.auburn.edu).

The authors are indebted to Barry G. Dorman and Tammy K. Dorman for technical assistance.

Support from Terrestrial Carbon Processes Program of the Environmental Sciences Division, US Department of Energy (Interagency Agreement No. DE-AI05-95ER62088) is gratefully acknowledged.

[Haworth co-indexing entry note]: "Implications of Elevated CO_2-Induced Changes in Agroecosystem Productivity." Prior, S. A. et al. Co-published simultaneously in *Journal of Crop Production* (Food Products Press, an imprint of The Haworth Press, Inc.) Vol. 8, No. 1/2 (#15/16), 2003, pp. 217-244; and: *Cropping Systems: Trends and Advances* (ed: Anil Shrestha) Food Products Press, an imprint of The Haworth Press, Inc., 2003, pp. 217-244. Single or multiple copies of this article are available for a fee from The Haworth Document Delivery Service [1-800-HAWORTH, 9:00 a.m. - 5:00 p.m. (EST). E-mail address: docdelivery@haworthpress.com].

(ambient and twice ambient) for three growing seasons. Elevated CO_2 increased soybean and sorghum yield by 53% and 17% increase, respectively; reductions in whole plant water use were also greater for soybean than sorghum. These findings suggest that increasing CO_2 could improve future food security, especially in soybean production systems. Elevated CO_2 increased aboveground residue production by > 35% for both crops; such shifts could complement conservation management by increasing soil surface cover, thereby reducing soil erosion. However, increased residue could negatively impact crop stand establishment and implement effectiveness during tillage operations. Elevated CO_2 increased total belowground dry weight for both crops; increased root proliferation may alter soil structural characteristics (e.g., due to increased number and extent of root channels) which could lead to increases in porosity, infiltration rates, and subsequent soil water storage. Nitrate leaching was reduced during the growing season (due to increased N capture by high CO_2-grown crops), and also during the fallow period (likely a result of altered decomposition patterns due to increased C:N ratios of the high CO_2-grown material). Enhanced crop growth (both above- and belowground) under elevated CO_2 suggests greater delivery of C to soil, more soil surface residue, and greater percent ground coverage which could reduce soil C losses, increase soil C storage, and help ameliorate the rise in atmospheric CO_2. Results from this study suggests that the biodegradability of crop residues and soil C storage may not only be affected by the environment they were produced in but may also be species dependent. To more fully elucidate the relationships between crop productivity, nutrient cycling, and decomposition of plant materials produced in elevated CO_2 environments, future studies must address species effects (including use of genetically modified crops) and must also consider other factors such as cover crops, crop rotations, soil series, tillage practices, weed management, and regional climatic differences. *[Article copies available for a fee from The Haworth Document Delivery Service: 1-800-HAWORTH. E-mail address: <docdelivery@haworthpress.com> Website: <http://www.HaworthPress.com>]*

KEYWORDS. Global change, carbon dioxide, sorghum, soybean, yield, residue, roots

INTRODUCTION

The global environment is changing with the rise in atmospheric CO_2 concentration (Keeling and Whorf, 1994) and the process is expected to continue into the future (Bolin et al., 1986). This rise can be attributed mainly to fossil fuel burning and land use change associated with industrial and/or population expansion (Houghton, Jenkins, and Ephraums, 1990). The most discussed consequence of the rise in atmospheric CO_2, along with other greenhouse trace gases, is a predicted shift in the Earth's climate. Aside from this debate, vegetation will be directly affected by the increase in atmospheric CO_2, the essential substrate of photosynthesis.

Since CO_2 is a primary input for crop growth, there is interest in how the rise in atmospheric CO_2 concentration will affect highly managed agricultural systems. Research has shown major plant responses, including increased growth and yield, increased water use efficiency (Rogers and Dahlman, 1993; Amthor, 1995), increased photosynthetic capacity (Huber, Rogers, and Israel, 1984; Radin et al., 1987; Bowes, 1991; Lawlor and Mitchell, 1991; Long and Drake, 1992), decreased respiration (Bunce, 1990; Amthor, Koch, and Bloom, 1992; Mousseau, 1993; Wullschleger, Ziska, and Bunce, 1994), and changes in plant structure (Pritchard et al., 1999). Relative to aboveground processes, CO_2 effects on crop root systems have received less attention despite their importance in attaining essential soil resources (i.e., water and nutrients) (Rogers, Runion, and Krupa, 1994). Crops have often shown increases in root dry weight under CO_2-enriched conditions (Chaudhuri et al., 1986; Del Castillo et al., 1989; Chaudhuri, Kirkham, and Kanemasu, 1990; Rogers et al., 1992) and in many cases, the largest proportion of the extra biomass produced as a result of elevated CO_2 is found belowground (Bazzaz, 1990; Rogers, Runion, and Krupa, 1994; Wittwer, 1995; Rogers et al., 1996). Findings suggest that whole plant nutrient uptake and nutrient utilization efficiency are increased while nutrient tissue concentration and nutrient uptake efficiency decline under elevated CO_2 (Rogers, Runion, and Krupa, 1994). However, most CO_2 studies have been conducted with containerized plants (i.e., confined rooting volume) in controlled environments which may obscure responses (above- and belowground) that would occur in the field (Sionit et al., 1984; Arp, 1991; Thomas and Strain, 1991).

Improved predictions on how changes in the global environment will impact agroecosystems will depend on obtaining realistic field data. Consequently, many current efforts focus on in-ground CO_2 studies utilizing open top chambers (OTC) and free-air CO_2 enrichment (FACE) (Allen et al., 1992). Recent field work has shown that elevated CO_2 can increase above- and belowground biomass (Mauney et al., 1994; Prior et al., 1994b; Kimball et al., 1995), alter root morphology (Prior et al., 1995) and the root system's capacity to explore soil volume through shifts in fine root distribution patterns (Prior et al., 1994a, 1994b; Weschsung et al., 1995, 1999), and induce changes in residue quality which alter soil carbon (C) and nitrogen (N) dynamics (Torbert, Prior, and Rogers, 1995; Henning et al., 1996; Torbert et al., 1996; Prior et al., 1997c). However, the extent of these CO_2-induced changes can be highly species dependent. The quantity and quality of crop residues produced under elevated CO_2 are important factors influencing soil C storage patterns (Torbert et al., 1997; Torbert et al., 2000). Furthermore, soil C storage in agroecosystems can be altered since they are very sensitive to management practices (e.g., conservation practices, tillage systems, and cropping systems) (Kern and Johnson, 1993; Potter et al., 1997, 1998; Torbert, Prior, and Reeves, 1999). The capability of soil to act as a sink for C storage in CO_2-enriched agroecosystems is a highly relevant issue since the potential for C storage in agricultural soils is of special interest in the current climate change policy debate. This has resulted from the possibility of developing CO_2 sequestration credits for land use changes to meet the CO_2 emission limits proposed by the Kyoto Protocol.

The effect of elevated CO_2 on grain yield and the amount of crop residue left in the field may depend on the differential effect of CO_2 on crop species utilized in agroecosystems. There have been few CO_2 studies with C_3 and C_4 crops grown concurrently under the same field conditions; these two photosynthetic types are known to respond differently to elevated CO_2 both with regard to C metabolism and water use (Rogers et al., 1983; Rogers, Thomas, and Bingham, 1983; Amthor, 1995). Due to differences in CO_2 utilization during photosynthesis, plants with a C_3 photosynthetic pathway often exhibit greater growth response relative to those with a C_4 pathway (Bowes, 1993; Poorter, 1993; Amthor, 1995; Amthor and Loomis, 1996; Rogers et al., 1997). For C_4 species, the CO_2-concentrating mechanism at the site of ribisco often limits their response to elevated CO_2. However, both C_3 and C_4 species do exhibit improved plant water relations under CO_2 enrichment due to decreased stomatal conductance and increased water use efficiency (Eamus and Jarvis, 1989; Rogers et al., 1983; Rogers, Thomas, and Bingham, 1983). For C_3 crops, the greater increase in biomass production coupled with improved plant water relations may impart a more competitive advantage over C_4 crops in a future CO_2-enriched world. This difference in response could become important with regard to future management decisions. In the current study, soybean [*Glycine max* (L.) Merr.] (a N-fixing C_3 crop) and grain sorghum [*Sorghum bicolor* (L.) Moench.] (a C_4 crop) were grown in a large outdoor soil bin under two atmospheric CO_2 concentrations; ambient and twice-ambient CO_2 levels were selected since atmospheric CO_2 may increase to 700 $\mu L\ L^{-1}$ within the next 100 years (Houghton, Callander, and Varney, 1992). The study design offered the opportunity to make a direct statistical comparison of C_3 and C_4 crop species under field conditions over a multi-year period. Our goal was to evaluate the effects of changing CO_2 level on biomass production for soybean and grain sorghum and the implications of these findings as they relate to food security issues, residue management, and belowground processes.

MATERIALS AND METHODS

Soybean ('Stonewall') and grain sorghum ('Savanna 5') were chosen as test crops to represent legume and non-legume crop species, respectively. Plants were grown from seed to maturity in open top field chambers at two atmospheric CO_2 concentrations (ambient and twice-ambient) for three growing seasons (1992-1994). The experimental site was an outdoor soil bin (2-m deep, 6-m wide, and 76-m long) located at the USDA-ARS National Soil Dynamics Laboratory, Auburn, AL, USA (32.6 °N, 85.5°W). The bin contained a Blanton loamy sand (loamy, siliceous, thermic Grossarenic Paleudult) that had been fallow for over 25 years prior to 1992 (Batchelor, 1984). Initial levels of phosphorus (8 kg ha^{-1}) and potassium (14 kg ha^{-1}) were in the 'very low' range. Cation exchange capacity averaged 2.45 cmol$_c$ kg^{-1}, and soil pH averaged 4.7. The initial level of organic matter averaged 5.0 g kg^{-1} and total N was 0.06 g kg^{-1}. A detailed description of the soil status prior to initiation of the study, fertilizer and lime amendments, and soil analysis results have been reported previously (Reeves et al., 1994).

The open top field chambers were constructed of a structural aluminum frame (3-m in diameter by 2.4-m in height) covered with a PVC film panel (0.2 mm thickness) similar to that described by Rogers, Heck, and Heagle (1983). Carbon dioxide was supplied from a 12.7 Mg liquid CO_2 receiver through a high volume dispensing manifold and the atmospheric CO_2 concentration was elevated by continuous injection of CO_2 into plenum boxes. Air was introduced into each chamber through the bottom half of each chamber cover which was double-walled; the inside wall was perforated with 2.5-cm diameter holes to serve as ducts to distribute air uniformly into the chamber. Three chamber volumes were exchanged every minute. Carbon dioxide concentrations were continually monitored (24 hr day^{-1}) using a time-shared manifold with samples drawn through solenoids to an infrared CO_2 analyzer (Model 6252, LI-COR, Inc., Lincoln, NE). Values were continuously recorded every 15-30 minutes for each chamber, depending upon whether or not an additional CO_2 study was on line. In 1992, the mean seasonal daytime CO_2 concentrations were 357.4 ± 0.1 (SE) and 705.0 ± 0.3 µL L^{-1} for ambient and enriched plots, respectively. In 1993, the mean CO_2 concentrations were 364.0 ± 0.2 and 731.7 ± 0.4 µL L^{-1}. In 1994, the mean CO_2 concentrations were 359.0 ± 0.1 and 706.9 ± 0.4 µL L^{-1}.

Seeds were sown in 6-m rows oriented across the width of the soil bin on 2 June, 5 May, and 6 May in 1992, 1993, and 1994, respectively. In 1994, sorghum plots were replanted in mid June because the first crop failed owing to root rot caused by moist cool soil. Soybean seeds were inoculated with commercial *Rhizobium* (Lipha Tech, Inc., Milwaukee, WI[1]) prior to planting. Plants were thinned for uniformity to a final density of 30 plants m^{-2} for soybean and 26 plants m^{-2} for sorghum. To ensure adequate plant establishment, fertilizer N was broadcast at a rate of 34 kg N ha^{-1} to both the grain sorghum and the soybean shortly after planting. In the grain sorghum, an additional 67 kg N ha^{-1} was applied 30 days after planting. All plots received ambient rainfall and were irrigated only when necessary to prevent drought-induced mortality; a drip irrigation system was used to uniformly distribute water throughout the bin. Total amounts of water received (rainfall + irrigation) were 623, 724, and 1001 mm for 1992, 1993, and 1994, respectively. Weeds were controlled manually. In the off season, weed control was both manually and by glyphosate (N-[phosphonomethyl] glycine) at a rate of 1.0 kg ai ha^{-1}. For three seasons, plants were grown as described above and managed using no-till practices.

The experiment used a split-plot design with three replications. Whole-plot treatments (plant species) were randomly assigned to half of each replication. Subplot treatments (CO_2 levels) were randomly assigned to two open top chambers (3 m diameter) within each whole-plot. Statistical analyses of data were performed using the mixed procedure of the Statistical Analysis System (Littell et al., 1996). A significance level of $P < 0.10$ was established a priori. Significant year effects were often observed for the measurements discussed due to the influence of different planting dates coupled with year-to-year weather variability, thus data were reported separately by year.

Plant material was collected at physiological maturity in all years. At each harvest, 12 and 16 plants were collected per chamber in 1992 and 1993/94. Leaf area was determined photometrically and dry weights of organ parts were determined after drying to constant weight at 55°C. In addition, estimates of root system biomass were calculated based on soil core (Prior and Rogers, 1992) and root extraction techniques (Bohm, 1979). Twelve root-soil cores (2.4 cm diameter, 30 cm length) were collected from each chamber. Roots were separated from soil with a hydropneumatic elutriation system (Gillison's Variety Fabrication, Inc., Benzonia, MI; Smucker, McBurney, and Srivastava, 1982). Organic debris was removed with tweezers and spring-loaded suction pipettes and root length was measured with a Comair Root Length Scanner (Hawker de Havilland, Victoria, Australia). Root weight was determined after drying samples at 55°C. The root extraction technique used a manual winch (Model 527, Fulton, Milwaukee, WI) mounted onto a portable metal tripod with a cable gripping tool (Model 72285K8, Klein Tools, Chicago, IL) attached to the plant stalk to break the roots from the soil; a scale (Model 8920, Hanson, Northbrook, IL) measured the peak force (load-kg $plant^{-1}$) required to uproot the plant (Prior et al., 1995). Root samples were collected from 12 and 16 plants per chamber in 1992 and 1993/94, respectively. Plants adjacent to uprooted plants were not sampled in subsequent measurements. After soaking in water, root samples were washed free of soil using a soft bristle brush, dried at 55°C, and weighed. Root dry weights from each root sampling method were expressed on an area basis and combined for an estimate of total belowground dry weight. Remaining plant stalks within each chamber were cut into 15 cm pieces using hedge clippers; aboveground non-yield residue, including 10% (by weight) of the seed yield, was added back to study plots to simulate normal farm operations (Prior et al., 1997b). Chambers were then removed during the fallow period, but their locations remained fixed and delineated by a permanent 3-m aluminum rings. Bird netting (1.6 cm by 1.9 cm openings; Dalen Products, Inc., Knoxville, TN) was placed over the entire soil bin to prevent movement of aboveground residue into or out of plots.

RESULTS AND DISCUSSION

In the context of future economic and environmental concerns, it is important to assess the response of crops managed under field conditions to reliably predict how agroecosystems will be altered in a future CO_2-enriched world. Most elevated CO_2 research has focused on crop plants, but the majority of these efforts have not been in-ground field studies (see reviews: Kimball, 1983; Rogers and Dahlman, 1993; Rogers, Runion, and Krupa, 1994; Strain and Cure, 1994). Increased C uptake and assimilation generally results in increased crop growth under CO_2-enriched conditions. Plants with a C_3 photosynthetic pathway often exhibit greater growth response relative to those with a C_4 pathway (Bowes, 1993; Poorter, 1993; Amthor, 1995; Amthor and Loomis, 1996; Rogers et al., 1997). The CO_2-concentrating mechanism utilized by C_4 species limits the response to CO_2 enrichment. For C_3 plants, positive responses are mainly attributed to competitive inhibition of photorespiration by CO_2 and the internal CO_2 concentrations of C_3 leaves (at current CO_2 levels) being less than the Michaelis-Menton constant of ribulose bisphosphate carboxylase/oxygenase (Amthor and Loomis, 1996). Furthermore, C_3 species exhibit improved plant water relations by reductions in stomatal apertures and leaf-level conductance under elevated CO_2 (Eamus and Jarvis, 1989; Rogers et al., 1983; Rogers, Thomas, and Bingham, 1983; Prior et al., 1991); C_4 plants also exhibit growth stimulation due to lowered conductance and increased water use efficiency (Rogers et al., 1983; Rogers, Thomas, and Bingham, 1983). A previous report from our field study demonstrated that elevated CO_2 decreased whole plant water use for both C_4 sorghum and C_3 soybean, but this reduction was greater for soybean (Dugas, Prior, and Rogers, 1997). Thus, part of the larger increase in water use efficiency for C_3 vs. C_4 plants with elevated CO_2 (Morrison, 1993) may be attributed to a greater decrease in whole plant water use and a greater increase in biomass production for C_3 crops.

In our study, differences between species were observed for most measured variables (Tables 1 and 2). Both crops exhibited increases in node number,

TABLE 1. Aboveground growth variables for sorghum (SG) and soybean (SB) grown under ambient (A) and CO_2-enriched (E) conditions in 1992, 1993, and 1994. Means and probabilities are shown.

Treatment	Height (cm)	Node Number	Stem Diameter (mm)	Leaf Area (cm^2)	Seed Number	Seed Mass Seed^{-1} (g)
1992 [a]						
A-SG	131.9	8.5	9.83	968.4	683.2	0.0273
E-SG	155.4	9.5	10.46	1468.4	756.1	0.0279
A-SB	81.5	13.1	5.57	1360.6	61.8	0.0758
E-SB	95.3	13.5	6.78	1637.0	86.7	0.0819
SPP[a]	0.0001	0.0002	0.0001	0.0279	0.0001	0.0009
CO_2	0.0006	0.0598	0.0166	0.0056	0.1681	0.0054
SPPxCO_2	0.1380	0.4342	0.3765	0.3426	0.4789	0.0103
1993						
A-SG	126.2	8.8	10.74	1333.1	1030.1	0.0233
E-SG	157.3	10.8	12.27	1850.5	1106.0	0.0241
A-SB	84.5	13.8	6.13	1660.7	70.2	0.0517
E-SB	95.2	15.0	7.09	1947.7	91.4	0.0612
SPP	0.0001	0.0001	0.0013	0.0364	0.0001	0.0020
CO_2	0.0003	0.0067	0.0002	0.0001	0.5024	0.0248
SPPxCO_2	0.0052	0.3446	0.0404	0.0132	0.7022	0.0417
1994						
A-SG	105.6	9.8	6.98	647.3	323.3	0.0302
E-SG	128.9	10.2	7.67	784.5	398.2	0.0290
A-SB	95.4	13.5	6.37	1636.4	61.3	0.0965
E-SB	106.8	14.1	7.08	1775.1	79.7	0.1166
SPP	0.0003	0.0001	0.0239	0.0001	0.0010	0.0001
CO_2	0.0002	0.0047	0.0001	0.0012	0.1442	0.0190
SPPxCO_2	0.0526	0.5494	0.8581	0.9771	0.3647	0.0126

[a] Values are Pr > F from mixed model analysis; SPP = main effect of species, CO_2 = main effect of CO_2 level, SPPxCO_2 = interaction.

basal diameter, and height under elevated CO_2 (Table 1). In general, elevated CO_2 increased sorghum and soybean stem biomass by > 45% (averaged across years) (Table 2). Leaf area and leaf dry weight were usually higher for soybean versus sorghum and CO_2-induced increases in these measures were due to larger leaves rather than a change in leaf number (data not shown). Elevated

TABLE 2. Above- and belowground dry weights ($g\ m^{-2}$) for sorghum (SG) and soybean (SB) grown under ambient (A) and CO_2-enriched (E) conditions in 1992, 1993, and 1994. Means and probabilities are shown.

Treatment	Stem	Leaf	Hull or Head[a]	Residue[b]	Seed	Belowground[c]
1992						
A-SG	253.5	110.3	89.2	453.0	472.5	130.6
E-SG	351.5	155.9	105.0	612.4	543.0	179.1
A-SB	244.1	148.7	127.5	409.8	142.9	199.9
E-SB	371.8	187.0	178.8	552.7	213.2	285.3
SPP[d]	0.7760	0.0077	0.0006	0.7817	0.0001	0.0001
CO_2	0.0003	0.0032	0.0070	0.0270	0.0105	0.0002
SPPxCO_2	0.4413	0.6938	0.0831	0.8611	0.9985	0.1211
1993						
A-SG	360.2	178.7	147.3	686.2	584.5	226.7
E-SG	540.4	250.6	161.8	952.8	672.5	271.7
A-SB	412.2	224.4	130.7	767.3	108.5	156.2
E-SB	663.1	304.9	193.4	1161.8	168.5	228.7
SPP	0.0074	0.0003	0.5012	0.0074	0.0001	0.0457
CO_2	0.0001	0.0001	0.0075	0.0001	0.0139	0.0054
SPPxCO_2	0.0506	0.5400	0.0569	0.0362	0.5707	0.2688
1994						
A-SG	171.9	76.8	63.0	311.8	245.5	120.7
E-SG	239.9	99.2	75.7	414.8	299.9	188.1
A-SB	480.1	235.2	131.7	847.0	177.8	196.7
E-SB	649.4	264.0	175.0	1088.5	273.2	280.8
SPP	0.0001	0.0001	0.0003	0.0001	0.1877	0.0591
CO_2	0.0001	0.0001	0.0083	0.0001	0.0040	0.0048
SPPxCO_2	0.0038	0.2653	0.0570	0.0015	0.1787	0.5848

[a] Soybean pod hull weight or sorghum head weight, minus seed.
[b] Sum of all aboveground non-yield components.
[c] Total root dry weight including nodules for soybean.
[d] Values are Pr > F from mixed model analysis; SPP = main effect of species, CO_2 = main effect of CO_2 level, SPPxCO_2 = interaction.

CO_2 increased soybean pod hull weight by ~ 40% (averaged across years), but had little effect on sorghum head weight (Table 2). Total non-yield residue production (i.e., stover) for both crops was increased by > 35% (averaged across years) due to elevated CO_2. Differences in seed number were observed between sorghum and soybean, but were not affected by CO_2 level. However, weight per seed increased for soybean under elevated CO_2 and soybean seed biomass was increased by 53% (averaged across years) compared to a 17% increase for sorghum (Table 2).

In the context of food security, the reported yield responses for these important crops is significant. The observed increase in soybean yield may have implications for major soybean production regions (FAO, 1996) located in the United States, South America (e.g., Brazil and Argentina), and Asia (e.g., China and India). Sustainability of sorghum production is also critical since it historically represents one of five major cereal crops used for food and is also important for animal consumption (Doggett, 1988; Bennett, Tucker, and Maunder, 1990). Major regions of sorghum production (FAO, 1996) are located in the United States, Mexico, Asia (e.g., China and India) and throughout Africa. Since sorghum is a major food staple for many developing countries (FAO, 1996), especially in semiarid regions (Doggett, 1988; Bennett, Tucker, and Maunder, 1990), it is essential to evaluate changes in the global environment that will affect productivity. Although reported increases in sorghum yield were substantially less than for soybean (Table 2), small yield shifts may be significant, particularly for semiarid regions of the world where reduced whole plant water use (Dugas, Prior, and Rogers, 1997) could help ameliorate periods of drought stress. Special emphasis should also be given to the reported positive shifts in non-yield biomass production in terms of future residue management considerations in these regions. For example, selecting planting and seed zone preparation implements that minimize disturbance of residue and underlying soil can lead to soil water conservation (Reicosky et al., 1999; Prior et al., 2000) which could be critical to successful seedling establishment in these semiarid regions. In order to improve food security and to alleviate poverty on the African continent, the African Conservation Tillage Network (ACT) has recently been established to promote the adoption of conservation tillage practices to ensure more sustainable use of the soil resources and to combat desertification (ACT, 2000). Findings from the current study suggest that CO_2-induced shifts in grain yield and crop water use could improve food security while increases in non-yield residue could complement conservation management efforts by ensuring greater soil surface coverage thereby promoting more soil water storage while preventing soil erosion losses. However, it is important to note that our work was conducted in a temperate region and more CO_2 research in this and other areas (e.g., semiarid and tropical regions) which evaluate regional crop management systems are needed before firm conclusions can be made.

Positive increases in non-yield residue inputs (Table 2) returned to the soil surface may impact implement effectiveness during tillage operations. In conventional tillage systems, the degree of residue cutting/burial (e.g., disc, chisel plow operations) may be altered by increased residue inputs resulting from

CO_2 enrichment. Such occurrences may require increasing tillage depth which could increase future field operational expenses (e.g., energy/fuel cost and tractor/equipment wear). However, current debate highlights the importance of modifying traditional tillage practices to promote sequestration of soil C in agroecosystems; management decisions that reduce tillage activities in favor of maintaining more soil surface residue and greater percent ground cover could reduce soil C losses, increase soil C storage, and help ameliorate the rise in atmospheric CO_2 (Follett, 1993; Lal et al., 1998a, 1998b). Adoption of such practices would also reduce water losses, erosion processes, and possibly improve overall soil quality (Reeves, 1997). In this context, land managers could realize some benefits by continuing to follow current operational schemes and accept less residue cutting/burial in conventional tillage systems.

In conservation tillage systems, these same CO_2-related advantages would exist, but to a greater extent due to lack of tillage and a higher accumulation of non-yield residue (Table 2). Previous work from our study showed a significant increase in percent ground cover under CO_2-enrichment (Figure 1; Prior et al., 1997b), but no significant effect on percent residue biomass recovery in litter bags was noted (Figure 2; Torbert et al., 2000). Measurement of mass losses from leaves and stems indicated a species effect which varied by tissue type; decomposition of soybean leaf tissue proceeded more rapidly than sor-

FIGURE 1. The average percent ground cover following an over-winter fallow period for sorghum and soybean (A) and for ambient and enriched CO_2 treatments (B). Main effect means are shown. Adapted from Prior et al. (1997b).

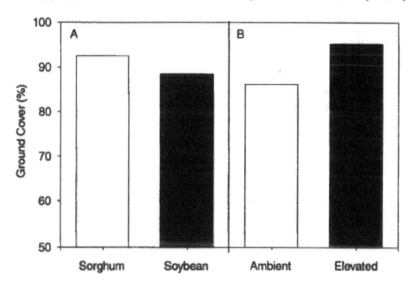

FIGURE 2. Recovery (%) of ambient and elevated CO_2-produced sorghum and soybean leaf (A) and stem (B) residue during an over-winter fallow period. Adapted from Torbert et al. (2000).

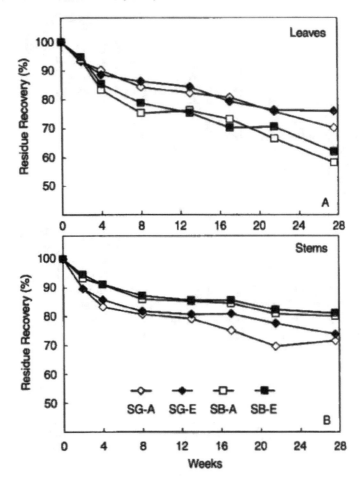

ghum, as would be expected with a lower C:N ratio; however, the opposite pattern was observed with stem tissue. Even though CO_2 level did not affect percent biomass recovery, greater production under elevated CO_2 resulted in more biomass remaining after the over-winter fallow period.

Some disadvantages associated with greater residue accumulation could occur in relation to crop stand establishment. Planter designs used in current conservation systems often have problems handling high amounts of residue during seed bed preparation/planting (e.g., clogged planters) resulting in poor stands (Phillips, 1984; Throckmorton, 1986). Precluding significant improvements in planter design, CO_2-induced increases in residue production could exacerbate this problem, especially in systems with additional residues from cover crops. Increased residue coverage of the soil surface can also reduce soil temperatures below optimum for seed germination, thereby delaying stand development (Erbach et al., 1986; Unger, 1986; Potter, Morrison, and Torbert, 1996). Higher production of non-yield residue in a CO_2-enriched world may require land managers to select conservation tillage methods such as strip tillage as opposed to no-tillage during planting to overcome such problems.

Relative to aboveground responses, CO_2 effects on root systems have received less attention despite their importance in attaining essential soil resources and their residue contributions to soil organic matter. In the current study, total belowground dry weight was increased by 44 and 38% (averaged across years) for soybean and sorghum, respectively (Table 2). CO_2-induced increases in soybean and sorghum root biomass have been previously reported (Chaudhuri et al., 1986; Chaudhuri, Kirkham, and Kanemasu, 1990; Del Castillo et al., 1989; Rogers et al., 1992) and in many instances the largest proportion of the extra phytomass produced as a result of elevated CO_2 is found belowground (Bazzaz, 1990; Rogers, Runion, and Krupa, 1994; Wittwer, 1995). Fine root density patterns (both length and dry weight) were also assessed in the current study and the extent of CO_2-induced changes were found to be species dependent; elevated CO_2 had a much greater positive affect on soybean compared to sorghum (Figures 3 and 4).

Other field studies using FACE have shown that high CO_2 can increase belowground production (Prior et al., 1994b), alter plant root morphology (Prior et al., 1995), and increase the root system's capacity to explore soil volume through shifts in fine root distribution patterns (Prior et al., 1994a; Weschsung et al., 1999). Such CO_2-induced changes in rooting patterns may influence whole-plant nutrient dynamics, thus influencing crop performance when demand for nutrients and water is high. In general, whole plant nutrient uptake and nutrient utilization efficiency are increased under elevated CO_2, while nutrient tissue concentration and nutrient uptake efficiency are lowered (Rogers, Runion, and Krupa, 1994; Prior et al., 1998). Nutrient management decisions may also be impacted by CO_2-induced shifts in root distribution patterns which could alter nutrient stratification within the soil profile. This would be more likely in reduced-tillage systems compared to conventional tillage systems which exhibit a more homogeneous plow layer due to mixing of soil with residues and amendments (e.g., fertilizers and lime).

The quality of water moving in the hydrological cycle is critically important in agroecosystems. Positive CO_2-induced shifts in crop root systems may enhance the ability of plants to capture a greater proportion of available nutrients, thus reducing the leaching of nutrients, such as nitrates, into groundwater. The

FIGURE 3. The effect of CO_2 concentration (A = ambient; E = elevated) on sorghum (SG) and soybean (SB) root length density (RLD) in 1992, 1993, and 1994.

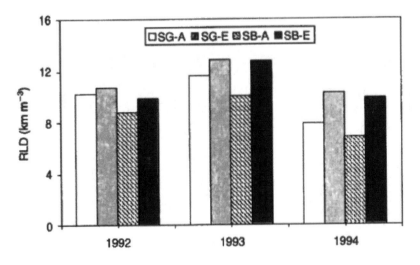

FIGURE 4. The effect of CO_2 concentration (A = ambient; E = elevated) on sorghum (SG) and soybean (SB) root dry weight density (RWD) in 1992, 1993, and 1994.

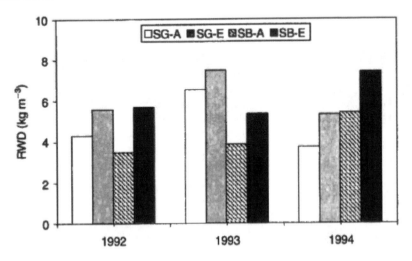

fate of N is a growing concern since nitrate contamination of groundwater is associated with potential health risks. Belowground N content is a balance of N in biomass, N loss processes such as nitrate leaching and denitrification, and N inputs through fertilizers, atmospheric deposition, and N_2 fixation by certain plant species. Nitrate leaching is dependent on the amount of nitrate in the edaphic environment and on the amount of water percolating through the soil profile. These factors may be altered by changes in atmospheric CO_2. Our study has previously demonstrated that elevated CO_2 will change both the C:N ratio of residue inputs to the soil (affecting the soil nitrate content; Torbert et al., 1996) and plant water relations (affecting water movement through the soil profile; Dugas, Prior, and Rogers, 1997). Results from two years of CO_2 enrichment in our study showed that nitrate leaching was reduced during the growing season because more N was captured by high CO_2-grown crops (Torbert et al., 1996). Furthermore, leaching was also reduced during the fallow period, which may be related to altered decomposition patterns resulting from the increased C:N ratio of residue. Nitrate leaching was generally higher for soybean compared to sorghum most likely due to higher N inputs to the soil from symbiotic N_2 fixation and lower residue C:N ratio. Furthermore, analysis of nitrate solution for ^{15}N content indicated that most of the N measured below the root zone originated from native N pools rather than from N-fertilizer application in both cropping systems. It is important to note that results from our study are indicative of leaching patterns associated with areas cropped with soybean and sorghum on a loamy sand soil under a no-till management system. Leaching patterns will likely vary with soil series and tillage practice. The impact of leaching on groundwater quality may be even more critical during disturbance events (e.g., plowing) which occur in conventional tillage systems.

Knowledge of changes in soil C due to elevated atmospheric CO_2 is essential to understanding global C cycling. Enhanced crop growth (both above- and belowground) under elevated CO_2 as reported in Table 2 suggests greater delivery of C to soil; extra C from elevated atmospheric CO_2 can enter the rhizosphere via residue decomposition, root growth, turnover, and exudation (Norby et al., 1987; Lekkerkerk, Van de Geijn, and Van Veen, 1990; Zak et al., 1993). Despite the well documented rise in atmospheric CO_2 concentration (Keeling and Whorf, 1994), not all C sinks are well defined, i.e., an estimated unknown sink of 1.4×10^{15} g C year^{-1} arises from the global C balance (Schimel et al., 1995). Crucial considerations in balancing the global C budget are that biospheric uncertainties are very large, that anthropogenic CO_2 is small relative to the natural exchange and abundance of C (Sundquist, 1993), and that description of the C cycle is incomplete (Bolin, 1981; Whipps, 1990). Although its specific identity has eluded detection, the sink is probably somewhere in the world's terrestrial plants and soils (Sundquist, 1993; Schimel, 1995). Soil plays a major role in the global accounting of C not only due to the large amount of C stored in soil (estimates of 1395 to 1636×10^{15} g) (Ajtay, Ketner, and Duvigneaud, 1979; Post, Emanuel, and King, 1992; Schlesinger, 1984), but also since annual soil flux of CO_2 to the atmosphere is 10 times that contributed by fossil fuel burning (Post et al., 1990). Enting and Pearman (1986) suggested that although in the past the biosphere has been a net C source, it is currently acting as a C sink. This is supported by estimates that the "pioneer agriculture effect in the USA" released some 60×10^{15} g C to the atmosphere from 1860 to 1890 (Wilson, 1978) which is 1.5 times the amount released by all industrial sources (mainly fossil fuel usage) prior to 1950. A portion of the terrestrial sink is likely the result of converting agricultural land back to natural or perennial vegetation (Post and Kwon, 2000). However, if the terrestrial biosphere has changed from a CO_2 source to a CO_2 sink, then agriculture, which accounts for fully 10% of all land on earth (Schlesinger, 1990), may play a pivotal role in global C sequestration (Cole et al., 1993; Kern and Johnson, 1993; Paustian et al., 1997; Lal et al., 1998a, 1998b). In a global context, agroecosystems are significant since approximately 1.3×10^{15} g of gross atmospheric CO_2 is removed by crops each year (Jackson, 1992) and soil C storage patterns in these systems are very sensitive to management practices (e.g., conservation practices, tillage systems, and cropping systems) (Kern and Johnson, 1993). All these factors combine to make the understanding of C cycling in soils of agroecosystems important, especially in the context of rising atmospheric CO_2.

The ability of terrestrial ecosystems to sequester additional C in soil from increasing levels of CO_2 in the atmosphere is highly debated. Schlesinger (1986, 1990) found little evidence for soil C storage and Lamborg, Hardy, and Paul (1984) have argued that increased soil microbial activity due to greater biomass C inputs in an elevated CO_2 environment (i.e., "the priming effect") would prevent accumulation of soil organic C. Alternatively, Goudriaan and de Ruiter (1983) proposed that increased soluble, easily decomposable C inputs (due to CO_2 enrichment) would accentuate soil microbial substrate preference mechanisms; that is preference for easily decomposable substrates would retard the decomposition of recalcitrant, structural plant debris and native soil organic matter resulting in an accumulation of soil organic matter. Experimental evidence with wheat (*Triticum aestivum* L.) grown under elevated CO_2 in a short-term growth chamber experiment (Lekkerkerk, Van de Geijn, and Van Veen, 1990) has supported the contentions of Goudriaan and De Ruiter (1983). Long-term field studies (at our laboratory and others) indicate that agroecosystems have the potential to sequester C from the atmosphere into the soil (Wood et al., 1994; Leavitt et al., 1994; Torbert, Prior, and Rogers, 1995; Henning et al., 1996; Prior et al., 1997c; Torbert et al., 1997). A 3-year study with cotton (*Gossypium hirsutum* L.) has suggested that soil C storage is more likely under non-limiting soil water conditions when CO_2 concentration is raised (Wood et al., 1994). Their findings indicated that factors other than total biomass input may affect soil C and N cycling; a possible explanation may be related to a differential effect of CO_2 and irrigation treatment on residue structure/composition which has altered decomposition patterns. In a similar study, an evaluation of soils after 2 years of wheat residue inputs indicated that more C storage may occur under elevated CO_2 for both irrigated and non-irrigated farm systems (Prior et al., 1997c). In our study, we observed that short-term CO_2 fluxes were greater for soybean under tillage or elevated CO_2; flux rates in the sorghum crop were affected by tillage, but they were not impacted by CO_2 level (Prior et al., 1997b). It is important to note that these short-term results were based on characterizing C losses associated with a simulated spring tillage event on microplots, thus, results should be viewed with caution when predicting long-term C turnover in agroecosystems. However, working in the same study using stable isotope techniques, Torbert et al. (1996) also noted differences in C storage patterns for sorghum and soybean after 2 years of CO_2 treatment. The high C:N ratio of sorghum residue slowed microbial decomposition resulting in increased new soil C, but CO_2-induced C storage occurred in the mineral fraction only. In comparison, the low C:N ratio of soybean residue promoted decomposition of new C inputs which reduced the decomposition of old C thereby increasing soil C storage. For a more thorough discussion on elevated CO_2 effects on residue decomposition as it relates to soil C and soil N interactions see Torbert et al. (2000). Collectively, the results suggest that the biodegradability of crop residue may not only be affected by the environment they were produced under but may also be species dependent, thereby accounting for differences in soil C storage patterns. To more fully elucidate the relationships between nutrient cycling and decomposition of plant materials produced in an elevated CO_2 environment, future studies must be concerned with crop species effects and must also consider the influence of other factors such as cover crops, crop rotations, soil series, tillage practices, and regional climatic differences.

The effects of additional residue input from elevated CO_2 on soil physical properties and their impact on soil C storage, has not been well studied. A more extensive residue mat should promote more favorable soil surface characteristics such as prevention of soil crusting. Minimizing soil crusting could enhance seedling emergence, water infiltration, soil water retention, and reduced soil erosional processes. This study clearly demonstrated that elevated CO_2 increased non-yield residue returned to the soil surface (Table 2) and percent ground cover following an over-wintering period (Figure 1; Prior et al., 1997b). Stabilization of the soil matrix by larger root systems under elevated CO_2 can be inferred from an increase in vertical root-pulling resistance (Figure 5); this may suggest reduced wind and water erosion on cropping systems located on highly erodible lands. Vertical root-pulling resistance was increased by

30% (averaged across years) for soybean and by 53% for sorghum (Figure 5). This finding is in general agreement with results reported for cotton (Prior et al., 1995). Positive shifts in crop root systems (Table 2; Figures 3 and 4) may alter soil structural characteristics (e.g., due to increased number and extent of root channels) which could lead to increases in aggregate stability, porosity, infiltration rates, and subsequent soil water storage. Changes in soil structure could possibly lead to increased rates of soil genesis (Brinkman and Sombroek, 1996). However, most of these hypothesized changes have yet to be examined in detail. A preliminary evaluation of soil physical characteristics indicated that soil structure was altered by elevated CO_2 in the soybean system only (Prior and Amthor, unpublished). In this case, the soil had lower bulk density values, more water stable aggregates, and exhibited positive shifts in saturated hydraulic conductivity, thereby suggesting that soil porosity had been increased under elevated CO_2. Such changes in the soil may be due to soybean residue quality (lower C:N ratio) in combination with a greater positive affect of elevated CO_2 on soybean fine root density patterns (both length and dry weight) compared to sorghum (Figures 3 and 4). Detailed examination of residue input (quality and quantity) in relation to soil C and N dynamics indicates that N availability exerts a strong influence on belowground decomposition processes (see review; Torbert et al., 2000) which may alter soil physical and chemical properties. Such shifts, in conjunction with root turnover, root exu-

FIGURE 5. The effect of CO_2 concentration (A = ambient; E = elevated) on sorghum (SG) and soybean (SB) vertical root-pulling resistance (RPR) in 1992, 1993, and 1994.

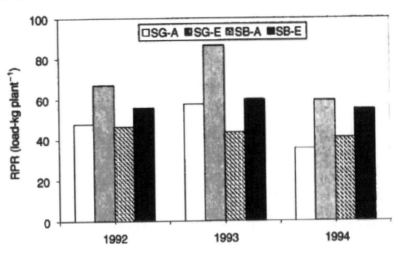

dation, and other biological activity (increased populations of microbes and soil fauna and fungi), may influence soil aggregate formation and nutrient cycling. CO_2-induced changes in soil physicochemical characteristics may lead to improvements in overall soil quality; such changes, however, will likely be dependent on crop species and management.

The direct effect of elevated CO_2 noted for crop productivity may also impact weed control management in agroecosystems. Some weed species may have competitive advantages under high CO_2 due to differential effects of CO_2 on growth which favor C_3 over C_4 weeds (Patterson, 1993). The reported increases in non-yield residue for soybean and sorghum cropping systems may suppress weeds (mulch effect) in conservation management systems. Although, the introduction of genetically modified crops (e.g., glyphosate tolerant soybean) into production systems represents another means to combat weeds, it is unknown if they will show the same growth responses to CO_2 as current day varieties. Weed management may be further complicated by response of weed species to herbicides under high CO_2 conditions. Ziska, Teasdale, and Bunce (1999) studied two of the world's worst weeds, representing a C_3 species, common lambsquarters (*Chenopodium album* L.), and a C_4 species, redroot pigweed (*Amaranthus retroflexus* L.), to a widely used postemergence herbicide (glyphosate) under conditions of elevated CO_2. They reported that current application rates could control *A. retroflexus*, but the economic cost of controlling *C. album* may increase under elevated CO_2 as standard herbicides rates were inadequate. These differential responses coincided with changes in stomatal conductance; however, changes in foliar absorption (and uptake) were not documented. Some studies have shown that elevated CO_2 may increase epicuticular wax deposition (based on SEM micrographs; Thomas and Harvey, 1983), while others have reported a decrease in wax density (Graham and Noble, 1996; Prior et al., 1997a) and changes in wax morphology (Prior et al., 1997a). Thus, one other possible mechanism explaining this finding might be related to changes in epicuticular waxes (quantity, composition, or morphology) which could alter permeability to chemicals including herbicides (Martin and Juniper, 1970; Von Wetsein-Knowles, 1993). If increases in commercial application rates of herbicides are required to control some weeds under high CO_2, it is unknown if tolerance levels of genetically modified crops (e.g., glyphosate tolerant soybean and cotton) are adequate since no information exists on responses of these altered crops to elevated CO_2.

Another unknown aspect of genetically modified material is how the introduction of such material might alter decomposition processes and microbial populations in the soil environment. Ellis, Thompson, and Bailey (1995) demonstrated that introduction of genetically modified microorganisms (as a seed dressing) did not disrupt the natural succession of microbial communities in a

231-day sugar beet (*Beta vulgaris*) study. Saxena, Flores, and Stotzky (1999) found that an insecticidal toxin was released into the rhizosphere of *Bt*-modified corn through root exudates and that this toxin remained biologically active in soil for at least 234 days; the impacts of such genetically altered plants on rhizosphere and soil microbial populations are unknown. Microbes are important for maintaining plant health and productivity and elevated CO_2 has been shown to affect soil microbial community composition and activity (Zak et al., 1993; Rice et al., 1994; Runion et al., 1994). However, limited research precludes drawing firm conclusions regarding the effects of elevated CO_2 on interactions of microbes with plants and plant material in soil; adding use of genetically modified plants and microbes into this scenario makes it even more difficult to predict how crop productivity might be affected by future farming practices under increasing levels of atmospheric CO_2.

The atmospheric CO_2 concentration has risen by 30% since the onset of the Industrial Revolution in the late 18th century; this increase may be the most significant change taking place on the earth today. No sector has more to lose or gain, in regard to global environmental change, than agriculture. Growth and yield of most plant species, including economically important crops, have been shown to increase under elevated CO_2. We found that yield response to elevated CO_2 was greater for soybean than for sorghum; however, the response of non-yield residues (including roots) of these contrasting (C_3 vs. C_4) crops was similar. Our findings suggest that increasing levels of atmospheric CO_2 could improve food security, soil physical properties, and groundwater quality. Increases in the non-yield components could have implications for residue management including farming practices to increase soil C sequestration and protect soil resources. Research will be required to fully understand the relationships between biomass production, nutrient cycling, and decomposition of residue produced in elevated CO_2 environments. Effects of CO_2 on crops grown under conservation tillage systems require further investigation. Future studies should address not only species effects, but must also consider how other factors (cover crops, crop rotations, soil series, tillage practices, and regional climatic differences) influence the response of agroecosystems to rising levels of atmospheric CO_2. Such factors must be considered due to the wide diversity of farm management systems that exist over an international scale. Evaluation of farming systems representative of underdeveloped countries must also be included to accurately assess how these regions will be impacted by the rise in atmospheric CO_2. Understanding the whole biological chain of events starting with transfer of C from air to leaf, transformation within the plant for growth and yield, return of plant residue to the soil, decomposition, C storage within soils of agricultural systems, and finally impacts of other environmental factors (e.g., nutrients and water) on these processes is necessary to optimize soil management for both agricultural production and C sequestration. Reducing uncertainty regarding the effects of rising atmospheric CO_2 is critical if the impacts of global change on agriculture and environmental quality are to be predicted.

NOTE

1. Trade names and products are mentioned solely for information. No endorsement by the USDA is implied.

REFERENCES

African Conservation Tillage Network-ACT. (2000). In: ACT World Wide Web site at <http://www.fao.org/landandwater/agll/consagri/home1.htm/>.

Ajtay G.L., P. Ketner, and P. Duvigneaud. (1979). Terrestrial primary production and phytomass. In *The Global Carbon Cycle*, eds. B. Bolin, E.T. Degens, S. Kempe, and P. Ketner, NY: John Wiley & Sons, pp. 129-181.

Allen, L.H., Jr., B.G. Drake, H.H. Rogers, and J.H. Shinn. (1992). Field techniques for exposure of plants and ecosystems to elevated CO_2 and other trace gases. *Critical Reviews in Plant Science* 11: 85-119.

Amthor, J.S. (1995). Terrestrial higher-plant response to increasing atmospheric $[CO_2]$ in relation to the global carbon cycle. *Global Change Biology* 1:243-274.

Amthor, J.S. and R.S. Loomis. (1996). Integrating knowledge of crop responses to elevated CO_2 and temperature with mechanistic simulation models: Model components and research needs. In *Carbon Dioxide and Terrestrial Ecosystems*, eds. G.W. Koch and H.A. Mooney, San Diego, CA: Academic Press, pp. 317-346.

Amthor, J.S., G.W. Koch, and A.J. Bloom. (1992). CO_2 inhibits respiration in leaves of *Rumex crispus* L. *Plant Physiology* 98:757-760.

Arp, W.J. (1991). Effects of source-sink relations on photosynthetic acclimation to elevated CO_2. *Plant, Cell and Environment* 14:869-875.

Batchelor, J.A., Jr. (1984). *Properties of Bin Soils at the National Tillage Machinery Laboratory, Publ. 218.* Auburn, AL: USDA-ARS National Soil Dynamics Laboratory.

Bazzaz, F.A. (1990). The response of natural ecosystems to the rising global CO_2 levels. *Annual Review of Ecology and Systematics* 21:167-196.

Bennett, W.F., B.B. Tucker, and A.B. Maunder. (1990). *Modern Grain Sorghum Production.* Ames, IA: Iowa State University Press.

Bohm, W. (1979). *Methods for Studying Root System, Ecological Series, Volume 33.* NY: Springer-Verlag.

Bolin, B. (1981). *Carbon Cycle Modeling: Scope 16.* NY: John Wiley.

Bolin, B., B.R. Doos, J. Jager, and R.A. Warrick. (1986). *Scope 29–The Greenhouse Effect, Climatic Change, and Ecosystems.* Chichester: John Wiley & Sons.

Bowes, G. (1991). Growth at elevated CO_2: Photosynthetic responses mediated through Rubisco. *Plant, Cell and Environment* 14:795-806.

Bowes, G. (1993). Facing the inevitable: Plants and increasing atmospheric CO_2. *Annual Review of Plant Physiology and Plant Molecular Biology* 44:309-332.

Brinkman, R. and W.G. Sombroek. (1996). The effects of global change on soil conditions in relation to plant growth and food production. In *Global Climatic Change and Agricultural Production, Direct and Indirect Effects of Changing Hydrological, Pedological, and Plant Physiological Processes*, eds. F. Bazzaz and W. Sombroek, NY: John Wiley & Sons, pp. 49-63.

Bunce, J.A. (1990). Short- and long-term inhibition of respiratory carbon dioxide efflux by elevated carbon dioxide. *Annals of Botany* 65:637-642.

Chaudhuri, U.N., R.B. Burnett, M.B. Kirkham, and E.T. Kanemasu. (1986). Effect of carbon dioxide on sorghum yield, root growth, and water use. *Agricultural and Forestry Meteorology* 37:109-122.

Chaudhuri, U.N., M.B. Kirkham, and E.T. Kanemasu. (1990). Root growth of winter wheat under elevated carbon dioxide and drought. *Crop Science* 30:853-857.

Cole, C.V., K. Paustian, E.T. Elliott, A.K. Metherell, D.S. Ojima, and W.J. Parton. (1993). Analysis of agroecosystem carbon pools. *Water, Air and Soil Pollution* 70:357-371.

Del Castillo, D., B. Acock, V.R. Reddy, and M.C. Acock. (1989). Elongation and branching of roots on soybean plants in a carbon dioxide-enriched aerial environment. *Agronomy Journal* 81:692-695.

Doggett, H. (1988). *Sorghum, Tropical Agriculture Series*. Singapore: Longman Publishers Ltd.

Dugas W.A., S.A. Prior, and H.H. Rogers. (1997). Transpiration from sorghum and soybean growing under ambient and elevated CO_2 concentrations. *Agricultural and Forestry Meteorology* 83:37-48.

Eamus, D. and P.G. Jarvis. (1989). The direct effects of increase in the global atmospheric CO_2 concentration on natural and commercial temperate trees and forests. *Advances in Ecological Research* 19:1-55.

Ellis, R.J., I.P. Thompson, and M.J. Bailey. (1995). Metabolic profiling as a means of characterizing plant-associated microbial communities. *FEMS Microbiology Ecology* 16:9-18.

Enting, I.G. and G.I. Pearman. (1986). The use of observations in calibrating and validating carbon cycle models. In *The Changing Carbon Cycle: A Global Analysis*, eds. J.R. Trabalka and D.E. Reichle, NY: Springer-Verlag, pp. 425-458.

Erbach, D.C., R.M. Cruse, T.M. Crosbie, D.R. Timmons, T.C. Kaspar, and K.N. Potter. (1986). Maize response to tillage-induced soil conditions. *Transactions in ASAE* 29:690-695.

Follett, R.F. (1993). Global climate change, U.S. agriculture, and carbon dioxide. *Journal of Production Agriculture*. 6:181-190.

Food and Agriculture Organization of the United Nations -FAO. (1996). *FAO Production Yearbook 1995, Volume 49, FAO Statistics Series No. 133*. Rome, Italy: FAO.

Goudriaan, J. and H.E. de Ruiter. (1983). Plant growth in response to CO_2 enrichment, at two levels of nitrogen and phosphorus supply. 1. Dry matter, leaf area, and development. *Netherlands Journal of Agricultural Science* 31:157-169.

Graham, E.A. and P.S. Nobel. (1996). Long-term effects of a doubled atmospheric CO_2 concentration on the CAM species *Agave deserti*. *Journal of Experimental Botany* 47:61-69.

Henning, F.P., C.W. Wood, H.H. Rogers, G.B. Runion, and S.A. Prior. (1996). Composition and decomposition of soybean and sorghum tissues grown under elevated atmospheric CO_2. *Journal of Environmental Quality* 25:822-827.

Houghton, J.T., G.J. Jenkins, and J.J. Ephraums. (1990). *Climate Change: The IPCC Scientific Assessment*. Cambridge: Cambridge University Press.

Houghton, J.T., B.A. Callander, and S.K. Varney. (1992). *Climate Change 1992: The Supplementary Report to the IPCC Scientific Assessment*. Cambridge: Cambridge University Press.

Huber, S.C., H.H. Rogers, and D.W. Israel. (1984). Effects of CO_2 enrichment on photosynthesis and photosynthate partitioning in soybean (*Glycine max*) leaves. *Physiologia Plantarum* 62:95-101.

Jackson, R.B. IV (1992). On estimating agriculture's net contribution to atmospheric carbon. *Water, Air and Soil Pollution* 64:121-137.

Keeling, C.D. and T.P. Whorf. (1994). Atmospheric CO_2 records from the sites in the SIO air sampling network. In *Trends '93: A Compendium of Data on Global Change, ORNL/CDIAC-65*, eds. T.A. Boden, D.P. Kaiser, R.J. Sepanski and F.W. Stoss, Oak Ridge, TN: The Carbon Dioxide Information Analysis Center, Oak Ridge National Laboratory, pp. 16-26.

Kern, J.S. and M.G. Johnson. (1993). Conservation tillage impacts on national soil and atmospheric carbon levels. *Soil Science Society of America Journal* 57: 200-210.

Kimball, B.A. (1983). Carbon dioxide and agricultural yield: An assemblage and analysis of 430 prior observations. *Agronomy Journal* 75:779-788.

Kimball, B.A., P.J. Pinter, Jr., R.L. Garcia, R.L. LaMorte, G.W. Wall, D.J. Hunsaker, G. Wechsung, F. Wechsung, and Th. Kartschall. (1995). Productivity and water use of wheat under free-air CO_2 enrichment. *Global Change Biology* 1:429-442.

Lal, R., J.M. Kimble, R.F. Follett, and C.V. Cole. (1998a). *The Potential of U.S. Cropland to Sequester Carbon and Mitigate the Greenhouse Effect*. Ann Arbor, MI: Ann Arbor Press.

Lal, R., J.M. Kimble, R.F. Follett, and B.A. Stewart (1998b). *Management of Carbon Sequestration in Soil*. Boca Raton, FL: CRC Lewis Publishers.

Lamborg, M.R., W.F. Hardy, and E.A. Paul. (1984). Microbial effects. In *CO_2 and Plants: The Response of Plants to Rising Levels of Atmospheric CO_2*, ed. E.R. Lemon, Washington, DC: Amer. Assoc. Adv. Sci. Selected Symp., pp. 131-176.

Lawlor, D.W. and R.A.C. Mitchell. (1991). The effects of increasing CO_2 on crop photosynthesis and productivity: A review of field studies. *Plant, Cell and Environment* 14: 807-818.

Leavitt S.W., E.A. Paul, B.A. Kimbal, G.R. Hendrey, J.R. Mauney, R. Rauschkolb, H. Rogers, K.F. Lewin, J. Nagy, P.J. Pinter, Jr., and H.B. Johnson. (1994). Carbon isotope dynamics of free-air CO_2-enriched cotton and soils. *Agricultural and Forestry Meteorology* 70:87-101.

Lekkerkerk, L.J.A., S.C. Van de Geijn, and J.A. Van Veen. (1990). Effects of elevated atmospheric CO_2-levels on the carbon economy of a soil planted with wheat. In *Soils and the Greenhouse Effect*, ed. A.F. Bouwman, NY: John Wiley & Sons, pp. 423-429.

Littell, R.C., G.A. Milliken, W.W. Stroup, and R.D. Wolfinger. (1996). *SAS System for Mixed Models*. Cary, NC: SAS Institute, Inc.

Long, S.P. and B.G. Drake. (1992). Photosynthetic CO_2 assimilation and rising atmospheric CO_2 concentrations. In *Crop photosynthesis: Spatial and Temporal Determinants*, eds. N.R. Baker and H. Thomas, NY: Elsevier, pp. 69-107.

Martin, J.T. and B.E. Juniper. (1970). *The Cuticles of Plants*. NY: St. Martin's Press.

Mauney, J.R., B.A. Kimball, P.J. Pinter, Jr., R.L. LaMorte, K.F. Lewin, J. Nagy, and G.R. Hendrey. (1994). Growth and yield of cotton in response to a free-air carbon dioxide enrichment (FACE) environment. *Agricultural and Forestry Meteorology* 70:49-67.

Morrison, J.I.L. (1993). Response of plants to CO_2 under water limited conditions. *Vegetation* 104/105:193-209.

Mousseau, M. (1993). Effects of elevated CO_2 on growth, photosynthesis and respiration of sweet chestnut (*Castanea sativa* Mill.). *Vegetation* 104/105:413-419.

Norby, R.J., E.G. O'Neill, W.G. Hood, and R.J. Luxmoore. (1987). Carbon allocation, root exudation and mycorrhizal colonization of *Pinus echinata* seedlings grown under CO_2 enrichment. *Tree Physiology* 3:203-210.

Patterson, D.T. (1993). Implications of global climate change for impact of weeds, insects and plant diseases. In *International Crop Science I*, ed. D.R. Buxton, Madison, WI: Crop Science Society of America, pp. 273-280.

Paustian, K., O. Andrén, H.H. Janzen, R. Lal, P. Smith, G. Tian, H. Tiessen, M. van Noordwijk, and P.L. Woomer. (1997). Agricultural soils as a sink to mitigate CO_2 emissions. *Soil Use and Management* 13:230-244.

Phillips, S.H. (1984). Equipment. In *No-Tillage Agriculture: Principles and Practices*, eds. R.E. Phillips and S.H. Phillips, NY: Van Nostrand Reinhold, pp. 254-269.

Poorter, H. (1993). Interspecific variation in the growth response of plants to an elevated ambient CO_2 concentration. *Vegetation* 104/105:77-97.

Post, W.M. and K.C. Kwon. (2000). Soil carbon sequestration and land-use change: Processes and potential. *Global Change Biology* 6:317-326.

Post, W.M., W.R. Emanuel, and A.W. King. (1992). Soil organic matter dynamics and the global carbon cycle. In *World Inventory of Soil Emission Potentials*, eds. N.H. Batjes and E.M. Bridges, Wageningen, The Netherlands: International Soil Reference Information Center, pp. 107-119.

Post, W.M., T.H. Peng, W.R. Emanuel, A.W. King, V.H. Dale, and D.L. DeAngelis. (1990). The global carbon cycle. *American Scientist* 78:310-326.

Potter, K.N., O.R. Jones, H.A. Torbert, and P.W. Unger. (1997). Crop rotation and tillage effects on organic carbon sequestration in the semi-arid southern Great Plains. *Soil Science* 162:140-147.

Potter, K.N., J.E. Morrison, and H.A. Torbert. (1996). Tillage intensity effects on corn and grain sorghum growth and productivity on a Vertisol. *Journal of Production Agriculture.* 9:317-390.

Potter, K.N., H.A. Torbert, O.R. Jones, J.E. Matocha, J.E. Morrison, and P.W. Unger. (1998). Distribution and amount of soil organic C in long-term management systems in Texas. *Soil and Tillage Research* 47:309-321.

Prior, S.A. and H.H. Rogers. (1992). A portable soil coring system that minimizes plot disturbance. *Agronomy Journal* 84:1073-1077.

Prior, S.A., S.G. Pritchard, G.B. Runion, H.H. Rogers, and R.J. Mitchell. (1997a). Influence of atmospheric CO_2 enrichment, soil N, and water stress on needle surface wax formation in *Pinus palustris* (Pinaceae). *American Journal of Botany* 84: 1070-1077.

Prior, S.A., D.C. Reicosky, D.W. Reeves, G.B. Runion, and R.L. Raper. (2000). Residue and tillage effects on planting implement-induced short-term CO_2 and water loss from a loamy sand soil in Alabama. *Soil and Tillage Research* 54:197-199.

Prior, S.A., H.H. Rogers, G.B. Runion, and G.R. Hendrey. (1994a). Free-air CO_2 enrichment of cotton: Vertical and lateral root distribution patterns. *Plant and Soil* 165:33-44.

Prior, S.A., H.H. Rogers, G.B. Runion, B.A. Kimball, J.R. Mauney, K.F. Lewin, J. Nagy, and G.R. Hendrey. (1995). Free-air CO_2 enrichment of cotton: Root morphological characteristics. *Journal of Environmental Quality* 24:678-683.

Prior, S.A., H.H. Rogers, G.B. Runion, and J.R. Mauney. (1994b). Effects of free-air CO_2 enrichment on cotton root growth. *Agricultural and Forestry Meteorology* 70:69-86.

Prior, S.A., H.H. Rogers, G.B. Runion, H.A. Torbert, and D.C. Reicosky. (1997b). Carbon dioxide-enriched agro-ecosystems: Influence of tillage on short-term soil carbon dioxide efflux. *Journal of Environmental Quality* 26:244-252.

Prior, S.A., H.H. Rogers, N. Sionit, and R.P. Patterson. (1991). Effects of elevated atmospheric CO_2 on water relations of soya bean. *Agriculture, Ecosystems and Environment* 35:13-25.

Prior, S.A., H.A. Torbert, G.B. Runion, G.L. Mullins, H.H. Rogers, and J.R. Mauney. (1998). Effects of CO_2 enrichment on cotton nutrient dynamics. *Journal of Plant Nutrition* 21:1407-1426.

Prior, S.A., H.A. Torbert, G.B. Runion, H.H. Rogers, C.W. Wood, B.A. Kimball, R.L. LaMorte, P.J. Pinter, and G.W. Wall. (1997c). Free-air carbon dioxide enrichment of wheat: Soil carbon and nitrogen dynamics. *Journal of Environmental Quality* 26:1161-1166.

Pritchard, S.G., H.H. Rogers, S.A. Prior, and C.M. Peterson. (1999). Elevated CO_2 and plant structure: A review. *Global Change Biology* 5:807-837.

Radin, J.W., B.A. Kimball, D.L. Hendrix and J.R. Mauney. (1987). Photosynthesis of cotton plants exposed to elevated levels of carbon dioxide in the field. *Photosynthesis Research* 12:191-203.

Reeves, D.W. (1997). The role of soil organic matter in maintaining soil quality in continuous cropping systems. *Soil and Tillage Research* 43:131-167.

Reeves, D.W., H.H. Rogers, S.A. Prior, C.W. Wood, and G.B. Runion. (1994). Elevated atmospheric carbon dioxide effects on sorghum and soybean nutrient status. *Journal of Plant Nutrition* 17:1939-1954.

Reicosky, D.C., D.W. Reeves, S.A. Prior, G.B. Runion, H.H. Rogers, and R.L. Raper. (1999). Effects of residue management and controlled traffic on carbon dioxide and water loss. *Soil and Tillage Research* 52:153-165.

Rice, C.W., F.O. Garci, C.O. Hampton, and C.E. Owensby. (1994). Soil microbial response in tallgrass prairie to elevated CO_2. *Plant and Soil* 165:67-74.

Rogers, H.H. and R.C. Dahlman. (1993). Crop responses to CO_2 enrichment. *Vegetation* 104/105: 117-131.

Rogers, H.H., G.E. Bingham, J.D. Cure, J.M. Smith, and K.A. Surano. (1983). Responses of selected plant species to elevated carbon dioxide in the field. *Journal of Environmental Quality* 12:569-574.

Rogers, H.H., W.W. Heck, and A.S. Heagle. (1983). A field technique for the study of plant responses to elevated carbon dioxide concentrations. *Air Pollution Control Association Journal* 33:42-44.

Rogers, H.H., C.M. Peterson, J.M. McCrimmon, and J.D. Cure. (1992). Response of soybean roots to elevated atmospheric carbon dioxide. *Plant, Cell and Environment* 15:749-752.

Rogers, H.H., S.A. Prior, G.B. Runion, and R.J. Mitchell. (1996). Root to shoot ratio of crops as influenced by CO_2. *Plant and Soil* 187:229-248.

Rogers, H.H., G.B. Runion, and S.V. Krupa. (1994). Plant responses to atmospheric CO_2 enrichment with emphasis on roots and rhizosphere. *Environmental Pollution* 83:155-189.

Rogers, H.H., G.B. Runion, S.V. Krupa, and S.A. Prior. (1997). Plant responses to atmospheric CO_2 enrichment: Implications in root-soil-microbe interactions. In *Advances in Carbon Dioxide Effects Research. ASA Special Publication No. 61,* eds. L.H. Allen, Jr., M.B. Kirkham, D.M. Olszyk and C.E. Whitman, Madison, WI: ASA, CSSA, and SSSA, pp. 1-34.

Rogers H.H., J.F. Thomas, and G.E. Bingham. (1983). Response of agronomic and forest species to elevated atmospheric carbon dioxide. *Science* 220:428-429.

Runion, G.B., E.A. Curl, H.H. Rogers, P.A. Backman, R. Rodriguez-Kabana, and B.E. Helms. (1994). Effects of free-air CO_2 enrichment on microbial populations in the rhizosphere and phyllosphere of cotton. *Agricultural and Forestry Meteorology* 70:117-130.

Saxena, D., S. Flores, and G. Stotzky. (1999). Transgenic plants: Insecticidal toxin in root exudates from *Bt* corn. *Nature* 402:480.

Schimel, D.S. (1995). Terrestrial ecosystems and the carbon cycle. *Global Change Biology* 1:77-91.

Schimel D., I.G. Enting, M. Heimann, T.M.L. Wigley, D. Raynaud, D. Alves, and U. Siegenthaler. (1995). CO_2 and the carbon cycle. In *Climate Change 1994: Radiative Forcing of Climate Change and an Evaluation of IPCC IS92 Emissions Scenarios,* eds. J.T. Houghton, L.G. Meira-Filho, J.P. Bruce, H. Lee, B.A. Callander, and E.S. Haites, Cambridge: Cambridge University Press, pp. 35-71.

Schlesinger, W.H. (1984). Soil organic matter: A source of atmospheric CO_2. In *The Role of Terrestrial Vegetation in the Global Carbon Cycle,* ed. G.M. Woodwell, NY: John Wiley, pp. 111-127.

Schlesinger, W.H. (1986). Changes in soil carbon storage and associated properties with disturbance and recovery. In *The Changing Carbon Cycle: A Global Analysis,* eds. J.R. Trabalka and D.E. Reichle, NY: Springer-Verlag, pp. 194-220.

Schlesinger, W.H. (1990). Evidence from chronosequence studies for a low carbon-storage potential of soils. *Nature* 348:232-234.

Sionit, N., H.H. Rogers, G.E. Bingham, and B.R. Strain. (1984). Photosynthesis and stomatal conductance with CO_2-enrichment of container and field-grown soybeans. *Agronomy Journal* 65:207-211.

Smucker, A.J.M., S.L. McBurney, and A.K. Srivastava. (1982). Quantitative separation of roots from compacted soil profiles by the hydropneumatic elutriation system. *Agronomy Journal* 74:500-503.

Strain, B.R. and J.D. Cure. (1994). *Direct effects of atmospheric CO_2 enrichment on plants and ecosystems: An updated bibliographic data base, ORNL/CDIAC-70.* Oak Ridge, TN: The Carbon Dioxide Information Analysis Center, Oak Ridge National Laboratory.

Sundquist, E.T. (1993). The global carbon dioxide budget. *Science* 259:934-941.

Thomas, J.F. and C.N. Harvey. (1983). Leaf anatomy of four species grown under continuous CO_2 enrichment. *Botanical Gazette* 144:303-309.

Thomas, R.R. and B.R. Strain. (1991). Root restriction as a factor in photosynthetic acclimation of cotton seedlings grown in elevated carbon dioxide. *Plant Physiology* 96:627-634.

Throckmorton, R.I. (1986). Tillage and planting equipment for reduced tillage. In *No-Tillage and Surface-Tillage Agriculture: The Tillage Revolution*, eds. M.A. Sprague and G.B. Triplett, NY: John Wiley & Sons, pp. 59-91.

Torbert, H.A., S.A. Prior, and D.W. Reeves. (1999). Land management effects on nitrogen and carbon cycling in an Ultisol. *Communications in Soil Science and Plant Analysis* 30:1345-1359.

Torbert, H.A., S.A. Prior, and H.H. Rogers. (1995). Elevated atmospheric carbon dioxide effects on cotton plant residue decomposition. *Soil Science Society of America Journal* 59:1321-1328.

Torbert, H.A., S.A. Prior, H.H. Rogers, W.H. Schlesinger, and G.L. Mullins. (1996). Elevated atmospheric carbon dioxide in agro-ecosystems affects groundwater quality. *Journal of Environmental Quality* 25:720-726.

Torbert, H.A., H.H. Rogers, S.A. Prior, W.H. Schlesinger, and G.B. Runion. (1997). Effects on elevated atmospheric CO_2 in agro-ecosystems on soil carbon storage. *Global Change Biology* 3:513-521.

Torbert, H.A., S.A. Prior, H.H. Rogers, and C.W. Wood. (2000). Elevated atmospheric CO_2 effects on agro-ecosystems: Residue decomposition processes and soil C storage. *Plant and Soil* 224:59-73.

Unger, P.W. (1986). Wheat residue management effects on soil water storage and corn production. *Soil Science Society of America Journal* 50:764-770.

Von Wetsein-Knowles, P.M. (1993). Waxes, cutin, and suberin. In *Lipid Metabolism in Plants*, ed. T.S. Moore, Boca Raton, FL: CRC Press, pp.127-166.

Weschsung, G., F. Weschsung, G.W. Wall, F.J. Adamsen, B.A. Kimball, R.L. Garcia, P.J. Pinter, Jr., and Th. Kartschall. (1995). Biomass and growth rate of a spring wheat root system grown in free-air CO_2 enrichment (FACE) and ample moisture. *Journal of Biogeography* 22:623-634.

Weschsung, G., F. Weschsung, G.W. Wall, F.J. Adamsen, B.A. Kimball, P.J. Pinter, Jr., Th. Kartschall, R.L. Garcia, and R.L. LaMorte. (1999). The effects of free-air CO_2 enrichment and soil water availability on spacial and seasonal patterns of wheat root growth. *Global Change Biology* 5:519-529.

Whipps, J.M. (1990). Carbon Economy. In *The Rhizosphere*, ed. J.M. Lynch, NY: John Wiley, pp. 59-97.

Wilson, A.T. (1978). Pioneer agriculture explosion and CO_2 levels in the atmosphere. *Nature* 273, 40-41.

Wittwer, S.H. (1995). *Food, Climate, and Carbon Dioxide: The Global Environment and World Food Production.* Boca Raton, FL: CRC Press.

Wood, C.W., H.A. Torbert, H.H. Rogers, G.B. Runion, and S.A. Prior. (1994). Free-air CO_2 enrichment effects on soil carbon and nitrogen. *Agricultural and Forestry Meteorology* 70:103-116.

Wullschleger, S.D., L.H. Ziska, and J.A. Bunce. (1994). Respiratory responses of higher plants to atmospheric CO_2 enrichment. *Physiologia Plantarum* 90:221-229.

Zak, D.R., K.S. Pregitzer, P.S. Curtis, J.A. Teeri, R. Fogel, and D.L. Randlett. (1993). Elevated atmospheric CO_2 and feedback between carbon and nitrogen cycles. *Plant and Soil* 151:105-117.

Ziska, L.H., J.R. Teasdale, and J.A. Bunce. (1999). Future atmospheric carbon dioxide may increase tolerance to glyphosate. *Weed Science* 47:608-615.

Weed Biology, Cropping Systems, and Weed Management

Douglas D. Buhler

SUMMARY. Weeds pose a recurrent threat to agricultural productivity in both industrialized and developing countries. Weeds respond dynamically to all cropping practices, and therefore, the design and function of cropping systems plays a central role in the composition of weed communities. The unique and challenging nature of weed communities requires more integrated approaches to weed management than are currently being employed by most growers. Integrating weed management with cropping system design and application may be an effective approach to diversifying weed management systems. Each crop-weed system is a unique mix of genetics and biology and will respond dynamically to changes in management practices. Practices such as crop rotation, tillage, cover crops, and fertility management modify weed populations. The challenge is to integrate these and other practices with the best available control tactics to generate integrated management systems. Cropping system design provides an excellent framework for developing and applying integrated approaches to weed management because it allows for new and creative ways of meeting the challenge of managing weeds. Weed science must integrate the theories and application of weed man-

Douglas D. Buhler is Chair and Professor, Department of Crop and Soil Sciences, Michigan State University, 286 Plant and Soil Sciences Building, East Lansing, MI 48824-1325 (E-mail: buhler@msu.edu).

[Haworth co-indexing entry note]: "Weed Biology, Cropping Systems, and Weed Management." Buhler, Douglas D. Co-published simultaneously in *Journal of Crop Production* (Food Products Press, an imprint of The Haworth Press, Inc.) Vol. 8, No. 1/2 (#15/16), 2003, pp. 245-270; and: *Cropping Systems: Trends and Advances* (ed: Anil Shrestha) Food Products Press, an imprint of The Haworth Press, Inc., 2003, pp. 245-270. Single or multiple copies of this article are available for a fee from The Haworth Document Delivery Service [1-800-HAWORTH, 9:00 a.m. - 5:00 p.m. (EST). E-mail address: docdelivery@haworthpress.com].

agement into cropping system design based on the unique characteristics of weed communities and the available weed management options. *[Article copies available for a fee from The Haworth Document Delivery Service: 1-800-HAWORTH. E-mail address: <docdelivery@haworthpress.com> Website: <http://www.HaworthPress.com>* © *2003 by The Haworth Press, Inc. All rights reserved.]*

KEYWORDS. IPM, integrated weed management, weed biology, weed ecology

INTRODUCTION

Weeds and weed problems are anthropocentric terms applied to various species and populations of plants (Ghersa et al., 1994; Harlan and de Wet, 1965). While many definitions have been proposed, plants are generally considered weeds when they are undesirable in a particular setting. Weeds typically originate when wild species adapt or are introduced to agricultural lands, when cultivated species escape domestication and persist as weeds, and when new weeds are generated due to hybridization and introgression between crop and wild species (Baker, 1974; Salisbury, 1961).

Despite the extensive use of technology and human labor, weeds continue to account for substantial economic costs and crop yield losses in both industrialized and developing countries. In the United States, weeds and weed control have an estimated annual economic cost of more than $15 billion (Bridges, 1994). In many developing countries the relative costs are even greater and hand labor for weed control consumes up to half of the total labor demand for crop production (Akobunudu, 1991). Akobunudu (1987) estimated that weeds reduce crop yields by 5% in the most developed countries, 10% in the less developed countries, and 25% in the least developed countries. Because weeds pose a recurrent and nearly ubiquitous threat to crop productivity and profitability, and because weeds respond dynamically to a wide range of cropping practices, the design and function of cropping systems should play a central role in weed management (Liebman and Gallandt, 1997).

Concern over the economic costs, environmental impacts, and long-term efficacy of current practices has reinforced the need for broader approaches to weed management (Buhler, 1999; Wyse, 1992). Herbicides are important tools for weed control and have improved production efficiency and facilitated reduced tillage production systems. Because of their effectiveness, herbicides and tillage are the dominant practices in many production systems. While the efficacy of these practices is evident, they may also lead to environmental contamination, human health problems, and soil erosion. In addition, weeds persist by adapting to production practices and by developing resistance to herbicides (Heap, 1999).

Crop and their associated management practices create a matrix of resource conditions, mortality events, and stresses that regulate the ability of various weed species to survive and proliferate (Liebman and Ohno, 1998). Management practices and crop characteristics that create this matrix interact with all phases of the weed life cycle. Weed population and community dynamics are affected by tillage systems (Buhler, 1995) and seedbed preparation (Mohler, 1993), residues of previous crops (Putnam and DeFrank, 1983; Einhelig and Rasmussen, 1989), cover crops (Teasdale, 1998), fertilizer application (Freyman, Kowalenko, and Hall, 1989; Dyck and Liebman, 1994), and manure, compost, and other organic amendments (Bloemhard et al., 1992; Conklin et al., 1998; Kennedy and Kremer, 1996). Weed and crop resource use, competitive interactions, growth, and reproduction are governed by factors such as planting density and spatial patterns (Minotti, 1991; Teasdale, 1995), fertility management (Alkamper, 1976; Tollenaar et al., 1994), crop species and cultivar (Callaway, 1992; Pester, Burnside, and Orf, 1999), and rotational sequence of crops (Crookston et al., 1991; Karlen et al., 1994; Liebman and Ohno, 1998). Once weeds are established with a crop, plants may be injured or killed by post-plant tillage (Buhler and Gunsolus, 1996; Mulder and Doll, 1994), grazing (Dowling and Wong, 1993; Thomsen et al., 1993), herbicides (Burnside et al., 1986; Schweizer and Zimdahl, 1984), or pathogens and insects (Harley and Forno, 1992; Charudattan and DeLoach, 1988).

Basic principles of plant ecology govern the behavior of weeds within cropping systems. Resource availability and utilization define competition both within and between crop and weed species. The competitive interactions of greatest concern in cropping systems are those that consume resources that could be used for crop growth and yield (Kropff, 1993). However, interactions among plants are not limited to resource competition. The release of phytotoxins into the environment and subsequent growth reduction of susceptible plants, known as allelopathy, is another important component of plant-to-plant interactions (Rice, 1995). Competition and allelopathy are distinct mechanistically, but are usually difficult to distinguish under field conditions. The term interference is often used to describe the combined effect of resource competition and allelopathy (Gliessman, 1986).

Crops are usually planted at a density and spatial pattern chosen to facilitate field operations and maximize yield potential (Stoskopf, 1985). Because the goal is total yield rather than maximum yield per individual plant, intraspecific competition commonly limits the yield of individual crop plants, demonstrating that even under weed-free conditions supplies of many resources are at suboptimal levels (Liebman and Gallandt, 1997). With weeds added to the system, the intensity of resource competition is intensified.

While resources almost always become limiting at some point in the growing season, the resources unused by the crop, especially soon after crop planting or after crop harvest, contribute greatly to the ability of weeds to infest crop land. In annual crops, large areas of bare soil with an ample supply of light, water, and nutrients are usually available to weedy vegetation for a significant portion of the growing season. Since soil devoid of vegetation is not a natural condition (Harper, 1977), nature moves to fill this void, commonly with plants considered weeds. Thus, it can be argued that the most basic condition that allows weeds to persist is the availability of resources not used by the crop (Harper, 1977; Radosevich, Holt, and Ghersa, 1997) and periods of excess resources may be more important in defining weed communities than periods of limiting resources. The majority of the research on resource dynamics has focused on resource competition after a weed population has become established, with comparatively little attention paid to the conditions that allow the weeds to exist in the first place.

When the impact of cropping systems on weeds is considered, temporal aspects of crop growth and development, crop harvest, tillage, and all other disturbances are critical components. For example, the ability of crop rotation to reduce weed densities is commonly attributed to differences in planting, harvesting, and tillage dates among crop species (Liebman and Ohno, 1998). Chancellor (1985) concluded that the season of planting the crop was one of the most important factors determining the species composition of the weed community.

Continuous production of a single crop species, or species with the same life cycle that are produced using the same set of management practices, exposes weeds to a consistent pattern of selection pressure. Such practices shift the composition of weed populations and communities toward species and genotypes best adapted to succeed under a recurring set of production conditions (Ghersa et al., 1994). The resulting weed communities are often difficult to control and highly competitive in the production system. The objective of this review is to examine the impacts of cropping practices on weeds and discuss ways in which these practices may be used in weed management systems.

CROPPING PRACTICES AND WEEDS

Crop Rotation

Crop rotation is often identified as a critical cultural component of weed management (Brust and Stinner, 1991; Leighty, 1938; Liebman and Gallandt, 1997). A substantial body of descriptive research has documented effects of crop rotation on weed population dynamics (i.e., Barberi, Silvestri, and Bonari,

1997; Cavers and Benoit, 1989; Loeppky and Derksen, 1994; Schweizer, Lybecker, and Zimdahl, 1988; Warnes and Andersen, 1984). Liebman and Ohno (1998) summarized the results of 25 test crop by rotation combinations for which comparisons with appropriate monoculture systems were possible. In 19 cases, weed plant density in rotation was less than in monoculture, higher than monoculture in 2 cases, and equivalent with monoculture in 4 cases. In the 12 cases where weed seed data were compared with monocultures of the component crops, weed seed density in the soil with crop rotation was lower in 9 cases and equivalent in 3 cases. Yields of test crops were higher in rotation than in monoculture in 9 of 12 cases where the crop yield was reported. These results support the conclusion of Leighty (1938) that "weed problems are likely to be least on farms where crop diversification is practiced and most severe on farms devoted for one reason or another to a single crop."

The regeneration niche of individual weed species can be disrupted and the buildup of adapted weed species reduced by rotating crops. Maximizing differences in planting dates and growth periods, tillage practices, competitive characteristics, and weed control practices increase the effectiveness of crop rotation. For example, downy brome (*Bromus tectorum* L.) density remained relatively constant when winter wheat (*Triticum aestivum* L.) was rotated with oilseed rape (*Brassica napus* L.), but downy brome density increased rapidly when wheat was grown continuously (Blackshaw, 1994a). Schreiber (1992) observed that giant foxtail (*Setaria faberi* Herrm.) density was greatest in continuous corn (*Zea mays* L.), intermediate in a corn/soybean [*Glycine max* (L.) Merr.] rotation, and lowest in a corn/soybean/winter wheat rotation.

Forage crops are also an important part of many crop rotations and offer diverse mechanisms to suppress weeds through competition, mowing, and grazing (Gill and Holmes, 1997; Liebman and Davis, 2000). In addition, if the forage is a perennial species, soil disturbance is eliminated for several years and the seed bank of annual species may decline due to the elimination of seed production at a time of continued seed germination, predation, and decay.

While quantitative responses of weeds to crop rotation have been documented, much less is known about the ecological mechanisms by which crop rotations affect weed communities (Jordan et al., 1995). Weeds respond to crop rotation when specific environmental conditions created by crop rotations affect weed demography and subsequent population dynamics (Liebman and Janke, 1990). Rotations may affect demography of plant survival, propagule production, or propagule survival and germination in the soil. An understanding of these processes could be an important aspect for developing crop rotations that benefit weed management with minimum economic cost and interference with other aspects of crop production.

Intercropping

Intercropping is one of the most widely available and inexpensive methods for increasing crop production per unit area of land (Plucknett and Smith, 1986). There is considerable evidence that the simultaneous culture of two or more crops on the same piece of land will produce a greater yield than a monoculture of any of the component crops (Barker and Francis, 1986). In regions of the world such as Latin America, Asia, and Africa, intercropping is the dominant cropping method (Harwood and Price, 1976; Okigbo and Greenland, 1976). While most commonly practiced on small farms with minimal mechanization or chemical technology, intercropping need not be restricted to such situations (Jagtap and Adeleye, 1999).

Most of the mechanical and chemical methods of weed control in monoculture can be used in intercropping (Carruthers et al., 1998; Liebman, 1988). Therefore, I will focus the discussion on how the changes in biological and ecological processes of intercropping systems affect weeds and weed management. Competitive suppression of weeds can take a very different form with intercropping than in crop monocultures. Increasing the complexity of a cropping system by interplanting species of differing growth forms, phenologies, and physiologies can create different patterns of resource availability to weeds, especially light (Ballare and Casal 2000). Because a more diverse crop population can capture resources more efficiently, these resources may be converted to a crop yield rather than leaving them available for weed growth. Thus, understanding how intercrops compete with weeds requires a thorough understanding of the growth and development characteristics of the crop and weed species in the intercropping system. Because resource availability is key to weed occurrence (Harper, 1977; Radosevich, Holt, and Ghersa, 1997), increasing resource utilization through intercropping may provide unique opportunities for weed management.

Cover and Smother Crops

Cover crops are included in cropping systems for long- and short-term improvements in soil fertility and crop performance. Long-term benefits are derived from reduced soil erosion, improved soil quality, and increased soil organic matter (Power, 1996). Short-term responses are the result of changes in radiation balance, soil temperature and moisture, nutrient availability, runoff and infiltration, crop establishment, and pest populations. When cover crops are used for weed control, the goal is to replace an unmanageable weed population with a manageable cover crop (Teasdale, 1998). This is accomplished by managing the cover crop to preempt niches previously available to weeds. There are at least two major types of cover crops that can be used for weed control: (1) off-season cover crops and (2) smother crops (a cover crop grown during part or all of the cropping season). When using off-season cover crops, the goal is to produce sufficient plant residue and/or allelochemicals to create an unfavorable environment for weed germination and establishment. When using a smother crop, the goal is usually to displace weeds from the harvested crop through resource competition.

In temperate regions, winter annual species are often used as off-season cover crops. These winter annual species are planted in late summer or autumn, become established before winter, and produce most of their biomass during the spring prior to planting a summer crop (Teasdale, 1998). The most appropriate species varies by region due to growth requirements and winter hardiness (Holderbaum et al., 1990; Johnson et al., 1998). Winter annual cover crops are most effective in areas where there is a sufficient establishment and growth period in the autumn and soil moisture is not a major limiting factor for crop growth in the spring.

There are many reports in the literature of variable effects of off-season cover crops on weed population dynamics (e.g., Creamer et al., 1996; Mohler and Teasdale, 1993; Putnam and DeFrank, 1983; Smeda and Weller, 1996). Because cover crops are not always effective in reducing weed densities, we need to develop a better understanding of the mechanisms by which cover crops change weed population dynamics, including how the growth of cover crops and the subsequent degradation of their residues change weed/soil interactions in multi-year cropping systems. As Teasdale (1998) concluded, "attention should be focused on defining the impact of cover crops on important rate-defining steps in the life cycle of weeds. This knowledge will help characterize how to use cover crops most effectively to disrupt the succession of important weed species."

Another approach to using cover crops for weed control is using a cover crop that remains alive for part or all of the cropping season, usually referred to as a smother crop or smother plant (Buhler, Kohler, and Foster, 1998; DeHaan et al., 1994). Competition for resources is usually the major mechanism of weed suppression by smother crop species. After conducting research using yellow mustard (*Brassica hirta* Moench) as a smother crop in corn, DeHaan et al. (1994) proposed that an ideal spring-seeded smother crop for weed control in corn in the North Central United States would have rapid seedling emergence under cool soil conditions, horizontal leaf angle, leaf size of 2 by 3 cm, rooting depth of 2.5 cm, maximum height of 10 cm, a life cycle of 5 weeks or less, and produce nondormant seed.

Buhler, Kohler, and Foster (1998) found that Berseem clover (*Trifolium alexandrinum* L.), annual medics (*Medicago* spp.), or yellow mustard provided 20 to 90% weed suppression in corn and soybean. However, weed suppression and crop yields were highly variable among smother crop species and locations. Rapid and uniform establishment of the smother crop was essential for smother crops to gain a competitive advantage over weeds. Smother crops represent a promising approach to weed control and the relationships among allelopathy, resource competition, weed suppression, and crop yield in smother crop systems warrant further study.

Tillage Systems and Cultivation

Some form of soil disturbance is a component of virtually all cropping systems. Since tillage affects the environment where weeds survive, weed communities are influenced by all forms of tillage (Buhler, 1995). Tillage for seed bed preparation can reduce densities of annual weed populations, especially if planting is delayed to allow weed seed germination prior to the final tillage operation (Buhler and Gunsolus, 1996; Gunsolus, 1990).

Tillage buries crop residues and alters the characteristics of the surface soil, regulating the germination environment of seeds and other propagules by reducing soil surface cover (Aase and Tanaka, 1987), altering soil temperature and moisture patterns (Al-Darby and Lowery, 1987; Johnson, Lowery, and Daniel, 1984), and altering weed seed distribution in the soil (Staricka et al., 1990; Yenish, Doll, and Buhler, 1992). The physical disruption caused by tillage (Pareja and Staniforth, 1985) and the variation in temperature requirements for germination among weed species (Egley and Williams, 1991; Wiese and Binning, 1987), in combination with tillage-induced changes in soil properties greatly influences weed population dynamics in crop production systems.

Tillage may also influence weed seed predation. Brust and House (1988) found that tillage reduced seed predation compared with untilled soil. In contrast, Cardina et al. (1996) found no difference in velvetleaf (*Abutilon theophrasti* Medikus) seed predation between tilled and untilled plots. To further confuse the matter, Cromar, Murphy, and Swanton (1999) found the highest predation in no-tillage and moldboard plowed environments and lowest predation in chisel-plowed environments.

In addition to affecting the growth environment of weeds, tillage systems impact the weed control options available to producers and the efficacy of those options. Certain herbicides may not be used in no-tillage systems because of the need for mechanical incorporation into the soil after application. Herbicide efficacy may also be altered by the presence of plant residues on the soil surface and changes in weed population dynamics (Buhler, 1995). With less tillage and more plant residues on the soil surface, mechanical weed control operations may become less effective (Springman et al., 1989). However, interrow cultivation in ridge-tillage systems is not hindered by surface residues (Buhler, 1992; Forcella and Lindstrom, 1988) and it has been shown that combining interrow cultivation with reduced herbicide rates can provide weed control similar to full-rate herbicide treatments in conservation tillage systems (Buhler et al., 1995; Mulder and Doll, 1993).

Trends in weed population dynamics in conservation tillage identified in the central United States included increased densities of annual grass, small-seeded broadleaf, winter annual, and perennial species; decreased densities of large-seeded broadleaf species; and reduced efficacy of soil-applied herbicides (Buhler, 1995). However, it is important to note that location, weather patterns, the type of conservation tillage system used, and weed control practices regulate responses in individual fields. Derksen et al. (1993) found that an increased association of perennial and annual grass species did not generally occur with conservation tillage in Saskatchewan, Canada. Wind-dispersed species and volunteer crops were most commonly associated with conservation tillage and summer annual dicots with conventional tillage. More important, species responded differently among sites or within a site over time. This indicated that weed control practices and other cultural practices used on the fields interacted with tillage to regulate weed populations.

Crop Competitiveness

Enhancing the ability of crops to compete with weeds is an attractive approach to improving weed management systems (McWhorter and Barrentine, 1975; Pester, Burnside, and Orf, 1999). Increasing crop competitiveness to weeds can be accomplished through both improved crop management practices and plant breeding efforts.

Enhancing the ability of a crop to compete with weeds can be accomplished by providing the best possible environment for crop growth combined with practices that reduce the density and/or vigor of the weeds. Practices such as narrow row spacing, increased plant density, appropriate time of planting, and fertility management are capable of shifting the competitive balance to favor crops over weeds (Buhler and Gunsolus, 1996; Malik, Swanton, and Michaels, 1993; Teasdale, 1995). Decreasing row spacing and/or increasing crop plant densities has been shown to increase competitiveness of many crops (i.e., Hauser and Buchanan, 1982; Malik, Swanton, and Michaels, 1993; Stoller et al., 1987; Teasdale, 1995; Teasdale and Frank, 1983). Teasdale (1995) found that growing corn in 38-cm-wide rows with increased density (compared with 76-cm-wide rows) improved weed control and reduced herbicide requirements. Malik, Swanton, and Michaels (1993) found differences in competitiveness among white bean (*Phaseolus vulgaris* L.) cultivars and that decreasing row spacing decreased yield losses due to weeds. In an extensive review of the effects of cultural practices on soybean/weed interactions, Stoller et al. (1987) concluded that soybean cultivar selection, row spacing, plant density, planting date, crop rotation, tillage, and herbicides can all be used to maximize the ability of soybean to compete with weeds. In addition to affecting crop growth, cultural practices also affect the nature of the weed population (Buhler and Gunsolus, 1996). For example, delaying planting 14 days reduced weed densities in soybean 25 to 90%, depending on weed species. These reductions in weed densities also increased the effectiveness of mechanical weed control treatments.

Differential weed competitive ability has been documented among commonly grown cultivars of several important crop species and several authors have suggested that weed management could be improved through crop breeding (i.e., Callaway, 1992; Challaiah et al., 1986; Lanning et al., 1997; Pester, Burnside, and Orf, 1999). It should be feasible to breed crop cultivars that are genetically superior competitors with weeds through crop tolerance to weeds (maintain yield in presence of weeds) or crop interference with weeds (suppress growth of weeds). Once we understand the genetics of crop tolerance and/or competitiveness, methods used for developing these cultivars will depend on the type of environment in which they will be grown (Martinez-Ghersa, Ghersa, and Satorre 2000). If the goal is to grow the crop in association with highly variable weed populations, plastic phenotypes will be required to respond to the variation in time and space. If the crop is to be grown in association with a uniformly competitive cover crop, a more rigid phenotype will be desired.

Characteristics commonly associated with crop competitiveness with weeds included rapid germination and root development, rapid early vegetative growth and vigor, rapid canopy closure and high leaf area index, profuse tillering or branching, increased leaf duration, and greater plant height (Callaway, 1992; Pester, Burnside, and Orf, 1999). In rice (*Oryza sativa* L.), leaf area index and biomass production early in the growing season were the traits most closely associated with competitiveness against weeds (Garrity, Movillon, and Moody, 1992). In dry bean (*Phaseolus vulgaris* L.), leaf area index and leaf size accounted for 73% of the total variation in weed biomass production (Wortmann, 1993). The most consistent conclusion among many studies has been that vigorous growth characteristics enhance weed competitiveness by reducing light quantity and quality beneath the crop canopy.

In addition to crop yield responses, enhanced crop competitiveness may also reduce the reproductive capacity of weeds and reduce weed problems in subsequent years. For example, cheat (*Bromus secalinus* L.) produced more seed when grown with semi-dwarf than tall cultivars of winter wheat (Koscelny et al., 1990). Blackshaw (1994b) concluded that planting a more competitive cultivar of winter wheat resulted in higher wheat yields, less downy brome seed, and lower weed infestations in subsequent crops. Given the importance of resource availability to weed establishment and persistence, reducing resource availability to weeds through crop interference may provide a useful component to weed management.

Edaphic Factors

The interaction of edaphic factors and the occurrence of specific weed species is an area of considerable speculation. Weeds as indicator species is often discussed, but poorly documented. Some have gone as far as to postulate that "weeds are witness of man's failure to master the soil . . . they only indicate our errors and Nature's corrections. Weeds want to tell a story, they are Nature's means of teaching, and their story is interesting" (Pfeiffer, 1970). Others have suggested that since most weeds are colonizing, ephemeral species that respond to soil disturbance and resource availability, soil properties have a secondary influence on weed community composition of annual species (Smith, 1970). However, perennial species are subjected to the specific conditions in a locality for several years and may be affected more by soil conditions than annual species. In general, it appears that within the range of adaptation, most weed species can be found in soils with widely differing characteristics. This adaptability may be an essential characteristic of a successful weed species (Zimdahl, 1993).

While mechanistic relationships may be difficult to quantify, it is likely that soil properties play some role in regulating weed population dynamics. Soil conditions in agricultural fields are usually maintained within a relatively narrow range to promote crop productivity, thus reducing the potential variability to which weeds may be exposed compared to that of natural systems (Harper, 1977). However, significant variation in soil properties may exist both within and among fields in a narrow geographic area (Mulla, 1993) and these differences may be correlated to weed occurrence. At the whole field scale, crop type and clay content helped explain variation in weed species presence and abundance in 316 fields planted to eight crops over three years in Denmark (Andreasen, Streibig, and Haas, 1991). In Finland (Ervio et al., 1994), the occurrence of the 16 most common weeds in 706 cereal and vegetable fields were related to soil properties including soil type, pH, and calcium concentration. In Saskatchewan, the abundance of weed species separated into community groups correlated with soil zones and the associated climatic gradient (Dale, Thomas, and John, 1992).

At the level of an individual field, Dieleman et al. (2000) identified associations between soil properties and weed species abundance. However, variations in the associations over years were substantial and were attributed to differences in agronomic and weed control practices for different crops, as well as the stochastic environmental variation from year to year. Swamp smartweed (*Polygonum amphibium* L.) was associated with high phosphorous and clay content and low sand content, while ivy-leafed speedwell (*Veronica hederifolia* L.) was correlated to sand content (Hausler and Nordmeyer, 1995). Extreme conditions such as flooding, salinity, alkalinity, and acidity also affect weed occurrence (Holm et al., 1977).

While specific soil conditions have been associated with weed infestations, it should also be recognized that these same soil conditions may reduce the vigor of the crop, making the crop less competitive with weeds. Therefore, the weeds associated with a specific soil condition may be a secondary effect related to crop vigor rather than a weed response to soil conditions (Buhler, Liebman, and Obrycki, 2000). In general, we have a poor understanding of the interactions of soil properties, crop vigor, and weed infestations. A better understanding of these relationships would help explain much of the spatial variability in weed populations. These patterns are becoming of increasing importance as site-specific weed management is developed and implemented (Mortensen, Dieleman, and Johnson, 1998).

Fertility Management

Management of soil fertility may create spatial and temporal changes in the soil environment that may affect weeds and weed management. Over the years, numerous studies and reviews on the effects of fertilization on crop/weed interactions have generated conflicting conclusions (e.g., Alkamper, 1976; Kirkland and Beckie, 1998; Tilman et al., 1999; Vengris, Colby, and Drake, 1955). There is little question that adding fertilizer to the soil can benefit weeds as much or more than the crop, but specific responses vary depending on the nature of the weed/crop association. For example, Carlson and Hill (1986) observed that nitrogen fertilization increased wheat yield in the presence of 3 plants m^{-2} of wild oat (*Avena fatua* L.), but the same fertilizer treatment reduced wheat yield at wild oat densities of 8 to 50 plants m^{-2}.

The impact of weeds may be reduced by management strategies that maximize the uptake of nutrients by crops and minimize the availability of nutrients to weeds (DiTomaso, 1995). Applying fertilizer 5 cm below the soil surface in every second interrow space reduced weed biomass by 55% and weed density by 10% while increasing barley (*Hordeum vulgare* L.) grain yield by 28% compared with a broadcast application (Rasmussen, Rasmussen, and Petersen, 1996). Weed density, biomass, and N uptake was 20 to 40% less and wheat yield was 12% more where fertilizer was banded beside the crop row compared with broadcast application (Kirkland and Beckie, 1998). The addition of N fertilizer was detrimental to green foxtail [*Setaria viridis* (L.) Beauv.] regardless of placement because of enhanced crop growth. In soils characterized by acidic conditions and low P availability, liming and P applications favored the growth and yield of the barley more than it favored weed growth (Légère,

Simard, and Lapierre, 1994). Other methods to alter the relative availability of nutrients to crops and weeds include timing of fertilizer applications (Anderson, 1991), altering nutrient sources (DeLuca and DeLuca, 1997), or altering nutrient availability using materials such as nitrification inhibitors (Teyker, Hoelzer, and Liebl, 1991).

A more fundamental method of manipulating the relative uptake of nutrients by crop and weeds may be through enhancing the mechanisms and kinetics of mineral uptake by crop plants. In almost all cases where nutrient concentrations in crops and associated weeds were compared, accumulation of nutrients in the weeds exceeded the levels in the corresponding crop (see DiTomaso, 1995). Therefore, to maximize nutrient uptake by crops in competition with weeds, we need to develop a better understanding of the mechanisms for nutrient fluxes in the roots of weed and crop plants. Managing fertilizer application to benefit the crop may not only increase nutrient uptake by the crop, but likely will improve the competitiveness of the crop for other resources that might otherwise be available for weeds (Anderson et al., 1998; Kirkland and Beckie, 1998).

Organic Amendments/Weed Suppressive Soils

Organic matter amendments have long been used to enhance soil fertility, improve soil structure, and recycle waste products. In addition to enhancing soil properties, amending soils with organic amendments may affect weed seed survival, emergence, growth, and reproduction. Organic matter amendments alter temporal patterns of nutrient availability, especially for nitrogen and phosphorous. Compared to pulsed application of synthetic fertilizers, composted manure (DeLuca and DeLuca, 1997) and fresh plant residues and manures (Gallandt et al., 1998) release nutrients more slowly over a longer period of time. Because germination and early growth of many weed species are strongly dependent on soil nutrient concentrations (DiTomaso, 1995; Karssen and Hilhorst, 1992), shifts in the timing of nutrient availability of organic amendments compared with synthetic fertilizers may affect weed density, emergence timing, and community composition. These responses may be similar to those discussed in the section on fertility management.

Organic matter amendments are a source of non-nutrient compounds that may affect plant growth. Some of these compounds are growth-inhibiting, whereas others are growth-promoting. Depending on its age, decomposition status, and time of application, manure and other organic materials may release short chain organic acids, phenols, ammonia, and other organic compounds at concentrations high enough to be phytotoxic (Ozores-Hampton et al., 1999). Mature compost can serve as a source of growth-stimulating substances. Valdrighi et al. (1996) reported that compost-derived humic acids increased the fresh weight of chicory (*Cichorium intybus* L.) plants by as much as 350%.

Organic matter amendments also contain persistent forms of organic carbon that may affect soil physical properties, such as water-holding capacity (Serra-Wittling, Houot, and Barriuso, 1996), temperature and thermal conductivity (Al Kayassi et al., 1990), and aggregate stability and porosity (Guidi, Pagliai, and Giachetti, 1981). Direct effects on weeds could include changes in moisture and temperature and its related effects on timing of germination and seedling development (Mester and Buhler, 1991; Wiese and Binning, 1987) and aggregation and porosity effects on the abundance of suitable regeneration niches (Gallandt, Liebman, and Huggins, 1999).

Organic matter amendments may also increase soil microbial biomass and activity, and change the incidence and severity of soil-borne diseases of weeds and crops. Fraser et al. (1988) found that amending soil with beef (*Bos* spp.) manure increased soil microbial biomass, respiration rate, dehydrogenase activity, and fungal and bacterial population densities. Conklin et al. (1998) reported that wild mustard (*Brassica kaber* L.) seedlings grown in soil amended with compost and red clover (*Trifolium pratense* Sibth.) residue were smaller and had a higher incidence and severity of *Pythium* infection than seedlings grown in soil receiving ammonium nitrate fertilizer; corn seedlings were unaffected by soil amendment treatments.

Kennedy and Kremer (1996) suggested that it might be possible to develop farming practices that created "weed suppressive soils" in which microbial community composition and activity are altered in ways that would lead to depletion of the weed seed bank, reduced probabilities of weed seedling establishment, and reduced weed growth and competitive ability. They suggested that this may be accomplished by managing residue and microbial activity with methods such as no-tillage to establish an area of increased seed decay potential within the residue zone. In a summary of recent research on biological control of weeds in Europe, Muller-Scharer, Scheepens, and Greaves (2000) also addressed this approach by concluding that one of the major routes to developing biological weed control systems should be an ecological approach. This approach is based on a better understanding of the interactions among the crop, the weed, the natural antagonist, and the environment, which must be managed in order to maximize the spread and impact of an indigenous antagonist on the weed.

Site-Specific Management

Site-specific agriculture has been defined as "an information and technology based agricultural management system to identify, analyze, and manage spatial and temporal variability within fields for optimum profitability, sustainability, and protection of the environment" (Robert, Rust, and Larson, 1994). This concept has direct application to weed management because of the spatial and temporal heterogeneity of weed populations across agricultural landscapes (Cardina, Sparrow, and McCoy, 1996; Johnson, Mortensen, and Martin, 1995). The spatial and temporal variation in weed populations are the result of the many interactions between plants and their environment as discussed elsewhere in this review.

While it is evident that weeds are not uniformly distributed across fields, most weed control practices are applied uniformly. This uniform application of inputs over the nonuniform weed population has been identified as an important source of inefficiency in weed management (Cardina, Johnson, and Sparrow, 1997). Large portions of crop fields are often below threshold densities when the average field density is above the threshold (Cardina, Sparrow, and McCoy, 1995; Johnson, Mortensen, and Martin, 1995). Cardina, Sparrow, and McCoy (1996) found that with a threshold of 10 weed plants m^{-2}, about 40% of a field did not require treatment one year, but 90% of the same field required treatment in another year. In a simulation analysis, Johnson, Mortensen, and Martin (1995) found that as the threshold level was increased, a larger proportion of the field would not require herbicide application. However, the relationship between mean density and percent weed-free area was not consistent.

One method of dealing with weed patchiness is to develop methods to detect or map weeds and use that information to spatially direct herbicide application. Optical reflectance and image analysis are two approaches that have been used for real-time sensing of weeds to operate "patch sprayers" (Woebbecke et al., 1993). Another approach is to link remote sensing technology, weed control recommendation models, and herbicide application with global positioning systems for delivering herbicides only to those areas with weed infestations that exceed threshold levels (Medlin et al., 2000). Both of these approaches have shown potential, but do not yet have the capabilities to reliably detect low densities of weeds within crop canopies.

Site-specific management of weeds involves both new concepts of weed biology and new technology. Principles of weed management and biology will need to be applied in a more precise fashion, with as much attention to where control practices are applied as to what is applied and when it is applied. Since efforts to manage weeds uniformly has lead to patchy weed distributions, it will also be important to understand how nonuniform management will change the nature of weed populations.

DEVELOPING WEED MANAGEMENT SYSTEMS

To this point, this paper has reviewed the influence of individual components of cropping systems on weed biology and management. While these are important, the central challenge of developing effective weed management systems is the integration of the options and tools that are available to make the cropping system unfavorable for weeds and to minimize the impact of the weeds that survive. None of the practices discussed in this paper should be considered as more than part of a total weed management strategy. No single weed management tactic has proven to be the "silver bullet" to eliminate weed problems, and given the nature of weed communities, we should not expect one to be developed in the near future. The best approach may be to integrate cropping system design and all available weed control strategies into a comprehensive weed management system that is environmentally and economically viable over the long term. Such an approach fits under the rubric of integrated pest management.

The term "integrated pest management" or IPM first appeared in the literature in 1967 (Smith and van den Bosch, 1967) and has its root in the concept of integrated control (Stern et al., 1959). While many definitions of IPM have been proposed over the years, they all contain two key elements: (1) the use of multiple control tactics and (2) the integration of knowledge of pest biology into the management system (Bottrell, 1979; Pedigo, 1995). As such, developing integrated management strategies for weeds calls for broader approaches that move beyond control of existing weed populations (Liebman and Gallandt, 1997; Navas, 1991). Integrating cropping system design and weed science could lead to systems that best utilize resources, diversify the selection pressure on weed communities, and provide producers with a broader range of management options.

The progression from a focus on tools targeted at individual weed populations at a point in time to the adoption of a holistic approach to crop and weed management will require analysis, theory, and information to support implementation at the cropping system and ecosystem levels (Cardina et al., 1999; Hall, 1995). Integrated weed management must be developed within the context of the entire cropping system with the farm and the surrounding area being considered as part of a larger ecosystem (Buhler, 1999; Wyse, 1992). A better understanding of the factors that affect ecosystem health, population dynamics of the weeds, and weed and ecosystem response to management practices is needed (Liebman and Gallandt, 1997). We need to integrate cropping systems and IPM concepts to include comprehensive theories that include all management variables, building on the foundations provided by the theories and practices of plant ecology, population management, plant protection, and cropping systems. As new cropping systems are considered, the impact on weed population dynamics and weed management options should be considered in the design phase, not as an afterthought.

The complexity and diversity of weed communities demand more integrated approaches if we are to break the cycle of applying a control tactic until it causes a community or a population shift followed by a new control tactic that results in the next shift. The ecological properties of weed communities may provide opportunities to reduce weed problems by implementing more integrated management systems than those currently employed (Buhler, Liebman, and Obrycki, 2000). These properties include: (1) weeds do not move and thus are very responsive to soil conditions, be they soil-applied herbicides, soil chemical and physical properties, or temperature regimes created by cover crops and tillage practices; (2) weeds spend a large amount of time in a quiescent, noncompetitive growth stage from which they emerge when the proper environmental and physiological conditions occur or are manipulated in particular ways; (3) many weeds, especially annual species, are poor competitors with established vegetation and require open space for establishment; and (4) in most cases, management practices regulate the weed population of parcel of land and problems do not rapidly migrate over wide geographic areas. All of these properties interact with the elements of cropping systems and generate the potential for modifying weed communities. Most of these concepts are not new, but have been lost in many of the simple cropping systems currently being practiced.

CONCLUSIONS

The propagule pools characteristic of weed communities provide unique management opportunities. Because of this characteristic, cropping systems should incorporate more practices that affect propagule production, survival, and the propagule/seedling transition. Approaches might include: (1) physical and/or chemical methods (cover crops, pathogens, predators, or soil disturbance) to reduce the success of the propagule/seedling transition to truncate population peaks; (2) reduction of the number of safe sites available for seedling establishment through manipulation of soil conditions and crop characteristics; (3) reduction of propagule production per weed plant by increasing the competitiveness of the crop, allelopathy, or weed suppressive soils; (4) reduction of propagule deposition and survival by physical means or by encouraging seed predation and degradation; (5) promotion of cropping system diversity through crop and tillage rotation, cover crops, intercropping, and other practices; (6) management of resources and soil conditions to increase crop competitiveness and increase the capacity of the cropping system to tolerate weeds; and most important; and (7) development of methods to combine multiple tactics in an integrated multi-year cropping system.

While new tactics and approaches for weed management need to be developed, we also need to understand the fundamental elements of agroecosystems and develop better monitoring and assessment methods. We also need better methods for gathering feedback on the effectiveness of control tactics and the relationship of weeds with edaphic factors because of the importance of the reproductive output of surviving weeds on future weed populations.

Over the coming decades agriculture will continue to struggle with weeds and management-induced population shifts. The use of broad-spectrum herbicides on transgenic crops will only change the selection pressures compared with previously used herbicides and we can expect that weed communities will continue to adapt. There may also be some surprising challenges as a result of the new tactics that genetic engineering may provide producers.

The connections between weed population biology and cropping systems are of increasing importance to the future of agriculture. Each crop-weed system is a unique mix of genetics and biology, and research to elucidate the mechanisms regulating these systems might expand the potential approaches to weed management. The need to understand more about weeds and their relationships to other components of agricultural systems may become more acute as development of new herbicides becomes more difficult for technologic, economic, and social reasons. Weed science must increase its efforts to explore new ways of managing weeds and slowing weed adaptation to existing control tactics, realizing that weed management is a continuous process, not an end product.

The integration of weed management as a component of cropping systems presents a major challenge to weed science. Weed scientists must take a larger role in guiding cropping systems research and implementation at the state, regional, and national levels. Cropping system design provides an excellent framework for expanded approaches to weed management because it allows for new and creative ways of meeting the challenge of managing weeds.

REFERENCES

Aase, J. K. and D. L. Tanaka. (1987). Soil water evaporation comparisons among tillage practices in the northern great plains. *Soil Science Society of America Journal* 51:436-440.

Al-Darby, A. M. and B. Lowery. (1987). Seed zone soil temperature and early corn growth with three conservation tillage systems. *Soil Science Society of America Journal* 51:768-774.

Akobunudu, I. O. (1991). Weeds in human affairs in sub-Saharan Africa: Implications for sustainable food production. *Weed Technology* 5:680-690.

Akobunudu, I. O. (1987). *Weed Science in the Tropics: Principles and Practices.* Chichester, UK: Wiley.

Alkamper, J. (1976). Influence of weed infestation on effect of fertilizer dressings. *Pflanzenschutz-Nachrichten Bayer* 29:191-235.

Al Kayassi, A. W., A. A. al Karaghouli, A. M. Hasson, and S. A. Beker. (1990). Influence of soil moisture content on soil temperature and heat storage under greenhouse conditions. *Journal of Agricultural Engineering Research* 45:241-252.

Anderson, R. L. (1991). Timing of nitrogen application affects downy brome (*Bromus tectorum*) growth in winter wheat. *Weed Technology* 5:582-585.

Anderson, R. L., D. L. Tanaka, A. L. Black, and E. E. Schweizer. (1998). Weed community and species response to crop rotation, tillage, and nitrogen fertility. *Weed Technology* 12:531-536.

Andreasen, C., J. C. Streibig, and H. Haas. (1991). Soil properties affecting the distribution of 37 weed species in Danish fields. *Weed Research* 31:181-187.

Ballare, C. L. and J. J. Casal. (2000). Light signals perceived by crop and weed plants. *Field Crop Research* 67:149-160.

Baker, H. G. (1974). The evolution of weeds. In *Annual Review of Ecology and Systematics*. Palo Alto, CA: Annual Reviews, pp. 1-24.

Barberi, P., N. Silvestri, and E. Bonari. (1997). Weed communities of winter wheat as influenced by input level and rotation. *Weed Research* 37:301-313.

Barker, T. C. and C. R. Francis. (1986). Agronomy of multiple cropping systems. In *Multiple Cropping Systems*, ed. C. A. Francis. New York: Macmillan, pp. 161-182.

Blackshaw, R. E. (1994a). Rotation affects downy brome (*Bromus tectorum*) in winter wheat (*Triticum aestivum*). *Weed Technology* 8:728-732.

Blackshaw, R. E. (1994b). Differential competitive ability of winter wheat cultivars against downy brome. *Agronomy Journal* 86:649-654.

Bloemhard, C. M. J., M. W. M. F. Arts, P. C. Scheepens, and A. G. Elema. (1992). Thermal inactivation of weed seeds and tubers during drying of pig manure. *Netherlands Journal of Agricultural Science* 40:11-19.

Bottrell, D. R. (1979). Integrated pest management: definition, features, and scope. In *Integrated Pest Management*. Washington, DC: Council on Environmental Quality, U.S. Government Printing Office, pp.19-26.

Bridges, D. C. (1994). Impacts of weeds on human endeavors. *Weed Technology* 8:392-395.

Brust, G. E. and G. J. House. (1988). Weed seed destruction by arthropods and rodents in low-input soybean agroecosystems. *American Journal of Alternative Agriculture* 3:19-25.

Brust, G. E. and B. R. Stinner. (1991). Crop rotation for insect, plant pathogen, and weed control. In *CRC Handbook of Pest Management in Agriculture*, ed. D. Pimentel. Boca Raton, FL: CRC Press, pp. 217-236.

Buhler, D. D. (1999). Expanding the context of weed management. *Journal of Crop Production* 2:1-8.

Buhler, D. D. (1995). Influence of tillage systems on weed population dynamics and management in corn and soybean production in the central USA. *Crop Science* 35:1247-1257.

Buhler, D. D. (1992). Population dynamics and control of annual weeds in corn (*Zea mays*) as influenced by tillage systems. *Weed Science* 40:241-248.

Buhler, D. D., J. D. Doll, R. T. Proost, and M. R. Visocky. (1995). Integrating mechanical weeding with reduced herbicide use in conservation tillage corn production systems. *Agronomy Journal* 87:507-512.

Buhler, D. D. and J. L. Gunsolus. (1996). Effect of date of preplant tillage and planting on weed populations and mechanical weed control in soybean (*Glycine max*). *Weed Science* 44:373-379.

Buhler, D. D., K. A. Kohler, and M. S. Foster. (1998). Spring-seeded smother plants for weed control in corn and soybean. *Journal of Soil Water Conservation* 53:272-275.

Buhler, D. D., M. Liebman, and J. J. Obrycki. (2000). Theoretical and practical challenges to an IPM approach to weed management. *Weed Science* 48:274-280.

Burnside, O. C., R. S. Moomaw, F. W. Roeth, G. A. Wicks, and R. G. Wilson. (1986). Weed seed demise in soil in weed-free corn (*Zea mays*) production across Nebraska. *Weed Science* 34:248-251.

Callaway, M. B. (1992). A compendium of crop varietal tolerance to weeds. *American Journal of Alternative Agriculture* 7:169-180.

Cardina, J., G. A. Johnson, and D. H. Sparrow. (1997). The nature and consequences of weed spatial distribution. *Weed Science* 45:364-373.

Cardina, J., H. M. Norquay, B. R. Stinner, and D. A. McCartney. (1996). Postdispersal predation of velvetleaf (*Abutilon theophrasti*) seeds. *Weed Science* 534-539.

Cardina, J., D. H. Sparrow, and E. L. McCoy. (1996). Spatial relationships between seedbank and seedling populations of common lambsquarters (*Chenopoduim album*) and annual grasses. *Weed Science* 44:298-308.

Cardina, J., D. H. Sparrow, and E. L. McCoy. (1995). Analysis of spatial distribution of common lambsquarters (*Chenopoduim album*) in no-till soybean (*Glycine max*). *Weed Science* 43:258-268.

Cardina, J., T. M. Webster, C. P. Herms, and E. E. Regnier. (1999). Development of weed IPM: levels of integration for weed management. *Journal of Crop Production* 2:239-267.

Carlson, H. L. and J. E. Hill. (1986). Wild oat (*Avena fatua*) competition with spring wheat: effects of nitrogen fertilization. *Weed Science* 34:29-33.

Carruthers, K., Q. Fe, D. Cloutier, and D. L. Smith. (1998). Intercropping corn with soybean, lupin and forages: weed control by intercrops combined with interrow cultivation. *European Journal of Agronomy* 8:225-238.

Cavers, P. B. and D. L. Benoit. (1989). Seed banks in arable land. In *Ecology of Soil Seed Banks*, eds. M. A. Leck, V. T. Parker, and R. L. Simpson. London, UK: Academic Press, pp. 309-328.

Challaiah, R. E. Ramsel, G. A. Wicks, O. C. Burnside, and V. A. Johnson. (1983). Evaluation of the weed competitive ability of winter wheat cultivars. *North Central Weed Science Society Proceedings* 38:85-91.

Chancellor, R. J. (1985). Changes in the weed flora of an arable field cultivated for 20 years. *Journal of Applied Ecology* 22:491-501.

Charudattan, R. and C. J. DeLoach. (1988). Management of pathogens and insects for weed control in agroecosystems. In *Weed Management in Agroecosystems: Ecological Approaches*, eds. M. A. Altieri and M. Liebman. Boca Raton, FL: CRC Press, pp. 245- 264.

Conklin, A. E., M. S. Erich, M. Liebman, and D. H. Lambert. (1998). Disease incidence and growth of wild mustard seedlings in red clover and compost amended soil. *Agronomy Abstracts* 90:279.

Creamer, N. G., M. A. Bennett, B. R. Stinner, and J. Cardina. (1996). A comparison of four processing tomato production systems differing in cover crop and chemical inputs. *Journal of the American Society of Horticultural Science* 121:559-568.

Cromar, H. E., S. D. Murphy, and C. J. Swanton. (1999). Influence of tillage and crop residue on postdispersal predation of weed seeds. *Weed Science* 47:184-194.

Crookston, R. K., J. E. Kurle, P. J. Copeland, J. H. Ford, and W. E. Lueschen. (1991). Rotation cropping sequence affects yield of corn and soybean. *Agronomy Journal* 83:108-113.

Dale, M. R. T., A. G. Thomas, and E. A. John. (1992). Environmental factors influencing management practices as correlates of weed community composition in spring seeded crops. *Canadian Journal of Botany* 43:1319-1327.

DeHaan, R. L., D. L. Wyse, N. J. Ehlke, B. D. Maxwell, and D. H. Putnam. (1994). Simulation of spring-seeded smother plants for weed control in corn (*Zea mays*). *Weed Science* 42:35-43.

DeLuca, T. H. and D. K. DeLuca. (1997). Composting for feedlot manure management and soil quality. *Journal of Production Agriculture* 10:235-241.

Derksen, D. A., G. P. Lafond, A. G. Thomas, H. A. Loeppky, and C. J. Swanton. (1993). Impact of agronomic practices on weed communities: tillage systems. *Weed Science* 41:409-417.

Dieleman, J. A., D. A. Mortensen, D. D. Buhler, C. A. Cambardella, and T. B. Moorman. (2000). Identifying associations among site properties and weed species abundance. I. Multivariate analysis. *Weed Science* 48:567-575.

DiTomaso, J. M. (1995). Approaches for improving crop competitiveness through the manipulation of fertilization strategies. *Weed Science* 43:491-497.

Dowling, P. M. and P. T. W. Wong. (1993). Influence of preseason weed management and in-crop treatments on two successive wheat crops: 1. Weed seedling numbers and wheat grain yield. *Australian Journal of Experimental Agriculture* 33:167-172.

Dyck, E. and M. Liebman. (1994). Soil fertility management as a factor in weed control: the effect of crimson clover residue, synthetic nitrogen fertilizer, and their interaction on emergence and early growth of lambsquarters and sweet corn. *Plant and Soil* 167:227-237.

Egley, G. H. and R. D. Williams. (1991). Emergence periodicity of six summer annual weed species. *Weed Science* 39:595-600.

Einhelig, F. A. and J. A. Rasmussen. (1989). Prior cropping with grain sorghum inhibits weeds. *Journal of Chemical Ecology* 15:951-960.

Ervio, R., S. Hyvarinen, L. Ervio, and J. Salonen. (1994). Soil properties affecting weed distribution in spring cereal and vegetable fields. *Agricultural Science in Finland* 3:497-504.

Forcella, F. and M. J. Lindstrom. (1988). Weed seed populations in ridge and conventional tillage. *Weed Science* 36:500-502.

Fraser, D. G., J. W. Doran, W. W. Sahs, and G. W. Lesoing. (1988). Soil microbial populations and activities under conventional and organic management. *Journal of Environmental Quality* 17:585-590.

Freyman, S., C. G. Kowalenko, and J. W. Hall. (1989). Effect of nitrogen, phosphorous, and potassium on weed emergence and subsequent weed communities in south coastal British Columbia. *Canadian Journal of Plant Science* 69:1001-1010.

Gallandt, E. R., M. Liebman, S. Corson, G. A. Porter, and S. D. Ullrich. (1998). Effects of pest and soil management systems on weed dynamics in potato. *Weed Science* 46:238-248.

Gallandt, E. R., M. Liebman, and D. Huggins. (1999). Improving soil quality: implications for weed management. *Journal of Crop Production* 2:95-121.

Garrity, D. P., M. Movillon, and K. Moody. (1992). Differential weed suppression ability in upland rice cultivars. *Agronomy Journal* 84:586-591.

Ghersa, C. M., M. L. Roush, S. R. Radosevich, and S. M. Cordray. (1994). Coevolution of agroecosystems and weed management. *Bioscience* 44:85-94.

Gill, G. S. and J. E. Holmes. (1997). Efficacy of cultural control methods for combating herbicide-resistant *Lolium rigidum*. *Pesticide Science* 51:352-358.

Gliessman, S. R. (1986). Plant interactions in multiple cropping systems. In *Multiple Cropping Systems*, ed. C. A. Francis. New York: Macmillan, pp. 82-95.

Guidi, G., M. Pagliai, and M. Giachetti. (1981). Modification of some physical and chemical soil properties following sludge and compost applications. In *The Influence of Sewage Sludge Application on Physical and Biological Properties of Soils*, eds. G. Catroux, P. L. L'Hermite, and E. Suess. Dordrecht, Netherlands: D. Reidel Publishing, pp. 122-136.

Gunsolus, J. L. (1990). Mechanical and cultural weed control in corn and soybeans. *American Journal of Alternative Agriculture* 5:114-119.

Hall, R. (1995). Challenges and prospects of integrated pest management. In *Novel Approaches to Integrated Pest Management*, ed. R. Reuventi. Boca Raton, FL: Lewis Publishers, pp.1-19.

Harlan, J. R. and J. M. J. de Wet. (1965). Some thoughts about weeds. *Economic Botany* 19:16- 24.

Harley, K. L. S. and I. W. Forno. (1992). *Biological Control of Weeds: A Handbook for Practitioners and Students*, Melbourne: Inkata Press.

Harper, J. L. (1977). *The Population Biology of Plants*. London, UK: Academic Press.

Harwood, R. R. and E. C. Price. (1976). Multiple cropping in tropical Asia. In *Multiple Cropping*, eds. R. I. Papendick, P. A. Sanchez, and G. B. Triplett. Madison, WI: American Society of Agronomy Special Publication 27, pp. 11-40.

Hauser, E. and G. A. Buchanan. (1982). Production of peanuts as affected by weed competition and row spacing. *Alabama Agricultural Experiment Station Bulletin 538*. 35 pp.

Hausler, A. and H. Nordmeyer. (1995). Impact of soil properties on weed distribution. In *Proceedings of a Seminar on Site Specific Farming*, SP-report 26, ed. S. E. Olesen. Amsterdam: Danish Institute of Plant and Soil Science, pp. 186-189.

Heap, I. M. (1999). International Survey of Herbicide Resistant Weeds. Herbicide Resistance Action Committee and Weed Science Society of America. *Internet <www.weedscience.com>*.

Holderbaum, J. F., A. M. Decker, J. J. Meisinger, F. R. Mulford, and L. R. Vough. (1990). Fall-seeded legume cover crops for no-tillage corn in the humid east. *Agronomy Journal* 82:117-124.

Holm, L. G., D. L. Plucknett, J. V. Pancho, and J. P. Herberger. (1977). *The World's Worst Weeds: Distribution and Biology*. Honolulu, HI: University Press of Hawaii.

Jagtap, S. S. and O. Adeleye. (1999). Land use efficiency of maize and soyabean intercropping and monetary returns. *Tropical Science* 39:50-55.

Johnson, G. A., D. A. Mortensen, and A. R. Martin. (1995). A simulation of herbicide use based on weed spatial distribution. *Weed Research* 35:197-205.

Johnson, M. D., B. Lowery, and T. C. Daniel. (1984). Soil moisture regimes of three conservation tillage systems. *Transactions of the American Society of Agricultural Engineers* 27:1385-1390.

Johnson, T. J., T. C. Kaspar, K. A. Kohler, S. J. Corak, and S. D. Logsdon. (1998). Oat and rye overseeded into soybean as fall cover in the Midwest. *Journal of Soil and Water Conservation* 53:276-279.

Jordan, N., D. A. Mortensen, D. M. Prenzlow, and K. Curtis Cox. (1995). Simulation analysis of crop rotation effects on weed seedbanks. *American Journal of Botany* 82:390-398.

Karlen, D. L., G. E. Varvel, D. G. Bullock, and R. M. Cruse. (1994). Crop rotations for the 21st century. *Advances in Agronomy* 53:1-45.

Karssen, C. M. and H. W. M. Hillhorst. (1992). Effect of chemical environment on seed germination. In *Seeds: The Ecology of Regeneration in Plant Communities*, ed. M. Fenner. Wallingford, UK: CAB International, pp. 327-348.

Kennedy, A. C. and R. J. Kremer. (1996). Microorganisms in weed control strategies. *Journal of Production Agriculture* 9:480-485.

Kirkland, K. J. and H. J. Beckie. (1998). Contribution of nitrogen fertilizer placement to weed management in spring wheat (*Triticum aestivum*). *Weed Technology* 12:507-514.

Koscelny, J. A., T. F. Peeper, J. B. Solie, and S. G. Solomon. (1990). Effect of wheat (*Triticum aestivum*) row spacing, seeding rate, and cultivar on yield loss from cheat (*Bromus secalinus*). *Weed Technology* 4:487-492.

Kropff, M. J. (1993). General introduction. In *Modelling Crop-Weed Interactions*, eds. M. J. Kropff and H. H. van Laar. Wallingford, UK: CAB International, pp. 1-7.

Lanning, S. P., L. E. Talbert, J. M. Matrin, T. K. Blake, and P. L. Bruckner. (1997). Genotype of wheat and barley affects light penetration and wild oat growth. *Agronomy Journal* 89:100-103.

Légère, A., R. R. Simard, and C. Lapierre. (1994). Response of spring barley and weed communities to lime, phosphorus and tillage. *Canadian Journal of Plant Science* 74:421-428.

Leighty, C. E. (1938). Crop rotation. In *Soils and Men, Yearbook of Agriculture 1938*. Washington, DC: United States Department of Agriculture, pp. 406-430.

Liebman, M. (1988). Ecological suppression of weeds in intercropping systems: a review. In *Weed Management in Agroecosystems: Ecological Approaches*, eds. M. A. Altieri and M. Liebman. Boca Raton, FL: CRC Press, pp. 198-212.

Liebman, M. and A. S. Davis. (2000). Integration of soil, crop and weed management in low- external-input farming systems. *Weed Research* 40:27-47.

Liebman, M. and E. R. Gallandt. (1997). Many little hammers: ecological management of crop-weed interactions. In *Ecology in Agriculture*, ed. L. Jackson. New York: Academic Press, pp. 291-341.

Liebman, M. and R. R. Janke. (1990). Sustainable weed management practices. In *Sustainable Agriculture in Temperate Zones*, eds. C. Francis, C. B. Flora, and L. D. King. New York: John Wiley and Sons, pp. 111-143.

Liebman, M. and T. Ohno. (1998). Crop rotation and legume residue effects on weed emergence and growth: applications for weed management. In *Integrated Weed and*

Soil Management, eds. J. L. Hatfield, D. D. Buhler, and B. A. Stewart. Chelsea, MI: Ann Arbor Press, pp. 181-221.

Loeppky, H. A. and D. A. Derksen. (1994). Quackgrass suppression through crop rotation in conservation tillage systems. *Canadian Journal of Plant Science* 74:193-197.

Malik, V. S., C. J. Swanton, and T. E. Michaels. (1993). Interaction of white bean cultivars (*Phaseolus vulgaris* L.) cultivars, row spacing, and seeding density with annual weeds. *Weed Science* 41:62-68.

Martinez-Ghersa, M. A., C. M. Ghersa, and E. H. Satorre. (2000). Coevolution of agricultural systems and their weed companions: implications for research. *Field Crops Research* 67:181-190.

McWhorter, C. G. and W. L. Barrentine. (1975). Cocklebur control in soybeans as affected by cultivars, seeding rates, and methods of weed control. *Weed Science* 23:386-390.

Medlin, C. R., D. R. Shaw, P. D. Gerard, and F. E. LaMastrus. (2000). Using remote sensing to detect weed infestations in *Glycine max*. *Weed Science* 48:393-398.

Mester, T. C. and D. D. Buhler. (1991). Effects of soil temperature, seed depth, and cyanazine on giant foxtail (*Setaria faberi*) and velvetleaf (*Abutilon theophrasti*) seedling development. *Weed Science* 39:204-209.

Minotti, P. L. (1991). Role of crop interference in limiting losses from weeds. In *CRC Handbook of Pest Management*, Volume 2, ed. D. Pimentel. Boca Raton, FL: CRC Press, pp. 359-368.

Mohler, C. L. (1993). A model of the effects of tillage on emergence of weed seedlings. *Ecological Applications* 3:53-73.

Mohler, C. L. and J. R. Teasdale. (1993). Response of weed emergence to rate of *Vicia villosa* Roth and *Secale cereale* L. residue. *Weed Research* 33:487-499.

Mortensen, D. A., J. A. Dieleman, and G. A. Johnson. (1998). Weed spatial variation and weed management. In *Integrated Weed and Soil Management*, eds. J. L. Hatfield, D. D. Buhler, and B. A. Stewart. Chelsea, MI: Ann Arbor Press, pp. 293-309.

Mulder, T. A. and J. D. Doll. (1994). Reduced input corn weed control: the effects of planting date, early season weed control, and row crop cultivator selection. *Journal of Production Agriculture* 7:256-260.

Mulder, T. A. and J. D. Doll. (1993). Integrating reduced herbicide use with mechanical weeding in corn (*Zea mays*). *Weed Technology* 7:382-389.

Mulla, D. J. (1993). Mapping and managing spatial patterns in soil fertility and crop yield. In *Soil Specific Crop Management*, eds. P. C. Robert, R. H. Rust, and W. E. Larson. Madison, WI: American Society of Agronomy, pp. 15-26.

Muller-Scharer, H., P. C. Scheepens, and M. P. Greaves. (2000). Biological control of weeds in European crops: recent achievements and future work. *Weed Research* 40:83-98.

Navas, M. L. (1991). Using plant population biology in weed research: A strategy to improve weed management. *Weed Research* 31:171-179.

Okigo, B. N. and D. J. Greenland. (1976). Intercropping systems in tropical Africa. In *Multiple Cropping*, eds. R. I. Papendick, P. A. Sanchez, and G. B. Triplett. Madison, WI: American Society of Agronomy Special Publication 27, pp. 63-101.

Ozores-Hampton, M., P. J. Stoffella, T. A. Bewick, D. J. Cantliffe, and T. A. Obreza. (1999). Effect of age of composted MSW and biosolids on weed seed germination. *Compost Science and Utilization* 7:51-57.

Pareja, M. R. and D. W. Staniforth. (1985). Soil-seed microsite characteristics in relation to seed germination. *Weed Science* 33:190-193.

Pedigo, L. P. (1995). Closing the gap between IPM theory and practice. *Journal of Agricultural Entomology* 12:171-181.

Pester, T. A., O. C. Burnside, and J. H. Orf. (1999). Increasing crop competitiveness to weeds through crop breeding. *Journal of Crop Production* 2:31-58.

Pfeiffer, E. E. (1970). *Weeds and What They Tell*. Kimberton, PA: Bio-Dynamic Farming and Gardening Association.

Plucknett, D. L. and N. J. H. Smith. (1986). Historical perspectives on multiple cropping. In *Multiple Cropping Systems*, ed. C. A. Francis. New York: Macmillan, pp. 20-39.

Power, J. F. (1996). Cover crops. In *1997 McGraw-Hill Yearbook of Science and Technology*, ed. S. B. Parker. New York: McGraw-Hill, pp. 124-126.

Putnam, A. R. and J. DeFrank. (1983). Use of phytotoxic plant residues for selective weed control. *Crop Protection* 2:173-181.

Radosevich, S., J. Holt, and C. Ghersa. (1997). *Weed Ecology: Implications for Management*, 2nd edition. New York: John Wiley and Sons.

Rasmussen, K., J. Rasmussen, and J. Petersen. (1996). Effects of fertilizer placement on weed in weed harrowed spring barley. *Acta Agriculturae Scandinavica* 46:192-196.

Rice, E. L. (1995). *Biological Control of Weeds and Plant Diseases: Advances in Applied Allelopathy*. Norman, OK: University of Oklahoma Press.

Robert, P. C., R. H. Rust, and W. E. Larson. (1994). Preface. In *Site-Specific Management for Agricultural Systems*, eds. P. C. Robert, R. H. Rust, and W. E. Larson. Madison, WI: American Society of Agronomy.

Salisbury, E. J. (1961). *Weeds and Aliens*. London, UK: Collins.

Schreiber, M. M. (1992). Influence of tillage, crop rotation, and weed management on giant foxtail (*Setaria faberi*) population dynamics and corn yield. *Weed Science* 40:645-653.

Schweizer, E. E., D. W. Lybecker, and R. L. Zimdahl. (1988). Systems approach to weed management in irrigated crops. *Weed Science* 36:840-845.

Schweizer, E. E. and R. L. Zimdahl. (1984). Weed seed decline in irrigated soil after six years of continuous corn (*Zea mays*) and herbicides. *Weed Science* 32:76-83.

Serra-Wittling, C., S. Houot, and E. Barriuso. (1996). Modification of soil water retention and biological properties by municipal solid waste compost. *Compost Science and Utilization* 4:44-52.

Smeda, R. J. and S. C. Weller. (1996). Potential of rye (*Secale cereale*) for weed management in transplanted tomatoes (*Lycopersicon esculentum*). *Weed Science* 44:596-602.

Smith, A. G. (1970). The influence of Mesolithic and Neolithic man on British vegetation: a discussion. In *Studies in the Vegetational History of the British Isles*, ed. D. Walker and R. G. West. Cambridge, UK: Cambridge University Press, pp. 81-96.

Smith, R. F. and R. van den Bosch. (1967). Integrated control. In *Pest Control: Biological, Physical, and Selected Chemical Methods*, eds. W. W. Kilgore and R. L. Doutt. New York: Academic Press, pp. 295-340.

Springman, R., D. Buhler, R. Schuler, D. Mueller, and J. Doll. (1989). Row crop cultivators for conservation tillage systems. *University of Wisconsin Extension Publication A3483*. 6 pp.

Staricka, J. A., P. M. Burford, R. R. Allmaras, and W. W. Nelson. (1990). Tracing the vertical distribution of simulated shattered seeds as related to tillage. *Agronomy Journal* 82:1131-1134.

Stern, V. M., R. F. Smith, R. van den Bosch, and K. S. Hagen. (1959). The integrated control concept. *Hilgardia* 29:81-101.

Stoller, E. W., S. K. Harrison, L. M. Wax, E. E. Regnier, and E. D. Nafziger. (1987). Weed interference in soybeans. *Reviews of Weed Science* 3:155-181.

Stoskopf, N. C. (1985). *Cereal Grain Crops*. Reston, VA: Reston Publishing.

Teasdale, J. R. (1995). Influence of narrow row/high population corn (*Zea mays*) on weed control and light transmission. *Weed Technology* 9:113-118.

Teasdale, J. R. (1998). Cover crops, smother plants, and weed management. In *Integrated Weed and Soil Management*, eds. J. L. Hatfield, D. D. Buhler, and B. A. Stewart. Chelsea, MI: Ann Arbor Press, pp. 247-270.

Teasdale, J. R. and J. R. Frank. (1983). Effect of row spacing on weed competition with snap beans. *Weed Science* 31:81-85.

Teyker, R. H., H. D. Hoelzer, and R. A. Liebl. (1991). Maize and pigweed response to nitrogen supply and form. *Plant and Soil* 135:287-292.

Thomsen, C. D., W. A. Williams, M. Vayssieres, E. L. Bell, and M. R. George. (1993). Controlled grazing on annual grassland deceases yellow starthistle. *California Agriculture* 47:36-40.

Tilman, E. A., D. Tilman, M. J. Crawley, and A. E. Johnston. (1999). Biological weed control via nutrient competition: potassium limitation of dandelions. *Ecological Applications* 9:103-111.

Tollenaar, M., S. P. Nissanka, A. Aguilera, S. F. Weise, and C. J. Swanton. (1994). Effect of weed interference and soil nitrogen on four maize hybrids. *Agronomy Journal* 86:596-601.

Valdrighi, M. M., A. Pera, M. Agnolucci, S. Frasinetti, D. Lundardi, and G. Vallini. (1996). Effects of compost-derived humic acids on vegetable biomass production and microbial growth within a plant (*Cichorium intybus*)-soil system: a comparative study. *Agriculture, Ecosystems, and Environment* 58:133-144.

Vengris, K., W. G. Colby, and M. Drake. (1955). Plant nutrient competition between weeds and corn. *Agronomy Journal* 47:213-216.

Warnes, D. D. and R. N. Andersen. (1984). Decline of wild mustard (*Brassica kaber*) seeds in soil under various cultural and chemical practices. *Weed Science* 32:214-217.

Wiese, A. M. and L. K. Binning. (1987). Calculating the threshold temperature of development for weeds. *Weed Science* 35:177-179.

Woebbecke, D. M., G. E. Meyer, K. Von Bargen, and D. A. Mortensen. (1993). Plant species identification, size, and enumeration using machine vision techniques on near-binary images. In *Proceedings of the SPIE Conference on Optics in Agriculture and Forestry. Volume 1836*, ed. J. A. DeShazer. Boston, MA: SPIE, pp. 208-219.

Wortmann, C. S. (1993). Contribution of bean morphological characteristics to weed suppression. *Agronomy Journal* 85:840-843.

Wyse, D. L. (1992). Future of weed science research. *Weed Technology* 6:162-165.

Yenish, J. P., J. D. Doll, and D. D. Buhler. (1992). Effects of tillage on vertical distribution and viability of weed seed in soil. *Weed Science* 40:429-433.

Zimdahl, R. L. (1993). *Fundamentals of Weed Science*. New York: Academic Press.

Cropping Systems
and Integrated Pest Management:
Examples from Selected Crops

K. R. Barker

C. Sorenson

SUMMARY. Cropping systems have been central to managing associated pests for centuries. This treatment focuses on the history, concepts, and the integration of available Integrated Pest Management (IPM) tools/ strategies into cropping systems. Pest assessments/diagnoses, IPM-decision-making aids, and examples of pest management in selected crops/ cropping systems (wheat, soybean, corn, cotton, potato, and strawberry) as well as emerging opportunities and challenges are discussed. The evolving philosophy of IPM and the recently renewed emphasis on ecologically based pest management address the fact that significant levels of predation and/or parasitism are desirable insofar as they promote diversity and sustainability of agroecosystems. Thus, cropping systems are beginning to focus on soil and crop health as well as specific IPM and production goals. Although extensive efforts have been directed toward modeling the many interactions between crops, associated pests and the

K. R. Barker is affiliated with the Department of Plant Pathology, North Carolina State University, Box 7616, Raleigh, NC 27695-7616 USA.

C. Sorenson is affiliated with the Department of Entomology, North Carolina State University, Box 7630, Raleigh, NC 27695-7630 USA.

[Haworth co-indexing entry note]: "Cropping Systems and Integrated Pest Management: Examples from Selected Crops." Barker, K. R.. and C. Sorenson. Co-published simultaneously in *Journal of Crop Production* (Food Products Press, an imprint of The Haworth Press, Inc.) Vol. 8, No. 1/2 (#15/16), 2003, pp. 271-305; and: *Cropping Systems: Trends and Advances* (ed: Anil Shrestha) Food Products Press, an imprint of The Haworth Press, Inc., 2003, pp. 271-305. Single or multiple copies of this article are available for a fee from The Haworth Document Delivery Service [1-800-HAWORTH, 9:00 a.m. - 5:00 p.m. (EST). E-mail address: docdelivery@haworthpress.com].

environment, the general implementation of a systems approach to integrated crop and pest management remains to be accomplished. *[Article copies available for a fee from The Haworth Document Delivery Service: 1-800-HAWORTH. E-mail address: <docdelivery@haworthpress.com> Website: <http://www.HaworthPress.com> © 2003 by The Haworth Press, Inc. All rights reserved.]*

KEYWORDS. Agroecosystem management, biological control, chemical control

INTRODUCTION

Cropping systems have been central to managing some crop pests for centuries. An early cropping system that evolved specifically to avoid the low yields associated with the potato cyst nematodes (*Globodera rostochiensis*) in Peru included rotation and fallow (Bridge, 1996). The concept of utilizing multiple management tactics was pioneered by Kühn (1888) when he tested fumigation, rotation, and cover crops as means of controlling the sugar beet cyst nematode (*Heterodera schachtii*) (see Campbell, Peterson, and Griffith 1999). However, the term Integrated Pest Management (IPM) was first used by Smith and van den Bosch (1967) in regards to controlling insect pests. The modern concept of IPM evolved from the initial concept of "Integrated Control," as developed by Stern et al. (1959).

Many definitions of IPM have been offered. A useful definition of IPM adopted by the National Coalition on Integrated Pest Management is as follows: "A sustainable approach to managing pests by combining biological, cultural, physical, and chemical tools in a way that minimizes economic, health, and environmental risks" (CAST 2003a). That definition, however, loses some of the aspects contained in earlier concepts of IPM, including the use of decision rules based on ecological principles, economic/social considerations, and a multidisciplinary approach (Rabb and Guthrie, 1970; Kennedy, 2000). Two recent publications, "Ecologically Based Pest Management" (EBPM) (National Research Council, 1996) and "Ecological Management of Weeds" (Liebman, Mohler, and Staver, 2001) stressed the importance of ecological considerations in pest control. Widespread concerns about the detrimental impact of pesticides on the environment and related health issues were responsible in large part for the development of the concept of IPM (Kennedy, 2000). In fact, the publication of the book "*Silent Spring*" by Rachael Carson in 1962 ignited widespread discussion/debate on the real and potential hazards of pesticides. This still ongoing dialogue includes scientists in many disciplines, environmentalists, and policy makers. *Silent Spring* contributed much to the development of alternatives to pesticides for pest management purposes, augmented global interests in developing cropping systems that limit crop pests, and added much to the environmental movement. Diverse publications continue to re-visit Carson's 1962 landmark treatment (van Emden and Peakall, 1996; Waddell, 2000).

Several related concepts interface with IPM, including "Integrated Crop Management" (ICM), or Integrated Farming Systems (IFS) as used in a number of countries (El Titi and Ipach, 1989), "Good Agricultural Practices" (GAP), and "Sustainable Agriculture." "ICM is a whole farm management strategy that uses IPM as a component, minimizes waste and pollution, safeguards natural farm assets, enhances energy efficiency and manages crop profitability" (Carroll, 2000). ICM systems are dynamic and encompass the latest research and technology, based on expert advice and experience. GAP is defined as "efficient production of good-quality food, feed, and fiber while maintaining natural resources and optimizing crop inputs to minimize environmental impacts and ensure responsibility for the health and safety of farmers" (Carroll, 2000). The concepts of IPM, ICM, and GAP, while sharing considerably in overall goals, provide much of the infrastructure of Sustainable Agriculture (Carroll, 2000).

Fundamental to effective IPM programs is the development of appropriate pest management strategies and tactics that best interface with cropping system-pest situations. A pest management strategy is the overall plan to eliminate or minimize the pest problem. A pest management tactic, in contrast, is a method used to implement a given strategy (Barker and Koenning, 1998; Pedigo, 1999). Depending on the type of pest, some of the primary management strategies include: exclusion, eradication; reduction of pest population numbers; reduction of crop susceptibility; combination of reduced population numbers and crop susceptibility; or do nothing (Agrios, 1997; CAST, 2003a). As discussed later, a first critical step in IPM is to secure site-specific information on pest incidence/levels and soil nutrient status. In addition, an understanding of the life cycles or history and ecology of the pests as well as the epidemiology of diseases is helpful in the formulation of IPM strategies and tactics.

Additional aspects of the evolving IPM concept included: the influence of the total agroecosystem on pest problems; the recognition that only certain levels of pest infestations caused crop damage; and the philosophy that multiple tactics or methods should be used to manage single pests or pest complexes (Kennedy, 2000). Because plant pathogens and weeds have very divergent characteristics versus those of insects, IPM has often been viewed as an activity among entomologists. While population levels and economic thresholds often can be determined for insect infestations, the rather different natures of many plant pathogen, weed, and nematode communities frequently require very different assessments as discussed herein.

Estimated annual yield losses due to a wide array of pests on food and fiber crops amount to 40% in the U.S.A. and 48% worldwide (Agrios, 1997). Worldwide estimates of yield losses to the various pests range from 5 to 12% annually for plant-parasitic nematodes (Koenning et al., 1999), 12% for insects and 12% for plant pathogens, and 10% for weeds (Agrios, 1997). Recent estimates on dollar losses to weeds in the U.S.A. include $4 billion annually in direct losses and $6 billion for herbicides (Liebman, Mohler, and Staver, 2001). The magnitude of these losses emphasizes the crucial importance of effective pest management. These economic losses also have instilled into growers a high sensitivity for risks associated with crop pests. In earlier years, for many growers, limiting these risks meant high pesticide usage as crop insurance. In some countries such as the United States, crop insurance is being viewed as a means of minimizing associated production risks–including risks sometimes encountered with minimal pesticide use. In addition, IPM certification may eventually enable private insurers to offer crop protection insurance products (CAST, 2003a).

Pest population assessments, in addition to being central in minimizing the use of unnecessary pesticides, are used in decision making related to cropping systems, including appropriate crop rotations, selection of pest-resistant cultivars, or a combination of pest-management strategies/tactics. Field histories related to pests are especially important for soilborne organisms such as plant-parasitic nematodes and soil-inhabiting insects as well as fungi, bacteria, and weeds. With the advent of precision agriculture, site-specific treatments of weed infestations and other crop pests can be utilized to further limit use of pesticide applications to mini-sites in given fields where they are needed (CAST, 2003a; Renner, Swinton, and Kells, 1999). Other new technologies, especially those related to genetic engineering, identification and populations-assessment methodologies, and integrated systems are having profound impacts on IPM and related cropping systems. For example, more than 50% of the soybeans (*Glycine max* L. Merr.) and cotton (*Gossypium hirsutum* L.) currently produced in the United States are herbicide-tolerant lines. The increased use of Bt-transgenic cotton from 1995 to 1998 resulted in a reduction of some 1.1 million liters of insecticides on that crop (CAST, 2003a).

The objectives of this article are to provide a synopsis of the concepts of IPM, the integration of available strategies and tactics into selected cropping systems, and related challenges. In the U.S., progress in the development and implementation of IPM tactics in suitable cropping systems has been heavily impacted by government polices and programs. The EPA in 1993 renewed support by setting a goal that IPM practices would be implemented on 75% of the nation's cropland by the year 2000; this goal included the reduction of pesticide use and associated risks (CAST, 2003a; GAO, 2001). In 1994, the USDA announced a related initiative to facilitate achieving the 75% goal by 2000 through research, outreach, and education (Benbrook et al., 1996; CAST, 2003a; Fernandez-Cornejo and Jans, 1999; GAO, 2001; Merrigan, 2000). The objectives of the UDSA 1994 IPM Initiative grouped various related farming practices (similar to those listed in Tables 1 and 2) into four categories: Pre-

TABLE 1. Cropping systems and other strategies and tactics used in IPM

Management practice[a]	Options	Utility per pest group[b]			
		Insects	Weeds	Pathogens	Nematodes
Crop Rotations	Continuous	0/--	+/--	0/--	0/--
	2-year	++/--	+/--	+/-	+/-
	≥ 3-year	++/--	+/-	++/-	++/-
Fallow	-	?	--	++/-	++/-
Cover crops	-	+/--	+/0	+/0	++/0
Refugia	-	+/0	NA?	0/-	NA
Tillage	No-Till	+/--	+/--	0/--	+/-
	Chisel	0/-	0/-	+/-	+/-
	Moldboard	+/-	+/-	++/-	-?
Planting Date	Early	0/--	+/-	+/0	+/0
	Late	0/--	+/-	+/-	+
Plant Population	Low	0/-	-	+/-	NA
	High	0/-	+	+/-	NA
Row Width	Narrow	+/0	+/0	0/--	NA
	Conventional	0	0/-	+/0	NA
Cultivation	Early	0	+/-	0	+/-
	Late	0	+/-	0	+/0
Pesticides	Foliar	++/--	++/-	++/-	+/NA
	Seed	?	++/0?	++/0	+/0
	Soil	++/NA	++/0?	++/NA	++/-
Sanitation	-	+/0	+/0	++/0	++/0
Resistant varieties	-	++/NA	++/NA[c]	++/-	++/-
Biological controls	-	++/NA	+/NA	+/NA	+/NA
Soil amendments	-	+/0?	+/0?	+/0	+/0
Field size and Borders	Small/borders	+/-	NA	0/--?	NA
	Large w/o borders	0/-	NA	0/--?	NA

[a] In part, adapted from Cavigelli et al. (2000).
[b] Symbols codes as follows:
   0 = No or little effect on pest risks
   + = Limits pest risks
   ++ = Greatly limits pest risks
   - = Slight increase in pest risks
   − = Strong increase in pest risks
   NA = Not applicable.
   ? = Definitive information lacking
[c] Herbicide tolerant crop cultivars can greatly limit weed risks.

TABLE 2. Pest management practices, field crops, 1996 (adatped from Fernandez-Cornejo and Jans, 1999).

Cultural techniques are the leading pest management practice for field crops

Item	Corn	Soybean	Cotton	Fall potato	Winter wheat	Spring wheat	Durum wheat
				Percent of planted acres			
Biological Techniques							
Considered beneficial insects in selected pesticides	8	5	52	29	10	4	12
Purchased and released beneficial insects	*	*	*	0	*	*	0
Used pheromone lures to control pests	na	*	7	2	*	1	0
Used *Bacillus thringiensis* (Bt)[2]	2.4	1.6	4.1	*	*	0	0
Cultural Techniques							
Adjusted planting or harvesting dates[3]	5	6	25	7	19	11	13
Used mechanical cultivation for weed control	51	29	89	86	na	na	na
Used a no till system	19	33	na	na	3	4	7
Crop rotations[4]							
Continuous[5]	18	11	67	2	42[11]	14	10
Rotation with other row crops[6]	54[8]	63[9]	15	2	2	2	0
Other[7]	28	26	18	96[10]	56[12]	83[13]	90[14]
Pesticide Efficiency							
Alternated pesticides to control pest resistance	31	28	41	69	13	38	32
Monitoring							
Used pheromone lures to monitor pests[1]	1	*	33	3	*	4	1
Used soil biological testing to detect pests such as insects, diseases, or nematodes	2	3	9	46	2	0	0

[1] For corn, pheromone lures were used to monitor black cutworm.
[2] Percent of insecticide-treated acres for Bt.
[3] Adjust planting dates only for corn.
[4] Crop rotations include 3 years 1994, 1995, and 1996. Column crop heading is for crop planted in 1996.
[5] The same crop was planted in 1994, 1995 and 1996.
[6] A crop sequence, excluding continuous same crop, where only row crops (corn, soybeans, sorghum, cotton, and peanuts) were planted for three consecutive years.
[7] Other excludes continuous same crop and rotation with row crops and includes fallow or idle.
[8] 49 percent of corn-planted acres were in rotation with soybeans.
[9] 56 percent of soybean-planted acres were in rotation with corn.
[10] 26 percent of potato-planted acres were fallow in 1994 and 1995, and 70 percent were in rotation with other crops or fallow in 1994 or 1995.
[11] Continuous same crop for winter wheat were for two years 1995 and 1996, for winter wheat planted in fall 1994 and winter wheat planted in fall 1995.
[12] 40 percent of winter-wheat-planted acres were fallow in fall 1994 and had winter wheat in fall 1995.
[13] 23 percent of spring-wheat-planted acres were fallow in 1994 and had spring wheat in 1995, and 60 percent were in rotation with other crops or fallow in 1994 or 1995.
[14] 24 percent of durum-wheat-planted acres were fallow in 1994 and had durum wheat in 1995, and 66 percent were in rotation with other crops or fallow in 1994 or 1995.
na = not available or not applicable. * Less than 0.5%. (Original source: NASS/ERS 1996 ARMS survey.)

vention, avoidance, monitoring, and suppression (PAMS) (GAO, 2001). For example, avoidance practices encompass cropping system components such as rotation, adjusting planting dates, and use of pest-resistant crop varieties. The percentages of acerages for various field crops utilizing these diverse IPM practices in 1996 are summarized in Table 2 (after Fernandez-Cornejo and Jans, 1999). Overall, the USDA has estimated that some IPM practices have been followed in about 76% of U.S. cropland, but biologically-based IPM practices were often less than 20% (GAO, 2001). A second major federal program now affecting IPM in the U.S. is the 1996 Food Quality Protection Act (FQPA) (CAST, 2003a; EPA, 1996; Ragsdale, 2000). This act will limit the use of a number of pesticides and likely foster the implementation of IPM, as discussed briefly under "Emerging Opportunities and Challenges."

IPM TOOLS, STRATEGIES, AND TACTICS

Pest Assessments and Diagnostics

Reliable identification of pests and diagnosis of related problems is a cornerstone for minimizing losses caused by insects, weeds, and plant pathogens. Assessment and monitoring of pest populations and/or damage as well as yield losses are becoming increasingly important to growers as they strive to limit production costs and unnecessary application of pesticides. In addition to the classical methods of pest identification, recently deployed molecular- and computer-based diagnostic technologies now offer much to IPM. Today, advanced DNA sensors and sensor arrays for direct, genetic analysis of pathogens are under development for human pathogens (Henkens et al., 2000). Already in use in some cropping systems, Global Positioning Systems (GPS) and Geographical Information Systems (GIS) offer a means of identifying site-specific pest-management needs within given fields.

Insect Infestations

Monitoring of insect infestations generally interfaces closely with the use of economic thresholds (Figure 1) as well as regulatory programs. However, assessments of insect populations in given insect-crop combinations often go beyond scouting and diagnosis by a skilled specialist (Pedigo, 1999). For example, methods for detecting pesticide-resistance in insects include biochemical, immunochemical, molecular, and bioassay procedures (Roe et al., 2000). This area of IPM, while especially important in detection and monitoring of insecticide resistance, is crucial to augmenting the durability of many pesticides. Due to the extremely dynamic nature of many insect pest populations, routine systematic sampling over substantial portions of the growing season ("scouting")

FIGURE 1. Relationships of crop injury/damage and economic damage caused by insects. (A) Effects of cotton boll set of plant capacity and insect injury (Adapted from Smith and van den Bosch, 1967); (B) General relationships between the damage boundary and the gain threshold (Adapted from Pedigo, 1999).

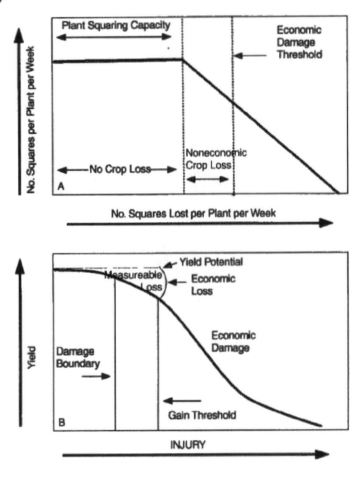

are better developed and more widely used for arthropods than for other pests (Kennedy, 2000).

Information on insect activity is fundamental to their management. This includes surveillance data on their invasion of crops, long-distance migration, local movement, feeding, and reproduction (Kennedy and Storer, 2000; Pedigo, 1999; Southwood, 1978). Surveillance of insect and other pest activities receives much effort via local, state, and national government agencies. In the

U.S.A., the Animal and Plant Health Inspection Service (APHIS) monitors for introductions and spread of invasive pests.

Weeds

Agricultural fields usually are infested with large numbers of seeds of highly diverse weed species which sharply contrasts to the dynamic spatial dispersal of most other plant pests (Liebman, Mohler, and Staver, 2001). Thus, weeds pose an ongoing and obvious threat to crops for which growers may use a combination of cultural and herbicide-management strategies/tactics to limit yield losses. Assessment of the weed-seed banks in given fields can be very effective, but costs involved may be very high (CAST, 2003a). Although such data are essential for utilizing the concepts of an economic threshold or action thresholds, the costs for assessing weed seed kinds and numbers may be prohibitive even for small units. As a result, scouting fields for weed seedlings in most crops is not done in a systematic manner. Since given fields typically contain five or more dominant weeds with rather specific image signatures that differ from the crop, efforts to automate weed seedling counts by remote sensing with specialized cameras have met with some success. Site-specific application of herbicides of weeds, as determined by soil sensors, also is under evaluation (Hall, 2000). The concept of "time-density-equivalent" also offers a new approach for weed assessments over a growing season (CAST, 2003a; Gunsolus et al., 2000).

Pathogen Detection and Diagnoses

Plant-disease assessments often encompass crop-pathogen field histories, assessments of environmental conditions that favor or limit epidemics, and in some instances data on likely disease progression. A "disease-progress curve" for a specified pathogen-crop cultivar is thus developed. A number of plant parasites frequently may be encountered on one plant or crop although only one pathogen is causing the primary disease problem. In this regard, "detection" concerns establishment of the presence of a particular target organism within a sample, whereas "diagnosis" involves identification of the nature and cause of an observed disease problem (Louws, Rademaker, and deBruijn, 1999). Both detection and diagnosis are very important for effective IPM programs.

The choices of diagnostic procedures for plant pathogens have increased greatly in recent years. Immunological diagnostic tests are now available for viral and bacterial pathogens, and a number of tests are becoming available for fungal pathogens (Stewart, 2000). Laboratory-oriented tests also have been developed for plant-parasitic nematodes. ELISA (enzyme-linked immunosorbent assay) is the primary method utilized for the detection and diagnosis of virus problems on a number of crops, especially for greenhouse ornamentals (Dinesen and van Zaayen, 1996). Detection kits also are available commercially for given groups of viruses such as the potyviruses. Still, bioassays using susceptible hosts often are needed to confirm negative ELISA results. A number of polymerase chain reaction (PCR)-based protocols such as rDNA-based PCR, and others, have been devised and adapted to enhance both the detection and identification of plant pathogenic bacteria (Louws, Rademaker, and debruijn, 1999). These protocols have been utilized widely in research programs and are resulting in the development of fundamental information on the ecology and population dynamics of bacterial pathogens, but remain to be applied in IPM programs. Additional protocols used for identifying bacteria include immuno-fluorescence and staining, the ELISA procedure, and immunobinding assay (Dinesen and van Zaayen, 1996). Currently, the use of these methods is limited largely to research programs and diagnostic laboratories. However, commercial ELISA kits are now available for fungi such as *Septoria nodorum, S. tritici* (Stewart, 2000), *Rhizoctonia solani,* and *Fusarium, Pythium* and *Phytophthora* species.

Nematode Diagnostics

Most plant-parasitic nematodes have very limited capacity for active dispersal. Therefore, field history and soil characteristics along with population data and damage or economic thresholds serve as the key parameters for their management. Assessment of population data and diagnoses of related problems rely on soil and tissue sampling, a range of extraction procedures, and identification by various methods (Barker and Davis, 1996). Nematode morphology and differential hosts have been the primary basis for identification of nematode species and host races. Protocols that use immunology, protein electrophoresis, isoelectric focusing, and a range of DNA-based protocols are still limited to laboratory and research programs (Barker and Davis, 1996; Fleming and Powers, 1998). For IPM purposes, the presence of nematode-specific symptoms (galled roots for root-knot nematodes or root lesions for lesion nematodes) and signs of nematodes (the pear- or round-shaped cysts of the cyst nematodes, etc.) as well as the typical spotty growth patterns of a given crop in an infested field are good indicators of nematode problems. Precise identifications of nematode species, however, continue to be done primarily by nematode advisory programs and research laboratories.

Elimination of Pathogens

The importance of pathogen-free plant material is sometimes overlooked in modern as well as subsistence agriculture. Still, interest in this area is increasing due to the need to reduce pesticide usage in response to environmental and health issues. Meristem culture, which has been in use for almost 50 years, has been employed on many crops as a method of eliminating virus infections (Dinesen and van Zaayen, 1996). This procedure may also free plants from bacteria, fungi, viroids, phytoplasmas, and nematodes. Although meristem culture has been used intensively on ornamentals, the technology has much potential for other crops. For example, Pesic-van Esbroeck and associates (personal communication) have a major ongoing program at North Carolina State University in which this protocol is utilized to free strawberry (*Fragaria chiloensis* Duchesne), sweet potato (*Ipomoea batatas* (L.) Lam.), and other crops from associated virus pathogens. This procedure also is invaluable for limiting virus problems on potato (*Solanum tuberosum* L.).

Decision-Making Aids

As indicated earlier, effective IPM depends on accurate diagnosis of pest problems, quantification of the infestations, and information on related population dynamics, or dispersal of pests over time (CAST, 2003a) as well as decision rules (Kennedy, 2000). Thus, general decision-making aids include pest-population monitoring; field histories; various models, including those based on experience, crop-loss models for given pests, and more comprehensive mathematical models; and weather-based advisories. Information on the rate of overwintering of soilborne insects and nematode pests is essential for management decisions before the establishment of susceptible crop. Fields with high infestations of given fungi or other soilborne pests frequently are planted to non-host crops, especially where no other effective management tactic is available or where treatment costs are excessive. For some of the highly damaging vegetable root rots, avoiding high-risk fields has saved the industry millions of dollars over the last 3 decades (CAST, 2003a).

The compilation and interpretation of pest-population data, beneficial organisms, distribution of weeds, field histories and the appearance and development of plant diseases serve as the infrastructure of many IPM programs. Although labor and data intensive, the resulting regional and national IPM programs provide improved pest control as well as reductions in pesticide usage, and enhanced profitability for the grower. IPM service-advisers generally provide a range of services, including pre-season pest and soil-nutrient assessments, and crop monitoring during the growing season for pests/diseases, nutrient status, collection of weather data, and application of these data via pest-prediction models to aid in management decisions (CAST, 2003a). An additional pre-harvest sampling for some pests and other variables can facilitate estimations of yield and the need for monitoring for post-harvest, or storage, pest problems. Reliable weather data and their interpretation serve as important variables for prediction of the appearance and development of a number of crop pests. This information is applied via models that then impact management decisions. Weather-based IPM advisories are available for field crops such as peanuts (*Arachis hypogea* L.) (Phipps, Deck, and Walker, 1997), potato, and certain vegetables (CAST, 2003a; Main et al., 2001).

While population levels and economic thresholds (Figure 1) can be determined for many insect infestations, most threats posed by plant-pathogen infestations are assessed by symptomology, environmental/weather conditions, and field history (Campbell and Benson, 1994). Projections of disease development can be offered via various "models," including those driven by environmental parameters, especially temperature and relative humidity. "Disease-progress curves" (Figure 2) that characterized disease on specific cultivars are useful in that regard.

Choice of appropriate crop rotations, selection of pest-resistant cultivars, or a combination of pest-management tactics/strategies are often based on pest diagnostics. In addition to resources such as traditional pest-resistant cultivars, cultural practices, and pesticide options, other new technologies, especially those related to genetic engineering, and precision agriculture are having an impact on IPM and related cropping systems.

SPECIFIC IPM TOOLS USED IN CROPPING SYSTEMS

A synopsis of the wide range of cropping components and other practices for managing crop pests, including their relative effectiveness, is given in Table 1. Cropping components include: rotation, fallow, cover crops, manipulating pest refugia, tillage, row width, type of cultivation, soil amendments, and field size and borders. These and other practices such as pesticide usage, plant populations, sanitation, and resistant varieties are discussed as follows.

Rotation/Cultural Practices, and Habitat Management

The growing interest in IPM and sustainable cropping systems has generated renewed emphasis on cultural methods of pest management (Cavigelli et al., 2000; Kennedy and Sutton, 2000; Madden, 1992). As indicated in Table 1, many cultural practices have proven to be effective for managing a wide range of pests. Key facets of cultural pest control include polyculture, crop diversification, destruction of residual roots of certain crops with a perennial growth habit, minimal tillage, and biological/environmental manipulation that lead to biological diversity (Altieri, 1994). Polyculture may include the use of "refugia" to maintain and enhance populations of natural enemies of some arthropod pests (Altieri, 1994; Landis, Wratten, and Gurr, 2000). Crop rotation provides for diversity in time and space and often is the preferred means of management for soilborne pests such as plant-parasitic nematodes. The benefits of rotation

FIGURE 2. Disease progress curves. (A) Comparison of single-cycle (one generation, usually soilborne parasite) and multi-cycle (multiple generations, usually foliage parasites) plant pathogens; (B) Rapid disease development induced by a virulent pathogen on a qualitatively resistant crop versus minimal disease with an avirulent pathogen (in large part, adapted from Cavigelli et al., 2000).

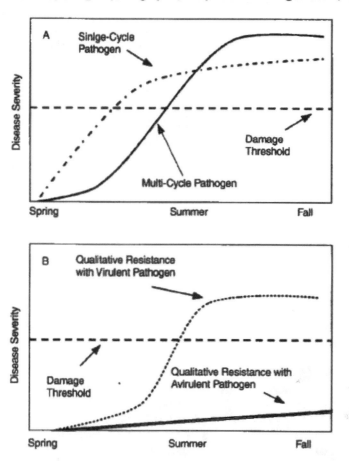

are derived from the destruction of given crop pathogens and often other pests by natural enemies or "sanitizing organisms" while other unrelated crops are grown (Cook and Veseth, 1991). Rotation, however, may be of limited value for pests that have a wide host range or are highly mobile. Furthermore, the need to rotate crops varies with given pest populations and location. Corn (*Zea mays* L.) can be grown continuously in some locations, whereas soilborne pests such as nematodes can cause serious yield losses to this crop in some regions (Barker and Koenning, 1998). In contrast, soybean (*Glycine max* L.), a crop highly susceptible to nematodes and other soilborne pathogens, is routinely rotated with corn and other crops to prevent related yield losses and to prolong the durability of resistant cultivars. Many intensively managed crops that are susceptible to numerous pathogens and nematodes often encounter severe disease and other pest problems under monoculture. Rotating crops such as tobacco (*Nicotiana tabacum* L.) with fescue (*Festuca arundinacea* L.) limits pest activity, improves soil structure and water-holding capacity, and enhances crop yield. Plants directly antagonistic to certain pests also can be used in crop rotations. *Crotalaria* species may function as trap crops for root-knot nematodes, and the African marigold (*Tagetes erecta* and other *Tagetes* species) provide excellent nematode control under certain conditions (Barker and Koenning, 1998). As discussed under potato, effective rotations are central to nematode and pest management of this high-value crop.

Effective use of cultural practices in a crop management system requires a considerable database (CAST, 2003a). Features of the cropping system that can be manipulated to retard pest activity while enhancing biological control should be central. This situation may call for the integration of a range of management practices such as rotation, minimal tillage, soil amendments, and the use of a cover crop (El Titi and Ipach, 1989; Zwart et al., 1994). To provide optimum returns, information on the impact of such management practices on beneficial organisms and crop pests is essential.

The use of cover crops or green manure crops between primary crops offers many benefits (Table 3). When utilized as green-manure or cover crops, a number of selected clovers, velvet bean [*Mucuna deeringiana* (Bort.) Merr.] and joint vetch (*Aeschynomene americana* L.) have considerable potential for augmenting soil health and crop production (Barker and Koenning, 1998; Kloepper et al., 1992). Benefits of using these legumes encompass enhanced soil nitrogen (N), promotion of soil populations of plant-growth-promoting rhizobacteria, and direct or indirect negative effects on a spectrum of plant pests (Magdoff and Van Es, 2000). A range of legume cover crops, including castor bean (*Ricinus communis* L.), sward bean [*Canavalia ensiformis* (L.) D.C. Bean], and velvetbean greatly enhance the numbers of plant-growth promoting rhizobacteria such as *Burholderia cepacia* and *Pseudomonas gladioli*. Both cyst and root-knot nematode populations are suppressed when soybean followed any of these cover crops (Kloepper et al., 1992). Certain rhizobacteria may induce systemic acquired resistance to foliage pathogens such as *Pseudomonas syringae* pv. *lacrymans* and *Colletotrichum orbiculare* on cucumber as well as provide some nematode control (Wei, Kloepper, and Tuzun, 1996). Other soilborne bacteria designated as "deleterious rhizobacteria," including certain strains of *Pseudomonas flourescens*, have potential as biological controls of weeds (Liebman, Mohler, and Staver, 2001).

TABLE 3. Effects of selected grass and legume cover crops on soil and associated pests/beneficials (adapted from Bowman, Shirley, and Cramer, 1998).

Cover crop	Loosen soil	Allelopathic	Benefits/risks per pest group[a]				Beneficial insects
			weeds	pathogens	Nematodes	insects	
Annual ryegrass	++++[a]	++	++++/----	++/-	++/-	-	+
Barley	+++	+++	+++/---	+/---	+/---	---	++
Oats	+++	+++	++++/0	++/-	•/--	--	0
Rye	++++	++++	++++/----	++/-	++/-	--	+
Wheat	+++	+	+++/-	+/---	+/---	---	+
Buckwheat	+++	+++	++++/----	•/0	+/-	-	++++
Sorghum-Sudan	++	++++	++++/--	+++/0	+++/-	-	++
Berseem clover	+++	+	+++/0	•/-	•/--	--	++
Cowpeas	+++	•	+++/0	•/-	•/--	--	+++
Crimson clover	++	+	++/--	++/-	+/----	----	+++
Field peas	+++	+	+++/0	+++/---	++/--	--	+++
Hairy vetch	+++	++	+++/--	++/0	+/--	--	++++
Medics	+	+	+++/--	++/0	++/---	---	+
Red clover	++	++	++/--	+/-	+/---	---	+++
Sweet clovers	++++	+	++/---	+/0	+/---	---	+++
White clovers	+++	++	+++/--	•/-	•/--	--	++

a Symbols for benefits

  • = Poor

  + = Fair

  ++ = Good

  +++ = Very good

  ++++ = Excellent

Symbols for increased pest risks

  0 = rarely becomes a problem

  - = occasionally a problem

  -- = can be a minor problem

  --- = can be a moderate problem

  ---- = can be a major increased pest risks

Various tillage and cropping options may differentially affect soil organisms and crop pests (Table 1). In an IFS, beneficial earthworms made up 17.6% of the total soil microflora/fauna whereas they may be absent in a conventional system (Zwart et al., 1994). Bacteria comprised 94% of the biomass in the conventional production systems versus 75% in the integrated system. Figure 3 shows that IFS that included altered soil tillage, a clover cover crop, organic manure, and reduced pesticide usage, resulted in highly suppressed populations of two nematode species (*Heterodera avenae* and *Ditylenchus dipsaci*) on cereals versus those in conventional cropping (El Titi and Ipach, 1989). For health management of crops such as wheat (*Triticum aestivum* L.), a key goal is to limit tillage as much as possible (Cook and Veseth, 1991).

FIGURE 3. Relative infestation levels by the cereal cyst nematode, *Heterodera avenae*, in integrated and conventional farming systems [expressed as the mean number of eggs + juveniles per 250 cc soil; differences are highly significant, according to Wilcoxon Test ($P = 96\%$)] from 1984 onwards (Adapted from El Titi and Ipach, 1989).

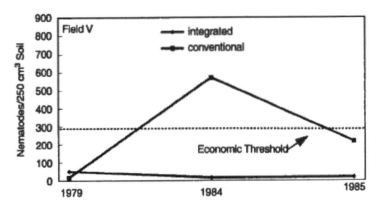

Soil organic matter (SOM) can play a multi-faceted role in IPM and cropping systems. Magdoff and Van Es (2000) concluded that good SOM is the very foundation for a sustainable and thriving agriculture. Objectives in all cropping systems should include an annual goal of returning as much or more organic matter to the soil than is withdrawn (Cook and Veseth, 1991). Bowman, Shirley, and Cramer (1998) provide general ratings on the capacity of various cover crops to suppress the activity of weeds, pathogens, nematodes, and insects (Table 3). Certain cover crops such as buckwheat (*Fagopyrun esculentum* Moench) and hairy vetch (*Vicia villosa* Roth) are very effective in attracting beneficial insects. Most grass cover crops serve to "loosen" the soil but can have allelopathic effects on subsequent crops. Although legume cover crops provide many benefits, including increased soil N, some can enhance pest/pathogen problems (Table 3). In addition to increased biological activity and diversity in the soil, organic matter improves soil quality through increased aggregation, pore structure, tilth, and water-holding capacity (Magdoff and van Es, 2000). These many benefits, although not providing a high level of control of all crop pests, are undoubtedly partly responsible for the rapid annual growth of 20 to 30% in the organic production industry (USDA, 2001).

Host-Plant Resistance/Tolerance

Crop resistance, carried in the seed or propagation materials, is an ideal tactic in IPM with the deterrent to the pest being inherited. The heritable trait that limits the normal host-pest interaction restricts associated damage and yield loss, compared to susceptible varieties (CAST, 2003a). Since crop resistance involves little or minimal input costs, it is an ideal tactic for managing a wide range of pests, including bacteria, fungi, viruses, insects, mites, nematodes, mammalian herbivores, and other plants (through allelopathy) (see Tables 1 and 3). Although pest-crop resistance requires minimal management input, its durability is heavily dependent on the manner in which it is deployed (CAST, 2003a). Continuous use of a given resistance source in the same fields often results in the development of biotypes of pests that overcome the impact of the resistance genes. Except for this important problem, resistance has a low-management requirement as compared to many management-intensive IPM tools. In addition to the classical pest resistance developed through traditional plant-breeding programs, genetically engineered crop lines or cultivars with resistance or tolerance to major pests are being widely deployed as indicated earlier. Examples of these types of resources are discussed under given cropping systems.

Biological Control

Among numerous definitions of biological control, Hoy (2000) provided a succinct description: "Biological control of arthropod pests and weeds is a method of pest management that employs parasitoids, predators, pathogens, and entomophilic nematodes (natural enemies) to reduce pest populations." A broader concept of biological control was given by Cook (1987): "The use of natural or modified organisms, genes, or gene products, to reduce the effects of undesirable organisms (pests) and to favor desirable organisms such as crops, trees, animals and other beneficial insects, and microorganisms." Biological-control products are also being used in IPM. They are differentiated from biological-control agents as follows: "Biological-control organisms are living organisms that can be used to manage arthropods (mites and insects), weed, and plant (bacteria, fungi, viruses, and nematodes) pests and pathogens"; "genes or gene products derived from living organisms that kill, disable, or otherwise regulate the behavior of plant pests are biological-control products" (National Research Council, 1996). Our increasing understanding of the basis for biological control is bringing new pest-control products to the market as well as advancing ecologically-based IPM (Kennedy and Sutton, 2000).

Pesticides

Pesticides have traditionally been an important component of cropping systems. Ecologically-based IPM is our most promising option for reducing the negative effects of pesticides on our environment while contributing to the sustainability of our societies (CAST, 2003a). Still, pesticides continue to be essential to feed and protect the ever-increasing world population. IPM programs have helped stabilize pesticide usage in the U.S. since the mid-1980s, whereas their use in rapidly growing economies such as those of the Pacific Rim is increasing at about 8% per year (CAST, 2003a). The niche for pesticides in crop production differs greatly with the cropping system, associated pests per region or site, weather and economics. Examples are given in the cropping systems presented herein, but usage continues to evolve as new IPM options are developed, changes in pest situations, and changes government regulations (CAST, 2003a; Kennedy and Sutton, 2000; Main et al., 2001).

EXAMPLES OF PEST MANAGEMENT IN MAJOR CROPS OF THE U.S.

Cropping systems vary along a continuum from very low intensity, low input, high acreage systems to high intensity, high input, low acreage systems. Pest management efforts and inputs vary along this continuum as well. In general, pest management in low value, extensive cropping systems must rely on inexpensive management strategies/tactics (i.e., many cultural practices and host-plant resistance). This situation results from the equilibrium points of many pests in these systems routinely falling well below economically justifiable pesticide-based, thresholds. Pest management in high value crops generally is far more intensive because the value of the crop dictates lower tolerance for production loss and therefore lower thresholds. Since pesticides are typically highly efficacious but often expensive (both in terms of direct costs of application and collateral impact) and temporary in effect, the intensity of pesticide use may be a strong indicator of the intensity of the overall pest management program in a given commodity. However, the potential for much greater economic return in high-value crops also provides opportunity to integrate high intensity, non-pesticidal tactics as well.

In the following, we discuss examples of cropping systems along the continuum from low value, extensively farmed crops to high-value, intensively farmed crops and the typical pest management options and considerations in each. Alternatively, crops intensively managed in one region may be grown with minimal inputs elsewhere. For example, potatoes are grown with minimal inputs in much of the developing world, while European and North American production is intensive. Likewise, wheat in the U.S. and Australia receives few pesticide inputs, whereas production in Europe usually is greatly intensified.

Wheat

Wheat occupies approximately 26 million ha in the U.S. This area includes winter wheat, durum, and spring wheat. Average yields in the U.S. in 1998 and

1999 were approximately 2,350 kg ha^{-1} with an average gross return of about $247.00 ha^{-1} (USDA/NASS, 1999a; 2000a). Significant weed pests of wheat include both grasses and broadleaves. Tillage is an important management tactic for all weeds, although demands for soil conservation now limit its use on wheat (Cook and Veseth, 1991). About 30-40% of area under wheat is treated with one or more herbicides annually; the most frequently used material is 2,4-D, which provides control of broad-leaved weeds (USDA/NASS, 2001c). Important insect pests of wheat in the U.S. include the Hessian fly (*Mayetiola distructor* Say), the Russian wheat aphid (*Diuraphis noxia*) and other aphids, the cereal leaf beetle (*Oulema melanopus* L.) in eastern North America, and chinch bug (*Blissus leucopterus* Say). Typically less than 10% of the area under wheat receives an application of insecticide over the course of the growing season. Insect management in wheat relies on non-insecticidal tactics. Hessian fly management is dependent on two main strategies: host-plant resistance and the use of a "fly-free" planting date (ensuring that most adult flies have died by the time the crop emerges). Cereal leaf beetle management over much of the area where it occurs depends on introduced parasitoids.

Infectious wheat diseases may be caused by bacteria, fungi, nematodes, viruses, and flowering plants (Wiese, 1987). Some pathogens can cause more than one disease on this crop. For example, *Fusarium graminearum* and related species, in addition to inducing seedling blights and foot rots, also cause a major disease, "scab," on wheat. This disease often is especially severe when wheat follows corn as the pathogen overwinters in corn stalks. The emerging spores are windborne and infect heads of wheat at flowering. While this disease is common in the Great Lakes states and other regions, it is limited to fields receiving pivot irrigation in the Pacific Northwest (Cook and Veseth, 1991). In addition to suppressing wheat yields, infected grain may contain mycotoxins that impact the health of cattle.

Cook and Veseth (1991) provide a comprehensive, holistic approach to wheat health management or IPM. They document the benefits of crop rotation and maintaining SOM/soil structure. Additional principles for managing wheat include: an understanding of the production limits of the cropping system; using well-adapted, pest and disease resistant varieties; choosing high quality, disease/weed-free seed; minimizing environmental and nutrient stresses; maintaining or enriching populations of beneficial insects and microorganisms; and scouting for pests and treating, when necessary, with pesticides. While promoting rotations and augmented organic matter, these authors view tillage as a prime means of destroying organic matter and enhancing soil erosion. A combination of increased organic matter and reduced tillage can suppress pathogens such as the cereal cyst nematode (*Heterodera avenae*) (El Titi and Ipach, 1989) while enhancing populations of beneficial organisms, including earth worms (Zwart et al., 1994). With the foliage pathogens, however, host resistance continues to be invaluable in wheat. In the U.S.A., wheat crop-ping systems included rotations in about 83 to 90% of the area in 1995 (Table 2). To minimize the development of resistance, pesticides were alternated in about 30 to 38% of area under wheat (Table 2). An average of 0.25 kg ha^{-1} of pesticides was used in U.S. wheat in 1998. Of this, about 98% was herbicide (USDA/NASS, 1999c).

Soybean

Soybean is grown on approximately 28 million ha in the U.S. Average yields across the U.S. in 1998 and 1999 were approximately 2130 kg ha^{-1} with an average gross return of about $450.00 ha^{-1} (USDA/NASS, 1999a, 2000c). Weeds are significant pests of soybean, and shifts over the last 15-20 years to-wards reduced-tillage systems have increased reliance on herbicide use in soy-beans. Approximately 95% of the soybean acreage in the U.S. received at least one application of herbicide in the 1998 and 1999 growing seasons (USDA/NASS, 1999c, 2000c). The development of transgenic, glyphosate-resistant soybean cultivars has been a major development in weed management in this crop. Approximately 60 to 70% of the U.S. soybean crop is planted to these cultivars (S. R. Koenning, personal communication). Another significant cul-tural development over the last 15-20 years has been the shift towards narrow row spacings; narrower rows close the plant canopy quicker and give the crop a competitive advantage over weeds (Howe and Oliver, 1987).

While a large number of insect species feed on soybean, and occasionally cause economic damage (Higley and Boethel, 1994), two significant arthropod pests of soybean are the corn earworm (*Helicoverpa zea* Boddie) and the bean leaf beetle (*Cerotoma trifurcata* Foster). Insecticides currently play a very small role in soybean since less than 5% of the area under soybean in the nation received an insecticide application in the 1998 and 1999 growing seasons. This represents a rather dramatic shift from the situation 20-25 years earlier, when as much as 40-50% of the acreage may have been treated with an insecticide. Several factors contributed to the decline in insecticide use in soybean. The profitability of soybean has declined as prices dropped to less than 50% (tak-ing inflation into account) of what they were at their highest in the late 1970s. This has elevated the economic injury levels of many of the insects, previously regarded as serious soybean pests, to points well above the equilibrium points for those species. The bean leaf beetle is both an early season and mid-season defoliator and pod-feeder; however, because of prevailing economic condi-tions, only very high populations warrant treatment (Pedigo, Zeiss, and Rice, 1990). Furthermore, cultural strategies for managing the most significant soy-bean pests have been developed. The importance of the corn earworm dropped dramatically with the adoption of narrow row spacing and the increase in double-cropping. Narrow rows close the crop's canopy much earlier making soybean less attractive to ovipositing female earworm moths; double-cropping results in asynchrony between ovipositing moths and the most attractive phenological stage of the soybean plant (Bradley and Van Duyn, 1979).

An array of diseases caused by nematodes, fungi, bacteria, and viruses is well documented for soybean (Hartman, Sinclair, and Rupe, 1999). Among these pathogens, the soybean cyst nematode (*Heterodera glycines*) causes the greatest yield losses which amount to billions of dollars annually. Severe stunting of plants by high infestations of this nematode also can result in greater weed and insect problems on the crop (Alston et al., 1991). In recent years, disease and nematode management in soybeans have been pursued largely through the development of resistant cultivars and the identification of appropriate cultural tactics. Bradley and Duffy (1982) documented the economic value of disease resistance on this crop via an economic analysis. The cost of developing one cyst-root-knot nematode resistant cultivar "Forrest" was $1 million, whereas the return to growers (avoiding losses) over a 6-year period amounted to $401 million. Statistical models also may be useful for predicting soybean as well as corn yields for specific soils and regions (Garcia-Paredes, Olson, and Lang, 2000).

Key cropping systems for IPM in soybean in the U.S. include rotation (89%), no-tillage (33%), mechanical cultivation for weeds (29%), and alternating pesticides (28%) (Table 2). An average of 1.23 kg ha^{-1} of pesticides was used in U.S. soybean in 1998. Of this, 99% was herbicide (USDA/NASS, 1999c).

Corn

Field corn is planted on approximately 28.3 million ha in the U.S. Across the country, annual gross returns over the 1998 and 1999 growing seasons were approximately $630.00 ha^{-1} on yields of approximately 8,000 kg ha^{-1}. As in the earlier examples, weed competition can significantly compromise corn production (USDA/NASS, 1999a, 2000a). Weed management in corn historically relied on tillage and cultivation. However, over the last 30 years, herbicides have assumed a much larger role in weed management in corn. Currently, virtually all corn acreage receives at least one application of one or more herbicides; the most widely used active ingredient is atrazine (USDA/NASS, 1999c). Again, pressure to conserve agricultural soils has increased adoption of reduced tillage production systems. As with soybean, herbicide tolerant corn varieties are being developed and deployed.

The major arthropod pests of corn are the European corn borer, *Ostrinia nubilalis* Hubner and the corn rootworm complex composed primarily of the western corn rootworm, *Diabrotica virgifera*, and the northern corn rootworm,

Diabrotica longicornis barberi. Many other arthropods feed on corn and are occasional pests; however, these three drive corn pest management over much of the country. The European corn borer is an introduced moth found throughout the eastern half of the U.S. Over much of this area, two generations per year attack corn. In most areas, the first generation is well suppressed by host plant resistance bred into most hybrid corn lines (Chiang and Hudon, 1976). This resistance dissipates before the onset of the second generation. Second-generation management has been quite problematic due to the large size of the plant during this part of the growing season and the boring habits of the larvae. Recently, transgenic corn varieties containing *Bacillus thureingensis* (Bt) toxins effective against European corn borer have been developed; however, social concern over this technology may limit its utility (CAST, 2003a).

The corn rootworm complex has, until recently, been managed through a combination of cultural and chemical tactics. These two species are univoltine, and most populations have traditionally laid their eggs in cornfields. The eggs over-winter and hatch in the spring about the time corn roots become available. This life cycle has made this complex susceptible to management through rotation to a non-host crop. Within the last decade, however, populations of western corn rootworm have been identified which remain in diapause for two years (Krysan, Jackson, and Lew, 1984), and populations of north corn rootworms have been found which oviposit in soybean fields, the most frequent rotational crop (O'-Neal, Gray, and Smyth, 1999). Transgenic corn hybrids containing genes coding for rootworm-active Bt toxins have been developed and hold great promise for management of the rootworm complex if societal concerns over the technology are resolved.

Numerous diseases often affect the yield and quality of corn (White, 1999). Except for pathogenic nematodes, disease management in corn has depended on the identification of resistant or tolerant germplasm and incorporation through conventional plant breeding, together with cultural disease management strategies. Reliance on a limited pool of genetic material for hybrid development has, in the past, caused crises in corn production; the most notable example of widespread failure of this nature was the southern leaf blight, caused by *Cochliobolus heterostrophus* (Drechs.) Drechs. (*Bipolaris maydis* (Nisik.) Shoem.), epidemic of the mid-1970s. Although corn is somewhat tolerant to several nematode species, the lesion nematode (*Pratylenchus* spp.), stubby-root nematode (*Paratrichodorus* spp.), needle nematode (*Longidorus* spp.), and the sting nematode (*Belonlaimus* spp.) can inflict extensive damage, especially in sandy soils.

More recently, concerns over mycotoxin contamination have complicated corn production (CAST, 2003b). Aflatoxin in corn is most serious during years with droughts and other conditions favorable for the causal fungus, *Aspergillus* spp. Although prevailing environmental conditions in the southeastern U.S.

result in frequent aflatoxin problems, the impact of this toxin in midwestern corn would be much greater as 75% of U.S. corn produced in the North Central States. Thus, any major aflatoxin problems in that region can limit the availability of clean corn for export as well as domestic use. A comprehensive treatment of aflatoxins is forthcoming (CAST, 2003b). Fungicide use is negligible in corn.

Corn cropping IPM systems in the U.S. for 1996 included rotation (82%), mechanical cultivation (51%), and no-tillage (19%) (Table 2). An average of 3.16 kg ha^{-1} of pesticides was used in U.S. corn in 1998. Of this, about 93% was herbicide (USDA/NASS, 1999c).

Cotton

Cotton is produced annually on approximately 5.4 million ha in the U.S. Cotton is a frost-intolerant perennial managed as an annual, and production is concentrated in the southern states from Virginia through California. Annual gross returns to the producer over the 1998 and 1999 growing seasons were approximately U.S. \$830 ha^{-1} on production of ca. 690 kg ha^{-1} of lint (USDA/NASS, 1999a, 2000a). Since cotton is initially not very competitive and is typically grown in warm climate areas with either abundant rainfall or substantial irrigation, weed management is a significant production issue. On average, in excess of 95% of the crop receives at least one application of herbicide; typically, two or three herbicides will be applied. Weed management also typically involves two or three cultivations, although the advent of transgenic, herbicide resistant cotton varieties and the development of reduced tillage systems for cotton have reduced the frequency of cultivation.

Cotton fruits are present and are vulnerable to attack by arthropod pests for approximately 70 days during the growing season; this, together with the wide diversity of insects attacking cotton, elevates the importance of insect management. Significant arthropod pests in cotton include the bollworm complex [*Helicoverpa zea* (Boddie) and *Heliothis virescens* F.], the pink bollworm [*Pectinophora gossypiella* (Saunders)] (restricted to the Southwest), thrips (Thysanoptera), stink bugs [*Acrosternum hilare* (Say), *Nezara viridula* (L.), and others] and plant bugs (*Lygus* spp.), and in a rapidly declining area in the mid-south and Texas, the boll weevil [*Anthonomus grandis* (Boheman)].

Insecticide use has been reduced in those areas where the boll weevil has been economically eradicated. Foliar applications of insecticide have declined from about 12 per season to about 4 per season in these areas (Bacheler, 1991) (other factors, including the advent of plant growth regulators and short-season varieties, have also contributed to this decline). However, the eradication program has been and continues to be expensive. Approximately 6% of total insect management costs are for eradication and the maintenance of eradication in areas were the insect has been extirpated (Williams, 1999).

Insect management for cotton often includes advanced strategies and tactics. Computer programs such as TEXCIM, a software package that focuses on costs of pests, benefits of management tactics, and crop value is used in southern Texas. Frequent, systematic sampling of insect pests is routine in cotton, and much of this activity (42% in 1996) is conducted by paid crop consultants (Williams, 1999) (Table 2). Most cotton areas are examined at least weekly for insect pest problems. Foliar insecticide use is more intense in cotton than in any of the crops discussed earlier. An average of 4.3 foliar applications of insecticides was applied to the 1998 crop (Williams, 1999); approximately half of this targeted the bollworm complex. In addition, approximately 54% of the acreage received an at-planting application of insecticide, primarily for thrips control. Total average costs of insect pest management in 1998 were U.S. $155.80 per ha^{-1}.

While disease management is also an important consideration in cotton, many diseases are of limited, regional importance (Kirkpatrick and Rothrock, 2001). Key pathogens of cotton include seedling parasites, nematodes and wilt fungi. Cottonseed is treated with a fungicide prior to planting for control of seedling diseases, and a small fraction of the total area also receives a foliar fungicide. The Columbia Lance nematode (*Hoplolaimus columbus*), occurs only in the southeastern U.S. Other important cotton diseases are managed through a combination of cultural strategies, host plant resistance, and soil-applied chemicals.

Monoculture was the choice cotton-cropping system in 1996 (67%), and mechanical cultivation was used for weed control in 89% of the area (Table 2). With the widespread use of herbicide-tolerant cotton, mechanical cultivation likely has diminished. An average of 3.55 kg ha^{-1} of pesticides was used in U.S. cotton in 1998. Of this, approximately 40% was insecticide and 59% was herbicide. In addition, each hectare also received about 1.12 kg of other agricultural chemicals, primarily plant-growth regulators and harvest aids (USDA/NASS, 1999a, 1999c). To aid in limiting the development of resistance, pesticides were alternated in 41% of U.S. cotton in 1996 (Table 2).

Potato

Potatoes are grown on approximately 565,000 ha in the U.S. Annual gross returns to potato producers over the 1998-1999 growing seasons averaged approximately $4,900 per ha^{-1} on average production of 39,300 kg ha^{-1} (USDA/NASS, 1999b, 2000c). Because of the value of this crop and the impact of its many pests, comprehensive or holistic management-production plans are available (Rowe, 1993). This program includes guidelines in establishing long-term rotations at least a year before growing potatoes and following definitive practices at preplant, all growth stages of the crop, and at harvest and storage. Rigorous disease/insect scouting is interfaced with highly developed IPM Programs (CAST, 2003a; Rowe, 1993). Weed management in potatoes is achieved through fairly intense herbicide use and frequent mechanical cultivation through the first half of the growing season. Approximately 95% of the potato hectarage receives at least one application of a herbicide, and a substantial portion of the land receives more than a single herbicide (USDA/NASS, 1999b).

The most significant insect pests of potato are the green peach aphid [*Myzus persicae* (Sulzer)], in the northern production regions, and the Colorado potato beetle (*Leptinotarsa decemlineata* Say), in the eastern and central production regions. The green peach aphid is an important vector of viral diseases of potato, particularly potato leafroll virus (PLRV), the causative agent of net necrosis in potato tubers. Green peach aphid management is based on the use of systemic insecticides at planting and early season foliar insecticides to limit the incidence of this disease. Field sanitation and the use of virus-free tubers are also important components of the management of PLRV (CAST, 2003a).

The Colorado potato beetle is most significant in areas where two generations develop per year. While crop rotation and tillage can reduce Colorado potato beetle populations, management of this insect relies on the use of insecticides. Due to heavy insecticide use, some populations of the Colorado potato beetle have developed resistance to virtually all classes of insecticides (Forgash, 1985). On average, in excess of two foliar insecticide applications are made to potato crops annually for all foliar pests.

Disease management in potato is critically important; several fungal, nematode, and viral diseases can seriously compromise potato yield and quality. Two foliage diseases, early blight (*Alternaria solani*) and late blight (*Phytophthora infestans*) can defoliate potato plants, resulting in low yields and quality. Frequent monitoring is crucial in managing these diseases. A number of forecasting models [HYRE system, WALLIN system, BLIGHTCAST system], based on temperature and moisture and/or humidity, have been developed (Rowe, 1993). "Early dying" is a soilborne disease complex caused by the lesion nematode (*Pratylenchus penetrans*) and *Verticillium dahliae*. Late-summer soil sampling in fields preceding spring planting can help growers assay for infestation levels of both pathogens and plan cropping systems appropriate to the pest hazard. Options could involve rotations with a nonhost crop or fumigation (Rowe, 1993).

Sanitation and certification of disease-resistant planting stock are important components of potato disease management, as is vector management. However, fungicides are the key disease management tools in this crop, and crops may receive seven or more applications of these materials over the course of a growing season. An average of 16.9 kg ha^{-1} of pesticides was used in potatoes in the U.S. in 1998. Of this, about 82% was fungicide (USDA/NASS, 1999b). Pesticide rotation is used (69% in 1996–Table 2) to minimize the development of resistance. Rotation is followed in about 98% of potato hectarage (Table 2). Rotation 'crops' include Sudan grass hybrids (*Sorghum bicolor* L. Moench) which give good control of *M. chitwoodi* on potato, but are not effective for lesion nematodes (MacGuidwin and Layne, 1995). Hybrids of sorghum-Sudan grass also control *Meloidogyne chitwoodi* but these plants must be grazed with care as they synthesize high concentrations of dhurrin, which is toxic to cattle (Mojtahedi, Santo, and Ingham, 1993). A federal quarantine has been very effective in limiting the spread of the highly damaging golden nematode (*Globodera rostochiensis*) in the U.S. (Marks and Brodie, 1998). In addition to using a carefully developed rotation system and certified, high quality tubers of resistant cultivars, fields heavily infested with soilborne pathogens should be avoided (Rowe, 1993).

Strawberries

Strawberry is a perennial, herbaceous vine grown commercially as an annual. A total of approximately 19,400 ha of strawberries are grown for commercial fresh market sales and processing annually in the U.S. (many additional ha are planted for "pick-your-own," purchaser-harvest operations). Average yields on commercial plantings in 1999 were approximately 32,300 kg ha^{-1} with gross returns to producers of about \$34,600 ha^{-1}. Approximately 80% of commercial strawberry production in the U.S. is in California (USDA/NASS, 2000d).

Commercial cultivation of strawberry is extremely intensive due to the high value of the crop and demand for high quality. Because of these key issues, IPM plays a limited role in producing this crop as compared to potato and many field crops. Also, weed and pathogen management usually must be implemented as preventatives rather than eradicatives. Strawberry fields are typically fumigated prior to planting; the material of choice has been a mixture of methyl bromide and chloropicrin (Maas, 1998), but this chemical is being phased out through federation and international regulation (EPA, 2001). On most commercial fields, plastic sheet mulch is applied either before or after planting for moisture, weed/pest, and soil-temperature management. Drip or sprinkler irrigation is usually supplied; sprinkler irrigation systems may be installed purely for frost protection purposes. Supplemental N and other nutrients (in some areas) are supplied.

Strawberry crops are confronted by even more suites of pests than most other crops. While damage thresholds are not available for weeds (Maas, 1998), management of these pests is far more integrated in this crop than in any other we have addressed previously. Pest control is largely effected by the fumigation of planting beds, the implementation of the plastic mulch, some use of herbicides, and hand labor. The shift from perennial culture of strawberries to annual culture, including preventative treatments, also simplifies weed management.

The most important arthropod pests of strawberry are mites. The spider mites (*Tetranychus* spp.) are managed through intense monitoring and applications of miticides. The cyclamen mite [*Steneotarsonemus pallidus* (Banks)] is managed primarily through rotation and destruction of old production beds. Other arthropods, including several species of aphids, *Otiorynchus* root weevils, and several species of stem and foliage feeding beetles and moths infest strawberry plantings and are managed through the use of insecticides and rotation with other crops. Vacuum insect extractors have been used in strawberries as well (Vincent and Lachance, 1993). For the southeastern U.S., the imported fire ants (*Solenopsis invicta*) can pose problems due to their large mounds and the painful stings inflicted on pickers (Maas, 1998).

A number of disease agents threaten strawberry production, and, indeed, disease management drives much of the culture of the crop. While direct sampling for pathogens is not practical, disease diagnosis and pathogen identification along with an understanding of related etiology, epidemiology and life cycles of the pathogens are fundamental to producing profitable crops. Basic control practices for strawberry pathogens include: use of certified, disease-free planting stock; sanitation; host resistance where available, rotation, and chemical treatments. Soil fumigation is done in large part to manage all soil-inhabiting pests. Host plant resistance has also been heavily exploited, but resistance can be complicated by interactions between pathogens, the host, environment, and the presence of vectors for some parasites. Also, some cultivars have resistance to only certain races of given pathogens such as *Phytophthora fragariae* var. *fragariae* (Maas, 1998). Sanitation and certification of planting stock are also critically important in strawberry disease management as a number of pathogens, including viruses and nematodes, can be introduced with poor plants. Recently, meristem propagation techniques have been adopted to help insure disease free planting stocks.

An average of approximately 335 kg ha^{-1} of pesticides were applied on strawberries the U.S. in 1999 (USDA/NASS, 2000d). Of this total, approximately 98% were fumigants used for broad-spectrum pest control with almost all acreage in California receiving a preplant treatment of methyl bromide-chloropicrin (Maas, 1998). Currently, much effort is being directed toward the development of alternatives to this biocide. Potential detrimental effects of the high usage of pesticides on this crop have received only limited attention. More effective linkage of IPM and carefully developed sustainable cropping systems that include strawberry could minimize these negative effects. Presently, however, the short-term economic benefits of current IPM programs fail to justify the associated costs (Maas, 1998).

Several beneficial organisms often are important in strawberry production. For example, more than 130 species of Mycorrhizal fungi are associated with this crop. These obligate symbionts benefit their hosts by facilitating nutrient update and may provide a level of resistance or tolerance to some pathogens such as root-knot nematodes *Meloidogyne* spp. (Barker and Koenning, 1998). However, highly effective biological control agents for most strawberry pests remain to be identified and exploited.

EMERGING OPPORTUNITIES AND CHALLENGES

Much progress has been made in characterizing the many interactions between crops and associated pests, including the development of simulation and other models. This includes the characterization of pest interactions on given crops (Abawi and Chen, 1998; Alston et al., 1991). Nevertheless, one of the greatest shortcomings of current IPM programs is the limited use of a systems approach (Merrigan, 2000)–an area in need of intensive research. The integration across pest types for specific crops and the overall production system continues to be a major challenge for IPM.

The recent advances in the development of IPM tools through biotechnology and other technologies pose new opportunities as well as weighty questions for IPM and cropping systems in agroecosystems (Bridges, 2000; CAST, 2003a; Ellsbury et al., 2000; Kennedy and Sutton, 2000). Unknown effects of genetically modified organisms (GMOs) on nontarget pests and associated organisms remain to be determined. A number of U.S. genetically engineered or transgenic crops with agronomic crop improvements such as insect resistance and/or herbicide tolerance have been economically successful. However, as 'Starlink' corn proved recently, the deployment of transgenic crops requires a number of special and complex considerations, including guidelines for production, harvesting, storage and marketing and processing. For long-term stability of these new tools, biointensive IPM-copping systems approaches will be essential. This includes industry support, monitoring of pest communities, rotation systems, judicious pesticide applications where needed, and a truly integrated crop production-IPM system (Kennedy and Sutton, 2000).

In addition to integrated systems and transgenic organisms, precision farming tools offer an option for site-specific IPM (CAST, 2003a). While precision farming provides tools for application of the appropriate amounts of inputs at the ideal time and areas in given fields, the required information and equipment are not readily available for most pests/crops and production regions. As more biotechnology and information-intensive products and related information are used, crop-oriented industries likely will have a greater impact in the development of cropping and IPM systems. For example, industry-developed crop-pest-management packages or systems are becoming available (Carroll, 2000). The focus of these types of packages is on large-acreage, high-value crops. These developments will likely impact the traditional technology-transfer systems that are provided by governmental agencies and Universities.

Another rapidly developing phenomenon is the growth of "Organic Farming" as a food-production system in a number of countries. Since the Final Rule in the U.S. prohibits the use of GMOs and greatly restricts the use of traditional pesticides (USDA/AMS, 2001), will our new technologies eventually be accepted by the organic-products communities? Clearly, the growth rates of 20 to 30% annually for organically produced foods in the U.S. and Europe will pose ongoing challenges for adaptations in IPM-cropping systems.

Increased global travel and trade have accentuated a bioinvasion of our agroecosystems and natural habitats. This problem is so severe in the U.S. that a national "Invasive Species Council" was established via a Presidential Order in 1999 (CAST, 2003a). Invasive pests clearly are bringing new challenges to IPM and cropping as well as animal systems. For example, widespread development of aggressive invasive pests, especially weeds, often greatly suppresses the activity of the normal fauna/flora, including microbes, and thereby reduces the biodiversity so important in ecology-based IPM (Altieri, 1994).

The ongoing development of pesticide resistance in numerous types of pests constitutes another huge barrier in IPM. With more than 700 pests already having acquired resistance to pesticides (CAST, 2003a), cropping systems should include "resistance-management plans." Government polices currently are having an increasingly important impact on IPM and related cropping systems. The international phase-out of methyl bromide offers one of the greatest challenges encountered by growers and agricultural scientists (EPA, 2001). The impending loss of this biocide has necessitated the development of alternative treatments for the control of an array of crop pests as well as altered cropping systems in many instances. Growers in the U.S. are faced with an equally difficult problem with the implementation of the "Food Quality Protection Act (FQPA)" (EPA, 1996; Ragsdale, 2000). This act redefines how pesticides are regulated and the act may limit the availability of organophosphates and carbamate insecticides and other older classes of pesticides. The FQPA encourages the adoption of IPM, and thus will affect many cropping systems. Ragsdale (2000) suggested that the FQPA could result in decreased availability of a wholesome and affordable food. Still, the FQPA is especially relevant to small acreage crops such as fruits and vegetables that comprise much of the diets of children.

As suggested by a call for new solutions for pest control through "ecologically-based pest management," there is a need for increased emphasis on the ecological facets of IPM and crop-production systems (National Research Council, 1996). While "Ecologically Based Pest Management" should not replace the well-established concepts of IPM, the critical needs for greater focus on ecological and other environmental factors cannot be ignored. As a recent GAO (2001) report and other assessments (CAST 2003a, Kennedy and Sutton, 2000) indicate, many other needs must be accommodated to facilitate further development and use of IPM. One of the greatest challenges for IPM is the development of truly integrated IPM-crop-production systems.

REFERENCES

Abawi, G. S. and J. Chen. (1998). Concomitant pathogen and pest interactions. In *Plant and Nematode Interactions*, Agronomy Monograph 36, ed. K. R. Barker, G. A. Pederson, and G. L. Windham, Madison, WI: ASA, CSSA, and SSSA, pp. 135-158.

Agrios, G. N. (1997). *Plant Pathology*. Academic Press, New York, London.

Alston, D. G., J. R. Bradley, Jr., D. P. Schmitt, and H. D. Coble. (1991). Response of *Helicoverpa zea* (Lepidotera: Noctuidae) populations to the canopy development in soybean as influenced by *Heterodera glycines* (Nematoda: Heteroderidae) and annual weed population densities. *Journal of Economic Entomology* 84:267-276.

Altieri, M. A. (1994). *Biodiversity and Pest Management in Agroecosystems*. Binghamton, NY: Food Products Press.

Bacheler, J. S. (1991). Life without the boll weevil: The status of cotton insect pests in North Carolina following eradication. In *Proceedings of the Beltwide Cotton Conference*, Memphis, TN: National Cotton Council, pp. 615-617.

Barker, K. R. and E. L. Davis. (1996). Assessing plant nematode infestations and infections. *Advances in Botanical Research (Pathogen Indexing Technologies)* 23:103-136.

Barker, K. R. and S. R. Koenning. (1998). Developing sustainable systems for nematode management. *Annual Review of Phytopathology* 36:165-205.

Benbrook, C. M., E. Groth III, J. M. Halloran, M. K. Hansen, and S. Marquardt. (1996). *Pest Management at the Crossroads*. Yonkers, New York: Consumers Union.

Bowman, G., C. Shirley, and C. Cramer. (1998). *Managing Cover Crops Profitably* (2nd Ed.). Beltsville, MD: Sustainable Agriculture Network.

Bradley, E. B. and M. Duffy. (1982). The value of plant resistance to soybean cyst nematode: A case study of Forrest soybeans. Rep. No. AGES820919. USDA, Washington, DC: Natural Resour. Econ. Div., Economic Research Service.

Bradley, J. R. Jr. and J. W. Van Duyn. (1979). Insect pest management in North Carolina soybeans. In *World Soybean Research Conference II: Proceedings*, ed., F. T. Corbin, Boulder, CO: Westview Press, pp. 343-354.

Bridge, J. (1996). Nematode management in sustainable and subsistence agriculture. *Annual Review of Phytopathology* 34:201-225.

Bridges, D. C. (2000). Implications of pest-resistant/herbicide-tolerant plants for IPM. In *Emerging Technologies for Integrated Pest Management—Concepts, Research, and Implementation*, ed., G. G. Kennedy and T. B. Sutton. St. Paul, MN: APS Press, The American Phytopathological Society, pp. 141-153.

Campbell, C. L. and D. M. Benson. (1994). *Epidemiology and Management of Root Diseases.* Berlin, New York: Springer-Verlag.

Campbell, C. L., P. D. Peterson, and C. S. Griffith. (1999). *The Formative Years of Plant Pathology in the United States.* St. Paul, MN: APS Press, The American Phytopathological Society.

Carroll, B. N. (2000). Development in transition: Crop protection and more. In *Emerging Technologies for Integrated Pest Management–Concepts, Research, and Implementation,* ed., G. G. Kennedy and T. B. Sutton, St. Paul, MN: APS Press, American Phytopathological Society, pp. 339-354.

Carson, R. (1962). *Silent Spring.* Boston: Hughton Mifflin Company.

CAST (Council for Agricultural Science and Technology). (2003a). *Integrated Pest Management–Current and Future Strategies.* IPM Report 140. Ames, IA: Council for Agricultural Science and Technology.

CAST (Council for Agricultural Science and Technology). (2003b). *Mycotoxins: Risks in Plant and Animal Systems.* Report No. 139. Ames, IA: Council of Agricultural Science and Technologies.

Cavigelli, M. A., S. R. Deming, L. K. Probyn, and D. R. Mutch. (2000). *Michigan Field Crop Pest Ecology and Management.* Michigan State University Extension Bulletin E-2704, 108 pp.

Chiang, M .S. and M. Hudon. (1976). A study of maize inbred lines for their resistance to the European corn borer, *Ostrinia nubilalis* (Hübner). *Phytoprotection* 57:36-40.

Cook, R. J. [Chair]. (1987). *Report on the Research Briefing Panel on Biological Control in Managed Ecosystems.* Washington, DC: National Academy Press.

Cook, R. J. and R. J. Veseth. (1991). *Wheat Health Management.* St. Paul, MN: APS Press, The American Phytopathological Society.

Dinesen, I. G. and A. van Zaayen. (1996). Potential of pathogen detection technology for management of diseases in glasshouse ornamental crops. *Advances in Botanical Research (Pathogen Indexing Technologies)* 23:137-170.

Ellsbury, M. M., S. A. Clay, S. J. Fleischer, L. D. Chandler, and S. M. Schneider. (2000). Use of GIS/GPS systems in IPM: Progress and reality. In *Emerging Technologies for Integrated Pest Management-Concepts, Research and Implementation,* ed., G. G. Kennedy and T. B. Sutton, St. Paul, MN: APS Press, The American Phytopathological Society, pp. 419-438.

El Titi, A. and U. Ipach. (1989). Soil fauna in sustainable agriculture: Results of an integrated farming system at Lautenback, F.R.G. *Agriculture, Ecosystems and Environment* 27:561-572.

EPA–United States Environmental Protection Agency. (1996). The Food Quality Protection Act of 1996. Washington, DC: United States Environmental Protection Agency.

EPA–United States Environmental Protection Agency. (2001). U.S. Methyl Bromide Phaseout. Washington, DC: United States Environmental Protection Agency. <http://www.epa.gov/ozone/mbr/mbrqa.html>.

Fernandez-Cornejo, J. and S. Jans. (1999). *Pest Management in U.S. Agriculture.* Washington, DC: USDA ERS Agriculture Handbook No. 717.

Fleming, C. C. and T. O. Powers. (1998). Potato cyst nematode diagnostics: morphology, differential hosts, and biochemical techniques. In *Potato Cyst Nematodes,* ed., R. J. Marks and B. B. Brodie, Wallingford, UK: CAB International, pp. 91-113.

Forgash, A. J. (1985). Insecticide resistance in the Colorado potato beetle. In *Proceedings- XVII International Congress of Entomology*, ed., D. N. Ferro and R. H. Voss, Amherst, MA: Massachusetts Agricultural Experiment Station Research Bulletin 704, pp. 33-52.

GAO–United States General Accounting Office. (2001). *Agricultural Pesticides–Management Improvements Needed to Further Promote Integrated Pest Management*. Washington, DC: United States General Accounting Office (GAO). GAO-01-815, August 2001.

Garcia-Paredes, J. D., K. R. Olson, and J. M. Lang. (2000). Predicting corn and soybean productivity for Illinois soils. *Agricultural Systems* 64(3):151-170.

Gunsolus, J. L., T. R. Hoverstad, B. D. Potter, and G. A. Johnson. (2000). Assessing integrated weed management in terms of risk management and biological time constraints. In *Emerging Technologies for Integrated Pest Management–Concepts, Research and Implementation*, ed., G. G. Kennedy and T. B. Sutton, St. Paul, MN: APS Press, The American Phytopathological Society, pp. 373-399.

Hall, F. Q. (2000). Delivering new crop protection agents within an IPM environment. In *Emerging Technologies for Integrated Pest Management–Concepts, Research and Implementation*, ed., G. G. Kennedy and T. B. Sutton, St. Paul, MN: APS Press, The American Phytopathological Society, pp. 323-338.

Hartman, G. L., J. B. Sinclair, and J. C. Rupe. (1999). *Compendium of Soybean Diseases* (4th Ed.). St. Paul, MN: APS Press, The American Phytopathological Society.

Henkens, R., C. Bonaventura, V. Kazantseva, M. Moreno, J. O'Daly, R. Sundseth, S. Wegner, and M. Wojciechowski. (2000). Use of DNA technologies in diagnostics. In *Emerging Technologies for Integrated Pest Management–Concepts, Research and Implementation*, ed., G. G. Kennedy and T. B. Sutton, St. Paul, MN: APS Press, The American Phytopathological Society, pp. 52-66.

Higley, L. G. and D. J. Boethel. (1994). *Handbook of Soybean Insect Pests*. Lanham, MD: Entomological Society of America.

Howe, O. W. III and L. R. Oliver. (1987). Influence of soybean (*Glycine max*) row spacing on pitted morningglory (*Ipomoea lacunosa*) interference. *Weed Science* 35:185-193.

Hoy, M. A. (2000). Current status of biological control of insects. In *Emerging Technologies for Integrated Pest Management–Concepts, Research and Implementation*, ed., G. G. Kennedy and T. B. Sutton, St. Paul, MN: APS Press, The American Phytopathological Society, pp. 210-225.

Kirkpatrick, T. L. and C. S. Rothrock. (2001). *Compendium of Cotton Diseases* (2nd Ed.). St. Paul, MN: APS Press, The American Phytopathological Society.

Kloepper, J. W., R. Rodriguez-K'bana, J. A. McInroy, and R. W. Young. (1992). Rhizosphere bacteria antagonistic to soybean cyst (*Heterodera glycines)* and rootknot (*Meloidogyne incognita)* nematodes: Identification by fatty acid analysis and frequency of biological control activity. *Plant Soil* 139:75-84.

Kennedy, G. G. (2000). Perspectives on progress in IPM. In *Emerging Technologies for Integrated Pest Management–Concepts, Research and Implementation*, ed., G. G. Kennedy and T. B. Sutton, St. Paul, MN: APS Press, The American Phytopathological Society, pp. 2-11.

Kennedy, G. G. and N. P. Storer. (2000). Life systems of polyphagous arthropod pests in temporally unstable cropping systems. *Annual Review of Entomology* 45:467-493.

Kennedy, G. G. and T. B. Sutton. (2000). *Emerging Technologies for Integrated Pest Management–Concepts, Research, and Implementation*, ed., G. G. Kennedy and T. B. Sutton, St. Paul, MN: APS Press, American Phytopathological Society.

Koenning, S. R., C. Overstreet, J. W. Noling, P. A. Donald, J. O. Becker, and B. A. Fortnum. (1999). Survey of crop losses in response to phytoparasitic nematodes in the United States for 1994. *Journal of Nematology Supplement* 31(4S):587-618.

Krysan, J. L., J. J. Jackson, and A. C. Lew. (1984). Field termination of egg diapause in *Diabrotica* with new evidence of extended diapause in *Diabrotica barberi* (Coleoptera: Chrysomelidae). *Environmental Entomology* 13:1237-1240.

Landis, D. A., S. D. Wratten, and G. M. Gurr. (2000). Habitat management to conserve natural enemies of arthropod pests in agriculture. *Annual Review of Entomology* 45:175-201.

Liebman, M., C. L. Mohler, and C. P. Staver. (2001). *Ecological Management of Agricultural Weeds*. Cambridge, UK: Cambridge University Press.

Louws, F. J., J. L. W. Rademaker, and F. J. debruijn. (1999). The three D's of PCR-based genomic analysis of phytobacteria: Diversity, detection, and disease diagnosis. *Annual Review of Phytopathology* 37:81-125.

Maas, J. L. (1998). *Compendium of Strawberry Diseases*. (2nd Ed.). St. Paul, MN: APS Press, The American Phytopathological Society.

MacGuidwin, A. F. and T. L. Layne. (1995). Response of nematode communities to sudangrass and sorghum-sudangrass hybrids grown as green manure crops. *Journal of Nematology Supplement* 27 (4S):609-616.

Madden, J. P. [Chair]. (1992). *Beyond Pesticides–Biological Approaches to Pest Management in California*. Oakland, CA: ANR Publications, University of California.

Magdoff, F., and H. van Es. (2000). *Building Soils for Better Crops* (2nd Ed.). Burlington, VT: Sustainable Agriculture Network.

Main, C. E., T. Keever, G. J. Holmes, and J. M. Davis. (2001). Forcasting long-range transport of mildew spores and spread of plant disease epidemics. APS net. St. Paul, MN: The American Phytopathological Society.

Marks, R. J. and B. B. Brodie. (1998). *Potato Cyst Nematodes*. Wallingford, UK: CAB International.

Merrigan, K. A. (2000). Politics, policy, and IPM. In *Emerging Technologies for Integrated Pest Management–Concepts, Research and Implementation*, ed., G. G. Kennedy and T. B. Sutton, St. Paul, MN: APS Press, The American Phytopathological Society, pp. 497-504.

Mojtahedi, H., G. S. Santo, and R. E. Ingham. (1993). Suppression of *Meloidogyne chitwoodi* with sudangrass cultivars as green manure. *Journal of Nematology* 25:303-311.

National Research Council. (1996). *Ecologically Based Pest Management–New Solutions for a New Century*. Washington, DC: National Academy Press.

O'-Neal, M. E., M. E. Gray, and C. A. Smyth. (1999). Population characteristics of a western corn rootworm (Coleoptera: Chrysomelidae) strain in east-central Illinois corn and soybean fields. *Journal of Economic Entomology* 92:1301-1310.

Pedigo, L. P. (1999). *Entomology and Pest Management* (3rd Ed.). Upper Saddle River, NJ:Prentice Hall.

Pedigo, L. P., M. R. Zeiss, and M. E. Rice. (1990). Biology and management of the bean leaf beetle in soybean. In *Proceedings of the 1990 Crop Production and Protection Conference.* Ames, IA: Iowa State University Extension Service.

Phipps, P. M., S. H. Deck, and D. R. Walker. (1997). Weather-based crop and disease advisories for peanuts in Virginia. *Plant Disease* 81:235-244.

Rabb, R. L. and F. E. Guthrie. (1970). *Concepts of Pest Management.* Raleigh, NC: North Carolina State University.

Ragsdale, N. N. (2000). The impact of the food quality protection act on the future of plant disease management. *Annual Review of Phytopathology* 38:577-596.

Renner, K. A., S. M. Swinton, and J. J. Kells. (1999). Adaptation and evaluation of the WEEDSIM weed management model for Michigan. *Weed Science* 47:338-348.

Roe, R. M., W. D. Bailey, G. Gould, C. E. Sorenson, G. G. Kennedy, J. S. Bacheler, R. L. Rose, E. Hodgson, and C. L. Sutula. (2000). Detection of resistant insects and IPM. In *Emerging Technologies for Integrated Pest Management–Concepts, Research and Implementation*, ed., G. G. Kennedy and T. B. Sutton, St. Paul, MN: APS Press, The American Phytopathological Society, pp. 67-84.

Rowe, R. C. (1993). Potato Health Management. St. Paul, MN: APS Press, The American Phytopathological Society.

Smith, R. F. and R. van den Bosch. (1967). Integrated control. In *Pest Control–Biological, Physical, and Selected Chemical Methods*, ed., W. W. Kilgore and R. L. Doutt. New York and London: Academic Press, pp. 295-340.

Southwood, T. R. E. (1978). *Ecological Methods*, 2nd ed., New York, NY: Chapman and Hall.

Stern, V. M., R. F. Smith, R. van den Bosch, and K. S. Hagen. (1959). The integrated control concept. *Hilgardia* 29:81-101.

Stewart, S. J. (2000). Pathogen detection and pesticide use. In *Emerging Technologies for Integrated Pest Management–Concepts, Research and Implementation*, ed., G. G. Kennedy and T. B. Sutton. St. Paul, MN: APS Press, The American Phytopathological Society, pp. 85-93.

USDA/AMS. (2001). *The National Organic Program.* USDA–Agricultural Marketing Service, Washington, DC: United States Department of Agriculture/Agricultural Marketing Service. RIN: 0581-AA40. <http://www.ams.usda.gov/nop/>.

USDA/NASS. (1999a). *Agricultural Chemical Usage:1998.* Field crops summary. U.S. Government Printing Office, Washington, DC. 141 pp.

USDA/NASS. (2000a). *Agricultural Chemical Usage:1999.* Field crops summary. U.S. Government Printing Office, Washington, DC. 120 pp.

USDA/NASS. (2001a). *Agricultural Chemical Usage:2000.* Field crops summary. U.S. Government Printing Office, Washington, DC. 120 pp.

USDA/NASS. (1999b). *Agricultural Chemical Usage:1998.* Vegetable summary. U.S. Government Printing Office, Washington, DC. 300 pp.

USDA/NASS. (1999c). XIV: Fertilizers and pesticides. 14 pp. In *Agricultural Statistics* 1998. U. S. Government Printing Office, Washington, DC.

USDA/NASS. (2000c). XIV: Fertilizers and pesticides. 14 Pp. In *Agricultural Statistics* 1999. U. S. Government Printing Office, Washington, DC.

USDA/NASS. (2000d). *Agricultural Chemical Usage:* 1999. Fruit crops summary. U.S. Government Printing Office, Washington, DC.

van Emden, H. F. and D. B. Peakall. (1996). *Beyond Silent Spring: Integrated Pest Management and Chemical Safety.* London and New York: Chapman & Hall.

Vincent, C. and P. Lachance. (1993). Evaluation of a tractor-propelled vacuum device for management of tarnished plant bug (Heteroptera: Miridae) populations in strawberry plantations. *Environmental Entomology* 22:1103-1107.

Waddell, C. (2000). *And No Birds Sing: Rhetorical Analyses of Rachel Carson's Silent Spring.* Carbondale and Edwardsville: Southern Illinois University Press.

Wei, G., J. W. Kloepper, and S. Tuzun. (1996). Induced systemic resistance to cucumber diseases and increased growth by plant growth-promoting Rhizobacteria under field conditions. *Phytopathology* 86:221-224.

Wiese, M. V. (1987). *Compendium of Wheat Diseases* (2nd Ed.). St. Paul, MN: APS Press, The American Phytopathological Society.

White, D. G. (1999). *Compendium of Corn Diseases* (3rd Ed.), St. Paul, MN: APS Press, The American Phytopathological Society.

Williams, M. (1999). Cotton insect losses. In *Proceedings of the Beltwide Cotton Conference.* Memphis, TN: National Cotton Council, pp. 785-808.

Zwart, K. B., J. Bloem, S. L. G. E. Burgers, L. A. Bouman, L. Brussaard, G. Lebbink, W. A. M. Didden, J. C. Y. Marinissen, J. J. Vreeken-Buys, and P. C. de Ruiter. (1994). Population dynamics in the below-ground food webs in two different agricultural systems. *Agriculture, Ecosystems and the Environment* 51:187-198.

Conceptual Model
for Sustainable Cropping Systems
in the Southeast:
Cotton System

Harry H. Schomberg
Joe Lewis
Glynn Tillman
Dawn Olson
Patricia Timper
Don Wauchope
Sharad Phatak
Marion Jay

Harry H. Schomberg is Ecologist, United States Department of Agriculture, Agricultural Research Service, J. Phil Campbell, Sr. Natural Resource Conservation Center, Watkinsville, GA.

Joe Lewis, Glynn Tillman, and Dawn Olson are Research Entomologists, and Patricia Timper is Research Plant Pathologist, United States Department of Agriculture, Agricultural Research Service, Crop Protection and Management Research Unit, P.O. Box 748, Tifton, GA 31793.

Don Wauchope is Research Chemist, United States Department of Agriculture, Agricultural Research Service, Southeast Watershed Research Laboratory, P.O. Box 946, Tifton, GA 31793.

Sharad Phatak is Professor of Horticulture, University of Georgia, College of Agricultural and Environmental Science, Department of Horticulture, Tifton Campus/ Coastal Plain Station, Tifton, GA 31793.

Marion Jay is Field Facilitator for Communities in Schools of Georgia, Inc., Tifton, GA 31791.

Address correspondence to: Harry H. Schomberg, USDA-ARS-JPCS, Natural Resource Conservation Center, 1420 Experiment Station Road, Watkinsville, GA 30677 (E-mail: hhs1@arches.uga.edu).

[Haworth co-indexing entry note]: "Conceptual Model for Sustainable Cropping Systems in the Southeast: Cotton System." Schomberg, Harry H. et al. Co-published simultaneously in *Journal of Crop Production* (Food Products Press, an imprint of The Haworth Press, Inc.) Vol. 8, No. 1/2 (#15/16), 2003, pp. 307-327; and: *Cropping Systems: Trends and Advances* (ed: Anil Shrestha) Food Products Press, an imprint of The Haworth Press, Inc., 2003, pp. 307-327. Single or multiple copies of this article are available for a fee from The Haworth Document Delivery Service [1-800-HAWORTH, 9:00 a.m. - 5:00 p.m. (EST). E-mail address: docdelivery@haworthpress.com].

SUMMARY. Many small to mid-size family farms face an economic and ecological crisis due to the changing face of agricultural production. Increasing production costs and lower revenues are causing many producers to leave the farm. Rural communities face economic hardships due to declining farm numbers and continued loss of the brightest youths who often seek employment in urban areas. Small to mid-size family farms and rural communities can be sustainable if economic and environmental risks are recognized and solutions developed that reach all members of the farm and rural communities. Our project focuses on the involvement of farmers, scientists, and other stakeholders to enhance understanding of sustainable principles at the farm level and extend awareness of the central components to sustainability of rural communities. Conservation tillage with cover crops is being used to modify pest pressures, reduce chemical inputs, improve soil productivity and reduce environmental risks to producers, the community and the environment in cotton (*Gossypium hirsutum* L.) production systems. Preliminary results indicate that reductions in use of pesticides can be achieved due to enhanced presence of beneficial insects. Cotton offers the best opportunity to enhance the understanding and use of sustainable practices in ecologically-based farming systems because of its predominance in southern farm enterprises. Farmer participation and understanding is being facilitated through the participation of the farmer based Georgia Conservation Tillage Alliance. To achieve greater outreach and broaden community participation within the region we are involving at-risk rural youth through the Communities in Schools of Georgia program. Outreach includes the use of traditional and newer internet based technologies through the development of databases and expert systems that allow, farmers, ranchers, and community members an opportunity to evaluate economic and environmental effects of alternative production practices at local and regional scales. Through interactions with existing federal, state, and private organizations we are encouraging expansion of these sustainable approaches regionally. *[Article copies available for a fee from The Haworth Document Delivery Service: 1-800-HAWORTH. E-mail address: <docdelivery@ haworthpress.com> Website: <http://www.HaworthPress.com> © 2003 by The Haworth Press, Inc. All rights reserved.]*

KEYWORDS. Nutrient management, pest management, economic models, environmental index, insects, cover crops, rural sustainability

INTRODUCTION

Sustainability as a Concept in Agricultural Systems

Since World War II, modern agricultural practices and farm policy have affected a shift in US agriculture from localized diverse production systems that included draft animals, legumes, and animal manures to systems that depend on machines, fossil fuels, chemical fertilizers, and pesticides (Hildebrand and Russell, 1996). This shift in technologies has created a landscape of centralized homogeneous cropping systems that rely less and less on interdependent components. Although highly productive, decreasing diversity of production within a local area has contributed to destabilization of many rural economies (Olson and Francis, 1995). Many farming enterprises in the US face increasing costs of production while commodity prices continue to decline due to global market forces. This economic imbalance contributes to loss of producers and erosion of the stability of agriculturally based rural communities.

Because of this economic imbalance, producers are more and more in need of environment specific technologies aimed at improving productivity and economic sustainability. Until recently, the general consensus was that sustainable production practices do not lead to improved net economic return, especially at small to mid size scales. Sustainable production practices that focus on reduced input costs are perceived to cost more in labor and management and may not be credited for the increased production per unit of land area that can be achieved with intensive small systems. The question remains whether sufficient net income can be generated on enough acreage to support a family farm unit through the use of intensive systems. The necessity of producer focus on short-term economic viability, along with commodity based government policies continue to limit acceptance of practices that could improve long-term environmental and economic sustainability. In addition, greater support from research, extension, financial institutions, risk management professionals, governmental bodies, and local leaders is needed to effect change (Lewis and Jay, 2001).

Need for Community Support of Sustainable Principles and Practices

Land stewardship has long been recognized for sustaining productivity. Certain inherent principles of natural ecosystems when applied to farms and communities enable them to maintain balance and minimize negative effects of adverse disturbances. These strengths are: *interdependence*–components rely on each other for energy and cycling of materials; *self-sufficiency*–minimal import of resources; *self-regulation*–feedback loops maintain balance within certain bounds; *self-renewal*–perpetuation through effective reproduction, defense, and other strategies; *efficiency*–minimal waste, i.e., recycling; and diversity/versatility C insures ability to cope with cycles of fluctuating conditions. Sustainability of small and limited resource family based farm operations depends on applying these core principles to develop systems where solutions to problems are "built-in" and renewable (i.e., crop rotation, cover crops, intercropping, and integrated animal-crop systems) (Lewis et al., 1997b).

Over the past 50 years, spectacular short-term solutions for problems such as soil nutrition, weeds, insects, and plant diseases have been achieved through scientific research. On the near-term, dosages and costs of these therapeutic solutions were nominal. Thus, the practice of monoculture and high-input agriculture surged as yields per acre quadrupled (Odum, 1989). With availability of these tools, emphasis on inherent, self-renewing regulators such as biodiversity, natural enemies of pests, and recycling of nutrients generally fell by the wayside and sometimes resulted in secondary negative effects, i.e., water pollution, wildlife injury, and soil erosion.

Current production practices depend on large inputs to maintain yields, thus placing US producers at risk for economic disaster which is often overcome only through emergency farm payments. To effectively change the current system will take multiple levels of interaction among producers, scientists, educators, economists, politicians and other stakeholders. One way to incorporate holistic sustainable management principles applicable to problems in rural communities and agriculture is through interdisciplinary, on-farm research and demonstrations, partnered with a broad, community-based educational outreach program. A successful approach will require collaborative interactions among existing federal, state, and private organizations so that their individual strengths can be drawn upon to insure expansion to other regions.

On Farm Research to Promote Sustainable Practices

Farmers are justifiably reluctant to adopt new technologies before seeing convincing tests and demonstrations under farming conditions similar to their own (Rzewnicki, 1991). This reluctance often results from limited producer involvement in technology development. Separation of research priority setting from actual agricultural production often results in development of inappropriate technologies that require significant end-user modification. Producers become the ultimate integrators of site-specific management systems based on their knowledge of current technologies, available resources, and environmental conditions. The current system of technology transfer increases the economic risks associated with adoption of new practices and limits early adoption to the most innovative and usually larger producers. This often inhibits adoption by limited resource or small farm producers.

Contributions of scientists become more important and more difficult as the need for integrating regional and site-specific factors increases. However, the site-specific applicability of data from on-farm research helps facilitate technology transfer to other regional farmers. Participatory research encourages synergism among scientists and farmers working together to design, implement, and evaluate research (Wuest et al., 1999). Including farmers ensures identification of high priority problems and potential solutions, aids in design and implementation, and improves interpretation of results and recommendations (Hildebrand and Russell, 1996). Farmer participation provides greater insight into how new technologies will be applied and provides a more robust evaluation due to the broader, more variable, and unpredictable range of environmental conditions (Rzewnicki, 1991; Wuest et al., 1999). Farmers also become "scientists" in learning to critically analyze their farms and self-initiate on-farm research activities. On-farm research/demonstrations and shared learning experiences help to facilitate major paradigm shifts both with producers and in research.

Historic Perspective on Cotton (Gossypium hirsutum L.)

Cotton played a significant role in the economic welfare of the south from the time of colonial settlement in the late 1700s until the boll weevil (*Anthonomous grandis grandis*, Boheman) caused significant declines in yields and increases in production costs during the early 1900s (Haney, Lewis, and Phatak, 1996). The long history of row crop production, predominantly cotton, and intensive tillage practices were responsible for extensive soil erosion and loss of soil productivity in the region. Trimble (1974) estimated that 15 to 30 cm of soil were lost on sloping soils of the region from 1865 to 1920. Much of the soil loss is attributed to lack of crop rotation that resulted in 50 to 75 years of continuous cotton. Arrival of the boll weevil could be heralded as an important stimulus for diversification and change at the farm, community, and regional scales in the south (Haney, Lewis, and Phatak, 1996).

A new era of cotton dominance in the south has emerged due to the success of the Boll Weevil Eradication program. Production increased from 3.7 million hectares in 1989 to 4.7 million hectares in 1998 (CTIC, 1998). This increase has not occurred without risks. Intensive tillage practices like fall plowing followed by winter fallow and spring discing are practiced on over 85% of the cotton grown in the south (CTIC, 1998). Most of this cotton is grown on land that is not rotated to other crops. These practices leave soils vulnerable to the intensive rain and wind that continue to cause erosion and loss of soil productivity. In addition to environmental problems, recent increases in per unit cost of inputs and drops in prices have contributed to reduced farm profitability for cotton farmers. Cotton prices declined from $2.53 kg^{-1} in 1995 to less than $1.10 in 2001 (Shurley, 2001) while farm expenditures from 1993 to 1998 increased 14% (USDA/NASS, 1999).

The expanded production of cotton, success of the Boll Weevil eradication program, and continued availability of economic support to producers from Loan Deficiency Payments (US government support of cotton prices) makes cotton an ideal crop on which to base a project for promoting sustainable practices such as the use of cover crops, conservation tillage, and integrated pest management (IPM). Producers are more familiar with IPM principles due to the success obtained in the Boll Weevil program and should be willing to try new and innovative approaches for reducing pesticide and other chemical inputs.

A SUSTAINABLE COTTON PRODUCTION SYSTEM FOR THE COASTAL PLAIN

The foundation components of the system are the use of conservation tillage and cover crops to manage insect habitat so as to enhance the presence of beneficial insects, and also improve nutrient and water availability. Previous work with cotton growers in south Georgia has shown that cotton grown in strip-killed crimson clover (*Trifolium incarnatum* L.) using reduced tillage improves soil health, cuts tillage and insecticide costs, and reduces fertilizer inputs by 56 to 67 kg ha^{-1} (Haney, Lewis, and Phatak, 1996; Lewis, Haney, and Phatak, 1996). One producer reported a savings of \$300 ha^{-1} on inputs and yields of 7.4 bales ha^{-1} of cotton compared to 3 bales ha^{-1} in his conventional system (Reed et al., 1997). Increases were observed in beneficial insect numbers and duration of presence in the fields. For many producers switching to a system that relies on reduced off-farm inputs will require planning, management, and time to implement (Stark et al., 1999). However, interactions among system components must be better understood to increase applicability to a wider area (Lewis et al., 1997c). We are working with producer members of the Georgia Conservation Tillage Alliance (GCTA) to evaluate these practices on small to mid-size farms in areas of rural Georgia in an effort to expand adoption of sustainable practices.

Role of Reduced Tillage and Cover Crops

Conservation tillage reduces the number of operations required to prepare a field for a crop thus reducing field traffic, labor and fuel costs, machinery needs and time (Liu and Duffy, 1996). In addition, reduced tillage practices can increase soil productivity due to influences on surface soil organic matter and water infiltration/availability (Bruce et al., 1995). Accumulation of organic matter with reduced tillage is attributed to a reduction in the rate of organic matter decomposition.

Cover crops are grown primarily to protect the soil from erosive forces and usually are not harvested. Use of green manure crops to increase biomass inputs back to soil has long been known to be a sound agronomic practice (Reeves, 1994). When used with conservation tillage, cover crops provide many of the benefits attributable to green manure crops. Besides protecting soil against erosion, they improve soil structure, enhance soil fertility, suppress pests, enhance soil quality, conserve soil moisture, protect water quality, and help safeguard personal health (Reeves, 1994). In addition to the physical effects, cover crops reduce runoff and erosion through effects on soil structure. Microorganisms decomposing crop residues produce compounds that increase aggregate stability which is only sustained through continuous inputs of new organic matter (Kladivko, 1994). Cover crops thus serve as a source for organic matter input.

Using cover crops with conservation tillage can restore soil productivity of degraded soils through increases (or reduced losses) in soil organic matter (Bruce et al., 1995; Franzluebbers, Langdale, and Schomberg, 1999). Soil organic matter supports the abundant diversity of organisms important in decomposition and nutrient cycling, serves as a source of plant nutrients through release of organic N, S, and P, and supplies inorganic nutrients through its cationic exchange capacity and chelation reactions (Schomberg, Ford, and Hargrove, 1994). Reduced tillage practices can result in organic matter increases of up to 2.3 Mg ha^{-1} yr^{-1} (Reicosky et al., 1995) depending on the rate at which biomass is added minus the rate at which erosion and biological oxidation are removing organic matter. Effectiveness of cover crops to increase biomass input will depend on how well the cover crop is adapted to the area and management variables like planting date, fertility, and killing date (Reeves, 1994).

Availability of N to a subsequent crop is directly influenced by cover crop residue effects on N mineralization and immobilization and/or through N fixation by legume cover crops. The N value of legumes can range from 30 to 180 kg N ha^{-1} depending on growing conditions and type of legume (Frye et al., 1988; Hargrove, 1986; Stute and Posner, 1995). Availability of N to a crop during the growing season can be 20 to 40% greater following a legume than following rye (*Secale cereale* L.) (Schomberg, 1998). Scavenging of N remaining in the soil profile by gramineous cover crops reduces loss of leached N up to 60% compared to no cover crop (Meisinger et al., 1991). Legume and grass cover crop mixtures can improve nutrient conservation through complementarity of residue chemical composition that affects decomposition and N mineralization rates thus leading to greater synchrony of nutrient availability to the following crop. Grasses conserve soil N (uptake) and impede release of N due to slower decomposition while legumes increase available N through N fixation and rapid decomposition supplying N early in the growing season (Rannells and Wagger, 1996). A complex mixture of crimson clover, hairy vetch (*Vicia villosa* Roth), rye and barley (*Hordeum vulgare* L.) provided significant inputs of N (220 to 360 kg N ha^{-1}) in a low input system for tomato (*Lycopersicon esculentum* Mill.) production and suppressed weeds as well as a herbicide system in Ohio (Creamer et al., 1996). Greater diversity in a mixture can provide greater resilience to climatic and biological adversity because of growth compensation by individual components of the mixture.

Habitat Management

Habitat management as a pest management tool is an ecologically-based strategy aimed at designing agroecosystems to support populations of natural enemies of pest species (Altieri and Whitcomb, 1979; Altieri, Martin, and Lewis, 1983). The well-known S-shaped curve of growth through time illustrates the sequential progression of natural ecosystems with growth beginning slowly, rapidly increasing, and leveling off thereafter (e.g., Flint and Van den Bosch, 1981). Conventional monoculture agroecosystems typically operate in the linear portion of this curve where large oscillations in species occur until the latter part of the growing season when increasing interactions tend to stabilize the oscillations. Conventional monoculture agroecosystems seldom reach the plateau of the S-curve as chemical inputs often remove or debilitate many species and annual removal of biomass forces growth to start over each year. Habitat management through conservation tillage and cover crops as well as other types of field landscaping (e.g., field borders, hedgerows, adjacent crops) help promote more year-round natural enemy-pest-species interactions by providing alternate prey or hosts, reproductive sites, and shelter from adverse conditions for natural enemies of pests. These landscape effects on natural-enemy-pest interactions suggest a potentially high utility as a pest management tool (Landis, Wratten, and Gurr, 2000) but more information on species-specific interactions of targeted pests and natural enemies are needed to facilitate the design of appropriate landscapes.

Studies of cotton arthropod pests and their natural enemies in conservation tillage and cover crop systems in the south-southeast have been conflicting. Generally ground-dwelling beneficial species are higher in conservation tillage cotton with and without cover crops compared to conventional tillage (Blumberg and Crossley, 1982; Sullivan and Smith, 1993; Haney et al., 1995; Lewis, Haney, and Phatak, 1996), while cutworm (*Noctuidae* sp.) pest populations are higher in reduced tillage and legume cover crop systems than in conventional tillage systems where crop residues are incorporated into the soil (Guthrie et al., 1993; Leonard et al., 1993; Sullivan and Smith, 1993; Turnock, Timlick, and Palaniswamy, 1993). However, no consistent patterns in significant pest populations and plant-dwelling beneficial species or in cotton yields have been reported (Leser, 1995; Ruberson, Phatak, and Lewis, 1997; Stapel et al., 1998). In some studies, increases in aphid (*Aphis* sp.) populations and decreases in heliothine eggs and plant-dwelling beneficial species were correlated with higher numbers of predacious fire ants (*Solenopsis invicta*) in conservation tillage cover crop systems (Leser, 1995; Ruberson et al., 1995;

Stapel et al., 1998). Differences among cover crop species, years, field histories and locations, and surrounding landscape contribute to the conflicting results of these studies. Longer-term studies may provide a better understanding of how various cover crops, reduced tillage, and other landscape factors affect arthropod pests and beneficial species, and how these translate into plant protection over time.

Habitat management also offers the potential to activate inherent mechanisms for suppressing plant-parasitic nematode populations (i.e., promoting the presence of nematophagous organisms like nematode-parasitic fungi and predaceous nematodes) (Stirling, 1991). Plant-parasitic nematodes feed on plant roots and are major pests of many crops including cotton. In Georgia, cotton yield losses from nematodes were $25 million and the cost of control was $11 million in 1998 (Williams-Woodward, 1999). The southern root-knot nematode [*Meloidogyne incognita* (Kofoid and White) Chitwood] and the reniform nematode (*Rotylenchulus reniformis*) are the most widespread and damaging plant-parasitic nematodes in cotton production. These nematodes reproduce on a wide range of plant species, including most winter cover crops. Moreover, no agronomically acceptable cotton varieties exist with resistance to southern root-knot nematode or reniform nematode and growers have few choices for non host crops to rotate with cotton. Alternative nematode management options are needed but the effects of most cropping practices on natural enemies of nematodes are unknown. Conventional tillage may displace natural enemies from the area of greatest nematode activity and expose them to upper layers of soil where their survival is diminished by desiccation and ultraviolet irradiation. Rotation with non-host plants, such as Bahia grass (*Paspalum notatum* Flügge), reduces nematode populations as well as populations of its natural enemies. Several well-documented cases indicate nematode-suppressive soils can develop in response to continuous planting of a host crop (Stirling, 1991). Year-round plant growth has the potential to increase populations of plant-parasitic nematodes because of the extended presence of nematode susceptible crops (cotton and cover crops); however, this may also lead to a buildup of host-specific natural enemies that consistently suppress nematode populations below damaging levels.

Environmental Impact

Pesticide losses in run-off are reduced with conservation tillage and cover crops because less water leaves the field. Conservation tillage promotes a change in soil physical properties while cover crops help slow the rate of water moving at the surface thus increasing infiltration. This reduced run-off has caused concern that there is greater potential for groundwater contamination from pesticides or nitrate (Fawcett, Christensen, and Tierney, 1994). Preferential flow through macropores, which may be more prevalent with no-till, can allow water and dissolved solids or suspended sediments to by-pass upper layers of soil. Although preferential flow through macropores can allow rapid transport of water and certain pesticides a few feet deep in the soil, it is not clear that this process can deliver pesticides to deeper depth (Fawcett, Christensen, and Tierney, 1994). Pesticides that move deeper into the soil have been found to diffuse into the soil matrix and are no longer subject to preferential flow (Gish et al., 1991; Gish, Helling, and Mojasevic, 1991).

Concern has also been raised that adoption of conservation tillage practices increases use of herbicides and insecticides with greater potential for contamination of the environment. While adoption of conservation tillage can change weed and insect problems and the types of herbicides used, total usage of pesticides has not changed when farmers convert to conservation tillage (Hanthorn and Duffy, 1983; Fawcett, 1987; Bull et al., 1993; Day et al., 1999). Day et al. (1999) evaluated pesticide use by producers in the major corn (*Zea mays* L.) and soybean (*Glycine max* L. Merr.) production areas of the US for 1990, 1993, and 1995. Combining the conclusions of their study with previous studies (Hanthorn and Duffy, 1983; Bull et al., 1993) indicated that as tillage moves from conventional systems to conservation tillage to no-till, herbicide use per hectare tends to increase. This increase in the no-till system was mostly related to the need for a burn-down herbicide. For ridge-till and mulch-till systems rates were not much different from those used in conventional systems. Statistical analysis of insecticide application rates showed that conventional tillage used more insecticide than no-till and about the same as mulch-till and ridge-till (Day et al., 1999). Measured changes in quantities of pesticides over time did not reflect quality changes that occurred (i.e., newer and more potent pesticides entering the market often require lower application rates). Future effects on the environment depend on the inherent toxicity of the active ingredients and characteristics that affect persistence as well as management strategies developed to reduce acquisition of resistance by target pests.

Economics (Farm and Rural Communities)

System benefits and costs of alternative management strategies are being evaluated at the farm level to determine optimum combinations of cover crops, crop rotation, and pest management that sustain revenues. Consideration must not only be given to the potential for increasing returns and reducing volatility due to changes in productivity but also to the environmental benefits of reduced fertilizer and pesticide inputs. Likelihood of producers adopting sustainable management strategies will depend on their expected future change in yields and associated economic volatility. Possible tradeoffs between short-term returns versus long-term sustainability can only be addressed to a limited degree with data that are now available.

We are developing a set of indices to evaluate how changes in system components affect long-term viability. These indices will be used in a general procedure for determining a "sustainability" score for different practices. They will also provide a useful measure of the contribution of farms to sustainability of communities and geographic areas, and an objective, numeric basis for conservation or environmental protection planning and payment programs. One index focuses on pesticide effects on density and diversity of pest and beneficial species over time and how these interact to affect production. An environmental impact index incorporates exposure and toxicity ratings for (a) terrestrial species in the field: non-target/biodiversity impacts on agroecosystems and (b) potential for agrochemical transport to aquatic ecosystems and impacts on indicator species. An index of soil quality is being used to determine an economic value of system effects on soil productivity. And a wildlife index describes the relative economic and environmental benefits of alternative crop management scenarios to producers and rural communities.

ON-FARM RESEARCH DEVELOPING A SUSTAINABLE COTTON PRODUCTION SYSTEM

Focusing on cotton as the base system, because of its prevalence throughout the south, we are working to achieve a more sustainable production system that will reduce pesticide, fertilizer, and fuel inputs through adoption of conservation tillage (minimizing tillage intensity and frequency) and cover crops to add diversity, fix N, and provide habitat to beneficial insects. In addition, the system encourages diversification to include other cash crops and livestock and extend the basic principles of sustainability to other crop production systems in the region. Work on six farms in two areas of the state began in the fall of 2000.

Our research plots focus on the use of conservation tillage to enhance soil quality factors such as increasing soil surface cover and organic matter content at the soil surface. Both factors are important for improving water infiltration and water-use efficiency. We are comparing cover crop mixtures (clovers plus rye) for biomass production and insect habitat. Cotton is planted into killed strips of cover crops. The remaining live strips serve to prolong the presence of insect habitat. The combined results of tillage and cover crop management should help to reduce inputs of fertilizer and pesticides and also help with water management thus reducing costs of cotton production.

Combining traditional field days, newer internet-based education, and extension-led outreach, producers, educators, and civic and community leaders will be exposed to holistic ecologically-based tools to foster sustainability at the farm and community levels. The support of a strong farmer-based conservation tillage alliance, i.e., GCTA, has been instrumental in helping to develop the project and provide contacts from its 200 plus members and four regional subchapters. Through its monthly newsletter and internet page, GCTA provides an effective conduit for disseminating information from the project. On-farm field day demonstrations in cooperation with GCTA and workshops on sustainability at GCTA's annual meetings provide effective means for transfer of information. Involvement of broader community components such as financial institutions, risk management professionals, governmental bodies, and community leaders is being targeted to help develop a sense of the need for sustainable practices at the community level (Lewis and Jay, 2001).

Internet-Based Technology Transfer

An internet-based system in which whole-farm economic analysis is combined with agronomic and horticultural knowledge and environmental impact analysis is being developed to extend the project's activities to a much broader audience (Figure 1). Numerous frameworks or approaches to whole farm planning are possible and are being explored by various groups in the US and in other countries (Freyenberger, Janke, and Norman, 1997). Janke and Freyenberger (1997) considered applicability of these approaches to range from user friendly to not likely to be used at all. They identified the Ontario Environmental Farm Plan as thorough but complex; the Farm-A-Syst checklist approach as providing a snapshot at a certain point in time but does not promote ongoing monitoring; PLANETOR, a computer based system, allowed more what if evaluations; and the Minnesota Land Stewardship Project incorporated several monitoring tools that encourage interaction among farm families and researchers.

The system under development in this project is based on interactive (farmer as well as scientist initiated) technology transfer and knowledge transfer. Methods to assess economic and environmental benefits of management practices are used to provide researchers and individuals a way to compare sustainable alternatives based on research results, whole-farm economic analysis (Lamb, Davidson, and Butts, 1992), and environmental indexing (a relative ranking of the environmental impacts of an agricultural management practice, see below).

Multiple interfaces will allow farmers, extension-research specialists, and the general public access to database information. Participating producers will manage records of their own farm through a password protected internet access. This producer data is aggregated to maintain privacy for research, analysis, policy, and community purposes. Expert system functionality will be used to provide knowledge exchange for alternate crops and production practices using input from farmers and specialists which will also provide direct link-

FIGURE 1. Internet Based System for Information Transfer and Agroecosystem Analysis.

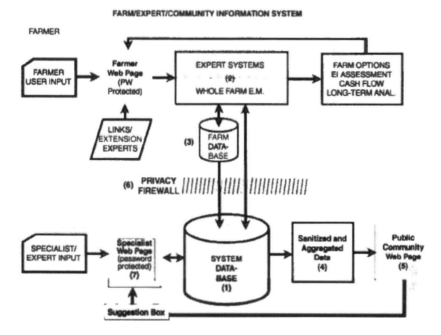

(1) System Database: Input data for whole-farm analysis, environmental impact index calculator, irrigation and any other expert systems used. Spatial information to calculate aggregate watershed information such as water use, chemicals and nutrients applied, added in future versions.

(2) Whole-Farm Economic Model/Expert Systems Management Assistant Model Suite: Accessed via the Farmer-Client Web Page using a password, farmer may either download suite for offline run or run online and construct private database of input data within system. Suite provides economic analysis, options, and answers to "what-if" questions, long-term economic viability of choices and cash flow analysis. Environmental index calculations for each field/management/crop combination.

(3) Client Farm Database: Private space provided in the system for clients to build a "permanent" database describing his/her farm operations and financial data.

(4) Sanitized and Aggregated Data: A "sanitizer filter" is used to remove the identity of individual farms (to protect individual farm operations and farmer privacy) and aggregate the data based on farm type, county, and region to provide data for community planners, environmental agencies, public planning, conservation payment system structuring, and public information web page.

(5) Public/Community Information Web Page: Internet based technology transfer providing a description of the project, how to participate, services provided, and data access. Using database information and expert systems, economic and environmental index calculations can be developed for alternative farm enterprises by anyone.

(6) Privacy Firewall: Separates client program suite and farmer database from public- and expert-access parts of system; sanitizes and disconnects private data for aggregation.

(7) Specialist/Farmer Input Page: Specialists (agronomists, ecologists, economists, entomologists, and others) create and maintain system databases through this interface. Public/farmers can provide information via a "suggestion box." Specific database areas are the responsibility of individual "authors," who have exclusive access to those fields.

ages between farmers and specialists. The system will allow evaluation of production alternatives for community planning and watershed environmental assessment as well as on-farm information for producers.

Environmental Impact: Quantifying Relative Risk Reduction

Because there is little knowledge relating off-site actual ecological impacts to specific practices on farms, research groups in the US and elsewhere are using the concept of relative risk as an initial approach to defining this aspect of the sustainability of practices (Bockstaller, Girardin, and van der Werf, 1997; Lewis et al., 1997a; Lukk, Tindall, and Potts, 1995; Newman 1995; van der Werf and Zimmer, 1998). Although this approach has mainly been used for comparing pesticides with each other and with alternative pest management practices, it can in principle be extended to agronomic practices such as use of herbicide resistant crops and application of animal waste in cropping systems. For pesticides, a weighted relative environmental impact "risk index" is calculated by combining indicator species, human toxicity and exposure data obtained in many cases from risk assessment data used for pesticide registration.

Initially the conceptual model uses a simplified version of the index developed by Kovach et al. (1992) which was developed to determine the relative environmental impact of pesticides in conventional, IPM, and organic systems of apple (*Pyrus malus* L.) production. Their index combines a relative risk calculation for "ecological," "consumer," and "farm worker" components using such indexes as dermal toxicity, fish toxicity, leaching and runoff potential, etc., for relative hazard, and using application rate as a surrogate for exposure.

Initially we will neglect the "farm worker" part of this index because it is less well characterized than consumer and ecological risk. However, as our experience with this process grows it will be added. The form of the resulting simplified Environmental Index Quotient (EIQ) is thus

$$EIQ = [(C \bullet ((S+P)^2) \bullet Sy) + L] + [(F \bullet R) + (D \bullet ((S+P)^2) \bullet 3 + (Z \bullet P \bullet 3) + (B \bullet P \bullet 5)]^2$$

Where the first and second terms in square brackets compute relative consumer risk and ecological risk, respectively. The components are: C = mammalian chronic toxicity, S = soil half-life, P = plant surface half-life, Sy = plant sorption potential, L = leaching potential, F = fish toxicity, R = runoff potential, D = bird toxicity, Z = bee toxicity, and B = beneficial arthropod toxicity. Each of these individual factors is in itself an index scaled in order to weigh properly in the calculation. For example, "toxicity to beneficial arthropods" can have values from 1 to 5 assigned to "low impact" through "high impact," respectively.

The arbitrariness and subjectiveness of this calculation is obvious. However, it provides a conceptual framework, which will require adjustment and modification with experience. We plan to include some measures of sustainability including biodiversity and soil microbiological impacts in the field and will refer to the resulting index as a relative sustainability index.

Economics

The basic economic unit of agriculture is the whole farm. Our approach is to analyze economic returns of practices within a single-owner "family farm" unit, assuming fixed land and water resources. Analyses will evaluate a variety of production options of varying relative environmental/ecological sustainability (as indicated by the relative sustainability index described above) available to the producer. A detailed model for whole-farm economic analysis has been developed by Lamb, Davidson, and Butts (1992), in a spreadsheet format which allows short- and long-term analysis of the profitability of each practice on each field.

Ultimately, if an economic comparison of practices is possible which includes "external" environmental costs, it seems in principle possible to develop a defensible system for reimbursing farmers when they are faced with economic versus environmental decisions (Prato and Wu, 1995; Kozloff, Taff, and Wang, 1992). In the short-term, the internet based system should help producers understand costs of production and improve economics of their current operations. In the longer-term, our hypothesis is that the environmental impact analyses of alternative choices will help quantify short-term costs to producers of adjustments required for conversions to more sustainable systems. Showing policy makers that producers deserve additional economic compensation through some kind of conservation or environmental payments to help them through the conversions would contribute to viability of agricultural communities. The goal of developing the model is to provide a sound basis for evaluating productivity and environmental risks associated with production systems and provide an economic basis for that evaluation. Conceivably, the model could allow for comparisons among conventional and sustainable practices in evaluating farm qualification for loans and participation in government programs.

Expanding the Concept of Sustainability to Rural Communities

Rural communities have been challenged with the same social ills impacting urban environments. These include large school drop-out rates for adults and teens, teen pregnancies, and juvenile arrests. For poor and minority populations these risk factors escalate dramatically. Often the best and brightest rural youth who typically complete their education at the university level rarely return to their rural communities. A greater proportion of youth that remain in the community are teen mothers and school drop-outs. Therefore, educational opportunities that demonstrate sustainable principles must be provided early on (e.g., Middle School) and continued through life. Understanding these principles can play an important role in development of stewardship responsibilities in the community.

A unique part of our approach to bringing sustainability principles to rural communities is through participation of the Communities in Schools of Georgia (CISG) program (www.cisnet.org). This program is designed to improve education by teaching kids how to help themselves. Taking a holistic view, the program seeks to combine the benefits of specialization and modern technology. Hands-on or applied learning techniques, which CISG has found to be effective for engaging youth who are most at risk of dropping-out of school, are used to present sustainability issues. Through hands-on service learning, youth identify an important social issue, plan an activity to address the issue, implement the plan and then reflect on the learning as the plan is implemented and concluded. Application of this method to engaging youth with sustainable farming practices allows rural youth to reconnect with their heritage while learning key components of safeguarding natural resources.

CONCLUSION

Economic and environmental sustainability of family based small-farms in the southern US depends on the development and promotion of integrated systems of crop and farm management. Most producers in the region are interested in protecting natural resources and being good land stewards, but are also economically motivated. Producers are increasingly interested in knowing the effects of management decisions on their immediate environment including soil health (Brock, 1999), water quality, and wildlife. A set of indices that allow an objective measure of the benefits and costs of alternative management strategies in sustainable agroecosystems will help evaluate economic returns of production as well as the environmental benefits of reduced run-off and inputs. These indices also provide a measure of a farm's contribution to sustainability of the community and geographic area: information needed for conservation or environmental protection planning and useful in determining payments to farms with high sustainability indices. Long-term benefits are potentially greater for researchers, producers, and society.

At this point our on field efforts have just begun and preliminary results from the 2001 season are encouraging. Producer involvement has presented real world problems that the researchers would not have faced on small plot scales such as planting problems, and cover crop management problems. Insect pressures have been reduced in some cases by the treatments with some of the producers surprised by the positive effects. Greater communication between researchers and producers is needed to clearly define the role of each group and expectations during the research process. As we continue through the project and put more of the concepts into practice we envision the expansion of the practices to surrounding farms and communities.

Support for sustainable agriculture requires expansion of the concepts within rural communities which can be accomplished by targeting youth (the future rural community leaders). Although youth evolvement has yet to be achieved, we are encouraged by the continued support and encouragement of the Communities in Schools of Georgia participants. By engaging rural youth to understand the complex interactions occurring within agroecosystems, we can help them understand and safeguard local resources as well as reconnect them with their rural heritage.

REFERENCES

Altieri, M.A., P.B. Martin, and W.J. Lewis. (1983). A quest for ecologically based pest management systems. *Environment Management* 7:91-100.

Altieri, M.A. and W.H. Whitcomb. (1979). The potential use of weeds in the manipulation of beneficial insects. *HortScience* 14:12-18.

Blumberg, A.Y. and D.A. Crossley. (1982). Comparisons of soil surface arthropod populations in conventional tillage, no tillage and old field systems. *Agroecosytems.* 8:247-253.

Bockstaller, C., P. Girardin, and H.M.G. van der Werf. (1997). Use of agro-ecological indicators for the evaluation of farming systems. *Journal of Agronomy* 7:261-270.

Brock, B.G. (1999). Rx for soil quality. In *Soil Quality and Soil Erosion,* ed. R. Lal, Boca Raton, FL: CRC Press, pp. 169-172.

Bruce, R.R., G.W. Langdale, L.T. West, and W.P. Miller. (1995). Surface soil degradation and soil productivity restoration and maintenance. *Soil Science Society of America Journal* 59:654-660.

Bull, L., H. Delvo, C. Sandretto, and B. Lindamood. (1993). Analysis of pesticide use by tillage system in 1990, 1991, and 1992 corn and soybeans. *Agricultural Resources: Inputs Situation and Outlook, AR 32.* Economic Resource Service, U.S. Dept. Agriculture, Washington, DC, pp. 41-54.

Creamer, N.G., M.A. Bennett, B.R. Stinner, and J. Cardina. (1996). A comparison of four processing tomato production systems differing in cover crop and chemical inputs. *Journal of the American Society for Horticultural Science* 121:559-568.

CTIC. (1998). National Crop Residue Management Survey. Conservation Technology Information Center, West Lafayette, IN: Purdue University. <http://www.ctic.purdue.edu/CTIC/CTIC.html>.

Day, J.C., C.B. Hallahan, C.L. Sandretto, and W.A. Lindamood. (1999). Pesticide use in U.S. corn production: does conservation tillage make a difference? *Journal of Soil and Water Conservation* 54:477-484.

Fawcett, R.S. (1987). Overview of pest management for conservation tillage. In *Effects of Conservation Tillage on Groundwater Quality: Nitrates and Pesticides,* eds. T.J.

Logan, J.M. Davidson, J.L. Baker, and M.R. Overcash, Boca Raton, FL: Lewis Pub., pp. 19-37.

Fawcett, R.S., B.R. Christensen, and D.P. Tierney. (1994). The impact of conservation tillage on pesticide runoff into surface water: a review and analysis. *Journal of Soil and Water Conservation* 49:126-135.

Flint, M.L. and R. Van den Bosch. (1981). *Introduction to Integrated Pest Management.* New York, NY: Plenum Press.

Franzluebbers, A.J., G.W. Langdale, and H.H. Schomberg. (1999). Soil carbon, nitrogen, and aggregation in response to type and frequency of tillage. *Soil Science Society of America Journal* 63:349-355.

Freyenberger, S., R. Janke, and D. Norman. (1997). Indicators of Sustainability in Whole Farm Planning: Literature Review. Kansas Sustainable Agriculture Series. #2 Manhattan, KA: Kansas State University Agricultural Experiment Station and Cooperative Extension. <http://www.oznet.ksu.edu/sustainableag/publications/ksas2.htm>.

Frye, W.W., J.J. Varco, R.L. Blevins, M.S. Smith, and S.J. Corak. (1988). Role of annual legume cover crops in efficient use of water and nitrogen. In *Cropping Strategies for Efficient Use of Water and Nitrogen*, ed. W.L. Hargrove, Madison, WI: American Society of Agronomy, pp. 129-154.

Gish, T.J., A.R. Isensee, R.G. Nash, and C.S. Helling. (1991). Impact of preferential transport on water quality. *Transactions of the American Society of Agricultural Engineers* 34:1745-1753.

Gish, T.G., C.S. Helling, and M. Mojasevic. (1991). Preferential movement of atrazine and cyanazine under field conditions. *Transactions of the American Society of Agricultural Engineers* 34:1699-1705.

Guthrie, D.S., B. Hutchinson, P. Denton, J. Bradley, J.C. Banks, W. Keeling, C. Guy, and C. Burmester. (1993). Conservation tillage cotton. *Physiology Today* (9), 4 pp.

Haney, P.B., O. Stapel, D.J. Waters, W.J. Lewis, S.K. Diffe, and J.R. Ruberson. (1995). Dynamics of insect populations in a reduced-tillage, crimson clover/cotton system. Part II: Pitfall surveys. In *Proceedings of the Beltwide Cotton Production Research Conference*, Memphis, TN: National Cotton Council, pp. 817-821.

Haney, P.B., W.J. Lewis, and S. Phatak. (1996). Continued studies of insect population dynamics in crimson clover and refugia/cotton systems. Part II: Pitfall trap sampling. In *Proceedings of the Beltwide Cotton Production Research Conference*, Memphis, TN: National Cotton Council, pp. 1115-1119.

Hanthorn, M. and M. Duffy. (1983). Corn and Soybean Practices for Alternative Tillage Strategies. Publication number IOS-2 Economic Resource Service, U.S. Department of Agriculture, Washington, DC, pp. 14-17.

Hargrove, W.L. (1986). Winter legumes as a nitrogen source for no-till grain sorghum. *Agronomy Journal* 78:70-74.

Hildebrand, P.E. and J.T. Russell. (1996). *Adaptability Analysis: A Method for the Design, Analysis and Interpretation of On-Farm Research-Extension.* Ames, IA: Iowa State University Press.

Janke R. and S. Freyenberger. (1997). Indicators of sustainability. In *Whole Farm Planning: Planning Tools*, Kansas Sustainable Agriculture Series, Paper #3, Manhattan KS: Kansas State University Agricultural Experiment Station and Cooperative Extension.< http://www.oznet.ksu.edu/sustainableag/publications/ksas3.htm>.

Kladivko, E.J. (1994). Residue effects on soil physical properties. In *Managing Agricultural Residues*, ed. P. W. Unger, Boca Raton, FL: Lewis Publishers, pp. 123-142.

Kovach, J., C. Petzoldt, J. Degni, and J. Tette. (1992). A method to measure the environmental impact of pesticides. *New York's Food and Life Science Bulletin* No. 139, Geneva, NY: New York State Agricultural Engineer Society.

Kozloff, K., S.J. Taff, and Y. Wang. (1992). Microtargeting the acquisition of cropping rights to reduce nonpoint source water pollution. *Water Resources Research* 28: 623-628.

Lamb, M.C., J.I. Davidson Jr., and C.L. Butts. (1992). PNTPLAN, an expert systems whole-farm planning model designed to optimize peanut-based rotation decisions. *Proceedings of the American Peanut Research and Education Society* 24:37.

Landis, D.A., S.D. Wratten, and G.M. Gurr. (2000). Habitat management to conserve natural enemies of arthropod pests in agriculture. *Annual Review of Entomology* 45:175-201.

Leonard, B.R., R.L. Hutchinson, J B. Graves, and E. Burris. (1993). Conservation-tillage systems and early-season cotton insect pest management. In *Conservation-Tillage Systems for Cotton: A Review of Research and Demonstration Results from Across the Cotton Belt*, eds. M.R. McClelland, T.D. Valco, and R.E. Frans, Fayetteville, AK: Arkansas Agriculture Experiment Station, Special Report 160:80-85.

Leser, J.F. (1995). Conservation-tillage systems: southwest insect management. In *Conservation-Tillage Systems for Cotton: A Review of Research and Demonstration Results from Across the Cotton Belt*, eds. M.R. McClelland, T.D. Valco, and R.E. Frans, Fayetteville, AR: Arkansas Agriculture Experiment Station. Special Report 169:131-135.

Lewis, J.L., M.J. Newbold, A.M. Hall, and C.E. Broom. (1997a). Eco-rating system for optimizing pesticide use at farm level, Part 1: Theory and development. *Journal of Agricultural Engineering Research* 68:271-279.

Lewis, W.J. and M. Jay. (2001). *Ecologically-Based Communities Putting It All Together at the Local Level*. Kerr Center for Agriculture, Poteau, OK.

Lewis, W.J., J.C. van Lenteren, S.C. Phatak, and J.H Tumlinson III. (1997b). A total system approach to sustainable pest management. *Proceedings of the National Academy of Sciences of the United States of America* 94:12243-12248.

Lewis, W.J., P.B. Haney, R. Reed, and A. Walker. (1997c). A total systems approach for sustainable cotton production in Georgia and the southeast: first year results. In *Proceedings of the Beltwide Cotton Production Research Conference*, Memphis, TN: National Cotton Council, pp. 1129-1134.

Lewis, W.J., P.B. Haney, and S. Phatak. (1996). Continued studies of insect population dynamics in crimson clover and refugia/cotton systems. Part I. Sweep and whole plant sampling. In *Proceedings of the Beltwide Cotton Production Research Conference*, Vol 2, Memphis, TN: National Cotton Council, pp. 1108-1114.

Liu, S. and M.D. Duffy. 1996. Tillage systems and profitability: an economic analysis of the Iowa MAX program. *Journal of Production Agriculture* 9:522-527.

Lukk, K.J., J.L. Tindall, and D.F. Potts. (1995). A GIS application for assessment of non-point-source pollution risk on managed forest lands. In *Water Quality Modeling*, ed. C. Heatwole, St. Joseph, MI: American Society of Agricultural Engineers, pp. 491-502.

Meisinger, J.J., W.L. Hargrove, R.L. Mikkelsen, J.R. Williams, and V.W. Benson. (1991). Effects of cover crops on groundwater quality. In *Cover Crops for Clean Water*, ed. W.L. Hargrove. Ankeny, IA: Soil and Water Conservation Society, pp. 57-68.

Newman, A. (1995). Ranking pesticides by environmental impact. *Environmental Science and Technology* 29:324A-326A.

Odum, E.P. (1989). Bridging the four major "gaps" that threaten human and environmental quality. *Bridges* 1:135-141.

Olson, R.K. and C.A. Francis. (1995). Introduction. In *Exploring the Role of Biodiversity on Sustainable Agriculture*, eds. R.K. Olson, C.A. Francis, and S. Kaffa. Madison, WI: American Society of Agronomy, pp. 1-4.

Prato T. and S. Wu. (1995). Economic and water quality impacts of using alternative farming systems for claypan soil in the Goodwater Creek watershed, a stochastic programming analysis. In *Water Quality Modeling*, ed. C. Heatwole, St. Joseph, MI: American Society of Agriculture Engineers, pp. 503-520.

Rannells, N.N. and M.G. Wagger. (1996). Nitrogen release from grass and legume cover crop monocultures and bicultures. *Agronomy Journal* 88:777-782.

Reed, R., S. Phatak, A. Page, P.B. Haney, and W.J. Lewis. (1997). Conservation tillage in Coffee county cotton. In *Proceedings of the Beltwide Cotton Production Research Conference*, Vol. 1, Memphis, TN: National Cotton Council, pp. 621-622.

Reeves, D.W. (1994). Cover crops and rotations. In *Crops Residue Management*, eds. J.L. Hatfield and B.A. Stewart, Boca Raton, FL: Lewis Publishers, pp. 125-172.

Reicosky, D.C., W.D. Kemper, G.W. Langdale, C.L. Douglas Jr., and P.E. Rassmusen. (1995). Soil organic matter changes resulting from tillage and biomass production. *Journal of Soil and Water Conservation* 50:253-261.

Ruberson, J.R., S.C. Phatak, and W.J. Lewis. (1997). Insect populations in a cover crop/strip till system. In *Proceedings of the Beltwide Cotton Production Research Conference*, Memphis, TN: National Cotton Council, pp. 1121-1124.

Ruberson, J.R., W.J. Lewis, D.J. Waters, O. Stapel, and P.B. Haney. (1995). Dynamics of insect populations in a reduced-tillage crimson clover/cotton system. Part 1. Pests and beneficials on plants. In *Proceedings of the Beltwide Cotton Production Research Conference*, Memphis, TN: National Cotton Council, pp. 814-817.

Rzewnicki, P. (1991). Farmers' perceptions of experiment station research, demonstrations, and on-farm research in agronomy. *Journal of Agronomic Education* 20:31-36.

Schomberg, H.H. (1998). *In situ* N mineralization and N availability in a no-till cotton cover crop system. In *Proceedings 16th World Congress of Soil Science, Symposium 12 Indicators for Soil Fertility Recapitalization Efforts*, Montpellier, France (20-21 August). International Society of Soil Science (ISSS). (CDROM).

Schomberg, H.H., P.B. Ford, and W.L. Hargrove. (1994). Influence of crop residues on nutrient cycling and soil chemical properties. In *Managing Agriculture Residues*, ed. P.W. Unger, Boca Raton, FL: Lewis Publishers, pp. 99-121.

Shurley, D. (2001). Depressed US textile industry means tough times and marketing challenges for cotton farmers. University of GA Agricultural Economics Cotton Outlook. http://www.ces.uga.edu/Agriculture/agecon/outlook/cotton/cot01july.htm (verified August 10, 2001).

Stapel, J.O., W.J. Lewis, S.C. Phatak, and J.R. Ruberson. (1998). Insect pest management as a component of sustainable cotton production system. In *Proceedings of the Beltwide Cotton Production Research Conference*, Memphis, TN: National Cotton Council, pp. 1107-1111.

Stark, C.R. Jr., K.J. Bryant, J.R. Ruberson, S.C. Phatak, and W.J. Lewis. (1999). Economic comparisons between ecologically-based and traditional cotton pest management systems. In *Proceedings of the Beltwide Cotton Production Research Conference*, Vol. 1, Memphis, TN: National Cotton Council, pp. 348-349.

Stirling, G.R. (1991). *Biological Control of Plant Parasitic Nematodes: Progress, Problems and Prospects*. Wallingford, England: CAB International.

Stute, J.K. and J.L. Posner. (1995). Legume cover crops as a nitrogen source for corn in an oat-corn rotation. *Journal of Production Agriculture* 8:385-390.

Sullivan, M.J. and T.W. Smith Jr. (1993). Insect management in reduced-tillage southeastern cotton. In *Conservation-Tillage Systems for Cotton: A Review of Research and Demonstration Results from Across the Cotton Belt*, eds. M.R. McClelland, T.D. Valco, and R.E. Frans, Fayetteville, AR: Arkansas Agriculture Experiment Station, Special Report 160: 39-41.

Trimble, S.W. (1974). *Man-Induced Soil Erosion on the Southern Piedmont, 1700-1970*. Ankeny, IA: Soil Conservation Society of America.

Turnock, W.J., B. Timlick, and P. Palaniswamy. (1993). Species and abundance of cutworms (Noctuidae) and their parasitoids in conservation and conventional tillage fields. *Agriculture Ecosystems and Environment* 45: 213-227.

United States Department of Agriculture/National Agricultural Statistics Service (USDA/NASS). (1999). *Track Records: United States Crop Production*. USDA, Washington, DC.

van der Werf, H.M.G. and C. Zimmer. (1998). An indicator of pesticide environmental impact based on a fuzzy expert system. *Chemosphere* 36:2225-2249.

Wuest, S.B., D.K. McCool, B.C. Miller, and R.J. Veseth. (1999). Development of more effective conservation farming systems through participatory on-farm research. *American Journal of Alternative Agriculture* 14:98-102.

Williams-Woodward, J.L. (1999). *1998 Georgia Plant Disease Loss Estimates*. University of Georgia Cooperative Extension Service, Athens, GA, Pathology 99-102.

Cropping Systems
and Water Quality Concerns

D. Brook Harker
Brian McConkey
Helen H. McDuffie

SUMMARY. The impact of cropping systems on water quality is uncertain, and its interpretation depends heavily upon our definition of acceptable risk. As a means of determining net effect, both classical and precautionary approaches to assessing risk have their strengths and weaknesses. Relating the impact of cropping practices to human health outcomes can be particularly difficult. A variety of guidelines and standards are used to assess water quality, and recent methods for assessing water quality seek to incorporate more than water chemistry alone. An understanding of the derivation of water quality guidelines and standards is essential to their effective application, and meaningful interpretation.

In addressing water quality concerns, it is essential to first clarify that there is indeed a problem, and whether agriculture makes a significant

D. Brook Harker is affiliated with Agriculture and Agri-Food Canada, PFRA, 603–1800 Hamilton Street, Regina, SK, Canada, S4P 4L2 (E-mail: harkerb@em.agr.ca).

Brian McConkey is affiliated with Agriculture and Agri-Food Canada, Research Branch, SPARC, P.O. Box 1030, Swift Current, SK, Canada, S9H 3X2 (E-mail: mcconkeyb@em.agr.ca).

Helen H. McDuffie is affiliated with the Centre for Agricultural Medicine, 3614 Royal University Hospital, University of Saskatchewan, 103 Hospital Drive, Saskatoon, SK, Canada, S7N 0W8 (E-mail: McDuffie@sask.usask.ca).

[Haworth co-indexing entry note]: "Cropping Systems and Water Quality Concerns." Harker, D. Brook, Brian McConkey, and Helen H. McDuffie. Co-published simultaneously in *Journal of Crop Production* (Food Products Press, an imprint of The Haworth Press, Inc.) Vol. 9, No. 1/2 (#17/18), 2003, pp. 329-359; and: *Cropping Systems: Trends and Advances* (ed: Anil Shrestha) Food Products Press, an imprint of The Haworth Press, Inc., 2003, pp. 329-359. Single or multiple copies of this article are available for a fee from The Haworth Document Delivery Service [1-800-HAWORTH, 9:00 a.m. - 5:00 p.m. (EST). E-mail address: docdelivery@haworthpress.com].

http://www.haworthpress.com/store/product.asp?sku=J144
10.1300/J144v09n01_01

contribution. Compound interactions and modes of chemical movement can render this troublesome. Yet, because farmers live on the land and drink the water, they want to be among the first to know what is happening and to take appropriate action when problems are identified.

Agriculture must be proactive in addressing water quality concerns. However, effective land management strategies depend greatly upon regional differences and may be highly site-specific. Hence, it is best to apply a set of common sense concepts at the local level. Because soil and water degradation are closely related, practices first developed to help conserve the soil (i.e., crop rotations, reduced tillage, cover crops) may also tend to conserve water quality. As well, restricting the loss of agricultural inputs (e.g., fertilizer nutrients, pesticides) from off farmland, and reducing the amount of those that might be available to do so, can assist in effectively reducing potential pollution. Buffer zones are a promising means of using plants and wetlands as a filter towards intercepting escaping contaminants. *[Article copies available for a fee from The Haworth Document Delivery Service: 1-800-HAWORTH. E-mail address: <docdelivery@ haworthpress.com> Website: <http://www.HaworthPress.com> © 2003 by The Haworth Press, Inc. All rights reserved.]*

KEYWORDS. Water quality, risk assessment, Precautionary Principle, guidelines, standards, human health, soil conservation, best management practices, fertilizer, pesticides

UNCERTAINTY AND WATER QUALITY

The State of Our Water

Water quality is said to be deteriorating across North America and around the world, and agricultural cropping systems are increasingly cited as a key factor. Yet, it is frequently difficult to establish clear cause-and-effect relationships between agricultural operations and water quality concerns (Swader, Adams, and Meek, 1994). Nevertheless, the wide-spread nature and varying intensity of agricultural activities clearly reflect a risk that surface and ground waters may become unacceptably contaminated. Runoff, leachate, and airborne deposition from agricultural lands can contribute significant levels of organic matter (OM), mineral nutrients, sediment, agrochemicals, and pathogens to surface and groundwater supplies. Pollution might adversely affect water use and safety for human life and the entire ecosystem.

There is a public perception that water quality is worsening. A 1993 national opinion poll in Canada found that many residents saw water pollution as "the most imperative environmental problem" of the day (Angus Reid Group,

1993). A recent survey indicates that most Canadians (87%) still rate their drinking water as of "acceptable-to-high quality," yet an increasingly large minority (4% in 1992; 17% in 2000) "believe the food and agriculture sector is the worst polluter" of their water (Environics International, 2000). There are continuing, conflicting messages about the role of agriculture in degraded water quality. In the US, for example, it is said that agriculture is a major contributor to water pollution (Offutt, 1990), while on the Canadian prairies it has been argued that there is "very little evidence" that agriculture is negatively affecting water quality (Lindwall, 1992). Although the opinion of individual authors might change as new information becomes available, such statements indicate the range of thought that prevails today.

It is certain that agriculture is the cause of pollution in some localities ('hot spots'), but it is uncertain how representative many of these hot spots are of agriculture in general. A fundamental question is: "Can we have agriculture in the landscape and expect to have no net change in water quality?" To a large extent, the difficulty in assessing the impacts of agriculture on water quality hinges upon our basic understanding of acceptable risk, and our understanding of associated assumptions and relationships.

The objective of this paper is to discuss issues and attitudes regarding the impact of cropping systems on water quality, the application of water quality guidelines, and cropping system strategies that can reduce contamination of surface- and groundwaters.

Defining Acceptable Risk

A central reason for differences of opinion on the effect of cropping practices on water quality is that individual attitudes toward acceptable risk greatly affect how we interpret water quality information (Harker, Hill, and McDuffie, 1998). Some are prepared to accept a guideline approach to water quality–the concept that there are contaminant levels below which our lives and the health of the ecosystem are at reasonable risk. Others take a zero tolerance position and hold that no amount of unnatural substance or elevated nutrient or sediment concentration in the environment is acceptable. We are often being asked to choose between these two schools of thought.

Some people believe that zero risk is unreasonable, and call for experts and policy makers to explain why we should strive to attain an objective that may not be feasible. Black (1995) points out that all of nature uses water to absorb waste products and to transport nutrients. According to him, humankind should be able to (responsibly) do the same thing and he has stated that

> Policies advocating the zero discharge of pollutants are contrary to the role of water as a natural resource buffer and aim for a fundamentally unnatural goal.

Hrudey and Krewski (1995) questioned the validity of a zero tolerance point of view. Using conservative United States Environmental Protection Agency (USEPA) estimates, they calculated the hazard of lifetime exposure to one molecule a day of the most potent known carcinogen (TCDD). Their calculations indicate that exposing the entire world population to this smallest conceivable daily dose would not yield a single case of cancer. Hence, "Within a realistic concept of safety, there is a safe level of exposure . . ." to even the most toxic of carcinogens–and the concept of zero tolerance is rendered invalid. However, it is argued that the long-term health implications of exposure even to minute quantities of these chemicals, particularly their combined effects, are not well understood (Linton, 1997). Until such relationships are better known, society can never be sure of the risks.

It is important to consider the effects of water quality on not only human life, but on the ecosystem at large–including aquatic life and plant growth. Hence, the different uses of water can become a critical factor in defining acceptable risk. Plant growth, for example, may be affected by trace amounts of pesticide that do not appear to affect human health. In an Alberta study looking at water quality for human and livestock health and irrigation purposes, water deemed suitable for human and livestock use contained herbicide levels considered unsafe for irrigation of certain crops (CAESA, 1998).

Evaluating Risk

Risk assessment is at best an imprecise science, severely constrained by what Finkel (1996) of the US Occupational Health and Safety Administration (OHSA) calls "a dearth of qualified practitioners." He cautions that we must be careful not to ask more of risk assessment than it can deliver. Scherer (1990) points out that traditional approaches to risk assessment which involve probabilities, statistics and risk analysis, are not sufficient in the public mind. He says that technical and scientific problems are ultimately social problems, and that public reaction to risk assessment is based on a set of criteria that requires both technical and social solutions.

Sandman (1987) says the risk assessment criteria used by the public are more likely to focus on what he calls a dimension of 'outrage.' This outrage dimension is based on the integration of more than 20 factors that include

- *Fairness of risk*–accounts for individual proximity to a hazard, like a nuclear reactor.
- *Degree of control*–access to a private well as opposed to relying on public water supply.
- *Familiarity*–exposure to common automobile accidents versus less understood risks such as pesticide contamination.

We can add to these outrage factors, a notion of trespass—the concept that individuals do not like their water sources being arbitrarily contaminated by others, regardless of whether or not the contamination can be shown to be harmful.

In the end, risk analysis generally comes down to a matter of probabilities. Probabilities are usually based on historical data, their application assumes average conditions, and it is expected that past trends will continue. But there is often little historical data from which to extrapolate the probable effects of trace amounts of water-borne agrochemicals on human and ecosystem life, and toxicological findings from laboratory rats, and other tests, may not be applicable (Caldwell, 1996).

Known Hazards vs. the Unknown

We generally live longer, healthier lives today than ever before, and when it comes to predicting many hazards, we have excellent long-term records of many relationships. We can say with confidence, for example, that the probability of a North American adult over 35 years of age dying from a heart attack in any given year is 1:77 (Kluger, 1996). Similarly, the odds of a young adult (14-25 yrs) dying in a car crash are 1:3500. These represent significant odds when compared with the uncertainty of death or the long-term consequences of trace amounts of agrochemical in drinking water.

Many in society are preoccupied with the possibility that products of our own making may be harming us—causing cancer for example. Yet there are very potent natural carcinogens, and when it comes to human diet, a multi-disciplinary task force representing the National Academy of Sciences has concluded, "there is no clear difference between the potency of known naturally occurring and synthetic carcinogens. . . ." The investigators go on to say, "Current evidence suggests that the contribution of excess macro-nutrients and excess calories to cancer causation . . . outweighs that of individual food micro-chemicals, both natural and synthetic" (NAS, 1996). In short, eating too much is far more likely to cause cancer than the micro-chemicals in the food we eat.

In addition to toxicity and carcinogenic concerns regarding human-made chemicals, it is possible that some of these chemicals (e.g., pesticides) are interfering with the development of the reproductive system in the fetus and in children, i.e., acting as endocrine disruptors. This may be the result of interference with the estrogen receptor, yet recent evidence indicates that this mechanism might play only a minor role in purported effects. Hence, according to Foster (1998), "the focus on estrogen mimics [as a source of endocrine disruption] may be too simplistic and alternate mechanisms could be more relevant. . . ." The author further says there is still cause for concern, however, as human-made chemicals might induce reproductive problems some other way. He indicates that more comprehensive research is needed to quantify the effect of such factors as: stage of development during exposure, thyroid function, and specific chemical mixtures.

In matters analytical, we've become so smart that perhaps we're foolish–wherein "The scientific community's ability to detect chemicals is much more advanced than the understanding of the toxicology associated with such discoveries" (CAST, 1992). Compared with the reality of other hazards (e.g., heart attack, car crash deaths), perhaps future generations will look back on us as a society preoccupied with chasing molecules–while ignoring the more likely prospect of being hit by a bus.

WEIGHING THE EFFECTS

The Classical Science Approach

In today's classical science, the Null Hypothesis is the general principle of evaluation–that is, unless the probability of change is documented to be greater than that due to a specified likelihood from chance alone, no change is assumed to have taken place. The process of evaluation thus incorporated into the Null Hypothesis is similar to the 'innocent until proven guilty' dictum of criminal law. This systematic, conservative methodology is used because we are often fooled by apparent relationships. A set of coincident circumstances (e.g., trace levels of pesticides in the Great Lakes drinking water of mothers experiencing birth defects) by no means confirms that potential cause-and-effect relationships are in force. There are often too many other variables that might be responsible, and we must be cautious of reaching conclusions that cannot be substantiated.

The strength of the scientific method is that it demands proof. An important weakness is that it requires a way to separate effects, in order to identify statistical cause-and-effect relationships. This can be problematic for constituents that are widely distributed in water. For example, at the 1998 conference on Children's Health and Environment, the question arose (Bertel, 1998) as to how researchers might expect to find abnormal effects within a statistically normally distributed population, if the entire population was somehow uniformly subject to the same adverse effect (e.g., trace pesticides in drinking water). A further weakness of the Null Hypothesis is that there is often an implicit conservatism in favor of the status quo, since, typically, the probability of detecting a real change is less than the probability of accepting the hypothesis that there is no change at all (Cox, 1958).

The Precautionary Principle

Some think that when it comes to environmental matters, the Null Hypothesis should be abandoned in favor of the 'Precautionary Principle.' This principle holds that the environment is too complex to understand and too difficult to repair once adversely affected, and that we should assume all actions have a negative effect until society has irrefutable evidence to the contrary. A recent version of the precautionary principle states that

> When an activity raises threats to the environment or human health, precautionary measures should be taken, even if some cause-and-effect relationships are not fully established scientifically. (SEHN, 1998)

There is, however, a wide range of interpretation as to the stringency with which the precautionary principle ought to be applied. Proponents of a moderate interpretation hold that instead of asking questions like: "How safe is safe?" or "What level of risk is acceptable?"; society ought to be asking: "How much contamination can be avoided?" or "What are the alternatives to this product?" and "Is this activity really necessary?" This approach places much less emphasis on the traditional risk assessment and benefit-cost analysis of individual chemicals or products, while still allowing such techniques to be used to effectively compare alternatives.

Others argue that the precautionary principle is dangerous, that it is antagonistic towards sound science, having its roots largely founded in instinct and feeling (Mongoven, 1998). They warn that it threatens the entire chemical industry, wherein hundreds of new chemicals are marketed annually. At present, the release of a new chemical may not require environmental testing for specific scenarios, because negative environmental effects are not known to exist. In its harshest interpretation, the precautionary principle could require any new product to prove that it has no negative effect on any aspect of the environment—a virtually impossible task, akin to what has been termed 'Environmental McCarthyism' (Harker and McConkey, 2001).

The strength of the precautionary principle is that it emphasises environmental assessment on the basis of whole system analysis—an understanding of the parts by looking at the whole (Ashford and Miller, 1998). Its weakness is that once cause-and-effect relationships have been linked to apparent factors (however tenuously), there is a tendency towards the wholesale condemnation of any and all of the constituents of the suspected 'chemical soup.'

Relating Human Health to Cropping Practices

Attempting to link human health outcomes with the real or perceived contamination of drinking water as a result of cropping practices, poses complex and challenging issues (Raja 1996; McDuffie et al., 1998; McDuffie et al., 2003). These can be subdivided as follows:

- characterization and documentation of water contamination, and the water source
- characterization of the individual(s) in question
- methodological and design considerations
- statistical analysis and interpretation of the data.

Characterizing the possible contamination of drinking water from agricultural sources is difficult, because of the dynamic status of many contaminants. For example, concentrations of a target compound or microbe can vary dramatically over a short time period, or seasonally. As well, there is no requirement to systematically sample and analyse water from private sources, or for very small communities–hence a dearth of information exists for these sources. Radically changing weather conditions such as heavy rainfall or rapid snowmelt can also greatly influence water quality.

Complex relationships between water characteristics (e.g., pH and temperature) can influence whether or not substances are dissolved or precipitated, and the concentration of dissolved contaminants. Factors of this type can also influence the potential for interaction between microbes (by promoting or inhibiting growth) and other kinds of pollutants, such as agricultural or non-agricultural chemicals. Movement of water and contaminants through soil can be unpredictable, causing contaminants to appear in unexpected locations. In addition, water treatment practices may interact with natural or synthetic contaminants to produce new compounds.

Because of the diversity of water sources on private land (and in small rural communities), detailed information on the water source and the environmental conditions surrounding the water source (e.g., distance from grazing cattle, history of agrochemical use, age of wells), must be collected using standardized questionnaires. If cancer or other medical conditions with long latency are of interest, historical information on drinking water quality must be sought. As to specific sample analysis, financial constraints usually induce the researcher to choose among: a variety of microbial analyses; and the concentration and bioavailability of specific chemical elements (including nitrates, pesticides or other compounds)–known or suspected of causing adverse health effects.

Within the human population, definitions of what constitute an abnormal concentration of contaminant and of the human health risks associated with abnormalities vary among studies. In some studies, the concentration of each contaminant may be statistically treated as independent of the concentration of all other known contaminants–despite the fact that there may be some information to suggest contaminants occur in quantitative relationships, or that potential interactions exist among contaminants.

For many risk factors, the prevalence and levels of exposure are low. Accordingly, future studies need to better take into account: (a) large population studies–perhaps identifying 'hard-to-detect' synergistic and additive effects by examining extremely large segments of a population; (b) age relationships–recognizing that certain life stages (e.g., children, the aged) may be more susceptible, or may reflect consequences only evident later in life (Chance and Harmsen, 1998). All such non-chemical evaluations of water quality involve increasingly complex analyses and rely upon a science that is relatively new.

INDICATORS OF WATER QUALITY

If a guidelines approach to water quality is accepted, then a well-established set of chemical indicators is generally available for monitoring the state (or existing condition) of agricultural pollutants in surface- and groundwaters (Bonnis, 1997). These indicators include the measurement of various forms of oxygen, phosphorus, and nitrogen (N); as well as acidity, pesticides, heavy metals, bacterial contamination, turbidity, and salinity.

The Canadian Water Quality Guidelines

The *Canadian Water Quality Guidelines* rely heavily on a chemical approach to defining acceptable water–for drinking, recreation, irrigation, and other uses (CCREM, 1987, and updates). The Canadian guidelines are based on studies of individual chemicals, conducted in Canada and elsewhere, that are considered applicable to Canadian conditions.

The Canadian guidelines recognize that there is no hard and fast boundary between good and poor water quality. So maximum concentrations in the guidelines generally incorporate a safety factor at least 10 to 100 and even 1,000 times greater than test results indicate. Hence the *Canadian Drinking Water Guidelines* might be portrayed as a conservative approach to water quality, and themselves avow only that *continually* exceeding them "*may*, in some instances, be capable of inducing deleterious effects on health. . . ." (FPSDW, 1987).

European Economic Community (EEC) Thresholds

Some view the Canadian and similar guidelines as a 'far too liberal' approach to water quality assessment. This may cause them to favor a more stringent limit on guidelines, such as the threshold levels established in the *European*

Commission's 1980 Drinking Water Directive (80/778/EEC). EEC thresholds include categories that cover

- *Individual pesticides*–Maximum Admissible Concentration (MAC) is 0.1 µg L^{-1} (Compare to CDWQG,[1] i.e., 5 µg L^{-1} for atrazine; 100 µg L^{-1} 2,4-D; 280 µg L^{-1} glyphosate)
- *Total pesticides*–combined MAC equals 0.5 µg L^{-1} (CDWQG–no combined MAC is specified)
- *Nitrate*–MAC 50 mg L^{-1} (25 mg L^{-1} reference level) (CDWQG–MAC is similar)
- *Phosphorus*–MAC 5000 µg L^{-1} (400 µg L^{-1} reference level) (CDWQG–no MAC is specified).

From a cause-and-effect standpoint, it can be argued that it does not make sense to treat all pesticides the same–as though similar concentrations of different contaminants are equally harmful. Yet, because of the uncertainty surrounding risk analysis, the simplified EEC standard is politically attractive and can even lead to reduced monitoring requirements, as regulators no longer have to search for hard-to-find contaminants but only need find 0.5 µg L^{-1} of total pesticides. Hence, such a guideline is attractive.

A Less Stringent, Intermediate Approach?

At a recent water quality workshop in Quebec City, Canada, a somewhat intermediate approach to the Canadian guidelines and EEC thresholds was proposed. On the Canadian prairies, for example, with the variety of application technologies currently being used, even for pesticides applied at application rates of 500 to 1000 g ha^{-1} (compared to several new pesticides, which are applied at rates less than 50 g ha^{-1}), concentrations in surface waters seldom exceed 5 µg L^{-1}. Perhaps, then, society ought to evaluate the stewardship of pesticide use in terms of whether residue levels reflect current application technologies, rather than whether they exceed an often much higher drinking guideline (e.g., 100 µg L^{-1} for 2,4-D). As well, since water quality guidelines have not yet been established by Health Canada for some pesticides, and because new pesticides are continually coming onto the market, a limit that mirrors good management practices could well be a simple precautionary means of addressing pesticide use in general (Allan Cessna, personal communications).

Other More Comprehensive Indicators

Chemical indicators are only one way of evaluating the state of water quality, and there is growing recognition that, "Traditional emphasis on chemical

---

[1]CDWQG = *Canadian Drinking Water Quality Guidelines*; 1 µg L^{-1} = 1 ppb.

indicators of water quality must be supplemented by more comprehensive indicators . . ." (Young, Dooge, and Rodda, 1994).

These indicators ought to be based on the total properties of a water body–including the physical, chemical, biological, radiological and ecological parameters of the water, and of the material it carries. As such, the health of biologic communities may be an indirect but effective indicator of the state of water quality. Biological indicators can be used to demonstrate the combined effects of pollutants, changes in habitat, and other impacts that chemical monitoring alone does not reveal. Variations on this kind of measurement include

- *Aquatic indicator species*–that reflect the relative health of a 'most-sensitive' species
- *Biologic diversity*–of species within the aquatic ecosystem
- *Functional diversity*–a sufficient range of species to perform the normal ecosystem functions of primary production, decomposition, etc.

APPLYING GUIDELINES AND STANDARDS

Contaminants in water are under increasing scrutiny–as they should be. The pervasive tendency has been to change guidelines and standards to match the lower level to which natural and unnatural' substances may be detected in water. But the reasons for doing so appear to have less to do with specific toxicities than they do with a fundamental uncertainty and distrust of the underlying rationale.

Aquatic Guidelines

Take, for example, the blanket application of aquatic guidelines–which are increasingly recommended to protect water quality overall. One reason for using aquatic guidelines is that people are uncomfortable with the prospect that significantly higher levels of contaminant are generally allowed for drinking water than for other guidelines. Another, is the belief that aquatic organisms are the most sensitive on the ecological scale, hence all waters ought to be protected to that level of sensitivity–even ground waters because they might discharge into surface waters.

The underlying issues hearken back to whether drinking water guidelines are adequate in the first place, and whether different guidelines (aquatic vs. drinking) ought to be separately applied to surface and ground waters. Until this dilemma is resolved, the merit behind systematically applying aquatic-use guidelines to all water quality will be muddied (Harker, Hill, and McDuffie, 1998).

The Nitrate Guideline

The necessity of clearly understanding the rationale behind a guideline in order to properly interpret its significance is amply illustrated in the nitrate-nitrogen (NO_3-N) standard for drinking water–10 mg L^{-1}. (Our purpose is not to attack the standard, but to better understand its overall interpretation.) If, for example, 10 mg L^{-1} of NO_3-N is judged to be unsafe, does that mean that 5 mg L^{-1} is half way to unsafe, or that 20 mg L^{-1} is cause for outright alarm?

Early research in the 1940s (Comly, 1945) indicated that a high nitrate level in drinking water was associated with the nitrite that causes 'blue baby' syndrome, an occasionally fatal condition. The conservative guideline of 10 mg L^{-1} NO_3-N was set for all drinking water–based on results showing apparent toxicity in some infants at greater than 60 mg L^{-1}. But well waters contaminated with nitrate may also be contaminated with bacteria, and subsequent research has repeatedly raised the possibility that blue baby syndrome may be principally due to bacteria alone (Cornblath and Hartmann, 1948; Hanukoglu and Danon, 1996), as discussed in detail by Avery (1999). Still, evidence for nitrate involvement remains (Tanase, Iacob, and Beldescu, 1998) and the debate continues.

Understanding the origin of the nitrate guideline and the debate surrounding it, helps us to keep the guideline in perspective. What then is the significance to human health of finding NO_3-N concentrations of 5, 10 or even 20 mg L^{-1} in groundwater? Perhaps very little–unless there's a well-documented trend to steadily increasing levels, and these levels are likely to persist at concentrations well above the guideline.

Phosphorus Levels

Phosphorus (P) is undoubtedly the greatest water quality concern on the Great Plains (prairies) of western Canada. Concentrations as little as 0.01 to 0.05 mg L^{-1} of dissolved inorganic P can represent eutrophic conditions (promoting excessive growth of aquatic vegetation) (Sosiak, 1997). To illustrate the extent of excessive nutrients already existing in surface water in the Prairie region, typical concentrations of dissolved inorganic P in farm dugouts (surface ponds used primarily for farmstead water) are 0.1 to 0.5 mg L^{-1}–about 10 times the concentration associated with eutrophic conditions (Corkal and Peterson, 1994).

In response to problems of excessive P in surface water, the province of Alberta, Canada, adopted an interim water quality objective of 0.05 mg L^{-1} of total P for surface waters (CAESA, 1998). However, most total P is unavailable for biologic growth, so this objective is clearly conservative. In fact, the guideline is so conservative that very few streams in Alberta have total P below that concentration, including those almost devoid of agricultural activity and human habitation (Anderson et al., 1998). If the interim water quality objective was applied indiscriminately, it would be not only unachievable, but would require scarce resources that might be better used elsewhere. Fortunately, Alberta has taken the approach of concentrating on specific problems first (Sosiak, 1997).

The origin of excessive P in surface water on the Prairies is easily understood. The naturally fertile prairie soils contain in the order of 1000 kg total P ha^{-1} in the 10 cm tillage layer (Sadler and Stewart, 1974), although most of that P is biologically unavailable, or only very slowly available (Selles, McConkey, and Campbell, 1999). Typical runoff from Alberta crop land would be less than 100 mm $year^{-1}$, and the driest two-thirds of crop land would have less than 25 mm of annual runoff (Environment Canada, 1978). So losses of as little as 50 mg of total P ha^{-1}—or less than 0.01% of the total P near the land surface—would bring the concentration of total P in runoff above the 0.05 mg L^{-1} objective. To make matters worse, evaporation almost invariably exceeds precipitation on the prairies. As water evaporates from lakes and reservoirs, the concentration of P increases. Not surprisingly, there is evidence that many shallow prairie lakes and ponds were eutrophic long before European settlement (Mitchell and Trew, 1992).

One way that agriculture is seeking to address this issue, is to clarify soil P limits—the amount of P that a soil can accumulate before it might be expected to negatively impact on surface water quality. Phosphorus limits acknowledge that differing soils have differing capacities to store P, and recognize that the distance of a specific soil from a waterway can significantly affect the relative hazard that it might contribute to degraded water quality. A comprehensive soil P limits study in Alberta is looking at various aspects of this issue. Research and workshop findings there are being weighed against case study results from jurisdictions in Michigan, Wisconsin, Texas, and The Netherlands (Olson et al., 2001).

Impact of Different Standards on Interpretation–An Example

Because of differences in the way standards are sometimes added to or applied, even findings from similar studies in the same geographic region can appear to be contradictory. For example, a fairly recent PFRA (Prairie Farm Rehabilitation Administration) study into the effects of nonpoint-source agricultural activities on water quality (Harker et al., 1997) came to the conclusion that there was, "no significant body of evidence to indicate the wide-spread contamination of surface and ground water on the prairies." Yet within a year of this report, a comprehensive study (CAESA, 1998) in the prairie province of Alberta (also funded in part by PFRA), found that, "Current agricultural management practices on many farms are not adequate to sustain water quality. . . ." How could this be?

The prairie-wide PFRA study represented findings on agricultural lands in general, for both surface and groundwater conditions. On the other hand, the CAESA findings, where negative, relate primarily to streams and small lakes associated with runoff prone landscapes, and apply largely to surface waters only. As well, conclusions in the prairie-wide report are reflective of Canadian Water Quality Guidelines per se, whereas much of the CAESA conclusions rely heavily upon a set of interim Alberta objectives for aquatic life–some of the most stringent in the nation. Hence, although both reports took a guidelines approach to their analysis, their conclusions may apply to differing landscapes and a different application of water quality guidelines.

QUANTIFYING AGRICULTURAL CONTRIBUTIONS

A full range of agriculture-related water quality problems are encountered around the world (Bonnis, 1997). In order to be assured that the effects of agricultural pollution on the environment are neither minimized nor exaggerated, a number of factors must be carefully considered.

Clarifying the Problem

Effective water quality solutions require a clear understanding of the nature and sources of the problem–starting with whether the threat is real or is only perceived to be so. Not that reality is any more important than perception in defining a problem (MacAlpine and Nguyen, 1993), but answering that simple question is often overlooked.

Where a water quality problem is known to exist, the cause of the problem should first be determined–because agriculture may not be a significant contributor. For example, in the US, sediment loading from agricultural lands is said to be "the most pervasive nonpoint pollutant" (Gomez, 1995), and many people assume this condition therefore applies in Canada. But on the Saskatchewan River system, which drains much of the Canadian prairies, silt loading attributed to agricultural lands is said to be relatively insignificant when compared to the erosion occurring naturally as a result of stream base-flow (Carson and Associates, 1990).

In order to ensure a substantial likelihood of project success, a practical solution is required–one that takes into account various social and technological aspects of benefit-cost; and that allows residents to live and work within the watershed while accomplishing environmental goals. Where practical, efforts at monitoring and mitigating water quality should focus on existing and potential hot spots as early indicators of developing problems. These are lands where difficulties are likely to be experienced first, because of the large volume or timing of nutrient and pesticide applications, or the nature of the soils and landscapes.

Interactions, Protocols, and Compounding Effects

A fundamental understanding of soil-water interactions involves the realization that the characteristics of a particular agrochemical, such as its solubility, adsorption, and persistence, can significantly affect the way it will react with surrounding biophysical conditions. Hence, the fate of pesticides and nutrients can be influenced by changes to the soil micro-environment caused by weather and farming practices. Choice of tillage practice, residue management, and related decisions can result in "unique combinations of aeration, water availability, temperature distribution and availability of substrates"–all of which will affect the movement of agrochemicals (Power, 1994).

Given the small concentration of agrochemical often involved, sampling and analysis protocol can play a major role in whether or not a residue is found, and at what levels. For example, shallow groundwater may need to be sampled weekly during the growing season, otherwise peak nitrate levels will be missed (Chang and Entz, 1996), and large seasonal and spatial variations can occur in the herbicide levels of shallow groundwater (Hill et al., 1996). As well, some pesticides may be absorbed by the PVC casing of sampling wells or by the plastic bottles used to store samples (Hill et al., 1995). Similarly, detecting pesticides in soil and plants requires careful sampling, considering that pesticides move within plants, and pesticides in the soil partition between soil minerals, organic materials, and water (Kookana and Simpson, 2000).

Significant amounts of pesticides can be transported by the atmosphere, in the vapor phase and in association with fine particles, for distances of metres to thousands of kilometres (Hawthorne et al. 1996; Goolsby et al. 1997; Rawn et al., 1999a). In some cases, the presence of herbicides in water cannot be attributed to cropping practices in the local watershed, but are due to atmospheric deposition from elsewhere (Muir and Grift, 1995; Rawn et al., 1999b). Although this contamination is a result of cropping practices somewhere, addressing such long-range pollution pathways can require regional, continental, or even global actions.

Potential agricultural effects can be further complicated by factors such as

- *Difficulty in tracing* the origin and level of nonpoint-source losses
- *Magnitude of sources* including the large numbers of farms, soil differences, and multiple management styles
- *Time lag* between application and response relationships

- *Poorly understood interactions* between many agricultural and environmental factors.

Finally, unpredictable and unusual weather events are generally the most important factor driving water pollution from agricultural lands, and typically overwhelm the effects of agricultural cropping practices (Randall and Mulla 2001; Power, Wiese, and Flowerday, 2001).

Farmers Are Concerned

It is evident that for a given region, even within the agricultural community, a consensus of opinion on water quality is difficult to derive and may not exist. In the past, a type of moral suasion was used with landowners to promote environmental quality and sustainable agriculture, with varying success (Huang and LeBlanc, 1994). More recently, the application of economic incentives to this end has been controversial amongst advocates, target groups, and social scientists (Dubgaard, 1991).

Nevertheless, because farmers live on the land and drink the water there, they want to be among the first to understand what is happening and to seek appropriate solutions. As well, many farmers have increasingly come to "realize that balancing . . . agricultural growth with a clean environment is a vital part of doing business in today's world marketplace" (CAESA, 1998). It has been suggested that basing policies solely on the idea that farmers are only motivated by profit maximization, needs to be re-examined and replaced with policies that acknowledge their concern for the environment (Prato and Hajkowicz, 2001).

CROPPING STRATEGIES

Proactive Agriculture

The agriculture industry must continue in its proactive role towards reducing the potential to degrade water quality. Attention to environmental impact remains prominent, and agriculture needs to be assertive in identifying and quantifying specific problems, taking action where needed because, "Denying that agriculture is having any impact will not be accepted. Agriculture must be proactive to avoid future regulation" (Hicks, 1992). It has been suggested that people "need to get over the shock of low levels of herbicides being detected. . . ," and focus their attention on identifying those worst case scenarios where herbicide concentrations may occasionally approach limits specified in water quality guidelines (Hill et al., 1995). But such opinion can vary over time, and might lean towards a more precautionary approach. Sharpley et al. (1994) have cautioned that a proactive approach to water quality must consider the inherent vulnerability of the land to agrochemical loss, as well as the sensitivity of impacted waters to degradation.

Landowners need to increasingly demonstrate that responsibility for land use and accountability for water quality go hand in hand with property rights (Wayland, 1990). North American society is spending billions of dollars annually to protect and restore the quality of rivers, lakes, and streams. Critics often acknowledge that the most sensible, cost-effective approach to water quality degradation may well be a reliance on the farm community to devise and implement pollution control. Yet, they have repeatedly warned that ever increasing regulation is imminent and that action is needed now (Offutt, 1990).

Best Management Principles

In many cases the agricultural community has decreased its potential impact on watercourses and groundwater systems by significantly reducing reliance on farm chemicals, and through advances in sustainable land management practices. Specific land management strategies to this end are often packaged as so-called 'best,' 'better,' or 'beneficial' management practices (BMPs). But results from such practices can be highly site-dependent, with effects varying widely between regions. Hence, in seeking to protect water quality, it is useful to center on some common sense concepts in three main areas of focus (Bernard et al., 2000):

- *Land management practices*–utilize cropping and tillage strategies such as crop rotations (including summer fallow), reduced tillage, cover crops, and shelterbelts–to reduce soil erosion, runoff, and leaching losses from agricultural lands.
- *Managing inputs*–control the amount, type, and timing of agricultural inputs (e.g., fertilizers, manures, and pesticides)–such that immediate applications and unused amounts at the end of the growing season are not at high risk of entering surface and ground waters through runoff, leaching or atmospheric deposition.
- *Buffer zones*–use vegetative strips (e.g., grassed or treed borders) and natural or artificial impoundments (e.g., ponds, wetlands) to intercept and retain contaminants that may have escaped from farmland–before they enter streams and lakes.

Land Management Practices

Soil and water degradation are closely connected, as eroding soil particles can cause increased turbidity, sedimentation, nutrient loading (organic and mineral forms), and pesticide concentrations in receiving waters (Gaynor and

McTavish, 1981; Cessna et al., 1994; Larney, Cessna, and Bullock, 1999). Hence, many practices first developed to help conserve soil through controlling wind and water erosion, have great potential towards reducing contamination of water from nonpoint agricultural sources (Bernard et al., 2000). These can reduce the movement of nutrients, pesticides, and bacteria that are in solution or attached to eroding soil particles. A careful balance between the requirements of soil erosion control and those for the protection of water quality may be required.

Crop Rotations

Crop rotations are often used for purposes of: soil conservation; moisture storage; nutrient recycling; weed, insect, and disease control. Periodically inserting a leguminous forage crop into a rotation, for example, can reduce runoff and erosion (Ghidey and Alberts, 1997). As well, crop rotations can reduce the overall need for nutrient and pesticide use in the first place, and thereby decrease the hazard that contaminants will leave farmland.

Summer fallow (leaving the soil bare) might be included as part of a cropping rotation in order to conserve soil moisture, or to address a weed or disease problem. Under fallow conditions, weeds are often controlled through frequent tillage passes, but this can break down organic matter (OM) and release N, degrade the soil, and threaten water quality. Findings on Prince Edward Island, for example, show that including a fallow period immediately after the breaking of forage lands can lead to substantial leaching losses of stored soil N in the form of nitrate (Milburn and Richards, 1991).

A Saskatchewan study found that nitrate and sediment loadings in snowmelt runoff from previously fallowed fields was higher than from previously cropped fields. Concentrations of non-sediment phosphorus were the same (Nicholaichuk and Read, 1978).

Reduced Tillage

The effects of reduced tillage on water quality are complex. Reducing the frequency and intensity of tillage can enhance soil residue cover and soil structure. Thus runoff, soil erosion and nutrient losses are often less extensive under reduced tillage and no-till than under conventional tillage. The higher OM content in no-till fields tends to filter the coarse soil particles from runoff, thereby enhancing the ratio of fine to coarse soil particles in runoff (Bernard et al., 1992). But Ontario experience has found that this combination can bring about increased concentrations of dissolved nutrients, particularly P, in surface runoff (Pesant, Dionne, and Genest, 1987). Such increased nutrient loadings might even be sufficient to offset the benefits of an otherwise reduced volume of runoff. Nevertheless, when it comes to snowmelt, no-till might actually increase runoff, because the standing crop residue in no-till can trap much more snow, hence more water is available for runoff than from tilled fields (McConkey, unpublished data).

Because no-till produces a moister soil, it is likely to result in increased soil OM and N, especially where soils are adequately fertilized (Campbell et al., 2001). This increased OM and nutrient content enhances microbial and enzyme activity, hence pesticides are more effectively degraded under no-till conditions. The higher OM content on the surface of soils can also increase the sorption of those pesticides having a strong affinity for organic materials. Still, research findings are mixed, depending on regional conditions (Locke and Bryson, 1997).

Work in Quebec has shown that reduced tillage can enhance infiltration and leaching—resulting in the hazard that nitrate and pesticides (or manure-derived bacteria) might enter tile drainage or groundwater (Karemangingo, 1998). Under irrigation, Elliot et al. (2000) found that no-tillage increased the leaching of the more water soluble herbicides compared with tilled soil. Under no-till, weeds cannot be controlled with tillage, thus increased herbicide use (e.g., chem fallow) or the use of differing herbicide strategies (each with inherent water quality hazards) may be required. Additional research is needed to clarify the effect of reduced tillage on water quality.

Cover Crops and Shelterbelts

Cover crops can substantially reduce the impacts of runoff, soil erosion, and nutrient loss, by providing an extra protective crop canopy over the soil. They include intercrops that are seeded between or underseeded within the major crop; green manures that are most often planted after the harvest of the main crop; and permanent cover where sensitive lands are permanently seeded (or remain seeded for many years) to grass and forage stands.

Cover crops can use up the excess nutrients that remain in soils or retard their movement, and so restrict their loss by runoff or leaching (Milburn, MacLeod, and Sanderson, 1997). When cover crops are eventually plowed under or intentionally killed with herbicides (in late fall or early spring), they will release their nutrients for possible use by subsequent crops. However, the timing of such events can be critical to the nitrate content of tile drainage water (MacLeod, 2000). Results from British Columbia indicate that specific winter cover crops can be used to provide good short-term N supplies to spring seeded crops, and thereby reduce the potential for N leaching losses (Odhiambo and Bomke, 2000).

Where summer fallow is a common practice, such as on the semi-arid prairies, strip-cropping uses alternating, narrow fields of cropped and fallow land to reduce the initiation of wind erosion and trap subsequent wind-blow particles. A similar action occurs where crop strips are planted on the contour to control water erosion. Permanent cover practices provide for both of these functions on a field-scale basis. In the end, the risk is reduced that soil particles and associated OM and nutrient loads will enter surface waters.

Shelterbelts (planted rows of trees and shrubs) provide general shelter from the wind and can protect vulnerable soils from wind erosion. As such, they reduce the potential for wind-blown soils to enter surface waters. As well, during winter, shelterbelts can trap and store significant volumes of snow in compact drifts, and thereby slow down peak localized runoff in the spring (Tabler, 1985; Kort, 1992). There is some evidence that belts of trees might also filter airborne pesticides before they can deposit in surface waters (Porskamp, Michielsen, Huijmans, 1994).

Managing Inputs

Restricting the movement of agricultural inputs from off farmland (e.g., fertilizer nutrients, pesticides), and reducing the amount of inputs that might be available to do so, are two ways of reducing potential water pollution. This includes assuring that input amounts match those required for production, and that excessive residues are not left at the end of the cropping cycle. In short this means, applying the right amount, of the right product, in the right place, and at the right time (PFRA, 2000a).

Nutrients and Bacteria

A close match between the constraints imposed by natural environmental factors (e.g., climate, soil texture, drainage) and realistic yield objectives, will help reduce the hazard of nutrient losses to leaching and runoff. On the Prairies, for example, water is frequently the limiting factor to crop production—hence high-rates of fertilizer application may be under-utilized.

An effective nutrient management plan allows for balancing crop nutrient needs with possible nutrient sources, thereby hedging against losses. This requires a knowledge of how much nutrient (e.g., N, P, and K) is already present in the soil, and thereafter warrants (Bernard et al., 2000)

- a careful accounting of contributions from other nutrient sources
- adding nutrients in smaller, sequential amounts where practical
- preventing the buildup of N, P and other nutrients within the soil.

Soil testing provides a means of determining N, P, and K level in the soil, but 40% of farmers still do not regularly soil test (Statistics Canada, 1996)–at least in part because of periodic discrepancies between soil test recommendations and apparent yield response. Emerging technologies such as precision agriculture are a promising means of trying to better compensate for these differences (Nolin, 2002), and can be used in conjunction with an Integrated Pest Management (IPM) plan, as discussed later in the text.

Work in Quebec has shown that interactions between cropping practices and variations within cropping practices (i.e., crops planted, tillage used, nutrient sources), can have a great impact on changes within the P fraction of a fine-textured soil (Zheng et al., 2001). In another study, strong interactions between soil carbon and N cycles tended to stimulate the mineralization of barley (*Hordeum vulgare* L.) straw when slurry N from liquid hog manure was added to soils, whereas straw added to soils already treated with pig slurry had the opposite effect (Chantigny, Rochette, and Angers, 2001).

Livestock manure can help improve soil tilth and fertility, but the nutrients, OM, bacteria and salts within manure can contaminate surface and ground waters. The nutrient content of manure varies greatly with the animal species (e.g., hogs vs. cows), animal feed type (nutrient load), and system of manure storage (open vs. closed). Safe storage allows farmers to retain manure supplies until circumstances are suitable for application. Findings in British Columbia confirm that when manure is applied in winter and crops are unable to use the nutrients, leaching and runoff account for significant nutrient losses (van Vliet, Zebarth, and Derksen, 1999). In an increasing number of jurisdictions, application of manure to frozen ground is restricted. On the Canadian prairies, heavy, repeat applications of cattle manure should generally be based on the P and not the N content of the manure, otherwise there can be an unacceptably high buildup of P in the soil over time, as manure has a relatively low N:P ratio. Modelling work in Texas has shown that moving from N-based to P-based application of manures can significantly reduce the hazard of P loss, at moderate cost (Osei et al., 2000). It is essential to remember that the main source of agriculturally-derived bacteria in water sources comes from livestock manure. Bacterial losses in runoff can be large when the manure is not incorporated into the soil, and largest if rainfall (or snowmelt) occurs shortly after the exposed manure is applied (Grando, 1996).

Some pathogenic bacteria, such as the virulent enterohemorrhagic *E. coli* O157:H7, require less than 100 bacteria to cause serious illness. Nevertheless, most such faecal pathogens from animal waste (e.g., dogs, cattle, deer) die relatively quickly in the soil. Two or three months can be sufficient to reduce bacterial counts to negligible levels, with half lives in the order of two or three days to a week. Bacteria survive longest under moist conditions, within fine soil sediments, and at cool temperatures (Mibiru, Coyne, and Grove, 2000).

Pesticides

Pesticide application might best be incorporated as part of an IPM plan, wherein pest control is sought through means that are environmentally friendly, yet economically viable. This can significantly reduce overall pesticide use and hence the risk of water contamination. IPM requires the collection of accurate information on pest numbers; identification of critical (economic and action-triggering) thresholds; and taking appropriate control action where required.

IPM incorporates crop management practices in four main areas (PFRA, 2000b):

- *Physical controls* that include barriers, trap crops, grazing, and adjusting planting location/timing to destroy or evade pests
- *Cultural controls* are geared to reduce persistent pest problems and include crop rotations, cultivation, and seeding practices to decrease crop vulnerability
- *Biological controls* that use natural or introduced pest enemies to regulate or suppress pest organisms
- *Chemical controls* that include conventional pesticides and other chemicals

Tillage and method of application can have a significant impact on pesticide losses. Nebraska studies on furrow-irrigated corn (*Zea mays* L.) indicate that runoff losses of atrazine were less from no-till and ridge-till (24-17%) than from disk-tilled fields. Subsequent model simulations indicated that for a given tillage practice, pre-emergent incorporation and pre-emergent banding of atrazine were the most effective means of reducing long-term losses (Gorneau et al., 2000).

Pesticide characteristics such as solubility, rate of degradation, and volatility, all affect the potential that pesticides might be transported into water. A significant mechanism of movement is atmospheric transport and deposition (Cessna, 2002). Contamination can also directly occur from spills or improper spraying practices. Where pesticides are required, it is best to choose formulations that are target specific, have low persistence and toxicity, and are low in vapor pressure (less tendency for wind drift) and have a low leaching potential. Application guidelines (including rate and appropriate weather conditions) should be followed–applying neither too much, nor too little as this may lead to pest resistance. The application method should allow for a targeted application, rather than generalized in-field distribution (Bernard et al., 2000).

An additional step in effective chemical control is to regularly (between applications of different chemicals) calibrate application equipment. According to Statistics Canada (1996), 76% of producers applying pesticides in 1995 operated their own sprayers. Of these, most (68%) calibrated their equipment at the beginning of the season, but only 16% re-calibrated equipment between applications of different chemical.

Buffer Zones

Despite the best preventative techniques, some contaminants will escape from agricultural lands. Buffer zones (buffer strips, riparian areas, wetlands) are a final way of retarding flow and reducing contaminant movement into surface water. With the exception of wetlands, buffer zones generally consist of grasses, shrubs or trees. Some buffer zones act best at filtering out sediments when the flow of water into them is shallow and uniform (Norris, 1993), but lab simulations indicate that field slope can be far more important than flow rate *per se* (Jin and Römkens, 2001). Sediment particle size and vegetation density are also limiting factors. In addition, trees and shrubs along waterways help to anchor the soil and reduce erosion (Geyer et al., 2000).

Buffer strips are fairly narrow bands of permanent vegetation (natural or planted) between agricultural lands and waterways. Efficiencies of sediment and nutrient removal can vary widely, and depends, for example, upon the source of irrigation runoff water (Tate et al., 2000). Some studies indicate that buffer strip efficiency increases with the relative width of the buffer zone (Daniels and Gilliam, 1996). But other lab tests have found that buffer strips act more like a barrier than a filter, wherein the great majority of sediment is deposited in front of the buffer strip, as the water slows before entering it (Ghadiri, Rose, and Hogarth, 2001).

The ability of buffer strips to trap N and P depends upon the form of the nutrient, since soluble N and P are less efficiently trapped than particulate forms. Over time, however, buffer zones might actually accumulate P and become a source of soluble P themselves. Plant growth in riparian buffers can be an important means of using nitrates in subsurface flow, although riparian conditions may transform some nitrate into greenhouse gasses.

Case studies have shown that wetlands (as buffer zones) can be a very effective means of reducing N and P in agricultural drainage waters. Their efficiency, however, depends upon the size of wetland in relation to the drainage area and water retention time within the wetland (Woltemade, 2000). Although P can be trapped in ponds and wetlands, it might be subsequently released during high flow conditions in the spring or fall. Pesticides can also be retained to some extent by buffer strips (Mickelson and Baker, 1993) and wetlands. Tests in Mississippi have shown that vegetation in drainage ditches can effectively sorb pesticides from storm water runoff moving from agricultural lands (Moore et al., 2001).

The ability of some riparian areas to act as buffer zones has sometimes been degraded–often due to prolonged or intensive cattle access. Signs of riparian degradation include: abundance of weeds and non-native plant species; absence of tree saplings (due to over-grazing); large areas of bare ground and slumping banks; and high sediment loading in the stream (PFRA, 2000c). One way of allowing riparian lands to regenerate, is by only briefly but intensively grazing them on a seasonal basis. Wisconsin studies have shown that such methods are as effective as the planting of artificial buffer strips towards rehabilitating trout streams (Lyons et al., 2000). Where high risk or severely disturbed areas are involved, fencing may be required.

CONCLUSION

Assessing the potential impact of agriculture on water quality is a difficult issue. An improved understanding of current uncertainties and complex relationships requires the continuing integration of knowledge from many perspectives and disciplines. Specifically designed, multi-disciplinary studies are needed to elucidate the often-contradictory nature of the elements that govern water quality. This should lead to the further development and hopefully judicious application of improved protective measures. Meanwhile, a number of common sense concepts that are presently being used, might be increasingly applied to a wide range of cropping strategies—with the clear aim to maintain or improve water quality. Through the incorporation of findings from multi-disciplinary dialogue and research, with practical and effective field management technique, both private landowners and society at large have reason to anticipate that good water quality can go hand in hand with productive agriculture.

REFERENCES

Anderson, A.-M., D.O. Trew, R.D. Neilson, N.D. MacAlpine, and R. Borg. (1998). Impacts of agriculture on surface water quality in Alberta. Part II: Provincial stream survey. Lethbridge, AB: CAESA, Alberta Agriculture, Food, and Rural Development.

Angus Reid Group. (1993). The Reid Report, Vol. 8, No. 6 (June). p 16.

Ashford, N.A. and C.S. Miller. (1998). *Chemical Exposures: Low Levels and High Stakes.* 2nd ed. New York, NY: Van Nostrand Reinhold/Wiley.

Avery, Alexander A. (1999). Infantile methemoglobinemia: Reexamining the role of drinking water nitrates. Environ Health Perspect 107:583-586.

Bernard, C., C.F. Drury, G.L. Fairchild, L.J. Gregorich, M.J. Goss, D.B. Harker, P. Lafrance, B. McConkey, J.A. MacLeod, T.W. Van der Gulik, L.J.P. van Vliet, and A. Weersink. (2000). Protecting water quality. In *The Health of Our Water–Toward Sustainable Agriculture in Canada*, ed. D.R. Coote and L.J. Gregorich, Ottawa, ON: AAFC, pp. 91-104.

Bernard, C., M.R. Laverdière, and A.R. Pesant. (1992). Variabilité de la relation entre les pertes de césium et de sol par érosion hydrique. *Geoderma* 52: 265-277.

Bertel, R. (1998). Environmental Influences On the Health of Children. Symposium papers, First International Conference on Children's Health and Environment, Amsterdam, The Netherlands.

Black, P.E. (1995). The critical role of "unused" resources. American Water Resources Association. *Water Resources Bulletin* 31: 589-592.

Bonnis, G. (1997). Agriculture and water quality. In *Agri-Environmental Indicators: Stocktaking Report.* OECD, Paris: Joint Working Party of the Committee for Agriculture and the Environment. 27 p.

CAESA (Canada-Alberta Environmentally Sustainable Agriculture). (1998). *Agricultural Impacts on Water Quality in Alberta–An Initial Assessment.* Canada-Alberta Environmentally Sustainable Agriculture Agreement. Edmonton, AB: Alberta Agriculture.

Caldwell, M. (1996). Beyond the lab rat. *Discover Magazine.* May 1996. pp. 70-75.

Campbell, C.A., F. Selles, G.P. Lafond, and R.P. Zentner. (2001). Adopting zero tillage management: Impact on soil C and N under long-term crop rotations in a thin Black Chernozem. *Canadian Journal of Soil Science* 81:139-148.

Carson and Associates, M.A. (1990). Off-Farm Sediment Impacts in the Saskatchewan River Basin. For Environment Canada, Inland Waters Directorate, Saskatchewan District. 87 p.

CAST (Council for Agricultural Science and Technology). (1992). *Water Quality: Agriculture's Role.* Task Force Report No. 120. 103 p.

CCREM (Canadian Council of Resource and Environment Ministers). (1987). *Canadian Water Quality Guidelines.* Ottawa, ON: Environment Canada.

Cessna, A. (2002). Pesticides–Balancing the risks. Abstracts, Effects of Agricultural Activities on Water Quality–A CCME-sponsored workshop, Jan. 31-Feb.1, 2002. Quebec City, QC.

Cessna, A., J.A. Elliot, L.A. Kerr, K.B. Best, W. Nicholiachuk, and R. Grover. (1994). Transport of nutrients and post-emergence-applied herbicides during corrugation irrigation of wheat. *Journal of Environmental Quality* 23:1038-1045.

Chance, G.W. and E. Harmsen. (1998). Children are different: Environmental contaminants and children's health. *Canadian Journal of Public Health:* S9-S13.

Chang, C. and T. Entz. (1996). Nitrate leaching losses under repeated cattle feedlot manure applications in southern Alberta. *Journal of Environmental Quality* 25:145-153.

Chantigny, M.H., P. Rochette, and D.A. Angers. (2001). Short-term C and N dynamics in a soil amended with pig slurry and barley straw: a field experiment. *Canadian Journal of Soil Science* 81:131-137.

Comly, H.H. (1945). Cyanosis in infants caused by nitrates in well water. *Journal of the American Medical Association* 129:112-116.

Corkal, D. and H.G. Peterson. (1994). Prairie surface water quality initiative. Proceedings of the National Conference on Drinking Water, 6th, Victoria B.C., 16-18 Oct. 1994, pp. 1-13.

Cornblath, M. and A.F. Hartman. (1948). Methemoglobinemia in young infants. *Journal of Pediatrics* 33:421-425.

Cox, D.R. (1958). *Planning of Experiments.* New York, NY: John Wiley and Sons.

Daniels, R.B. and W. Gilliam. (1996). Sediment and chemical load reduction by grass and riparian filters. *Soil Science Society of America Journal* 60:246-251.

Dubgaard, A. (1991). The Danish Nitrate Policy in the 1980s. Statens Jordbrugsøkonomiske Institut, København. Rapport nr. 59. 45 p.

Elliot, J.A., A.J. Cessna, W. Nicholiachuk, and L.C. Tollefson. (2000). Leaching rates and preferential flow of selected herbicides through tilled and untilled soil. *Journal of Environmental Quality* 29:1650-1656.

Environics International. (2000). Canadian Opinion on Drinking Water in the Aftermath of Walkerton. Environics International *e*Flash Report. August 2000 [cited 20 February 2002]. Available from World Wide Web: (www.environicsinternational. com/).

Environment Canada. (1978). Hydrological Atlas of Canada. Cat No. EN37-26/1978, Ottawa, ON: Environment Canada.

Finkel, A.M. (1996). Who's exaggerating? *Discover Magazine*, May 1996. pp. 48-54.

Foster, W.G. (1998). Endocrine disruptors and development of the reproductive system in the fetus and children: Is there a cause for concern? *Canadian Journal of Public Health*: S37-S41, S52.

FPSDW (Federal-Provincial Subcommittee on Drinking Water). (1987). *Guidelines for Canadian Drinking Water Quality*. Minister of National Health and Welfare. No. H48-10/1987E. 20 p.

Gaynor, J.D. and D.C. McTavish. (1981). Movement of granular simazine by wind erosion. *Hort Science* 16:756-757.

Ghidey, F. and E.E. Alberts. (1997). Plant root effects on soil erodibility, splash detachment, soil strength, and aggregate stability. *Transactions of the American Society of Agricultural Engineers* 40:129-135.

Geyer, W.A., T. Neppl, K. Brooks, and J. Carlisle. (2000). Woody vegetation protects streambank stability during the 1993 flood in central Kansas. *Journal of Soil and Water Conservation* 55:483-486.

Ghadiri, H., C.W. Rose, and W.L. Hogarth. (2001). The influence of grass and porous barrier strips on runoff hydrology and sediment transport. *Transactions of the American Society of Agricultural Engineers* 44:259-268.

Goolsby, D.A., E.W. Thurman, M.L. Pomes, M.T. Meyer, and W.A. Battaglin. (1997). Herbicides and their metabolites in rainfall: Origin, transport, and deposition pattern across the midwestern and northeastern United States. *Environmental Science and Technology* 31:1325-1333.

Gomez, B. (1995). Assessing the impact of the 1985 farm bill on sediment-related nonpoint source pollution. *Journal of Soil and Water Conservation* 50:347-377.

Gorneau, W.S., T.G. Franti, B.L. Benham, and S.D. Comfort. (2000). Reducing long-term atrazine runoff from south-central Nebraska. *Transactions of the American Society of Agricultural Engineers* 44: 45-53.

Grando, S. 1996. Effets de Deux Modes d'Epandage de Lisier de Porc sur la Qualité de l'Eau de Ruissellement. Memoured de Fin d'Etudes, ENITA de Bordeau, France (in French).

Hanukoglu A. and P.N. Dannon. (1996). Endogenous methemoglobinemia associated with diarrheal disease in infancy. *Journal of Pediatric Gastroenterology and Nutrition* 23:1-7.

Harker, D.B., B.D. Hill, and H.H. McDuffie. (1998). The Risk Agriculture Poses to Water Quality–Factors affecting our interpretation of findings. International Conference on Children's Health and Environment, 1st, Abstracts and Posters. E-WE-FP1-1 (p. 490).

Harker, D. B., K. Bolton, L. Townley-Smith, and B. Bristol. (1997). A Prairie-Wide Perspective of Nonpoint Agricultural Effects on Water Quality. Regina, SK: Prairie Farm Rehabilitation Administration (PFRA), Agriculture and Agri-Food Canada (AAFC). A98-3/2-1997E. 85 p.

Harker, D.B. and B. McConkey. (2001). Environmental McCarthyism and the Precautionary Principle. In *Sustaining the Global Farm*, ed. D.E. Scott, R.H. Mohtar, and G.C. Steinhardt. Selected Papers from the 10th International Soil Conservation Organization Meeting, May 24-29, Purdue University and the USDA-ARS National Soil Erosion Research Laboratory, pp. 106-111.

Hawthorne, S.B., D.J. Miller, P.K.K. Louie, R.D. Butler, and G.G. Mayer. (1996). Vapor-phase and particulate-associated pesticides and PCB concentrations in eastern North Dakota air samples. *Journal of Environmental Quality* 25:594-600.

Hicks, R. (1992). Alberta's environmental legislation and agricultural impacts on surface and groundwater quality. Proceedings of Agricultural Impacts on Surface and Groundwater Quality. Lethbridge, AB. pp. 92-93.

Hill, B., C. Chang, J.J. Miller, J. Rodvang, and N. Harker. (1995). Herbicides do leach in shallow Alberta groundwater. Proceedings of Agricultural Impacts on Water Quality. CAESA. Red Deer, AB. pp. 172-176.

Hill, B.D., J.J. Miller, C. Chang, and S.J. Rodvang. (1996). Seasonal variation in herbicide levels detected in shallow Alberta groundwater. *Journal of Environmental Science and Health*, B 31(4):883-900.

Hrudey, S.E. and D. Krewski. (1995). Is there a safe level of exposure to a carcinogen? *Environmental Science and Technology* 29:370A-375A.

Huang, W.-Y. and M. LeBlanc. (1994). Market-based incentives for addressing nonpoint water quality problems: A residual nitrogen tax approach. *Review of Agricultural Economics* 16:427-440.

Jin, C.-X. and M.J.M Römkens. 2001. Experimental studies of factors in determining sediment trapping in vegetative filter strips. *Transactions of the American Society of Agricultural Engineers* 44: 277-278.

Karemangingo, C. (1998). Évaluation des Risques de Pollution des Eaux sous Différents Systèmes de Production du Maïs-grain. Thèse de Doctorat. Université Laval, Montreal, Que. (In French).

Kluger, J. (1996). Risky business. *Discover Magazine*, May 1996, pp. 44-47, 82-83.

Kookana., R.S. and B.W. Simpson. (2000). Pesticide fate in farming systems: research and monitoring. *Communications in Soil Science and Plant Analysis* 31:1640-1659.

Kort, J. (1992). The effect of shelterbelts on snow distribution. Proceedings of Soils and Crops Workshop, Saskatoon, SK: University of Saskatchewan.

Larney, F.J., A.J. Cessna, and M.S. Bullock. (1999). Herbicide transport on wind-eroded sediment. *Journal of Environmental Quality* 28:1412-1421.

Lindwall, C.W. (1992). Future direction and key issues in surface and groundwater quality. Proceedings of Agricultural Impacts on Surface and Groundwater Quality. Lethbridge, AB. pp. 88-91.

Linton, J. (1997). Beneath the Surface–The State of Water in Canada. Canadian Wildlife Federation. Ottawa, ON. ISBN: 1-55029-98-3. p. 72.

Locke, M.A. and C.T. Bryson. (1997). Herbicide-soil interactions in reduced tillage and plant residue management systems. *Weed Science* 45:307-325.

Lyons, J., B.M. Weigel, L.K. Paine, and D.J. Undersander. (2000). Influence of intensive rotational grazing on bank erosion, fish habitat quality, and fish communities in southwestern Wisconsin trout streams. *Journal of Soil and Water Conservation* 55:271-276.

MacAlpine, N. and Q-T. Nguyen. (1993). Agricultural Nonpoint Source Pollution of Water: A literature survey from an Alberta perspective. Edmonton, AB: Conservation and Development Branch, AAFRD. 95 p.

MacLeod, J.A. (2000). Effect of timing and method of killing red clover forage on nitrate content of subsurface drainage water in Prince Edward Island. In *The Health of Our Water–Toward Sustainable Agriculture in Canada*, ed. D.R. Coote and L.J. Gregorich, Ottawa, ON: AAFC. p. 96.

McDuffie, H.H., K.M. Semchuk, R. Kerrich, A.J. Cessna, D.G Irvine, A. Senthilselvin, D.L. Ledingham, V. Juorio, P. Hanke, L.M. Hagel, M.L. Masley, J.A. Dosman, and M. Crossley. (1998). The Prairie Ecosystem Study (PECOS): Drinking water quality. *Epidemiology* 9(4) suppl 40 p.

McDuffie, H.H., K.M. Semchuk, M. Crossley, A. Senthilselvan, A.M. Rosenberg, L. Hagel, A. Cessna, D. Irvine, and D.L. Ledingham. (2003). The Prairie Ecosystem Study (PECOS): From Community to Chemical Elements, the Essential Role of Questionnaires. In *Managing for Healthy Ecosystems*, ed. D.J. Rapport, W.L. Lasley, D.E. Rolston, N.O. Nielsen, C.O. Qualset and A.B. Damania, Boca Raton, FL: CRC/Lewis Press. pp. 1169-1182.

Mibiru, D.N., M.S. Coyne, and J.H. Grove. (2000). Mortality of *E. Coli* in two soils. *Journal of Environmental Quality* 29:1821-1825.

Mickelson, S.K. and J.L. Baker. (1993). Buffer strips for controlling herbicide runoff losses. American Society of Agricultural Engineers/Canadian Society of Agricultural Engineers Meeting, Spokane, WA. Paper No. 93-2804.

Milburn, P. and J.E. Richards. (1991). Annual nitrate leaching losses associated with agricultural land use in New Brunswick. Annual Meeting of Canadian Society of Agricultural Engineers, July 29-31, Fredericton, N.B. Paper No. 91-109.

Milburn, P., J.A. MacLeod, and S. Sanderson. (1997). Control of fall nitrate leaching from early harvested potatoes on Prince Edward Island. *Canadian Agricultural Engineering* 39:263-271.

Mitchell, P. and D. Trew. (1992). Agricultural runoff and lake water quality. Proceedings of Agricultural Impacts on Surface and Groundwater Quality, Lethbridge, AB. pp. 73-79.

Mongoven, J.O. (1998). In *Hileman, Bette*. 1998. Precautionary Principle. Chemical and Engineering News, Feb. 9, 1998. pp. 16-18.

Moore, M.T., E.R. Bennett, C.M. Cooper, S. Smith Jr., F.D. Shields Jr., C.D. Milam, and J.L. Farris. (2001). Transport and fate of atrazine and lambda-cyhalothrin in an agricultural drainage ditch in the Mississippi Delta, USA. *Agriculture, Ecosystems and Environment* 87:309-314.

Muir, D.C.G. and N.P. Grift. (1995). Fate of herbicides and organochlorine insecticides in lake waters. In *International Congress of Pesticide Chemistry Options 2000*, ed. N.N. Ragsdale, P.C. Kearney, and J. R. Plimmer, Washington, DC: American Chemical Society. pp. 141-156.

NAS (National Academy of Sciences). (1996). *Carcinogens and Anti-Carcinogens in the Human Diet–A Comparison of Naturally Occurring and Synthetic Substances.* Washington, DC: National Academy Press.

Nicholaichuk, W. and D.W.L. Read. (1978). Nutrient runoff from fertilized and unfertilized fields in western Canada. *Journal of Environmental Quality* 7:542-544.

Nolin, M.C. 2002. Impacts of precision agriculture on water quality. Abstracts, Effects of Agricultural Activities on Water Quality–A CCME-Sponsored Workshop, January 31-February 1, 2002. Quebec City, QC.

Norris, V. (1993). The use of buffer zones to protect water quality: a review. *Water Resources Management* 7:257-272.

Odhiambo, J.J.O. and A.A. Bomke. (2000). Short term nitrogen availability following overwinter cereal/grass and legume cover crop monocultures and mixtures in south coastal British Columbia. *Journal of Soil and Water Conservation* 55:347-354.

Offutt, S. (1990). Agriculture's role in protecting water quality. *Journal of Soil and Water Conservation* 45:94-96.

Olson, B.M., T. Martin, R. Wright, A.-M. Anderson, A.E. Howard, and A. Sevderus. (2001). Soil phosphorus limits for agricultural land in Alberta. Final report for the Canada Adaptation and Rural Development (CARD) Fund. Lethbridge, AB: Soil Phosphorus Limits Project. 52 p.

Osei, E., P.W. Gassman, R.D. Jones, S.J. Pratt, L.M. Hauck, L.J. Beran, W.D. Rosenthal, and J.R. Williams. (2000). Economic and environmental impacts of alternative practices on dairy farms in an agricultural watershed. *Journal of Soil and Water Conservation* 55:466-472.

Pesant, A.R., J.L. Dionne, and J. Genest. (1987). Soil and nutrient losses in surface runoff from conventional and no-till corn systems. *Canadian Journal of Soil Science* 67:835-843.

PFRA (Prairie Farm Rehabilitation Administration). (2000a). *Nutrient Management Planning.* Water Quality Matters Factsheet Series, ed., C. Hilliard and S. Reedyk, Regina, SK: AAFC.

PFRA (Prairie Farm Rehabilitation Administration) (2000b). *Pest Management and Water Quality.* Water Quality Matters Factsheet Series, ed., C. Hilliard and S. Reedyk, Regina, SK: AAFC.

PFRA (Prairie Farm Rehabilitation Administration) (2000c). *Riparian Area Management.* Water Quality Matters Factsheet Series, ed., C. Hilliard and S. Reedyk, Regina, SK: AAFC.

Porskamp, H.A.J., J.M.P.G. Michielsen, and Ir. J.R.M. Huijmans. (1994). The reduction of the drift of pesticides in fruit growing by wind-break. Dienst voor Landbouwkundig Onderzoek, Instituut voor Milieu-en Agritechniek, Rapport 94-29. Wageningen, 27 p.

Power, J.F. (1994). Understanding the nutrient cycling process. *Journal of Soil and Water Conservation* 49:16-23.

Power, J.F., R. Wiese, and D. Flowerday. (2001). Managing farming systems for nitrate control: a research review from management systems evaluation areas. *Journal of Environmental Quality* 30:1866-1880.

Prato, T. and S. Hajkowicz. (2001). Comparison of profit maximization and multiple criteria models for selecting farming systems. *Journal of Soil and Water Conservation* 56:52-55.

Raja, M. (1996). Water Contamination as a Risk Factor for Non-Hodgkin's Lymphoma. Thesis towards a B.Sc. Med. degree, Saskatoon, SK: College of Medicine, University of Saskatchewan.

Randall, G.W. and D.J. Mulla. (2001). Nitrate-nitrogen in surface waters as influenced by climatic conditions and agricultural practices. *Journal of Environmental Quality* 30:337-344.

Rawn, D.F.K., T.H.J. Halldorson, B.D. Lawson, and D.C.G. Muir. (1999a.) A multi-year study of four herbicides in air and precipitation from a small prairie watershed. *Journal of Environmental Quality* 28:898-906.

Rawn, D.F.K., T.H.J. Halldorson, W.N. Turner, R.N. Woychuk, J.-G. Zakrevsky, and D.C.G. Muir. (1999b). A multi-year study of four herbicides in surface water of a small prairie watershed. *Journal of Environmental Quality* 28:906-917.

Sadler, J.M. and J.W.B. Stewart. (1974). *Residual Fertilizer Phosphorus in Western Canadian Soils: A Review*. Saskatoon, SK: Saskatchewan Institute of Pedology. Publication No. R136.

Sandman, P.M. (1987). Communicating risk: Some basics. *Health and Environment Digest* 1:3-4.

Scherer, C.W. (1990). Communicating water quality risk. *Journal of Soil and Water Conservation* 45:198-200.

SEHN (Science and Environmental Health Network). (1998). In *Hileman, Bette*. 1998. Precautionary Principle. Chemical and Engineering News, Feb. 9, 1998. pp. 16-18.

Selles, F., B.G. McConkey, and C.A. Campbell. (1999). Distribution and forms of P under cultivator and zero tillage for continuous and fallow-wheat cropping systems in the semi-arid Canadian prairies. *Soil and Tillage Research* 51:47-59.

Sharpley, A.N., S.C. Chupta, R. Wedepohl, J.T. Sims, T.C. Daniel, and K.R. Reddy. (1994). Managing agricultural phosphorus in protection of surface waters: Issues and options. *Journal of Environmental Quality* 23:437-431.

Sosiak, A. (1997). The Pine Lake restoration project. Proceedings of Agricultural Impacts on Water Quality. 21-22 Feb. 1995, Red Deer, AB. pp. 132-137.

Statistics Canada. (1996). Farm Inputs Management Survey, 1995. Catalogue No. 21F0009XPE. Ottawa, ON.

Swader, F.N., L.D. Adams, and J.W. Meek. (1994). USDA's water quality program--Environmentally sound agriculture. Proceedings of the 2nd Environmentally Sound Agriculture. Orlando, FL. pp. 86-92.

Tabler, R.D. (1985). Ablation rated of snow fence drifts at 2300 metres elevation in Wyoming. Proceedings of Western Snow Conference, Boulder, CO.

Tanase, I., I. Iacob, and N. Beldescu. (1998). GIS for Exposure of Well-Water Nitrate. International Conference on Children's Health and Environment, 1st, Abstracts and Posters. E-WE-FP1-3 (p. 491).

Tate, K.W., G.A. Nader, D.J. Lewis, E.R. Atwill, and J.M. Connor. (2000). Evaluation of buffers to improve the quality of runoff from irrigated pastures. *Journal of Soil and Water Conservation* 55:473-477.

van Vliet, L.J.P., B.J. Zebarth, and G. Derksen. (1999). Risk of Surface Water Contamination from Manure Management on Corn Land in South Coastal British Columbia. Agassiz, BC: Pacific Agri-Food Research Centre, Research Branch AAFC.

Wayland, R. (1990). What progress in improving water quality? *Journal of Soil and Water Conservation* 48:261-266.

Woltemade, C.J. (2000). Ability of restored wetlands to reduce nitrogen and phosphorus concentrations in agricultural drainage water. *Journal of Soil and Water Conservation* 55:303-309.

Young, G.J., J.C.I. Dooge, and J.C. Rodda. (1994). Protection of water resources, water quality and aquatic ecosystems. In *Global Water Resource Issues*. Cambridge, UK: Cambridge University Press. ISBN 0 521 461537. pp. 73-86.

Zheng, Z., R.R. Simard, J. Lafond, and L.E. Parent. (2001). Changes in phosphorus fractions of a Humic Gleysol as influenced by cropping systems and nutrient sources. *Canadian Journal of Soil Science* 81:175-183.

The Role of Precision Agriculture in Cropping Systems

Bradley Koch
Rajiv Khosla

SUMMARY. Precision agriculture is a new and developing discipline that incorporates advanced technologies to enhance the efficiency of farm inputs in a profitable and environmentally sensible manner. Yield monitoring and variable rate application are the most widely used precision technologies. Versatile guidance systems utilizing the global positioning system (GPS) and management zone approaches are also being developed to further increase productivity by reducing error, cost, and time. These technologies provide tools to quantify and manage variability existing in fields across an array of cropping systems. A review of precision farming technologies that are currently being used in the United States and around the world is presented in this article. *[Article copies available for a fee from The Haworth Document Delivery Service: 1-800-HAWORTH. E-mail address: <docdelivery@haworthpress.com> Website: <http://www.HaworthPress.com> © 2003 by The Haworth Press, Inc. All rights reserved.]*

Bradley Koch is Graduate Research Assistant, and Rajiv Khosla is Assistant Professor of Precision Farming, Department of Soil and Crop Sciences, Colorado State University, Fort Collins, CO.

Address correspondence to: Rajiv Khosla, Assistant Professor of Precision Farming, Department of Soil and Crop Sciences, C6 Plant Sciences Building, Colorado State University, Fort Collins, CO 80523 (E-mail: rkhosla@lamar.colostate.edu).

[Haworth co-indexing entry note]: "The Role of Precision Agriculture in Cropping Systems." Koch, Bradley, and Rajiv Khosla. Co-published simultaneously in *Journal of Crop Production* (Food Products Press, an imprint of The Haworth Press, Inc.) Vol. 9, No. 1/2 (#17/18), 2003, pp. 361-381; and: *Cropping Systems: Trends and Advances* (ed: Anil Shrestha) Food Products Press, an imprint of The Haworth Press, Inc., 2003, pp. 361-381. Single or multiple copies of this article are available for a fee from The Haworth Document Delivery Service [1-800-HAWORTH, 9:00 a.m. - 5:00 p.m. (EST). E-mail address: docdelivery@haworthpress.com].

10.1300/J144v09n01_02

KEYWORDS. Global positioning system, geographic information systems, yield monitoring, variable rate application, parallel swath bar, management zones

INTRODUCTION

As we venture into a new century of advanced information technology, agriculture has taken on a whole new approach filled with an array of exciting, sophisticated and useful concepts. Centuries ago the horse and plow were the only implements in farming, today agricultural implements are controlled via computers and located via satellites. Known as precision agriculture, or site-specific farming, we are using spatial information technology (Thrikawala et al., 1999) such as the global positioning system (GPS), and geographic information systems (GIS) to make precise management decisions in different cropping systems throughout the world. A cropping system can be defined as a multi-crop rotation in a particular area undergoing crop production. The majority of current research in precision agriculture is focused on one crop of interest, but as conclusions are made about particular crops this new discipline of site-specific farming will undergo research throughout a multitude of cropping rotations.

Precision agriculture is defined as the art and science of utilizing advanced technologies for enhancing crop production while minimizing potential environmental pollution (Khosla, 2001). This technology recognizes the inherent spatial variability that is associated with most fields under crop production (Thrikawala et al., 1999). Once the in-field variability (both, spatial and temporal) is recognized, located, quantified, and recorded, it can then be managed by applying farm inputs in specific amounts and at specific locations (Khosla, 2001).

Application of farm inputs on specific locations is achieved by farming equipment that utilizes GPS receivers. The GPS receivers secure satellite signals from a trilateration of at least four (of the 24) US military satellites orbiting 20,200 kilometers above the earth. In addition, there is a differential GPS (DGPS) receiver, which is capable of receiving another signal (i.e., "differential signal") from an earth-based network of stations such as US Coast Guard beacons or transmitters. Differential transmitters are at known stationary locations, thus the DGPS receiver is able to correct systematic errors (i.e., satellite clock errors, orbital errors, ionospheric and atmospheric errors, multi-path errors, receiver errors, induced errors or selective availability) associated with GPS signals. This allows the precision farming equipment to be accurate within 1 m on the ground, making this technology very site-specific. In addition to standard DGPS, the introduction of real-time-kinematic (RTK) DGPS can accomplish dynamic accuracy within 20 cm and static accuracy within 1 cm of the actual location. However, RTK differential systems are costly and require the user to construct a private base station and radio link, making standard DGPS the most practical system for precision farming at this time (Strombaugh and Shearer, 2000).

Precision farming yields a threefold advantage. First, it provides the farmer useful information, that can influence their use of seed, fertilizer, chemicals, irrigation, and other farm inputs. Second, economics are optimized by enhanced efficiency of farm inputs. Finally, by varying the amount of farm inputs (fertilizers, pesticides, and irrigation) used for crop production, and applying those inputs exactly where they are needed, the environment is sustained (Strombaugh and Shearer, 2000; Fleming, Westfall, and Bausch, 2001; Fleming et al., 2000).

Precision farming technologies today are being studied and adopted for varied cropping systems. Besides the traditional crops, i.e., corn (*Zea mays* L.), soybean (*Glycine max* L.), wheat (*Triticum aestivum* L.), and barley (*Hordeum vulgare* L.), precision farming practices are now being implemented in potato (*Solanum tuberosum* L.), onion (*Allium cepa*), tomato (*Lycopersicon lycopersicum*), sugar beet (*Beta vulgaris* L.), forages, citrus, grape (*Vitis* spp.), and sugarcane (*Saccharum* spp.) (Heacox, 1998). Practices of yield monitoring, variable-rate fertilizer and chemical application, variable-rate seeding, and parallel swath navigation, are being studied and used throughout a variety of crop production systems. Furthermore, different ways of determining quantity and location of crop inputs are being proposed and studied. Site-specific grid sampling is now being compared to the production level management zone approach to identify the most profitable method of determining crop-input application.

Yield monitoring, variable-rate application, parallel swath navigation, and crop-input determinations are the most important aspects of precision agriculture. The objective of this article is to review these aspects and present the current role of precision farming in cropping systems research.

YIELD MONITORING

Yield monitoring and mapping are key elements of site-specific farming and they were the most widely used components of precision farming initially (Heacox, 1998). Yield monitoring offers the most intensive measure of spatial yield variability that exists in farm fields allowing producers to assess how management skills and environmental factors effect crop production (Stombaugh and Shearer, 2000). This assessment provides direct and valuable feedback to the farmer enabling them to make better management decisions (Pelletier and

Upadhyaya, 1999). Such feedback includes but is not limited to: instantaneous yield and moisture documentation, creation of yield and moisture maps, digitally flagged pest documentation and organization of data by year, farm, field, load, and crop. Yield monitoring over time creates a unique GIS database that assists farmers to easily identify yield variability within a field, to make better variable-rate decisions, and create a history of spatial field data (Doerge, 1999a). Yield monitoring has become a common practice in traditional grain crops and corn-soybean rotation systems. This technology is being researched and commercialized for other crops such as potato, onion, sugar beet, tomato, hay, citrus, grape, and sugarcane.

Impact plates, optical volumetric measurements, radiometric techniques, and continuous weighing methods are some of the common practices used for grain yield monitoring. However, for bulkier products such as potatoes, onions, and tomatoes, there are fewer options: (1) bulk weighing/drop bucket; (2) total weight; or (3) the use of belt weighing methods related to the ones invented by Campbell, Rawlins, and Han (1994) and Hofman et al. (1995) for potatoes, sugar beets, and tomatoes. Pelletier and Upadhyaya (1999) have developed and researched a continuous weigh-type monitor consisting of a three-idle weigh-bridge, an angle transducer, a belt speed sensor, and a DGPS receiver that successfully measures and maps tomato yield. This device, installed on a harvester, is used to measure yield variability and produce yield maps.

Over the past few years, installation of yield monitors on commercial potato harvesters are increasing in North America and Europe. The idea is to assess the spatial variability in potato yields and relate this variability to yield limiting factors in the soil and landscape (DeHaan et al., 1999). Reports have indicated that after successful calibration of potato yield monitor; measured yields have been within 5% accuracy (DeHaan et al., 1999; Rawlins et al., 1994; Godwin and Wheeler, 1997). DeHaan et al. (1999) reported that analysis of yield monitoring data has quantified differences and shown yield benefits associated with no-tillage or conservation tillage as compared to conventional tillage. This technology has also proven to be valuable in determining the relationship between potato yield and the level of soil degradation.

Impact-plates, volumetric, and radiation-based sensing devices have been commonly used and found to be suitable for combine harvesters (Stombaugh and Shearer, 2000). Unfortunately, this technology is not reliable in monitoring forage crops due to challenges of texture and moisture variability in forages. Forage crops falsify impact measurement, vary in density, and require an intense radiation source for monitoring (Wild and Auernhammer, 1999). Wild, Auernhammer, and Rottmeirer (1994) reported a round bale yield monitoring system that uses load cells on the tongue and strain gauges on the axles of the baler to measure the total weight of the baler and bale. This method of weighing bales in the round baler has proved to be an acceptable method of yield determination and errors less than 1% were established. Although these weights of round bales provide yield-monitoring data, yield maps are limited to the size of the area from where the bale was produced (Wild and Auernhammer, 1999).

Extensive yield monitoring research is being conducted in the area of fruit production. Schueller et al. (1999) developed a simple system to produce reliable, low-cost yield maps for hand-harvested citrus. Using a commercial GPS receiver, individual harvest container locations were recorded to measure yield variations within a citrus block. In oranges (*Citrus sinensis* L. Osbeck) and various other citrus fruits, management practices are usually carried out on blocks that range from 2 to 100 ha. Traditionally, citrus management practices have been implemented uniformly within each block. However, soil and topographical differences, differences in pest infestations (nematodes), and variations in tree ages have all contributed to spatial variations within a block. The method used by Schueller et al. (1999) was to map the location of each harvest container as it was collected. This was achieved by the use of the Crop Harvest Tracking System (CHTS) developed by GeoFocus.[1] The CHTS uses a GPS receiver to record the container location each time the collection button is pushed on the container collector. The method proved to be efficient, reliable, and relatively inexpensive in mapping yield variation within citrus blocks. Initial yield maps have shown spatial yield variability to have a possible correlation to tree canopy size (Whitney et al., 1999). A regression model was produced plotting yield per hectare versus percent canopy ground cover. The results displayed an association between yield and tree canopy size with an r-value of 0.45 and a p-value of 0.0001. This association can visually be observed by dividing the aerial photograph of the citrus field into yield zones (Whitney et al., 1999). As citrus harvest becomes mechanized, individual tree yields will be mapped, and this will further enhance management decisions in citrus production (Schueller et al., 1999).

Researchers at Washington State University have developed a yield monitoring system in conjunction with HarvestMaster to examine how mechanical pruning and thinning influence yield and quality of juice grapes. The Harvest-Master-500[1] yield monitoring system was mounted on a mechanized grape harvester consisting of a conditioning and control unit, belt speed sensors, load cell sensors, an inclinometer, a DGPS receiver, and a hand-held computer. After the analysis of grape yield maps, it was concluded that variations across the vineyard were due to type of pruning technique and field variability (Wample, Mills, and Davenport, 1999).

Currently, there are no yield monitors commercially available for sugarcane (Cox, Harris, and Cox, 1999), but sugarcane yield monitoring is undergoing

1. Mention of a trade name neither constitutes endorsement of the equipment or products mentioned nor criticism of similar ones not mentioned by the authors or Colorado State University.

extensive research (Whitney et al., 1999). The University of Southern Queensland and DAVCO Farming in Australia have been exploring techniques of mass sensing, volume measurement, and measurement by power consumption. After rigorous field trials, they have selected and patented a direct mass measurement technique. Using GPS and ArcView, yield maps were produced. These yield maps displayed a significant amount of yield variability from 70 to 190 Mg ha^{-1}. This technique is still being assessed and analyzed (Cox, Harris, and Cox 1999).

Several examples of the use of yield monitoring were provided above in a vast array of row, vegetable, fruit, and specialty crops. Table 1 presents a variety of crops and their corresponding sensor types that are being used for yield monitoring purposes in North America and elsewhere. New ideas using yield monitors are being proposed and researched throughout all cropping systems. Commercial advancement of yield monitor research will definitely be significant in the years to come.

Errors in Yield Monitoring

Yield monitoring sensors measure grain mass or volume of a crop per unit area, while a combine is harvesting through a crop field. Measurement of such yield parameters is subject to many sources of error. Lag time, overlapping of data points, moisture content error, and velocity changes are common sources of error associated with yield monitoring, the first two are the most significant errors (Pierce et al., 1997). Most grain flow sensors are mounted on a harvester in a position to sense clean grain flow. This causes a distinct difference in harvester's location from the point of crop harvest and the location where the

TABLE 1. Crop type, sensor type, sensor location on harvester, state of development, and geographical location of various types of yield monitors.

Crop	Sensor Type	Sensor Location	State of Development	Geographical Location
Corn	IP, R, OV	Clean grain auger	C	Worldwide
Soybeans	IP, R, OV	Clean grain auger	C	Worldwide
Potatoes	LC	Under conveyor	C	N. America and Europe
Onions	LC	Under conveyor	C	N. America
Tomatoes	LC	Under belt	Exp	CA
Forages (hay)	LC, SG	Tongue and axle	Exp	Germany
Oranges	LC	Under truck bed	Exp	Florida
Grapes	LC	Under conveyor	Exp	WA, CA
Sugarcane	DMS	Crop intake area	Exp (pat.)	Australia

IP: Impact plate, R: Radiometric techniques, OV: Optical volumetric measurement, LC: Load cell, SG: Strain gauge, DMS: Direct mass sensor, C: Commercial, Exp: Experimental, Pat: Patented.

grain flow is measured (National Research Council, 1997). Lag time is the difference in time from the point of harvest to grain arrival at the sensor. Lag time causes significant errors that, at this point, cannot be completely removed from yield data (Pierce et al., 1997).

Overlapping of data is another significant source of data error that can occur in many ways. Movement of a yield monitor equipped harvester over already harvested areas, wheel-slippage while harvesting crops on rolling topography, and incorrect harvest widths are some common causes associated with overlapping of yield data points. All of these occurrences can either affect the data by over or under estimating yield (Pierce et al., 1997). Although yield monitoring is one of the most widely practiced components of precision agriculture, yield values should be used as a reference as there is much needed improvement in yield monitor accuracy (National Research Council, 1997).

VARIABLE RATE TECHNOLOGY (VRT)

Technology to vary the rate of farm inputs such as fertilizers, pesticides, and seeds is available and is being used with various cropping systems in North America and other parts of the world (Peterson and Wollenhaupt, 1996). Variable rate drills and planters, fertilizer spreaders, and sprayers are commercially available for VRT. Variable rate precision irrigation systems are also being currently studied and developed (D. Heermann, personal communications). Farm machinery equipped with VRT controllers typically have a DGPS receiver to identify the precise location of spatial variability in the field and automatically control the rate of application based on pre-derived input application maps. Integrated control systems have been developed that work across farm equipment so that they can be shared between combines, tractors, and variable-rate equipment. This allows a farmer to obtain a single, cost efficient system that can be implemented in many field operations (Dampney and Moore, 1999). However, because of the cost associated with this technology, the majority of farmers still rely on custom application when using VRT. There are various applications of VRT technology in site-specific cropping systems management. Some of the widely adopted applications of VRT are discussed below.

Site-Specific Nutrient Management

The most widely used form of VRT is variable-rate fertilizer application (Cambouris, Walin, and Simard, 1999). This practice has been used for the past several years and is being researched in a variety of cropping systems. It is well documented that spatial variability in soil properties across landscapes affects crop yield (Ortega, Westfall, and Peterson, 1997). Uniform application of fertilizers, therefore, can result in under-fertilization of certain parts of a field and over-fertilization in other areas (Frasier, Whittlesey, and English, 1999; Khosla, Alley, and Griffith, 1999). Under-fertilization may result in a yield loss and over-fertilization can be harmful to the environment (Cambouris, Walin, and Simard, 1999; Hammond, 1993). With the invention of VRT, it has become possible to manage soil nutrient variations throughout a field with pre-scription fertilizer applications.

Although variable-rate fertilization in corn and soybean is not a new concept, there is still a significant amount of on-going VRT research in this type of cropping system. Iowa State University's Departments of Agronomy and Statistics have been comparing uniform and variable-rate phosphorous (P) application in on-farm strip trials. Using differential GPS, yield monitors, and grid soil sampling, four strip-trials were conducted at four different farmer's fields. The P treatments consisted of a non-fertilized control, a uniform P rate, and a variable rate P determined by soil P tests that were done prior to planting. The grain yields were taken and recorded every second with combines equipped with yield monitors and DGPS receivers. It was concluded that yield corresponding to P fertilization methods varied among the trials (Mallarino et al., 1999). Variable-rate P reduced the amount of total P fertilizer applied on three out of four farm fields. However, a grain yield increase was observed only in one trial. These results indicate that variable-rate P fertilization in corn and soybean is a more economically efficient and environmentally prudent way of fertilizer distribution (Mallarino et al., 1999).

With the economic and environmental benefits of variable-rate fertilization, the adaptation to other, more non-traditional cropping systems is also undergoing extensive research. Vegetable crops, such as potatoes, demand high amounts of fertilizer and possess a great economic value. Significant correlations between yield and soil physical properties such as slope, cation exchange capacity, and water holding capacity have been measured (DeHaan et al., 1999). It has also been demonstrated that the spatial variability of P and potassium (K) contents in the soil affects yield and tuber quality (Cambouris, Walin, and Simard, 1999; Kunkel et al., 1991).

A study at the Soils and Crops Research Centre of Quebec, Canada investigated the efficiency of variable-rate application of P and K in potatoes. The experiment consisted of three P and K treatments: conventional, variable rate, and a control. A uniform rate was applied for conventional treatment, variable rates of P and K were applied in the VRT treatment by krigging soil test values to produce an application map, and no fertilizer was applied in the control. The first year results indicated that the VRT treatment produced similar yields as the conventional method. However, since lower amounts of fertilizer were applied to the VRT treatment, it reduced the input cost and enhanced the input use efficiency. In addition, the VRT treatment increased tuber quality as compared to that of the conventional method. This shows that there is an agronomic and economic payoff if the cost of the variable-rate application does not exceed the fertilizer savings (Cambouris, Walin, and Simard, 1999).

Besides vegetable and conventional crops, variable-rate fertilization is being explored in other crops. A case study involving sugarcane was conducted in Australia, where yield mapping and DGPS soil sampling were used as data-layers to determine a variable rate of fertilizer to be applied (Cox, Harris, and Cox, 1999). Soil properties coupled with previous year's yield maps to generate variable rate gypsum application maps on high sodic soils were used. An economic analysis of the study showed a $563 ha^{-1} benefit over five years when comparing VRT with standard uniform input application. Due to sugarcane's value and high input costs, the National Center of Engineering in Agriculture, Australia has proposed that this crop is a good candidate for precision farming practices.

Site-Specific Weed Management

For decades farmers have uniformly broadcast or band applied herbicide to decrease yield loss due to weed competition, reduce weed seed contamination in harvested grain, and improve crop harvestability (Johnson, Cardina, and Mortensen, 1997). In a century of increased concern over environmental issues and the need for higher input efficiency, uniform application of chemical herbicides may be replaced with a site-specific form of herbicide application. Pressure to reduce food, soil, and water contamination and increased herbicide costs have prompted the need for precision technologies to target herbicide application more accurately. Thus, providing a higher degree of optimization in herbicide use (Stafford and Miller, 1996).

It is documented that a degree of spatial variation exists in weed distributions (Johnson, Cardina, and Mortensen, 1997; Cardina, Sparrow, and McCoy, 1996; Johnson, Mortensen, and Gotway, 1996; Mortensen, Johnson, and Young, 1993; Dessaint, Chadoeuf, and Barralis, 1991). The spatial variability across an agricultural field in terms of weed density and species is due to factors such as seed dispersal mechanisms, physical and chemical soil properties, and past management practices (Johnson, Cardina, and Mortensen, 1997; Stafford and Miller, 1996). Therefore, these spatially distributed weed populations provide the use of new technologies that detect, describe, and manage these weed populations (Johnson, Cardina, and Mortensen, 1997).

Over-application of herbicides can result in environmental contamination, increased herbicide cost, and injured crop. Likewise, under-application can cause poor weed control resulting in yield losses (Johnson, Cardina, and Mortensen, 1997; Wilson et al., 1993). Three approaches have been proposed to manage spatial aggregation of weed populations to prevent the under- or over-application of herbicides (Johnson, Cardina, and Mortensen, 1997). The first approach involves the collection of spatially referenced weed populations carried out prior to the herbicide application. Digital application maps are generated corresponding to the aggregated weed populations to serve as a basis for the variable-rate herbicide spraying. Map-based variable herbicide application has resulted in herbicide reductions ranging from 7 to 69% (Stafford and Miller, 1996). The second approach instantaneously senses the presence of weeds and applies herbicides accordingly. The quantity of reflected light of a particular wavelength is measured using real-time weed detection sensors mounted on spraying equipment. These sensors use "optical contrast indices" to distinguish targeted weeds from the crop (Johnson, Cardina, and Mortensen, 1997). The information generated by the sensors is passed to a control system that turns the sprayer on or off accordingly. Guyer et al. (1986) suggested that the use of sensor based spot spraying would substantially reduce herbicide application in the US corn and soybean crops. Shearer and Jones (1991) reported herbicide reductions of 15% after sensors were used to control and activate spray nozzles. Christensen, Heisel, and Walter (1996) evaluated and reported variable herbicide application in cereal crops as a response to weed distributions. The study showed a 47% decrease in herbicide cost. The third approach manages spatially aggregated weed populations by varying herbicide rate according to soil physical and chemical properties. Soil properties are known to influence herbicide-plant interactions (Johnson, Cardina, and Mortensen, 1997; Sonon and Schwab, 1995). Herbicide rates could be varied according to variable soil properties such as organic matter, soil structure, and pH. Implementation of this concept would require algorithms combined with soil property maps to produce herbicide rates and variable application maps (Johnson, Cardina, and Mortensen, 1997). Other methods are being studied to vary herbicide based on in-field management zones that are delineated using stable physical and chemical soil properties (P. Westra, personal communications).

Site-Specific Planting

Another site-specific application is seed placement and planting rate. Site-specific planting involves the proper placement and population of seeds to achieve maximum yield and quality. There are known correlations among spatial variations in yield, crop quality, soil attributes, seed spacing, and plant population. Monitoring these variations provide an opportunity to plant various crops on a spatially selective basis (Rupp and Thornton, 1992; Hess et al., 1999; Cambouris, Walin, and Simard, 1999; Fleming et al., 2000).

A team of agronomists and engineers from the University of Idaho and the Idaho National Engineering and Environmental Laboratory has been researching and monitoring the seed placement of potatoes for the past several years. The primary objective of this study was to measure real-time seed placement and develop an analysis system to assist the growers in potato seed placement based on in-field spatial variability. It is essential to have a good stand of properly distributed plants to achieve high yield and quality in potato production. Improper seed distribution and skips cause varied plant growth and increased competition between individual plants (Hess et al., 1999; Holland, 1991; Rupp and Thornton, 1992). By obtaining seed spacing accuracy of 75% or higher, farmers can expect a 5 to 10% yield increase and a 20% improvement in crop quality. These improvements were obtained with essentially no cost increases (Hess et al., 1999; Harris, 1997; Holland, 1991, 1994). The monitoring system used in this study effectively proved that there are significant spatial deviations from the primary placement target when planting a potato crop. Spatial variations in soil types and soil conditions affected the performance of the planter's drive wheel.

Universities, extension centers, and engineering firms are not the only agencies that are conducting variable-rate planting research. Progressive farmers from the Great Plains to the midwestern corn-belt are also conducting on-farm, production level research. Nebraska growers have been utilizing automatic population rate controllers to vary corn populations between irrigated circles and non-irrigated corners. This is achieved by collecting position and guidance data on center pivots by means of GPS receivers. The data is stored on a data card and is put into a yield monitor mounted in the planter tractor. Electro-hydraulic population controllers are used to switch between seeding rates (Stombaugh and Shearer, 2000). The position information stored on the data card is sent to the population controller and seeding rate is controlled automatically. Irrigated circles are planted at a rate of about 67,950 seeds ha^{-1}, as the planter crosses into non-irrigated corners the rate is reduced to 44,475 seeds ha^{-1} (Wilcox, 2000b, 2001a). Growers in Iowa and Kansas are also studying variable-rate planting and making population changes based on observations. These changes in plant populations are based on "yield zones" which are derived by changes in soil tests, soil type, and past knowledge of the field (Fleming, Westfall, and Bausch, 2001).

Parallel Swath Navigation

A parallel swath bar, or light bar, is a guidance system that utilizes location information from GPS. This technology allows the user to map the field perimeter and have the guidance system automatically lay a set of parallel swaths between the boundaries. User inputs such as spacing between parallel lines and number of lines are usually required (Tyler, Roberts, and Nielsen, 1997). After these calibrations are made, if the operator deviates from parallel tracks, the swath bar will visually and/or audibly respond so a correction can be made (Wilcox, 2000b, 2001b). This guidance system allows a trained operator to drive straight while spraying pesticides, applying fertilizers, working at night, and almost any other instance where there is a need to drive in a parallel track (Wilcox, 2000b). With proper training on this guidance system such technological advances greatly increase the efficiency of farming operations.

The swath bars are excellent for guiding spraying in cornstalks or drilled beans (*Phaseolus* sp.) where foam makers are hard to follow (Holmberg, 2001). Newer systems are now displaying cursors, representing vehicle position, and fill color on the on-board computer screen for the area of the field that has been traveled over so an operator can be confident that all parts of the field were covered (Tyler, Roberts, and Nielsen, 1997). A convenient benefit of a parallel swath bar is that it can be easily moved between various pieces of equipment (Wilcox, 2001b). Grain producers ranging from the Midwest to the Pacific Northwest are implementing this versatile navigation system in large fields under crop production to optimize planting, tillage, and harvest patterns (Tyler, Roberts, and Nielsen, 1997). The precision of these swath bars relative to foam markers, tillage swaths, harvest patterns, and other guidance systems is only as accurate as the GPS receiver used. Preliminary results from an ongoing study at Ohio State University indicated that 95% of the time the accuracy of all GPS guidance system used in the study on a straight line was better than 0.51 m (R. Ehsani, personal communications). Although not commercially required, GPS receivers having sub-meter accuracy are recommended for optimum accuracy and efficiency of guidance systems.

GRID SAMPLING AND MANAGEMENT ZONES

Before crop inputs can be varied on a spatially selective basis, determination of application rates and location must be assessed. Therefore, different approaches of determining how much, when, and where crop inputs are to be distributed in the field are being proposed and researched. Developing accurate variable-rate application maps is a key element to implementing precision farming technology (Fleming and Westfall, 2001). Grid soil sampling was the very first approach used by researchers to make prescription application maps (Figure 1a). Since the early 1990s application maps have been developed for the use of variable-rate fertilizer application based on results from sparsely spaced grid soil sampling (Doerge, 1999b). Due to the spacing between sampling points, estimates of the soil test values in between the sample points must be interpolated. Kriging and Inverse Distance Weighing (IDW) are the most suited methods for interpolation (Doerge, 1999b). Figure 1b shows an interpolated soil tests results map using IDW interpolation technique. Grid soil sampling when performed at a scale that produces a spatially dependent data yields a more precise spatial information compared to whole field composite sampling. In addition, it detects spatial features previously unaware and ignored about a field (Doerge, 1999b; Ferguson et al., 1999). However, the accuracy of interpolation technique depends on unique spatial properties of each data set and may produce different predicted surface maps depending on the interpolation technique used. Furthermore, the scale at which grid sampling is performed plays a significant role in capturing the spatial variability and can lead to incorrect interpretation of existing variability in the field. Several different scales of grid soil sampling and their comparison have been discussed in the literature (Anderson-Cook et al., 1999; Wollenhaupt and Wolkowski, 1994; Bullock et al., 1994). Regardless of the technique used, grid soil sampling on small grids (i.e., 0.4 ha or less) is time consuming and cost intensive (Gotway, Ferguson, and Hergert, 1996). Furthermore, grid soil sampling for nitrogen fertilizer recommendations needs to be conducted each year and for every field that will undergo variable N fertilizer application (Khosla, 2001).

The goal of crop input determination is to collect the information needed to make accurate application maps at the lowest possible cost. The profitability

FIGURE 1. a, b, c, d. An example of grid soil sampling, IDW interpolation of soil test results, bare soil imagery, and management zone delineation on an experimental field.

a. Non-aligned grid soil sampling.

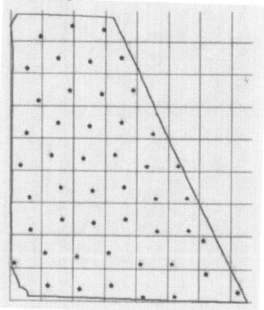

FIGURE 1 (continued)

b. Interpolated soil test results using IDW interpolation technique.

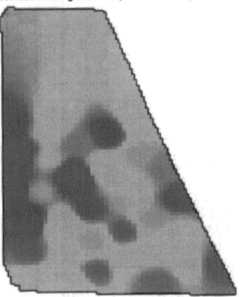

c. Bare soil imagery of an experimental field.

d. Management zones delineation based on bare soil imagery.

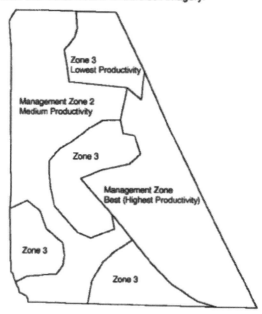

potential for site-specific fertilizer management is significantly enhanced if the initial means of forming application maps are inexpensive (Peterson and Wollenhaupt, 1996). There has been an increasing need for a method of managing crop input variability that is less time, labor, and cost intensive, as well as one that remains stable for several years. Ongoing research trials conducted by scientists from Colorado State University, USDA-ARS, and various other agencies in different parts of the country have established a more economically feasible system that divides farm fields into different regions referred to as production level management zones (Khosla, 2001; Fleming et al., 2000; Fleming, Westfall, and Bausch, 2001).

Production level management zones are defined as homogenous sub-regions of a field that have similar yield limiting factors (Doerge, 1999b; Khosla and Shaver, 2001). The current technique of delineating management zones in western US includes three GIS data layers, i.e., bare soil imagery, topography, and farmer's experience. Figure 1c presents an example of bare-soil imagery used for delineating management zones on an experimental field. This type of management zone delineation, however, is limited to conventional tillage.

Using this system of delineating management zones, a field can be divided into three different zones: high, medium, and low, based on the productivity potential of these areas. These management zones provide the grower with an opportunity to optimize their fertilizer applications. Areas of the field that have a high yield potential would receive a high fertilizer rate, medium productivity areas would receive a medium rate, and low productivity areas would receive the lowest rate. Figure 1d presents an example of delineated management zone on an experimental field. Varying the rate in this manner reduces the overall amount of fertilizer applied while maintaining or increasing the grain yields (Khosla and Shaver, 2001). Managing in-field spatial variability by management zones reduces the amount of crop input application, maximizes input efficiency, maintains or increases grain yields, reduces environmental impact, and enhances farm profitability (Khosla, 2001).

CONCLUSION

Precision farming technologies are being researched and implemented in a multitude of cropping systems. It is evident from the review of the literature presented herein that precision agriculture is successful in its role of enhancing crop production while minimizing environmental impact. Utilizing the GPS, precision agriculture recognizes and quantifies the inherent spatial variability in fields and manages this variability by applying inputs at specific amounts, when and where they are needed.

There are various elements of precision farming that play their own important role in cropping systems. Yield monitoring is being used to measure the yield variability of corn, soybean, potato, tomato, onion, sugar beet, hay, orange, grape, sugarcane and has become the most widely used component of precision farming. Besides yield variability, crop variety comparisons, yield damage reports, and field efficiency are being assessed with the use of yield monitoring systems.

Likewise, technologies to vary the rate of farm inputs such as seed, fertilizers, chemicals, and irrigation are being researched and are becoming available to a variety of cropping systems throughout the world. Varying the rate of fertilizers to correlate with the spatial variability of essential nutrients has become one of the most common practices of precision agriculture. Scientists and farmers alike have been monitoring placement and seeding rates. By conducting research in site-specific planting, input costs may be decreased and yields and quality may be increased.

Parallel swath navigation provides a guidance system that utilizes GPS to navigate equipment and reduce over application and skips. This new technology allows the day and/or night operation of applying fertilizers and chemicals without worry of costly skips or overlaps. It also provides better yield monitor accuracy in drilled soybean harvest.

Developing accurate application maps is a key element when implementing precision farming technologies. Several approaches are undergoing evaluation to determine which is the most efficient and profitable method of determining crop input application. Grid soil sampling is now being compared to the various management zone approaches. Although grid soil sampling has offered advantages in the past, it has proved to be time, labor, and cost intensive. Research in applying inputs to meet the production potential of individual management zones has indicated increase in input use efficiency, enhanced farm profitability, and reduced environmental impact.

Most importantly, throughout all types of cropping systems, the producers need to recognize, research, and implement these precision technologies and management practices at an on-farm production level. The future of farming needs profitability with minimal environmental impact. Precision agriculture technologies can help achieve these goals.

REFERENCES

Anderson-Cook, C., M.M. Alley, R. Noble, and R. Khosla. (1999). Accuracy of P and K fertilizer recommendations for two Mid-Atlantic Coastal Plain fields. *Soil Science Society of America Journal* 63:1740-1747.

Bullock, D.G., R.G. Hoeft, P. Dorman, T. Macy, and R. Olson. (1994). Nutrient management with intensive soil sampling and differential fertilizer spreading. *Better Crops with Plant Food* 78 (4):10-12.

Cambouris, A.N., M.C. Walin, and R.R. Simard. (1999). Precision management of fertilizer phosphorous and potassium for potato in Quebec, Canada. In *Proceedings of the 4th International Conference on Precision Agriculture*, St. Paul, MN, ed., P.C. Robert, R.H. Rust, and W.E. Larson, Madison, WI: American Society of Agronomy, Crop Science Society of America, and Soil Science Society of America, pp. 847-858.

Campbell, R.H., S.L. Rawlins, and S. Han. (1994). Monitoring methods for potato yield mapping. ASAE Paper 94-1584. St. Joseph, MI: American Society of Agricultural Engineers.

Cardina, J., D.H. Sparrow, and E.L. McCoy. (1996). Spatial relationships between seedbank and seedling populations of common lambsquarters (*Chenopodium album*) and annual grasses. *Weed Science* 44:298-308.

Christensen, S., T. Heisel, and A.M. Walter. (1996). Patch spraying in cereals. In *Proceedings of the 2nd International Weed Control Congressional*, Copenhagen, Denmark.

Cox, G., H. Harris, and D. Cox. (1999). Application of precision agriculture to sugar cane. In *Proceedings of the 4th International Conference on Precision Agriculture*, St. Paul, MN, ed., P.C. Robert, R.H. Rust, and W.E. Larson, Madison, WI: American Society of Agronomy, Crop Science Society of America, and Soil Science Society of America, pp. 753-766.

Dampney, P.M.R. and M. Moore. (1999). Precision agriculture in England: Current practice and research-based advice to farmers. *In Proceedings of the 4th International Conference on Precision Agriculture*, St. Paul, MN, ed., P.C. Robert, R.H. Rust, and W.E. Larson, Madison, WI: American Society of Agronomy, Crop Science Society of America, and Soil Science Society of America, pp. 661-674.

DeHaan, K.R., G.T. Vessey, D.A. Holmstrom, J.A. MacLeod, J.B. Sanderson, and M.R. Carter. (1999). Relating potato yield to the level of soil degradation using a bulk yield monitor and differential global positioning systems. *Computers and Electronics in Agriculture* 23:133-143.

Dessaint, F., R. Chadoeuf, and G. Barralis. (1991). Spatial pattern analysis of weed seeds in cultivated soil seedbank. *Journal of Applied Ecology* 28:721-730.

Doerge, T. (1999a). Defining management zones for precision farming. *Crop Insights* 8(21):1-5.

Doerge, T. (1999b). Yield monitors create on and off-farm profit opportunities. *Crop Insights* 9(14):1-3.

Ferguson, R.B., G.W. Hergert, J.S. Schepers, and C.A. Crawford. (1999). Site-specific nitrogen management of irrigated corn. In *Proceedings of the 4th International Conference on Precision Agriculture*, St. Paul, MN, ed., P.C. Robert, R.H. Rust, and W.E. Larson, Madison, WI: American Society of Agronomy, Crop Science Society of America, and Soil Science Society of America, pp. 733-744.

Fleming, K. and D. Westfall. (2001). Fertilizer application by management zone. *Colorado State University Agronomy Newsletter* 21(1):16-18.

Fleming, K.L., D.G. Westfall, D.W. Wiens, and M.C. Brodahl. (2000). Evaluating farmer defined management zone maps for variable rate fertilizer application. *Precision Agriculture* 2:201-215.

Fleming, K.L., D.G. Westfall, and W.C. Bausch. (2001). Evaluating management zone technology and grid soil sampling for variable rate nitrogen application. In *Proceedings of the 5th International Conference on Precision Agriculture*, Madison, WI: American Society of Agronomy, Crop Science Society of America, and Soil Science Society of America, CD Rom.

Frasier, W.M., N.K. Whittlesey, and M. English. (1999). Economic impacts of irrigation application uniformity in controlling nitrate leaching. Presented at the American Society of Civil Engineering, International Water Research Engineering Conference, Seattle, WA.

Godwin, R.J. and P.N. Wheeler. (1997). Yield mapping by mass accumulation rate. Presented at 1997 American Society of Agricultural Engineers Annual International Meeting, 10-14 August 1997, Minneapolis, MN, Paper No. 971061.

Gotway, C.A., R.B. Ferguson, and G.W. Hergert. (1996). The effects of mapping scale on variable-rate fertilizer recommendations for corn. In *Proceedings of the 3rd International Conference on Precision Agriculture*, St. Paul, MN, ed., P.C. Robert, R.H. Rust, and W.E. Larson, Madison, WI: American Society of Agronomy, Crop Science Society of America, and Soil Science Society of America, pp. 321-330.

Guyer, D.E., G.E. Miles, M.M. Schreiber, O.R. Mitchell, and V.C. Vanderbilt. (1986). Machine vision and image processing for plant identification. *Transactions of the American Society of Agricultural Engineers* 29:1500-1506.

Hammond, W.W. (1993). Cost analysis of variable fertility management of phosphorus and potassium for potato production in central Washington. In *Proceedings of the Soil Specific Crop Management: A Workshop on Research and Development Issues*, ed. P.C. Robert, R.H. Rust, and W.E. Larson, Madison, WI: American Society of Agronomy, Crop Science Society of America, and Soil Science Society of America, pp. 213-219.

Harris, J. (1997). Planting and seed-handling tips given. *Potato Grower of Idaho* March:8-9.

Heacox, L. (1998). Precision Primer. *American Vegetable Grower* 46(6):2-4.

Hess, J.R., J.M. Svobada, R.L. Hoskinson, D.W. Hempstead, and W.B. Jones. (1999). Spatial potato seed piece placement monitoring system. In *Proceedings of the 4th International Conference on Precision Agriculture*, St. Paul, MN, ed., P.C. Robert, R.H. Rust, and W.E. Larson, Madison, WI: American Society of Agronomy, Crop Science Society of America, and Soil Science Society of America, pp. 653-660.

Hofman, A.R., S. Panigrahi, B. Gregor, and J. Walker. (1995). In *Field Monitoring of Sugarbeets*. Paper 95-2114. St. Joseph, MI: American Society of Agricultural Engineers.

Holland, S. (1991). Planter performance: A key to high yield and quality. In *Spud Topics*, A Washington State Industry Report. Potato Country, USA, 34:24.

Holland, S. (1994). Potato planting for precise seed spacing. In *Spud Topics*, A Washington State Industry Report. Potato Country, USA, 36:21.

Holmberg, M. (2001) The guiding lights. *Successful Farming*, February 2001: bonus page.

Johnson, G.A., J. Cardina, and D.A. Mortensen. (1997). Site-specific weed management: Current and future directions. In *The State of Site-Specific Management for Agriculture*, ed., F.J. Pierce and E.J. Sadler, Madison, WI: American Society of Agronomy, Crop Science Society of America, and Soil Science Society of America, pp. 131-147.

Johnson, G.A., D.A. Mortensen, and C.A. Gotway. (1996). Spatial and temporal analysis of weed seedling populations using geostatistics. *Weed Science* 44:704-710.

Khosla, R. (2001). Zoning in on Precision Ag. *Colorado State University Agronomy Newsletter* 21(1):2-4.

Khosla, R., M.M. Alley, and W.K. Griffith. (1999). Soil-specific nitrogen management on Mid-Atlantic Coastal Plain soils. *Better Crops with Plant Food* 83(3):6-7.

Khosla, R. and T. Shaver. (2001). Zoning in on nitrogen needs. *Colorado State University Agronomy Newsletter* 21(1):24-26.

Kunkel, R., C.D. Moodie, T.S. Russel, and N. Hjolstad. (1991). Soil heterogeneity and potato fertilizer recommendations. *American Potato Journal* 48:153-173.

Mallarino, A.P., D.J. Wittry, D. Dousa, and P.N. Hinz. (1999). Variable-rate phosphorus fertilization: On-farm research methods and evaluation for corn and soybean. In *Proceedings of the 4th International Conference on Precision Agriculture*, St. Paul, MN, ed., P.C. Robert, R.H. Rust, and W.E. Larson, Madison, WI: American Society of Agronomy, Crop Science Society of America, and Soil Science Society of America, pp. 687-696.

Mortensen, D.A., G.A. Johnson, and L.J. Young. (1993). Weed distribution in agricultural fields. In *Soil Specific Crop Management*, ed., P.C. Robert, R.H. Rust, and

W.E. Larson, Madison, WI: American Society of Agronomy, Crop Science Society of America, and Soil Science Society of America, pp. 113-124.

National Research Council. (1997). *Precision Agriculture in the 21st Century: Geospatial and Information Technologies in Crop Management.* Washington, DC: National Academy Press, pp. 31.

Ortega, R.A., D.G. Westfall, and G.A. Peterson. (1997). Variability of phosphorus over landscapes and dryland winter wheat yields. *Better Crops with Plant Food* 81(2):24-27.

Pelletier, G. and S.K. Upadhyaya. (1999). Development of a tomato load/yield monitor. *Computers and Electronics in Agriculture* 23:103-117.

Peterson, T. and N. Wallenhaupt. (1996). Considerations in mapping soil properties or what about this grid sampling. *Crop Insights* 6(19):1-10.

Pierce, F.J., N.W. Anderson, T.S. Colvin, J.K. Schueller, D.S. Humburg, and N.B. McLaughlin. (1997). Yield mapping. In *The State of Site-Specific Management for Agriculture*, ed. F.J. Pierce and E.J. Sadler, Madison, WI: American Society of Agronomy, Crop Science Society of America, and Soil Science Society of America, pp. 211-243.

Rawlins, S.L., G.S. Campbell, R.H. Campbell, and J.R. Hess. (1994). Yield mapping of potatoes. Report of US Department of Energy, Idaho Field Office Contract DE-AC07-76IDO1570. Pullman, WA: Department of Crops and Soil Science, Washington State University.

Rupp, J.N. and R.E. Thornton. (1992). Seed placement for yield, quality. *Spudman* May: 14-32.

Schueller, J.K., J.D. Whitney, T.A. Wheaton, W.M. Miller, and A.E. Turner. (1999). Low-cost automatic yield mapping in hand-harvested citrus. *Computers and Electronics in Agriculture* 23:145-153.

Shearer, S.A. and P.T. Jones. (1991). Selective application of post-emergence herbicides using photoelectrics. *Transactions of the American Society of Agricultural Engineers* 34:1661-1666.

Sonon, L.S. and A.P. Schwab. (1995). Absorption characteristics of atrazine and alachlor in Kansas soils. *Weed Science* 43:461-466.

Stafford, J.V. and P.C.H. Miller. (1996). Spatially variable treatment of weed patches. In *Precision Agriculture*, Proceedings of the 3rd International Conference, Minneapolis, MN, ed., P.C. Robert, R.H. Rust, and W.E. Larson, Madison, WI: American Society of Agronomy, Crop Science Society of America, and Soil Science Society of America, pp. 465-474.

Stombaugh, T.S. and S. Shearer. (2000). Equipment technologies for precision agriculture. *Journal of Soil and Water Conservation* 55:6-11.

Thrikawala, S., A. Weersink, G. Kachanoski, and G. Fox. (1999). Economic feasibility of variable-rate technology for nitrogen on corn. *American Journal of Economics* 81:914-927.

Tyler, D.A., D.W. Roberts, and G.A. Nielsen. (1997). Location and guidance for site-specific management. In *The State of Site-Specific Management for Agriculture*, ed., F.J. Pierce and E.J. Sadler, Madison, WI: American Society of Agronomy, Crop Science Society of America, and Soil Science Society of America, pp. 161-181.

Wample, R.L., L. Mills, and J.R. Davenport. (1999). Use of precision farming practices in grape production. In *Proceedings of the 4th International Conference on Precision Agriculture*, St. Paul, MN, ed., P.C. Robert, R.H. Rust, and W.E. Larson, Madison, WI: American Society of Agronomy, Crop Science Society of America, and Soil Science Society of America, pp. 897-905.

Whitney, J.D., W.M. Miller, T.A. Wheaton, M. Salyoni, and J.K. Schueller. (1999). Precision farming applications in Florida citrus. *Applied Engineering in Agriculture* 15:399-403.

Wilcox, J. (2000a). What's the population? *Successful Farming* 98(3):44-46.

Wilcox, J. (2000b). Technology for Crops: Parallel tracking guides the way, day or night. *Successful Farming*, 98(3):Bonus page.

Wilcox, J. (2001a). Ready for a dual: Site-specific farming can be as simple as planting at two rates. *Corn Farmer*, 99(4A):20.

Wilcox, J. (2001b). A versatile high-tech toolbox. *Corn Farmer* 99(4A):32.

Wild, K. and H. Auernhammer. (1999). A weighing system for local yield monitoring of forage crops in round balers. *Computers and Electronics in Agriculture* 23:119-132.

Wild, K., H. Auernhammer, and J. Rottmeier. (1994). Automatic data acquisition on round balers. St. Joseph, MI: American Society of Agricultural Engineers Paper 94-1582.

Wilson, J.P., W.P. Inskeep, P.R. Rubright, D. Cooksay, J.S. Jacobson, and R.D. Snyder. (1993). Coupling geographic information systems and models for weed control and groundwater protection. *Weed Science* 7:255-264.

Wollenhaupt, N.C. and R.P. Wolkowski. (1994). Grid soil sampling. *Better Crops* 78(4):6-9.

Trends in Decision Support Systems for Cropping Systems Analysis: Examples from Nebraska

R. M. Caldwell

SUMMARY. Over the past 20 years, decision support systems (DSS) have been applied to a number of important problems in cropping systems analysis. Seven interrelated information technologies are currently driving new opportunities for decision support systems: global positioning systems (GPS), sensors for direct field measurements, variable rate application technology, remote sensing, data communication systems, geographic information systems (GIS), and personal computers. These technologies were used in three modeling examples designed to develop DSS at three different levels: competition between individual plants in an uneven stand (modeled at hourly time steps and sub-meter resolution), historical analysis of county-level yields (100 year sequence

R. M. Caldwell is Cropping Systems Specialist, 306 Keim Hall, Department of Agronomy and Horticulture, University of Nebraska–Lincoln, Lincoln, NE 68583 USA (E-mail: RCALDWELL1@UNL.edu).

Data for the analysis of plant-to-plant competition is from an experiment conducted with W. K. Russell, Department of Agronomy and Horticulture, University of Nebraska–Lincoln. Weather data for the yield map simulation was provided by mPower[3], Inc. Aaron Schepers provided the aerial photography and Angela Mittan assisted with map processing.

A grant from the Nebraska Soybean Board contributed to development of the software and data sets illustrated in this paper.

[Haworth co-indexing entry note]: "Trends in Decision Support Systems for Cropping Systems Analysis: Examples from Nebraska." Caldwell, R. M. Co-published simultaneously in *Journal of Crop Production* (Food Products Press, an imprint of The Haworth Press, Inc.) Vol. 9, No. 1/2 (#17/18), 2003, pp. 383-407; and: *Cropping Systems: Trends and Advances* (ed: Anil Shrestha) Food Products Press, an imprint of The Haworth Press, Inc., 2003, pp. 383-407. Single or multiple copies of this article are available for a fee from The Haworth Document Delivery Service [1-800-HAWORTH, 9:00 a.m. - 5:00 p.m. (EST). E-mail address: docdelivery@haworthpress.com].

at multi-kilometer resolution), and yield map modeling (daily time step for a 64 ha field). In each case, a sequential analysis of model predictions helped identify additional data or algorithms needed to model the system at its appropriate scale. Continued progress in system modeling combined with projected growth in computer power, near-term improvements in remote sensing and precision farming equipment, and new developments in the automation of data exchanges over the Internet should all contribute to expanded use of DSS for cropping systems analysis in the future. *[Article copies available for a fee from The Haworth Document Delivery Service: 1-800-HAWORTH. E-mail address: <docdelivery@ haworthpress.com> Website: <http://www.HaworthPress.com> © 2003 by The Haworth Press, Inc. All rights reserved.]*

KEYWORDS. Aerial photography, crop modeling, extensible markup language, Internet, variable rate application technology, yield mapping, *Zea mays* L.

INTRODUCTION

Our basic understanding of decision support predates the era of personal computers. Simon (1960), drawing on Dewey (1910), described three phases in decision-making: (1) identification of situations calling for a decision ("intelligence activity"), (2) creation and analysis of the array of possible decision options ("design activity"), and (3) selection of one or a few options for action ("choice activity"). While not dismissing the value of traditional techniques, Simon (1960) described problems observed with traditional techniques and presented how computers were transforming methods for both programmed (i.e., routine and repetitive) and nonprogrammed (i.e., novel and unstructured) decisions making. By the start of the 1980s the field of systems analysis had adopted simulation models as a primary means for evaluating production systems, and the definition of a "decision support system" became tied to computers: "Decision support systems are interactive computer-based systems that help decision makers utilize data and models to solve unstructured problems" (Sprague and Carlson, 1982).

In 1989 the International Benchmark Sites Network for Agrotechnology Transfer (IBSNAT) released version 2.1 of the Decision Support System for Agrotechnology Transfer (DSSAT) (IBSNAT, 1989; IBSNAT, 1993). DSSAT provides users with an integrated software package for: storing and retrieving soil profile parameters; entering observed weather data and generating simulated weather sequences for use in risk analysis; storing "genetic coefficients" for the key growth and development characteristics that differ by cultivar; entering the soil moisture and nutrient contents measured at the beginning of the experiment, and recording details on how an experiment was managed, such as tillage types, dates and depths, fertilizer types, dates and rates, planting date and depth, genotype, plant population, row spacing, and irrigation dates and rates. Once an experiment's minimum data set of soil, weather, genotype, and management information is recorded, the user can run the appropriate crop model to simulate the treatments in the experiment, view graphical displays of simulated data (e.g., transpiration, leaching, plant development and dry matter partitioning, and stress effects due to temperature, water deficiency or nitrogen (N) deficiency), and compare simulated results with experimental observations. Once run, the user can perform a sensitivity analysis by modifying one or more weather, soil, genotype, or management inputs and then re-running the simulation. Changes in the simulated results reflect the sensitivity of the system to the particular input(s). DSSAT also permits competing management strategies to be simulated using long-term weather sequences. The results of those simulations are then compared using stochastic strategy analysis–a method that helps identify the economic risks associated with the strategies (Dent, 1993). Through sensitivity analysis and stochastic strategy analysis, computer power is made available to the user for phase two of Simon's decision-making process: the creation and analysis of possible decision options (Simon, 1960).

DSSAT case studies illustrate the types of analyses possible with decision support systems (DSS). Published applications of DSSAT include: land evaluation (Alves and Nortcliff, 2000), yield loss assessment due to plant disease (Luo et al., 1997, 1998a, 1998b), evaluation of crop breeding strategies (Wortmann, 1998), analysis of spatial variability within fields (Engel et al., 1997; Thornton, Booltink, and Stoorvogel, 1997), analysis of optimum management strategies based on regional combinations of soil and weather (Lal et al., 1993; Simane, Wortmann, and Hoogenboom, 1998), assessment of development impacts on water resources (Luijten, Knapp, and Jones, 2001), and the analysis of crop yield and possible management strategies subject to global climate change conditions (Alexandrov, 1999; Amien et al., 1999; Cuculeanu, Marica, and Simota, 1999; Alexandrov and Hoogenboom, 2000; Yao et al., 2000; Guerena et al., 2001). Papers on the design and applications of DSSAT, including DSSAT version 3 (Tsuji, Uehara, and Balas, 1994), have been published in a series of books on systems approaches for agricultural development (Teng and Penning de Vries, 1992; Penning de Vries, Teng, and Metselaar, 1993; Goldsworthy and Penning de Vries, 1994; Teng et al., 1997; Kropff et al., 1997; Tsuji, Hoogenboom, and Thornton, 1998). Those books also provide reports and case studies from groups using other crop models (e.g., INTERCOM: Lindquist and Kropff, 1997; APSIM: McCown et al., 1994), groups concentrating on models for natural resource conservation (e.g., ITOPE: Boerboom,

Flitcroft, and Kanemasu, 1997), and groups using rule-based systems for decision support (e.g., AGFADOPT: Robotham, 1997; GOSSYM-COMAX: Boone, Kikusawa, and McKinion, 1997). The use of crop models in field research and technology transfer, including DSSAT, APSIM, GOSSYM, GLYCIM/GUICS, and RZWQM, was reviewed by Ma, Ahuja, and Howell (2002).

Early in the history of crop modeling, researchers were responsible for collecting practically all the experimental data needed to develop and test their models, including measurement of soil parameters, weather records, and crop growth and yield data. That is changing; accessibility of data for crop modeling is improving, as are the opportunities for use of DSS software in crop management. Seven interrelated technologies are driving these improvements: global positioning systems (GPS), sensors for direct measurements, variable rate application technology, remote sensing, data communication systems, geographic information systems (GIS), and personal computers.

A GPS combines a network of signal transmitters (satellite-based and/or ground-based, at known positions), signal receiver(s), microprocessor for processing the signals and calculating position, and a datalogger for recording positions (Hofmann-Wellenhof, Lichtenegger, and Collins, 2001). The system measures the time required for the GPS signal to travel from each transmitter to the receiver, estimates the distances traveled, and then solves for the position that best fits the distances.

Sensors that make direct field measurements (i.e., measurements made with contact) include: rainfall, solar radiation, and temperature sensors; probes used to measure apparent soil electrical conductivity (Hartsock et al., 2000; Jaynes, 1996); sensors used on planters to monitor seeding rate (Lang, 1999); and sensors used on harvesters to record yield based on the flow rate of the harvested commodity (Perez-Munoz and Colvin, 1996), the moisture content (Zoerb, Moore, and Burrow, 1993), and the ground speed of the harvester. Sensors installed on harvesters with a GPS unit enable farmers to create yield maps, which has been one of the key developments in the emergence of precision agriculture.

Variable rate application technology (Sawyer, 1994) allows farmers to adjust the application rate of an amendment while the applicator is moving, creating an opportunity for site-specific management. Commercial variable rate applicators are now available for essentially all types of crop inputs: fertilizers, lime, irrigation, pesticides, and seed.

Remote sensing is the indirect measurement of objects (i.e., without contact), usually from satellite, aircraft, or tractor platforms (for overviews see Schowengerdt, 1997; and Rees, 2001). Sources of remote sensing data for agriculture are expanding with the availability of the IKONOS satellite in 1999 (Franklin, Wulder, and Gerylo, 2001) and the planned availability of QuickBird 1 and OrbView-3 satellite data. These satellites have spatial resolutions from

0.37 to 1 m^2 for panchromatic data (i.e., gray scale) and 5.8 to 16 m^2 resolutions for multispectral data (i.e., color images with data for blue, green, red, and near-infrared spectral bands). Even with the availability of the high-resolution satellite images, film photography and digital imaging from aircraft will continue to be important for remote sensing, at least in the near-term (Baltsavias, 2000). New systems for tractor-based remote sensing are also becoming available. Raun et al. (2002) demonstrated the ability to perform site-specific applications of N fertilizer at a resolution of 1 m^2. Their system used an optical sensor system to measure an index of crop status. The index was then used to estimate the site-specific N need and the variable rate application system for delivering the fertilizer. Nitrogen use efficiency of wheat (*Triticum aestivum* L.) was improved 15% by the in-season N management system compared to a uniform preseason application of fertilizer. With hardware operating at 1000 cycles s^{-1}, the system has the spatiotemporal resolution that may enable sensing and fertilization of individual corn (*Zea mays* L.) plants (W. Raun and M. Stone, personal communications, 2001).

Data communication systems combine a physical infrastructure for transmitting data with protocols for interpreting the data. Important systems for agricultural data include manual transfers of data using solid-state memory cards and card readers; serial communications over cable and wireless connections, the Controller Area Network (CAN) data bus protocol (Bosch, 1991), the emerging International Organization for Standardization (ISO) 11783 protocol for use on farm equipment (Lang, 1999; Munack and Speckmann, 2001); and the Internet. Remotely sensed data and weather records are now routinely transmitted over the Internet for use in DSS (Bastiaanssen et al., 2001; Growth Stage Consulting, 2002; mPower[3], 2002; Righetti and Halbleib, 2000; Thomas and Gubler, 2000; Welch et al., 1999).

A GIS is an integrated set of: computer hardware for input, output and storage of spatial data; software for database management, data analysis and automated cartography (Chou, 1997); and trained personnel for organizing and guiding GIS applications (ESRI, 1994). GIS systems in agriculture enable the creation of prescription maps for variable rate applications through an analysis of data from remote sensing, GPS systems, yield mapping combines, soil sensors, and soil and plant analyses.

The current generation of personal computers (e.g., processor speeds of 1 gHertz or faster, with hard drive capacities of 40 gigabytes or more) provides farmers with the processing power needed to run a variety of DSS on their farms, including systems that employ GIS processing of the input data and output results. In the Great Plains, farmer adoption of computers is higher for less experienced producers, indicating that farmers may use information technologies to compensate for a lack of practical management experience (Ascough et al., 1999).

Even with the availability of powerful personal computers and the improvements in farmer access to data and models on the Internet, there is no evidence that simulation-based decision support has been adopted by a significant fraction of US grain producers. Those crop management issues that might be aided with a decision support system (i.e., selection of: tillage types, dates and depths; fertilizer types, dates and rates; planting date and depth; genotype; plant population; row spacing; and irrigation dates and rates) remain solved, in most cases, using traditional decision-making techniques. The possible reasons for this lack of adoption are numerous, and the barriers to adoption are likely to vary by geographical region. In the case of Nebraska, we have experience with Cooperative Extension programs organized to encourage the use of DSS for cropping systems analysis. Local farmers, Certified Crop Advisors, Extension Educators, researchers, managers and support staff from large agribusinesses, and independent crop consultants have participated in workshops designed to teach PCYield (Welch et al., 1999) and DSSAT (Tsuji, Uehara, and Balas, 1994). Based on observations of software use by the workshop participants and based on discussions with them on the practical use of the software in their work environments, two basic types of problems were identified as barriers to the adoption of existing DSS: lack of input data at the spatial and temporal resolutions required by the management problem, and lack of models for the dynamics that operate at those spatial and temporal resolutions. In this paper three case studies are used to illustrate those two types of problems: (1) an analysis of competition between individual corn plants, as a basis for evaluating high-precision in-season N fertilization strategies, (2) a historical analysis of dryland crop production at the western edge of the US Corn Belt, as a basis for evaluating changes in genetics, management, and the environment, and (3) yield map simulation, as a basis for evaluating strategies for site-specific management. Research methods for handling incomplete data sets and adding new features to crop models are also illustrated. Based on the examples, plus Kurzweil's projections for growth in computer power (Kurzweil, 1999), trends in DSS for cropping systems analysis and the changes needed in agronomic research and education programs over the coming decades to realize the potential benefits of DSS are also discussed.

MODELING COMPETITION AMONG INDIVIDUAL PLANTS

Corn yield can be reduced by lack of uniformity in plant spacing and time of emergence (Nafziger, Carter, and Graham, 1991). Traditionally there has been little a farmer could do about a poor stand other than replant. Given the cost of additional seed and the potential yield loss due to missing the optimum planting date, DSS software has been developed to assist farmers in making replant decisions (Heiniger et al., 1997). An alternative to replanting is to manage the crop in a way that would overcome some of yield loss expected from lack of uniformity and would save on fertilizer costs by reducing the rate to match localized plant needs. Raun et al. (2002) demonstrated the ability to fertilize wheat based on localized plant needs. The hardware in their sensor-plus-applicator system has a spatiotemporal resolution that may allow sensing and fertilizing individual corn plants (M. Stone and W. Raun, personal communications, 2001), providing corn producers with new options for managing problems in plant establishment.

In theory, the effect of variability in plant spacing and plant emergence will be a function of competition between individual plants, with the overall intensity of competition being a function of the supply of photosynthetically active radiation, soil water, and soil nutrients. Relative to plants grown in the normal spacing, a plant bordering a gap (i.e., spot with one or more missing plants) has more light, water, and nutrients available and therefore less of a chance for interplant competition. Farmers using an in-season fertilizer management strategy based on plant sensing (Raun et al. 2002) can save fertilizer by reducing fertilization of the gaps. Lack of uniformity in emergence is a more difficult problem. The optimum amount of in-season fertilizer to apply a plant depends on the plant's yield potential. A plant emerging much later than its neighbors may be subject to interplant competition so severe as to make the plant barren. In that case, no in-season fertilizer application is warranted; the position should be treated as a gap. A plant emerging slightly later than its neighbors poses a more difficult management problem. Currently there is no general rule for deciding when in-season plant fertilization is warranted. There may be circumstances in which a late-emerging plant would benefit from an in-season fertilizer application if the fertilizer would boost height growth and protect the plant from barrenness.

In order to develop a decision support system for in-season plant fertilization, a modeling project designed to predict individual plant performance subject to the influences of variability in emergence time and spacing, along with environment-specific effects of weather, soil and management was initiated. To handle competition for light among individual plants, the light interception model of Norman and Welles (1983) was added to JanuSys (Caldwell and Fernandez, 1998). The Norman and Welles (1983) model represents the foliage of each plant contained within an ellipsoid. The model was generalized using object-oriented programming in JanuSys so that the user can define any number of plants at any position relative to each other, planted on a surface defined by a gridded Digital Elevation Model (DEM). Partitioning of direct and diffuse light to sky positions is handled as in CropSys (Caldwell and Hansen, 1993). Simulations of competition for light can be performed at any spatial and temporal resolution, within the limits of the computer. Backscattering of light and the consideration of multiple wavebands, as in the full Norman and Welles (1983) model, are not currently included. Competition for water and N (nitrate and ammonium) follows the logic in CropSys (Caldwell and Hansen, 1993) generalized from two competitors to a user-defined number of competitors.

Experience modeling intercrop systems (Caldwell, Pachepsky, and Timlin, 1996) led us to suspect that the plant model used for corn in JanuSys (CERES-Maize, from DSSAT v2.1; IBSNAT, 1989; see Caldwell and Fernandez, 1998) would be unable to simulate some of the effects of interplant competition on plant development for late emerging plants. In order to test that hypothesis and evaluate priorities for model improvement, an experiment was conducted at the University of Nebraska–Lincoln East Campus. The study was machine planted 5 May 2000 using seeding rates above the target population densities. Stands were hand thinned to the target density on 2 June 2000. Workers were instructed to thin out the most unhealthy plants first, and then thin to create spacings as uniform as possible. Plant height (measured from the base of the plant to the tallest leaf tip when foliage is extended upward) and the width of the top leaf (i.e., most recent leaf with blade arched downward) were measured on 8 June 2000. Height and leaf width of nine border plants were also measured in the field and then harvested, measured for leaf area in the lab, and analyzed for the linear regression of total leaf area on the product of plant height and leaf width ($R^2 = 0.97$; n = 9). Leaf area of each plant on 8 June 2000 was estimated from the regression equation. Frequent observations of tasseling, silking, and pollen shed were made on each plant, normally every one or two days. At maturity, the ear of each plant was harvested, dried, and shelled to measure grain yield per plant. Seed was counted by machine to get seed set per plant and mass per seed. The positions of each plant within the plots were measured and recorded.

Table 1 contains the results from a systematic analysis of data from one field plot (Pioneer Brand hybrid 33J56; 5 plants m^{-2} target density; not irrigated), illustrating the predictive power of the model at four levels of calibration. In the initial model, Level 0, plants were assumed to emerge at the same time and follow the same growth in plant dimensions: ellipsoidal height and width. The position of each plant, as measured in the plot, was the only input to the model influencing predictions of plant-to-plant variability. While the basic function of the light interception model worked (i.e., predicted light interception was higher for plants whose neighbors were far away versus plants whose neighbors were close), the Level 0 model was a poor predictor of plant leaf area or seed yield. To remove the influence of errors in dimensional growth from the analysis, plant-specific calibrations were performed on the model's height growth parameter (i.e., height increment per phyllochron at leaf number 12), the phyllochron interval (i.e., thermal time per leaf), and a new emergence delay factor (i.e., unitless rate multiplier) used to account for the early-season

TABLE 1. Sequential modification and calibration of CERES-Maize inside JanuSys (Caldwell and Fernandez, 1998) based on the predictions of individual plants within a single field plot (n = 24). Results are the fraction of variance in the observed values explained by the model (R^2) for plant height and total leaf area 44 days after planting, and seed yield at maturity.

Level in modification/calibration sequence	Variable tested		
	Height	Leaf area	Seed yield
		R^2	
0. Initial model	0.0	0.12	0.08
1. Height, emergence and tasseling dates	0.99	0.73	0.16
2. Leaf area growth	0.99	0.83	0.44
3. Silk emergence delay, seed-set sensitivity	0.99	0.83	0.71

height differences. These calibrations improved the predictions of plant leaf area (Level 1, Table 1), but created only a small improvement in seed yield prediction. Adding a leaf area growth modifier (i.e., a unitless rate modifier, calculated as a function of photosynthate available relative the photosynthate required to fill the potential new leaf area) improved both leaf area predictions and seed yield predictions (Level 2, Table 1). Predictions of early-season light interception (Figure 1) were strongly affected by the combination of calibration factors for emergence delay and leaf area development in Level 2 of the analysis. At Level 3 of the analysis, seed yield predictions were improved by introducing a silk emergence delay factor (calibrated to cause predicted silking to match observed silking) along with a seed-set sensitivity factor (unitless modifier that reduces or increases the CERES estimate of kernel abortion; calibrated to match observed and predicted kernel number). Figure 2 shows the relationship between observed and predicted seed yields.

Given the poor performance at the lowest three levels of model calibration, Table 1, Level 3 capabilities will be required before JanuSys can be used in a decision support system for in-season corn plant fertilization. The model will need validated algorithms for effects of interplant competition on dimensional growth, leaf area growth, silking emergence delay, and seed-set sensitivity, as well as the effects of in-season N applications on those processes. Recent recommendations on improvements to CERES-Maize (Lizaso, Batchelor, and Adams, 2001; Lizaso et al., 2003) should help guide algorithm development, though models of individual plant performance in uneven stands may need approaches that are qualitatively different from those used to model the average plant in uniform stands.

FIGURE 1. (a) Leaf area index and (b) light transmitted to the soil surface above grid cells (85 mm × 85 mm) predicted within a single field plot of corn, 18 days after planting, using parameters at Level 2 (Table 1) in the sequence of model calibration.

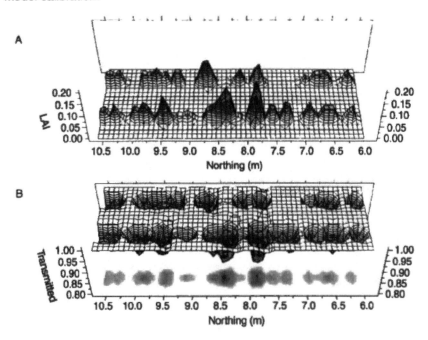

FIGURE 2. Observed versus predicted seed yield per plant at maturity for 24 plants measured within a single field plot of corn (see Figure 1), using parameters at Level 3 (Table 1) in the sequence of model calibration.

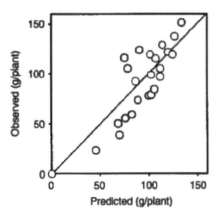

MODELING HISTORICAL TRENDS IN COUNTY-LEVEL YIELD

In 1963, grain sorghum (*Sorghum bicolor* L.) produced a higher average yield across Nebraska than corn (NASS, 1997). Sorghum dominated coarse grain production in parts of the region during the 1960s, but since that time farmers have increased dryland corn acreage relative to grain sorghum (NASS, 1997). There are a number of possible reasons for this shift in acreage. Farmers may have been planted corn on their best fields and sorghum on their worst, leading to an incorrect impression that corn is inherently higher yielding than sorghum. Genetic improvement of corn hybrids may have outpaced that of sorghum. Some farmers claim that their reduced tillage systems (now common in the region) have increased the water holding capacity of their soils, making corn production higher yielding and more dependable. Above average rainfall during the 1990s may have benefited farmers who grew corn more than farmers who grew sorghum. Given the history of drought in the region, at the fringe of the US Corn Belt, there is some concern that the shift away from sorghum may predispose dryland farmers to severe crop losses when drought reoccurs.

Crop models have been used to evaluate seasonal variation in crop yields as recorded in agricultural statistics (Hodges et al., 1987: four years for the US Corn Belt; Moen, Kaiser, and Riha, 1994: 30 years for a Crop Reporting District). In order to evaluate factors underlying the shift from dryland sorghum to corn and develop a decision support system for drought mitigation, we developed a simulation-based approach for analyzing historical data for grain production in Nebraska. Cass County was selected as a test location because historically there has been significant acreage of both corn and sorghum. Weather records dating back to the 1800s were obtained through the High Plains Climate Center for eight stations in eastern Nebraska: Albion, Ashland, Auburn, Crete, David City, Seward, Syracuse, and Weeping Water. Solar radiation was estimated from temperature based on a region-specific algorithm developed at the Center (Ken Hubbard, personal communication, 2000). During simulations, weather data for each grid position was interpolated from the eight stations using the spatial weather model in JanuSys (Caldwell and Fernandez, 1998). Soil profile descriptions for the simulations were developed by performing a grid conversion of the county's Soil Survey Geographic (SSURGO) database map (published 5 March 1999 by the U.S. Department of Agriculture, Natural Resources Conservation Service) using ArcView 3.1 (resolution: 0.02° latitude by 0.02° longitude, resulting in 396 grid positions inside the county). The map unit symbol obtained at each grid position was used to identify the corresponding profile description data in the Map Unit Interpretation Record. A standard, model-ready profile description was modified so that the water holding capacity, bulk density, pH, and organic carbon matched the Map Unit Interpretation Record, resulting in a model-ready soil profile map for the county.

A simulation of Cass County yields using parameters for a modern corn production system (7 plants m^{-2}; 12 May planting date; genetic coefficients (Jones and Kiniry, 1986) P1: 200; P2: 0.5; P5: 685; G2: 750; G3: 8.25; no N deficiency), applied without modification to the entire weather sequence, produced a poor fit to the recorded yields ($R^2 = 0.13$; statewide recorded yield was used 1897 to 1926; county records used 1927 to 1997). Model predictions were too high for years at the beginning of the period. The highest yield predicted using modern parameters occurred in 1915: 1152 g m^{-2}. The reasons for the poor fit are obvious: changes in management and genotypes over the last 100 years have increased yields. Those changes must be taken into account if we are to analyze the effect of drought on county-level yields and use DSS to mitigate future droughts.

To account for historical changes in management and genotypes, data input routines in JanuSys were modified to recalculate for each year in the 100-year simulation. Three parameters were modified as functions of time: population density (a key management variable known to have increased with time), a general stress factor for photosynthesis (to cover effects not included in the model, such as weeds, insects, diseases, and general soil fertility), and the seed-set sensitivity factor introduced during development of Level 3 of the plant competition study (to account for one possible component of genetic improvement; see Table 1). Table 2 provides the parameters for the three modifiers used in the case study. At each year of the simulation, the modifiers were calculated in JanuSys and the result multiplied by the modern values: 7 plants

TABLE 2. Relative change (unitless) in management and genotype parameters for the historical analysis of corn yields from Cass County, Nebraska, Figure 3. Parameters for years not shown were calculated by linear interpolation inside JanuSys (adapted from Caldwell and Fernandez, 1998).

	Parameter adjusted		
Year	Population density	Ps factor	Seed-set sensitivity
1850	0.20	0.20	1.3
1920	0.20	0.18	1.3
1940	0.20	0.18	1.3
1960	0.27	0.25	1.3
1980	0.80	0.90	1.0
2000	1.00	1.00	0.9

m^{-2} for population density, 1.0 for the general stress factor for photosynthesis (Ps factor), and 1.0 for the seed-set sensitivity factor.

Figure 3 illustrates results from the analysis. Using estimated historical trends in the three management and genotypic variables, model predictions accounted for 80% of the variation in the recorded yields. The model predicted yields above the statewide yields (i.e., prior to 1927) due in part to the higher rainfall expected in Cass County versus the rest of the state. Model predictions are likely to improve with the addition of multiple planting dates per year, following the approach of Moen, Kaiser, and Riha (1994).

MODELING YIELD MAPS

A grain flow monitor and a GPS are relatively low-cost additions to a new combine, therefore yield maps have become one of the most common sources of map data available to farmers for their fields. Use of the yield maps in decision-making varies from farmer to farmer. On-farm research, using production equipment to lay out plots and yield mapping to record treatment performance, is one approach being developed. For example, the Nebraska Soil Fertility Network (Caldwell and Peterson, 2002) and the Nebraska Soybean and Feed Grains Profitability Project (NSFGPP, 2002) are assisting farmers with the design and analysis of mapped experiments in Nebraska.

For a number of the experiments in the Nebraska Soil Fertility Network (Caldwell and Peterson, 2002) the possible treatment effects appear overwhelmed by factors that vary according to landscape position. Yields in some maps are highly correlated with drainage patterns. Cooperator feedback on a 64 ha experiment confirmed that flood damage in the lower sections of the field was the likely cause of fertilized plots yielding less than unfertilized plots. By random chance, intermittent ponding of water was more common in the fertilized plots than the controls.

In order to evaluate effects of landscape position on yield mapped research and to develop tools for use of yield maps in decision support, we developed a set of software tools for simulating yield maps with JanuSys. Figure 4 illustrates a case study in which N fertilizer was applied through a center pivot irrigation system as an experimental treatment on half of a 32 ha soybean field (experimental design: a radial randomized complete block with two treatments and four replicates) in 2000. Along with the yield map (Figure 4a) we collected color infrared aerial photography (Figure 4b), apparent soil electrical conductivity using a Veris™ probe (Figure 4c), and bare soil imagery from the public Digital Orthophoto Quarter Quadrangle (Figure 4d).

For spatial crop modeling, JanuSys requires a three-dimensional soil profile description. Topographic data for the soybean experiment was obtained from

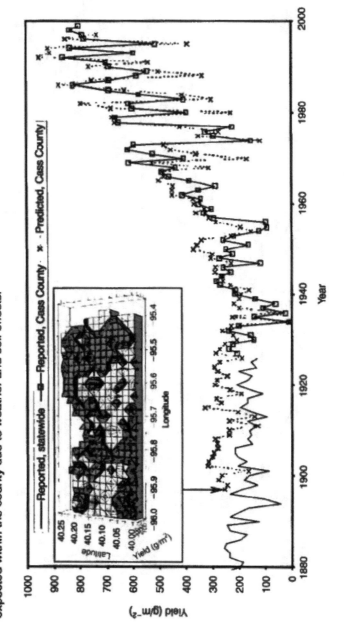

FIGURE 3. Analysis of historical trends in corn yields, Cass County, Nebraska. Predictions are the average across the year's simulated map (n = 396). Inset: yield map predicted for 1897, illustrating the spatial variability expected within the county due to weather and soil effects.

FIGURE 4. Yield map simulation in support of an on-farm trial of in-season N fertilization of soybean, using: (A) observed yield map, for testing model predictions, (B) a vegetation index from an aerial photograph (based on green and near infrared reflectances), for calibrating the site-specific general stress factor for photosynthesis, (C) apparent soil electrical conductivity maps, for calibrating soil profile depth and water holding capacity, (D) bare soil imagery, for calibrating topsoil organic carbon content, (E) soil classification map, for initial soil profile characterization data, (F) digital elevation model, for setting parameters in the overland flow and ponding model, and (G) map of fertilizer treatment applied through the center pivot irrigation system. See (G) for the spatial scale used in all the maps.

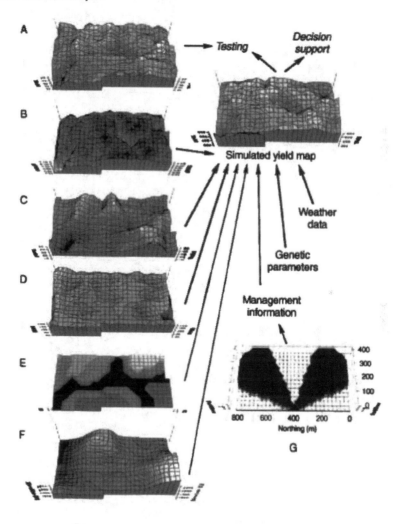

public DEM data, downloaded from the Internet. JanuSys reads the DEM and calculates from it parameters for the overland flow and ponding model. A soil profile description in either DSSAT v2.1 (IBSNAT, 1989) or DSSAT v3 (Tsuji, Uehara, and Balas, 1994) format is required for each region being simulated. In the absence of a certified digital version, the soil classification map for the field was redrawn from the published soil survey using ArcView version 3.2. The DEM and the soil classification map were gridded at a 30 m by 30 m spacing (Figure 4e and 4f). The soil classification map unit of each grid cell was associated with its profile description in the DSSAT v3 soil file (SOIL.SOL, Tsuji, Uehara, and Balas, 1994).

Crop simulation based on the soil survey map units did not produce a predicted map with spatial patterns matching the observed map (Figure 4a). Yield variation in the predicted map was dominated by the differences in the soil profile descriptions associated with the map units. A smaller amount of variation was created within the map units by the effects of overland flow and ponding. We concluded that the soil survey map was not drawn at a scale appropriate for the yield map simulation.

In order to reflect the continuous variation in soil properties across the field, the six original soil profile descriptions (one for each map unit) were replaced by a single composite profile description. Three types of map data were then used to modify that soil profile map for each grid cell in the simulation. The reflectance in the bare soil image, Figure 4d, was used to modify the topsoil (0 to 0.3 m) organic carbon content. The apparent soil electrical conductivity maps were used to modify the water holding capacity in each soil layer using data from either the shallow (Figure 4c) or deep probe for soil depths above 0.75 m or below 0.75 m, respectively. The vegetation index from the color infrared image (calculated as $(I - G)/(I + G)$, where I is the infrared reflectance and G is the green reflectance) was used to adjust the general stress factor for photosynthesis. In each case, the soil parameter being modified was assumed to be a linear function of the map data.

After calibrating the soil profile parameters based on the map data, the resulting simulated yield map (Figure 4) matched the basic patterns found in the observed yield map (Figure 4a). Yields predicted at the border of the field were poorly correlated to the observations (due presumably to physical damage to plants and compacted soils where the tractors make their turns), so they were dropped from the comparison of observed versus predicted in Figure 5. JanuSys accounted for 56% of the variance in the yield observations summarized in the 30 m by 30 m grid (n = 338; Figure 5), and 99% of the variance in the quintile averages (n = 5). In low yielding spots the model predictions tended to be too high (note observed yields less than 220 g m^{-2}, Figure 5) and predictions for the high yielding spots tended to be too low (note observed yields greater than 300 g m^{-2}). Further calibration of the soil profile parame-

FIGURE 5. Observed versus predicted soybean yield from the yield map simulation study (Figure 4). Yield values from the border of the field were removed from this graph. The simulated yields accounted for 56% of the variation in the observed yields.

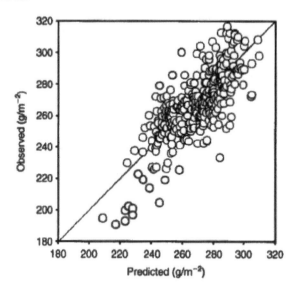

ters could eliminate that systematic error relative to the 2000 yield map. The farmer plans to provide his 2002 yield map, giving us the opportunity to test JanuSys under a new set of weather and management conditions. If JanuSys is able to account for the variation in the 2002 yield map without additional calibration, the farmer should gain confidence for using yield map simulations in decision making.

BUILDING A CAPACITY FOR DECISION SUPPORT

Over the last 100 years, computer power has grown at faster than an exponential rate (Figure 6; Kurzweil, 1999). This period spans a sequence of technologies from mechanical adding machines through vacuum-tube-based computers to the current age of processors built from silicon-based integrated circuits. Because new computing technologies are in development based on entirely new physical systems, Kurzweil (1999) argues that computers are likely to continue their faster-than-exponential increase in power over the next 100 years. If so, agricultural professionals entering the workforce now will be able to use, at some point in their career, personal computers operating at speeds

10^6 times faster than those today. Using models similar in complexity to those used today, the computational power will exist to simulate crop growth and development on every square meter of a farm.

Whether or not farmers use DSS resolved to every square meter on their farm depends on agronomic research and education programs that build a capacity for decision support. Remote sensing, from satellite-based, airplane-based, and tractor-based platforms, is already providing us with high resolution data on crop performance and soil condition. The most informative data source in the yield mapping case study, Figure 4, was the color infrared aerial photograph. Problem areas were clearly visible in the photograph (Figure 4c). Observed yields tended to be lower in the problem areas, though the data points from the yield map did not show as clear a boundary around the areas as the aerial photograph. Personal observations of the boundary around the spots revealed that within a span of as little as 3 m, topsoil texture changed from almost pure sand to the silt loam typical of the rest of the field, and visual indicators of soybean health changed from severely stressed and stunted to apparently stress-free. The grid cell size of 30 m by 30 m used for yield map modeling may be too coarse for effective modeling. In the future, we will not be forced to aggregate field data to 30 m grid cells just because of limitations in computer power. Decision support systems will be able to operate on spatial databases and models resolved to the nearest meter.

How data are shared is a key element in building capacity for decision support. The weather data used for the yield map simulation in this paper (Figure 4) was obtain through a commercial Internet information service (mPower[3], 2002) and downloaded to our personal computer in model-ready format. Initial attempts to put models to work for decision support required the farmer to build their own system for recording, transmitting, and processing weather data into model-ready format. That task, in retrospect, was too difficult for most farmers and a major barrier to adoption of decision support technologies. In the future we are likely to see steady improvements in the accessibility of weather, soil, and genotype data, along with cost data needed for economic analyses. For example, eXtensible Markup Language (XML) standards for field operations data (MapShots, 2002) can help automate the storage and retrieval of the management data necessary for yield map simulation and interpretation. XML is also being used for automated Internet services. Those technologies hold the promise of improving the linkages between public and private sectors by simplifying the exchange and interpretation of data. The technologies also may increase the level of competition between those offering the same types of information services, allowing farmers to more easily compare decision support tools for their cropping systems.

Over the last 40 years, with all the dramatic changes in agriculture (Figure 3) and computer technology (Figure 6), the core issues of decision support have not changed beyond the vision Simon (1960) articulated

> Organizations will still be constructed in three layers; [1] an underlying system of physical production and distribution processes, [2] a layer of programmed (and probably largely automated) decision processes for governing the routine day-to-day operation of the physical system, and [3] a layer of nonprogrammed decision processes (carried out in a man-machine system) for monitoring the first-level processes, redesigning them, and changing parameter values. (p. 49)

Much of the discussion of decision support in agriculture, and precision agriculture generally, has concentrated on Simon's third layer: the "nonprogrammed decision processes" in which farmers participate as the key decision makers. Less discussion has been dedicated to Simon's second layer (Simon, 1960). New tractor-based systems are being developed for in-season sensing of crop nutrient needs, at very fine spatial resolutions, and application of fertilizer (Raun et al. 2002). To be successful, these systems will need to solve a number of problems at Simon's layer two (Simon, 1960). Manual manipulation of data sets is unfeasible for these systems due to the short time frames required for

FIGURE 6. Growth in computer processing power, estimated as computations per second (CPS) per $1000 of cost (data, shown as open symbols, from Kurzweil, 1999; dollars adjusted for inflation to 1998 value; solid line: regression prediction; dashed lines: 95% confidence interval for the prediction).

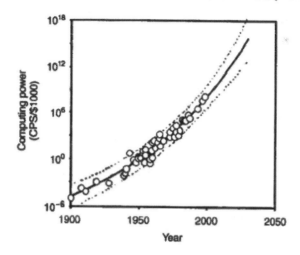

in-season management and due to the expense and availability of the labor capable of doing manual GIS and image processing operations. Much of the data processing and communication needs to be automated. Once these "programmed . . . decision processes" are created for the tractor-based systems, a whole new set of decision support needs (i.e., layer three) will be created. Farmers will need software, and consultants, to help them choose from among the different possible strategies for managing in-season fertilizer applications. GIS and image processing specialists are likely to be best employed as part of layer three activities, especially in creating improvements to the systems through testing and redesign. Failure to simplify and ultimately automate agriculture's layer-two processes may be one of the key reasons decision support systems have had such a low adoption rate in grain crop production.

CONCLUSION

While researchers have successfully applied DSS to a number of problems in cropping systems analysis, farmer use of the software appears to be low, at least for grain crop production in the US Corn Belt. Improvements in seven interrelated technologies should expand opportunities for the development and use of DSS: GPS, sensors for direct measurements, variable rate application technology, remote sensing, data communication systems, GIS, and personal computers. These technologies open up new decision support possibilities that work at finer spatiotemporal resolutions, integrate across more years and larger areas, and become more closely tied to the specific data sources found in precision agriculture. In examples operating at those three scales, a series of model calibrations revealed requirements for the crop models designed to operate at those scales. In the analysis of competition among corn plants, essentially all aspects of plant growth and development need to be modeled accurately if we are to predict individual plant yields subject to variability in plant spacing and time of emergence. In the analysis of historical corn yields for Cass County, Nebraska, historical trends in management and genotypes must be included in the analysis if we are to isolate the effect of drought on the yields. In the case of yield map modeling, the available soil survey data was not at the proper scale for describing soil variability in the simulation. A better approach was to calibrate soil profile characteristics using data sources associated with precision agriculture, such as apparent soil electrical conductivity maps and aerial photography. The limitations in the existing models and data sets, as illustrated in the examples, do not represent barriers to developing decision support systems but rather indicate the types of problems that should naturally arise as we analyze new types of systems at different scales. The cycle of model testing and improvement illustrated in the history of DSSAT (IBSNAT,

1989; Tsuji, Uehara, and Balas, 1994) and APSIM (McCown et al. 1996) should continue during the development of new systems models and applications of DSS. Long-term projections of growth in computer power, near-term projections of improvements in remote sensing and precision farming equipment, new developments in the automation of data exchanges over the Internet, and continued progress in crop modeling research all contribute to projecting an expanded use of DSS for cropping systems analysis.

REFERENCES

Alexandrov, V. (1999). Vulnerability and adaptation of agronomic systems in Bulgaria. *Climate Research* 12:161-173.

Alexandrov, V.A. and G. Hoogenboom. (2000). The impact of climate variability and change on crop yield in Bulgaria. *Agricultural and Forest Meteorology* 104:315-327.

Alves, H.M.R. and S. Nortcliff. (2000). Assessing potential production of maize using simulation models for land evaluation in Brazil. *Soil Use and Management* 16:49-55.

Amien, I., P. Redjekiningrum, B. Kartiwa, and W. Estiningtyas. (1999). Simulated rice yields as affected by interannual climate variability and possible climate change in Java. *Climate Research* 12:145-152.

Ascough, J.C. II, D.L. Hoag, W.M. Frasier, and G.S. McMaster. (1999). Computer use in agriculture: an analysis of Great Plains producers. *Computers and Electronics in Agriculture* 23:189-204.

Bastiaanssen, W.G.M., R.A.L. Brito, M.G. Bos, R.A. Souza, E.B. Cavalcanti, and M.M. Bakker. (2001). Low cost satellite data for monthly irrigation performance monitoring: benchmarks from Nilo Coelho, Brazil. *Irrigation-and-Drainage-Systems* 15:53-79.

Baltsavias, E.P. (2000). Is film scanning needed in the era of digital airborne and high-resolution spaceborne sensors? Future of photogrammetric scanners and film scanning. *Geomatics Info Magazine* 14:36-39.

Boerboom, L.G.J., I.D. Flitcroft, and E.T. Kanemasu. (1997). In *Applications of Systems Approaches at the Farm and Regional Levels, Volume 1*, ed. P. Teng, M.J. Kropff, H.F.M. ten Berge, J.B. Dent, F.P. Lansigan, and H.H. van Laar, Dordrecht, the Netherlands: Kluwer Academic Publishers, pp. 31-43.

Boone, M.Y.L., M. Kikusawa, and J.M. McKinion. (1997). Crop models and precision agriculture. In *Applications of Systems Approaches at the Field Level, Volume 2*, ed. M.J. Kropff, P. Teng, P.K. Aggarwal, J. Bouma, B.A.M. Bouman, J.W. Jones, and H.H. van Laar, Dordrecht, the Netherlands: Kluwer Academic Publishers, pp. 189-199.

Bosch, R. (1991). *CAN Specification Version 2.0.* Stuttgart, Germany: Robert Bosch GmbH.

Caldwell, R.M. and J.M. Peterson. (2002). Nebraska Soil Fertility Network [Online]. Available at http://southeast.unl.edu/caldwell/fertnet/overview.htm (verified 28 February 2002).

Caldwell, R.M. and A.A.J. Fernandez. (1998). A generic model of hierarchy for systems analysis and simulation. *Agricultural Systems* 57:197-225.

Caldwell, R.M. and J.W. Hansen. (1993). Simulation of multiple cropping systems with CropSys. In *Systems Approaches for Agricultural Development*, ed. F.W.T. Penning de Vries, P. Teng, and K. Metselaar. Dordrecht, The Netherlands: Kluwer Academic Publishers, pp. 397-412.

Caldwell, R.M., Y.A. Pachepsky, and D.J. Timlin. (1996). Current research status on growth modeling in intercropping. In *Dynamics of Roots and Nitrogen in Cropping Systems of the Semi-Arid Tropics*, ed. O. Ito, C. Johansen, J.J. Adu-Gyamfi, K. Katayama, J.V.D.K. Kumar Rao, and T.J. Rego, Kasuga, Tsukuba, Ibaraki, Japan: Japan International Research Center for Agricultural Sciences, pp. 617-635.

Chou, Y.H. (1997). *Exploring Spatial Analysis in Geographic Information Systems.* Santa Fe, NM: OnWord Press.

Cuculeanu, V., A. Marica, and C. Simota. (1999). Climate change impact on agricultural crops and adaptation options in Romania. *Climate Research* 12:153-160.

Dent, J.B. (1993). Potential for systems simulation in farming systems research. In *Systems Approaches for Agricultural Development*, ed. F.W.T. Penning de Vries, P. Teng, K. Metselaar. Dordrecht, The Netherlands: Kluwer Academic Publishers, pp. 325-339.

Dewey, J. (1910). *How We Think.* New York, NY: D.C. Heath & Company.

Engel, T., G. Hoogenboom, J.W. Jones, and P.W. Wilkens. (1997). AEGIS/WIN: a computer program for the application of crop simulation models across geographic areas. *Agronomy Journal* 89:929-928.

Environmental Systems Research Institute (ESRI). (1994). *Understanding GIS: The ARC/INFO Method, Version 7 for UNIX and OpenVMS.* Redlands, CA: ESRI.

Franklin S.E., M.A. Wulder, and G.R. Gerylo. (2001). Texture analysis of IKONOS panchromatic data for Douglas-fir forest age class separability in British Columbia. *International Journal of Remote Sensing* 22:2627-2632.

Goldsworthy, P. and F. Penning de Vries. (1994). *Opportunities, Use, and Transfer of Systems Research Methods in Agriculture to Developing Countries*, Dordrecht, The Netherlands: Kluwer Academic Publishers.

Growth Stage Consulting. (2002). Home page [Online]. Available at http://www.growthstage.com/ (verified 28 February 2002).

Guerena, A., M. Ruiz-Ramos, C.H. Diaz-Ambrona, J.R. Conde, and M.I. Minguez. (2001). Assessment of climate change and agriculture in Spain using climate models. *Agronomy Journal* 93:237-249.

Hartsock, N.J., T.G. Mueller, G.W. Thomas, R.I. Barnhisel, K.L. Wells, and S.A. Shearer. (2000). Soil electrical conductivity variability [CD-ROM computer file]. In *Proceedings of the Fifth International Conference on Precision Agriculture*, ed. P.C. Robert et al., Madison, WI: ASA, CSSA, and SSSA.

Heiniger, R.W., R.L. Vanderlip, J.R. Williams, and S.M. Welch. (1997). Developing guidelines for replanting grain sorghum. III. Using a plant growth model to determine replanting options. *Agronomy Journal* 89:93-100.

Hodges, T., D. Botner, C. Sakamoto, and J. Hays Haug. (1987). Using the CERES-Maize model to estimate production for the U.S. Cornbelt. *Agricultural and Forest Meteorology* 40:293-303.

Hofmann-Wellenhof, B., H. Lichtenegger, and J. Collins. (2001). *Global Positioning System: Theory and Practice.* Heidelberg, Germany: Springer-Verlag.

International Benchmark Sites Network for Agrotechnology Transfer (IBSNAT). (1989). *Decision Support System for Agrotechnology Transfer V2.10 (DSSAT V2.10)*. Honolulu, HI: Department of Agronomy and Soil Science, College of Tropical Agriculture and Human Resources, University of Hawaii.

International Benchmark Sites Network for Agrotechnology Transfer (IBSNAT). (1993). *The IBSNAT decade*. Honolulu, Hawaii: Department of Agronomy and Soil Science, College of Tropical Agriculture and Human Resources, University of Hawaii.

Jaynes, D.B. (1996). Improved soil mapping using electromagnetic induction surveys. In *Proceedings of the Third International Conference on Precision Agriculture*, ed. P.C. Robert, R.H. Rust, and W.E. Larson, Madison, WI: ASA, CSSA, and SSSA, pp. 169-179.

Jones, C.A. and J.R. Kiniry. (1986). *CERES-Maize: A Simulation Model of Corn Growth and Development*. College Station, TX: Texas A&M University Press.

Kropff, M.J., P. Teng, P.K. Aggarwal, J. Bouma, B.A.M. Bouman, J.W. Jones, and H.H. van Laar. (1997). *Applications of Systems Approaches at the Field Level, Volume 2*. Dordrecht, The Netherlands: Kluwer Academic Publishers.

Kurzweil, R. (1999). *The Age of Spiritual Machines*. New York, NY: Viking Press.

Lal, H., G. Hoogenboom, J.P. Calixte, J.W. Jones, and F.H. Beinroth. (1993). Using crop simulation models and GIS for regional productivity analysis. *Transactions of the American Society of Agricultural Engineers* 36:175-184.

Lang, F. (1999). Development of a CAN bus based air seeder monitor. In *Proceedings of the Fourth International Conference on Precision Agriculture*, ed. P.C. Robert, R.H. Rust, and W.E. Larson, Madison, WI: ASA, CSSA, and SSSA, pp. 991-995.

Lindquist, J.L. and M.J. Kropff. (1997). Improving rice tolerance to barnyardgrass through early crop vigour: simulations with INTERCOM. In *Applications of Systems Approaches at the Field Level, Volume 2*, ed. M.J. Kropff, P. Teng, P.K. Aggarwal, J. Bouma, B.A.M. Bouman, J.W. Jones, and H.H. van Laar, Dordrecht, The Netherlands: Kluwer Academic Publishers, pp. 53-62.

Lizaso, J.I., W.D. Batchelor, and S.S. Adams. (2001). Alternate approach to improve kernel number calculation in CERES-Maize. *Transactions of the ASAE* 44:1011-1018.

Lizaso, J.I., W.D. Batchelor, M.E. Westgate, and L. Echarte. (2003). Enhancing the ability of CERES-Maize to compute light capture. *Agricultural Systems* 16(1):293-311.

Luijten, J.C., E.B. Knapp, and J.W. Jones. (2001). A tool for assessing the implications of development on water security in hillside watersheds. *Agricultural Systems* 70:603-622.

Luo, Y., P.S. Teng, N.G. Fabellar, and D.O. TeBeest. (1997). A rice-leaf blast combined model for simulation of epidemics and yield loss. *Agricultural Systems* 53:27-39.

Luo, Y., P.S. Teng, N.G. Fabellar, and D.O. TeBeest. (1998a). The effects of global temperature change on rice leaf blast epidemics: a simulation study in three agroecological zones. *Agriculture Ecosystems and Environment* 68:187-196.

Luo, Y., P.S. Teng, N.G. Fabellar, and D.O. TeBeest. (1998b). Risk analysis of yield losses caused by rice leaf blast associated with temperature changes above and below for five Asian countries. *Agriculture Ecosystems and Environment* 68:197-205.

Ma, L., L.R. Ahuja, and T.A. Howell. (2002). *Agricultural System Models in Field Research and Technology Transfer*, Boca Raton, FL: CRC Press.

MapShots. (2002). Field operations data model [Online]. Available at http://www.mapshots.com/fodm/ (verified 28 February 2002).

McCown, R.L., P.G. Cox, B.A. Keating, G.L. Hammer, P.S. Carberry, M.E. Probert, and D.M. Freebairn. (1994). The development of strategies for improved agricultural systems and land-use management. In *Opportunities, Use, and Transfer of Systems Research Methods in Agriculture to Developing Countries*, ed. P. Goldsworthy and F. Penning de Vries, Dordrecht, The Netherlands: Kluwer Academic Publishers.

McCown, R.L., G.L. Hammer, J.N.G. Hargreaves, D.P. Holzworth, and D.M. Freebairn. (1996). APSIM: a novel software system for model development, model testing, and simulation in agricultural systems research. *Agricultural Systems* 50: 255-271.

Moen, T.N., H.M. Kaiser, and S.J. Riha. (1994). Regional yield estimation using a crop simulation model: concepts, methods and validation. *Agricultural Systems* 46:79-92.

mPower³. (2002). Crop growth and yield models [Online]. Available at http://www.mpower3.com/Public/Info/pc_exMain.asp (verified 28 February 2002).

Munack, A. and H. Speckmann. (2001). Communication technology is the backbone of precision agriculture [Online]. *Agricultural Engineering International: The CIGR Journal of Scientific Research and Development. Vol. III*. Available at http://agen.tamu.edu/cigr/Submissions/Munack%20Invited%20Paper.pdf (verified 28 February 2002).

Nafziger, E.D., P.R. Carter, and E.E. Graham. (1991). Response of corn to uneven emergence. *Crop Science* 31:811-815.

Nebraska Agricultural Statistics Service (NASS). (1997). *Nebraska Crop and Weather Summary, 1959-1996*. Lincoln, NE: Nebraska Agricultural Statistics Service.

Nebraska Soybean and Feed Grains Profitability Project (NSFGPP). (2002). Nebraska Soybean & Feed Grains Profitability Project [Online]. Available at http://on-farmresearch.unl.edu/ (verified 12 March 2002).

Norman, J.M. and J.M. Welles. (1983). Radiative transfer in an array of canopies. *Agronomy Journal* 75:481-488.

Penning de Vries, F., P. Teng, and K. Metselaar. (1993). *Systems Approaches for Agricultural Development*. Dordrecth, The Netherlands: Kluwer Academic Publishers.

Perez-Munoz, F. and T.S. Colvin. (1996). Continuous grain yield monitoring. *Transactions of the American Society of Agricultural Engineers* 39:775-783.

Raun, W.R., J.B. Solie, G.V. Johnson, M.L. Stone, R.W. Mullen, K.W. Freeman, W.E. Thomason, and E.V. Lukina. (2002). Improving nitrogen use efficiency in cereal grain production with optical sensing and variable rate application. *Agronomy Journal* 94:815-820.

Rees, W.G. (2001). *Physical Principles of Remote Sensing*. Cambridge, England: Cambridge University Press.

Righetti, T.L. and M.D. Halbleib. (2000). Pursuing precision horticulture with the internet and a spreadsheet. *HortTechnology* 10:458-467.

Robotham, M.P. (1997). Modelling socioeconomic influences on agroforestry adoption using a rule-based decision support system. In *Applications of Systems Approaches at the Farm and Regional Levels, Volume 1*, ed. P. Teng, M.J. Kropff,

H.F.M. ten Berge, J.B. Dent, F.P. Lansigan, and H.H. van Laar, Dordrecht, The Netherlands: Kluwer Academic Publishers, pp. 153-166.

Sawyer, J.E. (1994). Concepts of variable rate technology with considerations for fertilizer application. *Journal of Production Agriculture* 7:195-201.

Schowengerdt, R.A. (1997). *Remote Sensing: Models and Methods for Image Processing.* San Diego, CA: Academic Press.

Simon, H. (1960). *The New Science of Management Decision.* New York, NY: Harper & Row, Publishers.

Simane, B., C.W.S. Wortmann, and G. Hoogenboom. (1998). Haricot bean agroecology in Ethiopia: definition using agroclimatic and crop growth simulation models. *African Crop Science Journal* 6:9-18.

Sprague, R.H. Jr. and E.H. Carlson. (1982). *Building Effective Decision Support Systems.* Englewood Cliffs, NJ: Prentice Hall, Inc.

Teng, P. and F. Penning de Vries, ed. (1992). *Systems Approaches for Agricultural Development.* London, England: Elsevier Applied Science.

Teng, P., M.J. Kropff, H.F.M. ten Berge, J.B. Dent, F.P. Lansigan, and H.H. van Laar. (1997). *Applications of Systems Approaches at the Farm and Regional Levels, Volume 1.* Dordrecht, The Netherlands: Kluwer Academic Publishers.

Thomas, C.S. and W.D. Gubler (2000). A privatized crop warning system in the USA. *Bulletin of the European and Mediterranean Plant Protection Organization (OEPP)* 30:45-48.

Thornton, P.K., H.W.G. Booltink, and J.J. Stoorvogel. (1997). A computer program for geostatistical and spatial analysis of crop model outputs. *Agronomy Journal* 89:620-627.

Tsuji, G.Y., G. Hoogenboom, and P.K. Thornton. (1998). *Understanding Options for Agricultural Production.* Boston, MA: Kluwer Academic Publishers.

Tsuji, G.Y., G. Uehara, and S. Balas. (1994). *DSSAT v3.* Honolulu, HI: University of Hawaii.

Welch, S.M., J.W. Jones, G. Reeder, M.W. Brennan, and B.M. Jacobson. (1999). PCYield: model based support for soybean production. In *Proceedings of the International Symposium Modelling Cropping Systems,* ed. M. Donatelli, C. Stockle, F. Villalobos, and J.M. Villar Mir, University of Lleida, Lleida, Catalonia, Spain: European Society for Agronomy, pp. 273-274.

Wortmann, C.S. (1998). An adaptation breeding strategy for water deficit in bean developed with the application of the DSSAT3 drybean model. *African Crop Science Journal* 6:215-225.

Yao, M., H. Lur, C. Chu, J. Tsai, M.H. Yao, H.S. Lur, C. Chu, and J.C. Tsai. (2000). The applicability of DSSAT model to predict the production of rice and to evaluate the impact of climate change. *Journal of Agricultural Research of China* 49:16-28.

Zoerb, G.C., G.A. Moore, and R.P. Burrow. (1993). Continuous measurement of grain moisture content during harvest. *Transactions of the American Society of Agricultural Engineers* 36:5-9.

A Framework for Economic Analysis of Cropping Systems: Profitability, Risk Management, and Resource Allocation

Carl R. Dillon

SUMMARY. A cropping system must be economically viable for it to be sustainable. This study focuses upon the economic components of profitability, risk management potential, and optimal resource allocation. Some of the primary aspects of economic analysis of cropping systems to a broad multidisciplinary audience of researchers, extension specialists, graduate students, and senior undergraduate students are addressed. Analytical economic tools are discussed including data requirements, advantages, and disadvantages of each method. A case study of a Henderson, Kentucky producer is presented as a primary example along with empirical applications from the literature. The importance of various types of economic analysis is demonstrated through the selection of different production decisions dependent on the level of economic analysis included. *[Article copies available for a fee from The Haworth Document Delivery Service: 1-800-HAWORTH. E-mail address: <docdelivery@haworthpress. com> Website: <http://www.HaworthPress.com> © 2003 by The Haworth Press, Inc. All rights reserved.]*

Carl R. Dillon is Associate Professor, Department of Agricultural Economics, 403 C.E. Barnhart Building, University of Kentucky, Lexington, KY 40546-0276 USA (E-mail: cdillon@ca.uky.edu).

The author wishes to thank the many fine agricultural scientists he has worked with on past research studies, two anonymous reviewers for their comments, and Anil Shrestha for his valuable input.

[Haworth co-indexing entry note]: "A Framework for Economic Analysis of Cropping Systems: Profitability, Risk Management, and Resource Allocation." Dillon, Carl R. Co-published simultaneously in *Journal of Crop Production* (Food Products Press, an imprint of The Haworth Press, Inc.) Vol. 9, No. 1/2 (#17/18), 2003, pp. 409-432; and: *Cropping Systems: Trends and Advances* (ed: Anil Shrestha) Food Products Press, an imprint of The Haworth Press, Inc., 2003, pp. 409-432. Single or multiple copies of this article are available for a fee from The Haworth Document Delivery Service [1-800-HAWORTH, 9:00 a.m. - 5:00 p.m. (EST). E-mail address: docdelivery@haworthpress.com].

KEYWORDS. Economics, sustainable cropping systems, profitability, risk management

INTRODUCTION

A cropping system must be profitable for it to be sustainable. However, an economically sustainable cropping system has an acceptable level of risk and optimal resource allocation associated with it for the level of profit achieved. This study demonstrates the relevance of three economic issues of an economically sustainable cropping system: profitability, risk management potential, and optimal resource allocation. Examples for each of these issues are provided. While sustainability entails a broad spectrum of goals beyond economics, such as agronomic and environmental sustainability, these issues are addressed in other articles. Consequently, economic sustainability is the focus of this study.

Some publications attempt to introduce agricultural scientists to basic principles of the economic analysis of cropping systems (CIMMYT, 1988; Dillon and Hardaker, 1993; Roberts and Swinton, 1996). These publications do not totally address the concerns of Duffy, Guertal, and Muntifering (1997) who suggest a need to discuss "the uses and limits of economic analysis" with natural scientists. The purpose of this paper is to build upon the existing literature by providing a discussion of the strengths and weaknesses of several of the primary analytical tools available for assessing the economic sustainability of cropping systems. A case study is provided to aid in comprehension of selected economic analytical tools given an example can often be an effective method of understanding a concept. A comparison is provided in this case example illustrating differences between methods, highlighting the importance of each while partially explaining why different alternative cropping systems can be best depending on the circumstances.

The specific objectives of this article are threefold: (i) to identify the economically relevant characteristics of an optimal sustainable cropping system; (ii) to discuss previous research regarding these aspects in providing a framework of consideration; and (iii) to provide a case study of economic analyses to illustrate the relevant issues. The intent of this paper is thereby to establish an analytical framework for an economic assessment of the sustainability of a cropping system.

This paper provides a methodology for the economic assessment of cropping systems, includes economics in a multidisciplinary analysis, raises awareness of the potential contributions of economic evaluation beyond profitability, and communicates data requirements dictated by various levels of economic analysis. Assessment of economic potential becomes critical in the appropriate analysis of the degree to which a cropping system is sustainable. Specifically, it provides a common unit of dollars that permits comparison across crops and alternative production practices for a given crop with differing yields depending on the levels of inputs used. The appropriate evaluation of the complexity of cropping systems as well as new and alternative production practices, innovations, and technologies may inherently require a multidisciplinary analytical framework, including economic study. A properly conducted economic analysis is not simply a "checkbook accounting" procedure wherein one multiplies crop price and yield followed by subtracting a few simply derived cash expenses. There are a host of economic tools that need to be considered in order to expand the usefulness and applicability of innovative cropping system ideas. In communicating these concepts and establishing an analytical framework for economic assessment, it is hoped that this paper establishes the data required for economic analysis and explains the need to include economic considerations at the design level of the project.

A producer's decision is based upon the underlying economic consequences of the potential courses of action being considered. The economic consequences in turn are determined by the underlying production responses. Therefore, agricultural economic results drive a producer's decisions while the physical agricultural relationships (e.g., agronomic, engineering, pathological) provide the foundation for the economic results. In today's economic environment, successful producers must consider the multifaceted aspects of their decisions, relying upon appropriate analyses that adequately reflect these facets. In establishing the economic analytical framework, background information discussing the types of empirical models and previous literature will be presented leading to an economic framework. An agronomic framework is then provided in order to lay the foundation for economic evaluation.

BACKGROUND INFORMATION

Empirical modeling of the economic potential of cropping systems is a diverse area of published research. Duffy, Guertal, and Muntifering (1997) provide an excellent reference for interdisciplinary teams of researchers investigating agricultural systems. Some of the philosophical issues including the complications and advantages of the interaction between economists and natural scientists (difference in writing styles, unrealistic expectations, mutual ignorance, and poor timing) are presented in their paper. Economic methods for the evaluation of financial and environmental impacts and sustainability is the focus of a study by Roberts and Swinton (1996). Alternative performance criterion and analytical methods are discussed in this paper which also provides a worthwhile review of literature. In presenting these alternative performance criterion, the authors also stress the importance of incorporating several performance indicators and including an assessment of risk. This type of "inclusive systems approach" is also identified as important by Hansen and Jones (1996) who stress the need for incorporating the dynamic, stochastic elements of a farming system in evaluating its potential sustainability. The economic issues of profitability, risk management potential and optimal input allocation, proposed here as being critical to consider, represent major elements of the theoretical systems simulation framework offered in their study. Likewise, a series of performance indicators (net income, recycling, diversity, and capacity) are assessed in evaluating sustainability through consideration of economic and other factors by Lightfoot et al. (1993).

There are, nonetheless, a multitude of issues that arise even within the consideration of economic sustainability of cropping systems itself. These issues include the feasibility of the system in terms of economic requirements such as meeting cash flow obligations and the performance of the system in terms of overall profitability and level of risk associated with the cropping system. Numerous examples of empirical applications of a host of analytical techniques employed in economic research exist in literature.

These techniques have been discussed conceptually, theoretically, and empirically. Basic description of enterprise and partial budgeting (Kay and Edwards, 1999; Boehlje and Eidman, 1984; Giles and Stansfield, 1990; Kadlec, 1985), whole farm planning and linear programming (Kay and Edwards, 1999; Taha, 1992; Paris, 1991; Beneke and Winterboer, 1973), stochastic dominance (Hadar and Russell, 1969; Hardaker, Huirne, and Anderson, 1997; Robison and Barry, 1987), risk modeling (Freund, 1956; Boisvert and McCarl, 1990; Hardaker, Huirne, and Anderson, 1997; Robison and Barry, 1987) and resource allocation (Kay and Edwards, 1999; Boehlje and Eidman, 1984; Giles and Stansfield, 1990; Kadlec, 1985) are prevalent in the literature. Theoretical underpinnings are also found (Beattie and Taylor, 1985; Chambers, 1988; Robison and Barry, 1987). Philosophical and historical perspectives (Jensen, 1977; Day, 1977; Day and Sparling, 1977) have also been presented. Nonetheless, the focus of this study is an introduction to these methods and a discussion of their possible value. Application of economic analysis permits the researcher to assess the degree to which a cropping system is economically sustainable. Consequently, an economic framework discussing some basic analytical tools will be presented next, including a discussion of strengths, weaknesses and data requirements of each technique. However, an economic analysis must include an agronomic framework in order to develop a sustainable cropping system. Therefore, an agronomic framework of this study's main example is discussed after the economic framework to set the stage for economic analysis of the case example.

ECONOMIC FRAMEWORK

While economic evaluation of a cropping system typically occurs after the agronomic experiment has been at least partially conducted, it is advisable to incorporate consideration of the desired economic analysis in the designing phase of the experiment (Duffy, Guertal, and Muntifering, 1997). This section will discuss some of the primary methods of assessing the economic sustainability of cropping systems. Basic data requirements, advantages, disadvantages, and examples are presented. Furthermore, the three issues of profitability, risk management, and resource allocation are discussed.

Profitability

Several economic analytical tools are available for assessing profitability. However, none of these are a "checkbook accounting" method. Proper evaluation of production expenses can be an involved and complicated undertaking. While other means of profitability assessment exist, the three primary methods applicable to cropping systems are enterprise budgets, partial budgets, and whole farm analysis.

Enterprise budgeting is one of the most basic production economic tools available. It is frequently used to estimate cost and returns from a given enterprise [i.e., a single production activity such as corn (*Zea mays* L.)]. It is relatively simple compared to the other methods but still provides a detailed and in-depth analysis. It estimates the gross returns, variable costs, and fixed costs of the enterprise. Considerable detail is obtained on expected yield, commodity price, input requirements (e.g., fertilizer and herbicide), and input prices. Consequently, a detailed explanation of machinery and operations is required to develop an enterprise budget. This entails defining the machinery complement in specific detail such as well as the timing of operations and the amount of inputs required. When coupled with appropriate economic engineering procedures such as the American Society of Agricultural Engineering Standards (2000a, 2000b), this data also permits calculation of fixed expenses such as depreciation, interest on investment, and insurance. An enterprise budget is a fairly straightforward method of comparing cropping system alternatives and provides excellent insight into determining the economic sustainability of a cropping system, although it may have to be coupled with net present value analysis when evaluating cropping rotations (e.g., Westra and Boyle, 1991). Enterprise budgeting can also possess expanded usefulness when break-even analysis is incorporated wherein the level of a variable at which neither a profit nor a loss is made is calculated (Dillon, 1993; Dillon and Casey, 1990; Kay and Edwards, 1999). Furthermore, while other techniques prohibit statistical analysis or direct comparison to farms dramatically different than those modeled, enterprise budgeting may reasonably be combined with statistical evaluation techniques in assessing the relative economic performance of cropping systems using methods familiar to the natural sciences (e.g., analysis of variance).

Despite the popularity and many advantages of enterprise budgets, they are incomplete because they fail to consider specific issues including whole farm interactions and risk. For example, the feasibility of performing all necessary machinery operations for the entire acreage being produced under various weather conditions is excluded in an enterprise budget analysis. Thus, it is possible that maximizing the profit for each enterprise will not maximize the profit for the overall farm operation. Additionally, neither fluctuations in yield from year-to-year nor a farmer's attitude toward risk is incorporated in this type of assessment.

Examples of the use of enterprise budgeting to assess the economic potential of cropping systems are widespread in the literature. For example, enterprise budgets have been used to evaluate the economic potential for inclusion of annual forages in a corn-soybean (*Glycine max* L. Merr.) rotation (Olson et al., 1991). Enterprise budgets in soybean cropping systems include focus upon crop rotations (Dillon et al., 1999b; Dillon et al., 1997), seedbed preparation and weed management (Popp et al., 2000), row spacing and weed management (Oriade et al., 1997), and impacts of wheat (*Triticum aestivum* L.) residue management on double-cropped soybean (Oriade, Dillon, and Keisling, 1999). Examination of wheat planting methods (Oxner et al., 1997) and deep tillage in rice (*Oryza sativa* L.) (Pearce et al., 1999) have also relied upon enterprise budget analysis.

Partial budgeting is an economic analytical tool closely related to enterprise budgeting. This method focuses only upon changes made to an existing system and consequently simplifies the procedure even further in that data requirements are lessened. For example, if two systems with the same pre-plant machinery operations are to be compared, only costs associated after planting is taken into consideration while comparing the two systems. The partial budget has the potential to expand the scope of analysis by including interactions ignored by enterprise budget. It should be noted, however, that cropping systems usually entail the alteration of many components thereby rendering a partial budget inappropriate. Furthermore, risk analysis is still excluded and a thorough whole farm analysis is still precluded. Partial budget analysis is more limited in the literature but examples can be seen in evaluating production alternatives in a corn silage cropping system (Cox, Cherney, and Hanchar, 1998), full season sorghum (*Sorghum vulgare* L.), sorghum double-cropped with winter rye (*Secale cereale* L.), and corn and soybean cropping system (Buxton, Anderson, and Hallam, 1999). Partial budgets have also been applied in vegetable production systems (Trumble, Carson, and Kund, 1997; Kelly et al., 1995).

The most complete and potentially accurate form of economic assessment may be obtained from a whole farm analysis. Such analysis is more involved and relies upon the use of computer optimization techniques including mathematical programming models. Whole farm analysis can offer the advantage of explicit incorporation of interactions among enterprises and competition for common resources. A common view of the farmer's decision-making environment is that of an individual who desires to maximize profits in a constrained setting with a limited amount of land, labor, and capital. Modeling a decision-making framework reflecting this constrained optimization setting requires the use of whole farm analysis.

In addition to data requirements similar to that of enterprise budget analysis, data regarding resource allotments must be obtained. Thus, the amount of land, labor, and capital available is needed. In order to reflect this, a "representative" farm is often modeled which possesses an average amount of resources available. The representative farm is consequently unlikely to exactly match any one producer in the study area but does depict the situation for someone with an average amount of land, labor, and capital available. Field conditions which limit time available for machinery operations dictate the need to consider suitable field days in modeling labor constraints appropriately. Furthermore, correctly modeling labor constraints entails consideration of the timing of operations and inclusion of a series of periodic (e.g., weekly) labor constraints. Simply stated, labor available in December will not help a farmer needing to plant corn in April. This is also why a greater level of detailed information may be required for the whole farm analysis; a schedule of operations including time to be performed will be needed while an enterprise budget could be developed with less accurate timing information. Capital may or may not be modeled but a sufficient amount of funds for the purchase of operating inputs must obviously be available. The greater amount of data needed and expertise required are the primary disadvantages of the whole farm analysis method.

Limited examples of whole farm analysis of cropping systems exist in the literature. Examples include the economic analysis of alternative production practices of corn, sorghum, wheat and cotton (*Gossypium hirsutum* L.) (Dillon, Mjelde, and McCarl, 1989); the incorporation of early maturing soybean varieties in a soybean, wheat, and sorghum operation (Casey et al., 1998); and adoption of conservation tillage in a wheat cropping system (Helms, Bailey, and Glover, 1987).

Whole farm analyses have also incorporated the important dynamic issues involved in cropping system evaluation. The consideration of risk and resource requirements across time in evaluating the sustainability of a cropping rotation can be captured in a dynamic framework. Roberts and Swinton (1996) discuss the importance of these issues. Also, long run analysis of such systems is ideal (e.g., Christenson et al., 1995; Schoney and Thorsen, 1986).

Risk Analysis

While consideration of profit maximization in choosing alternative cropping systems is an excellent beginning, an assessment of risk is also necessary. Farmers are concerned not only with expected profit but also the fluctuation of profits. Willingness and ability to overcome the risk associated with a cropping system is necessary for it to be sustainable. Producers face production, marketing, financial, legal, and personal risk. While farmers have differing willingness and ability to bear risk, most are willing to sacrifice some level of mean or expected profit to reduce the risk. This risk is often measured by the variability of profits. Thus, a risk averse producer may prefer a cropping system with a lower mean net returns but a lower standard deviation or CV (coefficient of variation) of net returns over a cropping system with greater net returns but a higher level of risk (greater standard deviation and CV). Risk analysis through such mechanisms as stochastic dominance and risk modeling is useful in risk-return tradeoff wherein a decision-maker can reduce risk but experiences reduced profits.

Stochastic dominance serves as a tool to evaluate risk-return tradeoffs. Ideally, cropping system studies should include a measurement of economic risk associated with each system. However, economic analysis of cropping systems and true risk analysis is often excluded. Net returns are stochastic and they involve uncertainty. Hence, stochastic dominance attempts to assess the dominance or greater performance of one option over another in a probabilistic sense (Hadar and Russell, 1969; Hanoch and Levy, 1969). For example, if variety A yields greater than a similar variety B under any circumstance, A clearly dominates B in terms of first degree stochastic dominance. Higher orders of stochastic dominance can be calculated while evaluating more complicated circumstances (e.g., if variety A outperforms variety B under certain weather scenarios but the reverse is true under other weather patterns).

Stochastic dominance offers the advantages of requiring minimal or no additional data from developing enterprise budgets. Proper risk analysis requires an adequate measure of the central tendency (e.g., mean net returns) and dispersion (e.g., standard deviation of net returns). Consequently, an adequate number of observations across years for a cropping system is needed to properly assess risk. Such is often unavailable thereby precluding risk analysis. However, enterprise budgets for even a few years may be sufficient to establish some insights into risk management potential through stochastic dominance. Stochastic dominance also eliminates alternatives on broad categories of risk attitude, providing widely applicable results. A set of hundreds of choices may be narrowed to a handful with the mere assumption that the farmer is risk averse (wants to reduce risk as opposed to just maximize profits).

Disadvantages of stochastic dominance include its potential inability to separate a unique choice from a set of alternatives. Nonetheless, it does often dramatically reduce the alternatives to a reasonable set for examination. Similarly, most forms of stochastic dominance are also unable to directly consider combinations of alternatives. Therefore, if a combination of planting half of a given maturity class and half of another maturity class would best reduce risk, stochastic dominance would not select such unless it was directly entered as an alternative with the corresponding calculated data. Given the powerful opportunity afforded by negative correlation, this can be a serious shortcoming. Specifically, if cropping system A performs relatively well under dry weather but poorly in wet weather while B is exactly opposite. Exploiting the negative correlation of the two with a combination can offer a considerable reduction in risk. While conscious calculation of the combinations as an alternative allows one to circumvent this shortcoming of stochastic dominance, convex stochastic dominance provides an excellent recent development of overcoming this problem (Hardaker, Huirne, and Anderson, 1997).

Because of the required number of annual observations, stochastic dominance applications in cropping system studies are more limited in the literature. Nonetheless, a number of studies have used this method to investigate issues such as precision fertilization (Lowenberg-DeBoer and Aghib, 1999), cover crops (Larson et al. 1998; Giesler, Paxton, and Milhollon, 1993), double-cropped soybean production (Harper et al., 1991; Oriade et al., 1997), irrigation (Epperson, Hook, and Mustafa, 1992), tillage (Segarra, Keeling, and Abernathy, 1991; Williams, Llewelyn, and Barnaby, 1990), and weed management (Khan, Donald, and Prato, 1996; Zacharias and Grube, 1984).

Just as the stochastic dominance procedure parallels enterprise budgeting, risk modeling parallels whole farm analysis. Risk modeling is often a type of whole farm modeling itself; data requirements are therefore similar with the added caveat of the typical need of sufficient annual observations to measure risk. Advantages of the method include the ability to ascertain the best cropping systems combination for the decision-maker's attitude of risk assumed. Furthermore, the technique provides the benefits associated with whole farm analysis. Disadvantages are also similar but include the need to assume or quantify the farmer's attitude toward risk. This can be overcome by running several solutions under a series of risk attitude levels, but presenting too many solutions can be cumbersome. Another disadvantage relates to the fragile theoretical underpinnings of some methods for modeling risk. However, recent developments have severely weakened this criticism, especially regarding the mean-variance method. For example, this mean-variance technique maximizes risk adjusted net returns where a penalty related to the variability of net returns is subtracted from the mean net returns (Freund, 1956; Markowitz, 1959). Other risk models exist and include MOTAD (minimization of total absolute deviations) developed by Hazell (1971) and Target MOTAD (minimization of deviations below a target income level) developed by Tauer (1983).

Examples of risk modeling of cropping systems in the literature include the use of mean-variance analysis of integrated pest management (Musser, Tew, and Epperson, 1981), alternative production practices (Thornton and Hoogenboom, 1994; Dillon, Mjelde, and McCarl, 1989), impacts of suitable field day availability on production decisions (Dillon, 1999a), variety selection (Flood, McCamley, and Schneeberger, 1985; Dillon, 1992), and irrigation (Pandey, 1990). Target MOTAD models have been used to investigate the sustainable nature of rotations (Novak, Mitchell, and Crews, 1990), drip irrigation in a vegetable cropping system (Prevatt et al., 1992), and alternative single and double-cropped systems (Burton et al., 1996). MOTAD has been employed in comparing cropping systems and mixed livestock and crop production systems (Held and Zink, 1982).

Resource Allocation

While embodied in an appropriate economic analysis, the notion of optimal resource allocation is worth discussing specifically. The optimal level of input application is a relevant issue to producers that is sometimes either ignored in cropping systems analysis or dictated on the basis of physical sciences without regard to underlying economic principles. Specifically, the application of an input (e.g., fertilizer or herbicide) with the objective of yield maximization is almost never the same level as required for profit maximization. The price of both the crop and the input need to be considered, but these are often excluded from the decision of how much to apply and are only reflected in the end results if at all. An economic decision requires that the input be applied to the point where the additional benefits (marginal value product or value of extra yield) is just equal to the additional costs (price of the input). An example is evident in the case example below wherein strategies with higher plant populations had higher mean yields but were not selected in the economic model because the greater sales from this additional yield did not justify the additional costs of seed. Accurate reflection of these economic rules for all inputs will help enable the valid test as to whether or not a cropping system is truly sustainable.

AGRONOMIC FRAMEWORK

The agronomic framework for this study is based upon a cropping system representing alternative production practices available to a commercial grain production operation in Henderson County, Kentucky. The agronomic experimental design and the resultant yield provide the foundation of the defining elements of a cropping system and identifies the primary data required for economic assessment. In addition, the feasibility of the cropping system requires data on days available for successful completion of machinery operations (i.e., suitable field days). Consequently, the agronomic framework will focus on three issues: the underlying production environment, crop yield, and suitable field days.

The Underlying Production Environment

The cropping system in this study considers three crops commercially important to the region (corn, soybean, and wheat) with four enterprises of corn, full season soybean, wheat, and double-cropped soybean with wheat. These crops are assumed to be produced with no-till systems under dryland conditions. The production practices include alternative planting dates, plant populations, and crop maturity groups. A wide range of planting dates for all the crops were incorporated into the analysis consistent with farmer practices in the region (Kentucky Agricultural Statistics Service, various years). However, in order to investigate the effects of earlier planting dates, the ranges were moved earlier by one to two weeks to reflect a trend towards earlier planting in the study area. Alternative plant populations and maturity groups were examined for the production of both corn and soybean. Expert opinion of agricultural producers, Kentucky Cooperative Extension Service specialists, and other researchers were sought in determination of the agronomic experimental design. The production system parallels the cropping system analyzed by Dillon (1999a) for the mean suitable field day scenario.

Crop yields from the production system being considered were estimated from biophysical simulation models. Biophysical simulation provides the underlying yield response functions needed under a wide range of production practices and weather conditions. While agronomic field trials or similar data are preferable, such were not available with this breadth of information and sufficient observations that will allow a rich and diverse series of production strategies for this study. The use of biophysical simulation, therefore, enables the agronomic experimental design to illustrate a broad spectrum of alternative production practices for the corn, soybean, and wheat production cropping system used a case example.

Weekly corn planting from March 29 through May 24 was modeled for nine planting dates. Early, medium, and late maturity classes were included for corn as well as low, medium, and high plant populations of 49,421; 59,305; 69,189 (20,000; 24,000; and 28,000) plants ha^{-1} (plants ac^{-1}), respectively. The CORNF model (Stapper and Arkin, 1980) was used in simulating corn yields for the 81 (9 planting dates times 3 maturity classes times 3 plant populations) production strategies.

Soybean planting was simulated for nine weekly intervals from April 26 through June 21. Three generically representative varieties of maturity group III, IV, and V (MG III, MG IV, MG V) soybean were incorporated. Additionally, six plant and row spacing combinations were included. Specifically, soybean row spacing of 23 cm or 9 in (with 6.5 and 9.8 plants m^{-1} or 2 and 3 plants ft^{-1}), 48.3 cm or 19 in rows (with 13.1 and 19.7 plants m^{-1} or 4 and 6 plants ft^{-1}, and 76 cm or 30 in rows (with 19.7 and 29.5 plants m^{-1} or 6 and 9 plants ft^{-1}) were used. The SOYGRO (Wilkerson et al., 1983) model was used to simulate soybean yields associated with the 162 full season soybean production alternatives considered.

The nine weekly wheat planting dates modeled ranged from September 27 to November 22. A single cultivar of wheat was assumed to be drilled at a population of 150 plants m^{-2} (14 plants ft^{-2}). While wheat is almost always double-cropped with some other crop in the study area, most frequently soybean, the option of single crop wheat was simulated as well. When double-cropped, wheat is assumed to be double-cropped with soybean that is planted ten days after wheat harvest. The double-cropped soybean plant and row spacing as well as maturity groups parallel those utilized for the full season soybean crop. The CERES model (Ritchie and Otter, 1985) was used for simulating wheat yields resulting from these nine strategies and was integrated with SOYGRO for the joint simulation of double-cropped wheat and soybean for the 162 double-cropped production practice combinations. This allowed soil moisture impacts to be properly reflected in generating yield estimations.

A representative medium depth silt loam soil type was assumed in the biophysical simulation models. The meteorological data used for the model were from Henderson (1978 to 1999). Because of the overlap of winter wheat, twenty-one seasons of estimated yield data were provided. In addition to the use of Henderson weather data, solar radiation data for Evansville, Indiana were used. Insufficient data from field or plot experiments is the reason that biophysical simulation models are used, thereby prohibiting rigorous statistical model validation. However, model validation was performed by comparing overall yield levels and responsiveness of yields to alterations in production practices. This entailed examination of Kentucky Agricultural Statistics for various years, comparison to Ohio Valley region of Kentucky Farm Business Management Association results (Morgan, 1998; Gibson, 1998) and discussions with experts. Overall, the yield responses appear reasonable.

Crop Yield

Descriptive summary statistics for crop yield from the biophysical simulation models are displayed in Table 1. Data for selected relevant production

TABLE 1. Summary descriptive statistics for crop yield across strategies and years by crop

Crop	Number of Observations	Mean (kg ha^{-1})	Standard Deviation (kg ha^{-1})	CV (%)	Maximum (kg ha^{-1})	Minimum (kg ha^{-1})
Corn	1701	6893	1087	15.77	11,420	4462
Full Season Soybean	3402	2796	751	26.86	4355	888
Double-Cropped Soybean	3402	1828	1177	64.36	3900	0
Wheat	189	3184	697	21.88	5236	1795

practices are presented later. Corn averaged about 6,900 kg ha^{-1} (110 bu ac^{-1}) across all strategies and years while average full season soybean yield was 2,800 kg ha^{-1} (42 bu ac^{-1}) compared double-cropped soybean at 1,800 kg ha^{-1} (27 bu ac^{-1}). Average wheat yield was 3,200 kg ha^{-1} (47 bu ac^{-1}). Corn yield was the least variable with a coefficient of variation (CV) of 15.77% while double-cropped soybean yield was the most variable with a CV of 64.36%. As might be expected, double-cropped soybean failed to produce under some conditions.

Summary descriptive statistics for crop yield in selected cropping systems are displayed in Table 2. Earlier planting of corn and full season soybean generally yielded greater than later plantings, but possessed greater levels of relative yield variability. It is also noteworthy that, while the strategies selected in the economic analysis as discussed later were among the highest yielding, they did not represent the absolutely highest mean yields of any options. This is a result of economic considerations and conveys the need to conduct economic analysis when evaluating degree to which a cropping system is sustainable.

Suitable Field Day Results

Biophysical simulation was also used to determine the number of days suitable for fieldwork per week. Historical weather data was used in a soil moisture simulation for the medium depth silt loam conditions considered under a modified procedure discussed by Dillon, Mjelde, and McCarl (1989). Available field time was then calculated by multiplying the average number of workable field days per week by 12 working hours per day for 2.56 persons which is the average number of persons working on a commercial grain farm in the region (Morgan, 1998). Identification of a day unsuitable for fieldwork involved three criteria: (i) if it rained three consecutive days, both the third and

TABLE 2. Summary descriptive statistics for crop yields across years by selected strategy

Crop	Planting Date	Maturity Class	Number of Observ.	Mean (kg ha^{-1})	Standard Deviation (kg ha^{-1})	CV (%)	Max (kg ha^{-1})	Min (kg ha^{-1})
Corn	March 29	Medium	21	6587	651	9.88	7686	4775
	April 5	Late	21	7797	1047	12.43	10,243	5768
	April 12	Late	21	7779	1035	13.30	10,210	6130
	May 17	Late	21	7698	1029	13.36	9856	6009
	May 24	Late	21	7721	994	12.88	10,017	6125
Soybean	April 26	MG3	21	3069	530	17.28	3783	1738
		MG4	21	3106	581	18.70	3947	1715
	May 3	MG3	21	3013	525	17.41	3718	1951
		MG4	21	3048	587	19.23	3834	1901
	May 10	MG3	21	2977	515	17.28	3714	2022
	June 21	MG3	21	2802	842	30.04	3604	1275
		MG4	21	2811	854	30.37	3650	1273

the following day were considered unsuitable, (ii) if the soil moisture of the top 10 cm (3.9 in) was 80% or greater on a given day, then that day was not considered a field day, and (iii) if it rained 0.38 cm (0.15 in) or more on a given day, then that day was not considered a field day. The average number of days available per week under the weather conditions examined for Henderson was 5.5 with a standard deviation of 2.6.

A CASE EXAMPLE
OF ECONOMIC SUSTAINABILITY EVALUATION

The case study employed for this project serves as an example of whole farm analysis. The production decision model is formulated as a mathematical programming model wherein a producer attempts to maximize the net returns above specified costs (including all variable costs and relevant fixed costs). Land charges, property taxes, returns to management, and overhead labor were excluded because of the complexity of estimating these values and the assumption they would be constant. Notably, these need to be covered for economic sustainability but were excluded because of tractability and practicality. The four enterprises of corn, full season soybean, wheat, and double-cropped soybean with wheat were incorporated as decision variables in the model. Constraints included land, suitable field days by week, capital, and crop rotational constraints. Corn represented half of the planted land and full season soybean, wheat, and double-cropped soybean with wheat represented the other half. A mathematical depiction of the model may be found in Dillon (1999a).

The data required to specify the production decision model were: (1) available land, (2) available field days, (3) available capital, (4) labor requirements, (5) input requirements and prices, (6) crop prices, and (7) crop yields. The hypothetical farm was assumed to be a commercial size grain operation with 546.33 ha (1350 ac). This is derived by rounding the average number of tillable acres for an Ohio Valley grain farm of 1346 to 1350 ac or approximately 545 ha (Morgan, 1998). Suitable field days data results from biophysical simulation were used. Capital available was represented by the average capital available for an Ohio Valley grain farm at $225,000 (unpublished data).

The labor requirements per week, input prices, and input requirements ha^{-1} were taken from representative Tennessee no-till enterprise budgets (Gerloff and Maxey, 1998). Variable costs ha^{-1} for corn was $341.65, $352.89, and $364.16 ($138.26, $142.81, and $147.37 ac^{-1}) for low, medium and high populations. Full season (double-cropped) soybean had variable costs of $211.20 ha^{-1} and $222.47 ha^{-1} or $85.47 ac^{-1} and $90.03 ac^{-1} ($176.73 ha^{-1} and 188.00 ha^{-1} or $71.52 ac^{-1} and $76.08 ac^{-1}) for the two seeding rates while wheat had a variable cost of $259.51 ha^{-1}($105.02 ac^{-1}). Labor requirements were adjusted to weekly data and the timing of operations were shifted by planting date. Calculation of the mean simulated harvest dates allowed for adjustment of harvest time by maturity class. Loan deficiency payments were added to the 1994-1998 Kentucky average season prices for each crop with $0.1071 kg^{-1} ($2.72 bu^{-1}) for corn, $0.2451 kg^{-1} ($6.67 bu^{-1}) for soybean and $0.1260 kg^{-1} ($3.43 bu^{-1}) for wheat (Kentucky Agricultural Statistics Service, 1999).

The economic results for the case study example are presented in Table 3. Profit maximizing net returns above specified costs will be discussed here while risk reducing management strategies will be discussed under the section on risk analysis. The average net returns across all weather conditions is about $211,090 annually. This represents returns to land, overhead labor and management, and excludes property taxes. The risk associated with the production strategies selected is depicted in the standard deviation of $63,931 which on a relative basis in the CV is 30.29%. Another indication of risk is presented in the minimum value across years of $65,522.

The optimal production strategies for the case study are given in Table 4. The profit maximizing production practices includes late April to early May planting of soybean over a period of two weeks. MG III (maturity group III) and MG IV soybean were planted to spread harvest dates in consideration of

TABLE 3. Net returns above specified costs results for profit maximization and risk reduction by level

Component	Profit Maximization	Risk Reduction Level[a]		
		Low	Medium	High
Mean ($)	211,090	207,193	205,284	199,940
Std. Dev. ($)	63,931	56,091	54,465	52,584
CV(%)	30.29	27.07	26.53	26.30
Min ($)	65,522	81,571	96,613	98,277
Max ($)	326,424	313,448	310,186	300,204
Percent of Profit Max (%)	100.00	98.15	97.25	94.72

[a]The risk reduction level represents the attitude of the producer to reduce risk. Low, medium, and high correspond to 60%, 70% and 85% significance levels, respectively. These in turn reflect the certainty of receiving or exceeding a maximized lower level confidence limit on net returns. Assuming a normal distribution of net returns, a 50% certainty exists at risk neutrality that the actual net returns will be at or higher than the expected net returns. With risk aversion, a higher percentage of certainty in net returns is required; therefore, a certainty parameter larger than 50% is necessary. McCarl and Bessler (1989) provide details.

TABLE 4. Production strategy results for profit maximization and risk reduction by level

Crop	Planting Date	Maturity Class	Production Level (ha)			
			Profit Maximization	Risk Reduction Level[a]		
				Low	Medium	High
Soybean	April 26	MG3	10.4	207.2	207.2	207.2
Soybean	April 26	MG4	207.2	10.4	0.0	0.0
Soybean	May 3	MG3	0.0	55.6	23.8	0.0
Soybean	May 3	MG4	55.6	0.0	0.0	0.0
Soybean	May 10	MG3	0.0	0.0	22.0	28.9
Soybean	June 21	MG3	0.0	0.0	0.0	37.1
Soybean	June 21	MG4	0.0	0.0	20.2	0.0
Corn	March 29	MED	0.0	0.0	0.0	35.0
Corn	April 5	LATE	167.4	63.1	17.9	0.0
Corn	April 12	LATE	105.8	0.0	0.0	0.0
Corn	May 17	LATE	0.0	42.9	87.9	70.8
Corn	May 24	LATE	0.0	167.4	167.4	167.4
Soybean Mean Yield (kg ha^{-1})			3094	3043	3026	3010
Corn Mean Yield (kg ha^{-1})			7790	7824	7824	7383

[a]The risk reduction level represents the attitude of the producer to reduce risk. Low, medium, and high correspond to 60%, 70% and 85% significance levels, respectively. These in turn reflect the certainty of receiving or exceeding a maximized lower level confidence limit on net returns. Assuming a normal distribution of net returns, a 50% certainty exists at risk neutrality that the actual net returns will be at or higher than the expected net returns. With risk aversion, a higher percentage of certainty in net returns is required; therefore, a certainty parameter larger than 50% is necessary. McCarl and Bessler (1989) provide details.

available field days during soybean harvest. Soybean was planted in 23 cm or 9 in rows. Later maturing variety of corn was planted from April 5 to April 12. A low corn population was selected in all cases. The mean soybean and corn yield for the farm was about 3100 kg ha^{-1} (46 bu ac^{-1}) and 7800 kg ha^{-1} (124 bu ac^{-1}), respectively. Certain planting dates and maturity group combinations of corn produced greater yields than the ones selected by the economic model. However, soybean yields were reduced because of limited planting and harvest times. This provides an example of the importance of whole farm analysis and the interaction between enterprises, including the competition for common resources.

A mean-variance model was specified and solved as an example of risk modeling for this study. The same whole farm model above was modified to examine the potential for reducing risk associated with fluctuating yields. The net returns descriptive statistic results are presented in Table 3. An opportunity for managing risk clearly exists. A typical pattern of initially substantial risk reduction is possible for a relatively low cost of expected net returns. Note that a low level of risk reduction experiences a drop in mean net returns to about $207,193 or about 98% of the profit maximizing, risk neutral case. The standard deviation drops by about $7,840 or over 12% to $56,091 while the CV decreases from 30.29% to 27.07%. The minimum net returns increases substantially from $65,522 to $81,571. Further reduction in risk may be achieved at greater reductions of mean net returns.

The risk-return tradeoff may be visually depicted in a mean-variance graph (Figure 1). This figure displays a traditional mean-variance frontier that shows mean net returns and risk as measured by variance of net returns (which are actually in squared dollars for mathematical purists). The four points labeled identify the values corresponding to Table 3. The profit maximizing point would be selected by farmers desiring to ignore issues of risk and strictly seek the greatest expected return. Farmers who wish to reduce risk may do so for a relatively small decline in average net returns initially (see the low risk reduction point). As greater risk reduction is undertaken, greater relative decreases in mean net returns are required, as displayed in the nonlinear curve of risk-return tradeoff (Figure 1).

Identification of the mechanism for accomplishing risk reduction is of equal importance to the determination of the degree of risk reduction. Table 4 displays the alteration in optimal cropping systems that give rise to the risk management strategy by risk reduction level. Greater reliance on MG III soybean and the further diversification of corn planting dates over two periods (early April and middle to late May) permits initial risk reduction. As the desire to reduce risk increases, additional diversification in soybean planting time and corn planted early (late March) using a medium maturing variety is undertaken.

FIGURE 1. Risk-return tradeoff

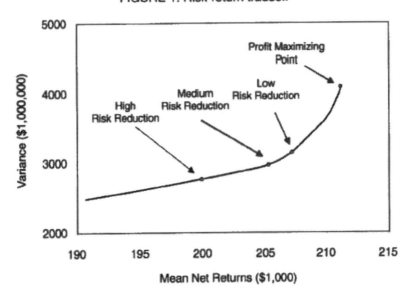

Economic analysis also can provide information regarding the use of limited resources. Land, labor, and capital are often cited as limiting resources and are now examined for the profit maximizing whole farm analysis above to provide an illustrative example for this study. The linear programming model provides a calculation of the value of an extra hectare of land to the representative Henderson producer using the optimal cropping system at about $381 ha^{-1} ($154 ac^{-1}) before considering returns to management and overhead labor. This value, known as dual or shadow price to programmers, reflects the marginal value product. In other words, this is the most the farmer would be willing to pay for additional land. Machinery operator labor is constraining in three weeks. The farmer would not be willing to pay for labor at other times based on model results because not all labor available is used. However, the values of labor for those three time periods constrained are $1.62, $5.13, and $8.59 hr^{-1} for weeks 35 (August 27 to September 2), 17 (April 23 to April 29), and 31 (July 30 to August 5), respectively. Cash flow feasibility and adequate capital funding is also required for an economically sustainable cropping system. The inclusion of a capital constraint in the model is designed to ensure this requirement. For the conditions modeled, sufficient operating capital was available with funds remaining that would presumably be invested by the farmer. Consequently, the representative Henderson farmer would be unwilling to borrow money for an operating loan under the conditions modeled.

CONCLUSIONS

This study identified three relevant economic characteristics important for a cropping system to be economically sustainable. These elements of profitability, risk management potential, and optimal resource allocation were discussed. Various techniques available for the analysis of cropping systems were presented along with a discussion of the data requirements, advantages, and disadvantages of each method. A case study of a Henderson, Kentucky producer was presented as an example. The importance of economics was demonstrated as evident in different production decisions on the basis of yield maximization versus profit maximization. The value of whole farm considerations as opposed to focus upon a single enterprise operation was demonstrated through different production decisions on the basis of profit maximization for a single enterprise versus the entire farm operation. Finally, the inclusion of risk attitude of the producer demonstrated different production decisions on the basis of profit maximization as opposed to considering the risk-return tradeoff.

Future research of cropping systems and their sustainability should include economic analysis. A wide variety of useful economic tools are available in providing worthwhile information to producers and fellow researchers in the search for the best cropping system. This information also expands the usefulness of cropping system research and enhances the likelihood of successful adoption by the producer. Inclusion of economic considerations at the design phase is also crucial for adequate and appropriate planning purposes. Indeed, economic analysis regarding the potential of innovative cropping systems is possible prior to the perfection of the systems. This can help in guiding researchers in the design of their research program given their limited time and funds for experimentation. Future economic cropping systems research should embody sound investigation of whole farm profitability coupled with the consideration of risk management potential and the implications of optimal resource allocation.

REFERENCES

American Society of Agricultural Engineers. (2000a). Agricultural machinery management. *Agricultural Engineering Standards* 47:344-349.

American Society of Agricultural Engineers. (2000b). Agricultural machinery management data. *Agricultural Engineering Standards* 47:350-357.

Beattie, B.B. and C.R. Taylor. (1985). *The Economics of Production.* New York, NY: John Wiley and Sons, Inc.

Beneke, R.R. and R. Winterboer. (1973). *Linear Programming Applications to Agriculture.* Ames, IA: Iowa State University Press.

Boehlje, M. and V. Eidman. (1984). *Farm Management.* New York, NY: John Wiley and Sons, Inc.

Boisvert, R.N. and B.A. McCarl. (1990). *Agricultural Risk Modeling Using Mathematical Programming.* Southern Cooperative Series Bulletin No. 356. Cornell University, New York.

Burton, R.O. Jr., M.F. Crisostomo, P.T. Berends, K.W. Kelley, and O.H. Buller. (1996). Risk/return analysis of double-cropping and alternative crop rotations with and without government programs. *Review of Agricultural Economics* 18:681-692.

Buxton, D.R., I.C. Anderson, and A. Hallam. (1999). Performance of sweet and forage sorghum grown continuously, double-cropped with winter rye, or in rotation with soybean and maize. *Agronomy Journal* 91:93-101.

Casey, W.P., T.J. Dumler, R.O. Burton, D.W. Sweeney, A.M. Featherstone, and G.V. Granade. (1998). A whole-farm economic analysis of early-maturing and traditional soybean. *Journal of Production Agriculture* 11:240-246.

Chambers, R.G. (1988). *Applied Production Analysis: A Dual Approach.* Cambridge, U.K.: Cambridge University Press.

Christenson, D.R., R.S. Gallagher, T.M. Harrigan, and J.R. Black. (1995). Net returns from twelve cropping systems containing sugar beet and navy beans. *Journal of Production Agriculture* 8:276-281.

CIMMYT. (1988). *From Agronomic Data to Farmer Recommendations: An Economics Training Manual.* Completely Revised ed. Mexico City: Centro Internacional de Mejoramiento de Maiz y Trigo (CIMMYT).

Cox, W.J., D.R. Cherney, and J.J. Hanchar. (1998). Row spacing, hybrid, and plant density effects on corn silage yield and quality. *Journal of Production Agriculture* 11:128-134.

Day, R.H. (1977). On economic optimization: A nontechnical survey. In *A Survey of Agricultural Economics Literature.* Volume 2, ed. L. Martin, St. Paul, MN: University of Minnesota Press. pp. 57-92.

Day, R.H. and E. Sparling. (1977). Optimization models in agricultural and resource economics. In *A Survey of Agricultural Economics Literature.* Volume 2, ed. L. Martin, St. Paul, MN: University of Minnesota Press. pp. 93-127.

Dillon, C.R. (1992). Microeconomic effects of reduced yield variability cultivars for soybeans and wheat. *Southern Journal of Agricultural Economics* 24:121-133.

Dillon, C.R. (1993). Advanced breakeven analysis of agricultural enterprise budgets. *Agricultural Economics* 9:127-143.

Dillon, C.R. (1999a). Production practice alternatives for income and suitable field day risk management. *Journal of Agricultural and Applied Economics* 31:247-261.

Dillon, C.R. and J.E. Casey. (1990). Elasticity of breakeven prices between agricultural enterprises. *Texas Journal of Agriculture and Natural Resources* 4:33-36.

Dillon, C.R., T.C. Keisling, R.D. Riggs, and L.R. Oliver. (1997). The profit potential of soybean production rotation systems in Arkansas. *Communications in Soil Science and Plant Analysis.* 28:1693-1709.

Dillon, C.R., J.W. Mjelde, and B.A. McCarl. (1989). Biophysical simulation in support of crop production decisions: A case study in the blacklands region of Texas. *Southern Journal of Agricultural Economics* 21:73-86.

Dillon, C.R., D.F. Rutherford, T.C. Keisling, and L.R. Oliver. (1999b). Economic assessment of irrigated and nonirrigated soybean cropping rotations on a clay soil. *Communications in Soil Science and Plant Analysis* 30:1165-1182.

Dillon, J.L. and J.B. Hardaker. (1993). *Farm Management Research for Small Farmer Development*. 2nd ed. FAO Farm Systems Management Series 6. Rome, Italy: Food and Agricultural Organization of the United Nations.

Duffy, P.A., E.A. Guertal, and R.B. Muntifering. (1997). The pleasures and pitfalls of interdisciplinary research in agriculture. *Journal of Agribusiness* 15:139-159.

Epperson, J.E., H.E. Hook, and Y.R. Mustafa. (1992). Stochastic dominance analysis for more profitable and less risky irrigation of corn. *Journal of Production Agriculture* 5:243-247.

Flood, M., F. McCamley, and K. Schneeberger. (1985). Mean variance efficiency as an approach to evaluate farmer adoption of crop technology. *North Central Journal of Agricultural Economics* 7:33-40.

Freund, R.J. (1956). The introduction of risk into a programming model. *Journal of the Econometric Society* 24:252-263

Gerloff, D.C. and L. Maxey. (1998). *Field Crop Budgets for 1998*. Knoxville, TN: Tennessee Agricultural Extension Service. AE and RD No. 30.

Gibson, C.D. (1998). *Enterprise Analysis, Ohio Valley Farm Analysis Group, 1993-1997*. Lexington, KY: University of Kentucky Cooperative Extension Service.

Giesler, G.G., K.W. Paxton, and E.P. Millhollon. (1993). A GSD estimation of the relative worth of cover crops in cotton production systems. *Journal of Agricultural and Resource Economics* 18:47-56.

Giles, T. and M. Stansfield. (1990). *The Farmer as Manager*. 2nd ed. New York, NY: CAB International.

Hadar, J. and W.R. Russell. (1969). Rules for ordering uncertain prospects. *American Economic Review* 59:25-34.

Hanoch, G. and H. Levy. (1969). The efficiency analysis of choices involving risk. *Review of Economic Studies* 36:335-346.

Hansen, J.W. and J.W. Jones. (1996). A systems framework for characterizing farm sustainability. *Agricultural Systems* 51:185-201.

Hardaker, J.B., R.B.M. Hiurne, and J.R. Anderson. (1997). *Coping with Risk in Agriculture*. New York, NY: CAB International.

Harper, J.K., J.R. Williams, R.O. Burton, Jr., and K.W. Kelley. (1991). Effect of risk preferences on incorporation of double-crop soybeans into traditional rotations. *Review of Agricultural Economics* 13:185-200.

Hazell, P.B.R. (1971). A linear alternative to quadratic and semivariance programming for farm planning under uncertainty. *American Journal of Agricultural Economics* 64:53-62.

Held, L.J. and R.A. Zink. (1982). Farm enterprise choice: Risk-return tradeoffs for cash-crop versus crop-livestock systems. *North Central Journal of Agricultural Economics* 4:11-19.

Helms, G.L., D.V. Bailey, and T.F. Glover. (1987). Government programs and adoption of conservation tillage practices on nonirrigated wheat farms. *American Journal of Agricultural Economics* 69:786-795.

Jensen, H.R. (1977). Farm management and production economics, 1946-70. In *A Survey of Agricultural Economics Literature*. Volume 1., ed. L. Martin, St. Paul, MN: University of Minnesota Press. pp. 3-89.

Kadlec, J.E. (1985). *Farm Management: Decisions, Operation, Control*. Englewood Cliffs, NJ: Prentice-Hall.

Kay, R.D. and W.M. Edwards. (1999). *Farm Management*. 4th Edition. New York, NY: McGraw-Hill Book Company.

Kelly, T.C., Y.C. Lu, A.A. Abdul-Baki, and J.R. Teasdale. (1995). Economics of a hairy vetch mulch system for producing fresh-market tomatoes in the mid-atlantic region. *Journal of the American Society of Horticultural Science* 120:854-860.

Kentucky Agricultural Statistics Service. *Kentucky Agricultural Statistics. Various years*. Louisville, KY: National Agricultural Statistics Service of USDA and Kentucky Department of Agriculture.

Khan, M.,W.W. Donald, and T. Prato. (1996). Spring wheat (*Triticum aestivum*) management can substitute for Diclofop for foxtail (*Setaria* spp.) control. *Weed Science* 44:362-372.

Larson, J.A., R.K. Roberts, D.D. Tyler, B.N. Duck, and S.P. Slinsky. (1998). Nitrogen fixing winter cover crops and production risk: A case study for no-tillage corn. *Journal of Agricultural and Applied Economics* 30:163-174.

Lightfoot, C., M.A.P. Bimbao, J.P.T. Dalsgaard, and R.S.V. Pullin. (1993). Aquaculture and sustainability through integrated resources management. *Outlook on Agriculture* 22: 143-150.

Lowenberg-Deboer, J. and A. Aghib. (1999). Average returns and risk characteristics of site specific p and k management: Eastern corn belt on-farm trial results. *Journal of Production Agriculture* 12:276-282.

Markowitz, H.M. (1959). *Portfolio Selection: Efficient Diversification of Investments*. New York, NY: John Wiley and Sons, Inc.

McCarl, B.A. and D. Bessler. (1989). Estimating an upper bound on the Pratt risk aversion coefficient when the utility function is unknown. *Australian Journal of Agricultural Economics* 33:56-63.

Morgan, R.D. (1998). *The Kentucky Farm Business Management Program 1997 Annual Summary*. Lexington, KY: University of Kentucky Cooperative Extension Service. Agricultural Economics-Extension Series No. 98-03.

Musser, W.N., B.V. Tew, and J.E. Epperson. (1981). An economic examination of an integrated pest management production system with a contrast between e-v and stochastic dominance analysis. *Southern Journal of Agricultural Economics* 13:119-124.

Novak, J.L., C.C. Mitchell, and J.R. Crews. (1990). Risk and sustainable agriculture: A target MOTAD analysis of the 92-year 'old rotation.' *Southern Journal of Agricultural Economics* 22:145-153.

Olson, K.D., N.P. Martin, D.R. Hicks, and M.A. Schmidt. (1991). Economic analysis of including an annual forage in a corn-soybean farming system. *Journal of Production Agriculture* 4:599-606.

Oriade, C.A., C.R. Dillon, and T.C. Keisling. (1999). Economics of wheat residue management in doublecrop soybean. *Journal of Production Agriculture* 12:42-48.

Oriade, C.A., C.R. Dillon, E.D. Vories, and M.E. Bohanan. (1997). An economic analysis of alternative cropping and row spacing systems for soybean production. *Journal of Production Agriculture* 10:619-624.

Oxner, M.D., C.R. Dillon, T.C. Keisling, and P. Counce. (1997). An agronomic and economic evaluation of commonly used wheat planting methods in the lower Mississippi River Delta. *Journal of Production Agriculture* 10:613-618.

Pandey, S. (1990). Risk-efficient irrigation strategies for wheat. *Agricultural Economics* 4 :59- 71.

Paris, Q. (1991). *An Economic Interpretation of Linear Programming.* Ames, IA: Iowa State University Press.

Pearce, A.D., C.R. Dillon, T.C. Keisling, and C.E. Wilson, Jr. (1999). Agronomic and economic effects of four tillage practices on salinity damage in rice. *Journal of Production Agriculture* 12:305-312.

Popp, M.P., L.R. Oliver, C.R. Dillon, T.C. Keisling, and P.M. Manning. (2000). Evaluation of seedbed preparation, planting method and herbicide alternatives for dryland soybean production. *Agronomy Journal* 92:1149-1155.

Prevatt, J.W., C.D. Stanley, P.R. Gilreath, and G.A. Clark. (1992). Return-risk analysis of adopting drip irrigation. *Applied Engineering in Agriculture* 8:47-52.

Ritchie, J.T. and S. Otter. (1985). Description and performance of Ceres–Wheat: A user oriented wheat yield model. In *ARS Wheat Yield Project*, ed. W.O. Willis, Washington, DC: USDA-ARS-38. pp. 159-175.

Roberts, W.S. and S. Swinton. (1996). Economic methods for comparing alternative crop production systems: A review of literature. *American Journal of Alternative Agriculture* 11:10-17.

Robison, L.J. and P.J. Barry. (1987). *The Competitive Firm's Response to Risk.* New York, NY: MacMillan Publishing Company.

Schoney, R.A. and T. Thorson. (1986). The economic analysis of extended crop rotations on Saskatchewan grain farms. *Canadian Farm Economics* 20:21-26.

Segarra, E., J.W. Keeling, and J.R. Abernathy. (1991) Tillage and cropping system effects on cotton yield and profitability on the Texas southern high plains. *Journal of Production Agriculture* 4:566-571.

Stapper, M. and G.F. Arkin. (1980). CORNF: A dynamic growth and development model for maize (*Zea mays* L.) College Station, TX: Texas Agricultural Experiment Station, Program and Model Documentation No. 80-2.

Taha, H.A. (1992). *Operations Research: An Introduction,* Fifth Edition. New York, NY: MacMillan Publishing Company.

Tauer, L. (1983). Target MOTAD. *American Journal of Agricultural Economics* 65:606-610.

Thornton, P.K. and G. Hoogenboom. (1994). A computer program to analyze single-season crop model outputs. *Agronomy Journal* 86:860-868.

Trumble, J.T., W.G. Carson, and G.S. Kund. (1997). Economics and environmental impact of a sustainable integrated pest management program in celery. *Journal of Economic Entomology* 90:139-146.

Westra, J.V. and K.J. Boyle. (1991). *An Economic Analysis of Crops Grown in Rotation with Potatoes in Aroostook County, Maine.* Maine Agricultural Experimental Station Bulletin 834. Department of Agricultural and Resource Economics, Orono, ME: University of Maine.

Wilkerson, G.G., J.W. Jones, K.T. Boote, K.T. Ingram, and J.W. Mishoe (1983). Modeling soybean growth for management. *Transactions of the American Society of Agricultural Engineers* 26: 63-73.

Williams, J.R., R.V. Llewelyn, and G.A. Barnaby. (1990). Risk analysis of tillage alternatives with government programs. *American Journal of Agricultural Economics.* 72:172-181.

Zacharias, T.P. and A.H. Grube. (1984). An economic evaluation of weed control methods used in combination with crop rotation: A stochastic dominance approach. *North Central Journal of Agricultural Economics.* 6:113-120.

Conceptual Framework
for Evaluating Sustainable Agriculture

Murari Suvedi
Christoffel den Biggelaar
Shawn Morford

SUMMARY. Evaluation of outcomes is arguably one of the most critical and challenging tasks of research and education programs in agriculture. Assessing research and education in sustainable agriculture is particularly complex because of the need to assess interactive social, environmental, and economic factors within farming systems. Definitions and indicators of sustainability vary widely among practitioners, and indirect and non-tangible factors are difficult to measure. No single evaluation

Murari Suvedi is Associate Professor, Department of Agriculture and Natural Resources Education and Communications Systems, 409 Agriculture Hall, Michigan State University, East Lansing, MI 48824.

Christoffel den Biggelaar is Assistant Professor, Department of Interdisciplinary Studies, 109 East Hall, Appalachian State University, Boone, NC 28608.

Shawn Morford is Socio-Economics Extension Specialist, Southern Interior Forest Extension and Research Partnership, 506 West Burnside Road, Victoria, BC, V8W 1M5 Canada.

This material is based upon work supported by the Cooperative State Research, Education, and Extension Service, U.S. Department of Agriculture under Cooperative Agreement No. 94-COOP-1-0809. Any opinions, findings, conclusions or recommendations expressed in this publication are those of the author(s) and do not necessarily reflect the views of the U.S. Department of Agriculture.

[Haworth co-indexing entry note]: "Conceptual Framework for Evaluating Sustainable Agriculture." Suvedi, Murari, Christoffel den Biggelaar, and Shawn Morford. Co-published simultaneously in *Journal of Crop Production* (Food Products Press, an imprint of The Haworth Press, Inc.) Vol. 9, No. 1/2 (#17/18), 2003, pp. 433-454; and: *Cropping Systems: Trends and Advances* (ed: Anil Shrestha) Food Products Press, an imprint of The Haworth Press, Inc., 2003, pp. 433-454. Single or multiple copies of this article are available for a fee from The Haworth Document Delivery Service [1-800-HAWORTH, 9:00 a.m. - 5:00 p.m. (EST). E-mail address: docdelivery@haworthpress.com].

model has been found to serve as an adequate tool for assessing investments in sustainable agriculture. This article describes challenges and issues in sustainable agriculture evaluation and proposes the incorporation
of elements of two frameworks: the Program Evaluation (Bennett's Hierarchy) and the Driving Force-State-Response (DSR) models. These
models help organize indicators and facilitate evaluation in light of the
complexity of sustainability. *[Article copies available for a fee from The
Haworth Document Delivery Service: 1-800-HAWORTH. E-mail address:
<docdelivery@haworthpress.com> Website: <http://www.HaworthPress.com>
© 2003 by The Haworth Press, Inc. All rights reserved.]*

KEYWORDS. Indicators, evaluation frameworks, monitoring and evaluation, assessment

INTRODUCTION

Sustainable agriculture as a practice and research priority has emerged in
response to widespread recognition of a need to balance food production with
environmental and social health (National Research Council, 1989; Pretty,
1995). The sustainability of increasingly mechanized, high-input, and specialized approaches is being questioned. Cropping systems are now viewed as
complex integrated farming systems involving economic, environmental, and
social factors (Edwards et al., 1990; Francis and Youngberg, 1990; Gliessman,
1997). Researchers and practitioners have begun to apply systems approaches
to scientific investigation and agricultural production that consider divergent
factors such as quality of life, regional economics, watershed health, and natural habitat. Cropping systems identified by characteristics such as on-farm inputs, diversity of products, local marketing, and Integrated Pest Management
(IPM) have begun to surface among the mainstream agricultural industry (Harwood, 1993; Madden and Chaplowe, 1997).

In the U.S., sustainable agriculture concepts began to emerge by the mid-
1980s. In 1980, the U.S. Department of Agriculture (USDA) released "Report
and Recommendations on Organic Farming" that defines organic farming as a
system that relies on crop rotations, crop residues, animal manures, legumes,
green manures, and off-farm organic mechanisms to maintain soil productivity
and control pests. The report was initiated in response to the finding that many
large and small-scale producers had begun developing unique systems for soil
and crop management, energy conservation, and pest control. Despite the existence of highly developed agricultural systems in the U.S., there has been
growing interest in alternative farming systems.

Sustainable agriculture practitioners use a range of technological and management activities to protect and enhance biological interactions and natural processes. These systems rely on both research-based and indigenous (local) knowledge. Sustainable agriculture is based on a range of practices, not on a prescribed set of techniques. It encompasses a philosophy that illustrates the dichotomy between reductionist/conventional and holistic/alternative approaches (Douglas, 1985; Batie and Taylor, 1991; Ikerd, 1996).

Institutional efforts to generate and disseminate sound and practical information about sustainable farming systems have accompanied the growing interest in these practices. In 1988, the U.S. Congress funded the Sustainable Agriculture Research and Education Program (SARE, formerly known as Low-Input Sustainable Agriculture or LISA) within USDA and approved a competitive grant program to enhance the capacity among research and extension organizations in the U.S.

The purpose of SARE is to expand knowledge and adoption of agriculture practices that are economically viable, environmentally sound, and socially acceptable. SARE is a competitive grant program with a regional orientation and leadership structure designed to provide research and education for the future economic viability of U.S. agriculture. At present, SARE is funded through the USDA-CSREES under Subtitle B, Chapter 1 of Title XVI of the Food, Agriculture, Conservation and Trade Act of 1990 (U.S. Congress 1990), amended in the Federal Agriculture Improvement and Reform Act of 1996 (U.S. Congress 1996).

Congress defined sustainable agriculture as

> An integrated system of plant and animal production practices having a site-specific application that will, over the long term, (a) satisfy human food and fiber needs; (b) enhance environmental quality and the natural resource base upon which the agriculture economy depends; (c) make the most efficient use of non-renewable resources and on-farm resources and integrate, where appropriate, natural biological cycles and controls; (d) sustain the economic viability of farm operations; and (e) enhance the quality of life for farmers and society as a whole. (Title XVI, Subtitle A, Section 1603)

A key challenge of the SARE Program has been how to assess the effectiveness, efficiency, and impact of investment in research and education. There is a constant and ongoing debate among agriculture professionals, policy makers and program evaluators about the evaluation criteria for sustainable agriculture and the development of indicators to measure sustainability in farm operations (Smoik, Dobbs, and Rickerl, 1995; Ikerd, 1996; Ikerd, Devino, and Traiyongwanich, 1996; Roberts and Swinton, 1996). Over the years, a number of models commonly used in the evaluation of agricultural research and extension programs have been developed and adapted for the assessment of sustainable agriculture. However, a host of issues and challenges limits the applicability of existing frameworks to evaluate SARE programs and projects. These issues and challenges have shaped our decisions about the evaluation design and implementation. Therefore, the objective of this paper is to describe challenges and issues in evaluation of sustainable agriculture, review existing models of evaluation, and to develop a framework of a conceptual model for evaluating sustainable agriculture. It is anticipated that professionals working for the development of sustainable cropping systems will benefit from this conceptual framework while planning, monitoring, and evaluating their programs.

CONCEPTS OF MONITORING AND EVALUATION

Horton, Peterson, and Ballantyne (1993) defined monitoring and evaluation as observing or checking on activities regarding their context, results, and impacts to ensure that implementation is proceeding according to plan; to provide a record of input use, activities and results; and to warn of deviations from initial goals and expected outcomes.

Monitoring provides information on project performance and allows managers to track whether progress is according to plan. It involves periodic recording, analysis, reporting and storage of data on research and management indicators (McLean, 1994). Managers track resources (e.g., funds, personnel, and supplies) and processes (e.g., research methods, occurrence of meetings, and program support). Monitoring allows for the comparison of progress against planned objectives, detection of deviations, and identification of bottlenecks and enables managers to take corrective actions while research is in progress (McLean, 1994). The information obtained can later be used for evaluation.

Evaluation involves judging, appraising, or determining worth, value, or quality of programs in terms of their relevance, effectiveness, efficiency, and impact. According to Rossi and Freeman (1985), evaluations are undertaken to: (1) judge the worth of ongoing programs and to estimate the usefulness of attempts to improve them; (2) assess the utility of innovative programs and initiatives; (3) increase the effectiveness of program management and administration; and (4) meet various accountability requirements. Evaluation can be done when a program is being planned (often called "needs assessments"), during programs, or after programs are completed.

A variety of methods may be used to evaluate programs and projects. In each case, the purpose is to provide managers, scientists or funding agencies with indications of program impacts. Two types of evaluations may be done after completion of a project: final evaluation and impact evaluation. In a final evaluation, the emphasis is on lessons to improve future program activities (Gapasin, 1993). Impact evaluation measures longer-term impacts of the program and the extent to which results have contributed to broader goals (Gapasin, 1993). Impact evaluations assess a variety of types of impacts, including economic, social, and environmental impacts (Peterson, 1993).

Peterson (1993) outlined benefits of impact evaluations for agricultural researchers and managers. They stated that impact evaluations provide valuable feedback from the agricultural community on the applicability of research results, evidence of the value of research to demonstrate to funding agencies, a basis for setting research priorities, information on which to base improvements, and information for discussions between researchers and policy makers.

INDICATORS AND THEIR ROLE IN EVALUATION

Indicators are variables that represent attributes (quality, characteristic, or property) of systems (Gallopin, 1997). They are observable phenomena that represent an intended and/or actual condition of situations, programs, or outcomes (Bennett and Rockwell, 1995). They are important tools of evaluation that allow managers to gauge change over time and space. Indicators should be simple, reliably determined, and relatively easy to verify (Lewandowski, Hardtlein, and Kaltschmitt, 1999). Frameworks that organize individual indicators or sets of indicators are useful for guiding the data collection process. They help evaluators communicate with decision-makers, aid in summarizing information from many different sources, and they can help identify information gaps. Frameworks help disperse reporting burdens by structuring data collection, analysis and reporting process across many issues and areas that pertain to sustainable development (UNEP-DPCSD, 1995).

Several authors discuss the function, characteristics, criteria for selecting, and concerns about indicators (Gallopin, 1997; Mortensen, 1997; Hart, 1995; and RIRDC, 1997). Based on those discussions, we propose the following criteria for selecting indicators for the evaluation of sustainable agriculture programs:

- Simple and unambiguous language
- Limited in number
- Relevant to the main objectives of the program
- Relevant to the overall objective of assessing progress towards sustainable development
- Feasible within the capacities of the organizations
- Conceptually well-founded
- Representative of international consensus, to the greatest extent possible

- Dependent on data that are available at reasonable cost, adequately documented, of known quality, and updated at regular intervals.

REVIEW OF EVALUATION FRAMEWORKS

To promote understanding of evaluation and to help organize indicators, scholars have developed several models for monitoring and evaluation. We reviewed seven frameworks for their applicability for evaluation of research and education in sustainable agriculture, as listed below. As we will discuss further, the last two models offered the most useful elements for assessing sustainable agriculture.

CIPP Evaluation Model (Mulholland, 1993)

Originally designed for evaluation of educational programs, this model is named after the four evaluation types that comprise it. *Context evaluation* assesses the contextual factors of programs, such as underlying problems; *Input evaluation* assesses factors relating to financial and human resources utilized; *Process evaluation* assesses program implementation; and *Product evaluation* assesses outcomes of the program. Although parts of this model are applicable to monitor sustainable agriculture programs, this model lacks elements needed to adequately document sustainability. The model does not track changes over time.

Development Agency Monitoring and Evaluation Model (Peterson, 1993)

This model is based on the work of international aid agencies and is concerned with accountability, impacts, operations, and lessons learned from past experience. This model mainly focuses on day-to-day project management rather than long-term impacts.

Economic Evaluation Model (Falconi, 1993)

This model applies economic principles such as benefit/cost analyses to evaluation. The existence of non-monetary benefits limits the utility of this model for sustainable agriculture, since many benefits are non-monetary and are not easily commensurable.

Logical Framework Model (Uribe and Horton, 1993)

The Logical Framework is a planning tool that can be used to help evaluate projects in the broader context of higher-level (such as national) goals. It helps to clarify logical links between project inputs and objectives, project activities and outputs, and broader purposes, and ultimate goal. It is particularly useful for planning activities, resources, and inputs required to meet objectives. One of the difficulties using this framework to evaluate research is that objectives may change as a result of research findings.

Research Project Management Model (Gapasin, 1993)

This model is used to translate research and education goals into individual projects for implementation, allowing managers to track progress and encourage project personnel to meet objectives on time and within budget. It can be a powerful tool for improving the effectiveness and efficiency of agriculture research and education but requires significant administrative resources.

Program Evaluation Model (Bennett, 1977)

Largely used by governments to evaluate social and educational programs, this model was originally designed for accountability purposes, but over time has broadened to include assessment of goals, strategies, technical and management processes, direct outputs and final outcomes (Horton, Peterson, and Ballantyne, 1993). With some limitations, this model has significant applicability for assessing sustainable agriculture investments.

The Driving Force-State-Response Model (OECD, 1997 and Mortensen, 1997)

This model is used extensively by many international development organizations to organize indicators of sustainable development. It is useful for considering multiple factors that influence social, economic, and environmental outcomes. Parts of this model also are highly applicable for assessing sustainable agriculture investments.

EVALUATION FRAMEWORK FOR THE SUSTAINABLE AGRICULTURE PROGRAMS

All but the final two frameworks we reviewed had limited applicability for assessing investments in sustainable agriculture. We found that elements of Program Evaluation and the Driving Force-State-Response model were the most suitable for evaluating research and education programs in sustainable agriculture. We combined elements of these models to develop and propose a hybrid conceptual framework for our evaluation. Details of these frameworks are as follows.

Program Evaluation Model (Bennett's Hierarchy)

Bennett's Hierarchy, and its recent version called the "Target of Program" model by Bennett and Rockwell (1995), has been used in various forms by the USDA Extension Service and state extension agencies to monitor and evaluate their programs and activities. In the Program Evaluation model (Figure 1), assessment is structured in a hierarchical fashion composed of seven levels as follows:

FIGURE 1. Program Evaluation Model (Bennett's Hierarchy) (adapted from Bennett, 1977 and Bennett and Rockwell, 1994).

Program evaluation hierarchy	Description
Level 7 End results	Measures the impact of the research and education on overall long-term goals
Level 6 Practice change	Measures changes in behavior of program participants such as adoption of new practices and management options
Level 5 KOSA change	Measures changes in Knowledge, Opinions, Skills, and Aspirations of program participants
Level 4 Reactions	Measures perceptions of program participants about research and education programs
Level 3 Outputs	Measures activities completed and products developed
Level 2 Activities	Measures what programs offer or do
Level 1 Inputs	Measures resources used in the programs (e.g., staff, budget, and time)

While the Program Evaluation model is useful for assessing inputs, activities, reactions, and KOSA changes (levels 1-5), we feel it is not rigorous enough to assess practice changes and end results (levels 6 and 7). For example, indicators are useful for tracking changes in practices and end results, but the program evaluation model does not provide for a logical organization of such indicators. The multi-faceted nature of sustainable agriculture necessitates multiple indicators in each of the three 'legs' supporting it (environment, society, and economy). One should be able to identify clear linkages between practice changes and end results within each of the components. Reviewing literature on the evaluation of sustainable agriculture and sustainable development, we found that the Driving Force-State-Response (DSR) model is widely used for the purpose.

The Driving Force-State-Response Model

Various versions of the DSR model are being used by international organizations for organizing information on sustainable development. The DSR model (Figure 2), adapted from Organization for Economic Cooperation and Development (OECD) (1997) and Mortensen (1997), contains the following components:

- *Driving force indicators* provide an indication of the causes of changes in the status of sustainable agriculture, and include both human activities and practices, and natural processes and patterns. Human-related driving forces include biophysical inputs and outputs at the farm level and economic and societal forces.
- *State indicators* provide an indication of the status or condition of sustainable agriculture, or a particular aspect of it, at a given point in time. The state of sustainable agriculture encompasses a wide range of elements that can be broadly organized into three categories: the state of the natural resources, the state of local economy, and the quality of life of the people in the community.
- *Response indicators* refer to the reaction by groups in society and policy makers to actual and perceived changes in the state of agricultural sustainability and to market signals. These indicators provide a measure of the willingness of a society and/or individuals to respond.

FIGURE 2. Relationships between indicators in the driving forces-state-response framework (adapted from OECD, 1997 and Mortensen, 1997).

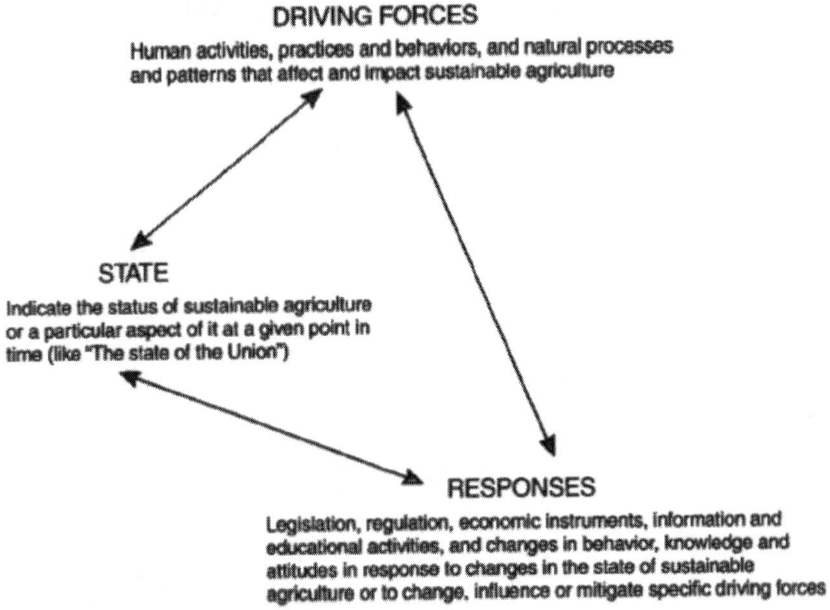

DRIVING FORCES

Human activities, practices and behaviors, and natural processes and patterns that affect and impact sustainable agriculture

STATE

Indicate the status of sustainable agriculture or a particular aspect of it at a given point in time (like "The state of the Union")

RESPONSES

Legislation, regulation, economic instruments, information and educational activities, and changes in behavior, knowledge and attitudes in response to changes in the state of sustainable agriculture or to change, influence or mitigate specific driving forces

INDICATORS AND MEASUREMENTS

In the first four levels of the model as depicted in Figure 3, i.e., inputs, activities, outputs, reactions, indicators focus on performance of research and educational programs during their implementation phase. For example, indicators of *inputs (Level 1)* could be

- program personnel and other resources devoted to the program
- the existence of clear objectives
- the existence of necessary resources (does the researcher have access to tools and other resources necessary to conduct the work?)

Measurement of input indicators would most appropriately occur at the proposal development stage.

Examples of *activity (level 2)* indicators

- number of workshops planned/held
- research methods used (is the researcher using appropriate research methods to address the research question?)

- number of research proposal submitted and funded
- number of demonstrations and field tours given

Measurement of activity indicators could occur through a site visit, as part of project monitoring.

Examples of *output (level 3)* indicators

FIGURE 3. Proposed program evaluation framework for sustainable agriculture

Feedback loops	Program evaluation hierarchy	Indicator examples
Response indicators: policy/legislation change, level of research and education support, consumer education	Level 7 End results	State indicators: soil nutrient levels, organic matter content of soil, biocontrol and Integrated Pest Management (IPM) practices, riparian buffer strips, species diversity, farm economic status, family quality of life
	Level 6 Practice change	Driving force indicators: level of fertilizer, cover crop, pesticide, and IPM use; markts, consumer demand, financial performance, and networking
	Level 5 KOSA change	Knowledge: Change in awarenss and understanding Opinion: Change in interest in ideas and practices Skills: Change in physical and verbal activities Aspirations: Change in plans for future actions
	Level 4 Reactions	Benefits, usefulness, and practicality of research and education programs as perceived by program participants
	Level 3 Outputs	Information on activities completed and products developed (such as number and type of articles, extension publications, presentations, recommendations formulated, demonstrations and field days)
	Level 2 Activities	Information on what the program offers (such as research projects funded/year; number of sustainable agriculture research projects funded through other sources)
	Level 1 Inputs	Information on resources used in the program (such as number of researchers and students involved, number of farmers involved in research and education and the value of their time)

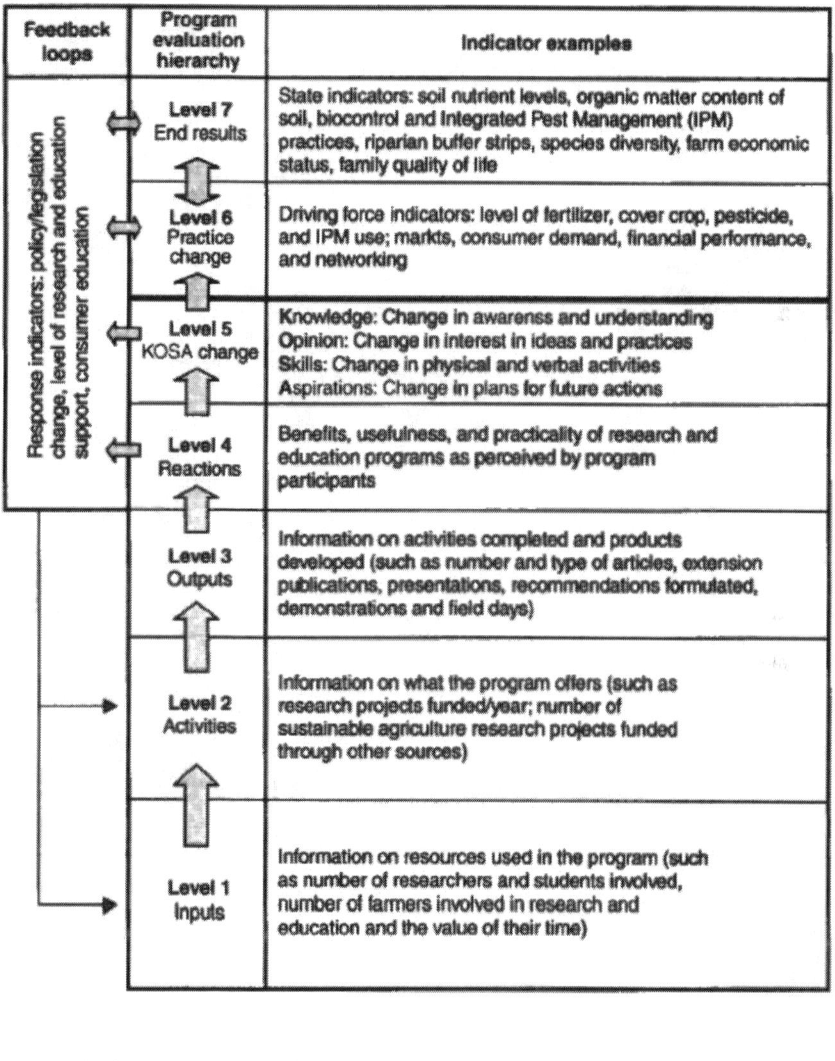

- percentage of target audience participating in programs
- appropriate use of communications tools
- number of articles published
- number of educational materials completed

Measurement of output indicators could occur through end-of-workshop questionnaires, review of articles published, and follow-up surveys and interviews to determine whether audience members perceived the message to be clear and appropriate.

Examples of *reaction (level 4)* indicators

- percentage of workshop participants who rated a workshop as "useful" or "very useful"
- perceptions by participants of strengths and weaknesses of a program
- verbal comments from participants during program regarding value of the information to them

Measurement of reactions could occur through end-of-workshop questionnaires, from observation of participants, and follow-up surveys.

Level five (KOSA) indicators represent changes in the intended audience relating to knowledge, skills, opinions, and aspirations as a result of exposure to the research and education. Examples of KOSA indicators

- further information sought by intended audience members
- increase in knowledge on topic by audience members
- perception by intended audience that they will likely use information gained from the program

Measurements for KOSA indicators could include pre- and post-workshop knowledge and skills tests and follow-up surveys to measure perceptions about participants' intentions.

Indicators for Assessing Sustainability (Levels 6 and 7)

Level 6 assesses the influence of sustainable agricultural research and education on farming practices. We used "driving force" indicators such as level of fertilizer, cover crop use, pesticide use, and IPM practices adopted. Level 7 assesses the level of sustainable agriculture being practiced overall. We used "state" indicators such as nutrient levels, soil organic matter content, biocontrol and IPM practices, as shown in Figure 3. These indicators help assess sustainable agriculture at a given point of time. "Response" indicators, such as policy/legislative changes and research and education funding levels, provide the feedback loop.

Indicators for assessing sustainable agriculture at levels 6 and 7 can be classified into seven interrelated categories. The following is a brief description of each category.

Agricultural Nutrient Use and Management

Crop growth depends on the presence of appropriate amounts of soil nutrients. Nutrients lost to crop production or other factors can be replaced through synthetic fertilizers or through on-farm inputs such as manure and compost. Nutrient management refers to practices to affect the efficiency and rate of nutrient use within the soil. Examples of indicators

- share of land that is regularly analyzed for nutrient content
- share of farms using a nutrient management plan
- area of land requiring less than normally recommended off-farm nutrient inputs

Pest Management

While pesticides have contributed significantly to agricultural production, their use for pest management in crop and animal production are studied because of potential and known environmental and health hazards associated with their use. Indicators for pest management include variables regarding pesticide use, use of alternative pest management practices, incidence of pests and diseases, as well as factors indicating farmer motivation to use alternative pest management practices. Examples of indicators

- participation in pesticide applicator refresher courses
- acres under IPM practices
- existence of emergency plan for pesticide use and storage

Agricultural Land Use and Conservation

While some agricultural land uses enhance conservation factors, agricultural land use patterns are also shown to negatively impact biodiversity, wildlife habitat and other environmental factors at a landscape level. Neave, Kirkwood, and Dumanski (1995) identified five objectives for sustainable agricultural land management in Canada (productivity, security, conservation, economic stability, and social acceptability). Since land use changes both positively and negatively affect sustainability, it is important to more clearly define the linkages between land use, land conservation practice, and environmental quality indicators. Examples of indicators

- share of land on which soil conservation practices (e.g., no-tillage, strip contouring, terracing) are adopted
- changes in total agricultural land area in relation to the total land area

Soil Quality and Management

Indicators reflecting the condition of the soil are important because soil is the link between the farm and the ecosystem level (Lewandowski, Hardtlein, and Kaltschmitt, 1999). A key issue in sustainable agriculture is the prevention of soil degradation resulting from erosion and land use. Assessing soil quality is difficult and costly because of the enormous variability in soils and because of the difficulty in isolating natural influences from human influences on soil quality. Parr et al. (1992) suggested the development of a soil quality index for various soil attributes to enable simulation and prediction. Examples of indicators

- soil organic matter content
- microbial and earthworm activity
- use of appropriate tillage practices

Agriculture and Biodiversity Levels

Development of suitable indicators for biodiversity in agriculture is complex because of the different levels at which biodiversity operates in agriculture. Diversity is considered within species, among species, and at the ecosystem level. Crop diversity reduces risk of large-scale pest infestation and minimizes the risk of domestic crop species encroaching natural habitats. Examples of indicators

- use of intercropping, cover crops, crop diversification, windbreaks, live fencing, hedgerows using a diversity of species
- presence of diversity of wild species (both flora and fauna)
- changes in ecosystem diversity (large scale habitats) associated with agriculture

Farm Financial Resources and Economic Performance

Developing meaningful indicators for sustainability regarding farm finances is complicated by the fact that financial and economic factors influence environmental factors to varying degrees. A farmer's financial management skills, adaptability, and risk aversion may influence his/her likelihood to use certain practices. Influences such as agricultural and land use policy and socio-cultural factors also affect practices. A farmer's access to financial resources such as off-farm income, capital, government support, and market returns also provide strong influences. Access to financial resources may influence farmers' attitudes toward environmental risk and regulation. Examples of indicators

- average rate of return on capital employed
- net farm and off-farm income
- average debt/equity ratio

Quality of Life

Sustainable agriculture depends on the social well being of those involved in the industry. Field experience shows that people who are stressed, isolated and unhappy, households that are dysfunctional, and communities that are disintegrating are unlikely to contribute to success in achieving a sustainable industry. Social indicators attempt to measure both sense of emotional and physical well being and a positive attitude about sustainability. Examples of indicators

- adequate access to information, skills, and resources needed or support networks
- ability and capacity to participate in social and public affairs
- sense of physical well-being
- satisfaction with family life
- understanding about what is required to achieve sustainability
- priority placed on self-education about sustainable practices

The Linkage Between Indicators

Farming systems are integrated webs with significant interplay between indicators. Changes in one indicator affect the status of others. Pest management indicators, for example, link to farm financial resources and economic performance, soil quality and management indicators. Biodiversity indicators, for another example, link to soil and water quality, crop productivity, economic performance, nutrient use and management, and pest management indicators. Understanding the importance of the linkages between indicators should help evaluators and managers to see the "whole picture" of sustainability.

LIMITATIONS OF APPLYING AN EVALUATION FRAMEWORK

Evaluators and managers should be aware of the limitations of applying an evaluation framework to evaluate sustainable agriculture. We describe these limitations as follows.

Lack of a shared definition of sustainability. Scholars do not share a single perspective on sustainability (Lewandowski, Hardtlein, and Kaltschmitt, 1999). Crews, Mohler, and Power (1991) consider sustainability to be a measure of a system's potential to endure. Richardson (1994) found that sustainability is not a state to reach, but a condition to maintain dynamically. White, Bradley, and Higley (1994) also saw sustainability as a general direction, rather than a fully defined objective or destination. Lack of a common definition can be attributed to different "world views" (Ikerd, 1996; Waltner-Toews, 1994). While there is substantial agreement that sustainable agriculture must be ecologically sound, there is less agreement about the economic and social and economic dimensions (Ikerd, 1996). Varying perceptions of sustainability makes it difficult to establish indicators of sustainability. Stakeholder processes to establish indicators often get "bogged down" or delayed, thereby acting as bottlenecks in the evaluation process.

Selected indicators may not sufficiently reflect the desired level of sustainability. While indicators enable us to see "the big picture" by revealing smaller pieces, they can also be misleading (Pomeroy, 1997). There is a risk that programs may target indicators rather than targeting the situation the indicators are intended to represent. One cannot assume that indicators can be transferred easily from one context to another; it is dangerous to generalize them over space, time, and different societies. It is easy to presume that one can "roll together" or aggregate indicators and that complex sets of data can be reduced to a simple statement or set of statistics. Information that is lost in aggregating the indicators may have been critical for revealing the "big picture."

Because of the seasonal nature of agriculture, certain management practices may be positive indicators at one point in time but may not have the same impact at another point in time (OECD, 1997). Because of the interaction of factors in a farming system, practices deemed sustainable might positively affect one component of the system while negatively affecting others. This makes aggregation at a state or regional level difficult, if not impossible. When aggregation is necessary or desirable, additional contextual information must be provided in order for indicators to be meaningful. This may involve a significant increase in the amount of data to be collected and analyzed.

Lack of ability to reliably demonstrate causal links between research and sustainability indicators. Because there are many non-research factors such as policy, markets, and cultural influences affecting farm management decisions, it is difficult to establish causal links between research recommendations and farm management decisions.

Since SARE research represents only a portion of the total amount of research-based information available to farmers, it may be impossible to isolate the impacts of SARE research from that of other research. To reliably claim causal linkages, evaluators would be forced to conduct costly fieldwork and analysis and wait several growing seasons to measure changes. Research managers, policy makers, and donors would be forced to endure a significant lag time between commencement of research and a time when impacts can be meaningfully assessed.

Where training has been considered an indicator of capacity building, it is difficult to make causal links because the impact of training on agricultural sustainability may be affected by other factors (Mabeza, 1993).

Difficulty of measuring indirect and non-tangible benefits, and problems of scale. One evaluation issue identified by Baidu-Forson (1996) is the intangible nature of many products available through sustainable agriculture. Benefits are often "collective"–not attributable to individuals, and thus difficult to measure.

The scale at which benefits are measured also influences the outcome of the evaluation. What one sub-region may gain from certain practices, for example, may not be significant when examining benefits at a larger landscape level. Herdt and Lynam (1992) looked at economic impact assessment of sustainable agriculture research and stated that such assessments require detailed micro-level data and cannot be carried out at the level of the market. As these authors write:

> ... The indirect benefits from the income streams generated by [sustainable agriculture] research will be virtually impossible to track through the market because, by their nature, the costs and benefits have external dimensions and so any market price effect will be spread across a diversity of different markets. Such research will have benefits to society in general, not only through lower direct costs of production, but also through lower external costs or at least through systems where increased external costs are off-set by lower internal costs–conditions which are difficult to measure. (7)

Moreover, conditions such as soil type, topography, and existing management practices heavily influence outcomes. Such heterogeneity introduces monitoring and sampling problems for data collection and aggregation of results.

Equity issues. A key question in assessing sustainability is "sustainable according to whom?" Who–and at what level–determines the criteria and indicators for evaluation? Once indicators are selected, how are they interpreted and who should do the interpreting? Ethical considerations arise when establishing indicators of sustainability to ensure that it is not only holders of influence who define sustainability, but those among the farm, community, and regional levels.

Difficulty of developing universally accepted indicators. The aggregation of indicators across regions causes them to lose meaning for specific locations. To be relevant to local areas, indicators must be locally defined. On the other hand, identifying indicators across regions is necessary for state and regional policy making. The process of identifying local indicators that involves stakeholders is time consuming and expensive.

Uncertainty regarding accountability for achievement of sustainable agriculture. Monitoring and evaluation are typically conducted to demonstrate program or policy impacts to a particular body that is held accountable for the outcomes of particular interventions. Given the multiple scales and variety of stakeholders within sustainable agriculture, no single body or group is accountable for ensuring that sustainable agriculture is achieved on a larger scale. This fact makes focusing an evaluation difficult for evaluators and managers, and as a result they can often be unclear about the purpose and methodology of the evaluation.

Variability in time frames for measuring sustainability. The time frame required by decision-makers and funders for program impact information may not be realistic for measuring progress toward sustainability. While some agricultural practices may lend themselves to short-term assessments, others may require longer periods of data collection to determine sustainability.

CONCLUSION

Among the challenges facing research and education managers and decision makers involved in sustainable agriculture is how to evaluate their investments in light of the integrated and dynamic nature of the systems. Research and education programs that attempt to balance environmental, economic, and social objectives are challenging to evaluate, and there are many challenges.

A common way to assess the performance of a system or program is through the use of indicators. The most important feature of indicators compared to other forms of information is their relevance to policy and decision making. While indicators are instrumental in gauging change over time and space, they have limited use for evaluating sustainable agriculture when they are considered in isolation from each other. Conceptual frameworks can help facilitate and organize indicators relevant to sustainable agriculture.

Evaluators must be aware that it may be difficult, if not impossible, to claim causal linkages using evaluation frameworks independently for assessing agricultural sustainability. Therefore, a framework that combines the elements of the Program Evaluation (Bennett's Hierarchy) and the DSR model (OECD, 1997 and Mortensen, 1997) offers the best tool for evaluating investments in sustainable agriculture.

Managers and evaluators must be careful not to make causal claims about their program in excess of what they can reliably defend. Indicators can help keep managers and decision makers clear about what program impacts they *can* make claims about, and what they can realistically measure given time and funding constraints. An iterative process involving key stakeholders to maintain a manageable indicator list enables managers to assess sustainability, support decision making, and satisfy reporting requirements.

Managers should recognize that selection of indicators is largely a political process that involves decisions relating to equity and scale. The question of "sustainable according to whom?" should not be ignored when selecting indicators. No evaluation process can "do it all": decision makers must make choices about what they want to (and reasonably can) influence with their programs and develop evaluation processes to measure those.

Managers and evaluators should appreciate the significance of the interplay among indicators of sustainable agriculture. Because changes in one indicator affect others, indicators should not be measured in isolation of one another. While other social scientists may choose to study individual aspects of sustainability, evaluators of investments in sustainable agriculture must consider all components as a system and measure them using a systems approach. Thus, we hope that professionals working for the development of sustainable cropping systems will find the conceptual model we have proposed in this paper to be helpful while planning, monitoring, and evaluating their programs.

REFERENCES

Baidu-Forson, J. (1996). Methodology for Evaluating Crop and Resource Management Technologies. In *Partners in Impact Assessment. Summary Proceedings of the ICRISAT/NARS Workshop on Methods and Joint Impact Targets in Western and Central Africa*, May 3-5, 1995, Sadore, Niger and May 9, 11-12, 1995, Samanko, Mali, ed., J. Baidu-Forson, M. C. S. Bantillan, S. K. Deborah, and D. D. Rohrbach, Patancheru, AP, India: ICRISAT.

Batie, S. S. and D. B. Taylor. (1991). Assessing the Character of Agricultural Production Systems: Issues and Implications. *American Journal of Alternative Agriculture* 6(4): 184-187.

Bennett, C. F. and S. K. Rockwell. (1995). *Targeting Outcomes of Programs (TOP): An Integrated Approach to Planning and Evaluation*. Draft. Lincoln, NE: Cooperative Extension, University of Nebraska.

Bennett, C. F. (1977). *Analyzing Impacts of Extension Programs*. Washington, DC: US Department of Agriculture Extension Service.

Crews, T. E., C.L. Mohler, and A. G. Power. (1991). Energetics and Ecosystem Integrity: The Defining Principles of Sustainable Agriculture. *American Journal of Alternative Agriculture* 6(3): 146-149.

Douglas, G. (1985). When Is Agriculture 'Sustainable'? In *Sustainable Agriculture & Integrated Farming Systems: 1984 Conference Proceedings*, ed., T. C. Edens, C. Fridgen, and S. L. Battenfield, East Lansing, MI: Michigan State University Press.

Edwards, C. L., R. Lal, P. Madden, R. H. Miller, and G. House. (1990). *Sustainable Agricultural Systems*. Ankeny, IA: Soil and Water Conservation Society.

Falconi, C. A. (1993). Economic Evaluation. In *Monitoring and Evaluating Agricultural Research*, ed., D. Horton, P. Ballantyne, W. Peterson, B. Uribe, D. Gapasin and K. Sheridan, Wallingford, Oxon, UK: CAB International, pp. 64-75.

Francis, C. A. and G. Youngberg. (1990). Sustainable Agriculture: An Overview. In *Sustainable Agriculture in Temperate Zones*, ed., C. A. Francis, C. B. Flora and L. D. King, New York: John Wiley & Sons, pp. 1-23.

Gallopin, G. C. (1997). Indicators and Their Use: Information for Decision-Making. In *Sustainability Indicators: Report of the Project on Indicators of Sustainable Development, Part One–Introduction*, ed., B. Moldan and S. Bilharz, New York, NY: John Wiley & Sons Ltd. pp. 13-27.

Gapasin, D. P. (1993). Research Project Management. In *Monitoring and Evaluating Agricultural Research*, ed., D. Horton, P. Ballantyne, W. Peterson, B. Uribe, D. Gapasin, and K. Sheridan, Wallingford, Oxon, UK: CAB International, pp. 155-161.

Gliessman, S. R. (1997). *Agroecology: Ecological Principles in Sustainable Agriculture*. Boca Raton, FL: Lewis Publishers.

Hart, M. (1995). *Guide to Sustainable Community Indicators*. Ipswich, MA: QLF/Atlantic Center for the Environment.

Harwood, R. R. (1993). A Look Back at USA's Report and Recommendations on Organic Farming. *American Journal of Alternative Agriculture* 8(4):150-154.

Herdt, R. W. and J. K. Lynam. (1992). Sustainable Development and the Changing Needs of International Agricultural Research. In *Assessing the Impact of International Agricultural Research for Sustainable Development, Proceedings from a symposium at Cornell University, Ithaca, NY*, June 16-19, 1991, ed., D. R. Lee, S. Kearl, and N. Uphoff. Ithaca, NY: Cornell International Institute for Food, Agriculture and Development.

Horton, D., W. Peterson, and P. Ballantyne. (1993). M&E Principles and Concepts. In *Monitoring and Evaluating Agricultural Research*, ed. D. Horton, P. Ballantyne, W. Peterson, B. Uribe, D. Gapasin, and K. Sheridan, Wallingford, Oxon, UK: CAB International, pp. 5-19.

Ikerd, J. (1996). Sustainable Agriculture: Do We Really Need to Define It? Presentation to the Michigan Agriculture Mega-Conference, January 12, 1996, Lansing, Michigan.

Ikerd, J., G. Devino, and S. Traiyongwanich. (1996). Evaluating the Sustainability of Alternative Farming Systems: A Case Study. *American Journal of Alternative Agriculture* 11(1):25-29.

Lewandowski, I., M. Hardtlein, and M. Kaltschmitt. (1999). Sustainable Crop Production: Definition and Methodological Approach for Assessing and Implementing Sustainability. *Crop Science* 39:184-193.

Mabeza, H. (1993). Training Evaluation. In *Monitoring and Evaluating Agricultural Research*, ed., D. Horton, P. Ballantyne, W. Peterson, B. Uribe, D. Gapasin and K. Sheridan, Wallingford, Oxon, UK: CAB International, pp. 180-186.

Madden, J. P. and S. G. Chaplowe. (1995). Introduction and Overview. In *For All Generations: Making World Agriculture More Sustainable*, ed., J. Patrick Madden, Glendale, CA: OM Publishing, pp. 3-37.

McLean, D. (1994). Principles of Project Monitoring and Evaluation. In *Management for Researchers*, ed., R. M. A. Lyons and F. Katepa-Mupondwa, Winnipeg, Manitoba, Canada: University of Manitoba, Department of Agricultural Economics and Farm Management, pp. 251-262.

Mortensen, L. F. (1997). The Driving Force-State-Response Framework Used by CSD. In *Sustainability Indicators: Report of the Project on Indicators of Sustainable Development* , ed., B. Moldan and S. Bilharz, New York, NY: John Wiley & Sons Ltd., pp. 47-53.

Mulholland, M. E. (1993). CIPP Evaluation Model. In *Monitoring and Evaluating Agricultural Research*, ed., D. Horton, P. Ballantyne, W. Peterson, B. Uribe, D. Gapasin, and K. Sheridan, Wallingford, Oxon, UK: CAB International, pp. 53-57.

National Research Council. (1989). *Alternative Agriculture*. Washington, DC: National Academy Press.

Neave, P., V. Kirkwood, and J. Dumanski. (1995). Review and Assessment of Available Indicators for Evaluating Sustainable Land Management. Technical Bulletin 1995-7E. Ottawa, Canada: Agriculture and Agri-Food Canada, Centre for Land and Biological Resources Research Central Experimental Farm.

Organization for Economic Cooperation and Development (OECD). (1997). *Environmental Indicators for Agriculture*. Paris, France: OECD.

Parr, J. F., R. I. Papendick, S. B. Hornick, and R. E. Meyer. (1992). Soil Quality: Attributes and Relationship to Alternative and Sustainable Agriculture. *American Journal of Alternative Agriculture* 7(1-2):5-11.

Peterson, W. (1993). Development-Agency M&E. In *Monitoring and Evaluating Agricultural Research*, ed., D. Horton, P. Ballantyne, W. Peterson, B. Uribe, D. Gapasin, and K. Sheridan, Wallingford, Oxon, UK: CAB International, pp. 58-64.

Pomeroy, A. (1997). Social Indicators of Sustainable Agriculture. In *Situation and Outlook for New Zealand Agriculture*. Wellington, New Zealand: Ministry of Agriculture and Forestry.

Pretty, J. N. (1995). *Regenerating Agriculture: Policies and Practices for Sustainability and Self-Reliance*. London: Earthscan Publication Ltd.

Richardson, R.H. (1994). *"Re: HRM Life Assessment/Goals–Answers, Long."* 7 October 1994, <http://www.sare.org/htdocs/hypermail/html-home/5-html/0374.html> (7 December 2001).

Rieper, O. and J. Toulemonde. (1997). *Politics and Practices of Intergovernmental Evaluation*. ed., O. Rieper and J. Toulemonde, New Brunswick, NJ: Transaction Publishers.

Roberts, W. S. and S. M. Swinton. (1996). Economic Methods for Comparing Alternative Crop Production Systems: A Review of the Literature. *American Journal of Alternative Agriculture* 11(1):10-17.

Rossi, P. H. and H. E. Freeman. (1985). *Evaluation: A Systematic Approach*. Beverly Hills: Sage Publications.

Rural Industries Research & Development Corporation (RIRDC). (1997). RIRDC Short Report No. 20: Developing Indicators for Sustainable Agriculture. Barton, ACT, Australia: RIRDC.

Smoik, J. D., T. L. Dobbs, and D. H. Rickerl. (1995). The relative sustainability of alternative, conventional and reduced-till farming systems. *American Journal of Alternative Agriculture* 10(1):25-35.

UNEP-DPCSD. (1995). The Role of Indicators in Decision-Making. Discussion paper prepared for UNEP for the Indicators of Sustainable Development for Decision Making Workshop, January 9-11, 1995, Ghent, Belgium.

Uribe, B. and D. Horton. (1993). Logical Framework. In *Monitoring and Evaluating Agricultural Research*, ed., D. Horton, P. Ballantyne, W. Peterson, B. Uribe, D. Gapasin, and K. Sheridan, ed., Wallingford, Oxon, UK: CAB International, pp. 113-119.

Waltner-Toews, D. (1994). Ecosystem Health: A Framework for Implementing Sustainability in Agriculture. In *Agroecosystem Health*, ed., N. O. Nielsen, University of Guelph, Guelph, Ontario, Canada, pp. 8-23.

White, D. C., J. B. Bradley, and L. G. Higley. (1994). Economics of Sustainable Agriculture. In *Sustainable Agriculture Systems*, ed., J. L. Hatfield and D. L. Karlen, Boca Raton: Lewis Publishers, pp. 229-260.

From Chemical Ecology to Agronomy:
Cropping Systems in the Humid Northeast

D. L. Smith
C. Costa
B. Ma
C. Madakadze
B. Prithiviraj
F. Zhang
X. Zhou

SUMMARY. Crop production in the northeastern North America poses unique challenges in that soils tend to be cool and wet in the spring, and there is only a brief period each summer when sufficient heat is obtained

D. L. Smith, C. Costa, B. Prithiviraj and X. Zhou are affiliated with the Plant Science Department, McGill University, Macdonald Campus, 21111 Lakeshore Road, Ste Anne de Bellevue, Quebec, Canada H9X 3V9.

B. Ma is affiliated with Agriculture and Agri-Foods Canada, Neatby Building, Central Experimental Farm, Ottawa, Ontario, Canada K1A 0C6.

C. Madakadze is affiliated with the Henderson Research Station, Private Bag 2004, Mazowe, Zimbabwe.

F. Zhang is affiliated with Bios Agriculture Inc., Macdonald Campus, McGill University, 21111 Lakeshore Road, Ste Anne de Bellevue, Canada H9X 3V9.

The authors thank the staff of the Lods Agronomy Research Station of McGill University, where most of their research was conducted.

The Intensive Cereal Management portion of this review constitutes Agriculture and Agri-Food Canada ECORC Contribution #11664.

http://www.haworthpress.com/store/product.asp?sku=J144
10.1300/J144v09n01_06

for the growth of the main crops produced in the area. This paper reviews seven areas of crop production and ecophysiology research in the context of the geo-climatic conditions of the extreme North American humid northeast. The seven areas are: (1) intercropping systems, (2) leafy reduced-stature corn, (3) intensive cereal management, (4) production of C_4 grasses, (5) development of a chronic injection system for physiology research, (6) legume-to-rhizobia signals and inhibition of soybean nodulation, and (7) rhizobia-to-legume signals and crop growth. Progress has been made in each area and there is now a longer-term need to integrate some of these findings at the cropping system level. *[Article copies available for a fee from The Haworth Document Delivery Service: 1-800-HAWORTH. E-mail address: <docdelivery@haworthpress.com> Website: <http://www. HaworthPress.com>* © *2003 by The Haworth Press, Inc. All rights reserved.]*

KEYWORDS. Intercropping, leafy reduced-stature corn, intensive cereal management, C_4 grasses, chronic injection, legume-rhizobia signals

INTRODUCTION

Crop Production Research in Southwestern Quebec–
The Extreme Humid Northeast

Adoption of agricultural crops and cropping systems in any geographical region is dictated by environmental, physical, social, and economic factors. Agriculture in eastern Canada faces significant obstacles related to climate and other aspects of the physical environment such as soils and landscape. In this paper, the St. Lawrence lowlands of Quebec will be used as an example of specific challenges dictating cropping systems in eastern Canada. Much of the soils in the area were deposited through marine sedimentation near the end of the last ice age, when the area was covered by the shallow Champlain sea (Parent and Occhietti, 1998). Thus the soils tend to be flat and heavy. Under natural conditions these soils drain very slowly in the spring, and would not be suitable for heavy vehicle traffic until some time in June during most years. This makes the area suitable only for perennial crops such as hay. However, from the 1960s to the 1980s, the provincial government provided subsidies for the installation of drainage tiles. Adequate drainage has turned large areas of marginal land into some of the most productive agricultural areas in the province of Quebec.

Precipitation is not a serious problem in most years as there is a complete soil recharge due to snow melt in the spring and, in most years, there is a reasonable level and distribution of rainfall during the season (Figure 1). How-

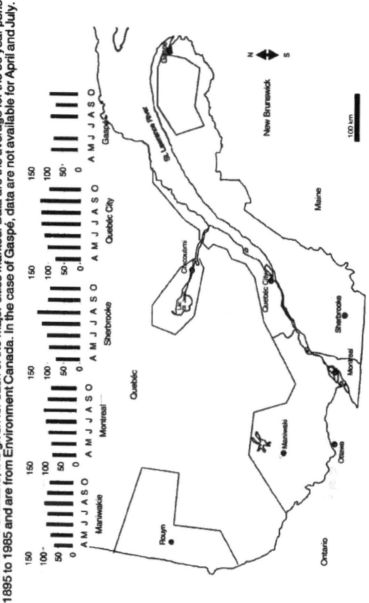

FIGURE 1. Southwestern Quebec, with the major agricultural areas indicated (bounded by straight lines, largely along the St. Lawrence River. Rainfall (mm) for the months of the growing season (A = April, M = May, J = June, J = July, A = August, S = September, O = October, in that order) is indicated in the graphs along the top. Where rainfall data is available, it is given for each of the major cities marked. Data are the average for the 90 year period, 1895 to 1985 and are from Environment Canada. In the case of Gaspé, data are not available for April and July.

ever, the growing-season in all areas of this region is short, with a limited amount of thermal-time available for crop development. While the area is quite suitable for the production of rapidly maturing small grain cereals, the production of longer-season crops, such as soybean (*Glycine max* L. Merr.) and corn (*Zea mays* L.), is more challenging. But, the yield level and/or quality attributes of these crops, however, have made their cultivation very desirable.

When hybrid corn was first introduced into the area during the 1960s, it was argued that the crop was much less well adapted to the region than the small grain cereals already being grown. However, the critics were proved wrong because corn yields were at least twice that of the small grain cereals in the region, making corn a very attractive crop to produce.

Soybean was introduced to Quebec in the mid-1970s, but its production was very limited until well into the 1980s. Although the yields of soybean are similar to those of the small grain cereals, the high protein and oil content of the crop makes it considerably more valuable. In addition, nitrogen (N) fixation by soybean makes it a viable rotation crop for corn and other cash crops.

For these reasons, there has been considerable impetus to adapt corn and soybean to the Quebec region. Both crops have received considerable attention in genetic improvement programs of the large seed companies in eastern Canada, often building on larger breeding programs further south. The short growing-season in eastern Canada led to the development of a very sensitive measure of thermal time, the corn heat unit (CHU) system (Brown and Bootsma, 1993). The CHU system provides a better measure of the thermal time required for maturity of corn and soybean genotypes than a simpler degree day system. In essence, the CHU system recognizes that growth, particularly during the daytime, is a quadratic, and not a simple linear function of thermal-time. The formula for calculation of CHU accumulated in a given day is:

$$CHU = \frac{9/5(T_{min} - 4.4) + 3.33(T_{max} - 10.0) - 0.084(T_{max} - 10.0)^2}{2}$$

where, T_{min} = the minimum temperature for each day
T_{max} = the maximum temperature for each day

A map of CHU distribution in the St. Lawrence lowlands and southern Ontario is found in Figure 2.

Corn is a "full season" row crop and it is often planted in early to mid-May in the St. Lawrence lowlands. The crop poses a threat to soil fertility and productivity because it does not germinate below 10°C, and develops slowly at temperatures below 20°C. During May and June soil and air temperatures in the region are frequently below 20°C. Thus the crop develops slowly during the spring, and, this coupled with the relatively high rainfalls for the area, make

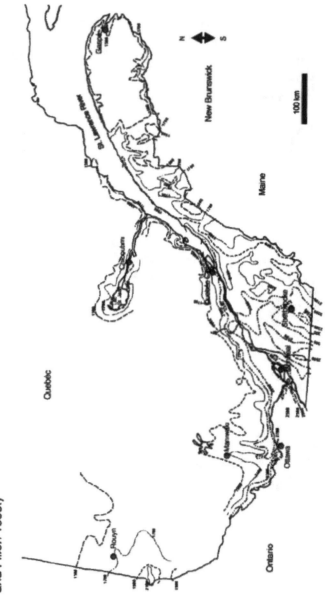

FIGURE 2. The distribution of corn head units in the agricultural areas of Quebec. Isobars do not extend into areas where levels of agriculture are minor. These areas exist because there are too few untis, or there is too little suitable soil, as is the case on most areas of the Canadian shield. (Adapted from Bootsma, Trembley and Filion 1999.)

the soils vulnerable to erosion. The crop is harvested late, often in late-October or November, when soils are again quite wet. Harvest is generally followed by fall tillage operations. Operation of heavy equipment on these soils can lead to compaction problems. These problems would seem to make no-till production an obvious choice. However, soils in no-till systems drain and warm up more slowly than under conventional tillage in the spring. In heavy soils and in years with cool early-season conditions, crop establishment in no-till systems can be delayed by several weeks. Given the already short-growing season, Quebec corn producers have thus been slow to adopt NT corn production.

Climate change will present significant challenges to crop production in this region. In general, the closer to the poles the greater the seasonal increases in temperature will be, and temperature increases have been greater for night temperatures than day temperatures (Karl et al., 1993). In northern locations, such as the St. Lawrence lowlands, the length of the growing season is usually between a time of sufficient average daily warmth in the spring and the first killing frost in the fall. Given that these frosts always occur at night and night temperatures are rising faster than day temperatures, the season will lengthen even faster than the average daily increase in temperature over the growing season. Corn producers in the area have been sensitive to the fact that seasonal temperatures have risen during the 1980s and 1990s, and that the CHU maps for the region, developed using data up to 1990, do not include data from the warmest decade on record. They are aware that hybrids that mature later have greater yield potentials and, as such, they often plant corn hybrids that are over the maturity rating (given in CHUs) for their areas. This has created problems in cooler summers. For instance, in 1996 and 2000 a substantial amount of the corn crop did not mature, and was simply plowed in. In 1996, producers with corn that did not mature were eligible for crop insurance. However, after 1996 the crop insurance program was altered such that producers growing hybrids outside the rating for their area would not receive payments for crop failure. In recent years the crop insurance program in this area has become less attractive, and so fewer producers are using it. However, those producers that do are wedded to a CHU map that is almost certainly out of date. One has to wonder if we have not, in this area, legislated lower yields for those who choose to use the crop insurance program.

A great deal of crop production in Quebec is tied to the production of dairy products as Quebec has the legal right to produce 48% of Canadian industrial milk (Anonymous, 1998). Corn and soybean are used as feeds for dairy herds in the region. Milk production in Canada, at present, is regulated through a quota system that makes milk production relatively lucrative for those who hold quota. Should the quota system be removed in the future, there may well be some reduction in the use of corn and soybean as feeds for dairy herds, and an increase in the use of grazing systems.

A system of subsidies, referred to as crop stabilization payments, is in place for the more widely produced crops in Quebec, except for hay and pasture crops. While this system removes some of the extreme variation in the annual earnings of farmers, and makes life for crop producers less precarious, it makes it more difficult to introduce new crops that, at least initially, do not qualify for stabilization payments.

Because of high yield potential of corn and the high quality (oil and protein) of soybean, these are now the most widely produced grain crops in southwestern Quebec. However, both of these crops are of tropical and subtropical origin and adapting them to the humid northeast continues to be a challenge. To retain a position of economic competitiveness, productivity needs to remain high. At the same time, there are environmental considerations, such as groundwater nitrate levels and greenhouse gas emissions, associated with crop production in this area. Higher crop productivity without increased environmental damage could be accomplished by utilization of innovative production systems, such as intercropping of corn and soybean. In addition, crops could be made more adapted to the region. For example, though the development of novel genotypes, such as leafy reduced-stature corn (develops light interception capability rapidly and reaches maturity quickly); or overcoming some of the climatic limitations, for instance, by allowing soybean to nodulate and fix N early in the growing season, in spite of low soil temperatures.

As the climate change issue grows, there will be a greater need to produce plant biomass that can act as an alternative source of high quality fiber, to reduce the pressure on standing forest carbon (C) stocks or as a source of material for biofuel production. New crops, such as the warm season C_4 grasses may help in this regard and their potential needs to be investigated. These crops may also serve as a source of forage for grazing livestock, such as dairy cows, during the warm summer period when the imported European grasses do poorly.

Although it has fallen during the past 15 years, there is still considerable production of small grain cereals in Quebec, largely barley (*Hordeum vulgare* L.) and oat (*Avena sativa* L.), with a smaller amount of wheat (*Triticum aestivum* L.). Sophisticated and intensive systems have been developed for the production of these crops under the maritime and modified maritime conditions of northern Europe. These systems have been very successful there. It is rational to evaluate the various inputs that make up these systems for small grain cereal production in Quebec.

Finally, we need to better understand the ultimate limitations on crop productivity, in the hope that these may be reduced or removed. While much of the work in the last 150 years has looked at the environmental limitations (soil fertility and pH, water availability, thermal conditions) and bringing together the best combinations of the existing genetic potentials (plant breeding). We have not previously examined the possibility of removing apparent intrinsic biological limitations. Work on biological constraints to N uptake by plants and to photosynthetic rates have the potential to direct us to new methods for increasing crop yields. Here we review data showing that root uptake of N is more limiting to the accumulation of protein in cereal seeds than is the uptake of N already inside the plant system by the seeds. We also show that the limitation of roots to uptake of soil N limits photosynthetic rates. Understanding this limitation and then removing it would allow higher crop yields and more efficient uptake of fertilizer N from the soil, leading to reduced soil N levels. In addition, we show that signal compounds involved in plant-microbe interactions have the ability to cause profound changes in overall plant metabolism, leading to increased photosynthetic rates and yields. Given that the major contributors to crop yield increases during the last 50 years (water availability, chemical fertilizers, increased harvest index) are largely exhausted, there is a growing need to look at the management of the basic biology. In short-season areas, such as the northeast, where the limitations to production are greater than in many other areas, the need for such biological interventions will be greatest.

In this paper we will summarize seven areas where our group has made research progress and, therefore, acquired some expertise, within the confines of the available conditions of the northeast.

INTERCROPPING SYSTEMS: CROP PRODUCTION WITH IMPROVED RESOURCE USE EFFICIENCY AND DIMINISHED NEGATIVE ENVIRONMENTAL EFFECTS

Intercropping has been a traditional crop production practice for millennia, providing sustainable farming systems with increased resource use efficiency. Intercropping has the potential to be more important to agriculture in the future because of its more efficient use of environmental and other resources than monocrop systems (Midmore, 1993; Innis, 1997). Research into intercropping systems in eastern Canada has demonstrated potential advantages in terms of resource use, reduced soil N residues after harvest, and better weed control.

Capture of Resources and Resource Utilization

The main advantage of intercropping is efficient resource capture (Innis, 1997). Martin et al. (1990, 1991) studied corn hybrid types (dwarf and tall corn), N rates (0, 60, and 120 kg ha^{-1}) and cropping systems (corn-soybean intercrop, corn and soybean monocrops) in the Ottawa area and demonstrated that dwarf corn with soybean intercrops, in a dry year, showed a consistent trend toward lower dry matter (DM) production and protein yield than a tall-corn intercrop at the same N level. The total silage protein concentration was 1.84% higher in intercrops than in the corresponding monocropped corn hybrid at the same N level over the two years of the study. Zhou et al. (1997a) studied corn intercropped with annual Italian ryegrass (*Lolium multiflorum* Lam.) and found that intercropping changed the allocation of the N taken up and DM production of corn plants more frequently at tasseling and mid-grain filling than other development stages, more in the leaves than other parts, and more in a wet year than that a dry year. When compared to monocropped corn, intercropping increased total DM production and N uptake by 2.9 Mg ha^{-1} and 64 kg N ha^{-1}, respectively, over the two years of this study. Their study emphasized the effect of climatic conditions and component crop growth/establishment on the good performance of intercropping systems in terms of DM production and N uptake.

Research on crop yield and yield components in intercropping systems can provide insights regarding crop competition and cropping system design. Zhou et al. (2000b) reported that corn grain yield was not affected by intercropped annual ryegrass as long as adequate nutrients and soil water were supplied. Furthermore, Carruthers et al. (2000) reported that intercropped forages, regardless of type, had no impact on yield and yield components of other intercrops [corn, soybean and lupin (*Lupinus albus* L.)]. They indicated that corn was the strongest competitor in such intercropping systems while lupin was a poor component crop for inclusion with corn. They concluded that lupin should not be considered as an intercropping component for eastern Canada, particularly if the goal of the cropping system was yield stability. On the other hand, if intercropped forage established rapidly and grew vigorously during early development, the yield reduction of another component crop could not be avoided. For instance, forage seeded into corn rows 10 days after corn emergence established better and accumulated more biomass than forage seeded 20 days after corn emergence (Abdin et al., 1997), but caused a 19% yield reduction in the corn component. They attributed this observation to the rapid forage establishment and vigorous early forage growth. Measurement of chlorophyll fluorescence [an indirect indicator for environmental stress (Krause and Weis, 1984)] by Abdin et al. (1998a) indicated that corn plants were more stressed when intercropped with early rapidly established forages than those seeded later. Similar results were reported by Exner and Cruse (1993) when corn was intercropped with forage legumes.

Generate and Convert Resources

The success of a legume and non-legume intercrop system greatly depends on how effectively N is fixed by the legume component, and more importantly, on subsequent transfer of N to the non-legume component. Nitrogen transfer has been observed in legume-cereal mixtures (Fujita et al., 1990; Ofosu-Budu, Noumura, and Fujita, 1995; Senaratne, Liyanage, and Soper, 1995), due to exudation of nitrogenous compounds by legume roots during growth (Virtanen, 1933) and decay of legume tissues after harvest (Simpson, 1965). The enhanced cereal growth in such cropping systems has been related to improved N nutrition, as a result of N transfer (Broadbent, Nakashima, and Chang, 1982) or the N "sparing" effect, by the legume (Herridge and Brockwell, 1988).

Understanding of indirect, direct or residual N transfer from the legume to a companion crop, or from a succeeding legume crop to a following non-legume, should lead to development of new, more N efficient and more productive legume-based intercropping systems. Using ^{15}N enrichment and dilution techniques, N transfer from soybean to corn was detected under both greenhouse (Martin, Voldeng, and Smith, 1991a) and field conditions (Martin, Voldeng, and Smith, 1991b) when the two were intercropped. In addition, Martin, Voldeng, and Smith (1991b) reported that N transfer was more pronounced in N-depleted soil, suggesting that there was less N transfer, and that it was less important to the non-legume component crop, when mineral N is plentiful (Eaglesham et al., 1981). The environmental factors affecting N transfer in corn-soybean intercrop systems were investigated by Martin et al. (1995). They found that N transfer from soybean to corn was most evident with P at 25% of the normal nutrient concentration, under full light at 200 mg N, or shade at 400 mg N. Their results suggest that N benefit under full light is highest and most readily detected where mineral N is limiting.

Mycorrhizal fungi are known to absorb and translocate soil N to their host plants (Ames et al., 1983), and to translocate N from donor to recipient plants. Mycorrhizal fungi have been found to increase crop grain yield and enhance forage legume growth, nodulation and N_2 fixation (Hamel et al., 1991a; Hamel and Smith, 1991), and P uptake (Hamel, Furlan, and Smith, 1991, 1992; Hamel and Smith, 1991). Thus application of mycorrhizal fungi in legume-non-legume intercrop systems may lessen the level of competition for resources between intercrop components by increasing the efficiency of resource utilization (Haynes, 1980). Hamel and Smith (1991, 1992) studied N transfer in field-grown soybean-corn mixtures, when mycorrhizal fungi were added by inoculation. Both studies found the inoculated mycorrhizal fungi enhanced growth in both corn and soybean under field conditions. They attributed this enhancement mainly to a better P uptake by mycorrhizal plants, and concluded that interspecific mycorrhizae-mediated N transfer may be limited. Hamel et al. (1991a) reported that nearly 60% of the ^{15}N-transfer from soybean to corn occurred during the three weeks following soybean physiological maturity and was enhanced by root and hyphal contact, and by inoculation with endomycorrhizal fungi (G. versiforme). The facilitation of plant-to-plant N transfer by mycorrhizal fungi was greatly limited by soil microorganism populations (Hamel et al., 1991b), the higher N transfer being associated with lower total microbial C. Maximizing the benefits of N transfer will depend on the associated crop species, the interactions of the crop with soil mycorrhizal-fungi, and soil N and P levels.

Reduced Soil N Residues

Under soil cropped with potato (*Solanum tuberosum* L.) in this area, the level of nitrate nitrogen (NO_3^- N) in the drainage water was up to 40 mg L^{-1}, which is well above the Canadian water quality guideline (10 mg NO_3^- N) (Madramootoo, Wiyo, and Enright, 1992). Concern about NO_3^- leaching and groundwater pollution has prompted investigation of alternative production systems that reduce soil NO_3^- leaching (Liang, Remillard, and Mackenzie, 1991; Madramootoo, Wiyo, and Enright, 1992). Intercropping may be a way to bring an aggressive N-absorbing component crop into corn production systems to more efficiently utilize excess N, and thus reduce the potential for soil NO_3^- leaching. A corn-annual ryegrass intercrop decreased the amount of soil NO_3^--N in the top 1 m of the soil profile by 47% (92.3 kg N ha^{-1}) when ryegrass was well established in the system (Zhou et al., 1997b). Furthermore, fall-plowed intercropped ryegrass increased NO_3^- concentration in the subsoil due to decomposition over the winter (Zhou et al., 1997b); a following crop would benefit from the N released through this decomposition.

Weed Control

Intercropping suppresses weeds through direct competition or allopathic effects (Midmore, 1993). Carruthers et al. (1998) suggested that the interseeding of cover crops between rows of the main crop becomes a possible alternative approach for weed control. They found that the biomass and density of the dicot weeds between corn rows were most effected by corn intercropped with soybean or lupin which, in some cases, reduced the need for weed control by 73-100%. Among the various intercrop systems tested, corn-soybean intercrops were more successful at reducing weed populations than corn-lupin intercrops in the humid northeast. The relative extent of weed control under the grasses and legume forages intercropped with corn greatly depend on the level of weed infestation, and the climatic and soil conditions (Abdin et al., 2000). In some cases, interrrow tillage or some herbicide application was still necessary; the cover crops can provide additional weed control but interrow tillage generally provided the bulk of it.

LEAFY REDUCED-STATURE CORN: RAPID DEVELOPMENT OF LEAF AREA AND EARLY MATURITY

Canadian corn yields have increased steadily in recent years (from 6.6 Mg ha^{-1} to 6.9 Mg ha^{-1} in Québec between 1990 and 1997), however, corn production for grain remains limited by environmental conditions, chiefly the short season, in north eastern North America. Earlier flowering corn genotypes grown under the restrictive growth and development conditions inherent to the Canadian climate are smaller in stature and have longer grain-filling periods, compared with the larger plant size and a shorter grain-filling periods of later flowering genotypes. Under Canadian climatic conditions, the former group is source limited and the latter is sink limited.

The discovery of the "Leafy" gene (*Lfy*) in 1971 (Shaver, 1983) offered a possible way to increase source size. At the same maturity, corn genotypes bearing the leafy trait have more leaves and leaf area above the ear, and are taller than conventional hybrids. A shorter vegetative period, longer grain-filling period, and higher yield potential were noted for *Lfy* than conventional genotypes, while reduced stature (*rd1*) types mature rapidly (Begna et al., 1997a, 1997b; Modarres et al., 1997b, 1997c, 1998), making the *Lfy* and *rd1* genotypes particularly well suited for short season areas. Most notable among corn genotypes bearing the *Lfy* trait are leafy reduced-stature (LRS, *Lfy rd1*) and leafy normal-stature (LNS, *Lfy*) genotypes. While the *Lfy* trait promotes faster leaf area development, the *rd1* trait allows rapid overall development and maturity.

The earliest studies of genotypes carrying the *Lfy* trait were focused on aspects of crop physiology (Modarres et al., 1997a, 1997b; Begna et al., 1997a, 1999, 2000; Dijak et al., 1999), mathematical characterization of canopy characteristics (Stewart and Dwyer, 1993, 1999), concentration of soluble sugars and starch and ear position and green leaf area duration (Dwyer et al., 1995). A few researchers have attempted to determine management requirements for *Lfy* hybrids, particularly their response to population density (Begna et al., 1999) and weed competition (Begna et al., 1997a, 1997b, 1999). The most noticeable differences between *Lfy*-bearing corn genotypes and conventional hybrids are their canopy architectures (Figure 3). The availability of scanner-based methods for root measurement systems has raised interest in studying root system characteristics of *Lfy*-bearing corn genotypes and the interactions of this trait with soil available N (Costa et al., 2000, 2001).

Development of Corn Genotypes with Leafy and Reduced-Stature Traits

While the *Lfy* trait was first reported in 1971, the recessive (Coe, Neuffer, and Hoisington, 1988) *rd1* trait was reported as early as 1957 (Nelson and Ohlrroge, 1957). Their rapid development makes the reduced stature genotypes particularly well suited for short season areas, such as eastern Canada (Daynard and Tollenaar, 1983). Corn genotypes bearing the *rd1* trait are 0.73 to 1.56 m tall (Modarres et al., 1997b; Dijak et al., 1999) and have exceptional stalk strength, leading to reduced lodging due to insects or wind (Stoskopf, 1985). Furthermore, they can be grown in narrow rows (Begna et al., 1997a)

FIGURE 3. Profiles showing architectures of non-leafy reduced-stature (NLRS), leafy reduced-stature (LRS), leafy normal-stature (LNS), and a conventional hybrid (Pioneer 3979, P3979).

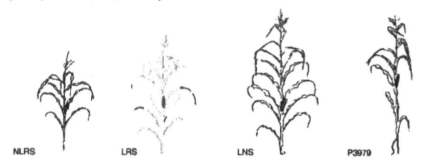

NLRS   LRS   LNS   P3979

which may, in turn, prevent birds from flying into the canopy, limiting bird damage to the edges of a stand (Hohenadel, 1984). The presence of the *Lfy* trait has also been shown to result in a greater allocation of photosynthate to the grain (Stoskopf, 1985).

The height, at silking, of LRS plants is between 0.84 m (Begna et al., 2000) and 1.50 m (Modarres et al., 1997b) (Table 1), at least 56% shorter than the average for conventional hybrids. LRS genotypes have been shown to be better suited to higher planting densities than their conventional counterparts (Begna et al., 1999). While plant population densities from 65 to 75 thousand plants ha^{-1} are generally recommended for conventional corn genotypes in Québec and Ontario, Canada (Russel, 1985; Tollenaar, 1991), LRS hybrids have better grain yields at higher plant population densities (9.8 at 100 thousand plants ha^{-1}; Begna et al., 1997b).

Number of Leaves

Plant height and total number of leaves above the ear are the characteristics most different between *Lfy*-bearing corn genotypes and their conventional counterparts (Figure 3). The *Lfy*-bearing corn genotypes produce, on average, a total of 17 or 18 leaves at silking, of which 7 to 10 are above the ear (Table 1; Shaver, 1983). This contrasts with 5 to 6 above-ear leaves for conventional hybrids. However, total and above-ear leaf numbers vary among *Lfy rd1* hybrids (Deng et al., 2001, unpublished data) and are affected by plant population density (Begna et al., 1999).

Leaf Area Index (LAI)

A number of studies have demonstrated that *Lfy*-bearing corn hybrids have LAIs (ratio of leaf area to ground area covered) about 20% greater than their

TABLE 1. Aboveground traits of leafy reduced-stature and conventional corn genotypes grown under field conditions.

| Trait | Corn genotype | | Soruce |
	Leafy	Conventional	
Plant height (cm)	84	.	Modarres et al., 1997b
	154	.	Dijak et al., 1999
	150	230	Begna et al., 2000
	143	219	Begna et al., 1999
	131	.	Modarres et al., 1998b
	177	258	Deng et al., 2001*
Total number of leaves	16.93	.	Modarres et al., 1997b
	16.10	12.50	Dwyer et al., 1998
	19.00	17.00	Deng et al., 2001*
Number of leaves above-ear	9.45	.	Modarres et al., 1997b
	8.00	.	Dijak et al., 1999
	7.80	5.30	Begna et al., 1999
	7.20	.	Modarres et al., 1998a
	7.30	.	Modarres et al., 1998b
	8.80	5.20	Dwyer et al., 1998
	8.80	6.60	Deng et al., 2001*
Total leaf area index	4.00	3.00	Begna et al., 1999
	5.31	4.04	Dwyer et al., 1998
	3.41	3.41	Deng et al., 2001*
Harvest index	0.47	0.44	Begna et al., 2000
	0.59	0.50	Begna et al., 1997a
	0.59	0.54	Begna et al., 1997b
	0.49	.	Modarres et al., 1998b
	0.53	0.58	Dwyer et al., 1998
	0.49	0.47	Deng et al., 2001*
Grain yield (Mg ha^{-1})	5.72	.	Modarres et al., 1997a
	9.80	.	Dijak et al., 1999
	7.93	8.62	Begna et al., 2000
	12.60	8.70	Begna et al., 1997a
	8.00	9.40	Begna et al., 1997b
	10.50	.	Modarres et al., 1998b
	9.90	10.10	Dwyer et al., 1998
	7.22	8.58	Deng et al., 2001*

* unpublished

conventional counterparts (Stewart and Dwyer, 1993, 1998; Begna et al., 1999) (Table 1).

High plant densities are often used to increase crop productivity (Hashemi-Dezfouli and Herbert, 1992). The use of *Lfy*-bearing corn genotypes to increase canopy leaf area may be equivalent to the use of high plant populations. Shading of lower leaves occurs at high plant population densities, lowering their photosynthetic rate (Ottman and Welch, 1989). The *Lfy* trait causes the plant to produce extra leaves above the ear without adding leaves below the ear (Shaver, 1983). The more upright above-ear leaves of *Lfy*-bearing corn genotypes minimize shading allowing for greater light penetration to lower leaves.

Harvest Index

Harvest index (ratio of grain DM to total aboveground biomass, HI) is positively correlated with grain yield. The average HI obtained for current conventional corn genotypes range from 0.44 to 0.58 (Begna et al., 2000; Table 1), and is affected by both genotype and climate. The average HI of *Lfy*-bearing corn genotypes tends to be greater than those of their conventional counterparts (Table 1) and especially LRS hybrids (Deng et al. 2001, unpublished data).

Grain Yield

Begna et al. (1997a) was the first to show higher grain yields in LRS hybrids than conventional counterparts of the same maturity group. One would expect this, given the morphological superiority of genotypes bearing the *Lfy* trait, as described above (Table 1). However, there has been no consistent trend over time, and while some of the more recently developed LRS hybrids have shown high yield potentials, many have not [(Costa et al., 2001(unpublished data); Deng et al., 2001 (unpublished data)]. These seemingly contradictory results are of great importance and present a challenge to our current understanding of the phenotypic consequences of the *Lfy* trait. Most work on the genetics of the *Lfy* trait has been conducted by private industry and so, results are not readily available.

Root Morphology and Fractal Dimension

The *Lfy* and *rd1* traits may have pleiotropic effects on root architecture, mirroring their effects on canopy structure (Figure 4). Several studies have characterized the rooting patterns of corn genotypes (Varney et al., 1991; Rosolem, Assis, and Santiago, 1994), but these studies were all conducted on hybrids that were similar in aboveground architecture. While the aboveground architectures of these *Lfy*-bearing and LRS hybrids have been well characterized (Stewart and Dwyer, 1993, 1999; Begna et al., 1997a; Modarres et al., 1997b; Dijak et al., 1999), their root morphology has been the subject of only one published study (Costa et al., 2000) and some unpublished observations (Table 2).

It has been suggested that the roots of *Lfy*-bearing and conventional hybrids were different in the field (Modarres, personal communication). Wang (personal communication), using the root analysis techniques described in Costa et

FIGURE 4. Root skeletons of corn gentoypes [leafy reduced-stature (LRS), leafy normal-stature (LNS) and Pioneer 3905 (P3905)] grown to 15 d post-emergence used for fractal dimension estimation.

TABLE 2. Previous observations at early and late developmental stages on total root length and fractal dimension of maize hybrid carrying the leafy trait vs. conventional hybrids.

Source	Experiment location and setup	Stage of development	Root length (km)		Fractal dimension	
			Leafy	Conventional	Leafy	Conventional
Modarres, pers. commun., 1996	Visual observation in the field	Harvest	Leafy > Conventional		——	——
Wang, pers. commun., 1997	Greenhouse	Tasseling	1.57	1.31	——	——
		Harvest	1.83	1.62	——	——
	Field	Tasseling	2.16	1.81	——	——
		Harvest	3.37	2.81	——	——
Costa et al., 2000*	Greenhouse	Silking	1.75	0.49	——	——
Costa et al., 2001 (unpublished data)	Greenhouse	15DFE	0.160	0.196	1.67	1.74

*Root surface area (m^2); LRS = 1.89, LNS = 2.60, P3905 (conventional) = 0.68.
† DFE = Days from emergence.

471

al. (2000), compared corn root length of field- and greenhouse-grown LRS and conventional (P3905) corn plants at both the tasseling and harvest stages (Table 2). Corn root length at late developmental stages was shown to be greater for the root systems of *Lfy*-bearing hybrids than those of their conventional hybrid counterparts (Table 1). Root length of LRS hybrids was 13-20% greater than that of conventional hybrids. At silking the root length of LRS and *Lfy* hybrids were 3.6- and 4.8-fold greater, respectively, than for their conventional counterparts.

One of the most novel recent research efforts on root systems of *Lfy* hybrids has been the characterization of their fractal dimension (FD). The description of root morphology by way of fractal analysis is a largely unexplored area initiated with the pioneering work of Tatsumi, Yamauchi, and Kono (1989). From a geometric perspective, the FD of a root system measures the complexity of its branching pattern. The degree of branching, the proportion of coarse/fine roots, and their total length are the structural components affecting the FD of a root system. Quantitative and qualitative characterization of the morphology of developing roots can use FD as a convenient single measure of root system complexity (Eghball et al., 1993), as has already been shown for crop canopy architecture (Foroutan-pour, Dutilleul, and Smith, 1999, 2000, 2001).

Thus, LRS corn hybrids have the potential for higher HI and yields. With careful selection, LRS hybrids with early maturity would be superior for production in short-season areas.

INTENSIVE CEREAL MANAGEMENT

Intensive cereal management (ICM) systems involve the use of high-yielding cultivars in narrower than conventional row spacing with high levels of fertilization (especially N), growth regulators and pesticides (Wiersma, Oplinger, and Guy, 1986). In the early 1970s, these systems caused great excitement in Europe. In North America, these systems were extensively researched and practiced on a small scale in the 1980s (Frederick and Marshall, 1985). In the following section, we discuss the feasibility of ICM systems under the growing conditions of northeastern North America.

ICM as a Package

In Quebec, Smith et al. (1993) found that under an ICM regime, grain and straw protein concentrations of spring barley were 5 to 20 g kg^{-1} higher, but HI was generally lower than under conventional crop management (CCM). Also, number of spikes m^{-2} was related to the interaction between the weather and ethephon (2-chloroethyl phosphonic acid) application. Dry conditions depressed plant growth and development, and few early tillers were formed on the main culm. Ethephon application exacerbated this situation, and also triggered late tillering and increased spike-bearing shoot numbers (Ma and Smith, 1991a, 1991b). Similarly, Caldwell, Mellish, and Norrie (1988) reported that application of ethephon, alone or in combination with chlormequat (2-chloroethyl trimethyl-ammonium chloride), increased tillering of barley in Maritime Canada. Intensive cereal management can increase the yield and protein levels of wheat (Shah et al., 1994). In the Maritimes, Caldwell and Starratt (1987) compared the individual and interactive effects of ICM inputs on spring wheat and found that N fertility increased grain protein levels but also induced foliar diseases, and yield response to the whole ICM package was inconsistent in the humid northeast.

Cultivar Selection

Decisions on the components or entire ICM package application must consider the genotype-by-environment (G × E) interaction (Ma and Smith, 1992a). Not only must yield and quality be assured, but resistance to disease and lodging must also be considered. The correct choice of cultivar is crucial for production of high quality milling oat (Humphreys, Smith, and Mather, 1994), bread quality wheat (Ayoub, Guertin, and Smith, 1995), and for profitable production of spring barley in eastern Canada (Bulman and Smith, 1994; Smith et al., 1993). For example, in a study to evaluate the response of spring wheat cultivars to N application, Ayoub et al. (1994b) found that all tested cultivars produced grain suitable for bread wheat quality. However, the N management required to produce this quality varied. Cultivars such as Katepwa and Columbus produced good quality grain for bread making at all N levels whereas acceptable bread wheat quality for Max could only be achieved with a split N application. In barley, Bulman, Zarkadas, and Smith (1994) found that cultivar differences in amino acid composition were not consistent over years. Higher rates of N fertilizer normally increase crude protein concentration (Ma et al., 1994; Bulman and Smith, 1993a, 1993b, 1993c, 1994) but might lower the nutritional value of the grain due to a reduction in the proportion of essential amino acids such as lysine and threonine (Bulman, Zarkadas, and Smith, 1994). In general, choice of superior cultivars adapted to the specific environment should be the first priority in a decision to design efficient and sustainable production systems for high yield and quality, while combating disease, lodging or other potentially stressful conditions.

Nitrogen Effect

Nitrogen fertility is a principal component of ICM (Bulman and Smith, 1993). Most ICM studies were designed to test yield and quality response to N with no shortage of the other essential nutrients.

Growth and yield of wheat is enhanced when plants are provided with mixtures of NO_3^- and NH_4^+ fertilizer, compared with either form alone (Wang and Below, 1992). Camberato and Bock (1990) attributed the positive effects of enhanced NH_4^+ supply to improved tiller development. Mixed-N-induced increases in tillering are likely due to an increased synthesis of cytokinins, stimulated by NH_4^+-N (Wang and Below, 1996), or a higher cytokinin/IAA ratio together with an increased GA_{1+3} level in the shoot (Chen et al., 1998). These changes in hormonal levels would enhance N uptake as well as improve root growth (Wang and Below, 1992).

It is clear that high N application can increase grain yield and protein concentration of spring barley (Leibovitch et al., 1992; Ma, Dwyer, and Smith, 1994) and wheat (Peltonen, 1992; Ayoub et al., 1994a). However, caution must be taken not to overfertilize the crop, because excessive N levels will lower N use efficiency (Ma and Dwyer, 1998), and cause potential N contamination of surface- and groundwater (Ayoub, MacKenzie, and Smith, 1995).

Plant Growth Regulators

Lodging is one of the most significant limitations to maximizing grain yield and quality in small grain cereals (Stoskopf, 1985). Lodging and grain yields of cereal crops are negatively correlated, while lodging is positively correlated with plant height (Wiersma, Oplinger, and Guy, 1986). Lodging can result in direct yield losses by reducing photosynthesis and indirect yield losses by promoting conditions conducive to disease, increased harvest difficulty and harvest losses (Stoskopf, 1985), and poor grain quality due to sprouting.

Since the discovery, in the early 1960s, of chlormequat as an anti-lodging agent, the use of plant growth regulators (PGRs) on cereal crops has offered a new managerial control of plant growth and development, potentially reducing yield losses (Green, 1986). Among synthetic PGRs, chlormequat and ethephon are the most commonly used in agricultural production. Both compounds have the ability to shorten and stiffen cereal culms, thereby reducing lodging losses in susceptible cultivars, particularly when they are grown under conditions of high fertility and favorable soil water conditions (Ma and Smith, 1991a, 1991b; Ma et al., 1992).

Apical development is a prerequisite to determine the optimum time for application of any ICM measure, especially PGR and fungicide (Ma and Smith, 1992a). Application of PGR, in addition to its effects on lodging control, may affect the development of reproductive structures in both main-culm and tiller apices. For example, application of chlormequat at Zadoks growth stage (ZGS) 13 (Zadoks, Chang, and Konzak, 1974) substantially delayed the development of the main-culm apex, resulting in enhancement of tiller number and adventitious root number, and decreasing the number of aborted spikelet primordia (Ma and Smith, 1991a). Early (ZGS 13) application of chlormequat to spring barley appears to synchronize development of spikelets along the rachis (Ma and Smith, 1991a) through partitioning of photoassimilates among plant organs (Ma and Smith, 1992b). Unlike chlormequat, ethephon significantly retards the development of embryonic spikes only at ZGS 39 (the early boot stage), which improves tiller survival (Ma and Smith, 1991a, 1992c) and tends to increase number of spikes m^{-2}, but had no effect on kernel weight and kernels per spike (Ma and Smith, 1992c). In general, the application of PGR reduces competition from dominant sinks, allowing the survival and greater development of more subordinate sinks.

Application of ethephon after heading increased grain yield (Ma and Smith, 1992e) while earlier application usually controlled lodging with no effect on spring barley yield (Ma and Smith, 1992d). Total plant N and DM of soft red winter wheat at anthesis was significantly increased by ethephon application (Van-Sanford et al., 1989). However, neither total N nor DM at maturity was altered by this treatment while remobilization of vegetative N was increased 13% by ethephon application. The effect of ethephon on the nature of cultivar differences in N use efficiency warrants further study (Ma, Leibovitch, and Smith 1994).

Pesticides

Yield promoting practices such as increased N input and use of narrower-row spacings create conditions more favorable for foliar disease development (Leibovitch et al., 1992). Growing disease-resistant cultivars has been the principal means of controlling foliar diseases of wheat and barley in the USA and Canada. However, pathogens can quickly develop the ability to overcome host resistance and integrated pest management, including fungicides, should be practiced to add long term stability to genetic resistance. Guy and Oplinger (1989) suggested that fungicide and PGR inputs should be routinely considered for use in soft winter wheat production. Results from an eight site-year study in southeast Kansas indicate that yield responses to higher N levels and foliar fungicide applications were environment and cultivar-dependent, thus the decision to use these practices should be based on cultivar selection and specific environmental conditions.

As expected, higher than recommended rates of N application, with or without fungicide and/or PGR, generally increased grain yields of winter wheat (Morris, Ferguson, and Paulsen, 1989), and spring barley (Ma et al., 1992), or grain protein concentration, especially when split N applications were used (Morris, Ferguson, and Paulsen, 1989; Ayoub et al., 1995a). However, the benefit of added N to grain yield can only be ensured with proper fungicide application (Leibovitch et al., 1992).

In summary, numerous studies have demonstrated that ICM, or at least some aspects of it, has the potential to increase grain yields and/or protein concentration of small grain cereals in short growing season areas. Compared to those reported in European studies, however, the average yield increases in our region are small and vary among years and sites. It appears that in the humid northeast, small grain cereal production will become more profitable not by the adoption of ICM *per se*, but by using ICM components that solve particular problems associated with each production system.

PRODUCTION OF C_4 GRASSES: FORAGE, BIOMASS, AND INDUSTRIAL FEED STOCK

The major forage grasses of North America are all cool season grasses imported from Europe and, typically, do not do well during the brief mid-summer period when continental conditions lead to high temperatures. However, a range of warm-season forage grass species is found in North America (Stubbendieck, Hatch, and Butterfield, 1992). These grasses utilize the C_4 photosynthetic pathway and are characterized by: (i) high optimal temperatures for photosynthesis, which creates forage production potential during the critical months of July and August, and (ii) high N and water use efficiencies for forage and biomass production (Jung et al., 1990). In North America, most of the research programs investigating these grasses are located in the southern US (Sanderson et al., 1996). The productivity and quality attributes of warm-season C_4 grasses have been little investigated in more northern latitudes where the low spring and fall temperatures limit their overall production.

DEVELOPMENT, PERSISTENCE, AND NUTRITIVE VALUE

Jacobson et al. (1984) found northern ecotypes of warm season grasses generally showed early growth in spring and matured early enough to avoid frost damage. Rate of morphological development in these grasses was determined largely by accumulated thermal-time (George et al., 1998; Gillen and Ewing, 1992). Madakadze et al. (1998a) found that, of 22 warm-season grasses evaluated in the humid northeast (southwestern Quebec) that cordgrass (*Spartina pectinata* L.) was the earliest to start spring regrowth. The earliest maturing entries were switchgrasses (*Panicum* spp.) and a big bluestem (*Andropogon gerardii* Vitman). Cordgrass was the latest maturing. Regrowth was predominantly from continued growth of biennial tillers in sandreeds (*Calamovilfa longifolia* (Hook) Scribn.), from rhizomes in cordgrass and from buds on stem bases for other species. Madakadze et al. (1998b) found that all nine switchgrass (*Panicum virgatum* L.) cultivars evaluated persisted through the three

TABLE 3. Frequently reported warm-season grass variables.

Material	Yield (Mg ha^{-1})	NDF (%)	ADF (%)	N Content (%)	Ash Content (%)	Location	Source
Switchgrass		77-86	46-54	0.4-1	4-7	HNE	Madakadze et al., 1998a
Switchgrass	9.8	81-87	51-57	0.3-0.9	4.7-6.6	HNE	Madakadze et al., 1998c
Switchgrass	10.6-12.2	85-87	65-67	0.5-2.5		HNE	Madakadze et al., 1999a
Switchgrass	6-11			0.5-1.5		HNE	Madakadze et al., 1999b
Kleingrass	6.6-11					SUS	Sanderson et al., 1996
Switchgrass	5.4-26					SUS	Sanderson et al., 1996
Buffalo grass	14.5					SUS	Sanderson et al., 1996
Big bluestem	3-9					NUS	Jung et al., 1990
Indiangrass	5.8-8.7					NUS	Jung et al., 1990
Switchgrass	7.7-12.3					NUS	Jung et al., 1990
Mix		72-84	42-52			NUS	Vona et al., 1984
Switchgrass					3-7	SUS	Agblevor et al., 1992
Big bluestem					1.9-5.7	SUS	Lanning and Eliuterius, 1987
Indiangrass					3.6-9.5	SUS	Lanning and Eliuterius, 1987
Switchgrass		66-76		0.2-2.5		SUS	Sanderson and Wolf, 1995

HNE = Humid Northeast; SUS = Southern US; NUS = Northern US.

years of a Quebec study and showed increases in tiller number from one year to the next. Plant heights increased at 1.9 to 2.8 cm d^{-1}. Ground cover ratings were as high as 85%. Switchgrass cultivars, and possibly cordgrass seem well suited for production in the humid northeast.

Warm-season grass species for short-season areas should have rapid initial leaf area development for high light interception, resulting in rapid growth and high yield (Muchow, Sinclair, and Bennett, 1990). High radiation use efficiency (RUE) values for aboveground biomass are also desirable, with values for other warm season grasses reported at: 1.25 g DM MJ^{-1} for *Pennisetum* spp. and 1.24 for *Saccharum* spp. (Woodard, Prine, and Bachrein, 1993); 1.40 for corn and 1.12 for sorghum [*Sorghum bicolor* (L.) Moench] (Muchow and Davis 1988). Madakadze et al. (1998c, 1998d) conducted a study to evaluate the performances of nine switchgrass populations in southwestern Quebec (Table 3). Maximum LAI ranged from 5.1 to 8. Overall average end of season biomass yield was 9.8 Mg ha^{-1}. The relationship between end of season yield and leaf area duration was linear. By early July the canopies were intercepting about 90% of the incoming light. The average switchgrass energy content was 17.4 MJ g^{-1} DM and total energy yields ha^{-1} ranged from 186 to 216 GJ. The

RUEs, during the near linear growth phase, based on photosynthetically active radiation values, were 1.9 to 2.2 g DM MJ^{-1}.

In the southern US biomass yields for warm-season grasses of over 40 Mg ha^{-1} have been obtained, with most values between 5.4 to 26 (Sanderson et al., 1996). Yields in northern US varied from 3 to 9 Mg ha^{-1} for big bluestem; 5.8 to 8.7 Mg ha^{-1} for Indiangrass [*Sorghastrum nutans* (L.) Nash] and 7.7 to 12.3 Mg ha^{-1} for switchgrass (Jung et al., 1990). Madakadze et al. (1999a) found that solid seeded switchgrass (three cultivars) had rates of DM accumulation during the 40-50 d near linear phases of growth were 175 to 191 kg DM ha^{-1} d^{-1} in southwestern Quebec. The end of season DM yields were 10.6 to 12.2 Mg ha^{-1}. Madakadze et al. (1999b) measured yields of 7, 9 and 11 Mg ha^{-1} when switchgrass was cut every 4 weeks, every 6 weeks, or left uncut, respectively. These values compare well with those reported from other locations in the northern US.

Acceptable levels of lignocellulose and other cell wall constituents vary according to end use. On the one hand, digestibility and intake of warm-season grasses decreases during development (Jung and Vogel, 1992), largely due to increased cell wall and lignin constituents. In contrast, high levels of lignocellulose are desirable for chemical and biofuel production (Trebbi, 1993); high levels of N and/or ash are not, as they reduce chemical output in thermochemical conversions (Agblevor et al., 1992). High ash levels also increase wear on industrial machinery and can also be fusible at high temperatures. Madakadze et al. (1999a, 1998a, 1998c) found that the lignocellulose content was high in 22 warm-season grasses, with neutral detergent fiber (NDF) and acid detergent fiber (ADF) ranging from 770 to 870 and 460 to 669 g kg^{-1}, respectively. Nitrogen contents ranged from 3.2 to 10 g kg^{-1} and ash contents ranged from 40 to 70 g kg^{-1}. Madakadze et al. (1999a, 1999b, 1999c) found that, for switchgrasses, N concentration of DM decreased curvilinearly from 25 g kg^{-1} at the beginning of the season to 5 g kg^{-1} DM at season end. Both ADF and NDF concentrations increased to a maximum early in the season, after which no changes were detected. Because of its composition and yield (Table 3) switchgrass has potential as a biomass crop in a short season environment.

Seed Germination

Uneven germination and slow seedling growth in spring results in both poor stand establishment and competition from cool season weeds. This has been attributed primarily to seed dormancy (Sanderson et al., 1996). Temperature affects germination through removal of seed dormancy (Benech-Arnold et al., 1989) and by determining rate of germination of non-dormant seeds (Jordan and Haferkemp, 1989). Reported values of minimum or base temperature (T_b)

for germination of the C_4 grasses vary between (Covell et al., 1986; Jordan and Haferkemp, 1989), and within species (Lawlor et al., 1990; Ellis et al., 1986). Madakadze et al. (2000) estimated T_b for four switchgrasses at 5.5 to 10.9°C, three big bluestem grasses at 7.3 to 8.7°C, two Indiangrasses at 7.5 and 9.6°C, and two prairie sandreeds at 4.5 and 7.9°C. Madakadze et al. (2000) and Madakadze, Stewart, and Smith (2001) found that seed conditioning (Hardegree, 1994) shortened germination time and increased total germination for some C_4 species.

Paper Production

Concerns regarding forest preservation and reduction of carbon monoxide and dioxide emissions associated with burning of agricultural residues (McCloskey, 1995; Wilson, 1996), and wood fiber shortages in areas like the North American west coast (Jacobs et al., 1996) have increased interest in non-wood fiber sources such as warm-season grasses. Non-wood fiber sources offer several advantages: (1) annual production systems, as opposed to the long growth cycles for wood; (2) lower production costs than annual crops as they do not have to be established each year; (3) lower lignin contents and easier delignification, requiring milder and faster cooking conditions (Paavilainen, 1994). Madakadze et al. (1999c) found that all warm-season grasses evaluated were easily pulped with a mild kraft process; pulp yields ranged from 44 to 51% and kappa numbers from 10 to 16. Klason lignin values ranged from 17.7 to 24%, typical for non-wood materials. The weight-weighted fiber length ranged from 1.29 to 1.43 mm. The unbeaten pulp freeness ranged from 275 to 411 mL. Tear indices were as high as 7.49 mN m^2 g^{-1} and burst indices as high as 5.7 kPa m^2 g^{-1}. The pulp from these grasses could provide good printability properties to paper.

In summary, the DM levels of switchgrass in the humid northeast is comparable with that in other locations in the northern US. With good N fertility, monthly grazing system produces high quality feed, while for hay, cutting every month and a half is better. The high fiber contents at the end of the season indicate good potential for fiber, biofuel, and paper production.

DEVELOPMENT OF A CHRONIC INJECTION SYSTEM AND INVESTIGATIONS OF CROP PHYSIOLOGY

Understanding the physiology of plants is key to understanding their growth, development, and agronomic performance in a given climate regime. In animal systems, much has been learned about the ability to withstand stresses and to function in a given set of conditions through the intravenous administration of materials affecting metabolism. Until recently, no such system has existed for plants. Foliar spray and root feeding have been the two most common methods used in the study of plant nutritional and metabolic requirements (Souza, Stark, and Fernandes, 1999; Kaya, Higgs, and Burton, 1999, Vieira et al., 1998). However, there are limitations for each of these methods, due to relatively slow movement of fed materials into the vascular systems of plants, slow absorbance, large amount of fed materials required, and dilution of fed material inside the plants.

More recently, stem perfusion/injection techniques have been developed for a range of plant species, e.g., wheat and barley (Ma, Dwyer, and Smith 1994), corn (Zhou and Smith, 1996), soybean (Abdin et al., 1998b), C_4 and C_3 weeds (Begna et al., 2002). The uptake rate by perfusion/injection techniques greatly depends on whether the plant has a solid (Zhou and Smith, 1996) or hollow stem (Ma and Smith, 1992a, 1992b, 1992c), the composition of the injected solution(s) (Ma, Dwyer, and Smith, 1994; Ma et al., 1992f) and the perfusion/injection site response to wounding (Zhou and Smith, 1996).

Infused and perfused N increased tissue N concentrations (Ma et al., 1995) but was not partitioned equally among plant tissues (Ma et al., 1998; Zhou et al., 1998). Infused and perfused N also increased seed protein concentrations (40%; Ma et al., 1992), even when soil N levels were saturating, indicating that uptake of N from the soil was more limiting to seed protein accumulation and uptake into the seed for both barley (Foroutan-pour, Ma, and Smith, 1997) and corn (Zhou et al., 1998). Interestingly, injection of ethephon (ethylene) also increased barley seed protein concentrations (Ma et al., 1994; Foroutan-pour, Ma, and Smith, 1995). Injected N could account for up to half the N in the plants (Ma et al., 1992). Perfusion of barley peduncles with 30 mM N increased net photosynthetic rates by 21% (Foroutan-pour, Ma, and Smith, 1997).

Sucrose injection reduced plant photosynthetic rates, causing premature leaf senescence in corn (Zhou et al., 1997b), but increased overall growth of corn (Zhou et al., 1997b), soybean (Zhou et al., 2000a), C_3 and C_4 weeds (Begna et al., 2002). Corn grain weight could be increased by as much as 30% (Zhou et al., 1999). Injected sucrose could account for as much as 60% of the total plant dry weight (Abdin et al., 1998b). Shade grown plants injected with sucrose could be as big as uninjected plants in full sun, indicating that sucrose injection could overcome limitations on plant growth due to shading (Begna et al., 2002). Morphological adaptations to shading occurred in both sucrose injected and water injected control plants, indicating that these changes were strictly due to light level, rather than sucrose availability (Begna et al., 2002). Zhou et al. (1997a) found that the highest grain yield was achieved by the plants injected with 150 g sucrose L^{-1}, but only under sufficient water supply, indicating that exogenous sucrose supplementation could not overcome reproductive failure due to water stress. Their study indicated that plant reproductive development after silking was limited more by the water availability than by assimilate supply, demonstrating that some overall plant response to water stress, perhaps mediated by hormonal signalling, was more import than carbohydrate supply. In addition, injected sucrose increased soybean nodule number and total dry weight (Abdin et al., 1998).

Injection of salicylic acid increased the photosynthetic rates of corn and barley plants, as did the combination of salicylic acid and sucrose (42%; Zhou et al., 1999). In corn, these treatments increased total DM and grain (9%; Zhou et al., 1999).

These techniques have become alternative approaches for researchers to study the effects of perfused/injected solutions on growth, morphology and physiology of plants under greenhouse (water stress, and shading) and field conditions.

IMPROVING LEGUME NODULATION UNDER NODULATION INHIBITORY CONDITIONS

Effects of Suboptimal Conditions on Symbiotic Nodulation in Northeastern North America

Suboptimal levels of many environmental factors can limit symbiotic nodulation and N fixation. Factors such as salinity, unfavorable soil pH, nutrient deficiency, mineral toxicity, the presence of soil mineral N, insufficient or excessive soil temperature, and plant diseases, impose limitations on the vigor of the host legume (Zhang and Smith, 2002). Environmental conditions that commonly inhibit the formation of legume root nodules in the humid northeastern part of North America are low root zone temperature (RZT), soil mineral N and water deficiency.

Soybean is a relatively new crop in Canada. As it originates from a tropical to sub-tropical region, its optimal symbiotic activity requires temperatures in the range of 25 to 30°C. In the range 25 to 17°C, the time between inoculation and onset of dinitrogen fixation is delayed by 2 to 3 days for each °C decrease. While, in the range 17 to 15°C each °C delays the onset of dinitrogen fixation by one week (Zhang and Smith, 1994; Zhang, Lynch, and Smith, 1995). All stages of nodule formation and function are affected by low RZT, but infection and early nodule development processes are the most sensitive (Zhang and Smith, 1994). Zhang and Smith (1995) found that low RZT decreases early infection and nodule initiation by disruption of the inter-organismal signal exchange. A recent study indicated that when N concentration in the rooting medium increased above 50 mg L^{-1}, both genistein and daidzein concentrations in root systems decreased (Zhang, Máce, and Smith, 2000). Application of up to 60 kg N ha^{-1} also affected total seasonal plant N_2-fixation by reducing $^{14}CO_2$ flux to nodules, and raising nitrogenase-linked respiration (Zahran, 1999). Water stress decreases the number of infection threads formed and inhibits nodulation. Following successful infection, decreased water supply can retard nodule development (Sprent and Sprent, 1990). Zahran, Karsisto, and Lindstorm (1994) showed that exposing rhizobia to osmotic stress resulted in alteration of bacterial membrane lipopolysaccharides, which are involved in the *Rhizobium*-host plant recognition process. Low pH generally decreases nodulation by increasing numbers of ineffective rhizobia (Date, 1988).

Improvement of Nitrogen Fixation by Manipulating the Exchange of Symbiotic Signals

An understanding of the mechanism(s) by which low RZT and mineral N inhibit legume nodulation and N fixation makes it possible to find a way to reduce its effects. Zhang and Smith (1995) demonstrated that preincubation of *Bradyrhizobium japonicum* cells with genistein increased soybean nodulation and N fixation at low RZT, and that at higher RZTs the efficacy of the genistein treatment diminished. Moreover, Zhang and Smith (1996, 1997) showed that incubation of *B. japonicum* with genistein, prior to application as an inoculant, or directly applied into the seed furrow at planting, increases soybean nodulation, N_2 fixation and total N yield, when field conditions are such that they would normally delay or inhibit nodulation. This was caused by a shortening of the time between inoculation and when the infection thread reached the bottom of the root hairs. Pan and Smith (2000a) showed that genistein treated *B. japonicum* 532C had higher levels of nodule occupancy than the untreated cells under greenhouse conditions. Paau et al. (1990) reported that adding soybean meal to the fermentation medium can alter the nodule readiness of the rhizobia and that this has an effect on the competitiveness of the inoculant strains.

Inoculation of soybean with *B. japonicum* that has been pre-activated with genistein and daidzein also improves plant nodulation and N fixation under levels of mineral N that are inhibitory to nodulation (Pan and Smith, 2000b). Pan and Smith (2000b) found that the plants receiving pre-incubated *B. japonicum* cells had more nodules, nodule weight, and plant N content, especially in a low N containing sandy soil and where 20 kg N ha^{-1} were added as mineral fertilizer. However, at higher levels of N application, genistein preincubation no longer influenced plant nodulation and N fixation.

Work conducted at McGill University, in the humid northeast, during the last 15 years has improved legume nodulation and N fixation under environmental conditions known to inhibit nodulation. This work was the first to clarify low soil temperature as a major limitation to soybean nodulation in this area, to show that effects on signal exchange were an important part of this inhibition, and that addition of specific signal compounds could help to over come this. Parallel phenomena have since been shown to exist for soybean nodule inhibition by high soil mineral N, low soil pH, and high soil salinity.

Plant Growth Promoting Rhizobacteria

Our understanding of rhizosphere biology has progressed with the discovery of a specific group of microorganisms, now called plant growth promoting rhizobacteria (PGPR), that can colonize plant roots and stimulate plant growth and development (Kloepper, Schroth, and Miller, 1980). Most of the identified strains of rhizobacteria occur within gram-negative genera, of which fluorescent pseudomonas are the most characterized, although strains in other genera, such as *Serratia* and *Azospirillum*, have also been reported. A number of publications have shown the beneficial effects of PGPR through direct plant growth promotion, disease suppression, associative N_2 fixation and improved access to soil nutrients (reviewed by Bashan and Holguin, 1997). Evaluation of 20 years of data from PGPR experiments showed that 60 to 70% of all field experiments were successful with significant yield increases ranging from 5 to 30%.

PGPR Associated with Legume Nodulation and Nitrogen Fixation

Some PGPR have beneficial effects on legume nodulation and N fixation; these have been designated as nodule promoting rhizobacteria (NPR). The mechanism by which these rhizobacteria improve legume nodulation is unknown. Recently, Zhang et al. (1996) found that inoculation of soybean plants with PGPR strains, together with *B. japonicum*, produced a wide range of effects. Some strains affected only nodulation or N fixation, whereas others affected both nodulation and N fixation. Photosynthesis was more sensitive to the application of PGPR, over a wide range of temperatures, than transpiration and stomatal conductance (Zhang et al., 1997). A field experiment showed that the onset of N_2 fixation by soybean plants receiving *B. japonicum* preincubated with PGPR was two to three days earlier than those receiving no PGPR treatment (Dashti et al., 1998), and this resulted in yield increases of 11% (Dashti et al., 1997).

RHIZOBIA-TO-LEGUME SIGNALS:
THE NITROGEN FIXING SYMBIOSIS AND BEYOND

Lipo-Chitooligosaccharides for Crop Production in the Northeast

Lipo-chitooligosaccharides (LCOs) or nod factors, are bacteria-to-plant signal molecules that play a key role in initiating the early events of the legume-rhizobia N-fixing symbiosis (Lhuissier et al., 2001). Extensive work on LCOs has revealed that members of the group are responsible for the host specificity of rhizobia (Perret, Staehelin, and Broughton, 2000). LCOs invoke multiple physiological responses in the host and non-hosts: root hair deformation (Spaink et al., 1991; Prithiviraj et al., 2000a), induction of nodulin genes essential for infection thread formation (Horvath et al., 1993), and cortical cell division (Schulaman et al., 1997). Purified nod factors *per se* are able to initiate complete nodule structures at submicromolar concentrations in some legume-rhizobia systems (Denarie and Cullimore, 1993). LCO reduces the salicylic acid (SA) level in alfalfa (*Medicago sativa* L.) roots and this might aid in the suppression of host defense responses, thus ensuring successful infection (Martinez-Abarca et al., 1998). Similar decreases in SA level in leaf tissues occurred when soybean plants were sprayed with LCO (Prithiviraj et al., 2000b). Coinoculation of soybean plants with *B. japonicum* enhanced root colonization of the mycorrhizal fungus *Glomus mosseae*, and similar results were observed when treated with highly purified nod factors [Nod NGR-V (MeFuc, Ac)] from *Rhizobium* sp. NGR 234 (Xie et al., 1995).

The growth promoting activity of various rhizobia on non-legumes have been investigated. Strains of *Rhizobium leguminosarum* bv. *trifolii* were isolated as endophytes of rice (*Oryza sativa*) and this association was shown to improve growth of rice under laboratory conditions and could improve nutrient uptake and grain yield of some rice cultivars under field conditions (Yanni et al., 1997; Biswas, Ladha, and Dazzo, 2000; Biswas et al., 2000). Recently, Prayitno et al. (1999) studied the association between rice and rhizobia using green fluorescent protein tagged *Rhizobium* strains and concluded that these bacteria were intimate epiphytic microorganisms that multiply on the root surface and produce plant growth stimulating molecules, which we think might be nod factors. Hungria and Stacey (1997) found that non-host plants could induce the nod genes of rhizobia. There are several reports of the induction of nod genes by non-host root exudates (Le Strange et al., 1990). Increased growth of corn and bean (*Phaseolus* sp.) in intercropping systems could be attributed to the reciprocal stimulation of *Rhizobium* and *Azospirillum* by root exudates of bean and corn (Hungria and Stacey, 1997).

In line with the above observations, we have found, over a decade of experimentation, that treatment of soybean seeds with *B. japonicum* cultures induced to produce LCOs through the addition of genistein markedly enhanced the germination and emergence of soybean and other crop plants under field conditions, when compared to un-induced cultures of *B. japonicum* or genistein alone. Collectively, these results suggest that some of the observed effects might be due to LCO present in the induced cultures. Systematic experiments in our laboratory have shown that the major LCO of *B. japonicum* enhanced the germination and early growth of a variety of crop plants in the model plant

Arabidopsis (Prithiviraj et al., 2000c). In cucumber (*Cucumis sativus* L.) and lettuce (*Lactuca sativa* L.) the germination promoting effect of LCO was more pronounced at a low temperature (15°C), suggesting that this class of compounds might be useful to alleviate the low temperature limitations imposed by short growing seasons in the northern latitudes, and may have a wider potential role in maintaining metabolic activity in the presence of stress conditions. Presoaking of seeds in LCO solutions induced rapid emergence of soybean, corn and cotton (*Gossypium hirsutum* L.) under field conditions. LCO also enhanced the early growth of corn and soybean in pot and hydrophonic cultures in the greenhouse. Irrigation of corn seedlings with LCO solution doubled leaf area, plant height, and root and shoot dry weight. Similarly, when 3-day-old seedlings of soybean and corn were grown in a hydrophonic system containing 10^{-7}, 10^{-9} or 10^{-11} M LCO the biomasses of both soybean and corn were enhanced. At 10^{-9} M and 10^{-7} M LCO, the soybean root biomass was 7-16% larger and roots were 34-44% longer than in the control. Further, spray application of LCO at submicromolar concentrations enhanced the photosynthetic rates of soybean, corn, rice, bean, canola (*Brassica campestris* L.), apple (*Pyrus malus* L.), and grapes (*Vitus vinifera* L.). On average there was 10-20% increase in the photosynthetic rate and this was concomitant with increases in stomatal conductivity. Under field conditions, spray application of LCO at concentrations of 10^{-6}, 10^{-8}, and 10^{-10} M resulted in increased soybean grain yields of up to 40%. Taken together, the results of our experiments suggest the possible use of this novel class of signal molecules in improving crop production especially in the short season areas like eastern Canada.

CONCLUSIONS AND FUTURE RESEARCH

The humid northeast faces a unique challenge for crop production due to the short growing season. There is clear potential for intercropping in the humid northeast to improve the production efficiency of corn and soybean. However, the challenge is to find a herbicide for weed control in such systems. The other major difficulty in wider adoption of intercrop systems is the inability of machinery to selectively harvest the individual crop components, either at different times or, when maturities coincide, synchronously. However, visual recognition by machines is an active area of research at this time and it may be that within a decade we will have sophisticated harvest equipment able to remove this last barrier to wide spread use of intercrops, with all of the attendant advantages.

There is further potential to adapt warm season crops, such as corn, soybean and warm season grasses to short-season northern areas. The LRS corn hybrids show the potential to develop light interception capacity quickly, mature rapidly and yield well. A careful and focused effort should be able to produce stable and highly productive LRS genotypes that are better suited to production in short-season areas. Use of signal compounds can improve soybean N fixation. This can be accomplished successfully by application of plant-to-bacteria flavonoid to rhizobial inoculants, or LCOs to soybean leaves. The latter also improves total photosynthetic rates and overall growth and yield. In the case of warm-season C_4 grasses, use of seed priming treatments can aid in stand establishment under cool spring conditions. Under northeastern conditions these grasses produce biomass at levels and qualities that make them viable for production of forage material, fiber for paper manufacture or biofuels.

The climate of the humid northeast does not allow all elements of ICM systems developed for production of small grain cereals in Europe to be applied in this region. However, with correct application some of the elements of these systems are effective, and this is particularly so when suitable cultivars are used.

It is now clear that, for at least some crops, the root uptake of soil N is an important barrier to crop production. Adding N to plants through stem injection increased photosynthetic rates and also the protein concentration in grain, resulting in potential for production of greater yields of grain with higher quality. In addition, more efficient removal of N from the soil will reduce the potential contamination of groundwater associated with crop production. This technology can further enhance crop production in this region.

Finding ways to increase inherent photosynthetic rates of crop plants is an important goal for the 21st century. There is a small body of published literature that suggests that the rates can be higher than those manifested, but the plants have homeostatic control processes in place (e.g., Dijak, Ormrod, and Smith, 1985). Our past successes have come through removing environmental limitations to expression of this inherent limit. The challenge now it to move this limit up. This is particularly important in short-season areas like northeastern North America. LCO applications seem to offer the first potential technology for use in this way.

Research challenges for the future include the production of intercrops of LRS corn and signal compound treated soybean. The altered root system of the LRS corn, coupled with the greater N fixation of the LCO treated soybean could lead to greater transfer of N from the legume to the cereal. If mycorrhizal inoculants could be added and the corn plant roots altered to reduce their resistance to N uptake this system could be made even more efficient. Mixtures of warm season North American and European grasses should be evaluated in livestock grazing systems as should methods for the production of warm season grasses for reduction of greenhouse gas emissions. There should also be an effort to look at warm-season grass intercrops, with delayed seeding, in corn production systems. All of these possibilities should be the focus of research in the near and more distant future, and have the potential to contribute to increased crop production with a reduced impact on the environment in the humid northeast.

REFERENCES

Abdin, O.A., B.E. Coulman, D. Cloutier, M.A. Faris, and D. L. Smith. (1997). Establishment, development and yield of forage legumes and grasses as cover crops in grain corn in eastern Canada. *Journal of Agronomy and Crop Science* 179:19-27.

Abdin, O.A., B.E. Coulman, D. Cloutier, M.A. Faris, X.M. Zhou, and D.L. Smith. (1998a). Yield and yield components of corn interseeded with cover crops. *Agronomy Journal* 90:63-68.

Abdin, O.A., X.M. Zhou, B. Coulman, D.C. Cloutier, M.A. Faris, and D.L. Smith. (2000). Cover crops and interrow tillage for weed control in short season corn (*Zea mays*). *European Journal of Agronomy* 12:93-102.

Abdin, O.A., X.M. Zhou, B.E. Coulman, D. Cloutier, M.A. Faris, and D.L. Smith. (1998b). Effect of sucrose supplementation by stem-injection on the development of soybean plants. *Journal of Experimental Botany* 40:2013-2018.

Agblevor, F.A., B. Rejaj, R.J. Evans, and K.D. Johnson. (1992). Pyrolytic analysis and catalytic upgrading of lignocellulosic materials by molecular beam mass spectrometry. In *Energy from Biomass and Wastes XVI*, ed., D.L. Klass, Chicago, IL: Elsevier Applied Science Publishers. pp. 767-795.

Ames, R.N., C.P.P. Reid, L.K. Porter, and C. Cambaradella. (1983). Hyphal uptake and transport of nitrogen from two ^{15}N-labelled sources by *Clomus masseae*, a vesicular-arbuscular mycorrhizal fungus. *New Phytologist* 95:381-396.

Anonymous. (1998). Statistiques. Producteur de Lait Quebecois. 18:40-42. (In French).

Ayoub, M., A. MacKenzie, and D.L. Smith. (1995). Evaluation of N fertilizer rate and timing and wheat cultivars on soil residual nitrates. *Journal of Agronomy and Crop Science* 175:87-97.

Ayoub, M., S. Guertin, and D.L. Smith. (1995). Nitrogen fertilizer rate and timing effect on bread wheat protein in eastern Canada. *Journal of Agronomy and Crop Science* 174:337-349.

Ayoub, M., S. Guertin, J. Fregeau-Reid, and D.L. Smith. (1994a). Nitrogen fertilizer effect on breadmaking quality of hard red spring wheat in eastern Canada. *Crop Science* 34:1346-1352.

Ayoub, M., S. Guertin, S. Lussier, and D.L. Smith. (1994b). Timing and level of nitrogen fertility effects on spring wheat yield in eastern Canada. *Crop Science* 34:748-756.

Bashan, Y. and G. Holguin. (1997). *Azospirillum*–plant relationships: environmental and physiological advances (1990-1996). *Canadian Journal of Microbiology* 43: 103-121.

Begna, S.H., L.M. Dwyer, D. Cloutier, L. Assemat, A. DiTommaso, X. Zhou, and D.L. Smith. (2002). Decoupling of light intensity effects on the growth and development of C_3 and C_4 weed species through sucrose supplementation. *Journal of Experimental Botany* 53:1935-1940.

Begna, S.H., R.I. Hamilton, L.M. Dwyer, D.W. Stewart, and D.L. Smith. (1997a). Effects of population density and planting pattern on the yield and yield components of leafy reduced-stature maize in short-season area. *Journal of Agronomy and Crop Science* 179:9-17.

Begna, S.H., R.I. Hamilton, L.M. Dwyer, D.W. Stewart, and D.L. Smith. (1997b). Effects of population density on the yield and yield components of leafy reduced-stature maize in short-season areas. *Journal of Agronomy and Crop Science* 178:103-110.

Begna, S.H., R.I. Hamilton, L.M. Dwyer, D.W. Stewart, and D.L. Smith. (1999). Effects of population density on the vegetative growth of leafy reduced-stature maize in short-season areas. *Journal of Agronomy. and Crop Science* 182:49-55.

Begna, S.H., R.I. Hamilton, L.M. Dwyer, D.W. Stewart, and D.L. Smith. (2000). Variability among maize hybrids differing in canopy architecture for above-ground dry matter and grain yield. *Maydica* 45:135-141.

Benech-Arnold, R.L., C.M. Ghersa, R.A. Sanchez, and P. Insautsi. (1989). Temperature effects on dormancy release and germination rate in *Sorghum halapense* L. Pers. seeds: a quantitative analysis. *Weed Research* 30:81-90.

Biswas, J.C., J.K. Ladha, and F.B. Dazzo. (2000). Rhizobia inoculation improves nutrient uptake and growth of lowland rice. *Soil Science Society of America Journal* 64:1644-1650.

Biswas, J.C., J.K. Ladha, F.B. Dazzo, Y.G. Yanni, and B.G. Rolfe. (2000). Rhizobial inoculation influences seedling vigor and yield of rice. *Agronomy Journal* 92:880-886.

Bootsma, A., G. Trembly, and P. Filion. (1999). Risk analyses of heat units available for corn and soybean production in Quebec. Technical Bulletin, ECORC Contribution No. 991396, 35 pages.

Broadbent, F.E., T. Nakashima, and G.Y. Chang. (1982). Estimation of nitrogen fixation by isotope dilution in field and greenhouse experiments. *Agronomy Journal* 74:625-628.

Brown, D.M. and A. Bootsma. (1993). Crop heat units for corn and other warm-season crops in Ontario. Ontario Ministry of Agriculture and Food, Factsheet No. 93-119, Agdex 111/31, 4 pp.

Bulman, P. and D.L. Smith. (1993a). Grain protein response of spring barley to high rates and post-anthesis application of fertilizer nitrogen. *Agronomy Journal* 85:1109-1113.

Bulman, P. and D.L. Smith. (1993b). Yield and yield component response of spring barley to fertilizer nitrogen. *Agronomy Journal* 85:226-231.

Bulman, P. and D.L. Smith. (1993c). Yield and grain protein response of spring barley to ethephon and triadimefon. *Crop Science* 33:798-803.

Bulman, P. and D.L. Smith. (1994). Post-heading nitrogen uptake, retranslocation, and partitioning in spring barley grain. *Crop Science* 34:977-984.

Bulman, P., C.G. Zarkadas, and D.L. Smith. (1994). Nitrogen fertilizer affects amino acid composition and quality of spring barley grain. *Crop Science* 34:1341-1346.

Caldwell, C.D. and C.E. Starratt. (1987). Response of Max spring wheat to management inputs. *Canadian Journal of Plant Science* 67:645-652.

Caldwell, C.D., D.R. Mellish, and J. Norrie. (1988). A comparison of ethephon alone and in combination with CCC or DPC applied to spring barley. *Canadian Journal of Plant Science* 68:941-946.

Camberato, J.J. and B.R. Bock. (1990). Spring wheat response to enhanced ammonium supply. II. Tillering. *Agronomy Journal* 82:467-473.

Carruthers, K., B. Prithiviral, Q. Fe, D. Cloutier, R.C. Martin, and D.L. Smith. (2000). Intercropping corn with soybean, lupin and forages: yield component responses. *European Journal of Agronomy* 12:103-115.

Carruthers, K., Q. Fe, D. Cloutier, and D.L. Smith. (1998). Intercropping corn with soybean, lupin and forages: weed control by intercrops combined with interrow cultivation. *European Journal of Agronomy* 8:225-238.

Chen, J.G., S.H. Cheng, W. Cao, and X. Zhou. (1998). Involvement of endogenous plant hormones in the effect of mixed nitrogen source on growth and tillering of wheat. *Journal of Plant Nutrition* 21:87-97.

Coe, E.H., M.G. Neuffer, and D.A. Hoisington. (1988). The genetics of corn. In *Corn and Corn Improvement* 3rd edition, ed., G.F. Sprague and J.W. Dudley, No. 8 Agronomy Series, Madison, WI: American Society of Agronomy. pp. 81-258.

Costa, C., L.M. Dwyer, C. Hamel, D.F. Muamba, L. Nantais, X.L. Wang, and D.L. Smith. (2001). Root contrast enhancement for measurement with optical scanner-based image analysis. *Canadian Journal of Botany* 79:23-29.

Costa, C., L.M. Dwyer, R.I. Hamilton, C. Hamel, L. Nantais, and D.L. Smith. (2000). A sampling method for measurement of large root systems with scanner-based image analysis. *Agronomy Journal* 92:621-627.

Covell, S., R.H. Ellis, E.H. Roberts, and R.J. Summerfield. (1986). The influence of temperature on seed germination rate in grain legumes. 1. A comparison of chickpea, lentil, soyabean and cowpea at constant temperatures. *Journal of Experimental Botany* 37:705-715.

Dashti, N., F. Zhang, R.K. Hynes, and D.L. Smith. (1997). Application of plant growth promoting rhizobacteria to soybean [*Glycine max* (L.) Merr.] increases protein and dry matter yield under short-season conditions. *Plant and Soil* 188:33-41.

Dashti, N., F. Zhang, R.K. Hynes, and D.L. Smith. (1998). Plant growth promoting bacteria accelerate nodulation and increase nitrogen fixation activity by field grown soybean [*Glycine max* (L.) Merr.] under short season conditions. *Plant and Soil* 200:205-213.

Date, R.A. (1988). Nodulation difficulties related to low pH. In *Current Perspectives in Nitrogen Fixation*, ed., A.H. Gibson and W.E. Newton, Canberra, Australia: Australian Academy of Sciences. pp. 216-262.

Daynard, T.B. and M. Tollenaar. (1983). Prospects for improving the productivity of early-maturing maize. *In* Communications au Coloque Physiologie du Mais, Royan, France, March 15-17. 1983. pp. 535-570.

Denarie, J. and J. Cullimore. (1993). Lipo-oligosaccharide nodulation factors: a new class of signalling molecules mediating recognition and morphogenesis. *Cell* 74: 951-954.

Dijak, M., A.M. Modarres, R.I. Hamilton, L.M. Dwyer, D.W. Stewart, D.E. Mather, and D.L. Smith. (1999). Leafy reduced-stature maize hybrids for short-season environments. *Crop Science* 39:1106-1110.

Dijak, M., D.P. Ormrod, D.L. Smith. (1985). Adaptation to supplementary incandescent radiation in growth chambers by seedlings of *Glycine max*. *Environ. Exp. Botany*. 25:375-384.

Dwyer, L.M., C.J. Andrews, D.W. Stewart, B.L. Ma, and J.A. Dugas. (1995). Carbohydrate levels in field-grown leafy and normal maize genotypes. *Crop Science* 35:1020-1027.

Eaglesham, A.R.J., A. Ayanaba, V. Ranga Rao, and D.L. Eskew. (1981). Improving the nitrogen nutrition of maize by intercropping with cowpea. *Soil Biology and Biochemistry* 13:169-171.

Eghball, B., J.R. Settimi, J.W. Maranville, and A.M. Parkhurst. (1993). Fractal analysis for morphological description of corn roots under nitrogen stress. *Agronomy Journal* 85:287-289.

Ellis, R.H., S. Covell, E.H. Roberts, and R.J. Summerfield. (1986). The influence of temperature on seed germination in grain legumes. II. Intraspecific variation in chick-pea (*Cicer arietinum* L.) at constant temperatures. *Journal of Experimental Botany* 37:1503-1515.

Exner, D.N. and R.M. Cruse. (1993). Interseeded forage legume potential as a winter ground cover, nitrogen source and competitor. *Journal of Production Agriculture* 6:226-231.

Foroutan-pour, K., B.L. Ma, D.L. Smith. (1997). Protein accumulation potential in barley seeds as affected by soil and peduncle-applied N and peduncle-applied plant growth regulators. *Physiologia Plantarum* 100:190-201.

Foroutan-pour, K., P. Dutilleul, and D.L. Smith. (1999). Advances in the implementation of the box-counting method of fractal dimension estimation. *Applied Mathematics and Computing* 105:195-210.

Foroutan-pour, K., P. Dutilleul, and D.L. Smith. (2000). Effects of population density and intercropping with soybean on fractal dimension of corn plant skeletal images. *Journal Agronomy and Crop Science* 184:89-100.

Foroutan-pour, K., P. Dutilleul, and D.L. Smith. (2001). Inclusion of the fractal dimension of leafless plant structure in the Beer-Lambert Law. *Agronomy Journal* 93:333-338.

Foroutan-pour, K., B.L. Ma, and D.L. Smith. (1995). Field evaluation of the peduncle perfusion technique. *Journal of Plant Nutrition* 18:1225-1236.

Frederick, J.R. and H.G. Marshall. (1985). Grain yield and yield components of soft red winter wheat as affected by management practises. *Agronomy Journal* 77:495-499.

Fujita K., S. Ogata, K. Matsumoto, K. Masuda, K.G. Ofosu-Budu, and K. Kuwata. (1990). Nitrogen transfer and dry matter production in soybean and sorghum mixed cropping system at different population densities. *Soil Science and Plant Nutrition* 36:233-241.

George, M.R., C.A. Raguse, W.J. Clawson, C.B. Wilson, R.I. Willoughby, N.K. McDougald. (1998). Correlation of degree-days with annual herbage yields and livestock gains. *Journal of Range Management* 41:193-197.

Gillen, R.L. and A.L. Ewing. (1992). Leaf development of native blue-stem grasses in relation to degree-day accumulation. *Journal of Range Management* 45:200-204.

Green, C.F. (1986). Modifications to the growth and development of cereals using chlorocholine chloride in the absence of lodging: a synopsis. *Field Crops Research* 14:117-133.

Guy, S.O. and E.S. Oplinger. (1989). Soft winter wheat cultivar response to propiconazole and ethephon. *Journal of Production Agriculture* 2:179-184.

Hamel, C. and D.L. Smith. (1991). Interspecific N-transfer and plant development in a mycorrhizal field-grown mixture. *Soil Biology and Biochemistry* 23:661-665.

Hamel, C. and D.L. Smith. (1992). Mycorrhizae-mediated ^{15}N transfer from soybean to corn in field-grown intercrops: effect of component crop spatial relationships. *Soil Biology and Biochemistry* 24:499-501.

Hamel, C., C. Nesser, U. Barrantes-Cartin, V. Furlan, and D.L. Smith. (1991b). Endomycorrhizal fungi species mediate ^{15}N transfer from soybean to maize in non-fumigated soil. *Plant and Soil* 138:41-47.

Hamel, C., U. Barrantes-Cartin, V. Furlan, and D.L. Smith. (1991a). Endomycorrhizal fungi in nitrogen transfer from soybean to maize. *Plant and Soil* 138:33-40.

Hamel, C., V. Furlan, and D.L. Smith. (1991). N_2-fixation and transfer in a field grown mycorrhizal corn and soybean intercrop. *Plant and Soil* 133:177-185.

Hamel, C., V. Furlan, and D.L. Smith. (1992). Mycorrhizal effects on interspecific plant competition and nitrogen transfer in legume-grass mixture. *Crop Science* 32:991-996.

Hardegree, S.P. (1994). Matric priming increases germination rate of great basin native perennial grasses. *Agronomy Journal* 86:289-293.

Hashemi-Dezfouli, A. and S.J. Herbert. (1992). Crops: intensifying plant density response of corn with artificial shade. *Agronomy Journal* 84:547-551.

Haynes, R.J. (1980). Competitive aspects of the grass-legume association. *Advances in Agronomy* 33:227-256.

Herridge, D.F. and J. Brockwell. (1988). Contribution of fixed nitrogen and soil nitrate to the nitrogen economy of irrigated soybean. *Soil Biology and Biochemistry* 20:711-717.

Hohenadel, A.B. (1984). Expanding the corn frontier. New dwarf hybrids may push the limits for corn production back to regions with 2000 heat units or less. *Country Guide*: March 1984. p. 34.

Horvath, B., R. Heidstra, M. Lados, M. Moerman, H.P. Spaink, J.C. Prome, A. van Kammen, and T. Bisseling. (1993). Lipo-chitooligosaccharide of *Rhizobium* induces infection-related early nodulin gene expression in pea root hairs. *Plant Journal* 4:727-733.

Humphreys, D.G., D.L. Smith, and D.E. Mather. (1994). Nitrogen fertilizer application and seeding date effects on oat grain milling quality. *Agronomy Journal* 86:836-843.

Hungria, M. and G. Stacey. (1997). Molecular signal exchanges between host plants and rhizobia: basic aspects and potential application in agriculture. *Soil Biology and Biochemistry* 29:819-830.

Innis, D. (1997). The nature of intercropping. In *Intercropping and the Scientific Basis of Traditional Agriculture*, ed., D. Innis, London, UK: Intermediate Technology Publications.

Jacobs, R., W. Pan, B. Miller, R.A. Allan, W.S. Fuller, and W.T. McKean. (1996). *Pacific Northwest Wheat Straw: A Glance at the Fiber Morphology*. Atlanta, GA: TAPPI Press.

Jacobson, E.T., D.A. Tober, R.J. Hans, and D.C. Darris. (1984). The performance of selected cultivars of warm-season grasses in northern prairie and plains states. In *The Prairie: Past, Present and Future, Proceedings of the 9th North American*

Prairie Conference, ed., G.K. Clambey and R.H. Pemble, Fargo, ND: Center for Environment Studies. pp. 215-221.

Jordan, G.L. and M.R. Haferkamp. (1989). Temperature responses and calculated heat units for germination of several range grasses and shrubs. *Journal of Range Management* 42:41-45.

Jung, G.A. and K.P. Vogel. (1992). Lignification of switchgrass (*Panicum virgatum*) and big bluestem (*Andropogon gerardii*) plant parts during maturation and its effect on fibre degradability. *Journal of Food and Agriculture* 59:169-176.

Jung, G.A., J.A. Shafer, W.L. Stout, and M.T. Panciera. (1990). Warm-season grass diversity in yield, plant morphology, and nitrogen concentration and removal in northeastern USA. *Agronomy Journal* 82:21-26.

Karl, T.R., P.D. Jones, R.W. Knight, G. Kukla, N. Plummer, V. Razuvayev, K.P. Gallo, J. Lindseay, R.J. Charlson, and T.C. Peterson. (1993). A new perspective on recent global warming: asymmetric trends of daily maximum and minimum temperature. *Bulletin of the American Meteorological Society* 74:1007-1023.

Kaya, C., D. Higgs, and A. Burton. (1999). Foliar application of iron as a remedy for zinc toxic tomato plants. *Journal of Plant Nutrition* 22:1829-1837.

Kloepper, J.W., M.N. Schroth, and T.D. Miller. (1980). Effects of rhizosphere colonization by plant growth promoting rhizobacteria on potato plant growth and yield. *Phytopathology* 70:1078-1082.

Krause, G.H. and E. Weis. (1984). Chlorophyll fluorescence as tool in plat physiology: II. Interpretation of fluorescence signal. *Photosynthesis Research* 5:139-157.

Lanning, F.C. and L.N. Eleuterius. (1987). Silica and ash in native plants to the central and southeastern of the United States. *Annals of Botany* 60:361-375.

Lawlor, D.J., E.T. Kanemasu, W.C. Albrecht III, and D.E. Johnson. (1990). Seed production environment influence on the base temperature for growth of sorghum genotypes. *Agronomy Journal* 82:643-647.

Le Strange, K.K., G.L. Bender, M.A. Djordjevic, B.G. Rolfe, and J.W. Redmond. (1990). The *Rhizobium* strain NGR234 nodD1 gene product responds to activation by the simple phenolic compounds vanillin and isovanillin present in wheat seedling extracts. *Molecular Plant-Microbe Interactions* 3:214-220.

Leibovitch, S., B.L. Ma, W.E. Maloba, and D.L. Smith. (1992). Spring barley responses to row spacing and fungicide triadimefon in regions with a short crop growing season. *Journal of Agronomy and Crop Science* 169:209-215.

Lhuissier, F.G.P., N.C.A. De Ruijter, B.J. Sieberer, J.J. Esseling, and A.M.C. Emons (2001). Time course of cell biological events evoked in legume root hairs by *Rhizobium* Nod factors: State of the art. *Annals of Botany* 87:289-302.

Liang, B.C., M. Remillard, and A.F. MacKenzie. (1991). Influence of fertilizer irrigation and no-growing season precipitation on soil nitrate-nitrogen under corn. *Journal of Environmental Quality* 20:123-128.

Ma, B.L., L.M. Dwyer, and D.L. Smith. (1994). Evaluation of peduncle perfusion for *in vivo* studies of carbon and nitrogen distribution in cereal crops. *Crop Science* 34:1584-1588.

Ma, B.L., L.M. Dwyer, S. Leibovitch, and D.L. Smith. (1995). Effect of peduncle-perfused nitrogen, sucrose, and growth regulators on barley and wheat amino acid composition. *Communications in Soil Science and Plant Analysis* 26:969-982.

Ma, B.L. and D.L. Smith. (1991a). Apical development of spring barley in relation to chlormequat and ethephon. *Agronomy Journal* 83:270-274.

Ma, B.L. and D.L. Smith. (1991b). The effects of ethephon, chlormequat chloride and mixtures of ethephon and chlormequat chloride applied at the beginning of stem elongation on spike-bearing shoots and other yield components of spring barley (*Hordeum vulgare* L.). *Journal of Agronomy & Crop Science* 166:127-135.

Ma, B.L. and D.L. Smith. (1992a). Apical development of spring barley under North American field conditions. *Crop Science* 32:144-149.

Ma, B.L. and D.L. Smith. (1992b). Growth regulator effects on above-ground dry matter partitioning during grain fill of spring barley. *Crop Science* 32:741-746.

Ma, B.L. and D.L. Smith. (1992c). Modification of tillering productivity in spring barley as affected by chlormequat and ethephon. *Crop Science* 32:735-740.

Ma, B.L. and D.L. Smith. (1992d). Chlormequat and ethephon timing and grain production of spring barley. *Agronomy Journal* 84:934-939.

Ma, B.L. and D.L. Smith. (1992e). Post-anthesis ethephon effects on yield of spring barley. *Agronomy Journal* 84:370-374.

Ma, B.L. and D.L. Smith. (1992f). New method for supplying substances to cereal inflorescence. *Crop Science* 32:191-194.

Ma, B.L., L.M. Dwyer, D.W. Stewart, C.J. Anderws, and M. Tollenaar. (1994). Stem infusion of field-grown maize. *Communications in Soil Science and Plant Analysis* 25:2005-2017.

Ma, B.L., L.M. Dwyer, T. Tollenaar, and D.L. Smith. (1998). Stem-infusion of nitrogen-15 to quantify nitrogen remobilisation in maize. *Communications in Soil Science and Plant Analysis* 29:305-317.

Ma, B.L., S. Leibovitch, W.E. Maloba, and D.L. Smith. (1992). Spring barley responses to nitrogen fertilizer and ethephon in regions with a short crop growing season. *Journal of Agronomy and Crop Science* 169:151-160.

Madakadze, C., B. Coulman, K. Stewart, P. Peterson, R. Samson, and D.L. Smith. (1998a). Phenology and tiller characteristics of selected cultivars of big bluestem and switchgrass in a short season area. *Agronomy Journal* 90:489-495.

Madakadze, C., B.E. Coulman, A.R. McElroy, K.A. Stewart, and D.L. Smith. (1998b). Evaluation of selected warm season grasses for biomass production in a short season area. *BioResource Technology* 65:1-12.

Madakadze, C., B.E. Coulman, P. Peterson, K. Stewart, R. Samson, and D.L. Smith. (1998c). Leaf area development, light interception and yield among switchgrass populations in a short season area. *Crop Science* 38:827-834.

Madakadze, C., K. Stewart, P. Peterson, B.E. Coulman, and D.L. Smith. (1999a). Switchgrass biomass and chemical composition for biofuel in eastern Canada. *Agronomy Journal* 91:969-701.

Madakadze, C., K. Stewart, P. Peterson, B.E. Coulman, R. Samson, and D.L. Smith. (1998d). Light intercepting, use efficiency and energy yield of switchgrass (*Panicum virgatum* L.) grown in a short season area. *Biomass and Bioenergy* 15:475-482.

Madakadze, I.C., T. Radiotis, J. Li, K. Goel, and D.L. Smith. (1999c). Kraft pulping characteristics and pulp properties of warm season grasses. *BioResource Technology* 69:75-85.

Madakadze, I.C., K.A. Stewart, P.B. Peterson, B.E. Coulman, and D.L. Smith. (1999b). Cutting frequency and N fertilization effects on yield and N concentration of switchgrass in a short season area. *Crop Science* 39:552-557.

Madakadze, I.C., K.A. Stewart, R.M. Madakadze, P.B. Peterson, B.E. Coulman, and D.L. Smith. (1999c). Field evaluation of the chlorophyll meter to predict yield and nitrogen concentration of switchgrass. *Journal of Plant Nutrition* 22:1001-1010.

Madakadze, I.C., K. Stewart, and D.L. Smith. (2001) Variation in base temperatures for germination and seedling growth in warm season grasses. *Seed Science Technology* 29:31-38.

Madakadze, I.C., R.M. Madakadze, K. Stewart, P.B. Peterson, B.E. Coulman, and D.L. Smith. (2000). Effect of preplant seed conditioning treatment on the germination of switchgrass (*Panicum virgatum* L.) seeds. *Seed Science Technology* 28: 403-411.

Madramootoo, C.A., K. Wiyo, and P. Enright. (1992). Nutrient losses through title drains from two potato field. *Applied Engineering Agriculture* 8:639-646.

Martin, R.C., A.R.J. Eaglesham. H.D. Voldeng, and D.L. Smith. (1995). Factor affecting nitrogen benefit from soybean [*Glycine Max* (L.) Merr. CV Lee] to interplanted corn (*Zea mays* L. CV CO-OP S259). *Environmental and Experimental Botany* 35: 497-505.

Martin, R.C., H.D. Voldeng, and D.L. Smith. (1991a). Nitrogen transfer from nodulating soybean to maize and nonnodulating soybean in intercrops: direct ^{15}N labelling methods. *New Phytologist* 117:233-241.

Martin, R.C., H.D. Voldeng, and D.L. Smith. (1991b). Nitrogen transfer from nodulating soybean to maize or to nonnodulating soybean in intercrops: the ^{15}N dilution method. *Plant and Soil* 132:3-63.

Martin, R.C., J.T. Arnason. J.D.H. Lambert, P. Isabelles, H.D. Voldeng, and D.L. Smith. (1989). Reduction of European corn borer (Lepidoptera: Pyralidae) damage by intercropping corn with soybean. *Journal of Economical Entomology* 82: 1455-1459.

Martin, R.C., H.D. Voldeng, and D.L. Smith. (1990). Intercropping corn and soybean for silage in a cool-temperate region: yield, protein and economic effects. *Field Crop Research* 23:295-310.

Martinez-Abarca, F., J.A. Herrera-Cervera, P. Bueno, J. Sanjuan, T. Bisseling, and J. Olivares. (1998). Involvement of salicylic acid in the establishment of the *Rhizobium meliloti*-alfalfa symbiosis. *Molecular Plant-Microbe Interaction* 11:153-155.

McCloskey, J.T. (1995). What about nonwoods? Proceedings of the TAPPI Global Fiber Supply Symposium. Atlanta, GA: TAPPI Press. pp. 95-106.

Midmore, D.J. (1993). Agronomic modification of resource use and intercrop productivity. *Field Crops Research* 34:357-380.

Modarres, A.M., R.I. Hamilton, L.M. Dwyer, D.W. Stewart, D.E. Mather, M. Dijak, and D.L. Smith. (1997a). Leafy reduced-stature maize for short-season environments: morphological aspects of inbred lines. *Euphytica* 96:301-309.

Modarres, A.M., R.I. Hamilton, L.M. Dwyer, D.W. Stewart, D.E. Mather, M. Dijak and D.L. Smith. (1997c). Leafy reduced-stature maize for short-season environments: morphological aspects of inbred lines. *Euphytica* 96:301-309.

Modarres, A.M., R.I. Hamilton, L.M. Dwyer, D.W. Stewart, M. Dijak, and D.L. Smith. (1997b). Leafy reduced-stature maize for short-season environments: yield and yield components of inbred lines. *Euphytica* 97:129-138.

Modarres, A.M., R.I. Hamilton, M. Dijak, L.M. Dwyer, D.W. Stewart, D.E. Mather, and D.L. Smith. (1998a). Plant population density effects on maize inbred lines grown in short-season environments. *Crop Science* 38:104-108.

Modarres, A.M., M. Dijak, R.I. Hamiltion, L.M. Dwyer, D.W. Stewart, D.E. Mather, and D.L. Smith. (1998b). Leafy reduced stature hybrid maize response to population density and planting patterns in a short growing season area. *Maydica* 43:227-234.

Morris, C.F., D.L. Ferguson, and G.M. Paulsen. (1989). Nitrogen fertilizer management with foliar fungicide and growth regulator for hard winter wheat production. *Applied Agricultural Research* 4:135-140.

Muchow, R.C. and R. Davis. (1988). Effect of nitrogen supply on the comparative productivity of maize and sorghum in a semi-arid tropical environment. II. Radiation interception and biomass accumulation. *Field Crops Research* 18:17-30.

Muchow, R.C., T.R. Sinclair, and J.M. Bennett. (1990). Temperature and solar radiation effects on potential maize yield across locations. *Agronomy Journal* 82:338-343.

Nelson, O.E. and A.J. Ohlrogge. (1957). Differential responses to population pressures by normal and dwarf lines of maize. *Science* 125:1200.

Ofosu-Budu, K.G., K. Noumura, and K. Fujita. (1995). N_2 fixation, N transfer and biomass production of soybean cv. Bragg or its supernodulating nts 1007 and sorghum mixed-cropping at two rates of N fertilizer. *Soil Biology and Biochemistry* 27: 311-317.

Ottman, M.J. and L.F. Welch. (1989). Planting patterns and radiation interception, plant nutrient concentration and yield in corn. *Agronomy Journal* 81:167-174.

Paau, A.S., M.L. Bennett, C.J. Kurtenbach, and L.L. Graham. (1990). Improvement of inoculant efficiency by strain improvement and formulation manipulation. In *Nitrogen Fixation: Achievements and Objective.* ed., P.G. Gresshoff, L.E. Roth, G. Stacey and W.E. Newton, New York, NY: Chapman and Hall. pp. 617-624.

Paavilainen, L. (1994). Fine paper from certain grass species. Paper presented at the Non-wood Fibres for Industry Conference. Pira International/Silsoe Research Institute Joint Conference. Pira International, Leatherhead, Surrey, UK.

Pan, B. and D.L. Smith. (2000a). Genistein preincubation of *Bradyrhizobium japonicum* cells improves strain competitiveness under greenhouse, but not field conditions. *Plant and Soil* 223:229-234.

Pan, B. and D.L. Smith. (2000b). Preincubation of *Bradyrhizobium japonicum* cells with genistein reduces the inhibitory effects of mineral nitrogen on soybean nodulation and nitrogen fixation under field conditions. *Plant and Soil* 223:235-242.

Parent, M. and S. Occhiett. (1988). Late Wisconsinan Deglaciation and Champlain Invasion in the St. Lawrence Valley, Quebec. *Geographie Physique et Quaternaire* 42:215-246.

Peltonen, J. (1992). Ear developmental stage used for timing supplemental nitrogen application to spring wheat. *Crop Science* 32:1029-1033.

Perret, X., C.H. Staehelin, and W.J. Broughton. (2000). Molecular basis of symbiotic promiscuity. *Microbiological and Molecular Biology Reviews* 64:180-201.

Prayitno, J., J. Stefaniak, J. McIver, J.J. Weinman, F.B. Dazzo, J.K. Ladha, W. Barraquio, Y.G. Yanni, and B.G. Rolf. (1999). Interactions of rice seedlings with bacteria isolated from rice roots. *Australian Journal of Plant Physiology* 26:521-535.

Prithiviraj, B., A. Solumenoev, X. Zhou, and D.L. Smith (2000a). Differential response of soybean (*Glycine max* (L.) Merr.) to lipo-chitooligosaccharide Nod *Bj* V ($C_{18:1}$ Me Fuc). *Journal of Experimental Botany* 51:2045-2051.

Prithiviraj, B., X. Zhou, A. Souleimanov, and D.L. Smith. (2000c). Nod Bj V (C18:1; MeFuc), a host specific bacteria-to-plant signal molecule, enhances germination and early growth of diverse crop plants. Abstract E6, 17th North American Conference on Symbiotic Nitrogen Fixation, 23-28 July 2000, Quebec, Canada. p. 38.

Prithiviraj, B., X. Zhou, A. Souleimanov, and D.L. Smith. (2000b). The lipo chitooligosaccharide Nod Bj V (C18:1;MeFuc) affects salicylic acid content of soybean (*Glycine max*) tissues. Abstract PE3, 17th North American Conference on Symbiotic Nitrogen Fixation, 23-28 July 2000, Quebec, Canada. p. 66.

Rosolem, C.A., J.S. Assis, and A.D. Santiago. (1994). Root growth and mineral nutrition of corn hybrids as affected by phosphorus and lime. *Communications in Soil Science* 25:2491-2499.

Russel, W.A. (1985). Evaluation for plant, ear and grain traits of maize cultivars representing seven years of breeding. *Maydica* 30:85-96.

Sanderson, M.A. and D.D. Wolf. (1995). Switchgrass biomass composition during morphological development in diverse environments. *Crop Science* 35:1432-1438.

Sanderson, M.A., R.L. Reed, S.B. McLaughlin, S.D. Wullschleger, B.V. Conger, D.J. Parrish, D.D. Wolf, C. Taliaferro, A.A. Hopkins, W.R. Ocumpaugh, M.A. Husley, J.C. Read, and C.R. Tischler. (1996). Switchgrass as a sustainable bioenergy crop. *BioResource Technology* 56:83-93.

Schulaman, H.R., A.A. Gisel, N.E. Quaedvlieg, G.V. Bloemberg, B.J. Lugtenberg, J.W. Kijne, I. Potrykus, H.P. Spaink, and C. Sautter. (1997). Chitin oligosaccharides can induce cortical cell division in roots of *Vicia sativa* when delivered by ballistic microtargeting. *Development* 124: 4887-4895.

Senaratne, R., N.D.L. Liyanage, and R.J. Soper. (1995). Nitrogen fixation of and N transfer from cowpea, mungbean and groundnut when intercropped with maize. *Fertilizer Research* 40:41-48.

Shah, S.A., S.A. Harrison, D.J. Boquet, P.D. Colyer, and S.H. Moore. (1994). Management effects on yield and yield components of late-planted wheat. *Crop Science* 34:1298-1303.

Shaver, D.L. (1983). Genetics and breeding of maize with extra leaves above the ear. In *Proceedings of the 38th Annual Corn and Sorghum Industry Research Conference*, Chicago, IL: American Seed Trade Association. pp.161-180.

Simpson, J.R. (1965). The transference of nitrogen from pasture legumes to an associated grass under several systems of managements in pot culture. *Australian Journal of Agricultural Research* 16:915-926.

Smith, D.L., B.L. Ma, S. Leibovitch, S. Lussier, and W.E. Maloba. (1993). Comparison of crop management effects on spring barley cultivars grown on three soil types in southwestern Quebec. *Canadian Journal of Plant Science* 73: 927-938.

Souza, S.R., E.M. Stark, and M.S. Fernandes. (1999). Foliar spraying of rice with nitrogen: effect on protein levels, protein fractions, and grain weight. *Journal of Plant Nutrition* 22:579-588.

Spaink, H.P., D.M. Sheeley, A.A.N van Brussel, J. Glushka, W.S. York, T. Tak, O. Geiger, E.P. Kennedy, V.N. Reinhold, and B.J.J. Lugtenberg (1991). A novel highly unsaturated fatty acid moiety of lipo-oligosaccharide signals determines host specificity of *Rhizobium*. *Nature* 354:125-130.

Sprent, J.I. and P. Sprent. (1990). *Nitrogen Fixing Organisms*. New York, NY: Chapman and Hall.

Stewart, D.W. and L.M. Dwyer. (1999). Mathematical characterization of leaf shape and area of maize hybrids. *Crop Science* 39:422-427.

Stewart, D.W. and L.M. Dwyer. (1993). Mathematical characterization of maize canopies. *Agricultural and Forest Meteorology* 66:247-265.

Stoskopf, N.C. (1985). Variability in cereals grains. In *Cereal Grain Crops*, Reston, VA: Reston Publishing Company, Inc.

Stubbendieck, J., S.L. Hatch, and C.H. Butterfield. (1992). *North American Range Plants*, 4th Edition. Lincoln, NE: University of Nebraska Press.

Tatsumi, J., A. Yamauchi, and Y. Kono. (1989). Fractal analysis of plant root systems. *Annuals of Botany* 64:499-503.

Tollenaar, M. (1991). Physiological basis of genetic improvement of maize hybrids in Ontario from 1959 to 1988. *Crop Science* 31:119-124.

Trebbi, G. (1993). Power-production options from biomass: the vision of a Southern European utility. *Bioresource Technology* 46:23-29.

Van-Sanford, D.A., J.H. Grove, L.J. Grabau, and C.T. MacKown. (1989). Ethephon and nitrogen use in winter wheat. *Agronomy Journal* 81:951-954.

Varney, G.T., M.J. Canny, X.L. Wang, and M.E. McCully. (1991). The branch roots of *Zea*. I. First order branches, their number, sizes and division into classes. *Canadian Journal of Botany* 67:357-364.

Vieira, R.F., C. Vieira, E.J.B.N. Cardoso, and P.R. Mosquim. (1998). Foliar application of molybdenum in common bean. II. Nitrogenase and nitrate reductase activities in a soil of low fertility. *Journal of Plant Nutrition* 21:2141-2151.

Virtanen, A.I. (1933). The nitrogen nutrition of plants. *Herbal Review* 1:88-91.

Vona, L.C., G.A. Jung, R.L. Reid, and W.C. Sharp. (1984). Nutritive value of warm season grass hays for beef and sheep: digestibility, intake and mineral utilization. *Journal of Animal Science* 59:1582-1593.

Wang, X. and F.E. Below. (1992). Root growth, nitrogen uptake, and tillering of wheat induced by mixed-nitrogen source. *Crop Science* 32:997-1002.

Wang, X. and F.E. Below. (1996). Cytokinins in enhanced growth and tillering of wheat induced by mixed nitrogen source. *Crop Science* 36:121-126.

Wiersma, D.W., E.S. Oplinger, and S.O. Guy. (1986). Environment and cultivar effects on winter wheat response to ethephon plant growth regulator. *Agronomy Journal* 78:761-764.

Wilson, R.A. (1996). Non wood fibres: precursors and criteria required for acceptance and expansion in pulp and paper markets. In *Conference Proceedings: Uses for Non-Wood Fibres–Commercial and Practical Issues for Papermaking. Vol. 1. Flax, Hemp and Cereal Straw*. Surrey, UK: Pira International. pp. 1-11.

Woodard, K.R., G.M. Prine, and S. Bachrein. (1993). Solar energy recovery by elephantgrass, energycane and elephantmillet canopies. *Crop Science* 33:824-830.

Xie, Z-P., C. Staehelin, H. Vierheilig, A. Wiemken, S. Jabbouri, W.J. Broughton, R. Vogeli-Lange, and T. Boller. (1995). Rhizobial nodulation factors stimulate mycorrhizal colonization of nodulating and non-nodulating soybeans. *Plant Physiology* 108:1519-1525.

Yanni, Y.G., R.Y. Rizk, V. Corich, A. Squartini, K. Ninke, S. Phillip-Hollongsworth, G. Orgambide, F. deBuijn, J. Stoltzfus, D. Buckley, T.M. Schmidt, P.F. Mateos, J.K. Ladha, and F.B. Dazzo. (1997). Natural endophytic association between *Rhizobium leguminosarum* bv. Trifolii and rice roots and assessment of its potential to promote rice growth. *Plant and Soil* 194:99-114.

Zadoks, J.C., T.T. Chang, and C.F. Konzak. (1974). A decimal code for the growth stages of cereals. *Weed Research* 14:415-421.

Zahran, H.H. (1999). *Rhizobium*-legume symbiosis and nitrogen fixation under severe conditions and in an arid climate. *Microbiology and Molecular Biology Reviews* 63:968-989.

Zahran, H.H., L.A.M. Karsisto, and K. Lindstorm. (1994). Alteration of lipopolysaccharides and protein profiles in SDS-PAGE of rhizobia by osmotic stress. *World Journal of Microbiology and Biotechnology* 10:100-105.

Zhang, F. and D.L. Smith. (1994). Effects of low root zone temperatures on the early stages of symbiosis establishment between soybean [*Glycine max* (L.) Merr.] and *Bradyrhizobium japonicum*. *Journal of Experimental Botany* 279:1467-1473.

Zhang, F. and D.L. Smith. (1995). Preincubation of *Bradyrhizobium japonicum* with genistein accelerates nodule development of soybean [*Glycine max* (L.) Merr.] at suboptimal root zone temperatures. *Plant Physiology* 108:961-968.

Zhang, F. and D.L. Smith. (1996). Inoculation of soybean [*Glycine max* (L.) Merr.] with genistein-preincubated *Bradyrhizobium japonicum* or genistein directly applied into soil and soybean protein and dry matter yield under short season conditions. *Plant and Soil* 179:233-241.

Zhang, F. and D.L. Smith. (1997). Application of genistein to inocula and soil to overcome low spring soil temperature inhibition of soybean nodulation and nitrogen fixation. *Plant and Soil* 192:141-151.

Zhang, F., D.H. Lynch, and D.L. Smith. (1995). Impact of low root zone temperatures in soybean [*Glycine max* (L.) Merr.] on nodulation and nitrogen fixation. *Environmental and Experimental Botany* 35:279-285.

Zhang, F., F. Máce, and D.L. Smith. (2000). Mineral nitrogen availability and isoflavonoid accumulation in the root system of soybean (*Glycine max* (L.) Merr). *Journal of Agronomy and Crop Science* 184:197-204.

Zhang, F., N. Dashti, R.K. Hynes, and D.L. Smith. (1996). Plant growth promoting rhizobacteria and soybean [*Glycine max* (L.) Merr.] nodulation and nitrogen fixation at suboptimal root zone temperatures. *Annals of Botany* 77:453-459.

Zhang, F., N. Dashti, R.K. Hynes, and D.L. Smith. (1997). Plant growth promoting rhizobacteria and soybean [*Glycine max* (L.) Merr.] growth and physiology at suboptimal root zone temperatures. *Annals of Botany* 79:243-249.

Zhang, F. and D.L. Smith. (2002). Interorganismal signaling in suboptimum environments: The legume-rhizobia symbiosis. *Advances in Agronomy* 76:125-161.

Zhou, X.M., A.F. MacKenzie, C.A. Madramootoo, and D.L. Smith. (1997b). Management practices to conserve soil nitrate in maize production system. *Journal of Environmental Quality* 26:1369-1374.

Zhou, X.M., O.A. Abdin, B. Coulman, D.C. Cloutier, M.A. Faris, and D.L. Smith. (2000a). Carbon and nitrogen supplementation to soybean through stem injection and its effect on soybean plant senescence. *Journal of Plant Nutrition* 23:605-616.

Zhou, X.M. and D.L. Smith. (1996). A new technique for continuous injection into stems of field-grown corn plants. *Crop Science* 36:452-456.

Zhou, X.M., C.A. Madramootoo, A.F. MacKenzie, and D.L. Smith. (1997a). Biomass production and N uptake by corn-ryegrass systems. *Agronomy Journal* 89:749-756.

Zhou, X.M., C.A. Madramootoo, A.F. MacKenzie, and D.L. Smith. (2000b). Corn yield and N recovery in water-table-controlled corn/ryegrass systems. *European Journal of Agronomy* 12: 83-92.

Zhou, X.M., C.A. Madramootoo, A.F. MacKenzie, and D.L. Smith. (1999). Effects of stem-injected plant growth regulators, with or without sucrose, on grain production, biomass, and photosynthetic activity of field-grown corn plants. *Journal of Agronomy and Crop Science* 183:103-110.

Zhou, X.M., C.A. Madramootoo, A.F. MacKenzie, and D.L. Smith. (1998). Distribution of ^{15}N-labeled urea injected into field-grown corn plants. *Journal of Plant Nutrition* 21:63-73.

Current and Potential Role
of Transgenic Crops in U.S. Agriculture

C. S. Silvers
L. P. Gianessi
J. E. Carpenter
S. Sankula

SUMMARY. Transgenic crop cultivars with resistance to insects, pathogens, and herbicides offer growers powerful new pest management tools. We reviewed the observed and potential farm-level impacts of transgenic cultivars, including those with regulatory approval and commercial availability as well as those still being researched and developed. Direct grower benefits, such as yield and production increases and decreased management costs, have led to rapid and extensive adoption of *Bt* corn and cotton, herbicide-resistant cotton, soybean, and canola, and virus-resistant papaya. Other transgenic crops, including *Bt* sweet corn and potato, and herbicide-resistant sugar beet and corn, have not been adopted despite strong agronomic and pest management performance, largely because growers fear there will be no market for their harvests. Despite inconsistent adoption of transgenic cultivars, demonstrated benefits of the technology encourage ongoing efforts to incorporate pest management traits into a wider variety of crops, including broccoli, tomato, lettuce, grape, citrus, pineapple, raspberry, peanut, wheat, barley, and rice. Po-

---

C. S. Silvers, J. E. Carpenter, and S. Sankula are Research Associates, and L. P. Gianessi is Senior Research Associate, National Center for Food and Agricultural Policy (NCFAP), 1616 P Street, NW, Washington, DC 20036.

[Haworth co-indexing entry note]: "Current and Potential Role of Transgenic Crops in U.S. Agriculture." Silvers, C. S. et al. Co-published simultaneously in *Journal of Crop Production* (Food Products Press, an imprint of The Haworth Press, Inc.) Vol. 9, No. 1/2 (#17/18), 2003, pp. 501-530; and: *Cropping Systems: Trends and Advances* (ed: Anil Shrestha) Food Products Press, an imprint of The Haworth Press, Inc., 2003, pp. 501-530. Single or multiple copies of this article are available for a fee from The Haworth Document Delivery Service [1-800-HAWORTH, 9:00 a.m. - 5:00 p.m. (EST). E-mail address: docdelivery@haworthpress. com].

tential impacts of these upcoming transgenic cultivars range from a decrease in weed management costs for lettuce and tomato growers, to the defense of stone fruit, grape, and citrus against devastating new pests. *[Article copies available for a fee from The Haworth Document Delivery Service: 1-800-HAWORTH. E-mail address: <docdelivery@haworthpress.com> Website: <http://www.HaworthPress.com> © 2003 by The Haworth Press, Inc. All rights reserved.]*

KEYWORDS. Agricultural biotechnology, pest management, economic impacts

INTRODUCTION

Advances in genetics and molecular biology have made it possible to identify a gene that codes for a specific trait in one organism, isolate and clone that gene, and incorporate it into the genome of another organism. This transfer of genes may occur regardless of relatedness of the source and donor species, or of linkage of the desirable trait with undesirable traits. The organism into whose genome a foreign gene, or transgene, has been incorporated is described as transgenic (Zaid et al., 1999). Transgenic technology is now being used experimentally on a myriad of crop species, transforming them with resistance to insects, pathogens, and herbicides in an effort to improve management of the pests that plague modern agriculture.

Several transgenic cultivars have full regulatory approval and are commercially available for planting by U.S. growers. Adoption of these transgenic cultivars has been rapid in some cases. In other cases, adoption has been minimal. Public debate over agricultural biotechnology has prompted many commodity buyers and processors across the country not to purchase food produced from genetically engineered plants, despite the potential benefits to growers shown in the field. The limited adoption of some transgenic varieties therefore stems from growers' uncertainty of finding a market for their harvested product rather than from poor agronomic or pest management performance. Despite inconsistent adoption, however, research continues to improve and expand the application of biotechnology in agriculture.

An understanding of the contributions, both realized and potentially forthcoming, of agricultural biotechnology for crop pest management is critical to the unfolding public discussion that surrounds it and, ultimately, will determine its future. The objective of this article is to provide brief descriptions of traits transferred to crop plants for resistance to insects, pathogens, and herbicides, and to discuss current adoption levels and farm-level impacts of available transgenic cultivars. Also reviewed are several transgenic crops under development, the agronomic pests they target, and projected farm-level impacts of their commercialization and adoption.

INSECTICIDAL CROPS

Insecticidal plants produced through biotechnology express traits derived from *Bacillus thuringiensis* (*Bt*), a species of soil-borne bacteria. Enclosed with the spores of *Bt* are protein crystals (Gill, Cowles, and Pietrantonio, 1992). When an insect ingests spores, the alkaline conditions in the insect gut dissolve the protein crystal, releasing protoxins that are then activated by specific enzymes in the gut. Activated toxins bind to cells lining the insect gut and disrupt the ionic balance within the cells by creating membranous pores. When the cells rupture, a hole is produced in the gut lining. The infected insect may suffer from paralysis of the gut or entire body, cease to feed, and starve to death. If paralysis does not occur, the insect will be killed by systemic infection after the *Bt* spore germinates and begins vegetative growth within its body.

There are several *Bt* varieties, and each produces one or more crystalline (Cry) proteins and protoxins (Gill, Cowles, and Pietranonio, 1992). Each *Bt* toxin has specific insecticidal activity against certain groups of insects. This specificity is based on characteristics of the *Bt* toxin itself, such as its chemical structure, and of the affected insect, such as the presence of toxin binding sites in the gut, as well as pH level and the digestive enzymes present. *Bt* varieties include *Bt* var. *kurstaki* and var. *morrisoni*, with activity against lepidopteran larvae; *Bt* var. *israelensis*, specific to larvae of mosquitoes and blackflies; *Bt* var. *aizawai*, with activity against wax moth larvae; and *Bt* var. *tenebrionis*, specific against coleopteran larvae (Swadener, 1994).

Toxins isolated from *Bt* varieties have been commercialized and applied as foliar insecticides for more than 40 years (Swadener, 1994). They dominate the biopesticides market and are widely used throughout the U.S., particularly in organic production. In 1981, a *Bt* gene encoding a Cry protein was cloned and successfully transferred to and expressed in another organism, the bacterium *Escherichia coli* (Schnepf and Whiteley, 1981). Within ten years, tomato (*Lycopersicon esculentum* L.), tobacco (*Nicotina tabacum* L.), and cotton (*Gossypium hirsutum* L.) plants had been transformed to express *Bt* Cry proteins (Fischhoff et al., 1987; Vaeck et al., 1987; Perlak et al., 1990), and *Bt* corn and potato plants were developed soon thereafter (Koziel et al., 1993; Perlak et al., 1993).

Bt Crops with Regulatory Approval

To date, *Bt* varieties for four crop plants have been commercialized: field corn (*Zea mays* L.), cotton, potato (*Solanum tuberosum* L.), and sweet corn.

Acres planted to *Bt* varieties for each crop, from the first year of commercialization through 2000, are listed in Table 1.

Field Corn

In the U.S., annual losses in production and management costs associated with European corn borer (ECB), *Ostrinia nubilalis* (Hubner), a major lepidopteran pest in field corn, total an estimated $1 billion (Mason et al., 1996). *Bt* field corn varieties were first introduced for planting by U.S. farmers in 1996 for control of ECB, and their adoption was swift (see Table 1). Adoption decreased in 2000, however, due to historically low ECB pressure in 1998 and 1999 (Gray and Steffey, 1999).

Because of difficulties inherent in scouting and insecticide application for ECB control, historically, few corn growers treat for the pest (Ostlie, Hutchison, and Hellmich, 1997; Mason et al., 1996; Fernandez-Cornejo and Jans, 1999). The main advantage of *Bt* field corn adoption, therefore, has been increased yields rather than decreased insecticide use.

Bt corn varieties provide a high level of protection from the corn borer, equal to and often greater than current insecticide options (Ostlie, Hutchison, and Hellmich, 1997: Walker, Hellmich, and Lewis, 2000; Barry et al., 2000). By protecting against the previously uncontrolled ECB, *Bt* corn plantings produce higher yields than unprotected corn varieties. Table 2 shows average estimates of per acre yield advantages conferred by *Bt* corn varieties for 1997-99, as compiled by Carpenter and Gianessi (2001). Despite increased yields in 1998 and 1999, the planting of *Bt* corn resulted, on average, in net losses for growers in those years because yield differentials, due to low ECB pressure, did not compensate for the cost of *Bt* varieties.

TABLE 1. Adoption of *Bt*-protected crops in the U.S.

	1996	1997	1998	1999	2000
	------------------ Percent of U.S. Acreage ------------------				
Bt Corn[1]	1	6	18	26	19
Bt Cotton[2]	12	18	23	32	39
Bt Potato[3]	1	2.5	< 4	< 4	2-3
Bt Sweet Corn[4]			1	4	< 1

[1]USEPA OPPBP (2000); USDA NASS (2000a)
[2]USDA AMS (1996-2000)
[3]USEPA OPPBP (2000); E. Owens (2000, Monsanto, personal communication)
[4]D. Warnick (2001, Syngenta, personal communication)
Adapted from Carpenter and Gianessi (2001)

TABLE 2. Aggregate production and revenue impacts of *Bt* Corn, 1997-1999.

	Yield Increase	Production Increase	Crop Value Increase	Costs	Net Gain (Loss)
	(bu/A)	(1,000 bu)	---------------- (millions) ----------------		
1997	11.7	55,832	$136	$47	$89
1998	4.2	60,606	$118	$144	($26)
1999	3.3	66,436	$126	$161	($35)

Adapted from Carpenter and Gianessi (2001)

Cotton

Among the major insect pests in U.S. cotton production are tobacco budworm [*Heliothis virescens* (Fabricius)], cotton bollworm [*Helicoverpa zea* (Boddie)], and pink bollworm [*Pectinophora gossypiella* (Saunders)]. More than any other cotton insect pests, these three lepidopterans reduce yields and drive insecticide use (Luttrell, 1994). In 1996, *Bt* cotton varieties that provided protection from bollworm/budworm were introduced, offering an alternative to insecticide-based management programs.

The adoption of *Bt* cotton varieties was extremely rapid in some states such as Alabama, where 65% of cotton acreage was planted to *Bt* varieties in 1996, the first year of commercialization (USDA AMS, 1996-2000). However, adoption was much slower in the two major cotton producing states of Texas and California, lowering national adoption rate statistics.

The impacts of the adoption of *Bt* cotton varieties include a reduction in yield losses due to the *Bt* target pests, reductions in insecticide use, and associated cost savings. In 12 of 16 states for which annual yield losses to insects are published (Williams, 1995-1999), yield losses due to *Bt* target pests have declined. An extreme illustration of this is seen in central Alabama, where the average yield loss due to bollworm/budworm was 55% in 1995, but was between 2 and 7% from 1996 to 1999.

The most substantial impact of the adoption of *Bt* cotton varieties has been a reduction in insecticide use. Comparing U.S. Department of Agriculture (USDA) pesticide use data (USDA NASS, 1996-2000) for six states from 1995 to 1999 indicates a dramatic reduction in the use of insecticides to control *Bt* target pests. By 1999, the amount of insecticide active ingredient applied had declined by 2.7 million pounds from 1995, and the number of insecticide applications had declined by 15 million acre-treatments.

Potato

Colorado potato beetle (CPB) [*Leptinotarsa decemlineata* (Say)] is the primary defoliating insect pest of potato in the U.S. (Ferro and Boiteau, 1993). Compounding CPB's threat as a pest is its facility for developing resistance to a broad range of insecticides (Forgash, 1985). *Bt* potato varieties, with protection against CPB, were introduced in 1996, but have been adopted on a very limited basis (Table 1). Adoption figures reflect combined adoption of the three types of *Bt* potato available: NewLeaf *Bt* potato, introduced in 1996; NewLeafPlus *Bt* potato with additional resistance to the potato leafroll virus (PLRV), introduced in 1999; and NewLeafY *Bt* potato with additional resistance to potato virus Y, also introduced in 1999. Both potato leafroll virus and potato virus Y are aphid-borne diseases that can cause significant yield losses in potato (Hooker, 1981).

Control of CPB populations on *Bt* potato is greater than on nontransformed potato plants treated with insecticides, and protection against defoliation is season long (Perlak et al., 1993; Reed et al., 2001). *Bt* potato varieties with protection against PLRV exhibit high resistance both in terms of visible symptoms and virus incidence (Thomas et al., 2000). Agronomic, horticultural, and tuber qualities of transgenic potato varieties also are comparable to those of the parental variety, Russet Burbank.

Pesticide use data from USDA (USDA NASS, 2000b) show insecticide use on approximately 90% of potato acreage in Washington, Oregon, and Idaho. Of the total two millions pounds of insecticide active ingredient applied annually, approximately two-thirds target CPB and aphid populations. The strong field performance transgenic potato varieties have demonstrated against CPB and PLRV suggest the primary impact of their adoption would be a decrease in insecticide use for CPB and aphid control, and a reduction in grower costs as well. An analysis by the U.S. Environmental Protection Agency (USEPA OPPBP, 2000) of national *Bt* potato plantings at peak adoption levels of 4% in 1999 calculated a $500,000 grower benefit. Adoption of *Bt* potato has since declined.

The low adoption rates of *Bt* potato varieties are due to a combination of factors. Early adoption of *Bt* varieties, prior to the introduction of stacked varieties with virus protection, was hindered by the continuing need by potato growers to control other insect pests in addition to the CPB, particularly the virus-transmitting aphids. Imidacloprid, a systemic insecticide effective against CPB, aphids, and other foliar-feeding insects (Boiteau, Osborn, and Drew, 1997), recently had been introduced and *Bt* varieties with protection against CPB only were unable to compete. By the time *Bt* potato stacked with virus resistance were introduced in 1999, their adoption was preempted by processors responding to refusal of fast food chains and other buyers to accept genetically modified potato products (Kilman, 2000).

Sweet Corn

Similarly, *Bt* sweet corn varieties with protection against lepidopteran insects, first introduced in 1998, have had little commercial success, due largely to fresh corn marketers' unwillingness to purchase ears of *Bt* varieties. Fresh sweet corn production in Florida is dependent on intensive insect management due to a high potential for lepidopteran insect damage, a low market tolerance for ear damage, and high crop value (Foster, 1989). Insecticide usage reports from USDA (USDA NASS, 1999) indicate an average of 12 insecticide applications per acre per season on Florida sweet corn. Sweet corn varieties with a *CryIA(b) Bt* gene exhibit a high level of resistance to fall armyworm [*Spodoptera frugiperda* (J. E. Smith)] and other economic lepidopteran pests of Florida sweet corn, even under high infestation levels (Lynch et al., 1999a, 1999b). In trials where no insecticide applications were made, the percentage of damaged ears in *Bt* sweet corn varieties was minimal, whereas up to 100% of non-*Bt* ears were damaged. In addition, within damaged ears, the number of kernels damaged was significantly lower in *Bt* varieties than in non-*Bt* varieties.

Bt sweet corn varieties do not, however, provide protection against non-lepidopteran insect pests, such as the corn silk fly (*Euxesta stigmatis* Loew), which may cause economic damage (Lynch et al., 1999b). Consequently, adoption of *Bt* fresh sweet corn varieties in Florida is not expected to eliminate insecticide applications altogether, but rather is expected to drop average per season applications by approximately 10.

With control of lepidopteran pests and the potential to drastically reduce pesticide use in Florida sweet corn, extensive adoption of *Bt* varieties might be expected on the 80% of the state's acreage where corn silk fly is not a primary pest. The main impact is expected to be a reduction in insecticide use of more than 100,000 pounds of active ingredient and the accompanying reduction in pest management costs of approximately $1.3 million (Gianessi et al., 2002). With current adoption of *Bt* sweet corn at less than 1% of the acreage (Table 1), however, these impacts on Florida sweet corn production remain hypothetical.

Bt Crops in Development

Despite the current lack of adoption of *Bt* sweet corn and potato varieties, the insect control benefits of incorporating *Bt* genes into crop plants continue to fuel research on its application in other crops, particularly in the academic sector. Discoveries of novel *Bt* toxins and the genes encoding them, and advances in gene expression techniques continue to increase the efficacy of *Bt*-protected plants and the range of insects targeted. The range of crop plants transformed with *Bt* genes expands with improvements in plant cell tissue culture and regeneration techniques. Among the many agronomic species to which transgenic *Bt* technology is being applied experimentally are soybean (*Glycine max* L. Merr.), eggplant (*Solanum melongena* L.), peanut (*Arachis hypogaea* L.), and broccoli (*Brassica oleracea* var. *italica* L.).

Soybean

In southern soybean production, the most damaging defoliating insects are velvetbean caterpillar (*Anticarsia gemmatalis* Hubner) and soybean looper (*Pseudoplusia includens* (Walker)) (Higley and Boethel, 1994). Other lepidopteran pests of economic significance include lesser cornstalk borer [*Elasmopalpus lignosellus* (Zeller)] and corn earworm [*Helicoverpa zea* (Boddie)]. Losses from velvetbean caterpillar alone in Georgia soybean may be as high as $2 million in combined damage and cost of control (McPherson, Hudson, and Jones, 1999). Researchers at the University of Georgia have transformed soybean with a synthetic *crylAc Bt* gene for protection against velvetbean caterpillar and lesser cornstalk borer (Walker et al., 2000). In addition to the transgene-induced insect resistance, natural lepidopteran-resistance traits from Japanese soybean varieties were bred into the *Bt* soybean for increased protection against corn earworm and soybean looper (Walker et al., 2002). Several lines with different combinations of *Bt* and natural resistance genes are being tested in the laboratory and in the field for a wide range of lepidopteran protection. If commercialized, southern soybean growers, who spend an estimated $10 million per year on insecticide use, are expected to benefit most through insecticide use reduction.

Eggplant

In eggplant, insecticide use is dominated by the systemic imidacloprid (USDA NASS, 2001b). As in potato, a primary pest in eggplant and a primary target of insecticide use is Colorado potato beetle (CPB) (Hamilton, 1995). Transgenic eggplants expressing a synthetic *crylIIA Bt* gene have been developed and field-tested (Hamilton et al., 1997; Jelenkovic et al., 1998). Presence of and feeding damage by CPB on *Bt* eggplant were significantly lower than on nontransgenic, untreated eggplants, and comparable to nontransgenic plants treated with imidacloprid. In variety field trials, *Bt* cultivars met commercial standards for preferred plant and fruit qualities (Kline and Garrison, 2000).

Peanut

Bt peanut lines also have been developed (Singsit et al., 1997), expressing a *CrylA(c)* gene with efficacy against lesser cornstalk borer (LCB) [*Elasmopalpus lignosellus* (Zeller)]. Larval LCB cause serious economic damage by feeding on parts of the plant at or just below the soil surface (Funderbunk and Brandenburg, 1995). In addition to direct damage, LCB feeding scars and wounds facilitate infestations of soilborne plant pathogens such as *Aspergillus* fungi (Lynch and Wilson, 1991). *Aspergillus* fungi are of major concern to the peanut industry because of aflatoxin production, especially in drought conditions that favor their growth. Peanut lots that are contaminated with *Aspergillus* are downgraded at a significant loss to the grower. An economic analysis, based on the non-drought years 1993-1996, estimated the average net cost to Georgia growers of downgrading due to *Aspergillus* contamination to be $1.7 million (Lamb and Sternizke, 2001). Experimental *Bt* peanut lines demonstrated high control of LCB and moderate to high control of two other peanut lepidopteran pests (Lynch, Singsit, and Ozias-Akins, 1995; Lynch and Ozias-Akins, 1998). If they prove to reduce *Aspergillus* and aflatoxin levels as well, they could prevent significant economic losses to contamination.

Broccoli

In broccoli, preventing contamination of heads by insects and their frass is as essential for the production of a marketable crop as is protection from feeding damage (Wyman and Oatman, 1977). Feeding by larvae of the diamondback moth (DBM) (*Plutella xylostella* Linnaeus) causes both significant damage as well as larval and frass contamination. The biology of DBM and its history of developing resistance to multiple classes of insecticides make its economic control a dynamic challenge (Shelton et al., 1993; Talekar and Shelton, 1993). In 1997, an outbreak of DBM in California broccoli overwhelmed standard insecticide-based control and resulted in crop losses estimated at more than $6 million (Shelton et al., 2000).

In laboratory assays, an experimental line of *Bt* broccoli producing a CryIC protein was protected against *Bt*-susceptible DBM, CryIA-resistant DBM, and DBM with moderate resistance to CryIC, as well as newly hatched cabbage looper [*Trichoplusia ni* (Hübner)] and imported cabbageworm larvae [*Pieris rapae* (Linnaeus)] (Cao et al., 1999). Broccoli varieties with CryIC have the potential to prevent the severe economic damages incurred during periodic DBM outbreaks such as the one in 1997, as well as reduce annual losses to lepidopteran feeding and contamination.

HERBICIDE-RESISTANT CROPS

A consistent limitation in crop weed management is the lack of cost-effective herbicides with broad-spectrum activity and no crop injury. Consequently, multiple applications of several herbicides are routinely used to control a wide range of weed species infesting agronomic crops (Wilcut et al., 1996). Weed management strategies tend to follow one of two patterns. The first relies on preemergence (PRE) herbicide applications made in response to expected weed infestations rather than made postemergence (POST) in response to weeds present. The other strategy employs a combination of POST herbicides carefully mixed, timed, and directed to avoid damage to the growing crop. Mechanical cultivation and hand weeding often are necessary supplements for controlling weeds not controlled by herbicide applications, especially in high value crops that can sustain high labor costs.

Crops transformed with resistance to a broad-spectrum, POST herbicide provide the basis for a simplified weed management program centering on one herbicide to control a wide spectrum of weeds with minimal crop damage. Qualitatively, ease of weed management is potentially increased because decisions about which herbicide active ingredients to apply, optimal timing, and placement of applications are reduced. Potential quantifiable impacts include a reduction in the number of herbicide active ingredients used for weed management, a reduction in the number of herbicide applications made during a season, yield increases due to improved weed management and less crop injury, and, presumably, a resultant increase in grower returns (Wilcut et al., 1996; Carpenter and Gianessi, 2001). To this end, transgenic crops have been developed that express resistance to one of three herbicides: glyphosate (N-(phosphonomethyl)glycine), glufosinate (2-amino-4-(hydroxymethylphosphinyl) butanoic acid), and bromoxynil (3,5-dibromo-4-hydroxybenzonitrile).

Herbicide Modes of Action and Mechanisms of Resistance

Glyphosate is a contact, non-selective, non-residual, systemic herbicide. It is effective against both annual and perennial weeds. The phytotoxic activity of glyphosate is due to its inhibitory effect on the enzyme 5-enolpyruvyl-shikimate-3-phosphate synthase (EPSPS), the key enzyme in synthesis of aromatic amino acids that are essential for several critical processes, including cell wall formation, hormone production, and energy transduction (Steinrucken and Amrhein, 1980). Inhibition of EPSPS by glyphosate creates a lethal chemical imbalance within the plant, by which the plant starves for the products of EPSPS and accumulates toxic levels of compounds normally metabolized by EPSPS.

There are several organisms with natural resistance to glyphosate that serve as sources for glyphosate-resistance traits. The gene for one such trait, a form of EPSPS that is not susceptible to glyphosate inhibition, was isolated from the soil bacterium *Agrobacterium* sp. strain CP4 and used to produce glyphosate-resistant crops (Padgette et al., 1996).

Glufosinate is a broad-spectrum herbicide that inhibits glutamine synthetase, a plant enzyme essential to the processing of accumulated ammonia into a form of nitrogen usable by plants (Bayer et al., 1972; Tachibana et al., 1986). Interfering with the activity of glutamine synthetase leads to toxic cellular accumulation of ammonia. But glufosinate's most phytotoxic effects are less direct. The inhibition of glutamine synthetase indirectly inhibits carbon (C) fixation, with cascading destructive effects that quickly kill the plant.

Glufosinate is a modified, synthetic version of a naturally occurring compound, bialaphos, which is produced by the soil bacterium *Streptomyces*. To avoid being poisoned by their own bialaphos production, *Streptomyces* species also produce an enzyme that detoxifies bialaphos. The detoxifying enzyme, phosphinothricin acetyl transferase (PAT), detoxifies glufosinate as well. In *Streptomyces hygroscopicus*, PAT is encoded by the *bar* gene. The *bar* gene has been isolated and used to produce glufosinate-resistant crops (Vasil, 1996).

Bromoxynil, a benzonitrile compound with herbicide activity against broadleaf plants, binds to a protein on the thylakoid membrane of plant chloroplasts, where energy transfers take place that drive C fixation in photosynthesis. By binding to the thylakoid membrane protein, bromoxynil disrupts its functioning and prevents photosynthesis from continuing (Buckland, Collins, and Pullin, 1973). In soil contaminated with bromoxynil, a bacterium, *Klebsiella ozaenae*, was found that uses bromoxynil as its only nitrogen source. The bacterium produces a nitrilase enzyme that specifically breaks down bromoxynil. The gene encoding the bromoxynil-specific nitrilase was isolated from *Klebsiella ozaenae* and used to develop bromoxynil-resistant cotton (Stalker et al., 1996).

Herbicide-Resistant Crops with Regulatory Approval

The first herbicide-resistant crop variety to be commercially introduced was bromoxynil-resistant cotton, in 1995. Adoption of bromoxynil-resistant cotton was slow and remains very limited (Table 3), although cotton and soybean varieties with glyphosate resistance introduced in the two subsequent years were adopted readily and their plantings remained high at more than 50% of cotton and soybean acreage in 2000. Both glyphosate and glufosinate-resistant canola (*Brassica napus* L.) are available in the U.S. and, combined, were planted on almost 50% of canola acreage in 2000. Glyphosate-resistant and glufosinate-resistant field corn varieties also are available, but planted on less than 10% of corn acreage. Sugar beet (*Beta vulgaris* L.) varieties with glyphosate resistance were granted regulatory approval in 1999, but have yet to be marketed.

Cotton

Cotton weed management involves POST herbicide applications directed at target weeds before their height approaches that of the cotton crop. Such herbicide applications are time-consuming and require special equipment and care-

TABLE 3. Adoption of herbicide-resistant crops in the U.S.

		1995	1996	1997	1998	1999	2000
Crop	Herbicide(s)	--------------- Percent of U.S. Acreage ---------------					
Cotton[1]	Bromoxynil	0.1	0.1	1.2	5.8	7.8	7.2
Cotton[1]	Glyphosate			4	21	37	54
Soybean[2]	Glyphosate		2	13	37	47	54
Field corn[3]	Glyphosate, Glufosinate		3	4.3	9	8	7
Sugar beet	Glyphosate					0	0
Canola[4]	Glyphosate, Glufosinate					31	47

[1]USDA AMS (1996-2000)
[2]K. Marshall (2000, Monsanto, personal communication)
[3]USDA ERS (2001)
[4]James (2001)

ful planning. Weed control programs based on glyphosate-resistant varieties are widely adopted by cotton growers because they rely on one herbicide to control a broad spectrum of weeds without crop injury, special equipment or careful timing, thereby providing a simpler alternative to directed herbicide applications. Bromoxynil resistance, only available in cotton, has had low national adoption rates due in large part to circumstances specific to bromoxynil. Bromoxynil registration limits its use in cotton to 1.3 million acres (less than 10% of national acreage), and, as a broadleaf herbicide, it is not effective against grasses. Finally, glyphosate resistance stacked with *Bt* traits has been commercialized in several cotton varieties, whereas availability of bromoxynil-resistant varieties stacked with *Bt* traits is severely limited, further reducing bromoxynil marketability.

In cotton, decreased complexity of herbicide usage due to adoption of herbicide-resistant varieties is reflected in observable shifts in herbicide use patterns since 1994. Surveys of pesticide use by USDA (USDA NASS, 1995-2000) show percentages of cotton acreage treated with several common herbicides, such as trifluralin [2,6-dinitro-*N,N*-dipropyl-4-(trifluoromethyl)benzenamine], fluometuron (*N,N*-dimethyl-*N'*-[3-(trifluoromethyl)phenyl]urea), and MSMA (monosodium salt of methylarsonic acid), have declined since 1994, while the percentages treated with bromoxynil and glyphosate have increased. The average number of herbicide applications per year in cotton also has declined, by

1.8 million between 1994 and 1998, and by 1.3 million between 1994 and 1999.

Overall, yields and economic returns among conventional and herbicide-resistant weed management programs in cotton are comparable, although a clear economic advantage to planting glyphosate-resistant cotton has been demonstrated in localized, single-year trials (Stark, 1997; Bloodworth et al., 1998). A more general advantage potentially provided by herbicide-resistant cotton, however, is the expansion of cotton production into areas where weed management was previously uneconomical due to limited POST control of troublesome weeds.

Soybean

Impacts of glyphosate resistance in soybean production have been similar to those in cotton, with an expansion of production into new areas and changes in herbicide use patterns. Glyphosate use in soybean has increased, as expected, while use of other herbicides has decreased (USDA NASS, 1996-2000). Imazethapyr (2-[4,5-dihydro-4-methyl-4-(1-methylethyl)-5-oxo-1H-imidazol-2-yl]-5-ethyl-3-pyridinecarboxylic acid) was the most widely used herbicide in soybean in 1995, applied to 44% of the national acreage. By 1999, its usage had declined to 16%. USDA NASS (1996-2000) statistics also show that the average number of herbicide applications made per acre has declined, although the use rate per acre per application has remained steady. Carpenter and Gianessi (2001) estimate glyphosate-resistant soybean have saved farmers $216 million in annual weed control costs and reduced the number of yearly herbicide applications by 19 million acre-treatments.

Herbicides used in conventional soybean cause crop injury, leading to significant yield loss (Vidrine, Reynolds, and Griffin, 1993), but reducing their rates to minimize crop injury may reduce weed control consistency, leading to potentially greater yield losses. However, no injury was reported on glyphosate-resistant soybean with glyphosate application (Elmore, 2001; Wait, Johnson, and Massey, 1999). The broad-spectrum activity of glyphosate reduces the need for tillage as a weed control practice. Consequently, another impact of glyphosate-resistant soybean plantings has been an expansion of no-tillage acreage (Conservation Tillage Information Center, 2001), which offers numerous benefits such as soil moisture conservation, reduction of soil compaction, improved water infiltration, and reduction in soil erosion.

An indirect impact of glyphosate-resistant crop varieties, particularly in soybean production, has been a lowering of other herbicide costs. Glyphosate-resistant soybean weed management programs were initially priced to be competitive with existing conventional programs. In an effort to remain competitive, manufacturers of other soybean herbicides responded by lowering their prices, in some cases by as much as 40%.

Canola

Although herbicide-resistant canola has been commercialized and widely planted in Canada since 1996, it was only introduced for planting in the U.S. in 1999 (Table 3). Its introduction in late spring of that year, after many canola growers had already purchased seed for the year, limited adoption. However, adoption doubled to almost 50% of the acreage in 2000.

Canola, a member of the mustard family, is not competitive in the seedling stage, but once established, is a good competitor with most weeds (Bergland and McKay, 1997). In addition to potentially causing significant yield reductions in establishing canola plantings (Kirkland, 1995; Davis et al., 1999), weed infestations can lead to seed contamination and problems in processing. Weeds in the mustard family tend to have high levels of erucic acid and glucosinolate, two undesirable compounds that reduce the quality of canola oil and feed quality of canola meal, respectively, leading to price discounts or rejection in the market (Kirkland, 1995). Canada thistle (*Cirsium arvense* L. Scop.), for which there are no economical herbicides available, is also troublesome in major canola producing regions. The cornerstone of weed management in canola is herbicide use, as the narrow row planting of canola inhibits use of cultivation. Despite use of available conventional weed control practices, an estimated 11% of canola yields are lost due to weed infestation (Kmec and Wiess, 1998).

Glyphosate provides a wider spectrum of control than other herbicides currently available for use in canola, including management of Canada thistle, mustard species, and ALS-resistant kochia (*Kochia scoparia* L. Schrad.) (Zollinger, 2001). With glyphosate-resistant canola varieties, growers can use one or two well timed glyphosate applications for effective weed control with no crop injury (Jenks, 1999; Endres et al., 2001). Estimated savings to growers include a reduction in herbicide use of approximately 0.7 pound acre^{-1} with a \$15 acre^{-1} reduction in herbicide costs, and a yield increase of up to 10% (Johnson et al., 2000; Canola Council of Canada, 2001).

Although canola with resistance to glufosinate is approved for planting in the U.S., there are few varieties available, and those that are available exhibit poor agronomic qualities. In addition, a weed control program based on glufosinate provides limited control of wild oat (*Avena fatua* L.), which is a major weed problem on more than 80% of canola acreage (Gregoire, 2000; Zollinger, 2001). Consequently, adoption of glufosinate-resistant canola varieties in the U.S. has been low, at an estimated 5% of total transgenic herbicide-resistant canola acreage.

Field Corn

Transgenic, herbicide-resistant field corn varieties have been available since 1996. Combined plantings of glufosinate-resistant varieties and glyphosate-resistant varieties have yet to exceed 10% of national field corn acreage (Table 3). Lack of foreign market acceptance of transgenic corn has restricted U.S. adoption to regions that produce corn solely for domestic use. Within those regions, the availability of atrazine, an inexpensive, broad-spectrum corn herbicide, has given growers little incentive to switch to herbicide-resistant varieties.

Sugar Beet

In sugar beet, competition from uncontrolled weeds can severely reduce production (Schweizer and Dexter, 1987). Particular weed problems and practices for their economic control differ among sugar beet production regions, but management programs largely consist of a combination of herbicides, cultivation, and hand weeding (Jacobsen et al., 2001). Growers in the U.S. typically make three to four herbicide applications each year, with each application consisting of multiple active ingredients. Extrapolating from USDA chemical use data (USDA NASS, 2001a), an average of 12 herbicide treatments are applied per year to each sugar beet acre, for an annual aggregate total of 1.3 million pounds active ingredient. Mechanical cultivation and hand weeding are employed for control of weeds, such as kochia, that are not adequately controlled by available herbicides.

Glyphosate-resistant sugar beet varieties were granted regulatory approval in 1999. Experimental data show weed suppression and subsequent yields in glyphosate-resistant sugar beet, treated with two glyphosate applications, were equivalent to or better than those in plots treated with standard herbicides (Morishita, Wille, and Downard, 1999; Dexter and Luecke, 2000). Weeds controlled included species most troublesome in conventional sugar beet fields, such as kochia, pigweed (*Amaranthus* sp.), common lambsquarters (*Chenopodium album* L.), foxtail (*Setaria* sp.), and wild oat.

Adoption of sugar beet varieties with glyphosate-resistance is expected to lead to a reduction in the number of herbicide applications applied per acre, as in soybean and cotton production. In addition, reductions are expected in cultivation and hand weeding practices. With fewer herbicide applications, fewer cultivation passes, and fewer hours of hand weeding, a significant reduction in weed control costs is expected. To date, however, herbicide-resistant sugar beet varieties have not been marketed or planted because major sugar buyers, including candy makers, will not buy sugar refined from transgenic beets; therefore, sugar beet processors will not purchase transgenic beets from growers (Kilman, 2001).

Herbicide-Resistant Crops in Development

Rice (*Oryza sativa* L.) varieties with resistance to glufosinate and varieties with resistance to glyphosate have been developed (Chambers and Childs, 1999). The main benefit of these varieties would be improved control of red rice (*Oryza sativa* L.), a major weed in Gulf Coast and Delta rice production. In California, where red rice is not as troublesome, the main impact would be improved management of weeds that have developed resistance to common herbicides. Weed management in rice currently relies on herbicide use and water management. Improved weed management in rice with herbicide-resistant varieties is expected to reduce herbicide and water use and soil erosion.

Glyphosate-resistant lettuce (*Lactuca sativa* L.) and strawberry (*Fragaria X Ananassa* Duch.) varieties (Nagata et al., 2000; Morgan and Baker, 1999), and glufosinate-resistant tomato varieties (S. Schroeder, personal communications) have been developed and tested for performance in field trials. In high value crops such as lettuce, tomatoes, and strawberries, there is little tolerance for weeds that interfere with harvest and reduce yields, or for crop injury due to herbicide applications. Tomato and lettuce production in California and strawberry production in the east are all similarly dependent on combinations of soil fumigants, multiple applications of numerous herbicides, and extensive use of cultivation and hand weeding, all of which may add up to several hundred dollars per acre in weed management costs (Agamalian, 1989; Lange and Orr, 1989; Ashton and Monaco, 1991; Mullen, 1997). If eventually commercialized and adopted, transgenic crops with herbicide resistance are expected to reduce herbicide applications, cultivation and hand weeding, and therefore significantly reduce production costs.

PATHOGEN-RESISTANT CROPS

The greatest advances in pathogen-resistant plants developed through biotechnology have been in the area of pathogen-derived resistance to plant viruses. In addition to *Bt* potato varieties with virus resistance, papaya (*Carica papaya* L.) and summer squash (*Cucurbita* spp.) varieties with transgenic virus resistance are commercially available in the U.S. Pathogen-derived resistance in general relies on the assumption that there are certain biochemical functions associated exclusively with the pathogen and upon which survival of the pathogen depends, but not survival of the host (Sanford and Johnston, 1985). By transforming the host plant with a gene integral to one of the pathogen's essential and exclusive functions, the expression of that gene by the host plant will interfere with the pathogen's essential process by upsetting the balance of related components.

The most common application of pathogen-derived resistance in commercial crop plants has been the use of viral coat protein genes, although use of truncated or entire viral replicase genes is increasing in the development of new pathogen-resistant crop plants (Kaniewski and Lawson, 1998). The specific mechanism for coat protein mediated resistance is not yet known, but experimental evidence suggests there may be several that involve interference with critical viral processes, including viron coating and uncoating, replication, post-transcriptional gene expression, and intercellular transport (Beachy, Loesch-Fries, and Tumer, 1990; Scholthof, Scholthof, and Jackson, 1993; Kaniewski and Lawson, 1998; Ravelonandro et al., 2000).

Biotechnology also is being used to develop plants with resistance to pathogens other than viruses, although none have reached the commercialization stage yet. A variety of antimicrobial and antitoxin defensive proteins have been identified in plants, animals, and microorganisms. Advances in genetics and transformation techniques make it possible to harness some of these natural defense mechanisms and incorporate them into crop plants that lack them (Broglie et al., 1993; Destefano-Beltran et al., 1993).

Potential impacts of transgenic plant pathogen resistance include a reduction in fungicide and bactericide use and a reduction in use of insecticides to prevent insect-vectored diseases. Many plant diseases for which effective management practices exist still incur economic losses, particularly during epidemic years. Transgenic pathogen resistance has the potential to provide a high level of protection consistent enough to prevent epidemic losses. Transgenic pathogen resistance could be particularly valuable in crops threatened by diseases for which there are few effective management practices, and in defending crops against devastation by emerging disease problems.

Pathogen-Resistant Crops with Regulatory Approval

Papaya

The development and adoption of transgenic papaya in Hawaii provides one of the most dramatic illustrations of the potential for crop improvement through biotechnology. Papaya ringspot virus (PRSV) is a limiting factor in papaya production worldwide, including Hawaii, where papaya is economically the second most important fruit crop (USDA NASS, 1992-2000). Transformation of papaya with a PRSV coat protein gene produced successful protection against Hawaiian strains of the virus (Fitch et al., 1992; Tennant et al., 1994). In the 1990s, Hawaiian papaya production was declining significantly due to a severe PRSV epidemic (Fitch, Moore, and Leong, 1998; USDA NASS, 1992-2000). During that time, two papaya cultivars with coat protein mediated PRSV resistance performed well in long-term field trials in Hawaii

(Lius et al., 1997; Fitch, Moore, and Leong, 1998), and were made available to growers in 1998.

Adoption of PRSV-resistant papaya varieties in Hawaii was rapid. In 2000, two years after their introduction, approximately 40% of total papaya acreage and 53% of bearing acreage in Hawaii was planted with one of the two PRSV-resistant cultivars (USDA HASS, 2000). Statewide production, which had fallen 45% from 1992 to 1998, rebounded by 35% from 1998 to 2000 (USDA NASS, 1992-2000).

Summer Squash

The other commercially available crop with transgenic pathogen resistance is summer squash. In 1998, transgenic summer squash varieties with coat protein mediated resistance to zucchini mosaic virus (ZMV), watermelon mosaic virus 2 (WMV 2), and cucumber mosaic virus (CMV) were introduced. These viruses, together with papaya ringspot virus (PRSV), make up a mosaic virus complex which is a limiting factor for summer squash production in the U.S.

Transgenic summer squash varieties with resistance to ZMV, WMV 2, and CMV offer several benefits to growers, including a prolonged growing season and higher yields of marketable fruit (Tricoli et al., 1995; Webb and Tyson, 1997; Schultheis and Walters, 1998). Varieties with resistance to only one or two of these viruses also increase marketable yields and grower returns (Arce-Ochoa et al., 1995; Clough and Hamm, 1995; Fuchs et al., 1998). Even in the absence of viral infections, the transgenic lines produce fruit that is equivalent to nontransgenic fruit in both quantity and quality (Webb and Tyson, 1997).

By lowering the risk of losses to viruses, transgenic squash plantings may also lead to a reduction in stylet oil and insecticide applications used for suppression of virus-vectoring aphids (Zitter and Ozaki, 1978). Transgenic virus protected varieties may also reduce regional disease incidence because, unlike susceptible varieties, they do not serve as virus reservoirs for further spread by aphids (Clough and Hamm, 1995).

Despite the benefits of virus-resistant summer squash observed in field trials, adoption has been low, estimated at between 5% and 10%. Low adoption rates may be due to the lack of protection against the fourth virus, PRSV. Also, higher adoption rates in localized areas may be masked by low national adoption estimates. There is a wide array of summer squash varieties, each adapted to local growing conditions and consumer preferences (Paris, 1996). The limited number of varieties that have been transformed for virus protection may restrict adoption to localized areas in which those varieties historically have been planted.

Pathogen-Resistant Crops in Development

Virus Resistance

Other agricultural crops have been transformed for pathogen-derived resistance to viruses but are still in experimental stages. Peanut lines with resistance to tomato spotted wilt virus are being developed (Yang et al., 1998; Magbanua et al., 2000). Tomato spotted wilt virus is a limiting factor in peanut production worldwide, costing Georgia peanut farmers alone as much as $43 million in losses during epidemic years (Bertrand, 1998). Tomato lines are being developed with resistance to whitefly-transmitted geminiviruses (Hou et al., 2000), which in one epidemic year caused a 20% yield reduction in Florida (Polston and Anderson, 1997). Citrus tristeza virus, already present in Texas, threatens to severely cripple the Texas citrus industry once its most effective vector, the brown citrus aphid (*Toxoptera citricida*), becomes established in the state. As a protective measure, citrus varieties with resistance to tristeza are in development (Yang et al., 2000). Transgenic raspberry cultivars are being field tested for resistance to raspberry bushy dwarf virus (Martin, Keller, and Mathews, 2001), an economically significant disease recently established in the Pacific Northwest (Daubeny, Freeman, and Stace-Smith, 1982; Martin, 1999). Also in field trials are stone fruit cultivars with resistance to plum pox virus. If established in the U.S., despite a multimillion dollar eradication program, plum pox could devastate the nation's $2 billion stone fruit industry (Ravelonandro et al., 1997; Scorza et al., 2000).

Bacteria Resistance

Several perennial crop species have been experimentally transformed for resistance to bacterial diseases and are at various stages of development and evaluation. Arrival of the glassy winged sharpshooter (*Homalodisca coagulata*) in California in 1989 and its efficient spread of Pierce's disease (*Xylella fastidiosa*) poses a serious threat to grape (*Vitis vinifera* L.) production throughout the state. More than 500 acres of grapevines have already been removed in southern growing regions due to the disease, and $10 million in local, state, and federal money has been appropriated to researching management of the vector/disease combination (Meadows, 2001). Biotechnology is under consideration as a way to impart disease resistance to currently produced commercial grape varieties without compromising the defining qualities of their berries. One approach is the transformation of grape with genes encoding antibacterial lytic proteins (Scorza and Gray, 2001). The natural Pierce's disease resistance found in muscadine grapes (*Vitis rotundifolia*) is also being investigated as a potential source of resistance for transfer to commercial cultivars through biotechnology (Meadows, 2001).

Outbreaks of fire blight (*Erwinia amylovora*) in apples (*Malus sylvestris* L. Mill.), each causing millions of dollars in damages, are increasing in frequency as fire blight populations develop resistance to the foliar antibiotics currently applied for their management (McManus and Jones, 1994). A variety of transgenic traits are being tested as sources of fire blight resistance, including antimicrobial proteins from animals such as a giant silkworm moth (*Hyalophora cecropia*), and pathogen-derived genes that elicit systemic acquired resistance to the disease (Abdul-Kader et al., 1999; Norelli et al., 1999; Aldwinkle et al., 2000). If successful, transgenic fire blight resistance could eliminate antibiotic use in orchards and reduce economic losses to the disease.

In Florida citrus, significant economic losses to citrus canker include yield reductions and downgrading of fresh market fruits to juice markets. The most recent efforts to eradicate citrus canker in Florida have led to the destruction of more than 1.5 million trees since 1995 (Schubert et al., 2001; Brown, 2001). Biotechnology is being used to investigate several mechanisms for citrus canker resistance, including expression of synthetic antimicrobial peptides and interference with pathogenic proteins produced by the canker bacterium (Gabriel and Hartung, 2000).

Fungi Resistance

Biotechnology also is being applied to develop cultivars with resistance to fungal pathogens, with the expected consequence of improved production and significantly lower use of soil fumigation and foliar fungicides. For example, an alfalfa (*Medicago sativa* L.) gene encoding for an antifungal protein has been used to transform potato for resistance to verticillium wilt (*Verticillium dahliae*), a primary cause of yield losses in U.S. potato production (Gao et al., 2000). Wheat (*Triticum aestivum* L.) genes encoding the antitoxin enzyme oxalate oxidase have been transferred to sunflower (*Helianthus annuus* L.) for resistance to the economic pest *Sclerotinia sclerotiorum* (Bazzalo et al., 2000).

Recent epidemics of scab (*Fusarium graminearum*) in the U.S. have caused billions of dollars in lost revenues to barley (*Hordeum vulgare* L.) producers due to severe yield reductions as high as 45% and price discounts from lowered grain quality and mycotoxin contamination (McMullen, Jones, and Gallenberg, 1997). As a consequence of these drastic economic losses, North Dakota barley acreage has fallen sharply, with many grain growers leaving farming altogether. In an effort to combat scab in barley, transgenic resistant varieties are being developed that express antitoxin genes from another *Fusarium* species and from a species of yeast, *Saccharomyces cerevisiae* (Wood et al., 1999).

Nematode Resistance

Transgenic technology is being used to develop nematode resistance in crop plants. Wild rice naturally produces cystatin, a proteinase inhibitor that interferes with nematode feeding and digestion (Atkinson et al., 1996). A commercial pineapple (*Ananas comosus* L. Merr.) variety has been transformed with a wild rice-derived cystatin transgene (Rohrbach et al., 2000) for resistance to the reniform nematode (*Rotylenchulus reniformis*), a primary pest of pineapple in Hawaii that is currently controlled with soil fumigants.

CONCLUSION

Transgenic crop varieties offer growers an additional choice in pest management, but not necessarily to the exclusion of all other management tactics. As always, growers rely on a variety of pest control practices for economic production of their crops, basing use decisions on the specific parameters of their particular crops. Biotechnology is another tool, the benefits, risks, and costs of which must be weighed as they are for other tools such as pesticides and cultural practices. To expect perfect performance from transgenic varieties sets the technology up for failure. Unrealistic expectations also threaten more conventional technologies by creating the false impression that growers will no longer need access to them. Access to an increasing number of pest management choices, biotechnology being only one, allows growers to tailor their practices as needed in response to varying weather and pest patterns, and thereby improves their chances of producing a profitable crop year after year.

Specific farm-level impacts of planting transgenic varieties vary, with the type and magnitude of impacts being dictated by the combination of crop, production region, pest targeted, and available management alternatives in each situation. For example, adoption of *Bt* corn has resulted in increased yields with little impact on insecticide use, whereas the main impact of *Bt* cotton has been a reduction in insecticide use. In general, however, the planting of transgenic crops with pest management traits may increase yield and production, or both, and reduce pesticide use. This has been demonstrated in the field by transgenic varieties with regulatory approval that have been commercialized and planted, and in field trials with approved varieties pending commercialization and adoption. Continued research to expand the application of agricultural biotechnology to a wider variety of pests and crops–including fruit, vegetable and grain crops, field and minor crops, and annual and perennial crops–is being driven by these already demonstrated impacts. Continued research in agricultural biotechnology also reflects the importance of these impacts, as well as confidence in the role of biotechnology in the future of crop protection.

REFERENCES

Abdul-Kader, A.M., J.L. Norelli, H.S. Aldwinkle, D.W. Bauer, and S.V. Beer. (1999). Evaluation of the *hrpN* gene for increasing resistance to fire blight in transgenic apple. *Acta Horticulturae* 489:247-250.

Aldwinkle, H., J. Norelli, S. Broan, T. Robinson, E. Borejsza-Wysocka, H. Gustafson, J.-P. Reynoird, and M.V. Bhaskara Reddy. (2000). Genetic engineering of apple for resistance to fire blight. *New York Fruit Quarterly* 8:24-26.

Agamalian, H.S. (1989). Lettuce (*Lactuca sativa*). In *Principles of Weed Control in California*, Fresno, CA: Thomson Publications, pp. 337-341.

Arce-Ochoa, J.P., F. Dainello, L.M. Pike, and D. Drews. (1995). Field performance comparison of two transgenic summer squash hybrids to their parental hybrid line. *Hortscience* 30:492-493.

Ashton, F.M. and T.J. Monaco. (1991). *Weed Science: Principles and Practice*. New York, NY: John Wiley and Sons, Inc.

Atkinson, H.J., P.E. Urwin, M.C. Clarke, and M.J. McPherson. (1996). Image analysis of the growth of *Globodera pallida* and *Meloidogyne incognita* on transgenic tomato roots expressing cystatins. *Journal of Nematology* 28:209-215.

Barry, B.D., L.L. Darrah, D.L. Huckla, A.Q. Antonio, G.S. Smith, and M.H. O'Day. (2000). Performance of transgenic corn hybrids in Missouri for insect control and yield. *Journal of Economic Entomology* 93:993-999.

Bayer, E., K.H. Guge, K. Hagele, H. Hogenmajer, S. Jessipow, W.A. Konig, and H. Zahner. (1972). Phosphinothricin and phosphinothricin-alanyl-alanin. *Helvetica Chimica Acta* 55:224.

Bazzalo, M.E., I. Bridges, T. Galella, M. Grondona, A. Leon, A. Scott, D. Bidney, G. Cole, J.-L. D'Hautefeuille, G. Lu, M. Mancl, C. Scelonge, J. Soper, G. Sosa-Dominguez, and L. Wang. (2000). *Sclerotinia* head rot resistance conferred by wheat oxalate oxidase gene in transgenic sunflower. In *Proceedings of the 15th Annual International Sunflower Conference*, Toulouse, France, pp. 65-65.

Beachy, R.N., S. Loesch-Fries, and N.E. Tumer. (1990). Coat protein-mediated resistance against virus infection. *Annual Review of Phytopathology* 28:451-74.

Bergland, D.R. and K. McKay. (1997). *Canola Production*. Fargo, ND: North Dakota State University Extension Service.

Bertrand, P.F. (1998). *1997 Georgia Plant Disease Loss Estimates*. Tifton, GA: The University of Georgia Cooperative Extension Service.

Bloodworth, K.M., D.B. Reynolds, D. Laughlin, and C.E. Snipes. (1998). A comparison of transgenic cotton weed control program efficacy and economics. *Proceedings of the Southern Weed Science Society* 51:51.

Boiteau, G., W.P.L. Osborn, and M.E. Drew. (1997). Residual activity of imidacloprid controlling Colorado potato beetle (Coleoptera: Chrysomelidae) and three species of potato colonizing aphids (Homoptera: Aphidae). *Journal of Economic Entomology* 90:309-319.

Broglie, K., R. Broglie, N. Benhamou, and I. Chet. (1993). The role of cell wall degrading enzymes in fungal disease resistance. In *Biotechnology in Plant Disease Control*, ed. I. Chet, New York, NY: Wiley-Liss, Inc., pp. 139-156.

Brown, K. (2001). Florida fights to stop citrus canker. *Science* 292:2275-2276.

Buckland, J.L., R.F. Collins, and E.M. Pullin. (1973). Metabolism of bromoxynil octanoate in growing wheat. *Pesticide Science* 4:149.

Canola Council of Canada. (2001). *An Agronomic and Economic Assessment of Transgenic Canola*. Winnipeg, MB, Canada.

Cao, J., J.D. Tang, N. Strizhov, A.M. Shelton, and E.D. Earle. (1999). Transgenic broccoli with high levels of *Bacillus thuringiensis* Cry1C protein control diamondback moth larvae resistant to Cry1A or Cry1C. *Molecular Breeding* 5:131-141.

Carpenter, J. and L. Gianessi. (2001). *Agricultural Biotechnology: Updated Benefit Estimates*. Washington, DC: National Center for Food and Agricultural Policy.

Chambers, W. and N. Childs. (1999). Herbicide-resistant varieties in commercial rice production: Implications for the future. In *Rice Situation and Outlook Yearbook*, Washington, DC: USDA Economic Research Service, pp. 24-26.

Clough, G.H. and P.B. Hamm. (1995). Coat protein transgenic resistance to watermelon mosaic and zucchini yellows mosaic virus in squash and cantaloupe. *Plant Disease* 79:1107-1109.

Conservation Tillage Information Center (2001). *Crop Residue Management Survey*. <www.ctic.purdue.edu/CTIC/CTIC.html>.

Daubeny, H.A., J.A. Freeman, and R. Stace-Smith. (1982). Effects of raspberry bushy dwarf virus on yield and cane growth in susceptible red raspberry cultivars. *Hortscience* 17: 645-647.

Davis, J.B., J. Brown, J.S. Brennan, and D.C. Thill. (1999). Predicting decreases in canola (*Brassica napus* and *B. rapa*) oil and meal quality caused by contamination by Brassicaceae weed seeds. *Weed Technology* 13: 239-243.

Destefano-Beltran, L., P.G. Nagpala, S.M. Cetiner, T. Denny, and J.M. Jaynes. (1993). Using genes encoding novel peptides and proteins to enhance disease resistance in plants. In *Biotechnology in Plant Disease Control*, ed. I. Chet, New York, NY: Wiley-Liss, Inc., pp. 175-189.

Dexter, A.G. and J.L. Luecke. (2000). Herbicides on Roundup Ready and Liberty Link sugarbeet, 1999. *1999 Sugarbeet Research and Extension Reports* 30:90-97.

Elmore, R.W. (2001). Glyphosate resistant soybean cultivar response to glyphosate. *Agronomy Journal* 93:404-407.

Endres, G.J., B.M. Jenks, J.R. Lukach, and M. Pauli. (2001). Weed management strategies with glyphosate-resistant canola. *Proceedings of the Western Society of Weed Science* 54: 75.

Fernandez-Cornejo, J. and S. Jans. (1999). *Pest Management in U.S. Agriculture*, Economic Research Service Agricultural Handbook No. 717. Washington, DC: USDA.

Ferro, D.N. and G. Boiteau. (1993). Management of Insect Pests. In *Potato Health Management*, ed. R.C. Rowe, St. Paul, MN: The American Phytopathological Society, pp. 103-115.

Fischhoff, D.A., K.S. Bowdish, F.J. Perlak, P.G. Marrone, S.M. McCormick, J.G. Nidermeyer, D.A. Dean, K. Kusano-Kretzmer, E.J. Mayer, D.E. Rochester, S.G. Rogers, and R.T. Fraley. (1987). Insect tolerant transgenic tomato plants. *Biotechnology* 5:807-813.

Fitch, M.M.M., R.M. Manshardt, D. Gonsalves, J.L. Slightom, and J.C. Sanford. (1992). Virus resistant papaya plants derived from tissues bombarded with the coat protein gene of papaya ringspot virus. *Biotechnology* 10:1466-1471.

Fitch, M., P. Moore, and T. Leong. (1998). Progress in transgenic papaya (*Carica papaya*) research: Transformation for broader resistance among cultivars and micropropagating selected hybrid transgenic plants. *Acta Horticulturae* 461:315-319.

Forgash, A. (1985). Insecticide resistance in the Colorado potato beetle. In *Proceedings of the Symposium on Colorado Potato Beetle, XVII International Congress of Entomology*, Massachusetts Agricultural Experiment Station, Bulletin No. 704, ed., D.N. Ferro and R.H. Voss, Amherst, MA: University of Massachusetts.

Foster, R.E. (1989). Strategies for protecting sweet corn ears from damage by fall armyworms in southern Florida. *Florida Entomologist* 72:146-151.

Fuchs, M., D.M. Tricoli, K.J. Carney, M. Schesser, J.R. McFerson, and D. Gonsalves. (1998). Comparative virus resistance and fruit yield of transgenic squash with single and multiple coat protein genes. *Plant Disease* 82:1350-1356.

Funderbunk, J.E. and R.L. Brandenburg. (1995). Management of insects and other arthropods in peanut. In *Peanut Health Management*, ed., H.A. Melouk and F.M. Shokes, St. Paul, MN: The American Phytopathological Society, pp. 51-58.

Gabriel, D. and J. Hartung. (2000). Discussion on genomics, resistance and GMOs: A moderated discussion session. *Proceedings of the International Citrus Canker Research Workshop*, June 20-22, Ft. Pierce, FL, p. 23.

Gao, A.-G., S.M. Hakim, C.A. Mittanck, Y. Wu, B.M. Woerner, D.M. Stark, D.M. Shah, J. Liang, and C.M.T. Rommens. (2000). Fungal pathogen protection in potato by expression of a plant defensin peptide. *Nature Biotechnology* 18:1307-1310.

Gianessi, L.P., C.S. Silvers, J.E. Carpenter, and S. Sankula. (2002). *The Potential for Biotechnology to Improve Crop Pest Management in the U.S.: 40 Case Studies*. Washington, DC: National Center for Food and Agricultural Policy.

Gill, S.S., E.A. Cowles, and P.V. Pietrantonio. (1992). The mode of action of *Bacillus thuringiensis* endotoxins. *Annual Review of Entomology* 37:615-636.

Gray, M. and K. Steffey. (1999). European corn borer populations in Illinois near historic low. *University of Illinois Pest Management and Crop Development Bulletin* 24.

Gregoire, T.D. (2000). Liberty for Weed Control in Canola. *2000 North Dakota Weed Control Research*, p. 7.

Hamilton, G.C. (1995). *A Comparison of Eggplant Grown Under Conventional and Biological Control Intensive Pest Management Conditions in New Jersey*, New Brunswick, NJ: Rutgers University Cook College.

Hamilton, G.C., G.L. Jelenkovic, J.H. Lashomb, G. Ghidiu, S. Billings, and J.M. Patt. (1997). Effectiveness of transgenic eggplant (*Solanum melongena* L.) against the Colorado potato beetle. *Advances in Horticultural Science* 11:189-192.

Higley, L.G. and D.J. Boethel. (1994). *Handbook of Soybean Insect Pests*, Lanham, MD: The Entomological Society of America.

Hooker, W.J. (1981). *Compendium of Potato Diseases*, St. Paul, MN: The American Phytopathological Society.

Hou, Y.-M., R. Sanders, V.M. Ursin, and R.L. Gilbertson. (2000). Transgenic plants expressing geminivirus movement proteins: Abnormal phenotypes and delayed infection by *Tomato mottle virus* in transgenic tomatoes expressing the *Bean dwarf virus* BV1 or BC1 proteins. *Molecular Plant-Microbe Interactions* 13:297-308.

Jacobsen, B.J., A.G. Dexter, L.J. Smith, and M.B. Mikkelson. (2001). Survey of pesticide use, IPM practices and pests of sugarbeet in the U.S.A. for 1998 and 1999. *Proceedings of the 32nd Biennial Meeting of the American Society of Sugarbeet Technologists.*

James, C. (2001). *Global Status of Commercialized Transgenic Crops: 2000.* Ithaca, NY: International Service for the Acquisition of Agri-Biotech Applications (ISAAA).

Jelenkovic, G., S. Billings, Q. Chen, J. Lashomb, G. Hamilton, and G. Ghidiu. (1998). Transformation of eggplant with synthetic *cryIIIA* gene produces a high level of resistance to the Colorado potato beetle. *Journal of the American Society of Horticultural Science* 123:19-25.

Jenks, B.M. (1999). Weed control in glyphosate, imidazolinone, and glyphosate-resistant canola. *Proceedings of the Western Society of Weed Science* 52:130-131.

Johnson, B., R. Zollinger, B. Hanson, E. Erikson, N. Riveland, R. Henson, and B. Jenks. (2000). Herbicide-tolerant and conventional canola production systems comparison. *2000 North Dakota Weed Research*, pp. 8-10.

Kaniewski, W. and C. Lawson. (1998). Coat protein and replicase-mediated resistance to plant viruses. In *Plant Virus Disease Control*, ed., A. Hadidi, R.K. Khetarpal, and H. Koganezawa, St. Paul, MN: APS Press, pp. 65-78.

Kilman, S. (2000). Monsanto's biotech spud is being pulled from the fryer at fast-food chains. *Wall Street Journal*, April 28.

Kilman, S. (2001). Food industry shuns bioengineered sugar. *Wall Street Journal*, April 27.

Kirkland, K.J. (1995). HOE 075032 for wild mustard (*Sinapis arvensis*) control in canola (*Brassica rapa*). *Weed Technology* 9:541-545.

Kline, W. and S. Garrison. (2000). 2000 Eggplant variety evaluations. *Proceedings of the 31st Annual Mid-Atlantic Vegetable Workers Conference*, November 8 and 9, Newark, DE, pp. 50-53.

Kmec, P. and M.J. Weiss. (1998). *Assessment of Pesticide Use for Canola.* Fargo, ND: North Dakota State University Department of Entomology.

Koziel, M.G., G.L. Beland, C. Bowman, N.B. Carozzi, R. Crenshaw, L. Crossland, J. Dawson, N. Desai, M. Hill, S. Kadwell, K. Launis, K. Lewis, D. Maddox, K. McPherson, M.R. Meghji, E. Merlin, R. Rhodes, G.W. Warren, M. Wright, and S.V. Evola. (1993). Field performance of elite transgenic maize plants expressing an insecticidal protein derived from *Bacillus thuringiensis*. *Biotechnology* 11:194-200.

Lamb, M.C. and D.A. Sternitzke. (2001). Cost of aflatoxin to the farmer, buying point, and sheller segments of the southeast United States peanut industry. *Peanut Science* 28(2):59-63.

Lange, A.H. and J.P. Orr. (1989). Tomatoes (*Lycopersicium exculentus*). In *Principles of Weed Control in California*, Fresno, CA: Thomson Publications, pp. 351-357.

Lius, S., R.M. Manshardt, M.M.M. Fitch, J.L. Slightom, J.C. Sanford, and D. Gonsalves. (1997). Pathogen-derived resistance provides papaya with effective protection against papaya ringspot virus. *Molecular Breeding* 3: 161-168.

Luttrell, R.G. (1994). Cotton pest management: Part 2. A U.S. perspective. *Annual Review of Entomology* 39:527-542.

Lynch, R.E. and D.M. Wilson. (1991). Enhanced infection of peanut, *Arachis hypogaea* L., seeds with *Aspergillus flavus* group fungi due to external scarification of peanut pods by the lesser cornstalk borer, *Elasmopalpus lignosellus* (Zeller). *Peanut Science* 18:110-116.

Lynch, R.E., C. Singsit, and P. Ozias-Akins. (1995). Efficacy of peanut containing the *Bt* gene for delta endotoxin against the lesser cornstalk borer. *1995 Proceedings of the American Peanut Research and Education Society* 27:35.

Lynch, R.E. and P. Ozias-Akins. (1998). Evaluation of peanut containing a *cryIA(c)* gene from *Bacillus thuringiensis* for activity against the lesser cornstalk borer, corn earworm, and fall armyworm. *1998 Proceedings of the American Peanut Research and Education Society* 30:38.

Lynch, R.E., B.R. Wiseman, D. Plaisted, and D. Warnick. (1999a). Evaluation of transgenic sweet corn hybrids expressing Cry1A(b) toxin for resistance to corn earworm and fall armyworm (Lepidoptera: Noctuidae). *Journal of Economic Entomology* 92:246-252.

Lynch, R.E., B.R. Wiseman, H.R. Sumner, D. Plaisted, and D. Warnick. (1999b). Management of corn earworm and fall armyworm (Lepidoptera: Noctuidae) injury on a sweet corn hybrid expressing a *cryIA(b)* gene. *Journal of Economic Entomology* 92:1217-1222.

Magbanua, Z.V., H.D. Wilde, J.K. Roberts, K. Chowdhury, J. Abad, J.W. Moyer, H.Y. Wetzstein, and W.A. Parrott. (2000). Field resistance to tomato spotted wilt virus in transgenic peanut (*Arachis hypogaea* L.) expressing an antisense nucleocapsid gene sequence. *Molecular Breeding* 6:227-236.

Martin, R.R. (1999). Raspberry viruses in Oregon, Washington, and British Columbia. *Acta Horticulturae* 505:259-262.

Martin, R.R., K.E. Keller, and H. Mathews. (2001). Engineered resistance to *Raspberry bushy dwarf virus* in red raspberry. *Phytopathology* 91: S58.

Mason, C.E., M.E. Rice, D.D. Calvin, J.W. Van Duyn, W.B. Showers, W.D. Hutchison, J.F. Witkowski, R.A. Higgins, D.W. Onstad, and G.P. Dively. (1996). *European Corn Borer: Ecology and Management*, North Central Regional Extension Publication 327. Ames, IA: Iowa State University.

McManus, P.S. and A.L. Jones. (1994). Epidemiology and genetic analysis of streptomycin-resistant *Erwinia amylovora* from Michigan and evaluation of oxytetracycline for control. *Phytopathology* 84:627-633.

McMullen, M., R. Jones, and D. Gallenberg. (1997). Scab of wheat and barley: A re-emerging disease of devastating impact. *Plant Disease* 81:1340-1348.

McPherson, R.M., R.D. Hudson, and D.C. Jones. (1999). Soybean insects. In *Summary of Losses from Insect Damage and Costs of Control in Georgia, 1997*, ed., G.K. Douce and R.M. McPherson, Athens, GA: University of Georgia College of Agriculture Experiment Station.

Meadows, R. (2001). Scientists, state aggressively pursue Pierce's disease. *California Agriculture* 55:8-11.

Morgan, A. and C. Baker. (1999). Production of herbicide tolerant strawberry through genetic engineering. *Proceedings of the Annual International Conference on Methyl Bromide Alternatives and Emissions Reductions*, p. 11.

Morishita, D.W., M.J. Wille, and R.W. Downard. (1999). Weed control in glyphosate resistant sugarbeet. *Proceedings of the Western Society of Weed Science* 52: 120.

Mullen, R.J. (1997). New weed control developments in tomatoes. *Proceedings of the 49th Annual California Weed Science Society*, Santa Barbara, CA, January 20-22, pp. 83-89.

Nagata, R.T., J.A. Dusky, R.J. Ferl, A.C. Torres, and D.J. Cantliffe. (2000). Evaluation of glyphosate resistance in transgenic lettuce. *Journal of the American Society of Horticultural Science* 125:669-672.

Norelli, J.L., E. Borejsza-Wysocka, M.T. Momol, J.Z. Mills, A. Grethel, H.S. Aldwinkle, K. Ko, S.K. Brown, D.W. Bauer, S.V. Beer, A.M. Abdul-Kader, and V. Hanke. (1999). Genetic transformation for fire blight resistance in apple. *Acta Horticulturae* 489:295-296.

Ostlie, K.R., W.D. Hutchison, and R.L. Hellmich. (1997). *Bt Corn and European Corn Borer*, North Central Regional Extension Publication 602. St. Paul, MN: University of Minnesota.

Padgette, S.R., D.B. Re, G.F. Barry, D.E. Eichholtz, X. Delannay, R.L. Fuchs, G.M. Kishore, and R.T. Fraley. (1996). New weed control opportunites: Development of soybeans with a Roundup Ready™ gene. In *Herbicide-Resistant Crops*, ed., S.O. Duke, Boca Raton, FL: CRC Press, Inc., pp. 53-84.

Paris, H.S. (1996). Summer squash: History, diversity, and distribution. *Horttechnology* 6:6-13.

Perlak, F.J., R.W. Deaton, T.A. Armstrong, R.L. Fuchs, S.R. Sims, J.T. Greenplate, and D.A. Fischhoff. (1990). Insect resistant cotton plants. *Biotechnology* 8:939-943.

Perlak, F.J., T.B. Stone, Y.M. Muskopf, L.J. Petersen, G.B. Parker, S.A. McPherson, J. Wyman, S. Love, G. Reed, D. Biever, and D.A. Fischhoff. (1993). Genetically improved potatoes: Protection from damage by Colorado potato beetles. *Plant Molecular Biology* 22:313-321.

Polston, J.E. and P.K. Anderson. (1997). The emergence of whitefly-transmitted geminiviruses in tomato in the western hemisphere. *Plant Disease* 81:1358-1369.

Ravelonandro, M., R. Scorza, J.C. Bachelier, G. Labonne, L. Levy, V. Damsteegt, A. Callahan, and J. Dunez. (1997). Resistance of transgenic *Prunus domestica* to plum pox virus infection. *Plant Disease* 81:1231-1235.

Ravelonandro, M., R. Scorza, A. Callahan, L. Levy, C. Jacquet, M. Monson, and V. Damsteegt. (2000). The use of transgenic fruit trees as a resistance strategy for virus epidemics: the plum pox (sharka) model. *Virus Research* 71:63-69.

Reed, G.L., A.S. Jensen, J. Riebe, G. Head, and J.J. Duan. (2001). Transgenic *Bt* potato and conventional insecticides for Colorado potato beetle management: Comparative efficacy and non-target impacts. *Entomologia Experimentalis et Applicata* 100:89-100.

Rohrbach, K.G., D. Christopher, J. Hu, R. Paull, and B. Sipes. (2000). Management of a multiple goal pineapple genetic engineering program. *Acta Horticulturae* 529: 111-113.

Sanford, J.C. and S.A. Johnston. (1985). The concept of parasite-derived resistance: Deriving resistance genes from the parasite's own genome. *Journal of Theoretical Biology* 113:395-405.

Schnepf, H.E. and H.R. Whiteley. (1981). Cloning and expression of the *Bacillus thuringiensis* crystal protein gene in *Escherichia coli*. *Proceedings of the National Academy of Sciences* 78:2893-2897.

Scholthof, K.G., H.B. Scholthof, and A.O. Jackson. (1993). Control of plant virus diseases by pathogen-derived resistance in transgenic plants. *Plant Physiology* 102:7-12.

Schubert, T.S., S.A. Rizvi, X. Sun, T.R. Gottwald, J.H. Graham, and W.N. Dixon. (2001). Meeting the challenge of eradicating citrus canker in Florida–again. *Plant Disease Reporter* 85:340-356.

Schultheis, J.R. and S.A. Walters. (1998). Yield and virus resistance of summer squash cultivars and breeding lines in North Carolina. *Horttechnology* 8:31-39.

Schweizer, E.E. and A.G. Dexter. (1987). Weed control in sugarbeets (*Beta vulgaris*) in North America. *Reviews of Weed Science* 3:113-133.

Scorza, R., A. Callahan, L. Levy, V. Damsteegt, K. Webb, and M. Ravelonandro. (2000). Post-transcriptional gene silencing in plum pox virus resistant transgenic European plum containing the plum pox potyvirus coat protein gene. *Transgenic Research* 1054:1-9.

Scorza, R.A. and D.J. Gray. (2001). Disease resistance in *Vitis*. United States Patent 6,232,528.

Shelton, A.M., J.A. Wyman, N.L. Cushing, K. Apfelbeck, T.J. Dennehy, S.E.R. Mahr, and S.D. Eigenbrode. (1993). Insecticide resistance of diamondback moth (Lepidoptera: Plutellidae) in North America. *Journal of Economic Entomology* 86:11-19.

Shelton, A.M., F.V. Sances, J. Hawley, J.D. Tang, M. Boune, D. Jungers, H.L. Collins, and J. Farias. (2000). Assessment of insecticide resistance after the outbreak of diamondback moth (Lepidoptera: Plutellidae) in California in 1997. *Journal of Economic Entomology* 93:931-936.

Singsit, C., M.J. Adang, R.E. Lynch, W.F. Anderson, A. Wang, G. Cardineau, and P. Ozias-Akins. (1997). Expression of a *Bacillus thruringiensis cry1A(c)* gene in transgenic peanut plants and its efficacy against lesser cornstalk borer. *Transgenic Research* 6:169-176.

Stalker, D.M., J.A. Kiser, G. Baldwin, B. Coulombe, and C.M. Houck. (1996). Cotton weed control using the BXN™ system. In *Herbicide-Resistant Crops*, ed. S.O. Duke. Boca Raton, FL: CRC Press, Inc., pp. 93-105.

Stark, C.R., Jr. (1997). Economics of transgenic cotton: Some indications based on Georgia producers. In *Proceedings of the Beltwide Cotton Conferences*, Memphis, TN: National Cotton Council, pp. 251-253.

Steinrucken, H.C. and N. Amrhein. (1980). The herbicide glyphosate is a potent inhibitor of 5-enolpyruvyl shikimic acid-3-phosphate synthase. *Biochemical and Biophysical Research Communications* 94:1207-1212.

Swadener, C. (1994). Insecticide fact sheet: *Bacillus thuringiensis* (B.t.). *Journal of Pesticide Reform* 14:13-20.

Tachibana, K., T. Watanabe, Y. Sekizawa, and T. Takematsu. (1986). Accumulation of ammonia in plants treated with bialaphos. *Journal of Pesticide Science* 11:33.

Talekar, N.S. and A.M. Shelton. (1993). Biology, ecology, and management of the diamondback moth. *Annual Review of Entomology* 38:275-301.

Tennant, P.F., C. Gonsalves, K.-S. Ling, M. Fitch, R. Manshardt, J.L. Slightom, and D. Gonsalves. (1994). Differential protection against papaya ringspot virus isolates in coat protein transgenic papaya and classically cross-protected papaya. *Phytopathology* 84:1359-1366.

Thomas, P.E., E.C. Lawson, J.C. Zalewski, G.L. Reed, and W.K. Kaniewski. (2000). Extreme resistance to *Potato leafroll virus* in potato cv. Russett Burbank mediated by the viral replicase gene. *Virus Research* 71:49-62.

Tricoli, D.M., K.J. Carney, P.F. Russell, J.R. McMaster, D.W. Groff, K.C. Hadden, P.T. Himmel, J.P. Hubbard, M.L. Boeshore, and H.D. Quemada. (1995). Field evaluation of transgenic squash containing single or multiple virus coat protein gene constructs for resistance to cucumber mosaic virus, watermelon mosaic virus 2, and zucchini yellow mosaic virus. *Biotechnology* 13:1458-1465.

U.S. Department of Agriculture, Agricultural Marketing Service. (1996-2000). *Cotton Varieties Planted* (multiple years). Washington, DC.

U.S. Department of Agriculture, Economic Research Service. (2001). Agricultural biotechnology: Adoption of biotechnology and its production impacts. In *Agricultural Biotechnology Briefing Room*, <www.ers.usda.gov/Briefing/biotechnology/chapter1.htm/>.

U.S. Department of Agriculture, Hawaii Agricultural Statistics Service. (2000). *Papaya Acreage Survey Results*. National Agricultural Statistics Service, Washington, DC.

U.S. Department of Agriculture, National Agricultural Statistics Service. (1992-2000). *Noncitrus Fruits and Nuts: Summary* (multiple years). Washington, DC.

U.S. Department of Agriculture, National Agricultural Statistics Service. (1995-2000). *Agricultural Chemical Usage: Field Crops Summary* (multiple years). Washington, DC.

U.S. Department of Agriculture, National Agricultural Statistics Service. (1996-2000). *Agricultural Chemical Usage: Field Crops Summary* (multiple years). Washington, DC.

U.S. Department of Agriculture, National Agricultural Statistics Service. (1999). *Agricultural Chemical Usage: 1998 Vegetables Summary*. Washington, DC.

U.S. Department of Agriculture, National Agricultural Statistics Service. (2000a). *Acreage*. Washington, DC.

U.S. Department of Agriculture, National Agricultural Statistics Service. (2000b). *Agricultural Chemical Usage: 1999 Field Crops Summary*. Washington, DC.

U.S. Department of Agriculture, National Agricultural Statistics Service. (2001a). *Agricultural Chemical Usage: 2000 Field Crops Summary*. Washington, DC.

U.S. Department of Agriculture, National Agricultural Statistics Service. (2001b). *Agricultural Chemical Usage: 2000 Vegetables Summary*. Washington, DC.

U.S. Environmental Protection Agency, Office of Pesticide Programs, Biopesticides, and Pollution. (2000). *Biopesticides Registration Action Document: Preliminary Risks and Benefits Sections, Bacillus thuringiensis Plant Pesticides*. Washington, DC.

Vaeck, M., A. Reybnaerts, J. Hofte, S. Jansens, M. DeBeucheleer, C. Dean, M. Zabeau, M. Van Montagu, and J. Leemans. (1987). Transgenic plants protected from insect attack. *Nature* 328:33-37.

Vasil, I.K. (1996). Phosphinothricin-resistant crops. In *Herbicide-Resistant Crops*, ed., S.O. Duke, Boca Raton, FL: CRC Press, Inc., pp. 85-91.

Vidrine, P.R., D.B. Reynolds, and J.L. Griffin. (1993). Weed control in soybean (*Glycine max*) with lactofen plus chlorimuron. *Weed Technology* 7:311-316.

Wait, J.D., W.G. Johnson, and R.E. Massey. (1999). Weed management with reduced rates of glyphosate in no-till, narrow row, glyphosate-resistant soybean (*Glycine max*). *Weed Technology* 13:478-483.

Walker, D.R., J.N. All, R.M. McPherson, H.R. Boerma, and W.A. Parrott. (2000). Field evaluation of soybean engineered with a synthetic *cryIAc* transgene for resistance to corn earworm, soybean looper, velvetbean caterpillar (Lepidoptera: Noctuidae), and lesser cornstalk borer (Lepidoptera: Pyralidae). *Journal of Economic Entomology* 93:613-622.

Walker, D.R., H.R. Boerma, J.N. All, and W.A. Parrott. (2002). Combining *CryIAc* with QTL alleles from PI 229358 to improve soybean resistance to lepidopteran pests. *Molecular Breeding* 9(1):43-51.

Walker, K.A., R.L. Hellmich, and L.C. Lewis. (2000). Late-instar European corn borer (Lepidoptera: Crambidae) tunneling and survival in transgenic corn hybrids. *Journal of Economic Entomology* 93:1276-1285.

Webb, S.E. and R.V. Tyson. (1997). Evaluation of virus-resistant squash varieties. *Proceedings of the Florida State Horticultural Society* 110:299-302.

Wilcut, J.W., H.D. Coble, A.C. York, and D.W. Monks. (1996). The niche for herbicide-resistant crops in U.S. agriculture. In *Herbicide-Resistant Crops*, ed., S.O. Duke, Boca Raton, FL: CRC Press, Inc., pp. 213-230.

Williams, M.R. (1995-1999). Cotton insect losses. In *Proceedings of the Beltwide Cotton Conference* (multiple years). Memphis, TN: National Cotton Council.

Wood, M., D. Comis, B. Hardin, L.C. McGraw, K. Barry, and K.B. Stelljes. (1999). Fighting fusarium. *Agricultural Research* 47:18-21.

Wyman, J.A. and E.R. Oatman. (1977). Yield responses in broccoli plantings sprayed with *Bacillus thuringiensis* at various lepidopterous larval density treatment levels. *Journal of Economic Entomology* 70:821-824.

Yang, H., C. Singsit, A. Wang, D. Gonsalves, and P. Ozias-Akins. (1998). Transgenic peanut plants containing a nucleocapsid protein gene of tomato spotted wilt virus show divergent levels of gene expression. *Plant Cell Reports* 17:693-699.

Yang, Z.N., I.L. Ingelbrecht, E. Louzada, M. Skaria, and T.E. Mirkov. (2000). *Agrobacterium*-mediated transformation of the commercially important grapefruit cultivar Rio Red (*Citrus paradisi* Macf.). *Plant Cell Reports* 19:1203-1211.

Zaid, A., H.G. Hughes, E. Porceddu, and F. Nicholas. (1999). *Glossary of Biotechnology and Genetic Engineering*, FAO Research and Technology Paper No. 7. Rome, Italy: Food and Agriculture Organization of the United Nations.

Zitter, T.A. and H.Y. Ozaki. (1978). Aphid-borne vegetable viruses controlled with oil sprays. *Proceedings of the Florida State Horticultural Society* 91:287-289.

Zollinger, R.K. (2001). *2001 North Dakota Weed Control Guide*. Fargo, ND: North Dakota State University Extension Service.

Problems and Perspectives
of Yam-Based Cropping Systems in Africa

Indira J. Ekanayake
Robert Asiedu

SUMMARY. Yams (*Dioscorea* spp.) constitute an important starchy
staple in sub-Saharan Africa (SSA) where food security for a growing
population is a critical issue. Mixed cropping in yam based systems is the
norm in the region and productivity of yams in these systems is below
potential. It is concluded that there is much scope for improvement of
yam based cropping systems in SSA in order to meet the needs of the region. The strategy of crop breeding to select yam varieties suitable for
various cropping systems must consider a truly multidisciplinary systems approach. Further manipulation must be made to tuber dormancy to
expand flexibility in field propagation in different cropping systems and
improve storage and marketing. The sustainability of yam-based cropping systems in SSA could improve if agronomic research was focused

Indira J. Ekanayake (E-mail: iekanayake@yahoo.com) is Agricultural Consultant,
and Robert Asiedu (E-mail: r.asiedu@cgiar.org) is Breeder and Project Coordinator,
International Institute of Tropical Agriculture (IITA), Oyo Road, Ibadan, Nigeria.

Address correspondence to the authors at: IITA, C/O L.W. Lambourn & Co., Caro-
lyn House, 26 Dingwall Road, Croydon CR9 3EE, England.

The authors acknowledge the various contributions of their colleagues N. Wanyera,
R. Carsky, and G. Tian of the International Institute of Tropical Agriculture (IITA) to
this manuscript.

This study was funded by IITA.

[Haworth co-indexing entry note]: "Problems and Perspectives of Yam-Based Cropping Systems in Af-
rica." Ekanayake, Indira J., and Robert Asiedu. Co-published simultaneously in *Journal of Crop Production*
(Food Products Press, an imprint of The Haworth Press, Inc.) Vol. 9, No. 1/2 (#17/18), 2003, pp. 531-558; and:
Cropping Systems: Trends and Advances (ed: Anil Shrestha) Food Products Press, an imprint of The Haworth
Press, Inc., 2003, pp. 531-558. Single or multiple copies of this article are available for a fee from The
Haworth Document Delivery Service [1-800-HAWORTH, 9:00 a.m. - 5:00 p.m. (EST). E-mail address:
docdelivery@haworthpress.com].

on strategies for improving soil fertility, weed and pest management including design of cropping systems and suitable rotations. *[Article copies available for a fee from The Haworth Document Delivery Service: 1-800-HAWORTH. E-mail address: <docdelivery@haworthpress.com> Website: <http://www.HaworthPress.com> © 2003 by The Haworth Press, Inc. All rights reserved.]*

KEYWORDS. Root and tuber crops, *Dioscorea* spp., cropping systems, food security, African agriculture, yam breeding

INTRODUCTION

Assurance of food security and expansion of income-generating opportunities continue to be important challenges as human population increases in sub-Saharan Africa (SSA). Yams, cassava (*Manihot esculenta* Crantz), banana and plantains (*Musa* spp.) are key starchy staples among the various food crops grown in SSA. Yams constitute a multi-species crop important for food, income, and sociocultural activities. The species show diversity of aerial and underground plant parts, geographical distribution, usage, modes of multiplication, and ploidy levels. The Guinea yams of African origin, *Dioscorea rotundata* Poir and *D. cayenensis* Lam., account for most of the yam production in Africa. Water yam (*D. alata* L.), of Asian origin, is next in volume of production but has the widest geographical distribution. West and Central Africa account for about 93% of the world's annual yam production of 38 million Mg year^{-1} (FAOSTATS, 2000). This dominant yam production zone stretches from the Côte d'Ivoire through Ghana, Togo, Benin, Nigeria, Cameroon, Gabon, Central African Republic, and the western part of the Democratic Republic of Congo. Yams have been cultivated historically in East and Southern Africa but production levels are very low. *Dioscorea alata*, *D. cayenensis*, and *D. burkiliana* are the predominant food yam species (Wanyera et al., 1996). *Dioscorea rotundata* is the most recent introduction from West Africa to this region. Virtually all production of yams is used for human consumption. Average per capita consumption figures for 1994 to 1998 (FAOSTATS, 2000) in this sub-region were highest in Benin: 129 kg year^{-1} that provided 5.6 grams of protein and 353 calories day^{-1}. Comparative figures for Nigeria, the world's leading producer of yams, were 88 kg year^{-1} providing 3.8 grams of protein and 241 calories day^{-1}. Ethiopia and Sudan are the major producers in East Africa. Yams that enter international market generally originate from the Caribbean islands, Jamaica being the leading exporter. Most of the yams produced in Africa are traded locally. Brazil leads the production in South America, while Japan accounts for most of the production in Asia. Yams are also important in the South Pacific islands especially in Papua New Guinea.

In most yam-producing countries, the importance of yams has been increasing with respect to food security and nutritional diversification, provision of employment and incomes, as well as alleviation of rural poverty (Baudoin and Lutaladio, 1998). Orkwor and Ekanayake (1998) reported that, in Nigeria, yams could constitute up to 32% of gross income derived from annual cropping. In response to increasing demand for yam, cropping systems and processing methods are undergoing development and innovation. For example, modest research efforts at the International Institute of Tropical Agriculture (IITA) in the last five years have been directed to improving agronomic aspects of yam-based systems while emphasizing yam breeding research that has been on-going since 1971 (Asiedu et al., 1992; IITA, 2001).

In 2000, nearly 4 million hectares were planted with yam throughout the world (FAOSTATS, 2000). More than 69% of this total area was located in Nigeria. The average yield in Nigeria was nearly 10-Mg ha^{-1}. West Africa dominated yam production and contributed to tripling of world yam production in ten years (FAOSTATS, 2000). Its growth rate reached 12% and its share of world production went from 80% to 92% in the same period (from 1985 to 1995). Nigeria's share of world production went from 40% in 1985-1987 to 74% in 1995-1997. The accelerated increase in production recorded in the past ten years was due to increases in yield at the rate of 4 to 5% year^{-1} and increases in area planted by 6 to 7% year^{-1}. The accelerated increase in yam production, however, has resulted in rapid decline of soil fertility and increase in weeds. In view of the biophysical, technical, and socioeconomic constraints that lead to degradation of the natural resource base and declining crop yields, the challenge in research and development is to identify or develop appropriate options for increasing productivity in a sustainable way through working with farmers and national scientists in the region. The objective of this paper is to describe the yam-based cropping systems, analyze the production constraints, and suggest measures to overcome these constraints in the SSA.

CROPPING SYSTEMS

Yams that were considered as a predominantly forest zone crop in the past are now found in the moist savannas of West Africa constituting 73% of the area cultivated. In Nigeria for instance, it has moved from the humid forest area into the Guinea savanna zones, where disease problems are less acute and where land is available (Manyong et al., 1996). Farmers in the Guinea Savanna zone indicated that profitable cropping systems must always include yam (Anchirinah et al., 1996). Yam has already become the traditional crop in areas of the *Lobi* ethnic group in Côte d'Ivoire, in the M'bé plain in Cameroon (Dumont, Hamon, and Seignobos, 1994), and in Pilimpikou region of Burkina Faso (Vernier, N'kpenu, and Orkwor, 2001).

Traditional Cropping Systems and Indigenous Knowledge of Farming in West Africa

There are limited reports on knowledge of yam cropping held by the traditional societies of West Africa, where yam is an important part of civilization (Degras, 1993; Vernier and Dumont, 1997). For example, the *Bariba* ethnic group in the Republic of Benin are familiar with the process leading to the production of cultivated germplasm from the wild savanna yams (Dumont, 1997). This process of domestication, described as intra-varietal adaptive diversification is reported to lead to the selection of new varieties.

Yam production is associated with the practice of clearing forests in nomadic agriculture and therefore exhibits inherent sustainability of production issues. The system of shifting cultivation has two advantages: first, chemical and physical fertilities are guaranteed; second, seed yam tubers have protection from excessive exposure to the sun, which may be harmful at emergence and at the end of the vegetative cycle (Vernier and Dumont, 1997). The main drawbacks to shifting cultivation are the limitations in the adoption of intensification practices (in particular, mechanization) and the increase in the workload for the harvest of yam product.

Yam growing requires particular skills in choosing the land, land preparation, preparation of planting tuber setts, choosing varieties, planting methods, staking, harvesting and post-harvest handling, including storage, packaging and transportation. In general little research or transfer of technologies has been done within SSA zone to improve yam cultivation techniques to overcome constraints to production, and particularly the production costs (Dorosh, 1988; Nweke et al., 1991; Baudoin and Lutaladio, 1998). As an example, yam farmers in SSA still depend largely on hoe-cutlass labor. Land clearing by slash and burn precedes manual clearing of burnt tree trunks and other vegetation. Between 20 and 30 days are needed for clearing. The manual preparation of the mounds for yams is labor intensive and requires twice the time needed for clearing. In general about 5,000 to 6,000 mounds are put per hectare. On the other hand a low density of mounds is used for growing double-harvest *D. rotundata*, probably as a precaution against the risk of hydromorphy in lowland soils (Dumont and Vernier, 1997a, 1997b). Large sized mounds are prepared for the production of ceremonial yams leading to lower mound densities.

Seed yams can be planted during the dry season when the soil is dry and before first rains unlike other tuberous crops such as cassava. This period also coincides with low labor demand. Once planted in dry soil, seed can wait for the rains, or produce a bare vine that is ready to put out its leaves when the rains come. The yam canopy develops rapidly during the rainy season and, depending on the variety, requires staking to support the foliage growth. The tuber growth of yam is heavily dependent on the available resources. The tubers have a large sink capacity and continue to grow and store food reserves for most of the year as long as growing conditions remain favorable. Yams have a tremendous yield potential and individual tubers may weigh as much as 20 to 30 kg. Tubers have a relatively long storage life (4-6 months at ambient temperature).

Gender division of labor in yam production is interesting to note (Nweke et al., 1991; Baudoin and Lutaladio, 1998). Solely men carry out land clearing and mound-making. Women normally purchase and select the seed yams for planting. Women and/or men do the cutting of yam tuber setts. Males prefer to cut the setts while the women carry the pieces to the farm and place them on mounds. Planting and weeding are carried out by both sexes, however, men do the physically demanding weeding of perennial and persistent weed species. Cutting of stakes and staking are done mostly by men whilst the training of vines are done by both sexes but mostly by women. Men do harvesting of yams, and women do the collection of yam tubers and marketing. In the absence of a man, a woman takes full control and participates in almost all farm operations.

Yam cultivation is no longer completely dependent on slash and burn practices (Dumont, 1997). This is because of the need to have production close to the main road network or land shortage. Yam requires fertile lands and is thus planted as the first crop on cleared land or after fallow periods. It is usually repeated in the third year of the rotation and sometimes even in the fifth year. Yam is grown for almost one full year after which the cultivation is shifted to another area. The land is rotated with crops such as cassava and/or vegetables or left fallow for two to five years. In the West African zone, settler farmers generally grow yams as sole crop commercially whereas, most indigenous farmers intercrop yams or farm for subsistence. Yams are grown on large commercial farms in the Northern Region (Vernier, N'kpenu, and Orkwor, 2001).

According to recent surveys in West Africa (Antwi, Adu-Mensah, and Asiedu, 2000; Some, Kam, and Ouedraogo, 1995; Aighewi, 1998) farmers plant several varieties of *D. rotundata* (the most preferred species) and *D. alata*. Farmer's preference for a variety, most importantly, depends on early maturity, market demand, larger tuber size, and suitability for 'fufu' or 'ampesi' (preparation of local dishes depending on the community). When yams are grown in monoculture, farmers keep separate fields of *D. rotundata* and *D. alata* due to the difference in the growth habits and maturity. *Dioscorea praehensilis* is either semi-domesticated or found in the wild as undergrowth in the forest areas of Eastern, Ashanti, Brong Ahafo, Central and Western Regions in Ghana where the trees are used by the yam vines as live stakes (Anchirinah et al., 1996). Some varieties of the *D. praehensilis* are found only under cocoa (*Theobroma cacao* L.) farms. Farmers however cultivate *D. praehensilis* mainly for home consumption.

Mixed Cropping Systems

Yam production has great potential as a monocrop or an intercrop under annual or perennial systems. It is cultivated as an intercrop in home gardens, an annual monocrop in shifting cultivation, or an annual intercrop in sedentary system with semi-perennials, perennials or as an intercrop in large plantations across the SSA zone. The annual and perennial systems are discussed below.

Annual Systems

Annual food crops are commonly combined with yam (Nweke et al., 1991). According to recent surveys in West Africa (Manyong et al., 1996; Manyong and Oyewole, 1997; Dumont and Vernier, 1997a; Vernier, N'kpenu, and Orkwor, 2001; Antwi, Adu-Mensah, and Asiedu, 2000) and in East Africa (Wanyera et al., 1996) yam is intercropped with other staples like cassava, cocoyam (*Xanthosoma sagittifolium* L.) and maize (*Zea mays* L.) with vegetables such as okra (*Hibiscus esculentus* L.), and pepper (*Capsicum annum* L.). (Okigbo, 1980). Yam in the home garden type of mixed cropping system is mostly grown with cowpea (*Vigna unguiculata* L.), egusi (*Citrillus lanatus* L. and *Cucumeropsis* sp.), okra and pepper (Okigbo and Greenland, 1976; Okigbo, 1980; Baudoin and Lutaladio, 1998). Other food crops that are commonly used in yam mixed systems are maize and cocoyam in the north and the forest area of Côte d'Ivoire, bambaranut (*Voandzeia subterranea* L.) in the west of Burkina Faso (Some, Kam, and Ouedreogo, 1995), and rice (*Oryza sativa* L.) (Baudoin and Lutaladio, 1998). In East Africa, annual food crops grown in combination with yam include cassava, maize, sweet potato (*Ipomeoa batatas* L.), beans (*Phaseolus vulgaris* L.), Irish potato (*Solanum tuberosum* L.), and other green vegetables (Wanyera et al., 1996).

In the cropping cycle there is often a delay between the planting of intercrops and yam. For example, sesame (*Sesamum indicum* L.) is always seeded before yam in the Southwest of Burkina Faso. On the other hand intercrops may also be planted after yam. For example, pearl millet [*Pennisetum glaucum* (L.) R. Br.] is planted after yam by the ethnic group Lobi in savanna zone of Côte d'Ivoire and cowpea after yam in the central area of Benin and Togo (Vernier, N'kpenu, and Orkwor, 2001). Yam/cassava intercropping is perhaps the most common cropping system followed by yam/sweet potato in African farms. However, there are regional differences in the dominant systems. In East Africa for example, yam/cocoyam (taro), yam/cocoyam/cassava, and yam/cassava/sweet potato are the predominant crop combinations (Wanyera et al., 1996).

Perennial Systems

Recently, attention has been drawn to yam production with perennial industrial crops because yams are grown for commercial purposes. It also appears that large-scale production reduces the number of potential intercrops with yam (Orkwor, 1990). One example in Central Africa is the Bamileke area in Cameroon, where yam and coffee production is frequently combined (Dumont, Hamon and Seignobos, 1994). In East Africa, yam is cultivated in banana-based home gardens and coffee plantations (Wanyera et al., 1996). It is inter-cropped with banana and fruit tress or with coffee and banana. It is noted that yams are preferred as an alternate crop by farmers in areas where perennial banana yield decline occurs or where coffee has failed (Kapinga, 1992). In order to reduce production costs, it is desirable to integrate yam into a perennial system as compared to the traditional slash and burn and fallow practiced in annual systems. Generally there has been a reduction in the length of the fallow period and commercial cultivation has been kept close to the main roads (Baudoin and Lutaladio, 1998). The introduction of a second yam crop in the rotation has become common practice among the Malinké ethnic group in Guinea (Dumont, 1993) and with the Duru of northern Cameroon (Dumont, Hamon and Seignobos, 1994). In the Republic of Benin it is not uncommon to see *kokoro* varieties used four or five times in two fallow periods (Vernier, N'kpenu, and Orkwor, 2001; Baudoin and Lutaladio, 1998).

Agroecological Distribution of Yam

The various species of yams are cultivated across the common yam growing agroecologies ranging from the forest zone to southern and northern Guinea savannas of West Africa. The varieties adapted to a given agroecological zone, however, may differ. Therefore, there is a high potential for expansion of yam cultivation to new areas on the basis of agroclimatic suitability in SSA alone (Jagtap, 2000). A substantial area (> 80%) of SSA, except for the greater horn and some parts of Central Africa, are climatically suited for yam culture. However, cultivation limits are set by soil fertility conditions, traditional preference for yams versus other starchy staples, and to a larger extent labor demand and costs.

Yams are more prevalent in the savannas than in the forest zones. For example, the Guinea savanna zone is an area with higher profit margins than the forest zone mainly due to lower labor demand associated with small mounds, less staking costs, easy land preparation, etc. (Nweke et al., 1991). Yams are generally cultivated in annual farming systems in the lowlands (< 800 m) of West and Central Africa. Limited cultivation occurs in the mid-altitudes of Nigeria and Cameroon. In East Africa, yam is grown at altitudes of 1100 m to 1750 m (Wanyera et al., 1996).

Yam Consumption

Tuber quality of yam for preparation of yam-based foods is a major criterion for acceptance of new varieties by farmers and consumers. The most preferred method of preparation of tubers from *D. rotundata* cultivars in West Africa is boiling and pounding into a thick paste ('pounded yam') which is then consumed with soup. Tubers may also be consumed directly after boiling or cooked into pottage with added protein sources and oils. Frying in oil or roasting are also important cooking methods. In some parts of the SSA region, the peeled tubers are dried and later ground into flour and stored in households or are marketed. Flour is normally mixed with water to make a paste for consumption. A few commercial products such as dry yam tuber flakes or flour from the tuber are marketed especially in Nigeria and Côte d'Ivoire, or are exported outside Africa.

CHALLENGES TO YAM PRODUCTION IN AFRICA

In many yam growing areas in Africa, the most serious constraints to production include high costs of planting material, labor (for field operations like land preparation, planting, staking, weeding and harvesting), and pest damage in the field and storage (Robin et al., 1984, Nweke et al., 1991; Baudoin and Lutaladio, 1998). Each of these constraints is discussed below.

Planting Material

Planting materials for the production of ware yams (tubers above 1 kg meant for market or home consumption) are derived from the edible portion, the tuber, which is expensive (at least 50% of production cost) and bulky to transport. Farmers often use small tubers from the previous harvest or plant cut portions of large tubers. Usually 10 to 30% of the annual harvest are reserved for planting. For early maturing varieties of *D. rotundata*, harvesting of tubers about two-thirds into the growing season without destroying the root system (this process is called 'milking') provides early yams for home consumption and market (Anchirinah et al., 1996). This also allows the regeneration of fresh and small tubers from the corm at the base of the vine. These small tubers are harvested at the end of the season and used as planting materials for the next season.

In the Malinké area, Upper Guinea (Dumont, 1993) farmers grow *D. rotundata* with and without milking. This practice of double harvesting helps provide a management solution to the high cost and production problems of generating adequate planting material. Only the second harvest of the first crop provides the planting material required. Yam is the first crop in the rotation immediately after the land is cleared. In the second year others crops (mainly cereals) are grown. In the third year yam is planted again. The multiplication ratio for seed yam production in the field is very low (less than 1:10) compared, for instance, to some cereals (1:300). The technique of rapid multiplication using small tuber pieces (for example 25 g-weight tuber is called a minisett) in field nurseries has been developed to accelerate the multiplication process. However, one cycle of multiplication still requires a year: seven to ten months of growth and development followed by two to four months of tuber dormancy. Meristem culture and micropropagation *in vitro* offer even higher multiplication rates and healthier propagules but this is yet to achieve viable commercial status for the crop. For bulbiferous cultivars of *D. alata* there is another option for propagation through the aerial tubers. The use of botanic seeds and vine cuttings for propagation require more research if they are to become viable alternatives in the future.

Labor

Many aspects of yam production—land preparation, planting, weeding, staking, and harvesting require considerable amounts of manual labor. Studies indicate that labor accounts for over 40% of yam production costs. Mechanization as an option to reduce manual labor is not common in yam cultivation. The only notable exception is mechanized land preparation in the Caribbean, Brazil, and French overseas territories (Vernier and Vasin, 1996).

Land Preparation

Yam cultivation in the SSA starts with land clearing. Land preparation for yam varies considerably depending on the region, the soil properties, and the purpose of the cultivation. The types of seedbeds for yams are mounds, flat seedbeds, or raised flat seedbeds. Cultivation for ware-yam is normally done on mounds or ridges. In the sedentarized farming systems in north Benin, ploughing with animals requires 3 to 5 days ha^{-1}, after which the mounds are made manually (Dumont, Hamon, and Seignobos, 1994). Farmers normally leave stumps during land clearing in the forest zone. Maize or sorghum (*Sorghum vulgare* L.) stems are left after harvesting in the savanna, which are later uses as stakes for growing yam vines.

The preference for large tubers in traditional yam producing districts of West Africa imposes heavy demands on the production system. For instance, for the production of 'ceremonial' yams (very large tubers used in traditional ceremonies) in Nigeria, very large mounds and tall stakes are essential (Orkwor and Asadu, 1998). In well-drained areas where hydromorphy is not critical (lower water tables) small mounds are made with associated lower labor costs (Nweke et al., 1991). Making of mounds and planting of *D. rotundata* are normally done in November/December before the minor rain ends in Ghana. Planting of *D. alata* is done later and could continue till March. The size of the mound depends on the variety and potential market value. On the average in Ghana, 1,500 mounds per acre are made but the range could be between 1,200 and 1,700 mounds acre^{-1} or around 4,000 ha^{-1} (Anchirinah et al., 1996). Nweke et al. (1991) reported up to 13,000 mounds ha^{-1} in the Guinea savanna zone of Nigeria.

Weeding

Weed management in yam fields is challenging due to its complexity in terms of weed species composition, density, and competitiveness. Weed competition during the first four months of yam growth may reduce yields by as much as 43%. In traditional mixed cropping systems in SSA, the critical period of weed-free requirements for yam (the first four months) is normally extended because of the range of component crops that require longer weed-free periods (Orkwor, 1990). Weeds in the yam growing areas are mostly grasses like *Imperata cylindrica* and others broad-leaves like *Euphorbia* spp. and *Chromolaena odorata*. A study in northeast Benin showed *Digitaria horizontalis, Commelina benghalensis,* and *Ipomoea eriocarpa* to be dominant weeds associated with yams (Ahanchede and Gasquez, 1997). The roots of *I. cylindrica* often pierce through the yam tuber thus exposing it to other pathogens (e.g., fungi) and reducing the market value. Although, *Chromolaena odorata* is also a big competitor for light as it grows taller and faster than yam. Some farmers claim that the presence of weeds in the field indicates high fertility status of the soil.

Yams emerge slowly when mixtures of heads, middle, and tail portions of tubers are used as planting setts. Emergence may range from two weeks after planting to two to three months depending on the physiological state of the planting material. Many varieties develop leaves slowly and do not cover the ground, permitting weeds to grow in the first four months. Furthermore, yams are combined with intercrop species and cultivars with varying growth requirements. This complicates the timing of application of herbicides (Orkwor et al., 1994). Under mixed farming systems, a safer and more reliable approach to using herbicides would be the application of pre-emergence herbicides or pre-planting soil-applied herbicides to control early weed emergence. In certain field conditions, for example, when yam is intercropped with maize, it is possible to use selective applications for the two crops, for example, pendimethaline + atrazine or metolachlor + atrazine (Marnotte and Téhia, 1986). Chemical weed control for yam production in SSA is not yet proven to give high economic returns on investments and relatively few farmers use chemicals unless targeted for commercial purposes.

Given the characteristics of yams and the traditional farming systems used, it is necessary to carry out several cycles of weeding, e.g., three to five times (depending on variety or species and time of harvest), before the yams are finally harvested (Anchirinah et al., 1996). In the forest zone, it is not possible to avoid hand hoeing for more than three months (Vernier, N'kpenu, and Orkwor, 2001). *Dioscorea rotundata* requires two to three weedings, as it is an early maturing variety, whereas *D. alata* requires more weedings due to its later maturity. Field operations are manually carried out with cutlass and hoe and in addition, 'mattock' is included for harvesting of yams.

Staking

The yam vine twines onto whatever supports it comes across. Where the crop is grown on stubble-burnt areas, the dead trees on the land function as natural stakes and the majority of the plants manage to wrap themselves around these stakes, sometimes with the help of a small rail placed by the farmer. Staking is less important where there is abundant solar radiation, for example, in the savannas as compared to the forest zone. The light saturation for photosynthesis is not critical for unstaked yam plants where solar radiation budget is higher. Experiments conducted in Nigeria revealed no significant increase in yield with staking north of latitude 8°30' (Hahn et al., 1987). Similarly, manual cultivation without staking is a common practice in many savanna regions in SSA (Benin, Nigeria). An intercropped yam field with no staking is shown in Figure 1a. When staking is practiced, the stakes are put near the mounds at planting or when the sett germinates, and the vines are trained onto the stake as the yam grows. Smaller and shorter stakes are used for individual mounds but usually many vines from different mounds are trained to one stake due to scarcity and cost of staking materials (Anchirinah et al., 1996). A single stake supporting four yam plants is shown in Figure 1b. Lack of staking material can limit the production of large yam setts that produce plants with extensive shoot systems.

Artificial staking is widely used in some regions. For example, in the forest belt near the coastline in Nigeria, southeast Nigeria, and the corresponding region in Cameroon, where staking material is available yams are staked up to 4 meters (Figure 2). Between 40 and 60 work days ha^{-1} are required for setting up the stakes in this zone. In the rest of West Africa, staking is normally limited to double-harvest of *D. rotundata* grown in the agroclimatic belt on the north-

FIGURE 1a. Intercropped yam field without staking.

FIGURE 1b. A single stake as support for four yam plants.

ern border of the distribution area of yams. It includes the Mbé plain in Cameroon, the northern Ghana, the Pilimpikou plain of Burkina Faso, and the Lobi area of Côte d'Ivoire. All varieties of yam may be staked but larger and taller stakes are used for *D. alata*. Figure 3a shows staked versus unstaked *D. alata* plants in the forest-savanna transition zone. In home gardens yams may trail along fences (Figure 3b).

The agroecology in which yam is cultivated is important in terms of staking requirements of the crop as well as the availability of staking materials. In a

FIGURE 2. Full height staking of yams in the forest agroecological zone.

humid climate, where clouds can greatly limit the number of hours of sunshine, staking improves photosynthesis of plants, prevents foliar diseases, and allows the cultivation of interim crops. In dry savanna areas, staking makes the first harvest easier and aids the production of seed yams. Staking yams using cereal stalks of previous crop in a Guinea savanna site is shown in Figure 4a. Prostrate yam varieties are grown with no staking. In a mid-altitude site in Uganda non-staked prostrate growth of yam is common (Figure 4b).

Live staking (Figure 5) is an option and farmers routinely use existing trees as support either in the forest belt or even in perennial plantations. Live-stakes may provide additional benefits for the yam crop. For example, the leguminous hedgerow specie *Gliricidia sepium* used as stakes for yams additionally provide mulch and nutrients compared to bamboo stakes (Otu and Agboola, 1994).

Mulching

Mulching or capping of mounds is done after planting. It is usually practiced for *D. rotundata* as it is often planted at the onset of the dry season (Anchirinah et al., 1996). Mulching of yams is common also in both West (Antwi, Adu-Mensah, and Asiedu, 2000) and East (Osiru and Hahn, 1994; Wanyera et al., 1996) Africa. In East Africa (mainly in Tanzania and Uganda) mulching materials used for yams are grasses and banana crop residues. In Uganda's perennial systems mulching is normally intended for the companion banana or coffee crop but the yam intercrop also benefits. Mulching helps

FIGURE 3a. Staked versus unstaked *Dioscorea alata* yam plants in the forest-savanna transition zone in Ibadan, Nigeria.

FIGURE 3b. Yam vines trailing along a fence in a home garden in Gambia.

conserve moisture in the soil and prevents overheating of the area around the sett, thus initiating ideal conditions, for sprouting and growth of yams. Mulching also prevents direct exposure of the tuber to sunlight. Mulching is reported to increase biomass and tuber weight in yams (Toyohara et al., 1997).

FIGURE 4a. Staking yams using cereal stalks of the previous crop in the Guinea savanna zone.

FIGURE 4b. Unstaked *Dioscorea alata* yams growing prostrate on the ground in Uganda.

Soil Amendments

Yams require rich soils with a high organic matter content. They have been reported to respond to high nutrient levels and fertilizer applications under various agronomic conditions (Irizarry, Goenaga, and Chardon, 1995). Yams also extract large quantities of nutrients from the soil. For example, a yam yield of

FIGURE 5. Use of live stake for yam support in Uganda.

29 Mg ha^{-1} removed 133 kg N, 10 kg P, and 85 kg K from the soil (Sobulo, 1972). According to reports on trends of resource management constraints in high intensity yam growing areas in Nigeria 72% of the fields had worsening soil fertility conditions whereas only 3% of the fields had improving fertility status (IITA, 1999). It is obvious that intervention to reverse this trend is needed in the yam-based systems.

Surveys have shown that farmers in West Africa often do not apply chemical fertilizers to yam as they believe these have detrimental effects on cooking (Anchirinah et al., 1996) and storage qualities of the tuber (Vernier, N'kpenu, and Asiedu, 2001). In East Africa, however, the farmers consider applying manure at the time of ridging or mounding as beneficial to yams. Farmers also noted that the use of compost has a residual positive effect on the following year's crop. Given the lack of fertilizer application, farmers have to grow the crop in deforested areas or after a long fallow period. Lack of soil amendments, as mentioned above, suggest that the long-term maintenance of soil fertility status in yam based systems is a critical issue.

Harvest and Storage

Milking or periodic removal of one or more tubers while keeping the plant intact (Figure 6a) is a relatively common practice in the yam growing zone.

The sizes and shapes of yam tubers and the nature of the soil have important influences on the ease and efficiency of harvesting. Yams are harvested when the leaves turn yellowish brown in the single harvest system. Harvested yams are either stored in the farmer's field or at home (Anchirinah et al., 1996). Care is needed while digging the yams out of the soil. Damage to the tubers should be avoided because injured areas will provide entry points for pests and pathogens. Figure 6b illustrates the difficult task of harvesting large size tubers.

Storage systems for yam differ widely, but it is essential for the storage barn to have adequate ventilation, shade, and protection from rain. On the field, the yams are stored by piling under shady trees or burying in the ground in pits or mounds and covering with soil and/or dry grass. Yams stored in underground pits can be kept for longer periods but once exposed to air they can deteriorate very quickly. *Dioscorea rotundata* can be stored for 2 to 5 months whereas, *D. alata* can be stored for 6 months or more depending on the variety. Postharvest losses and deterioration of cooking quality increase during long-term storage. The high moisture content (70 to 80%) makes them susceptible to microorganisms while in storage.

The major sources of yam quality loss during storage are due to respiration, transpiration, and sprouting. For the export market, as well as storage for in-country sales, a long dormancy period is preferable. Once dormancy is bro-

FIGURE 6a. Milking or removal of one or more yam tubers while keeping the plant intact.

FIGURE 6b. The laborious manual harvesting process of large sized *Dioscorea cayenensis* yams in Uganda.

ken, labor is required to ensure that vines are detached from tubers as they emerge and elongate. Failure to do so will lead to rapid loss in tuber quality.

Pest and Disease Problems

Yams are affected by many pests and pathogens that, either singly or in combination are responsible for suboptimal yields as well as deterioration of quality of tubers in storage. The major insect and nematode pests are yam beetles (*Heteroligus meles* Billb. and *H. Appius* Klug), yam leaf beetles (*Crioceris livida* Dalm. and *Lema armata* Fab.), yam crickets (*Gymnogryllus lucens* and *Brachytrypes membranaceus*), mealybugs (*Planococcus halli* Ezat and McConnel, *Pseudococcus brevipes* Ckll., *Rhizoecus angustus* James, and *Planococcus citri*), yam scale insects [*Aspidiella* (*Aspidiotus*) *hartii* Ckll.], yam tuber beetle (*Araecerus fasciculatus* Degeer), yam nematode [*Scutellonema bradys* (Steiner and Lehw) Andrassy], root-knot nematode (*Meloidogyne incognita* and *M. javanica*), and lesion nematode (*Pratylenchus brachyurus*). Yam anthracnose [*Colletotricum gloeosporioides* (Penz.) Penz & Sacc.], leaf spot (*Curvularia eragrostidis*), leaf blight (*Rhizoctonia solani*), fungal rot (*Fusarium monoliforme, Aspergillus niger, Botryodiplodia theobromae*), bacterial rot (*Erwinia* spp. and *Corynebacterium*), yam mosaic virus, and water yam virus are the most important diseases.

It is generally acknowledged that most pathological causes of losses in storage can be attributed to interaction of nematodes, fungi, and bacteria moderated by environmental factors such as temperature and humidity. Most losses originate from pre-harvest invasion or infection and/or damage, during harvest and transit. The practice of placing yam first in the cropping sequence and avoiding continuous yam cultivation in West Africa may have as much to do with avoidance of pest and pathogen build-up (nematodes, mealybugs, scales, etc.) as with loss in soil fertility *per se*. Use of geographical information system (GIS) to map yam growing regions infected with nematodes in West and Central Africa has revealed a concentration of infested fields in areas where the length of fallow was less than 4 years (Manyong and Oyewole, 1997). Hence, intensification of yam cultivation would benefit immensely from selection and breeding for host plant resistance to prevalent pests.

Socioeconomic, Institutional, and Policy Constraints

In general, large and uniform tubers (5-20 kg) are used for traditional ceremonies and rituals. Thus, consumers prefer large tubers. To produce large tubers, large planting setts or seed yams (1000-1500 g) are used to obtain the desired size of tubers; large mounds (1-2 m in height, with a diameter of 2-3 m) are made to accommodate large tubers in the ground; a wide plant spacing is used; and long stakes (up to 4 m) are used to support the growing yam vines and to lift them up from the ground. By doing so the yam canopy is more spread out for better light interception and aeration thus reducing foliar disease problems.

Yams are a major source of income for a wide range of smallholders, including women who are very active in the marketing of yams and yam products. The yam sector lacks innovative processing technologies that could reduce losses and create new value added products. Yams have generally suffered from lack of institutional arrangements and policy decisions related to production and marketing. In fact, government policies such as the ban on export of yams from some African countries are counterproductive. In many areas, poor infrastructures and poor access to markets are the major challenges to expansion in production of yams (Robin et al., 1984).

TECHNOLOGICAL OPTIONS
FOR SUSTAINABLE YAM PRODUCTION

Yams constitute a multi-species crop with great potential for increases in productivity in SSA. Some reviews in the past three decades have stressed the positive outlook for improved and sustained productivity in yam based systems (Hahn et al., 1987; Baudoin and Lutaladio, 1998; Quin, 2000). There is interest in yam as a valuable component of the farming and food systems in SSA. This interest is because of food security and household income provided by yams. Farmers in this zone are highly knowledgeable about their traditional cultivars, farming practices, and are willing to participate in experiments aimed at increasing yam productivity. Opportunities for farmer participatory technology development (yam breeding in particular), testing, and dissemination therefore are high in either West or East Africa. Increased productivity of yam-based systems is necessary to offset production costs, expand opportunities for income generation, and increase quantities consumed at low-income levels. This depends on the development and dissemination of high yielding yam varieties with good food and storage quality tubers, strategies for integrated control of pests and diseases in the field and during storage, strategies for soil and crop management suited to intensified cultivation in traditional production areas, reduced labor input, and the expansion of production. Cultivation of yams has not benefited from research and development efforts in a manner commensurate with their relative importance or the challenges facing the crop (Kabbaj, 1997).

Within the CGIAR, IITA holds the global mandate for research on yam-based systems with emphasis on Africa. Increasing on-farm productivity through better varieties is the *modus operandi*. New and more productive varieties of yams are needed to sustain productivity in Africa. Farmers have been relying on natural variation for their varietal selection but are challenged by the biophysical and socioeconomic environment, especially that of pest buildup. A combination of farmers' indigenous knowledge and the expertise of breeders in expanding genetic variability would be ideal for alleviating the current situation (Vernier and Dumont, 1997; Degras, Pierre, and Arnolin, 1996; Ekanayake, Asiedu, and Dixon, 2000). Improved seed yam propagation method has been developed (NRCRI, 1986) for production of high quality, low cost, and abundant planting material (Otoo et al., 1985). In collaboration with Nigeria's National Root Crops Research Institute (NRCRI), farmers in Nigeria, Benin, Ghana, Togo, and the Caribbean and Pacific Islands have adopted this propagation system.

Physiological Considerations

Yams have the greatest sink capacity among the tuber crops. Yam tubers grow and store food reserves throughout the growing season as long as conditions remain ideal for plant growth. Individual tubers may weigh as much as 20 to 30 kg and yield as high as 33 kg per stand. The yield limitation appears to be associated more with the photosynthetic capacity of the plant than with its ability to store food reserves in the tubers (Orkwor and Ekanayake, 1998). Improvement should, therefore, focus on increasing the photosynthesis capacity of the crop for example by encouraging early emergence. The vegetative development of yam or optimal leaf area development depends on the duration between emergence and summer solstice (Vandevenne and Castanié, 1985) or on the control of foliar diseases (Hahn et al., 1987).

Manipulation of tuber dormancy may also help in vegetative development. Researchers have been trying to get a better understanding of the physiology of yam especially on physiological maturity and manipulation of tuber dormancy. Gibberellic acid, irradiation, and refrigeration techniques have been developed for the extension of tuber dormancy. Technology is also available for manipulating yams into flowering and seeding. Most recently, scientists have been able to reduce the dormancy period of water yam from three to one month by hormone treatment (H. Shiwachi, Personal communication, 2000). This may enable rapid multiplication for yam breeding and allow farmers to grow water yams twice a year in certain areas.

Agronomic Perspectives

Research is needed on ways to improve soil fertility. Inexpensive methods of staking are required as farmers still prefer to stake despite the scarcity of staking materials. Yam production has always been labor-intensive and expensive (Nweke et al., 1991). Traditionally, hand labor has been used. However, more research is needed on mechanized weed control and harvesting in order to reduce labor costs and overcome labor scarcity. Intercropping and rotations are potential areas for sustainable yam cropping. Such practices could include legumes in the rotation, addition of organic matter, and development of efficient weed control methods. Similarly, impact of fallow systems on performance tuber quality (texture, taste, dry matter content, starch content, storability) of various yam varieties should be evaluated.

Cover crops can be used instead of leaving land fallow in yam-based systems (Obiagwu and Agbede, 1996). Cover crops protect the soil from erosion, improve the soil fertility, and help in control of weeds. Some of the promising legume cover crop species that are beneficial in yam-based cropping systems are *Mucuna* sp., *Pueraria* sp., and *Centrosema* sp. (Tian et al., 1999; IITA, 2000). Cover crops can also provide fodder for animals, while the animals provide organic manure to the crop-livestock systems (Tarawali and Peters, 1996). The impact of yam-based cropping and livestock rearing in the Guinea savanna zone in West Africa requires further experimentation.

Multistrata systems combining annual and perennial crops are being developed. Such systems mimic natural forest structures and provide several benefits. This system is proposed as an alternative to slash-and-burn agriculture as it provides ecological diversity to the farming system. Introduction of high yielding varieties of yams in such multi-strata systems is potentially a highly viable option.

Integrated Pest Management

The impact of cultural practices on pest and disease incidence/severity in yam fields is being studied (IITA, 2000). Agronomic practices play an important role in determining the extent of damage caused by pests and pathogens. The development of improved cropping practices that provide natural deterrents to pest and pathogen build-up is, therefore, essential for sustainable yam production (Hahn, 1992). Such practices include use of healthy seed yams; selection of flat well-drained soils, with few stones; effective crop rotations; use of cover crops; care in weeding, harvesting, and post-harvest handling, and adding organic amendments to the soil. Benefits of these practices include, for example, reduction of soil nematodes by *Aeschenamena histix*, *Pueraria phaseoloides*, and *Mucuna pruriens* cover crops (IITA, 2000); reduction in bruising and wounding of tubers during weeding and harvesting, thus reducing occurrence of tuber diseases; and suppression of yam nematode (*Scutellonema bradys*) by organic manure (Ikotun, 1989).

Plant Breeding

Reasonably good progress has been made in improving yam varieties, particularly in *D. rotundata*, by combining parental genotypes of reliable flowering and good agronomic attributes (Asiedu et al., 1998). However, major constraints to further efforts in conventional breeding presently exist. These constraints prevent full and timely exploitation of many sources of durable resistance and other elements of productivity in the available germplasm. Among technical constraints, insufficient diversity of yam germplasm material is a serious factor. The main obstacles encountered in sexual hybridization of yams for genetic improvement include the scarcity of flowering, poor synchronization of male and female flowering phases and lack of efficient pollination mechanisms. Advances have been made in studies of reproductive biology of yams at the Central Tuber Crops Research Institute (CTCRI), Trivandrum, India, and IITA, Ibadan, Nigeria (Sadik and Okereke 1975; Akoroda 1983, 1985; Abraham and Nair 1990; Bai and Ekanayake, 1998). However, further work is required to ensure that the desirable genetic diversity in non-flowering genotypes is exploited. When compared to other major food crops, very limited genetic studies have been conducted in the genus *Dioscorea*. The few yam breeding programs reported in the literature have relied on selections from land-races and hybridization of desired genotypes within and between species

(Sadik and Okereke, 1975; Doku, 1985; Abraham et al., 1986) without tapping the benefit of characteristics from complex polyploid species.

Improvements in screening methodologies are required to increase the efficiency and effectiveness of selection for resistance against pests and pathogens such as nematodes (*Scutellonema* sp., *Pratylenchus* sp., and *Meloidogyne* sp.), viruses (mosaic and shoe string, etc.), anthracnose, blight, leaf spots (*Colletotrichum* sp. and *Fusarium* sp.), tuber rots (*Aspergillus* sp., *Botryodiplodia* sp., and *Erwinia* sp.), and insects (beetles, mealy bugs, scales, etc.) (Akem and Asiedu, 1994).

On-going research in the SSA includes performance evaluation of local yam varieties in cropping systems, field evaluation of yam genotypes in a selection program, seedbed preparation (e.g., mounds and ridges), aspects of staking, fertilizer application, size of planting sett, plant density and arrangement, and plot size and border effects. By 1999, IITA held trust of 3190 yam accessions from both cultivated and wild species in its elite collection of yam germplasm (IITA, 2000).

Genotype × environment (G × E) interaction analysis is a tool used to a limited extent in assessing ecological adaptation on yams (Asiedu et al., 1998; Ekanayake, Asiedu, and Dixon, 2000; Apte, Karnik, and Patil, 1994). Recent studies using this analysis has shown that crop husbandry practices in yam greatly contribute to the variation in G × E effects (Quin, 2000). Breeding of yams for ecological adaptation and resistance is an ongoing activity at IITA. In Nigeria, planting materials of nine clones of *D. rotundata* were evaluated in 1997 by three collaborating national institutes under sole and mixed cropping in four agroecological zones. These zones represented the Forest, Forest/Savanna transition, Southern Guinea savanna, and Mid-Altitude savanna.

The successful development of diagnostics for yam viruses has made a vital contribution to the safe movement of yam germplasm (Asiedu et al., 1998). The micropropagation of virus-indexed clones is now a routine service activity at IITA. These safe-yam accessions have been distributed to 40 countries worldwide.

CONCLUSION

The sustainability of yam-based cropping systems in SSA could improve if agronomic research was focused on strategies for improving soil fertility, weed and pest management. Such strategies could include design of cropping systems and suitable rotations. Cover crops could replace fallows to protect soil erosion and replenish soil fertility. Further manipulation should be made to tuber dormancy to expand flexibility in field propagation and improve storage and marketing. Improvement and diversification of methods of yam propagation should be made to contribute to yield increases. Strategies should be developed to reduce labor requirement in yam cultivation. The above efforts should be complemented by diagnostic surveys to document farmers' perceptions of resource management constraints in yam-based systems; mapping the distribution of *Dioscorea* spp., cultivars, production practices, pests/diseases; characterizing resource use intensification in yam-based systems; and identifying new areas with potential for profitable yam production. The strategy of crop breeding to select varieties suitable for various cropping systems must also consider a truly multidisciplinary systems approach with focus on natural resource management (Izac and Sanchez, 2001).

IITA has been involved in the development of GIS coordinated maps for root and tuber crops in SSA. This information enables scientists to access valuable information on distribution of genotypes, species, pests, diseases, and other farming practices. Development of skilled human resources for yam research and improvement in national programs and the adoption of an end-user approach in the process of technology development are important elements that would ensure rapid progress towards achievement of sustainable yam-based systems. The increasing interest in regionalization of yam research offers good opportunities for high returns from focused and coordinated research and development efforts. An African Yam Network was formed in 1993, which brought together yam researchers and development workers on the African continent. Collaborative networks are important in yam research and development considering the limited number of active research on the crop. Finally methods of technology transfer to the end-users should be improved.

REFERENCES

Abraham, K. and S. G. Nair. (1990). Floral biology and artificial pollination in *Dioscorea alata*. *Euphytica* 48:45-51.

Abraham, K., S. G. Nair, M. T. Sreekumari, and M. Unnikrishnan. (1986). Seed set and seedling variation in greater yam (*Dioscorea alata* L.). *Euphytica* 35:337-343.

Aighewi, B. A. (1998). Seed yam (*Dioscorea rotundata* Poir) production and quality in selected yam zones of Nigeria. *Ph.D. Thesis*, Ibadan, Nigeria: University of Ibadan.

Ahanchede, A. and J. Gasquez. (1997). Weeds of rainfed crop fields in northeastern Benin. *Agriculture et Development* (special issue), pp. 17-23.

Akem, C. N. and R. Asiedu. (1994). Distribution and severity of yam anthracnose in Nigeria. In *Root crops for food security in Africa*, ed. M.O. Akoroda, Proceedings of the fifth triennial symposium of the International Society for Tropical Root Crops–Africa Branch, Kampala, Uganda, 22-28 November 1992.

Akoroda, M. O. (1983). Floral biology and hand pollination in white yam. *Euphytica* 32: 831-838.

Akoroda, M. O. (1985). Pollination management for controlled hybridization of white yam. *Scientia Horticulturae* 25:201-209.

Anchirinah, V., D. Ojha, R. Owusu-Sekyere, N. Ramnanan, Zhou-Sheng Kun, and S. K. Zhan. (1996). Production and marketing of yams in the forest/savanna transition zone of Ghana. Working document No. 53, Ottawa, Canada: IDRC. 99 p.

Antwi, A., J. Adu-Mensah, and R. Asiedu. (2000). The status of yam industry in Ghana: Research and development needs. Report, Kumasi, Ghana: Crop Research Institute (CRI).

Apte, U. B., A. R. Karnik, and V. H. Patil. (1994). G × E interactions in cassava and lesser yams. *Journal of Root Crops* 20:55-56.

Asiedu, R., M. Bokanga, A. G. O. Dixon, I. J. Ekanayake, S. Y. C. Ng, and K. Vijaya Bai. (1992). Cassava, sweet potato, yams and cocoyams. In *Sustainable Food Production in Sub-Saharan Africa. 1. IITA's Contributions: Chapter on Crop Improvement Division*, Ibadan, Nigeria: IITA. pp. 71-85.

Asiedu, R. A., S. Y. C. Ng, K. Vijaya Bai, I. J. Ekanayake, and N. Wanyera. (1998). Genetic Improvement. In *Food Yams: Advances in Research*, ed., G. C. Orkwor, R. Asiedu, and I. J. Ekanayake, Ibadan, Nigeria: IITA. pp. 63-104.

Bai, K. V. and I. J. Ekanayake. (1998). Taxonomy, morphology and flowering behavior of yams. In *Food Yams: Advances in Research*, ed., G. C. Orkwor, R. Asiedu, and I. J. Ekanayake. Ibadan, Nigeria: IITA. pp. 13-37.

Baudoin W. O. and N. B. Lutaladio. (1998). Yam cultivation and utilization for improved food security. *FAO's position paper*, Rome, Italy: FAO. 40 p.

Degras L. (1993). *The Yam. A Tropical Root Crop*. London, UK: MacMillan Press.

Degras, L., F. Pierre, and R. Arnolin. (1996). Some breeding aspects of yam (*Dioscorea* spp.). In *Tropical Tuber Crops: Problems, Prospects and Future Strategies*, ed., G. T. Kurup, M. S. Palaniswamy, V. P. Potty, G. Padmaja, S. Kabeeraththama, and S. V. Pillai. Delhi, India: Oxford and IBH Publ. pp. 146-159.

Doku, E. V. (1985). Sex expression and tuber yields of seedlings and clones derived from seedling tubers of white yam (*Dioscorea rotundata*). *Legon Agricultural Research Bulletin* 1:13-18.

Dorosh, P. (1988). The economics of root and tuber crops in Africa. *Resource and Crop Management Program Research Monograph No. 11*. Ibadan, Nigeria: IITA. pp. 31-48.

Dumont, R. (1993). Les ignames de Guinée. Aspects botaniques et techniques. Unité de Coordination des Recherches sur l'igname. Cotonou, Bénin: CIRAD/IITA. (In French)

Dumont, R. (1997). La production d'ignames dans un village bariba du Benin septentrional. *Carhiers de la Recherche Developpement* 43:35-51. (In French)

Dumont, R., P. Hamon, and C. Seignobos. (1994). Les ignames au Cameroun. Coll. *Repères–Cultures Annuelles*. Montpellier, France: CIRAD-CA, 80 p. (In French).

Dumont, R. and P. Vernier. (1997a). La domestication des ignames (*D. cayenensis-rotundata)* chez la population Bariba du Bénin. The domestication of yam (*D. cayenensis-D. rotundata*) within the Bariba ethnic group in the republic of Benin. In *Colloque Gestion des Ressources Génétiques des Plantes en Afrique des Savanes*. Bamako, Mali. 24-28 February 1997. p. 47-54. (In French)

Dumont, R. and P. Vernier. (1997b). L'igname en Afrique: des solutions transférables pour le développement. *Les Cahiers de la Recherché Developpement*, 44:115-120. (In French)

Ekanayake, I. J., R. Asiedu, and A. G. O. Dixon. (2000). Recent advances in root and tuber crops improvement research at the International Institute of Tropical Agriculture. In *Proceedings of the International Symposium on Tropical Root and Tuber Crops* ISOTUC III. January 19-22, Trivandrum, Kerala, India.

FAOSTATS (2000). *Crop Statistic of FAO*, URL: <http://fao.org> Rome Italy: FAO.

Hahn, S. K., D. S. O. Osiru, M. O. Akoroda, J. A. Otoo. (1987). Yam production and its future prospects. *Outlook on Agriculture* 16:105-110.

Ikotun, T. (1989). Diseases of yam tubers. *International Journal of Tropical Plant Diseases* 7:1-21.

IITA (International Institute of Tropical Agriculture) (1999). *Project 13. Improvement of Yam-Based Systems*, Annual Report 1998. Ibadan, Nigeria: IITA.

IITA (International Institute of Tropical Agriculture) (2000). Project 5 *Improvement of Yam-Based Systems*, Annual Report. Ibadan, Nigeria: IITA.

IITA (International Institute of Tropical Agriculture) (2001). IITA Annual Report. Ibadan, Nigeria: IITA.

Irizarry, H., R. Goenaga, and U. Chardon. (1995). Nutrient uptake and dry matter yield in the Gunang yam (*Dioscorea alata*) grown on an ultisol without vine support. *Journal of Agriculture of University of Puerto Rico* 79:121-130.

Izac, A. M. and P. A. Sanchez. (2001). Towards a natural resource management paradigm for international agriculture: the example of agroforestry research. *Agricultural Systems* 69:5-25.

Jagtap, J. J. (2000). The agroecological zones in sub-Saharan Africa. In *Genotype × Environment Interaction Analysis of IITA Mandate Crops in Sub-Saharan Africa*, ed., I. J. Ekanayake and R. Ortiz, Ibadan, Nigeria: IITA. pp. 182-194.

Kabbaj, O. (1997). The challenge of development and poverty reduction in Africa. In *Report CGIAR International Centers Week*, Washington, DC, USA: CGIAR Secretariat.

Kapinga, R. E. (1992). Root and tuber crops production and their constraints in Bukoba district. In *A Report: Tanzania/Netherlands Farming Systems Research Project*, Lake zone, Mwanza, Tanzania. 37 p.

Marnotte P. and K. E. Téhia. (1986). Tests d'herbicides pour la culture d'igname dans la région de Bouaké en 1983, 1984 et 1985. In *13ème Conférence COLUMA*. Versailles (France). 09-10 Déc. 1986. pp. 221-231. (In French)

Manyong, V. M. and B. Oyewole. (1997). Spatial patterns of biological constraints to cassava and yam production in West and Central Africa: Implications for technology development and transfer. *African Journal of Root and Tuber Crops* 3:15-21.

Manyong, V. M., J. Smith, G. K. Weber, S. S. Jagtap, and B. Oyewole. (1996). Macro-characterization of agricultural systems in West Africa: An overview. *Resource and Crop Management Monograph* No. 21, Ibadan, Nigeria: IITA.

NRCRI (National Root Crops Research Institute) (1986). *Annual Report NRCRI for 1985*, Umudike, Nigeria: NRCRI.

Nweke, F. I., B. O. Ugwu, C. L. A. Asadu, and P. Ay. (1991). Production costs in the yam-based cropping systems of southeastern Nigeria. *RCMP Research Monograph* No. 6, Ibadan, Nigeria: IITA.

Obiagwu, C. J. and O. O. Agbede. (1996). Interactive effects of food legume cover crops on the productivity of sandy soils of Benue river basins. *Journal of Sustainable Agriculture* 8:11-20.

Okigbo, B. N. (1980). A review of cropping systems in relation to residue management in the humid tropics of Africa. In *Organic recycling in Africa*. Rome, Italy: FAO. pp. 13-37.

Okigbo, B. N. and D. J. Greenland. (1976). Intercropping systems in Tropical Africa. In *Multiple Cropping: Proceedings of a Symposium*, ed., R. I. Papendick, P. A. Sanchez, and G. B. Triplett, Madison, WI: ASA Special Publication No. 27.

Osiru, D. S. O. and S. K. Hahn. (1994). Effects of mulching materials on the growth, development and yield of white yam. *African Crop Science Journal* 2:153-160.

Orkwor, G. C. (1990). Studies on the critical period of weed interference in yam (*Dioscorea rotundata* Poir) intercropped with maize (*Zea mays* L.), okra (*Abelmoschus esculentus* L. Moench), sweet potato and the biology of associated weeds. Ph. D. Thesis, Nsukka, Nigeria: University of Nigeria. 262 pp.

Orkwor, G. C. and C. L. A. Asadu. (1998). Agronomy. In *Food Yams: Advances in Research*, ed., G. C. Orkwor, R. Asiedu, and I. J. Ekanayake, Ibadan, Nigeria: IITA. pp. 105-141.

Orkwor, G. C. and I. J. Ekanayake. (1998). Growth and development. In *Food Yams: Advances in Research*, ed., G. C. Orkwor, R. Asiedu, and I. J. Ekanayake, Ibadan, Nigeria: IITA. pp. 39-62.

Orkwor, G. C., O. U. Okereke, F. O. C. Ezedinma, S. K. Hahn, H. C. Ezumah, and O. Akobundu. (1994). The response of yams (*Dioscorea rotundata* poir) to various periods of weed interference in an intercropping with maize (*Zea mays* L.), okra (*Abelmoschus esculentus* L. Moench), and sweet potato (*Ipomeae batatas* L. Lam). *Acta Horticulturae* 380:349-354.

Otoo, J. A., D. S. O. Osiru, S. Y. Ng, and S. K. Hahn. (1985). *Improved Technology for Seed Yam Production*. Ibadan, Nigeria: IITA.

Otu, O. I. and A. A. Agboola. (1994). The suitability of *Gliricidia sepium* in situ live stakes on the yield and performance of white yam (*Dioscorea rotundata*). *Acta Horticulturae* 380:360-366.

Quin, F. M. (2000). Genotype x environment analyses provide critical insights for improvement of IITA mandate crops. In *Genotype × Environment Interaction Analysis of IITA Mandate Crops in Sub-Saharan Africa*, ed., I. J. Ekanayake and R. Ortiz. Ibadan, Nigeria: IITA. pp. 1-5.

Robin, G., B. Clarke, H. Adams, S. Bellon, and M. Genthon. (1984). Introduction of clean *Dioscorea alata* planting material into small farm systems of Dominica. In *20th Annual Meeting of the Caribbean Food Crops Society*; St. Croix, US Virgin Islands; October 21-26, 1984.

Sadik, S. and O. U. Okereke. (1975). Flowering, pollen grain germination, fruiting, seed germination and seedling development of white yam, *Dioscorea rotundata* Poir. *Annals of Botany* 39:597-604.

Sobulo, R. A. (1972). Studies on white yam (*Dioscorea rotundata*). 1. Growth analysis. *Experimental Agriculture* 8:99-106.

Some, S., O. Kam, and O. Ouedraogo. (1995). Constrantes a la production del'gname au Burkina Faso. *Cahiers Agricultures* 4:163-169. (In French)

Tarawali, S. A. and M. Peters. (1996). The potential contribution of selected forage legume pastures to cereal production in crop-livestock farming systems. *Journal of Agricultural Sciences* 127:175-182.

Tian, G., G. O. Kolawole, F. K. Salako, and B. T. Kang. (1999). An improved cover crop fallow system for sustainable management of low activity clay soils of the tropics. *Soil Science* 164:671-682.

Toyohara, H., H. Kikuno, K. Irie, and F. Kikuchi. (1997). Effects of mulching on the growth of yam (*Dioscorea alata* L.) varieties introduced from tropical region. *Japanese Journal of Tropical Agriculture* 41:74-80.

Vandevenne, R. and D. Castanié. (1985). Croissance et développement de trois variétés d'ignames appartenant aux espèces *Dioscorea cayenensis, D. rotundata* et *D. alata* en Côte d'Ivoire. In *7th Symposium of the International Society for Tropical Root Crops (ISTRC)*, 1-6/7/1985, Gosier, Guadeloupe. (In French)

Vernier, P. and R. Dumont. (1997). L'igname en Afrique: des solutions transfearables pour le developpment. *Cahiers de la Recherche Developpement* 44:115-120. (In French)

Vernier, P. and D. Vasin. (1996). Recolte méchanique de l'igname en nouvelle-Caledonie. *Agriculture et Developpement* 10:56-60. (In French)

Vernier, P., K. E. N'kpenu, and G. C. Orkwor. (2001). Analyse comparée de la production d'igname pour la transformation en cossette au Bénin, Nigéria, et Togo: un exemple de sédentarisation de la culture de l'igname. In *Root Crops in 21st Century*, ed., M. O. Akoroda and J. M. Ngeve, proceedings of the 7th Symposium, International Society for Tropical Root Crops–Africa Branch (ISTRC-AB), October 1998, Cotonou, Republic of Benin. pp. 759-773. (In French)

Wanyera, N. M. W., R. Asiedu, R. Kapinga, and P. R. Speijer. (1996). Yams (*Dioscorea* spp.) in Tanzania and Uganda. *A survey report*. IITA, Ibadan, Nigeria and IITA-ESARC, Uganda.

Crop Technology Introduction
in Semiarid West Africa:
Performance and Future Strategy

John H. Sanders

Barry I. Shapiro

SUMMARY. The periodic food crises in agricultural production in semiarid sub-Saharan Africa have been the main public view of the subcontinent since the media perpetuates a Gloom and Doom perception. Actually the agricultural systems of sub-Saharan Africa have made some impressive achievements and there is potential to make more impact even in the more difficult regions where the poor are concentrated. In this paper the focus is on the successes, a strategy for introduction of new technologies in semiarid regions of West Africa, and supportive

John H. Sanders is Professor, Department of Agricultural Economics, Purdue University, 1145 Krannert Building, West Lafayette, IN 47907-1145 USA.

Barry I. Shapiro is Program Director, Natural Resource Management, ICRISAT, BP 320, Bamako, Mali.

The authors are grateful for the suggestions of John McIntire, Jock Anderson, and Tom Reardon in developing their ideas, and to various conference participants for comments on previous versions of this paper. Also excellent reviews from two anonymous reviewers prompted the authors to revise several important topics of technology development and policy support. Any remaining errors or omissions are the responsibility of the authors.

USAID supported much of the fieldwork reported in this article.

[Haworth co-indexing entry note]: "Crop Technology Introduction in Semiarid West Africa: Performance and Future Strategy." Sanders, John H., and Barry I. Shapiro. Co-published simultaneously in *Journal of Crop Production* (Food Products Press, an imprint of The Haworth Press, Inc.) Vol. 9, No. 1/2 (#17/18), 2003, pp. 559-592; and: *Cropping Systems: Trends and Advances* (ed: Anil Shrestha) Food Products Press, an imprint of The Haworth Press, Inc., 2003, pp. 559-592. Single or multiple copies of this article are available for a fee from The Haworth Document Delivery Service [1-800-HAWORTH, 9:00 a.m. - 5:00 p.m. (EST). E-mail address: docdelivery@haworthpress.com].

policy. More rapid technological change, with supportive policy, would have a substantial effect on rural poverty. *[Article copies available for a fee from The Haworth Document Delivery Service: 1-800-HAWORTH. E-mail address: <docdelivery@haworthpress.com> Website: <http://www.HaworthPress. com> © 2003 by The Haworth Press, Inc. All rights reserved.]*

KEYWORDS. Semiarid, West Africa, technological change, water retention, fertilizers

INTRODUCTION

Conventional wisdom is that agricultural technology development in sub-Saharan Africa (SSA) has been unsuccessful. Developed country donors are not convinced that returns to their investments in research have been sufficient (Lynam and Blackie, 1994); hence, financial support from donors for agricultural research in SSA has been declining in recent years (Pinstrup-Andersen, Lundberg, and Garrett, 1995). This attitude of pessimism regarding agricultural development potential for the region has been reinforced by declining food production per capita over the last three decades.

During most of this period high population growth has turned moderate agricultural production growth rates of approximately 2% annually (Larson and Frisvold, 1996) into declining per-capita trends. These moderate growth rates were achieved in a period of chronic drought as well as some acute drought periods (1968-1973; 1982-1984) and with macro and trade policy regimes generally adverse to the food production sector.

Economic policy changes combined with increased confidence in research systems from some successful technological development should make technology development even more successful in the new century. The stage is set for more rapid agricultural output increases even in semiarid areas.

We argue that the present highly pessimistic attitude toward the prospects for improving agriculture in semiarid West Africa is not warranted. This conclusion is based on a review of successes that have occurred over the last two decades, including substantial empirical evidence from farm-level studies carried out in four semiarid SSA countries, Mali, Burkina Faso, Niger, and Sudan (Vitale, 2001; Abdoulaye, 2002; Sanders, Shapiro, and Ramaswamy, 1996). Then modeling estimates of the potential impact of new technologies are reported for the two main semiarid agroecological zones and soil types, sandy loam soils in the Sudanian zone (using a Burkina Faso) and the sandy dune soils in the Sahelo-Sudanian zone (Niger).

The common elements in these successes and potential successes lead to a strategy for new technology development. This strategy builds upon what farmers are doing and then identifies other practices to further increase yields by employing a more complicated science-generated response to the same basic constraints. Emphasis is put on the public policy and investment support needed from national governments and the evolution of the private sector to facilitate and accelerate this new technology introduction.

We describe first the semiarid region and the food crops considered and then propose a strategy of technology development. We test this strategy by reviewing the technologies successfully introduced into different agroecological zones and then estimating the impact of and identifying the constraints to potential new technologies suggested by our strategy. Then the remainder of the paper considers issues related to putting into practice our strategy of technology development. First, we consider the viability of alternatives to these technologies. The following sections briefly review the economics of fertilizer, fertilizer policy and profitability, new product markets for the traditional cereals, and the evolution of the input markets. Then a section focuses on the impact on women of new technologies. Finally, future technological change and the implications for donors, national governments, and research policy are proposed in the conclusions.

THE REGION AND CROPS

The semiarid region is composed of two zones in West Africa, the Sudanian (600 to 800 mm of annual rainfall at 90% probability) and the Sahelo-Sudanian (350-600 mm; Figures 1 and 2). Most countries in West Africa with semiarid regions have some subhumid regions (Sudano-Guinean zone, 800-1100 mm) and there are complementary relations between agricultural development in the various regions. The World Bank and ICRISAT have included the Sudano-Guinean zone as semiarid. Since this latter region is the cotton (*Gossypium hirsutum* L.)-corn (*Zea mays* L.) zone, we prefer to put it in the subhumid category even though sorghum (*Sorghum bicolor* L. Moench.) and millet are still important crops. Our definition of semiarid then concentrates on where the drought-resistant cereals, sorghum and millet, are the principal crops. The concentration here is on the development of technological change for these two semiarid regions and on the traditional cereals (sorghum and millet) but some of the discussion will also be relevant to the subhumid zone. The figures show these agro-climatic zones across West Africa and then in more detail in Burkina Faso.

WHY FOCUS ON THE SEMIARID ZONE AND TRADITIONAL FOOD CROPS?

Contrary to conventional wisdom, technological change has been taking place in semiarid SSA. However, technology has been concentrated on the

FIGURE 1. Agro-climatic zones in semiarid West Africa (90 probability)

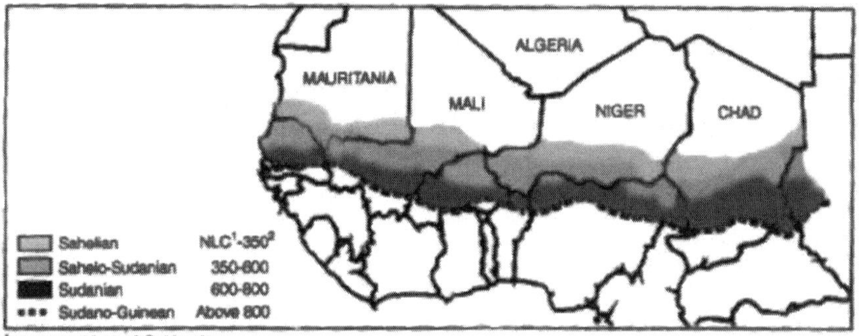

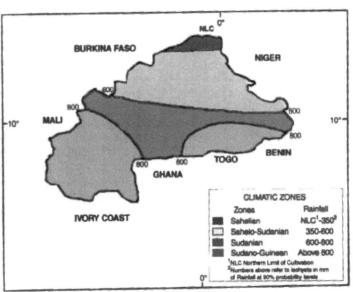

ALGERIA
MAURITANIA
MALI
NIGER
CHAD
Sahelian NLC[1]-350[2]
Sahelo-Sudanian 350-600
Sudanian 600-800
Sudano-Guinean Above 800

[1]Northern Limit of Cultivation
[2]mm of rainfall at 90% probability level
Adapted from Gorse and Steeds (1987).

FIGURE 2. Agro-climatic zones in Burkina Faso

NLC
0°
BURKINA FASO
NIGER
MALI
600
800
10°
800
800
BENIN
TOGO
GHANA
IVORY COAST
CLIMATIC ZONES
Zones Rainfall
Sahelian NLC[1]-350[2]
Sahelo-Sudanian 350-600
Sudanian 600-800
Sudano-Guinean Above 800
[1]NLC Northern Limit of Cultivation
[2]Numbers above refer to isohyets in mm of Rainfall at 90% probability levels

Adapted from Sanders, Shapiro, and Ramaswamy (1996).

main export crops, presently cotton and earlier peanuts (*Arachis hypogaea* L.) and corn (Sanders, Shapiro, and Ramaswamy, 1996). Among the food crops, corn is the main success story, with substantial corn-yield increases and rapid growth of input use (principally seed of new cultivars but also some fertilization). The area planted to the new corn varieties in SSA was 43% of the corn area in 1990 (Byerlee, 1996). But the gains to corn have been concentrated in subhumid regions. In the semiarid zone there have been substantial yield gains with corn but on the very small areas close to the household with higher fertility and better water retention due household wastes being dumped there and animal corralling there at night (Sanders, Shapiro, and Ramaswamy, 1996).

The economic environment for using higher inputs on most food crops has been adverse especially in semiarid regions with their higher risk of insufficient water at critical times. Moreover, most governments have maintained a series of policies to keep urban food prices down, thereby discouraging agricultural intensification. As a consequence, the levels of purchased inputs on food crops have been very low especially in semiarid regions.

In developing countries it is well recognized that with water and increased soil fertility there is a comparative advantage for semiarid and arid regions due to less disease and more sunlight. Hence, the highest crop yields in the world are regularly attained in these regions.

So if low cost strategies are available for moderately increasing water availability and soil fertility, then developing these regions has a high potential payoff. This potential is indicated by the differences between experiment station and on-farm yields for different cereals such as those developed recently in the analysis of Malian yield gaps (Figure 3; Vitale, 2001). Substantial gains in yields have reduced these gaps for corn and rice (*Oryza sativa* L.). Corn and rice have been grown predominantly in the semihumid or irrigated zones. Corn is often fertilized in the subhumid zone and in the irrigated regions rice is always fertilized. Even in the absence of potential for irrigation, there are a series of measures for increasing the availability of water in semiarid regions. With this increased water availability the response to fertilization is increased. With these two inputs (water and soil fertility amendments) there will be a large potential for new cultivars and these are the same combined inputs, fertilizers and new cultivars, that have resulted in yield increases for corn and rice over the last two decades.

Sorghum is the principal crop in the Sudanian zone and millet in the Sahelo-Sudanian region, but both are the predominant food grains found all over the semiarid region. In the higher rainfall, Sudano-Guinean zone, corn and cotton have the highest value. Nevertheless, even in this subhumid zone sorghum and millet are still the predominant crops with approximately half of the crop area in southern Mali (Vitale, 2001).

The focus of this paper is with reaching the potential of semiarid regions by providing moderately increased water and fertilization and extending the same yield increasing inputs to the traditional cereals (sorghum and millet) and other food crops [cowpeas (*Vigna unguiculata* L. Walp.)] that have been so successful with corn, rice, and cotton. New markets will need to be developed at the

FIGURE 3. Experiment station and average farmer yields in Mali for the principal cereals

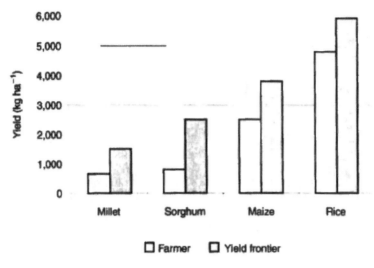

Adapted from Vitale (2001).

same time that new technologies are introduced or the improved agronomic potential of the semiarid zone will be used for other crop activities with more market potential.

A STRATEGY OF INTENSIVE AGRICULTURAL TECHNOLOGY DEVELOPMENT FOR SEMIARID WEST AFRICA

Sub-Saharan African countries have experienced increasing population pressure on land, causing a breakdown of the traditional rotation system of long fallow periods. Reduced fallow periods, without replacing soil nutrients with purchased inputs, result in declining crop yields. Farmers must then purchase inputs or increase their cultivated area to maintain production levels. With substantial between-season product price fluctuation and high input prices farmers often do not purchase inputs and then have to push crop cultivation into marginal areas, thus decreasing land for communal grazing. As cattle are sold or seasonally entrusted to herders, this source of organic fertilizer declines.

With rural population growth still rapid, and a high percentage of the rural population in agriculture, rural wage rates continue to be low. Reardon (1997) has pointed out the importance of non-farm work in determining incomes in some rural areas but also noted that there is substantial regional diversity in the availability of this activity. Moreover, the effect of non-farm job expansion on incomes has generally been overwhelmed by rapid population growth.

With agricultural labor productivity reduced by falling land quality, lack of capital investment in land-substituting inputs, and continuing rural labor-force growth, the value of labor relative to land deteriorates. Hence, in much of SSA farmers are observed making extremely labor-intensive investments in water-retention technologies, such as the dirt and rock bunds and the zaï (Mossi tradition from Burkina Faso–digging of small holes in the fields to retain water) in Burkina Faso and Niger, the animal traction ridges in Mali, terracing on the hillsides in Kenya and Rwanda, the use of a series of water retention techniques in Ethiopia including both bunds and a variation of the zaï (Shapiro and Sanders, 1992). These techniques are frequently augmented with manure. Extending the quantity and improving the quality of manure through corralling and compost heaps are other highly labor-intensive techniques also being rapidly introduced. These techniques only have small absolute yield effects because of the considerable soil degradation before their introduction. However, their introduction indicates responses by farmers and public agencies to the dual constraints of water availability and soil fertility.

In the semiarid regions these dual problems of low soil fertility, especially the principal nutrients, nitrogen (N) and phosphorus (P), and the lack of water availability at critical times for crop development, are hypothesized to be the principal factors leading to poor yields. Making water available when nutrient levels in the soil are very low generally has only a small yield response. Even at slightly higher soil nutrient levels, cereals will quickly deplete the available nutrients. Conversely, applying inorganic fertilizers without adequate water at the critical stages of plant development is economically risky because the response to fertilizer depends upon the availability of water. ·

One primary hypothesis of this paper is the necessity of resolving these two constraints simultaneously rather than trying to reduce the constraint discussion to one. Researchers tend to consider one input change at a time. Farmers prefer one input change so they can acquire experience from learning by doing before introducing other inputs (Byerlee and de Polanco, 1986). However, semiarid SSA is a harsh environment and one-input changes are not expected to have much effect on yields or the sustainability of the new technologies. The yield increase is greater and the risk is reduced when the two constraints are addressed simultaneously.

Others have emphasized the combination of organic and inorganic fertilizers (Breman and Sissoko, 1998). Since many farmers are already utilizing more intensively their organic matter (OM), it is more important now to emphasize the further addition of better water use and inorganic fertilizers. Nevertheless, in the sandy soils this addition of organic fertilizers can be critical and should be considered as a water retention technique since the sun cooked manure has minimal soil fertility help. Manure supply and quality also can be moderately increased with a series of very labor intensive measures mentioned above.

The appropriate techniques for responding to these dual constraints of water availability and soil fertility depend upon soil characteristics and the economic environment. In the heavier soils with more clay, the effects on crop yields of low and irregular rainfall are frequently aggravated by soil crusting. In these crusting soils, with serious infiltration problems, water-retention techniques become critical to reduce runoff and to use the available rainfall. Fortunately, there are numerous techniques for water conservation or retention (Shapiro and Sanders, 2002).

In the sandy dune soils found in most of the Sahelo-Sudanian zone, crusting exists but it is a less serious problem on these soils. However, rapid percolation of water through the soil can have the same effect as excessive runoff. The most commonly recommended water retention technique here is organic fertilizers especially crop residues and manure to better hold the water within the soil so that it is accessible to plant roots. However, moderate fertilizer use increases water use efficiency as well as providing plant nutrients (Shapiro et al., 1993). Apparently more plants and OM increase the capability of the sandy soils to retain water available to the roots.

A PRELIMINARY EVALUATION OF THE STRATEGY

Much of SSA is considered a land-surplus region, with seasonal-labor availability as the main constraint to increased output (Pingali, Bigot, and Binswanger, 1987). Animal traction is being introduced in some regions without yield-increasing technologies, such as the introduction of donkey traction in the Sudanian and Sahelo-Sudanian zones of Burkina Faso and oxen traction in the Gambia (McIntire, personal communication). However, the most pervasive and highest rates of animal traction diffusion are found in the high-rainfall Sudano-Guinean region where there are high-value cash crops and rapid introduction of new intensive technologies. In these regions such as the Sudano-Guinean zones of southern Mali and southwestern Burkina Faso, agriculture is more profitable and farmers acquire more capital and can buy the animals and the implements. The introduction of more intensive technologies on export crops, cotton and corn, also creates seasonal labor bottlenecks for critical operations (weeding, planting), thereby increasing the demand for animal traction (Sanders, Shapiro, and Ramaswamy, 1996; Jaeger, 1987).

The general failure to adopt animal traction, except where intensive, yield-increasing technologies were introduced, indicates the need for reevaluating this conventional analysis. The major thesis here is that the principal constraints to increasing agricultural output in semiarid West African agriculture are soil quality and water availability. They need to be simultaneously resolved. To evaluate this principal hypothesis of the paper we first do an overview of whether the new technologies being successfully introduced in semiarid West Africa are responding to these dual constraints (Table 1).

The region with the most rapid technological change, the Sudano-Guinean zone, has the highest rainfall. Here, the introduction of new cotton and corn cultivars has been combined with increasing levels of inorganic fertilizer. Other food crops, especially sorghum, benefit from being in the rotation. Since this is actually a subhumid region, there is normally sufficient rainfall and the focus can be on soil fertility and new cultivars. In many regions, the new corn cultivars have been introduced with minimal fertilization. Since the new cultivars mine the available nutrients, the importance of soil-fertility improvements to accompany new cultivar introduction is increasingly evident (CIMMYT, 1990).

Dikes to slow runoff have been rapidly introduced in the Sudanian and the Sahelo-Sudanian regions in Burkina Faso and Niger. In northern Burkina alone in the late '80s, 60,000 ha had been put into these dikes or bunds (World Bank, 1989, p. 98; Sanders, Negy, and Ramaswamy, 1990). These dikes are constructed on the contour at fairly wide intervals of 10 to 20 m, depending on the slope. The dikes retain some water and accumulate soil from higher up on the toposurface for 1 to 2 m behind them. This water retention is generally accompanied by the application of organic fertilizer (Wright, 1985). These combined techniques respond to both soil and water constraints (Table 1). The large inputs of labor required for constructing the bunds can be applied in the off-season when the opportunity costs of family labor are lower than during the crop season.

Another important innovation in both the Sudanian and the Sahelo-Sudanian zones has been shorter-season cultivars (sorghum, millet, and cowpeas). There has been diffusion of new cultivars from both farmers and experiment stations (Vierich and Stoop, 1990; Matlon, 1990; Coulibaly, 1987). With recurrent droughts since 1968, the payoff from early or short-season varieties for drought escape has increased. Improved farmer cultivars have been selected under low soil-fertility conditions. Therefore, early cultivars are a partial response to both soil and water constraints. This introduction of new cultivars alone results in minimal increases or even decreases in income in normal and better-rainfall years but raises yields and incomes in adverse rainfall seasons. This is only a temporary solution since early cultivars will need fertilizers or they will mine the soil nutrients. Unfortunately, earliness reduces the potential to respond to fertilizer during normal and good rainfall years.

TABLE 1. Technologies successfully introduced in the three principal agroecological regions of crop production in the semiarid tropics of West Africa

Zones	Rainfall Expected 90% Probability (mm)	Technologies	Responses to Principal Constraints	
			Water Availability	Soil Fertility
Sudano-Guinean	800–1100	New cotton and corn cultivars with inorganic fertilizers and improved agronomic practices	Sufficient rainfall in most years in this zone	Inorganic fertilizers used in the combined technology package
Sudanian	600–800	Contour dikes and organic fertilizers, zaï	Holds runoff water	Organic fertilizers
		Improved early cereal and cowpea cultivars	Earliness gives drought escape	Selected for low soil-fertility conditions
Sahelo-Sudanian	350–600	Contour dikes and organic fertilizers, zaï	Holds runoff	Organic fertilizers
		Early cereal and cowpea cultivars	Drought escape with earliness	Selected for low soil-fertility conditions
		Supplementary irrigation[a]	Full water control	Rice heavily fertilized

[a] Only small areas of supplementary irrigation (< 1 ha) provided by government to farmers. These are a type of income stabilization for dryland farmers.
Source: Adapted from Sanders, Shapiro, and Ramaswamy (1996).

Critics of the intensive water-retention technologies recommended above have emphasized the large increases in labor requirements (Matlon, 1990). This labor-use change is substantial. However, the soil degradation and the consequent fall in the marginal productivity of labor already required this increased amount of labor to produce the same level of output with less land. Thus, it is the degradation and the decreased value of farm-family labor with insufficient non-farm alternatives that encourage the increase in labor use. The technological change is a response to this demand for substitutes for the degraded land.

There is a literature on coping with adversity that emphasizes income diversification (Davies, 1996). Empirical studies of this strategy consistently shows that with higher resource bases farmers prefer to specialize more in agricultural activities and diversify less (Reardon, Matlon, and Delgado, 1988); hence, we emphasize reducing risk and increasing incomes within agriculture rather than just encouraging strategies allowing farmers to maintain very low but stable incomes on the basis of small commerce activities, migratory labor, remittances, and agricultural labor.

So this preliminary evaluation of the successfully introduced technologies into the semiarid regions appears to be consistent with our principal hypothesis of the dual constraints. The next phase of our examination will be to look at the potential effects of new technologies not yet widely diffused to respond to these two constraints. Farm models are constructed for the two principal zones to undertake this analysis.

WATER AND SOIL FERTILITY TECHNOLOGIES FOR THE SEMIARID REGION

The two main soil types in the Sudanian and Sahelo-Sudanian zones are the sandy clay loams (lixisols) and sandy dunes (arenosols) (Table 2). In much of the Sudanian zone, the clay content is higher; hence, crusting and runoff are common with continuous cultivation. Since runoff is the principal problem, water-retention techniques are a prerequisite to the application of fertilizer. In the more degraded regions, there are a number of techniques for retaining water common among farmers, including the earth and rock bunds and the zaï. Combining the water retention with manure severely degraded soils can be partially recovered and yields moderately increased, but labor inputs are large (Shapiro and Sanders, 2002).

The main characteristic of all these techniques is the use of large labor inputs undertaken outside the crop season. The return on labor is very low but these farmers are pressed to increase output after their soil resources have been degraded by the high population pressure, the disappearance of a frontier, and the lack of a sufficiently profitable environment for purchasing inputs.

TABLE 2. Distribution of the principal types of soils in West Africa

Pays	Arenosols	Lixisols	Nitisols	Acrisols	Ferrasols
Benin	-	80	6	-	-
Burkina Faso	6	46	-	-	-
Côte d'Ivoire	-	-	-	72	-
Gambia	-	-	25	-	-
Ghana	-	52	-	25	-
Guinea	-	-	-	22	13
Guinea-Bissau	-	55	5	-	18
Mali	15	-	-	-	-
Mauritania	10	-	-	-	-
Niger	30	-	-	-	-
Nigeria	13	34	14	5	-
Liberia	-	-	-	8	79
Senegal	30	-	6	-	-
Sierra Leone	-	-	-	-	70
Togo	-	59	11	-	-

Reprinted from Shapiro and Sanders (1998) with the permission from Elsevier.

Farmers' reactions to severe land degradation then indicate the directions to look for new technologies even before degradation. There are a series of water retention techniques that are more effective than the farmers' practices observed in the field. They need to be done within the crop season when there are many other demands on the time of the farm family hence they generally require the use of animal traction. Since these other techniques are commonly done before extreme degradation, their relative effects on yields are less than the techniques already being diffused (Shapiro and Sanders, 2002). However, their absolute yield effects are much larger. These new techniques include improved land preparation, ridging, and tied ridging (see Figure 4). They are generally combined with inorganic fertilizers.

To identify the potential adoption and income effects of these techniques farm modeling was undertaken for the tied ridges-inorganic fertilizer activities. Tied ridging consists of perpendicular ridges with a depression in the center where water collects rather than running off. The cereals are then grown on the ridges. The ridging can be done at planting, at first weeding, or at the second weeding, but the best results for yield increases in farm trials were at first

FIGURE 4. Corn under tied ridges soon after a rainstorm in Burkina Faso

Source: Sanders and McMillan (2001).

weeding (Sanders, Shapiro, and Ramaswamy, 1996; Sanders and McMillan, 2001). Their height and spacing vary with construction method (manual or animal tillage), farmer preference, soil type, and rainfall characteristics.

In the Sahelo-Sudanian zone, sandy dune soils predominate and soil crusting is less of a problem (Scott-Wendt, Hossner, and Chase, 1988). Rather overly rapid infiltration into these sandy, porous soils aggravates water and nutrient deficiencies. Organic fertilizers can provide an important water and nutrient retention technique here. Unfortunately, the soil fertility effect of manure left in the sun is minimal. Also soil temperatures become so high that the OM is burnt up within the soil during the crop season so it is difficult to build up the OM (Groot, Hassink, and Konè, 1998).

Fortunately, in these soils even moderate inorganic fertilizer use enables higher plant densities, which result in more OM and root development in these sandy soils. Hence, the addition of inorganic fertilizers increases the water-use efficiency of plants (ICRISAT, 1988; Reddy, 1988; Fussell and Serafini, 1985).

Because P is so deficient in these sandy soils or unavailable due to insolubility, a large yield response is generally obtained from the application of P fertilizer alone. ICRISAT/IFDC (International Fertilizer Development Center) trials have shown that, although there is not a response in poor rainfall years, there can be a significant economic carryover effect from P fertilizer (Jomini, 1990).

Advocates of using P fertilizer alone point out that this strategy reduces the costs and production risks associated with N fertilizer. Nitrogen deficiencies, however, soon limit growth when the plant increases its P use (Reddy and Samba, 1989). Nitrogen fertilizer will burn plants if there is insufficient soil moisture thereby compounding losses in poor rainfall years. Farmers can reduce the risk associated with N fertilizer use by applying it later in the season, thereby adjusting to the quantity of early season rainfall (Shapiro et al., 1993). Then complementing the inorganic fertilizer with the organic fertilizer will increase water and nutrient retention and increase microbial activity. The combination of the two has been shown to substantially increase yields in the Sahelian zone (Bationo and Mokwunye, 1991).

NEW TECHNOLOGY FOR THE SUDANIAN ZONE

The adoption of tied ridges made by hand increases farm income 12% when farmers plant all of the corn-compound land (0.15 ha) and half of their sorghum land (0.78 ha) using this new technique (Table 3, line 2). Another alternative is for farmers to use inorganic fertilizer without the tied ridges. The results show fertilizer would be used on the sorghum land, but the income increases would not be as great as those for tied ridges alone (Table 3). The increase in income from adding inorganic fertilizer to tied ridges is very small (+1%) compared with the effect of tied ridges alone (Table 3). However, this is a more stable, long-term solution because the major soil nutrients will not be mined. Here an annual programming solution is clearly insufficient given the importance of the dynamic effects.

TABLE 3. Effects of a water-retention device and new product markets on adoption of fertilizer on a representative farm in the Sudanian zone in Burkina Faso

Policy or Program	Fertilizer Use (ha)	Rainfed Crop Income (US$)	% Change Crop Income
Current practices	N/A	558	.
Tied ridges	N/A	624	12
Fertilizer	0.73	587	5
Tied ridges/fertilizer	0.73	632	13
New product markets (or price supports)	0.93	733	31

N/A: not available. Price supports in 5 could be either from the public sector or a local cooperative enabling the farmer to hold onto his product until the post-harvest price recovery.
Exchange rate: 273 FCFA/US$ (IMF, 1990).
Adapted from Shapiro and Sanders (1998).

Farm trial results demonstrate that combining tied ridges with fertilizer increases income and lowers risk. The combined technologies raise sorghum yields 50 to 100% across farms and over two years. Moreover, while 98% of the farms that used both inorganic fertilizer and tied ridges made profits in a poor-rainfall year, profits were made by only 76% who used inorganic fertilizer alone (Sanders, Shapiro, and Ramaswamy, 1996). So the combined technology is not only profit increasing but also risk reducing.

The programming predicts increasing area in tied ridges and inorganic fertilizers with either increasing implicit land costs (higher population pressure) higher output prices or lower input prices (Ramaswamy and Sanders, 1992). Introduction of tied ridging and fertilization has occurred where farmers have seen the results of this technology but diffusion has not been widespread (Sanders, Shapiro, and Ramaswamy, 1996). How do we explain the slow diffusion of this practice?

The more advanced water retention techniques including tied ridging need to be performed during the crop season when there are many other requirements for family labor. Hence, animal traction and a new implement is required. Four different types of animal traction ridgers have been developed in the Sahel and Ethiopia. Prototypes have been distributed in Burkina Faso and Ethiopia (Sanders, Shapiro, and Ramaswamy, 1996 for Burkina; Sanders and McMillan, 2001 for Ethiopia). In both cases farmers demanded adaptations and in Ethiopia the public research system (EARO) responded. A new version (Figure 5) responding to farmers' complaints was released in a year following farm trials and a private company has taken over the production and distribution from the national research center (Ethiopia report in Sanders and McMillan, 2001).

Supporting engineering services for adaptation of the ridgers to different soil types as well as the village-level development of some local blacksmith capacity for repair, appear to be necessary to accelerate the diffusion of new animal traction implements, such as the tied ridgers. In some developing regions, the private sector can be counted on to provide these adaptations and services. In the Punjab the rapid expansion of tubewells was facilitated by the growth of local village repair services. In the Sahel the public sector will undoubtedly have to participate in fostering an economic environment in which farmers make money and invest in food crop production, and entrepreneurs feel secure in making investments in the input industries. Specific repair services such as mobile repair units may even need the initial impetus from the public sector. Though there needs to be a plan from the onset to phase this into a private sector activity.

In the past, African governments have tried price supports to lower the risk of adopting new cereal technologies that involve cash inputs, but with little success. They have been successful with cotton. Hoping to control cereal mar-

FIGURE 5. Ethiopia-designed oxen-drawn tied ridger on the traditional plow

Source: Sanders and McMillan (2001).

keting, the public sector usually set cereal prices unrealistically high and tried to make purchases in most years. They could not sustain the drain on their treasuries. The model results show that if the Burkina Faso and other governments in semiarid Africa were to moderate cereal-price collapses when weather conditions are good, farmers would earn more money from the new cereal technologies and diffusion would be accelerated. For example, a 55 CFA kg^{-1} (French Community Franc–common currency utilized in eight French speaking countries of West Africa; here a value of $0.20 kg^{-1}) support price for sorghum and millet and improved agronomy (a combination of water retention and moderate inorganic fertilizer) would produce a household income of $733 as compared to $558 with traditional technologies at 45 CFA kg^{-1} (a 31% increase). With this slightly higher sorghum-millet price, the area in sorghum with tied ridges/fertilization increases to 0.93 ha (Table 2).

Innovative approaches are required to make price supports work. One approach could take advantage of the systems of regional storehouses constructed in recent years in developing countries to help mitigate the effects of drought and enhance food security. These facilities have been erected with donor assistance, and the donors have been filling them with imported food aid to be distributed to local populations in poor-rainfall years. African governments, however, could make targeted purchases of cereals in good-rainfall years at moderate support prices, store the grain, and then sell where needed in poor-rainfall years at prices that covered storage and transport costs. Assistance from donors might be required to get these programs started, but they could become self-sustaining. This would be more effective and sustainable than donors simply donating cereals in poor-rainfall years as part of famine relief.

Another alternative to support prices would be the development of local cooperatives, which would give the farmers advances at harvest time with the holding of his cereals. The farmer could afford to sell the grain later in the year after the post harvest price recovery. Then he would pay the coop for the storage function and the advance. The critical point here is the importance of more focus on obtaining better product prices and lower input costs as an essential part of the technology introduction process.

Given the previous failures, donors and national governments will be difficult to convince to engage in price support programs. A more viable alternative to increase the prices of the basic food products is the development of these cooperatives or even public support to the identification and response to new market opportunities. As economic growth proceeds, food grains become feed grains and the demand for cereals rapidly expands. There is substantial potential for the traditional cereals if they can assure quantity and quality to the feed mixers and thereby compete due to their lower transportation costs with the imported corn (Vitale, 2001; Vitale and Sanders, 2002). Similarly, there are a series of new food products with the processing of traditional foods that are being successfully introduced.

Some public support of research and consumer trials would undoubtedly facilitate this identification and development of new markets for traditional foods. A simultaneous approach to introducing technological change and expanding demand is expected to be more viable and supported by donors than any direct attempts by the public sector to support prices. The high returns to this combined approaches have been demonstrated for sorghum in Mali (Vitale and Sanders, 2002).

In some cases, productivity increase will lead to reduced output and area but this will free up resources for legumes and forages. After raising the productivity of the critical cereals the same techniques of increased water availability and higher soil fertility can increase the productivity of the animal production activity of these farms. The forages are used to feed animals and legumes can be used to improve the diets of farm households. Inorganic fertilizer use also increases crop residues of cereals and legumes and thus leads to more OM in these deficient soils.

NEW TECHNOLOGY FOR THE SAHELO-SUDANIAN ZONE

Since fertilization and higher plant densities have been shown to both raise soil fertility and increase water-use efficiency (Shapiro et al., 1993), the critical input to introduce into the region is higher levels of inorganic fertilizer. Where available, organic fertilization is a useful complement.

According to the model results improved shorter-cycle cultivars would be adopted, but not in combination with both N and P fertilizer (Table 4). Adoption of early varieties with higher cowpea planting densities increases expected rainfed-crop income 30% to $631, from $486 under traditional agronomic practices. The shorter-cycle cultivars provide some drought escape through earliness, making rainfed crop production more stable in lower-rainfall years, thus lowering the production risk. The coefficient of variation of cash income from rainfed crop production declines 13% with the introduction of the improved cultivar.

These model results are validated by adoption evidence from the field sites and other locations with similar conditions throughout the region (Lowenberg-DeBoer et al., 1992; Shapiro et al., 1993). Without additions to soil fertility, however, the use of improved cultivars will lead to further mining of the soil

TABLE 4. Effects of various technologies and development of improved marketing channels on adoption of fertilizer on a representative farm in the Sahelo-Sudanian zone in Niger

Policy or Program	Fertilizer Use (ha)	Rainfed Crop Income (US$)	% Change Crop Income
Current practices	N/A	486	-
Improved short-cycle cultivars	0	631	30
Phosphorus only	2.1	685	41
Long-cycle cultivars[a]	1.5	651	34
Input marketing improvements (10% cost reduction)	1.2	657	35

[a] Combined with both N and P fertilizers.
Exchange rate: 273 FCFA/US$ (IMF, 1990).
Source: Adapted from Shapiro and Sanders (1998).

nutrients so again the annual programming results are an inadequate technology planning device. When P fertilizer alone is available to the farmer, it raises total rainfed crop income 41% over current practices (Table 4).

The adoption of inorganic P fertilizer and shorter-cycle millet varieties on rainfed fields has been ongoing among farmers in the Dosso region of Niger since the mid-1980s. In interviews at one village site where on-farm testing was done over a 6-year period with 20 farmers, ICRISAT found that over 98% of all 150 farms in the village were not only using P, but both N and P fertilizer by the sixth year of the on-farm trials. Consumption of inorganic fertilizer in the village had increased from two tons of super-simple phosphate (SSP) to 115 tons of SSP, urea and compound N-K-P (Mokwunye and Hammond, 1992).

Has there been adoption of improved varieties and fertilizer at other sites in the same agroclimatic zone as the study site? While the improved shorter-cycle millet variety P3Kolo is widely used in the Niamey region, fertilizer has generally not accompanied the new cultivar. In some villages near the study site, however, farmers are making the zaï and are putting inorganic fertilizer into the holes. Furthermore, inorganic fertilizer from Nigeria and the improved shorter-cycle millet cultivar, CIVT, developed by INRAN, the national agricultural research institute of Niger, have also been widely used on farmers' fields since 1988 in villages in the Maradi Region, where the rainfall is higher and smuggled, low cost fertilizer is available from Nigeria (Lowenberg-DeBoer et al., 1992; Shapiro et al., 1993). Another recent study (Abdoulaye, 2002) documents the increasing use of inorganic fertilizers in the region of Niger where the IFDC/ICRISAT trials took place.

Although adoption of shorter-cycle cultivars is ongoing, they are unable physiologically to take advantage of normal or good season rainfall. If a mix of improved cultivars of different season length were available, farmers could adapt the type of cultivars they plant to the prevailing rainfall conditions. A longer-cycle cultivar with moderately higher yields than the local cultivars, would increase incomes 34% and result in the use of both N and P (Table 3). INRAN millet breeders have now begun selecting for late maturity to develop long-cycle cultivars that can respond more to fertilizer in better-rainfall years. A mix of long- and short-cycle cultivars is then a portfolio response to rainfall variability between years.

Several policy changes, including reductions in input prices, would also result in the introduction of inorganic fertilizer. After structural adjustment programs from the 1980s to the present there is minimal interest of donors in subsidies. So these price reductions of inputs would need to result from improvements in marketing services.

The market price of SSP was 50 CFA kg^{-1} in Niger from 1984 to 1988, and the price of urea was 65 CFA kg^{-1}. If prices could be reduced by 10% on N, there would be adoption of both N and P fertilizer on 1.2 ha of rainfed fields and increased rainfed-crop income of 35% over income levels under current practices (Table 3). Since the border with Nigeria reopened in 1988, subsidized fertilizer has been smuggled into Niger and is in use in areas contiguous to the border, especially in the Maradi region. The market price of the Nigerian fertilizer available in these border areas is lower than at the research sites in the Niamey region. The use of this subsidized fertilizer from Nigeria validates our model results on the potential impact of improved marketing on the introduction of the fertilizers. In other regions, improvements in transportation infrastructure or the formation of farmer cooperatives and then bulk purchases of fertilizer could both result in price reductions for fertilizer.

Moving farther north into the lower rainfall part of the Sahelo-Sudanian zone, none of the feasible policy changes or marketing innovations resulted in inorganic fertilizer introduction. Without fertilization, soils will be depleted with continuous cropping, hence in the future, this region appears to be inappropriate for crop production. The use of programming to define the potential regions, where fertilization will become profitable, appears to be a useful planning tool for defining regional research priorities between crop and agro-forestry production zones.

Crop research cannot resolve all income distribution issues especially in those sites with erratic and low rainfall, low soil fertility, and poor access to markets. In the northern section of the Sahelo-Sudanian zone, the development of agroforestry systems with improved grazing appears to be more appropriate than continuing efforts to improve crop production. Shifts to more extensive production systems will require out-migration of small farmers. So more profitable, non-farm alternatives or new agricultural regions will need to be sought to facilitate these adjustments. However, without fertilization, these farmers are not going to be able to increase their productivity in crop production.

This programming is then useful for developing research strategies for different regions based upon the future potential profitability of inorganic fertilizer. If inorganic fertilizer is not going to become profitable in crop production with foreseeable technological change and viable policy change, then we need to look for other activities besides crops. Otherwise incomes will continue to decline and soils degrade.

The model results for potential technology introduction and field observations of farmer practices are consistent with our strategy of technology development. Appropriate techniques need to be adapted to different soils, labor availability, and economic environments. However, the more intensive technologies clearly fit into the farmers' production systems, are profitable, and would be adopted according to model results in much of the semiarid zone.

The difficulty of simultaneously introducing at least two new inputs in the Sahelo-Sudanian zone (new varieties, fertilizer) or three in the Sudanian zone (water retention techniques, fertilizer, animal-traction implements) may explain previous failures in introduction of these higher-yielding water-retention/soil-fertility technologies.

ARE THERE ALTERNATIVES TO INORGANIC FERTILIZERS?

Numerous alternatives to imported inorganic fertilizer have been proposed over many years. Most propose to substitute locally produced materials–animal or crop by-products, or sophisticated production techniques, such as inoculation–for imported inorganic fertilizer. Many are claimed to be lower-cost or lower-risk.

Inorganic fertilizer can be risky because cash outlays can jeopardize the ability of farm households to make cereal purchases to feed family members when the rains are not adequate and crops fail. Also, input distribution systems often do not function efficiently and inorganic fertilizer can be available at the wrong times, lessening effectiveness.

Unfortunately, sources of organic fertilizer, such as manure, are severely limited in most of semiarid West Africa. For most farmers, sufficient manure is available only for small areas surrounding family compounds (Sanders, 1989; Williams, Powell, and Fernandez-Rivera, 1993; Sanders, Shapiro, and Ramaswamy, 1996). At the present time, most of the crop residues are removed for uses with higher economic value, such as feed, fuel, and building materials.

Problems associated with inorganic fertilizer use in sandy dune soils can include soil acidification and aluminum toxicity due to low soil OM (Scott-Wendt, Hossner, and Chase, 1988). Nevertheless, International Crops Research Institute for the Semi-Arid Tropics (ICRISAT) results consistently show that combining moderate levels of N-P-K fertilizer (45, 20, and 25 kg ha^{-1}, respectively, of pure nutrients) leads to higher yields on these sandy dune soils. With higher production, more crop residues can be left on the soil, raising the OM content. The potential problems of acidification and aluminum toxicity associated with inorganic fertilizers when there is insufficient OM can thus be avoided (ICRISAT, 1987; Groot, Hassink, and Konè, 1998).

On the sandy clay loam soils inorganic fertilization on sorghum has led to yield declines after 12 years. Over the longer term inorganic fertilizers also need to be supplemented on these heavier soils with organic fertilizers (Sanders, 1989; Nagy, Sanders, and Ohm, 1988). This Institut des Recherches d'Agriculture Tropicale/Institut National d'Investigation Agricole (IRAT/INIA) experiment assumed that despite more than doubling yields with inorganic fertilizer, farmers would not change their cultural practices and begin to incorporate some of the increased crop residues. Farmers presently use almost all of these residues and often burn what is left. With increased yields, farmers can could use part of the residues and leave the rest to decompose for the next crop year. It would be necessary either to keep animals from grazing these residuals or to find a method of incorporating them into the soil with animal traction.

With foreign exchange constraints discouraging the use of imported fertilizer, some Sahelian governments have been developing their own sources of natural rock phosphate. Research in Niger, meanwhile, indicates that high-quality rock phosphate from Tahoua, even when partially acidulated, would only be competitive with imported SSP if supplied to farmers at approximately 25% of the price of imported SSP (Jomini, 1990). This implies that at present efficiency levels, rock phosphate is unlikely to become competitive with the imported inorganic sources even after the recent devaluation of the CFA by 50% in 1994 (Mokwunye and Hammond, 1992). Furthermore, adequate costing of potential alternatives needs to include the additional labor, management, and extension costs of either labor or skill-intensive activities such as rock phosphate or inoculation (Sanders and Ahmed, 2001).

Environmentalists have also aroused concern among donors and development agencies about natural resource degradation in developing countries citing use of inorganic fertilizer as a contributing factor. The environmental impacts of fertilizer are assumed to be analogous to those in the developed countries where high levels of fertilizer use can pose a threat to the environment (Larson and Frisvold, 1996). However, in SSA only 5 to 10 kilograms of fertilizer are used per hectare of cropland (Bumb and Baanante, 1996; Larson and Frisvold, 1996), as compared to 60 kg ha^{-1} in India (Seckler, 1994). The very low levels of fertilizer use contribute to soil degradation by requiring the extension of cropping onto more marginal lands to maintain output once the fallow system breaks down. The traditional extensive cropping systems are not sustainable under increasing population pressure (Larson and Frisvold, 1996; Seckler, 1994). Per capita cereal production declined in SSA at an average annual rate of over 1% from 1961-1991 (Sanders, Shapiro, and Ramaswamy, 1996). Hence, the low levels or absence of inorganic fertilizer are a factor accelerating environmental damage in SSA.

At the present time, imported inorganic fertilizers are the only technically efficient and economically profitable way to overcome the soil-fertility constraints in the Sudanian and Sahelo-Sudanian zones for most regions and farmers. Some scientists have perpetuated this belief in alternatives to inorganic fertilizers. However, until substantial on-station and on-farm experimentation with alternatives is more successful, other measures of improving soil fertility must be considered as complements rather than as substitutes for imported inorganic fertilizers.

FERTILIZER POLICY

Governmental failures in West Africa to develop a fertilizer policy or to place inorganic fertilizer near the top of the list for foreign-exchange entitlements are apparently due to the lack of concern by governments and marketing parastatals with the profitability of agriculture (Lele, Christiansen, and Kadiresan, 1989; Larson and Frisvold, 1996). Public sector urban bias is consistently demonstrated in West Africa by cereal importation and subsidies whenever basic cereal prices increase above their long term normal prices.

Moreover, foreign technical advice often has contributed negatively to the problem. In the Sahelian countries researchers and others have been claiming that a local solution involving some combination of more manure, rock phosphate, grain-legume rotations, livestock-crop interactions, and low soil-fertility-tolerant cultivars would enable African governments to save foreign exchange and their farmers to avoid input purchases. Technicians in bilateral and multilateral programs, without evaluating the viability and economics of the alternatives to inorganic fertilizer, often pick up one or more of the arguments against using inorganic fertilizer. The technicians then reinforce the propensity of national government officials to put imported fertilizer at the bottom of the foreign exchange rationing list.

In the next decade, the availability and use of inorganic fertilizer will need to be rapidly increased. Inorganic fertilizer is a known technology with a substantial body of research behind it. Now it is time to utilize the clear advantages of inorganic fertilizer to increase food crop yields (Larson and Frisvold, 1996; Kuyvenhoven, Ruben, and Kruseman, 1998). In the sandy soils and other poorly buffered soils inorganic fertilizers will often need to be combined with organic fertilizers such as manure and crop residues (Byerlee and Heisey, 1993). Moreover, on-going research to find low-cost complementary activities to inorganic fertilizer and thereby reduce the costs of fertilization for farmers is an important research activity. However, the future potential of this research should not affect the present need for governments to pursue a development strategy of accelerating the diffusion of inorganic fertilizer.

PROFITABILITY OF INORGANIC FERTILIZER
AFTER STRUCTURAL ADJUSTMENT

Structural-adjustment programs have eliminated fertilizer subsidies and, with devaluation increased the prices of the imported inputs for agriculture, including inorganic fertilizer. However, these programs also raise prices of imported agricultural products and thus can encourage substitution of domestically produced for imported agricultural products. In the short run, the price elasticity of substitution between imported and traditional cereals has been shown to be very low (Reardon, 1993; Delgado and Reardon, 1992).

In the longer run, low-income consumers are expected to shift back to traditional cereals and other traditionally produced foods as price differentials continue over time between the imported and the domestically produced foods. Taste changes are expected to be a function of relative prices with some lags for adjustments.

The higher opportunity costs of the time of women in urban West Africa than in rural areas is an important factor explaining the shifts to the new cereals such as rice in urban areas. The traditional foods can also be adapted to respond to these demands. For example, both coucous of millet and millet flour are presently being sold in small plastic bags in urban areas of Bamako, Dakar, and are exported to Africans in Europe. This is still a very small part of consumption of these cereals but these sales are expanding rapidly. There are various on-going efforts to introduce traditional cereals and cowpeas in manufactured weaning foods (Vitale and Sanders, 2002). Public and private investments in the processing and preparation of the traditional cereals into easier-to-prepare food products not only responds to the changing preferences of women in the urban areas for food products that are easier to prepare but also is expected to enable some reclaiming of urban market share by the traditional cereals.

As imported food product become more expensive and population pressure makes land productivity increases essential to increase output, the prices of the domestic food products relative to fertilizers will increase. This has already been observed for the price of traditional cereals (sorghum and millet) relative to inorganic fertilizers in Mali in the late '90s (Sanders and Vitale, 1997). This increased incentive to use inorganic fertilizer will reverse the initial effects of the structural adjustment of reduced use of inorganic fertilizer observed in the late '80s and early '90s.

A positive effect from the higher inorganic fertilizer prices in the '80s and early '90s was the rapid introduction of practices to improve the quality and extend the quantity of organic fertilizers. At that time there was increasing diffusion of corrals, cutting crop residuals for beds in the corrals, compost heaps, and covering and watering the compost across the Sahel (Shapiro and Sanders, 2002; Sanders, Shapiro, and Ramaswamy, 1996). Since very large quantities of organic fertilizer are needed to provide adequate P and N, organic fertilizers was only a partial and inadequate substitute for inorganic fertilizer (McIntire, Bourzat, and Pingali, 1992; Seckler, 1994).

The on-going privatization of agricultural marketing will also benefit farmers by increasing competition and making the use of inorganic fertilizer more profitable. Historically, parastatal monopolies for cereal marketing in West Africa have been more concerned with maintaining low prices for urban areas and extracting a tax on agriculture for industrial development than with increasing the incentives for farmers to raise productivity and to make capital investments (Bates, 1981).

Public investment in transportation, communication and in facilitating the growth and strengthening of farmer organizations are important public activities to reduce the costs of the critical inputs especially inorganic fertilizers and new seeds. Various other policies to improve regional marketing and trade could also increase the product price share received by farmers (Davies, 1996). Since the private sector focuses on larger farmers, public policy oriented towards small farmers will be needed to improve the income distribution effects of marketing improvements.

INCREASING DEMAND FOR TRADITIONAL CEREALS

There is another critical problem for increasing the productivity of traditional food products. With an inelastic demand the prices of these products collapse with a good rainfall year and/or rapid technological change. Most countries in the West Africa have had official programs to maintain cereal prices that they were not able to implement. These parastatals (state financed and run companies performing roles usually undertaken by the private sector, in this case grain marketing) historically maximized employment and rarely had sufficient funds from the central government to support major food commodity price declines in good rainfall years. In contrast, developed countries do not let the prices of major food commodities collapse unless they have some other type of income support for farmers. The sub-Saharan countries need to find new approaches to the price collapse problem of the food crops in order to encourage the continuing introduction of new technologies.

With changes in milling technologies, sorghum can partially substitute for wheat in the flour mills. In northern Nigeria, one major beer company was producing sorghum seed for contracting with farmers. We noted above the introduction of new processed forms of the traditional cereals to sell to urban women in West Africa. A useful strategy is to for public policy to encourage with research and quality control the development of new processed domestic foods including weaning foods. Some sorghum-producing countries, such as Sudan, need to invest in the contacts and the quality requirements to expand their participation in the world sorghum market.

The big change is the shift from food to feed grains. Governments can encourage the development of a feed industry for goat- and sheep-fattening for years of good rainfall to moderate cereal price collapses in these years. Sheep fattening is already practiced for the high seasonal demands of religious holidays.

A much larger effect will result from the shifts in consumption patterns towards more animal products, fruits and vegetables with economic growth. When this occurs, the demand for poultry has grown in many countries at 8 to 10% annually and requires modern confinement and feeding technologies. This process can last two decades and includes large feed grain imports until countries can expand their domestic cereal production (Vitale and Sanders, 2002).

This dietary transformation has already begun among the middle and upper income sectors in several African countries including Senegal, Kenya, Botswana, and Zimbabwe (Vitale and Sanders, 2002). The public sector needs to begin now the process of developing farmer institutions, which could make advance contracts and insure quantity and quality of the traditional cereals (sorghum and millet) so that they could compete with imported corn at the local feed mixing facilities. These same institutions could also purchase inputs in bulk, do some extension services, and provide other marketing services to their farmers.

INPUT MARKETS AND AGRICULTURAL TECHNOLOGY

In the agricultural development process, the functioning of input markets—especially for seed, fertilizer, and credit—will become increasingly important. The private sector tends to be much more effective at responding to the need for quality control and to be able to develop more rapidly in response to rapid changes especially when adoption accelerates. But successful private sector development occurs primarily in middle and upper income countries and has not functioned as well in the agricultural input industries of low income countries as in most of West Africa. Hence there is a need for caution in the general advocacy of the reduction of the role of the public sector. Important public investments (transportation and communication infrastructure, research and extension support to small farmers for their public good aspects) and transitional assistance to facilitate the evolution of the private sector will all be necessary from an expanded and more efficient public sectors.

Illustrating the importance of the public sector in the transitional role is the recent Sudanese example in the production of a new sorghum hybrid. When the price for a new sorghum hybrid collapsed in the mid-'80s, two public agencies continued seed production as eight private firms left the industry. In the early '90s when the diffusion of the hybrid sorghum sales began accelerating again, the public sector had difficulty responding to the rapidly increasing demand for quality seeds and fertilizers. There are few incentives working effectively in the public sector to encourage bureaucrats to anticipate and respond to rapid demand shifts or to be sufficiently concerned with quality control. So in the '90s the private sector re-entered the market but it was the continuing seed production of the public sector in the late '80s that built up the demand sufficiently for the private sector to return (Sanders, Shapiro, and Ramaswamy, 1996). Otherwise the diffusion process would have halted in the mid-'80s.

Effective functioning of the public sector will often be necessary in this risky environment of technology development and introduction in West Africa. One difficult problem is knowing when to phase down and eliminate public sector activities and how to build a profitable environment for the private sector to take over. For instance, in seed production the public sector and NGOs often feel a pressure to keep seed prices low. To obtain private sector involvement an 8 to 10:1 ratio between seed and grain prices is sought by private companies in developed countries. So the public sector role in seed production needs to be on maintaining high quality seed production and convincing farmers that there are economic returns to higher quality seed even at a higher cost. The public sector needs to define a transitional role from the start of the public seed production process by gradually charging higher prices for seed and by following a time table for turning the seed production and distribution business over to the private sector.

A key public investment accelerating the entrance and competition of private firms in input and product marketing in agriculture is in roads and other infrastructure, including communications. Donor involvement will undoubtedly be necessary for these investments to improve the infrastructure supporting agriculture.

A continuing problem is that the private sector usually neglects small-farmers (Lynam and Blackie, 1994; Byerlee, 1996). The public sector can help offset some of the disadvantages of the smallholder sector by organizing cooperatives. These cooperative then will have more market power and lower transactions costs both necessary for more effective dealing with the private sector.

NEW TECHNOLOGIES AND THE WELFARE OF WOMEN

The stylized facts about the impacts of new technology on women are: (1) most of the technological innovations introduced into agriculture are obtained by men; (2) technology introduction increases the demand for the time of the women on the communal or family fields; and (3) new technology introduction can make women worse off due to the reduction of time they have to spend on their own fields.

One important question is how the within-family labor market functions. Some have argued that women's work on the communal fields is a responsibility of the marriage and not compensated. Or that all family members are fed and housed and in payment for this are expected to work on the farm.

However, there is a significant evolution in the farm when there is rapid technological change. As new income streams become available to the family and subsistence is no longer a predominant concern, what is the ability of the women to capture some of these increased income streams. Specifically, we first evaluated whether women were compensated for their increased labor on the communal fields with the introduction of new technologies. In the econometric analysis women's wages for the communal work are determined--as in other labor markets--by their opportunity costs, shifters such as increased technological change, and by their bargaining power (Lilja et al., 1996).

The significant technology coefficient indicates that women are receiving a share of these new income streams from technological change. The next question is the absolute size of wage payments and whether women are sufficiently compensated for their additional labor to offset the loss in home production from the reduction in private-plot effort. Unfortunately for most women, the increased time spent on family fields reduced their time available and returns from their private fields sufficiently to offset their higher wage payments. So their incomes were reduced by the technological change. Moreover, collaborating these results many women in the region expressed their discontent with their increased labor and decreased welfare resulting from technological change (Lilja and Sanders, 1998). Even though the welfare of women was apparently reduced by the introduction of new technologies considering only the production effects, the broader analysis of the changes in family consumption patterns with technological change still needs to be done to resolve this question of the net effect on the welfare of women from technological change.

Moreover, several institutional changes were observed, which had the apparent objective of obtaining a larger share of the increased income streams for the women from technological change. Note that there was not much difference between the wages paid to men and women workers after age and task levels were held constant. The problem is that the household head has been the primary beneficiary of most of the increased income streams. In East and southern Africa, there have been substantial out-migration and increasing shifts to farming by nuclear families so that at least income distribution is becoming more equal among men. Another notable institutional change in West Africa has been the village gender work groups, which perform seasonal agricultural tasks for fixed fees. These groups are increasingly paying their participants rather than saving the money for community functions, as was done historically with the proceeds (Lilja and Sanders, 1998; Sanders, Shapiro, and Ramaswamy, 1996). Since many seasonal jobs such as weeding and harvesting can have critical effects on yields, these gender groups have potential market power to negotiate a larger share of the profits.

The policy emphasis by donors to raise the income of women has been on increasing the access of women to input markets for their private field production especially extension, credit, and fertilizer. Another emphasis has been on improving the markets for the products of women and improving their access to land. If women have private plots, they can benefit from higher input levels and better markets for their rice, vegetables, and other products.

However, with the increasing land shortages, it is appropriate to stop focusing on the private plots of women as the primary instrument to increase their incomes. One-third of our sample of women did not have access to private land even though this region was in one of the higher-rainfall regions with more abundant land (Lilja and Sanders, 1998). Technological change will remain the main source of increased income so an emphasis on improving the institutions responsible for the continued development and diffusion of new technologies will benefit both sexes. In West Africa, the determinants of implicit wages paid for communal labor will be increasingly important. More attention needs to be put on increasing the opportunity costs and the bargaining power of women. Another important activity is the more rapid introduction of household technologies so that women have more time for off-farm activities as well as child rearing and taking care of themselves (Lawrence, Sanders, and Ramaswamy, 1999).

CONCLUSIONS

Soil-fertility improvement is a critical requirement in SSA and the gains from inorganic fertilizer are well documented. Fertilizer use in food crop production is expected to be profitable over a wide range of agroecological conditions. Unfortunately, there is resistance in some countries of West Africa to an extension emphasis on imported inorganic fertilizer. Some developing countries are actively trying to substitute domestic rock phosphate, manure, and other substances for imported inorganic fertilizers. This type of economic nationalism can slow the growth of their agricultural sectors.

The recent natural-resources emphasis of donors may be contributing to this delay in the promotion and diffusion of inorganic fertilizers. The dangers of chemicals in the environment of sub-Saharan agriculture are very different from those in countries that have attained high agricultural yields with high input levels (Bumb and Baanante, 1996). Environmental problems will be more significant in West Africa from the lack of intensification that results in further extension into marginal crop production regions (Larson and Frisvold, 1996). West Africa needs to take advantage of known technologies and to substantially increase food-crop yields now by facilitating the diffusion of imported inorganic fertilizers combined with water-retention techniques in semiarid regions.

The water-availability and soil-fertility constraints need to be given more attention by researchers and donors in semiarid regions. Region-specific agronomic research in semiarid regions is critical and will then raise the payoff to crop breeding programs. With water and fertilizer, semiarid zones shift from marginal to prime production areas as many regions in the world have demonstrated. The Mediterranean climates of California and Israel and the Australia semiarid regions have been turned into high-productivity areas, as has also occurred in some semiarid tropical regions of India and Sudan.

In most of West Africa, further expansion of irrigation is generally not economic; however, the potential for obtaining part of the benefits obtained with irrigation from water control and conservation has been demonstrated all over the world and various examples were given here. The problem then is to regionally identify the best methods to conserve water and to increase soil fertility and then to accelerate diffusion with public policy support to expand product markets so as to moderate the price collapses of traditional food crops resulting from the introduction of new technologies and/or good weather. In the absence of these demand expanding measures, diversification into alternative crops for niche markets and improved forages will be appropriate and the higher traditional food crop yields on reduced areas will facilitate this process.

For agricultural transformation to take place in semiarid SSA, i.e., the rapid evolution of input and product markets, farmers need to make money. The public sector will need to eventually phase itself out of these markets as the demand for inputs accelerates. West African governments will need to invest in infrastructure and in creating an economic environment (rule of contracts and law) to insure competition and protection for investments.

Moreover in the short run, there is still a need for governments to support the diffusion of new technologies. Innovative policy approaches are required to promote fertilizer adoption. These include investments and research to increase the demand for the traditional cereals (processed foods, animal feeds, alcoholic beverages), continuing research to develop cultivars that are more responsive to fertilizer, and infrastructure and other public investments to reduce the marketing costs of inputs and products.

There is an immediate need, meanwhile, for policy support to make more fertilizer available to farmers. Governments in semiarid West Africa need to make fertilizer imports a priority including the avoidance of overvaluation of the currency and the devolution of the fertilizer business to the private sector. Researchers and development agencies need to help extend this known and proven technology.

Plant breeders have been doing a good job in many countries of SSA but they cannot be expected to produce miracle varieties. Allowing them to concentrate on insect and disease resistances and higher yields is enough. The agronomic improvements need to come from increased input use (water and inorganic fertilizers) rather than new super varieties or the proposed substitutes for inorganic fertilizers.

REFERENCES

Abdoulaye, T. (2002). *Farm-Level Analysis of Agricultural Technological Change: Inorganic Fertilizer Use on Dryland in Western Niger.* PhD dissertation, Department of Agricultural Economics, West Lafayette, IN: Purdue University.

Bates, R.H. (1981). *Markets and States in Tropical Africa: The Political Basis of Agricultural Policies.* Berkeley, CA: University of California Press.

Bationo, A. and A.U. Mokwunye. (1991). Role of manures and crop residues in alleviating soil fertility constraints to crop production with special reference to the Sahelian zones of West Africa. *Fertilizer Research* 29:117-125.

Breman, H. and K. Sissoko. (1998) *L'intensification Agricole au Sahel* (in French). Paris, France: Éditions Karthala.

Bumb, B. and C. Baanante. (1996). The role of fertilizer in sustaining food security and protecting the environment to 2020. Food, Agriculture and the Environment Discussion Paper 17. Washington, DC: International Food Policy Research Institute.

Byerlee, D. and P. Heisey. (1993). Strategies for technical change in small-farm agriculture, with particular reference to Sub-Saharan Africa. In *Policy Options for Agricultural Development in Sub-Saharan Africa*, ed. N.C. Russell and C.R. Dowswell. Proceedings of workshop, Airlie House, VA, Aug. 23-25, 1992. Atlanta, GA: Global 2000, Inc.

Byerlee, D. (1996). Modern varieties, productivity, and sustainability: Recent experience and emerging challenges. *World Development* 24(4):697-718.

Bylerlee, D. and E.H. de Polanco. (1986). Farmers' stepwise adoption of technological packages: Evidence from the Mexican altiplano. *American Journal of Agricultural Economics* 68:519-28.

CIMMYT, in collaboration with IITA. (1990). *1989/1990 CIMMYT World Maize Facts and Trends: Realizing the Potential of Maize in Sub-Saharan Africa.* Mexico City, Mexico: CIMMYT.

Coulibaly, O.N. (1987). *Factors Affecting Adoption of Agricultural Technologies by Small Farmers in Sub-Saharan Africa: The Case of Improved Varieties of Cowpeas in Mali.* MS thesis. Department of Agricultural Economics, East Lansing, MI: Michigan State University.

Davies, S. (1996). *Adaptable Livelihoods: Coping with Food Insecurity in the Malian Sahel.* New York, NY: St. Martin's Press.

Delgado, C.L. and T. Reardon. (1992). Cereal consumption shifts and policy changes in developing countries: General trends and case studies from the West African semiarid tropics. In *Proceedings of the International Sorghum and Millet CRSP Conference*, ed. T. Schilling and D. Stoner, Publication No. 92-1, Lincoln, NE: University of Nebraska, INTSORMIL Management Entity Office, pp. 27-39.

Fussell, L.K. and P.G. Serafini. (1985). Crop associations in the semi-arid tropics of West Africa: Research strategies past and future. In *Appropriate Technologies for Farmers in Semi-Arid West Africa*, ed. H.W. Ohm and J.G. Nagy, West Lafayette, IN: Purdue University, International Programs in Agriculture.

Gorse, J.E. and D.R. Steeds. (1987). *Desertification in the Sahelian and Sudanian Zones of West Africa.* Technical paper 61. Washington, DC: World Bank.

Groot, I.J.R., J. Hassink, and D. Konè. (1998). Dynamique de la matière organique du sol. Ch. II.2.5 en *L'intensification Agricole au Sahel* (in French). Paris, H. Breman et K. Sissoko (eds.) France: Éditions KARTHALA.

ICRISAT. (1988). ICRISAT Sahelian Center Annual Re'Dort. 1987. Niamey, Niger: International Crops Research Institute for the Semi-Arid Tropics.

ICRISAT. (1987). *ICRISAT Sahelian Center Annual Re'Dort 1986*. Niamey, Niger:

IMF (International Monetary Fund). (1990). *International Financial Statistics, 1990*. Washington, DC: IMF.

Jaeger, W. (1987). *Agricultural Mechanization: The Economics of Animal Draft Powers in West Africa*. Boulder, CO: Westview Press.

Jomini, P. (1990). *The Economic Viability of Phosphorus Fertilization in Southwestern Niger: A Dynamic Approach Incorporating Agronomic Principles*. PhD Dissertation, Department of Agricultural Economics, West Lafayette, IN: Purdue University.

Kuyvenhoven, A., R. Ruben, and G. Kruseman. (1998). Technology, market policies and institutional reform for sustainable land use in southern Mali. *Agricultural Economics* 19:53-62.

Larson, B. and G.B. Frisvold. (1996). Fertilizers to support agricultural development in Sub-Saharan Africa: What is needed and why. *Food Policy* 21:509-525.

Lawrence, P.G., J.H. Sanders, and S. Ramaswamy. (1999). The impact of agricultural and household technologies on women: A conceptual and quantitative analysis in Burkina Faso. *Agricultural Economics* 20:203-214.

Lele, U., R.E. Christiansen, and K. Kadiresan. (1989). Fertilizer policy in Africa lessons from development programs and adjustment lending, 1979-87. MADIA Discussion Paper 5. Washington, DC: World Bank.

Lilja, N., J.H. Sanders, C.A. Durham, H. De Groote, and I. Dembèlè. (1996). Factors influencing the payments to women in Malian agriculture. *American Journal of Agricultural Economics* 78:1340-1345.

Lilja, N. and J.H. Sanders. (1998). Welfare impacts of technical change on women in southern Mali. *Agricultural Economics* 19:73-79.

Lowenberg-DeBoer, J., H. Zarafi, and M. Abdoulaye. (1992). *Enquête sur l' adoption des technologies Mil-Niébe á Kouka, Maïguero, Rigial et Kandamo* (in French). Publication 26F, Programme de Recherche Sur Les Systèmes de Production Agricole. Department de Recherches en Economie Rurale, Ministère de l' Agriculture, Niamey, Niger.

Lynam, J. and M.J. Blackie. (1994). Building effective agricultural research capacity. In *Agricultural Technology: Policy Issues for the International Community*, ed. J.R. Anderson. New York, NY: Cambridge University Press, pp. 106-134.

Matlon, P. (1990). Improving productivity in sorghum and pearl millet in semiarid Africa. *Food Research Institute Studies* 22:1-44.

McIntire J., D. Bourzat, and P.L. Pingali. (1992). *Crop-Livestock Interaction in Sub-Saharan Africa*. Washington, DC: World Bank.

Mokwunye, A.U. and L.L. Hammond. (1992). Myths and science of fertilizer use in the tropics. In *Myths and Science of Soils in the Tropics*, ed. R. Lal, P.A. Sanchez. Madison, WI: American Society of Agronomy and Soil Science Society of America.

Nagy, J.H., J.H. Sanders, and H.W. Ohm. (1988). Cereal technology interventions for the west African semi-arid tropics. *Agricultural Economics* 2:197-208.

Pingali, P.L., Y. Bigot, and H.P. Binswanger. (1987). *Agricultural Mechanization and the Evolution of Farming Systems in Sub-Saharan Africa*. Baltimore, MD: Johns Hopkins University Press.

Pinstrup-Andersen, P., M. Lundberg, and J.L. Garrett. (1995). *Foreign Assistance to Agriculture: A Win-Win Proposition*. 2020 Vision Food Policy Report. Washington, DC: International Food Policy Research Institute.

Ramaswamy, S. and J.H. Sanders. (1992). Population pressure, land degradation, and sustainable agricultural technologies in the Sahel. *Agricultural Systems* 40:361-378.

Reardon, T. (1993). Cereals demand in the Sahel and potential impacts of regional cereal protection. *World Development* 21:17-35.

Reardon, T. (1997). Using evidence of household income diversification to inform study of the rural nonfarm labor market in Africa. *World Development* 25:735:747.

Reardon, T., P. Matlon, and C. Delgado. (1988). Coping with household food insecurity in drought affected areas of Burkina Faso. *World Development* 16:1065-1074.

Reddy, K.C. (1988). *Strategies Alternatives pour la Production de Mil/Niebe Pendant l' Hivernage* (in French). Fascicule No. 1, INRAN, Niamey, Niger.

Reddy, K.C. and L. Samba. (1989). Farming systems research: Some suggestions for agronomists. Paper presented at the Regional Agronomists Training Workshop, Bamako, Mali, organized by SAFGFRAD/ICRISAT/USAID, Sept. 18-30.

Sanders, J.H. (1989). Agricultural research and new technology introduction in Burkina Faso and Niger. *Agricultural Systems* 30:139-154.

Sanders, J.H. (2000). The economic potential of the agricultural sector in the Sahel. In *Proceedings, 12th Danish Sahel Workshop*, January 3-5, Hanne Adriansen, ed. A. Reenberg and I. Nielson. Occasional Paper No. 11, pp. 39-51. Copenhagen, Denmark: Sahel-Sudan Environmental Research Institute.

Sanders, J.H. and M. Ahmed. (2001). Developing a fertilizer strategy for Sub-Saharan Africa. In *Sustainability of Agricultural Systems in Transition*, ASA Special Publication No. 64. Madison, WI: American Society of Agronomy, Crop Science Society of America, and Soil Science Society of America.

Sanders, J.H. and D.E. McMillan. (2001). *Agricultural Technology for the Semiarid African Horn. Vol. 2: Country Studies: Djibouti, Eritrea, Ethiopia, Kenya, Sudan, and Uganda*. IGAD/INTSORMIL/USAID-REDSO. Lincoln, NE: INTSORMIL, University of Nebraska.

Sanders, J.H. and J. Vitale. (1997). Institutional and economic aspects of fertilizer use in West Africa: Experiences and perspectives. Paper presented at Regional Workshop on Soil Fertility Management in West African Land Use Systems, University of Hohenheim and ICRISAT, Niamey, Niger.

Sanders, J.H., B.I. Shapiro, and S. Ramaswamy. (1996). *The Economics of Agricultural Technology in Semiarid Sub-Saharan Africa*. Baltimore, MD: Johns Hopkins University Press.

Sanders, J.H., J. Nagy, and S. Ramaswamy. (1990). Agricultural technologies for the Sahelian Countries: The Burkina Faso case. *Economic Development and Cultural Change* 39:1-22.

Scott-Wendt, J., L.R. Hossner, and R.G. Chase. (1988). Soil chemical variability in sandy ustalfs in semiarid Niger, West Africa. *Soil Science* 145: 414-419.

Scott-Wendt, J. (1989). Rejuvenation of desertified sandy soils in Sahelian West Africa. Proposal submitted to USAID/Mission to Niger.

Seckler, D. (1994). Nutrient mining and a fertilizer strategy for Sub-Saharan Africa. Mimeo. Morrilton, AR: Winrock International.

Shapiro, B.I. and J.H. Sanders. (1998). Fertilizer use in semiarid West Africa: Profitability and supporting policy. *Agricultural Systems* 56:467-482.

Shapiro, B.I. and J.H. Sanders. (2002). Natural-resource technologies for semiarid regions of Sub-Saharan Africa. In *Towards Sustainable Sorghum Production, Utilization, and Commercialization in West Central Africa*, ed. C.B. Barrett, F.M. Place, and A. Aboud, Tucson, AZ.: CAB International of North America, pp. 261-274.

Shapiro, B.I., J.H. Sanders, K.C. Reddy, and T. Baker. (1993). Evaluating and adapting new technologies in a high risk agricultural system–Niger. *Agricultural Systems* 42:153-171.

Vierich, H. and W.A. Stoop. (1990). Changes in west African savannah agriculture in response to growing population and continuing low rainfall. *Agriculture, Ecosystems, and Environment* 31:115-132.

Vitale, J.D. (2001). *The Economic Impacts of New Sorghum and Millet Technologies in Mali*. PhD dissertation, Department of Agricultural Economics, West Lafayette, IN: Purdue University.

Vitale, J.D. and J.H. Sanders. (2002). New markets and technological change for the traditional cereals in Sub-Saharan Africa: The Malian case. West Lafayette, IN: Purdue University, Department of Agricultural Economics, Staff paper.

Williams, T.O., J.M. Powell, and S. Fernandez-Rivera. (1993). Manure utilization, drought cycles, and herd dynamics in the Sahel: Implications for cropland productivity. Paper presented at the Nutrient Cycling Workshop, November 26-30, ILCA, Addis Ababa, Ethiopia.

World Bank, (1989). *Sub-Saharan Africa: From Crisis to Sustainable Growth*. Washington, DC: World Bank.

Wright, P. (1985). Water and soil conservation by farmers. In *Appropriate Technologies for Farmers in Semiarid West Africa*, ed. H.W. Ohm and J.G. Nagy. West Lafayette, IN: Purdue University, Department of International Programs.

Brazilian Agriculture:
The Transition to Sustainability

Robert M. Boddey
Deise F. Xavier
Bruno J. R. Alves
Segundo Urquiaga

SUMMARY. Brazil has a total area of 850 million ha, of which 90% is within the tropics. Historically, the system of exploitation of the land for agriculture and forestry was based on land clearing, cultivation for some years and then moving on to new areas. This process often left degraded areas behind, especially in the mountainous areas of the country. With

Robert M. Boddey, Bruno J. R. Alves, and Segundo Urquiaga are affiliated with Embrapa Agrobiologia, Caixa Postal 74.505, Seropédica, Rio de Janeiro, 23890-000, Brazil.

Deise F. Xavier is affiliated with Embrapa Gado de Leite, Rua Eugênio do Nascimento 610, Dom Bosco, 36038-330, Juiz de Fora, Minas Gerais, Brazil.

The authors would like to thank Alexander Resende and Claudia Sisti for contributing data included in this review, Verônica Reis for helping them to find some of the statistical data on sugarcane

The three authors from Embrapa Agrobiologia gratefully acknowledge research fellowships from the Brazilian National Research Council (CNPq).

This article is dedicated to the memory of Dr. Johanna Döbereiner, who died on 5th October 2000. Dr. Johanna dedicated almost 50 years of her life working on biological nitrogen fixation and its utilization in low input sustainable agricultural systems at the institute she founded, now known as Embrapa Agrobiologia. She strongly influenced and motivated all of the present authors and is sadly missed by all.

[Haworth co-indexing entry note]: "Brazilian Agriculture: The Transition to Sustainability." Boddey, Robert M. et al. Co-published simultaneously in *Journal of Crop Production* (Food Products Press, an imprint of The Haworth Press, Inc.) Vol. 9, No. 1/2 (#17/18), 2003, pp. 593-621; and: *Cropping Systems: Trends and Advances* (ed: Anil Shrestha) Food Products Press, an imprint of The Haworth Press, Inc., 2003, pp. 593-621. Single or multiple copies of this article are available for a fee from The Haworth Document Delivery Service [1-800-HAWORTH, 9:00 a.m. - 5:00 p.m. (EST). E-mail address: docdelivery@haworthpress.com].

modern agricultural technology, crops can be successfully grown in virtually any region of the country. The current challenge for Brazil is to feed its population and provide agricultural surpluses for the growing export markets, while preserving its rich and biologically-diverse native vegetation which still covers almost half of the country. The objective of this review is to trace the history of agricultural activity in this country, and to assess the sustainability of the cropping and pasture systems which today occupy the largest areas. At present approximately 50 million ha are under annual and perennial crops, while almost twice this area is under pastures. These pastures, predominately *Brachiaria* spp., are mostly in a degraded state due to lack of fertilization and over grazing. The various options available to recover these pastures or convert these areas for sustainable cropping are explained. In recent years, increasing proportions of soybean, wheat, and corn are produced under zero tillage which favors the conservation of soil organic matter. This not only radically reduces the risk of erosion but also increases the capacity of the soils to retain nutrients and water. Small holders who represent a considerable fraction of Brazil's food crop production, generally do not have access to fertilizers or other agricultural chemicals. They obtain very low yields and their farming practices exhaust the soil of nutrients. The Brazilian sugarcane industry is the largest in the world and recent changes in the management of this crop and its impact upon sustainability issues are also discussed. Brazil also has vast areas of degraded pastures and abandoned hillsides that can be used for agricultural expansion. This would prevent further destruction of native vegetation and its accompanying biological diversity for agriculture. *[Article copies available for a fee from The Haworth Document Delivery Service: 1-800-HAWORTH. E-mail address: <docdelivery@haworthpress.com> Website: <http://www.HaworthPress.com> © 2003 by The Haworth Press, Inc. All rights reserved.]*

KEYWORDS. Coffee, soybean, sugarcane, tropical pastures, zero tillage

INTRODUCTION

Brazil is the fifth largest country in the world with an area of 850 million ha. Unlike other large countries, the temperature and rainfall regimes combined with modern agricultural technology, permits that 90% of the area of Brazil could be used for cropping, forestry or pastures. At present, approximately 50 million ha are used for cropping (IBGE, 2001) and 100 million ha are occupied by planted pastures (Zimmer and Euclides-Filho, 1997). More than half of the country is covered by native vegetation, of which approximately two-thirds is the Amazon rain forest. The traditional model of agricultural activity in Brazil has been to clear and crop the land, often until the soils were severely depleted in nutrients and organic matter (OM). When the capacity of the soil to support cropping became severely diminished, the general rule was to abandon the land, often for use as rough pasture, and move onto to clear new areas. Brazil's population has increased from 70 million in 1960 to approximately 170 million today, and the areas of abandoned/degraded land now far exceed those used for productive agriculture. It is obvious that this traditional model of agricultural activity is no longer viable. If Brazil wishes to preserve its large areas of native vegetation and the diversity of fauna and flora within it, it must develop sustainable agricultural systems which produce high crop/pasture yields without exhausting the soil resource.

Sustainable agriculture has been defined as "the successful management of resources for agriculture to satisfy changing human needs while maintaining or enhancing the quality of the environment and conserving natural resources" (FAO, 1989). Many other definitions of "sustainable agriculture" have been proposed (for discussion of those most applied to tropical/developing countries see Reijntjes, Haverkort, and Waters-Bayer, 1992; Eswaran, Virmani, and Spivey, 1993; Scholes et al., 1994). All of these definitions encompass the five basic principles of soil management proposed as essential for sustainable agricultural production by Greenland (1975): (i) chemical nutrients removed by crops must be replenished, (ii) the physical condition of the soil must be maintained, (iii) there must be no build up of weeds, pests or diseases, (iv) there must be no increase in soil acidity or toxic elements, and (v) soil erosion must be controlled to be equal or less than that the rate of soil genesis.

More recent definitions have considered concerns such as agrochemical pollution, loss of biological diversity, lack of social equity, well-being and gender issues. No matter what definition of "sustainable agriculture" is adopted, the exploitation of land for crop and animal production in Brazil has, until recently, been a model of "non-sustainability."

In this review, we discuss the historical evolution of Brazilian agriculture and its degree of "sustainability" with respect to the two most traditional export crops, sugarcane (*Saccharum* spp.) and coffee (*Coffea robusta* L.). We then discuss those crops and pastures which today are most widely cultivated in Brazil, and changes in management that have occurred in recent years which have had significant positive and negative impacts upon the development of sustainable crop and animal production systems.

HISTORY OF AGRICULTURE IN BRAZIL

When the Portuguese commander Pedro Alvarez Cabral first sighted Brazil on Easter Sunday in the year 1500, the Atlantic coast of the land he christened

"Santa Cruz" was covered with a resplendent tropical forest which occupied approximately 100 million ha, an area 12 times that of Portugal. The population of the Tupi and other indigenous people at that time has been estimated as high as 9 individuals per km^2 in the Atlantic forest region (Dean, 1995). This suggests that the total population of this region was approximately 9 million, although most other authors give estimates between 3 and 6 million.

Until 40 years ago, when the country's capital was relocated from Rio de Janeiro to Brasília, it was the Atlantic forest region that hosted virtually all of Brazil's agricultural and pasture production. However, the history of the human devastation of this forest region, which to a large degree was the result of completely non-sustainable agricultural activities, has been recently described in detail by Dean (1995).

The Tupi people cultivated this region by cutting and burning the forest (slash and burn or swidden agriculture). In the 1st century or so after the first Europeans arrived in Brazil, the total population decreased partly because of incitement by the Europeans of warfare amongst the Tupi to yield slaves and, even more significantly, because of introduction of fatal European diseases amongst the indigenous peoples. Consequently, the pressure on the land was reduced. Later, when export crops were introduced and expanded, larger areas became more intensively used, the cycle of slash and burn was shortened, or the land was brought under permanent agriculture and became degraded.

Sugarcane and coffee were historically the most important export crops; but today, in terms of value and production they have been overtaken by soybean (*Glycine max* L. Merr.). However, in terms of area, planted pastures are the most important component of agricultural land-use. The following sections will discuss the issues surrounding the sustainability of these crop and pasture production systems.

SUGARCANE

The first important export crop introduced to Brazil was sugarcane. It was introduced from the island of Madeira by the Portuguese colonizing expedition of 1532 (Machado, da Silva, and Irvine, 1987). They founded the settlement of São Vincente on the coast of what is now São Paulo State and in the following year the first sugar mill was built. Sugar production was estimated to be approximately 10,000 Mg in 1600 and did not exceed 20,000 Mg until the start of the 19th century (Galloway, 1989). Most of the cane was initially planted in areas cleared from the forest, allowed to ratoon once or twice, and then replanted. This cycle of planting was repeated once or twice, then as soil fertility declined the field was abandoned and new areas of forest were cleared. It is estimated that until 1700 the area under sugarcane plantation and subsequently abandoned land may have totalled 100,000 ha. A further 120,000 ha may have been deforested for fuel for the mills (Dean, 1995). Because of the long period of cropping (10-15 years), very little of this area would have reverted to mature forest. The area probably would have remained as rough pasture or shrubs.

Sugarcane production expanded in the mid-19th century and more than 100,000 Mg sugar year^{-1} was produced. By the 1970s, sugarcane was planted on approximately 1.5 million ha and total sugar production was 5 to 6 million Mg year^{-1}. The government launched the "ProÁlcool" program in the mid-1970s to produce ethanol from cane juice as a biofuel. This program grew rapidly (Homewood, 1993) and today almost 5 million ha are planted to cane, over half of which is processed to produce approximately 13 billion liters of alcohol annually. Currently all gasoline in Brazil, contains 20 to 22% ethanol, and other octane-enhancing additives such as tetraethyl lead or MTBE (methyl *tert*-butyl ether) have been totally substituted. Furthermore, approximately 3 million vehicles are powered by hydrated (95%) ethanol. Details of the environmental advantages of this biofuel program have been given by Boddey (1993) and Macedo (1998).

Sugarcane is a C_4 graminaceous crop that produces a large amount of biomass. The average yield of sugarcane in Brazil is currently 67 Mg fresh cane ha^{-1} (equivalent to approximately 20 Mg dry matter), which is all taken to the mill. A further 10 Mg of senesced leaves may be left in the field as a trash blanket. This aboveground input of 10 Mg ha^{-1} of plant material, along with a similar quantity of plant root material (Souto et al., 1993), could help in maintaining soil organic matter (SOM) levels. However, the problem is that the trash is often burned off prior to harvest because the immense volume of residue makes manual harvesting very difficult and laborious.

Over the last century there has been considerable debate over the merits and demerits of burning cane. Cane grown in fields that have been burned must be taken to the mill within 48 h in order to prevent losses in sugar content. Traditionally burned cane was given a lower price at the mill. Some specialists lamented the detrimental effects of burning on SOM levels. However, others claimed that it was essential to control pests and diseases. Therefore, in the 19th century the fields were often burned after harvest (Valsechi, 1951). Pre-harvest burning was rarely practised prior to the 1940s, but as labor became more expensive, pre-harvest burning became common practice. However, over the last few years the heavy smoke and soot produced by the burning has encouraged the environmentalists, especially in the State of São Paulo, to lobby for the abolition of cane burning. Initially, there was a strong resistance by the plantation/mill owners ("Usineiros"), but with the introduction of mechanized green-cane harvesters, machine harvesting of unburned cane is expanding rapidly on flat land. However, mechanization is creating social problems in terms of rural unemployment because one harvesting machine replaces approximately 80 field workers. Today 14% of the cane is mechanically harvested and this proportion could increase to 60%. The remaining cane is grown on sloping land unsuitable for machine harvesting (CanaWeb, 2001).

Few long-term studies have been conducted to assess the impact of residue conservation on the sustainability of cane production. In 1983, a team at Embrapa-Agrobiologia initiated an experiment which investigated the effects of burning, and the application of nitrogen (N) fertilizer and distillery waste (vinasse) on cane yields and soil properties at a plantation in Pernambuco in NE Brazil. The results of the first cycle (first year crop + 7 ratoons) showed that residue conservation increased cane yield, crop N accumulation, and soil pH (Oliveira, Urquiaga, and Boddey, 1994). This experiment was then repeated on the same plots and continued until 1999 (first year crop + 5 ratoons). The effect of pre-harvest burning on cane yields was evident only after the second ratoon crop in both cycles (Figure 1). In the first cycle, for the last five ratoons, cane yields were 24% higher (12.5 Mg cane ha^{-1} $year^{-1}$) where trash was preserved than in plots where it was burned. In the second cycle, the plots where residues were conserved produced on average 45% more cane (16 Mg ha^{-1} $year^{-1}$). It was found that the conservation of cane residues had two main benefits: conservation of soil moisture and the preservation of SOM. The data from this trial showed that after 16 years the C content of the top 10 cm of soil under the burned plots was 1.16% compared to 1.33% where trash was conserved. In the 10-20 cm layer, these values were 1.10 and 1.28%, respectively, together representing a difference of 4.28 Mg C ha^{-1}, or a mean increase of 270 kg C ha^{-1} $year^{-1}$.

The annual vinasse addition (80 m^3 ha^{-1}) contained approximately 40 kg N, 200 to 300 kg K and 3.5 kg P. When the "ProÁlcool" program was initiated, huge quantities of vinasse were produced and almost all of it was disposed directly into local rivers. However, the State and Federal authorities forced the mill owners to recycle this liquid onto the fields because of the serious pollution problem it was causing. In response, the mill owners constructed long concrete canals to feed fields close to the distilleries, while tanker trucks carried the waste to outlying areas. Today virtually all vinasse is recycled, and thus little K fertilizer is required for sugarcane cultivation. However, fields further than 10 km from the mills generally require additional fertilizer to maintain soil K levels.

Mean annual N fertilizer use on sugarcane in Brazil is 60 kg N ha^{-1}. This is very low in comparison to most other cane-producing countries where annual N applications range from 150 to 300 kg N ha^{-1} (Urquiaga, Cruz, and Boddey, 1992). Although some N is returned in the vinasse, the burning of the cane leaves little or no N in the field. This suggests that the N balance associated with the crop is negative and that yields should fall within a few years due to N

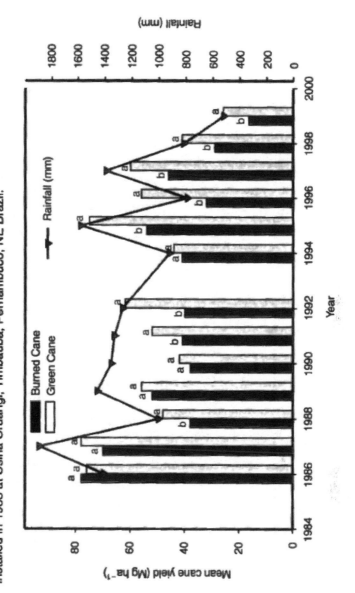

FIGURE 1. Effect of pre-harvest burning of sugarcane on cane yields (Mg cane ha^{-1}) from the first ratoon crop of the first cycle (1986) until the fifth ratoon crop of the second cycle (harvested 1999). Experiment installed in 1983 at Usina Cruangi, Timbaúba, Pernambuco, NE Brazil.

deficiency. That this does not commonly occur led some Brazilian scientists to suggest that sugarcane could benefit from N_2-fixing bacteria associated with the plants (Döbereiner, 1961; Ruschel, Henis, and Salati, 1975). Subsequent investigations led to the discovery of several hitherto-unknown N_2-fixing bacterial species that were found to colonize the interior of cane plants (Cavalcante and Döbereiner, 1988; Baldani et al., 1997). N balance and ^{15}N techniques have confirmed that Brazilian varieties of sugarcane can benefit considerably from contributions of biological N_2 fixation (BNF), in experimental conditions (Urquiaga, Cruz, and Boddey, 1992) and in commercial plantations (Oliveira, Urquiaga, and Boddey, 1994; Boddey et al., 2001).

Therefore, it can be concluded that conservation of the residues and recycling of vinasse can sustain highly productive sugarcane cultivation in Brazil. These processes also help in reducing fertilizer input and maintaining SOM levels.

COFFEE

Most people associate Brazil with coffee more than any other crop or export product. Coffee was introduced to Brazil in the late 18th century and it is estimated that production in 1790 was a little over 1 Mg of beans (Dean, 1995). Coffee is a crop that does not tolerate waterlogging but requires a wet climate (1300 to 1800 mm rainfall year^{-1}). It was soon found that the hill slopes in the regions near Rio de Janeiro were ideal for its cultivation. To provide nutrients for this crop, the practice adopted was to clear virgin forest with its thick layer of plant litter. In many other parts of the world coffee was grown under the canopy of larger trees, but in Brazil the forest was cleared and burned, and the seedlings were planted in the remaining ash. The result was that after one cycle of coffee (10 to 15 years) the soils were depleted of nutrients and unable to sustain coffee production. Soil erosion on the steep slopes was severe because it was customary to plant coffee in wide rows (4 to 5 m) and to clean all inter-row vegetation. Hence, new areas of virgin forest were cleared for renewal of the plantations. As Dean (1995) laments "Thus coffee marched across the highlands, generation by generation, leaving nothing in its wake but denuded hills." By the mid-19th century, Brazil produced 70% of the world's coffee. Dean (1995) estimates that in the 1st century (1788-1888) of coffee production, a minimum of 720,000 ha of virgin forest were destroyed along with burning of 3 million Mg of forest biomass. After most of the Paraíba valley inland from Rio had been devastated, coffee cultivation was moved to the hilly areas of Minas Gerais and São Paulo, and later to Espírito Santo. In the second half of the 20th century large areas of forest of northern Paraná (the State immediately to the south of São Paulo) were also cleared for coffee planting.

Today coffee occupies an area of 2.35 million ha making it Brazil's sixth largest crop on an area basis (Table 1). Brazil is still the world's largest coffee producer (1.83 million Mg out of world total of 7.09 million Mg) (FAO, 2001). Much of the coffee is still grown on hill slopes, but recently the crop is also being cultivated in relatively flat and well-drained areas in the Cerrado.

Huge amounts of lime, fertilizers and pesticides are generally applied to coffee. It is estimated at present that the mean annual addition of fertilizer is 407 kg ha^{-1} (Anonymous, 1998). The use of fertilizers has enabled the coffee plantations to be fixed in one location without having to clear new land. Therefore, the devastating cycle of forest clearing has stopped. However, the excessive use of agro-chemicals on this crop has created negative environmental effects. New insect pests and diseases such as "amarelinha" (*Xylella fastidiosa*) have become more difficult to control (Zambolim and Ribeiro de Vale, 2000).

Organically produced coffee is becoming an economically viable option, despite the lower yields, because of the higher price for the product and the lower input costs. There is also a movement to grow coffee under tall trees, as is the common practice in Central America and Colombia. However, the coffee bean yields are lower under such a system but there is a considerable recycling of nutrients from both the litter of the large trees and the coffee bushes themselves. This system protects the soil from erosion and greatly decreases nutrient loss. However, traditional high input production systems still dominate coffee growing in Brazil and at present there seems only slow progress towards management systems that are less polluting and more sustainable.

TABLE 1. Area and average yields of the major crops of Brazil in 2000.

Crop	Area (ha × 10³)	Yield (kg ha¹)
Soybean	13,618	2,400
Corn	11,614	2,730
Sugarcane	4,819	67,507*
Phaseolus/Vigna beans	4,318	696
Rice	3,671	3,036
Coffee (hulled beans)	2,347	1,240
Cassava	1,707	13,550
Oranges	1,008	126,985**
Cotton	809	2,368
Cacao	681	262

* kg milled cane ha^{-1}
** Orange yield in number of fruits ha^{-1}

SOYBEAN

The first report of soybean cropping in Brazil was in 1882 in the northeastern state of Bahia. Later, Japanese immigrants planted it in São Paulo as early as 1908 (Bonetti, 1981). However, it was grown mainly on an experimental basis until the 1940s. The first report of significant soybean production was from the State of Rio Grande do Sul in 1941 (total area planted 7,651 ha with a production of 9,146 Mg) (Medina, 1981). Expansion of the soybean area was initially slow. However, with a government campaign in the 1960s, the area under soybean rapidly increased because it was an ideal summer crop for rotation with wheat (*Triticum aestivum* L.) grown in winter. Soybean production since 1960 has increased steadily (Figure 2), spreading from Rio Grande do Sul, through Santa Catarina and Paraná and later into the tropical savanna regions of Mato Grosso, Goiais, Minas Gerais, western Bahia and most recently towards Amazonia, especially in the State of Mato Grosso (Table 2). Over the last five years, the area planted to soybean (13.5 million ha in 2000) has overtaken that of corn (*Zea mays* L.–11.6 million ha). Soybean now occupies more area than any other crop in Brazil and its products now constitute the country's foremost agricultural export. Today Brazil produces over 20% of the world's soybean crop (32.7 million Mg) second only to the USA (75.4 million Mg). Average soybean yields have reached 2,400 kg ha^{-1} which is only slightly lower than that of the USA (2,561 kg ha^{-1}) (FAO, 2001).

Soybean is invariably grown in rotation with other winter crops, and in the 1960s and early 1970s, the winter crop in rotation was usually wheat. Thus, the land had to be prepared twice a year for the rotation. The frequent ploughing and disking reduced SOM levels and often produced plough pans which restricted root growth and water acquisition by the crops (Juo and Lal, 1977; Guérif et al., 2001). In warm climates the loss of SOM due to intensive cultivation can be drastic within a few years, especially in sandy soils. For example, in the savanna region of western Bahia, where cultivation started in the 1980s, da Silva, Lemainski and Resck (1994) found that annual planting of soybean (with a dry season fallow) in virgin soils (Oxisols) with clay contents of less than 20%, reduced SOM content and cation-exchange capacity (CEC) of the soil by 50% within 3 and 5 years, respectively. In Paraná especially, after a few years of soybean/wheat rotation, crop yields began to decrease because of SOM depletion, and in many areas there were serious soil erosion problems (Gassen and Gassen, 1996).

In the early 1970s, a few innovative farmers with help from the Agronomic Institute of Paraná (IAPAR), the Embrapa wheat center, and the pesticide companies, started to experiment with zero tillage (ZT) agriculture. Initially there were many obstacles in ZT cultivation of soybean. Such obstacles included the lack of suitable direct drilling machines, and problems with weed

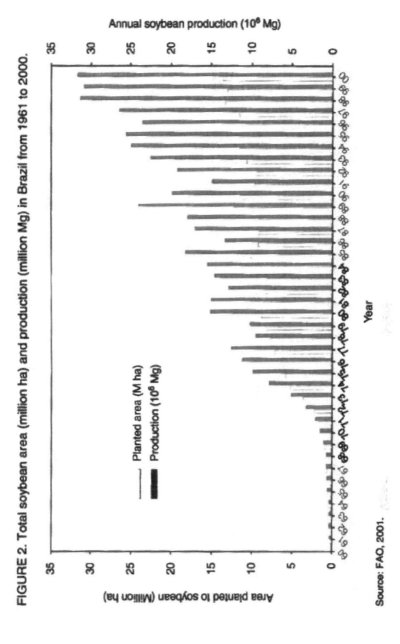

FIGURE 2. Total soybean area (million ha) and production (million Mg) in Brazil from 1961 to 2000.

Source: FAO, 2001.

TABLE 2. Area and average yield of soybean in different Regions/States of Brazil in 2000.

Region/State	Harvested area (ha)	Yield (kg ha^{-1})
North-East	846,776	2,419
Marahão	178,416	2,463
Piauí	40,004	2,524
Bahia	628,356	2,400
South East	1,134,564	2,313
Minas Gerais	600,054	2,391
São Paulo	534,510	2,225
South	6,098,532	2,054
Paraná	2,852,585	2,509
Santa Catarina	212,412	2,470
Rio Grande do Sul	3,033,535	1,593
Central West	5,528,377	2,794
Mato Grosso do Sul	1,103,301	2,261
Mato Grosso	2,897,728	3,022
Goiás	1,490,766	2,745
Federal District	33,582	2,733
TOTAL	13,608,000	2,399

Source: IBGE 2001.

and pest control. However, these barriers were gradually overcome with the development and distribution of appropriate machinery and chemicals. The first result of zero tillage was drastic reduction in soil erosion. A study by Mondardo (1981) on a wheat/soybean rotation in the Campos Gerais region north of Paraná (which can be regarded as the birthplace of ZT in Brazil) on a medium textured red podzolic soil (Typic Hapludult) with a 4% slope, showed that annual soil losses were almost 95% lower when ZT was used instead of conventional tillage (CT) (Table 3). It has been estimated in a recent review that in the areas where ZT had been introduced in Brazil, soil loss was reduced by 75% and water run off was lowered by 25% (de Maria, 1999).

Through the 1970s and 80s, the use of ZT spread slowly through the southern states of Brazil (Figure 3). To facilitate control of crop diseases and pests, the wheat/soybean rotations were diversified. Wheat was often replaced by other winter crops such as oats (*Avena sativa* L. or *A. strigosa* L.), oil radish (*Raphanus sativus* L. var. *oleifera*) or leguminous green manures (e.g., vetch–*Vicia sativa* L., lupins–*Lupinus albus* L.) and this substitution was encouraged by the low price of imported wheat. Oats or vetch/corn-wheat/soybean, wheat/soybean-oats/soybean-lupins/corn became typical 2 or 3 year rotations, respectively. In the early 1990s, the adoption of the ZT system rapidly increased

TABLE 3. Soil loss due to runoff in soybean-wheat rotation under conventional and zero tillage on a 4% slope in Bela Vista de Paraiso, North Paraná.

Crop	Conventional tillage	Zero tillage
Soybean	2,889	179
Wheat	267	0
Total	3,156	179
%	100	5.7

Soil Type: Typic Hapludult.
Adapted from Mondardo (1981).

and spread into the Cerrado region (Figure 3). The principal driving force behind this expansion is that far fewer mechanical field operations are required and this considerably reduces production costs while obtaining similar or higher crop yields.

Zero tillage associations with names such as "Clube da Minhoca" (the Earthworm Club) and "Amigos da Terra" (Friends of the Land) have been organized by enthusiastic farmers. These associations are very active in spreading the "gospel" of ZT. It is surprising that many farmers in these regions plant winter crops such as oats, oil radish, vetch or lupins which are not harvested for sale, but utilized solely to benefit the subsequent soybean or corn summer crop. Vetch and lupins are principally used prior to the corn crop and can often substitute as much as 40 to 80 kg ha^{-1} of fertilizer N (Amado, Mielnikzuck, and Fernandes, 2000). Oats and oil radish have deep penetrating roots which help break through plough pans and help to recover nutrients (principally N and P) that have leached down the profile. The capacity of oil radish or oats to recover soil N is such that they often accumulate more of this element than N$_2$-fixing green legumes (Lara-Cabezas, 1999).

In ZT the soil is left physically undisturbed and all crop residues are left on the soil surface. The increased OM on the soil surface has been shown to increase the microbial biomass and soil faunal populations (Colozzi Filho, Balota, and Andrade, 1999). Recently Embrapa researchers have compared the long-term effect of CT and ZT of different crop rotations on soil C stocks to a depth of 1 meter. The experiments consisting of three different rotations of 1, 2, and 3 years, respectively, were planted in 1986 at the Embrapa-Wheat center (Sisti, 2001).

The rotations were wheat/soybean, wheat/soybean-vetch/corn and wheat/soybean-oats/soybean-vetch/corn. Soil samples (0-100 cm) were taken in 1999 and evaluated for bulk density, and C and N content (Figure 4). The C stock in the soils, corrected for the different degrees of soil compaction

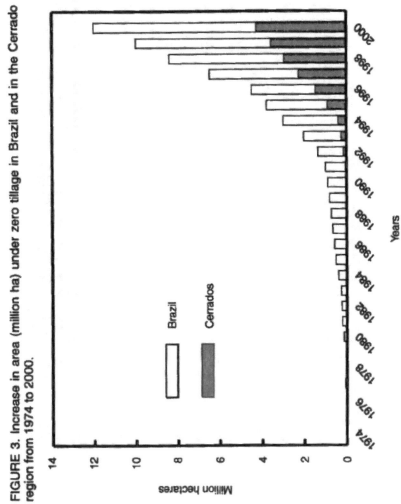

FIGURE 3. Increase in area (million ha) under zero tillage in Brazil and in the Cerrado region from 1974 to 2000.

Adapted from Plantio Direto no Cerrado—monthly magazine July 2000.

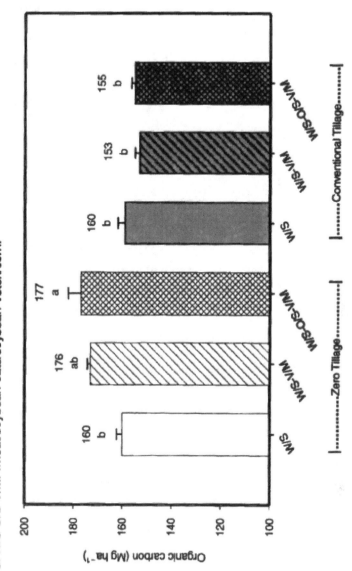

FIGURE 4. Effect of zero tillage versus conventional tillage on the change in stocks of soil carbon (0-100 cm) on three different crop rotations: 1. W/S: wheat/soybean, 2. W/S-V/M: wheat/soybean-vetch/corn, 3. W/S-O/S-V/M: wheat/soybean-oats/soybean-vetch/corn.

using the technique described by Neill et al. (1997), showed that the wheat/soybean rotation had maintained the same C stock under both CT and ZT. However, the introduction of a winter legume into the rotation promoted a greater accumulation of organic C, especially when combined with crops such as oats and corn that produced large amounts of high C:N ratio residues (both roots and trash).

Oxisols and Ultisols dominate much of the cropping regions of both southern Brazil and the Cerrado region. In these soils, the OM is responsible not only for aggregate stability and the water-holding capacity but also for a large proportion of the CEC. As ZT conserves SOM, it is to be expected that this system will favor long-term sustainable cropping. Generally the monitoring of C stocks is only possible in long-term studies, but as C:N ratios of SOM only vary within a very restricted range, short-term tracing of N fluxes can yield information on long-term trends. It can be concluded from a study by Zotarelli (2000) on an Oxisol that CT stimulates mineralization of the soil organic N to a level that inhibits N_2 fixation in initial stages of soybean growth. Thus, the total BNF contribution under CT is generally lower than under ZT. Also the inclusion of a winter legume in the rotation maintained soil N and C reserves.

Thus, the introduction and rapid expansion of the ZT system in Brazil has brought great advantages to the maintenance of soil fertility and integrity. Although, initially the system requires extra nutrients, principally N, to build up SOM (Boddey et al., 1997), the fertilizer requirement will subsequently decrease as a result of the increased SOM. Gassen and Gassen (1996) report that after some years, demand for P fertilizer is lower (up to 50%) under ZT because unlike CT the phosphate is not mixed into a large volume of P-fixing soil minerals. However, there are concerns about the widespread adoption of ZT systems because they generally require higher inputs of herbicides. The trend today is towards more specific herbicides that are less toxic to non-target life forms. As farmer skills improve, it can be expected that herbicide application rates will decrease. Thus, it is important that the great advantages that ZT offers for sustainable land use will not be marred by associated environmental pollution and health problems of farm workers.

PLANTED PASTURES

The total cropped area in Brazil is approximately 50 million ha, but additionally a huge area is also covered by pastures planted to grasses of African origin. A recent survey concluded that there were 49.5 million ha of such pastures in the Cerrado region alone (Sano, Barcellos, and Bezerra, 2000). Estimates for the Amazon region suggest that of the 47 million ha of forest cleared (Fearnside, 1997) approximately 20-24 million ha are under grazed pastures

(Neto and Dias Filho, 1995; Fearnside and Barbosa, 1998). The other main region where there are large areas of such pastures is the Atlantic forest area. São Paulo State (area 25 million ha) alone is reported to have 7 million ha of *Brachiaria*. The total area under these pastures in the states of Rio de Janeiro, Espirito Santo, Bahia, and other coastal areas of the Northeast is approximately 20 million ha. In total, therefore, Brazil has over 90 million ha (over 10% of the entire country) of pastures planted with African grasses, of which approximately 80% are of the three *Brachiaria* species; *B. decumbens*, *B. humidicola*, and *B. brizantha* (Macedo, 1995).

Many grass species of African origin were introduced to Brazil during the 18th century, mostly as straw bedding of the slave ships. These accidental transfers continued and by the late 20th century up to 40 different species of African grasses were found in Brazilian pastures (Parsons, 1972). The first species to be exploited for pasture were molasses grass (*Melinis minutiflora*), jaraguá (*Hyperrhenia rufa*), pangola or digit grass (*Digitaria decumbens*), and Guinea grass (*Panicum maximum*). When planted on newly deforested land or areas abandoned from cropping, these species can sustain animal production only for a year or two. In the Atlantic forest region today, the abandoned coffee lands and other deforested hillsides are generally covered by spontaneous invasion of these grasses. These grasses inhibit the return of pioneer forest plants and are prone to fire. Hence, there are problems of soil erosion especially when heavy rains follow dry periods. These degraded lands are not classified as pastures, but they cover hundreds of thousands of ha of hillsides that were once covered by tropical forest.

The new "miracle" grass *B. decumbens* was introduced in the 1970s (Macedo, 1995). This coincided with several important events such as the opening up of the Cerrado region, the acceleration of deforestation in Amazonia and the building of the Atlantic coastal highway (BR 101) from the north of Espirito Santo through the previously-untouched forested tablelands of the south of Bahia. In the Amazon, the newly deforested land was planted to Guinea grass which initially thrived on the nutrient-rich ash, but it was soon replaced by *B. decumbens* because of the persistence of *B. decumbens*. In the wet climate of the Amazon, however, spittle bug (e.g., *Deois* and *Zulia* spp.) became a serious problem and land owners favored *B. humidicola* over *B. decumbens* because it was less sensitive to this pest and to waterlogging (Neto and Dias Filho, 1995). In 1984, the Embrapa Beef Cattle center released the variety 'Marandú' of *B. brizantha* which was resistant to spittle bug attack and in recent years this variety has become popular especially in more humid areas.

Currently, just these three species (*B. decumbens*, *B. humidicola*, and *B. brizantha*) and only one genotype of each dominate the vast extensively grazed areas of Brazil. In the Cerrado, after the land was cleared and planted to a grain crop, usually dryland rice (*Oryza sativa*), *Brachiaria* grasses were planted and they thrived on the residual nutrients from the fertilizer applied to the grain crop. Live weight gains (LWG) of cattle were usually increased at least 10 fold on *Brachiaria* pastures compared to the native savanna grasses. However, productivity of the *Brachiaria* pastures eventually declined because no further fertilizer was applied and animal stocking rates were high. Although *Brachiaria* grasses were able to maintain satisfactory animal performance for longer periods than other grasses, their vigour decreased after about five years and animal production fell. Bare patches appeared in the sward whereas, in the Amazon the main characteristic of this pasture decline was the invasion by non-palatable "weed" species that finally reverted the pastures to secondary forest (Buschbacher, Uhl, and Serrão, 1988). Fearnside and Barbosa (1998) estimate that approximately 28% of the deforested area (~13 million ha) of Amazonia is covered by secondary forest derived from abandoned pastures. In the Cerrado region, areas of degraded pasture generally remain with very sparse vegetation cover and are susceptible to rain and wind erosion. No reliable estimates are available of the proportion of these pastures which can be considered "degraded," but it has been estimated to be between 60 and 80% (Macedo, 1995; Barcellos, 1996).

The process of pasture establishment followed by poor management leading to pasture decline, is similar to the non-sustainable land-use model adopted for coffee in the 19th century. However, most of the degraded *Brachiaria* pastures are not on steep land. In the Cerrado region, where there is only limited invasion by "weed" species, the process can be reversed by pasture recovery techniques. The soil can be decompacted by ploughing, a seed bed can be prepared, fertilized, and then reseeded. This process is expensive, and can cost approximately US$ 350 to 400 ha^{-1} (Kichel, Miranda, and Zimmer, 1998). However, production costs can be offset in some areas by using deep ploughing followed by fertilization with P and K, and planting of rice or corn under-sown with *Brachiaria*. Such a system ("Sistema Barreirão"–Kluthcouski et al., 1991) has been developed by researchers at the Embrapa Rice and Bean center and has found wide acceptance amongst farmers who have the necessary expertise, access to loans and to the planting and harvesting equipment.

Recent research has shown that this pasture decline is caused by the decrease in availability of N and P for grass growth. Experiments performed in the Cerrado region showed that recovery of grass growth was principally controlled by the addition of N and P (Oliveira et al., 2001). Primavesi and Primavesi (1997) in São Paulo obtained similar results on degraded *Brachiaria* pastures. The general consensus amongst farmers and researchers is that soil compaction is a major limiting factor for grass regrowth, but these results suggest that compaction may be because of pasture degradation rather than a cause (Arruda, Cantarutti, and Moreira, 1987; Carvalho et al., 1990). If this is true, then decompaction may not be necessary and the costs of pasture recovery could be significantly lowered. Reduced pasture recovery costs may encourage landowners to recover at least some of the vast areas of degraded pastures.

Recent investigations have shown the importance of N supply and demand and the effect of increased animal stocking rate on N fluxes in the soil/plant/animal system (Boddey et al., 2003; Cantarutti et al., 2002). The quantity of N recycled in the senescent plant tissues (litter) decreased as stocking rate and animal consumption increased (Rezende et al., 1999). There was a concomitant rise in the quantities of N excreted by the animal, and hence increased losses of N by volatilization and leaching. Studies conducted with ^{15}N-labelled cattle urine showed N losses were high and only 20 to 60% of the labelled N was recovered in the soil and plants in and adjacent to the urine patch (Ferreira et al., 1995, 2000). Losses were highest (over 70%) in areas without grass cover.

In some long-term experiments where *Brachiaria* pastures were amended with modest doses of P and K fertilizers (and animal stocking rate was regulated with respect to the amount of forage on offer), the pastures were found to be productive for many years. For example, Lascano and Euclides (1996) reported that after 16 years of careful management of a *B. decumbens* pasture at Carimagua in the Colombian Llanos (eastern savanna), annual animal LWG (~225 kg ha^{-1} year^{-1}) was the same as in the first few years of establishment. The pasture was fertilized every two years with 10, 13, 10, and 16 kg ha^{-1} of P, K, Mg, and S (but no N), respectively, and grazed at 1 animal unit (AU) ha^{-1} in the dry season and 2 AU ha^{-1} in the wet season. Similar results at a higher level of animal production (300 to 400 kg LWG ha^{-1} year^{-1}), were obtained over an 8 year period on continuously grazed *B. humidicola* at the Itabela site in the south of Bahia (Boddey et al. 1995). At both sites, there were no inputs of fertilizer N and no significant proportion of legumes in the pastures. Long-term studies have shown that in well managed pastures there is an accumulation of soil C, and at Itabela no significant loss of soil N (Fisher et al., 1994; Fisher, Thomas, and Rao, 1997; Tarré et al., 2001). ^{15}N dilution studies have confirmed that *B. decumbens*, *B. humidicola* and Guinea grass are able to obtain up to 40 kg N ha^{-1} year^{-1} from BNF (Boddey and Victoria, 1986; Miranda, Urquiaga, and Boddey, 1990).

It has been suggested that forage legumes should be introduced into such pastures to increase animal production and enhance pasture sustainability (Thomas, 1992; Boddey, Rao, and Thomas, 1996). Early attempts to establish mixed legume/grass pastures were not very successful because the legumes did not to persist for more than a year or two under grazing (Souto, 1992). This was mainly because the legume genotypes used in mixed tropical pastures in Australia were used. The soils in Australia, principally Queensland, have pH values close to 7.0 (Robertson, Myers, and Saffigna, 1993). The legumes that were successful in Queensland, when tested in Brazil were found not to be resistant to the high Al^{3+} availability and the low P availability. Hence, efforts were made by CIAT, and at various Embrapa centers and other research institutes in Brazil, to select from larger collections of legume genotypes more adapted to Brazilian conditions. This has resulted in the selection of certain varieties of *Stylosanthes guianensis*, *Arachis pintoi* ("forage groundnut"), and *Desmodium ovalifolium*. These varieties enhance animal production and persist in the sward in mixture with *Brachiaria* grasses.

In summary, careful regulation of stocking rates and adequate fertilization of P and K can help *Brachiaria* pastures support good animal weight gains for many years. At present, very few landowners apply fertilizer and pastures are over-grazed especially in the dry season. The introduction of forage legumes in the pasture can help in increasing animal yields and in maintaining long-term sustainability of the pastures. However, adoption of this technology by landowners is extremely slow. More progressive farmers are opting for ley cropping systems. Several Embrapa centers have developed viable and profitable systems based on typically four years of *Brachiaria* pasture followed by three to four years of rotational cropping, often based on ZT. Long-term studies are underway at several sites and preliminary evidence indicate that a few years of *Brachiaria* or Guinea grass pastures restore SOM levels and soil aggregate stability. However, at present only a tiny portion of the degraded pastures in the Cerrado and the Atlantic coastal region are being restored for sustainable pastures or for crop production.

SMALLHOLDER AGRICULTURE

The sections above discussed the sustainability of crops and pastures which occupy the greater portion of the agricultural area in Brazil. However, a large proportion of the main food crops, i.e., rice, beans (*Phaseolus vulgaris* L. and *Vigna unguiculata* L.), corn, and cassava (*Manihot esculenta* L.) are produced on small family farms. This review would not be complete without some discussion of smallholder agriculture. The production systems on these farms are diverse and only some generalizations about the sustainability of these systems can be made.

Brazil has the ninth largest gross domestic product amongst all countries in the world and a mean per capita income of ~US$ 5,000. Total grain production for the current 2000/2001 season is estimated to be 97 million Mg (IBGE, 2001). A part of its agricultural sector uses the latest technology such as ZT, central pivot irrigation, and even precision agriculture. On the other hand, 33 million people in Brazil are considered to be suffering from absolute hunger, of whom 50% are in the rural areas (INCRA/FAO, 2000). The increase in wealth of the commercial agricultural sector has elevated land prices and led to the expulsion of smallholders from family owned farms. The result has been mass migration to the cities, a process that started in the 1960s and continues today.

A recent detailed national survey was made of family agriculture in Brazil (INCRA/FAO, 2000). The survey divided family farms into four groups (A to D) in descending order of wealth generation based on the daily wage paid in the respective State. Under this classification there are 824,000 family properties in Group C and 1,916,000 in group D. Of the farmers in group C, 25% plant rice, 49% beans, 60% corn, and 29% cassava, for group D the values are 15, 45, 49 and 17% for these crops, respectively. Unfortunately, the INCRA survey does not provide the yields per ha of these crops. However, the IBGE (Brazilian Institute for Geography and Statistics) provides yields of these crops based on the total area grown on each property for 1996 (IBGE, 2001) (Figure 5). For example farms with more than 10 ha of corn had mean yields of 2.1 to 4.0 Mg ha^{-1}, while those with less than 5 ha produced less than 1.1 Mg ha^{-1} (Figure 5). The INCRA census data help in explaining the reason for these low yields on smaller farms because only 32.7% of group C and 24.4% of group D farms use any form of chemical fertilizer or lime. However, 26% of beans (and 62% of green pod "runner" beans), 35% of cassava, 8.5% of corn and 7.9% of rice produced in Brazil is produced on farms with less than 10 ha under these crops. Drought, pests, and diseases also limit crop growth on many farms. Similar to small farming units in sub-Saharan Africa, the nutrients are harvested in the produce and exported from the farm system, and are not replaced with fertilizers (Smaling, Nandwa, and Janssen, 1997). Many of the farms do not generate sufficient food/income for the family and the income is supplemented by younger members who send in funds from their jobs in the cities.

It is a challenge to improve the levels of productivity on these farms on a sustainable basis. The access of smallholders to credit facilities, fertilizers, and agricultural technical advice, will be necessary for the small farms to be sustainable. At present, many non-government organizations and farmers unions are working to improve these aspects for smallholders. There is also a movement in most parts of the country led by the radical landless group, "Movimento sem Terra" (MST), for land reform. The present government has made more progress in the distribution of land to the landless than any other in Brazil's history (INCRA, 2001). Since President Fernando Henrique Cardoso was elected in 1994, the government claims that it has settled 540,000 families on 19.8 million ha of land. The MST hotly disputes these figures but the pressure created by the, sometimes violent, land invasions and occupations of INCRA offices in the state capitals, has certainly kept the government continuously on

FIGURE 5. Total crop area on each farm property in Brazil planted to rice, common beans, cassava and corn and their corresponding yields.

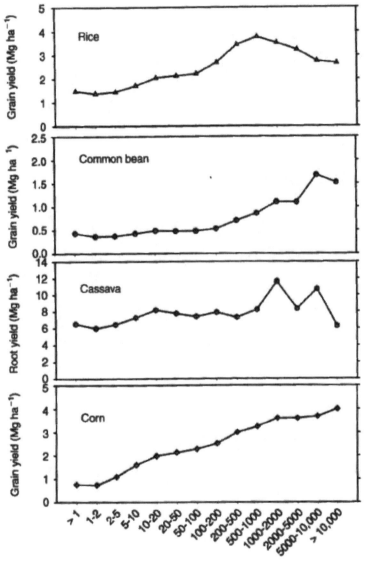

Area of land planted to crop on each property (ha)

Adatped from: IBGE, 2001.

the defensive. However, it remains to be seen how many of the settled farmers will be able to develop viable and sustainable production systems.

CONCLUSION

Brazil has huge areas of land degraded by heavy deforestation and by non-sustainable agricultural practices. Most of this destruction has occurred over the last 150 years. On the other hand, approximately 80% of the Amazon forest which covers half of the country is still intact and much of it in pristine condition. It is estimated that between 50 and 100 million ha of the Cerrado are relatively untouched. Approximately 50 million ha in Brazil are used for crop production, and there are 60 to 70 million ha of degraded pastures. These degraded pastures in the Cerrado, the Amazon, and the Atlantic coastal region should be considered as the true agricultural frontier of Brazil. Although these areas have already lost their biological diversity, soil fertility can be restored for agricultural production without excessive investments. Several technological options exist, for the conversion of these degraded lands to permanent cropping, ley cropping or productive pastures. However, such an effort requires motivation of the landowners, and incentives from the government to favor agricultural development and prevent the destruction of forest and other native vegetation. An incentive for the expansion of agriculture in Brazil may come with free international trade in agricultural products.

Brazil receives adequate rainfall for crop production because 90% of the country receives over 1000 mm of rainfall annually. With the introduction of ZT systems and ley cropping, truly sustainable crop production systems are being adopted on a wide scale. The agricultural research and development infrastructure is strong and sustainable systems for production of all of Brazil's major crops such as soybean, corn, sugarcane, *Phaseolus* beans, rice, and coffee have been developed.

For Brazil the major problem is not in the development of sustainable agricultural practices but in their adoption on a wide scale. The integration of annual crops into rotations under ZT has virtually eliminated soil erosion. At the same time this practice has increased fertilizer use efficiency which results in lower nutrient losses to the surrounding environment and water resources. The movement towards trash (residue) conservation in sugarcane is having similar results and in both cases adoption of these practices is increasing year by year. However, only a tiny proportion of the huge areas of degraded pastures are being recovered even though economically viable technologies exist to recover them for cropping or animal production. Similarly, very limited progress has been made to develop sustainable farming systems for smallholders. The adoption of sustainable systems for coffee production is extremely limited.

Brazil has larger areas of land suitable for agriculture than any other country in the world. This area could be tripled without further destruction of the Amazon forest, or other reserves of native vegetation. With fairer international trade, the demand for its agricultural products will certainly increase. It is to be hoped that, as this occurs, government policies will provide incentives for the utilization of the abandoned and degraded lands for the establishment of sustainable cropping systems. Such cropping systems that have already been developed by Brazilian farmers and researchers should be adopted, rather than destroying areas of native vegetation which are valuable reserves of biological diversity for future generations.

REFERENCES

Amado, T.J.C., J. Mielnikzuck, and S.B.V. Fernandes. (2000). Leguminosas e adubação mineral como fontes de nitrogênio para o milho em sistemas de preparo do solo. Revista Brasileira de Ciência do Solo 24:179-189 (in Portuguese).

Anonymous (1998). Lavoura Orgânica. *Manchete Rural* (Rio de Janeiro) 11:37-38 (in Portuguese).

Arruda, N.G.D., R.B. Cantarutti, and E.M. Moreira. (1987). Tratamentos físico-mecânicos e fertilização na recuperação de pastagens de *Brachiaria decumbens* em solos de tabuleiro. *Pasturas Tropicales* 9:36-39 (in Portuguese).

Baldani, J.I., L. Caruso, V.L.D. Baldani, S.R. Goi, and J. Döbereiner. (1997). Recent advances in BNF with non-legume plants. *Soil Biology and Biochemistry* 29: 911-923.

Barcellos, A. de (1996). Nitrogen cycling and sustainability of improved pastures in the Brazilian cerrados. In *Proceedings of the 1st International Symposium on Tropical Savannas*, eds. R.C. Pereira and L.C.N. Nasser, Planaltina, DF, Brazil: EMBRAPA-CPAC, pp. 130-136.

Boddey, R.M. (1993). "Green" energy from sugar cane. *Chemistry and Industry* (London) 17 May, No. 10, pp. 355-358.

Boddey, R.M. and R.L. Victoria. (1986). Estimation of biological nitrogen fixation associated with *Brachiaria* and *Paspalum* grasses using ^{15}N-labelled organic matter and fertilizer. *Plant and Soil* 90: 265-292.

Boddey, R.M., C. de P. Resende, J.M. Pereira, R.B. Cantarutti, B.J.R. Alves, E. Ferreira, M. Richter, G. Cadisch, and S. Urquiaga. (1995). The nitrogen cycle in pure grass and grass/legume pastures: Evaluation of pasture sustainability. In *Nuclear Techniques in Soil-Plant Studies for Sustainable Agriculture and Environmental Preservation*, Vienna, Austria: FAO/IAEA, pp. 307-319.

Boddey, R.M., I.M. Rao, and R.J. Thomas. (1996). Nutrient cycling and environmental impact of *Brachiaria* pastures. In *Brachiaria: The Biology, Agronomy and Improvement*, eds. J.W. Miles, B.L. Maass and C.B. do Valle, Cali, Colombia: CIAT, Publication 259, pp. 72-86.

Boddey R.M., J.C. de Moraes Sá, B.J.R. Alves, and S. Urquiaga (1997). The contribution of biological nitrogen fixation for sustainable agricultural systems in the tropics. *Soil Biology and Biochemistry* 29:787-799.

Boddey, R.M., J.C. Polidoro, A.S. Resende, B.J.R. Alves, and S. Urquiaga. (2001). Use of the ^{15}N natural abundance technique for the quantification of the contribution of N_2 fixation to grasses and cereals. *Australian Journal of Plant Physiology* 28:889-895.

Boddey, R.M., R. Macedo, R.M. Tarré, E. Ferreira, O.C. de Oliveira, C. de P. Rezende, R.B. Cantarutti, J.M. Pereira, B.J.R. Alves, and S. Urquiaga. (2003). Nitrogen cycling in *Brachiaria* pastures: The key to understanding the process of pasture decline? *Agriculture, Ecosystems and Environment* (in press).

Bonetti, L.P. (1981). Distribuição da soja no mundo. In *A Soja no Brasil*, eds. S. Miyasaka and J.C. Medina, Campinas, SP, Brazil: ITAL, pp. 1-6 (in Portuguese).

Buschbacher R., C. Uhl, and E.A.S. Serrão. (1988). Abandoned pastures in eastern Amazonia. II. Nutrient stocks in the soil and vegetation. *Journal of Ecology* 76: 682-699.

CanaWeb. (2001). www.jornalcana.com.br (Web site) February, 2001.

Cantarutti, R.B., R.M. Tarré, R. Macedo, G. Cadisch, C. de P. Rezende, J.M. Pereira, J.M. Braga, J.A. Gomide, E. Ferreira, B.J.R. Alves, S. Urquiaga, and R.M. Boddey. (2002). The effect of grazing intensity and the presence of a forage legume on nitrogen dynamics in *Brachiaria* pastures in the Atlantic forest region of the south of Bahia, Brazil. *Nutrient Cycling in Agroecosystems* 64:257-271.

Carvalho, S.I.D., L. Vilela, J.M. Spain, and C.T. Karia. (1990). Recuperação de pastagens degradadas de *Brachiaria decumbens* cv. Basilisk na região dos Cerrados. *Pasturas Tropicales* 12:24-28 (in Portuguese).

Cavalcante, V.A. and J. Döbereiner. (1988). A new acid-tolerant nitrogen-fixing bacterium associated with sugarcane. *Plant and Soil* 108:23-31.

Colozzi Filho, A., E.L. Balota, and D. de S. Andrade. (1999). Microrganismos e processos biológicos no sistema plantio direto. In *Inter-Relação Fertilidade, Biologia do Soil e Nutrição de Plantas*, eds J.O. Siqueira, F.M.S. Moreira, A.S. Lopes, L.R.G. Guilherme, V. Faquin, A.E. Furtini Neto, and J.G. Carvalho, Minas Gerais, Brazil: Federal University of Lavras, pp. 487- 508 (in Portuguese).

da Silva, J.E., J. Lemainski, and D.V.S. Resck. (1994). Perdas de matéria orgânica e suas relações com a capacidade de troca catiônica em solos da região do oeste Baiano. *Revista Brasileira de Ciência do Solo* 18:541-547 (in Portuguese).

Dean, W. (1995) *With Broadax and Firebrand: The Destruction of the Brazilian Atlantic Forest.* Berkeley, CA: University of California Press.

de Maria, I. C. Erosão e terraços em plantio direto. (1999). Boletim Informativo da Sociedade Brasileira de Ciência do Solo. 24:17-21 (in Portuguese).

Döbereiner, J. (1961). Nitrogen-fixing bacteria of the genus *Beijerinckia* Derx in the rhizosphere of sugar cane. *Plant and Soil* 15:211-216.

Eswaran, H., S.M. Virmani, and L.D. Spivey. (1993). Sustainable agriculture in developing countries: Constraints, challenges, and choices. In *Technologies for Sustainable Agriculture in the Tropics*, Madison, WI: American Society of Agronomy, Special Publication No. 56, pp. 7-24.

FAO (1989). *Sustainable Agricultural Production: Implications for International Agricultural Research*. Rome, Italy: Technical Advisory Committee, CGIAR, FAO Research and Technical Paper No. 4.

FAO (2001). FAOSTAT Agriculture data (www.apps.fao.org).

Fearnside, P.M. (1997). Amazonie: la déforestation repart de plus belle. La destruction de la forêt dépend étroitement des choix politiques. *La Recherche* 294:44-46 (in French).

Fearnside, P.M. and R.I. Barbosa. (1998). Soil carbon changes from conversion of forest to pasture in Brazilian Amazonia. *Forest Ecology and Management* 108:147-166.

Ferreira, E., A.S. Resende, B.J.R. Alves, R.M. Boddey, and S. Urquiaga. (1995). Destino do ^{15}N-urina bovina aplicado na superfície de um solo podzólico descoberto, ou sob cultura de *Brachiaria brizantha*. In *Anais XXXII Congresso Anual da Sociedade Brasileira de Zootecnia*. 17-21 Julho. pp. 109-110 (in Portuguese).

Ferreira, E., J.C.D. Santos, O.C. de Oliveira, R.M. Tarré, R. Macedo, C.H.B. Miranda, B.J.R. Alves, S. Urquiaga. (2000) A recuperação do nitrogênio da urina de bovinos por pastagens degradadas de *Brachiaria decumbens*. In *Proceedings of the International Symposium–Soil Functioning Under Pastures in Intertropical Areas*. Embrapa-Cerrados, Brasilia, 16-20 October CD-ROM (in Portuguese).

Fisher, M.J., I.M. Rao, M.A. Ayarza, C.E. Lascano, J.I. Sanz, R.J. Thomas, and R.R. Vera. (1994). Carbon storage by introduced deep-rooted grasses in the South American savannas. *Nature* 371:236-238.

Fisher, M.J., R.J. Thomas, and I.M. Rao. (1997). Management of tropical pastures in acid-soil savannas of south America for carbon sequestration in the soil. In *Management of Carbon Sequestration in Soil*, eds. R. Lal, J. Kimble, R.F. Follett, and B.A. Stewart. Boca Raton, FL: CRC Press, pp. 405-420.

Galloway, J.H. (1989) *The Sugar Cane Industry: An Historical Geography from Its Origins to 1914*. Cambridge, UK: Cambridge University Press.

Gassen, D.N. and Gassen, F.R. (1996). *Plantio direto, o caminho do futuro*. Aldeia Sul Editora, Passo Fundo, RS, Brazil. 207 p. (in Portuguese).

Greenland, D.J. (1975). Bringing the green revolution to the shifting cultivator. *Science* 190:841-844.

Guérif, J., G. Richard, C. Dürr, J.M. Machet, S. Recous, and J. Roger-Estrade. (2001). A review of tillage effects on crop residue management, seedbed conditions and seedling establishment. *Soil and Tillage Research* 61:13-32.

Homewood, B. (1993). Will Brazil's cars go on the wagon? *New Scientist* (9 January) 137:22-23.

IBGE (2001). Brazilian Institute for Geography and Statistics (www.ibge.gov.br) (in Portuguese).

INCRA (2001). October (www.incra.gov.br/reforma) (in Portuguese).

INCRA/FAO (2000). *Novo Retrato da Agricultura Familiar: O Brasil Redescoberto*. Joint Publication of the 'Instituto Nacional de Colonização e Reforma Agrária'– INCRA and FAO, Ministério do Desenvolvimento Agrário, Brasília, DF (February, 2000), 74 p. (in Portuguese).

Juo, A.S.R. and R. Lal. (1977). The effect of fallow and continuous cultivation on the chemical and physical properties of an alfisol in Western Nigeria. *Plant and Soil* 47: 567-584.

Kichel, A., C.H.B. Miranda, and A.H. Zimmer. (1998). Fatores de degrdção de pastagem sob pastejo rotacionado com ênfase na fase de implantação. In *Fundamentos do Pastejo Rotacionado–Anais 14 Simpósio Sobre Manejo da Pastagem*, Piracicaba. SP, Brazil. pp. 193-211 (in Portuguese).

Kluthcouski, J., A.R. Pacheco, S.M. Teixeira, and E.T.D Oliveira. (1991). *Renovação de pastagens do Cerrado com arroz. I. Sistema Barreirão*. Embrapa Rice and Bean Research Centre, Goiânia, GO, Brazil (Document 33) (in Portuguese).

Lara-Cabezas, W.A.R. (1999). Dinâmica do nitrogênio e estratégias de adubação nitrogenada no sistema plantio direto em solos do Cerrado. In *II Seminário sobre o sistema plantio direto na UFV*. ed. L.A.M. Cordeiro. Viçosa, MG, Brazil, Universidade Federal de Viçosa, pp. 117-156 (in Portuguese).

Lascano, C.E. and V.P.B. Euclides. (1996). Nutritional quality and animal production of *Brachiaria* pastures. In *Brachiaria: The Biology, Agronomy and Improvement* eds. J.W. Miles, B.L. Maass and C.B. do Valle, Cali, Colombia: CIAT, Publication 259, pp. 106-123.

Macedo, M.C.M. (1995). Pastagens no ecossistema Cerrados: Pesquisa para o desenvolvimento sustentável. In *Pastagens nos Ecossistemas Brasileiros: pesquisas para o desenvolvimento sustentável*, eds. R.P. de Andrade, A. de O. Barcellos and C.M.C. da Rocha, Sociedade Brasileira de Zootecnia, Viçosa, MG, Brazil, Universidade Federal de Viçosa, pp. 28-62 (in Portuguese).

Macedo, I. de C. (1998). Greenhouse gas emissions and energy balances in bio-ethanol production and utilization in Brazil (1996). *Biomass and Bioenergy* 14:77-81.

Machado, G.R., W.M. da Silva, and J.E. Irvine. (1987). Sugar cane breeding in Brazil: the Copersucar Program. In *Copersucar International Sugarcane Breeding Workshop*, Cooperativa de Produtores de Cana, Açúcar e Álcool do Estado de São Paulo Ltda, São Paulo, SP, Brazil. pp. 215-232.

Medina, J.C. (1981). Introdução e evolução da soja no Brasil. In *A Soja no Brasil*, eds., S. Miyasaka and J.C. Medina, Campinas, SP, Brazil: ITAL, pp. 17-20 (in Portuguese).

Miranda, C.H.B., S. Urquiaga, S., and R.M. Boddey. (1990). Selection of ecotypes of *Panicum maximum* for associated biological nitrogen fixation using the ^{15}N isotope dilution technique. *Soil Biology and Biochemistry* 22:657-663.

Mondardo, A. (1981) Manejo do solo e sistemas de cultivo. In *A Soja no Brasil*, eds., S. Miyasaka and J.C. Medina, Campinas, SP, Brazil: ITAL, pp. 354-363 (in Portuguese).

Neill, C., J. Melillo, P.A. Steudler, C.C. Cerri, J.F.L. Moraes, M.C. Piccolo, and M. Brito. (1997). Soil carbon and nitrogen stocks following forest clearing for pasture in the southwestern Brazilian amazon. *Ecological Applications* 7:1216-1225.

Neto, M.S. and M.B. Dias Filho. (1995). Pastagens no ecossistema do trópico úmido: Pesquisa para o desenvolvimento sustentável. In *Pastagens nos Ecosistemas Brasileiros: pesquisas para o desenvolvimento sustentável*, eds. R.P. de Andrade, A. de O. Barcellos, and C.M.C. da Rocha. Sociedade Brasileira de Zootecnia, Viçosa, MG: Brazil, Universidade Federal de Viçosa, pp. 76-93 (in Portuguese).

Oliveira, O.C. de, S. Urquiaga, and R.M. Boddey. (1994). Burning cane: the long term effects. *International Sugar Journal* 96 (1147):272-275.

Oliveira, O.C. de, I.P. de Oliveira, E. Ferreira, B.J.R. Alves, C.H.B. Miranda, L. Vilela, S. Urquiaga, and R.M. Boddey. (2002). Response of degraded pastures in the Brazilian Cerrado to chemical fertilisation. *Pasturas Tropicales* 23(1):14-18.

Parsons, J.J. (1972). Spread of African pasture grasses to the American tropics. *Journal of Range Management* 25:12-14.

Primavesi, O. and A.C.P. de A. Primavesi. (1997). Recuperação de pastagens degradadas, sob manejo intensiva, sem revolvimento de solo, e seu monitoramento. In *Proceedings of the III Simpósio Nacional de Recuperação de Áreas Degradadas (SINRAD)*, 18-24 May, Ouro Preto, MG, Brazil, pp. 150-155 (in Portuguese).

Reijntjes, C., B. Haverkort, and A. Waters-Bayer. (1992). *Farming for the Future: An Introduction to Low-External-Input and Sustainable Agriculture.* London, UK: Macmillan Education Ltd.

Rezende, C. de P., R.B. Cantarutti, J.M. Braga, J.A. Gomide, J.M. Pereira, E. Ferreira, R.M. Tarré, R. Macedo, B.J.R. Alves, S. Urquiaga, G. Cadisch, K.E. Giller, and R.M. Boddey. (1999). Litter deposition and disappearance in *Brachiaria* pastures in the Atlantic forest region of the south of Bahia, Brazil. *Nutrient Cycling in Agroecosystems* 54:99-112.

Robertson, F.A., R.J.K. Myers, and P.G. Saffigna. (1993). Carbon and nitrogen mineralization in cultivated and grasslands soils in subtropical Queensland. *Australian Journal of Soil Research.* 31:611-619.

Ruschel, A.P., Y. Henis, and E. Salati. (1975). Nitrogen-15 tracing of N-fixation with soil-grown sugar cane seedlings. *Soil Biology and Biochemistry* 7:181-182.

Sano, E.E., A.O. Barcellos, and H.S. Bezerra. (2000). Assessing the spatial distribution of cultivated pastures in the Brazilian Savanna. *Pasturas Tropicales* 23(3):2-15.

Scholes, M.C., M.J. Swift, O.W. Heal, P.A. Sanchez, J.S.I. Ingram, and R. Dalal. (1994). Soil fertility research in response to the demand for sustainability. In *The Management of Tropical Soil Fertility*, eds. P.L. Woomer and M.J. Swift, Chichester, UK: John Wiley, pp. 1-14.

Sisti, C.P.J. (2001). Influência de sistemas de preparo do solo e manejo de culturas sobre o estoque de carbono e nitrogênio do solo em diferentes condições agrícolas. MSc thesis. Seropédica, RJ, Brazil: Universidade Federal Rural do Rio de Janeiro (in Portuguese).

Smaling, E.A., S.M. Nandwa, and B.H. Janssen. (1997). Soil fertility in Africa is at stake. In *Replenishing Soil Fertility in Africa.* Madison, WI: Soil Science Society of America, Special Publication No. 51, pp. 47-61.

Souto, C.M., M.R. Romano, S. Urquiaga, and R.M. Boddey. (1993). Acumulação de matéria seca, N, P e K por cana de açúcar (cv. SP 70-1143). *Resumos do Congresso Brasileira de Ciência do Solo*, Goiânia, GO, 25-31 July. Sociedade Brasileira de Ciência do Solo, Goiânia, GO, Brazil. Vol. II. pp. 239-240 (in Portuguese).

Souto, S.M. (1992). Contribuição e persistência de leguminosas em pastagens tropicais. Embrapa-CNPBS, Seropédica, RJ, Brazil. Document No. 8. p. 20 (in Portuguese).

Tarré, R.M., R. Macedo, R.B. Cantarutti, C. de P. Rezende, J.M. Pereira, E. Ferreira, B.J.R. Alves, S. Urquiaga, and R.M. Boddey. (2001). The effect of the presence of a forage legume on nitrogen and carbon levels in soils under *Brachiaria* pastures in the Atlantic forest region of the South of Bahia, Brazil. *Plant and Soil* 234:15-26.

Thomas, R.J. (1992). The role of the legume in the nitrogen cycle of productive and sustainable pastures. *Grass and Forage Science* 47:133-142.

Urquiaga, S., K.H.S. Cruz, and R.M. Boddey. (1992). Contribution of nitrogen fixation to sugar cane: Nitrogen-15 and nitrogen balance estimates. *Soil Science Society of America Journal* 56:105-114.

Valsechi, O. (1951). *A queima da cana-de-açúcar e suas conseqüências.* "Tese de livre docência." Escola Superior de Agricultura Luiz de Queiroz, ESALQ, Piracicaba, SP, Brazil: University of São Paulo (in Portuguese).

Zambolim, L. and F.X. Ribeiro de Vale. (2000). Perdas na produtividade e qualidade do cafeeiro causadas por doenças bióticas e abióticas In *Café, Produtividade, Qualidade e Sustentabilidade.* ed. L. Zambolim, Viçosa, MG, Brazil: Universidade Federal de Viçosa, pp. 239-261 (in Portuguese).

Zimmer, A.H. and K. Euclides-Filho. (1997). Brazilian pasture and beef production. In *Proceedings of the International Symposium on Animal Production Under Grazing.* ed. J.A. Gomide, Viçosa, MG, Brazil: Universidade Federal de Viçosa, pp. 1-29.

Zotarelli, L. (2000). *Balanço de nitrogênio na rotação de culturas em sistema de plantio direto and convencional na região de Londrina-PR.* MSc Thesis, Department of Soil, Science, Universidade Federal Rural do Rio de Janeiro, Seropédica RJ, Brazil (in Portuguese).

Cropping Systems in Eastern Europe: Past, Present, and Future

Imre Molnar

SUMMARY. The evolution of cropping systems in Eastern Europe has been a long and gradual process. The objective of this paper is to provide a comprehensive review of the cropping systems in Eastern Europe before World War II, the present cropping systems, and speculate on the future systems. Before World War II, the crop-long term fallow, the grass-crop, and the field rotation were the prominent cropping systems. Crop rotations were adopted in the 16th-17th century and various modifications of the 'Norfolk' and the 'crop-grass' rotation systems were used. Cash cropping, industrial cropping, and monoculture were extensively used after World War II. Major political, social, and economic changes took place in Eastern Europe after the dissolution of the Soviet Union. Nationalized land was returned to rightful owners. The new owners, however, lacked the skills and financial resources for intensive crop production and this has resulted in drastic reductions in crop yields. Integrated crop production seems to be the most viable option for the East European countries because organic or biological cropping is financially too demanding. Considerable efforts are being made to design new cropping systems which emphasize reductions in mineral fertilizers and pesticides,

Imre Molnar is Professor, University of Novi Sad, Faculty of Agriculture, 21000 Novi Sad, Trg Dositeja Obradovica 8, Yugoslavia, and is also affiliated with the Institute of Field and Vegetable Crops, 21000 Novi Sad, Maksima Gorkog 30, Yugoslavia (E-mail: molnar@polj.ns.ac.yu).

[Haworth co-indexing entry note]: "Cropping Systems in Eastern Europe: Past, Present, and Future." Molnar, Imre. Co-published simultaneously in *Journal of Crop Production* (Food Products Press, an imprint of The Haworth Press, Inc.) Vol. 9, No. 1/2 (#17/18), 2003, pp. 623-647; and: *Cropping Systems: Trends and Advances* (ed: Anil Shrestha) Food Products Press, an imprint of The Haworth Press, Inc., 2003, pp. 623-647. Single or multiple copies of this article are available for a fee from The Haworth Document Delivery Service [1-800-HAWORTH, 9:00 a.m. - 5:00 p.m. (EST). E-mail address: docdelivery@haworthpress.com].

and conservation tillage. Eastern Europe may become important exporters of agricultural commodities by the year 2020. *[Article copies available for a fee from The Haworth Document Delivery Service: 1-800-HAWORTH. E-mail address: <docdelivery@haworthpress.com> Website: <http://www.HaworthPress. com> © 2003 by The Haworth Press, Inc. All rights reserved.]*

KEYWORDS. Crop-long term fallow, field rotation, Norfolk system, crop-grass system, integrated cropping, crop rotation, conservation tillage, genetically modified crops

INTRODUCTION

The term 'East European countries,' in this paper, will refer to the northern part of Yugoslavia and eastern Croatia (with the rivers Sava and Danube as the southern borders), Bulgaria, Romania, Moldavia, Hungary, Poland, the Czech Republic, Slovakia, and the countries in the European part of the former USSR (the Baltic states, Belarus, Russia, Ukraine, and the Caucasian states). Cropping systems in these countries evolved with the social system, agrarian policy, and the level of development of agricultural science and technology. The history of crop production in these countries has shown that each cropping system allowed for a more or less rational use of arable land over a certain period of time. As population and food demand increased, the prevalent system became inadequate and it had to yield to a new, more intensive system. The transition was gradual and the old system was replaced by a new, more efficient, and intensive system (Molnar, 1995).

Cropping systems were interlaced in time and space, and frequently similar systems were used on the same continent or in the same country. For example, in the beginning of the 20th century in Hungary, the prevailing crop-fallow system was combined with crop rotation and cash cropping. In Russia, however, the outdated crop-fallow system remained in use together with the subsequently implemented systems (Sipos, 1964). East European countries like other European countries located in the same geographic region shared common problems associated with cropping systems although certain dissimilarities occurred with variations in climate (humid-arid, maritime-continental) and altitude.

Grigg (1974) provided a historical review of agricultural cropping systems in the world. His review, however, did not cover the territory of Eastern Europe. Therefore, the objective of this paper is to provide an overview of the past, present, and future cropping systems in Eastern Europe. The discussion will focus on the cropping systems before World War II, cropping systems after World War II, present cropping systems, and cropping systems envisaged for the 21st century.

CROPPING SYSTEMS BEFORE WORLD WAR II

The Crop-Long Term Fallow System

The development of cropping systems in Eastern Europe did not essentially differ from that in Western Europe, except for certain periods. For example, in some parts of Western Europe that belonged to the Roman Empire, permanent settlements were established and crop rotation was adopted earlier than in Eastern Europe (Grigg, 1974). The first cropping system to come into practice in Eastern Europe was the 'crop-long term fallow system.' Characteristic features during the period when this cropping system was adopted were low population density, a nomadic way of life, and the use of primitive agricultural tools. In this cropping system, a limited acreage in the vicinity of a settlement was cultivated for a few years and then left fallow for reestablishment of the natural vegetation. The spacious steppe zones in the East European lowlands had ample agricultural land. The soils were rich in humus and plant nutrients, and these soils are still considered among the most fertile soils in the world. These soils were simple to cultivate; the ancient agriculturist only had to till and plant seeds. As a rule, monoculture cereal production prevailed. In humid mountain areas, agricultural land was obtained by clearing forest. The prevailing method was slash-and-burn followed by several years of cereal monoculture. Vegetative cover was burned, the ashes of which enriched the soil and reduced acidity. When crop yields declined after four to six years of cultivation, the land was abandoned and a new part of the forest was cleared and converted into agricultural land.

There were two variations of the 'crop-long term fallow system': the wild and the managed. The managed 'crop-long term fallow' system evolved from the wild system after the establishment of permanent settlements. The managed system differed from the wild system in that rotations were used in the cropland and grassland (meadow and pasture). Again, after several years of cultivation, the land was abandoned and converted to grassland when cereal yields decreased. This cropping system was very popular in Eastern Europe and it remained in use until the beginning of the 20th century in Russia. Two periods were clearly distinguished within this cropping system: cropping period (4-6 years) and fallow period (6-50 years). Weeds established first on the fallow land and other vegetation gradually developed. According to Sipos (1972), complete reestablishment of natural vegetation required 50 to 60 years. The 'crop-long term fallow system' was gradually replaced by a 'crop-short term fallow system' which gave rise to a new cropping system, the 'field rotation.'

The Field Rotation System

The 'field rotation system' was initially used in Eastern Europe in the beginning of the 7th century. This system became the dominant crop production system in the 10th and 11th centuries, brought about the diversification of agricultural production, and introduced the use of permanent fields. Meadows and pastures were relegated to less fertile areas. Available arable land was used more efficiently under this system than in the 'crop-long term fallow system.' The production area occupied 50-60%, and in some cases up to 80% of the total arable land.

Cereals were not included in long term grassland rotation because the fallow fields were regularly tilled (bare fallow) or were used for grazing. Cereal fields were divided into two or three tracts, cultivated for two or three years, and subsequently fallowed. The purpose of fallowing was to control weeds and to replenish soil fertility. The division of arable land into tracts was the beginning of early two- and three-crop rotations. The planting sequence of the two-field rotation was winter cereal-fallow. Only cereals such as wheat (*Triticum aestivum* L.), rye (*Secale cereale* L.), barley (*Hordeum vulgare* L.), and oat (*Avena sativa* L.) were grown in this system and they occupied 50% of the total area. This system was also known as the 'Mediterranean two-field rotation with fallow' (Mihalic, 1970). Later this system developed into a two-field rotation without a fallow period. A common rotation was winter cereal-spring cereal. Later, with the introduction of row crops [e.g., corn (*Zea mays* L.)], the common rotation was row crop-winter cereal. Both fields were cropped in this system, one-half with row crops and the other half with cereals.

After the discovery of America, corn was introduced as a manured row crop. Corn replaced fallow in the 17th century (Mándy, 1972) and considerably intensified the classical two-field rotation. Corn significantly increased the production capacity of the rotation because of its higher yields than cereals and because the manure applied to corn had a prolonged positive effect on cereal yields. Hoeing on a half of the production area contributed to weed control. This rotation gained popularity in the entire Pannonian Plain, Ukraine, Romania, Bulgaria, and the Caucasian states. In the Croatian and the Yugoslavian corn belts, this system known as the 'Balkan rotation' is still in use (Mihalic, 1970). This rotation had several advantages and disadvantages. Stojkovic (1951) pointed out that the two-field rotation ensured maximum production of a crop like wheat which could be stored and consumed for a long period. On the other hand, the corn-wheat rotation did not provide sufficient quality roughage and created several biological, organizational, and economic problems.

The next phase of the two-field rotation system was the substitution of corn with other row crops such as sugar beet (*Beta vulgaris* L.), potato (*Solanum tuberosum* L.), sunflower (*Helianthus annuus* L), hemp (*Cannabis sativa* L.), and the substitution of wheat with other cereals such as barley, oat, and rye. A special feature of this system was the cultivation of alfalfa (*Medicago sativa* L.) or red clover (*Trifolium pratense* L.) in separate fields outside of the rotation. After several years, the alfalfa or red clover field was cropped while the previously cropped fields were excluded from the rotation. Forage crops were thus included in the system and this was called the 'feudal three-field rotation system.' The original rotation that followed the winter cereal-spring cereal sequence was replaced by a winter cereal-winter cereal-spring cereal rotation. In this rotation, cereals occupied 100% of the area and there was no fallow period. This rotation was further modified by the introduction of row crops and annual legumes. This modification was known as the 'Slavonian three-field rotation' (Mihalic, 1970) and the system comprised of row crop (manured)-annual legume-winter cereal sequence.

After the discovery of America, in the late 17th and the early 18th century, potato was introduced as a row crop in the cooler parts of Eastern Europe while corn was introduced in the warmer parts (Mándy, 1972). In the Pannonian Plain, corn became the major row crop, vetch (*Vicia sativa* L.) the major annual legume, and winter wheat remained the major cereal. Thus, the three-field rotation was corn (manured)-vetch-winter wheat. The improved three-field rotation was an important step forward in improving the efficiency of the cropping system both biologically and agronomically. Inclusion of an annual legume in the rotation provided several benefits. For example, the legume improved the physical status and nitrogen content of soil, and its early harvest permitted time for soil preparation for planting winter wheat. It also contributed to higher yields of wheat and corn (Stojkovic et al., 1975). In the three-field rotation, corn could be replaced with other row crops such as sugar beet, sunflower, potato or even hemp whereas, vetch could be replaced with other annual legumes such as pea (*Pisum sativum* L.), soybean [*Glycine max* (L.) Merr.], bean (*Phaseolus vulgaris* L.) or lupine (*Lupinus* spp.), and wheat could be replaced with barley, oat or rye. The two- and three-field rotations are still in use in the East European countries.

The Crop Rotation System

The crop rotation system in Europe was introduced in Belgium and The Netherlands in the 16th-17th century to grow a greater array of crops necessary for industry and livestock production. Crop rotation became the dominant system in England and France in the 18th century (Martin, Leonard, and Stamp, 1976). Arthur Young (1741-1820) promoted the system in the counties of Norfolk and Suffolk. In the 19th century, the system was introduced in Germany.

This system was commonly referred to in the literature as the 'Norfolk four-course system.'

The original 'Norfolk system' had the following cropping sequence: row crop-spring cereal and red clover-red clover-winter cereal. Therefore, this cropping system was rightfully called the 'crop rotation system.' In the 'Norfolk rotation' cereals were grown in 50% of the area just like in the 'two-field rotation.' However, the difference was that row crops and legumes were also included in the 'Norfolk rotation.' This allowed quality feed to be produced for an increased number of livestock. The 'crop rotation system' had several beneficial effects. The increased numbers of livestock provided more manure, which was applied to row crops. The practice improved soil fertility and improved the yields of all crops in the rotation. Red clover improved soil structure and added nitrogen (N) to the soil. The expansion of the 'Norfolk rotation' on the continent, however, led to a phenomenon called 'soil exhaustion' that resulted in a rapid decline in soil fertility. This phenomenon was caused by red clover that was grown in the same field in short rotations instead of the normal six-year interval. The cause of the problem was identified and Albrecht Daniel Thaer (1752-1828) of Germany recommended the substitution of red clover with an annual legume or a row crop. Consequently, the original 'Norfolk rotation' was prevented from further expansion on the continent. Modified systems were developed using either annual legumes or row crops in the following sequences: row crop-spring cereal-annual legume-winter cereal, or row crop-spring cereal-row crop-winter cereal.

In the 17th and 18th centuries, after the discovery of America, potatoes were included in the modified 'Norfolk rotation' in the northern and mountain parts, and the arid regions of Eastern Europe (Mándy, 1972). The respective sequences were: sugar beet-spring barley-potato-winter wheat in the humid regions and corn-spring barley-vetch-winter wheat in the arid regions. In the northernmost and mountain regions of Eastern Europe, because of severe winters, spring barley and winter wheat were replaced with a winter hardy crop, i.e., winter rye. As the demand for red clover as a quality feed source increased, the 'Norfolk rotation' was expanded to include five or six fields, i.e., it evolved in the direction of long-term rotations. Long-term rotations combined rotation units that consisted of two, three, or four fields.

The Crop-Grass Management System

The 'crop-grass management system' was developed by the Russian academician Vilyams (1935). He believed that natural restoration of soil fertility in long-term fallow was a very slow process and suggested that growing mixtures of perennial grasses and legumes would accelerate the process of soil fertility restoration. He named this method the 'crop-grass management system' in which several years of annual crops were followed by a period of perennial grasses. According to Vilyams (1935), a stable crumby soil structure established favorable aeration, moisture and temperature regimes, and improved soil fertility. Some perennial grasses [e.g., timothy (*Phleum pratense* L.), perennial ryegrass (*Lolium perenne* L.), wheatgrass (*Agropyron* spp.)], and legumes [red clover, alfalfa, and esparcet (*Onobrychis viciaefolia* Scop.)] helped in forming a stable crumby soil structure. He further distinguished two basic types of the crop-grass rotation: the field crop type and the meadow-pasture type. The portion of grass mixtures was 20-30% in the former type whereas, it was 50-70% in the latter. An example of the field crop type nine-field rotation is: row crop-spring cereal (interplanted with clover and grasses)-grass mixture-grass mixture-row crop-spring cereal-row crop-annual legume-winter cereal. The rotation was often extended into a field rotation including row crop-annual legume-winter cereal. An example of the meadow-pasture type rotation is: row crop-grass-grass-grass-grass-grass-row crop-spring cereal-annual legume-winter or spring cereal.

Vilyams (1935) and his proponents dogmatically lauded this 'crop-grass management system' as the only cropping system suitable for the planned socialist economy. This system was hence legislated and promoted in the territories of the former USSR. In this system, wheat and fodder were produced within a singe rotation. This was considered as an advantage especially for large state farms which had limitless acreage suitable for crop production. This system ensured the production of roughage on a regular basis, stimulated the development of animal production, and increased the availability of manure for the crops. Criticisms about this system, however, were heard in the 1940s (Pryanisnikov, 1945; Tulaykov, 1963; Lőrinc, Sipos, and Sipos, 1978). It was pointed out that the beneficial effects of the 'crop-grass management system' lasted for only a year or two and the effectiveness of the system in restoring soil fertility was exaggerated. The authors further claimed that roughage could be produced only in regions with favorable growing conditions. In dry regions, mixtures of perennial grasses and legumes lowered cereal and row crop yields, and contributed little to the improvement of soil structure. This led to serious shortages in staple food crops. The critics emphasized that a certain cropping system should not be prescribed for an entire country or for a group of countries sharing the same social system. A cropping system should be based on the natural conditions of a location.

After World War II, attempts were made to introduce the 'crop-grass system' in the newly-formed socialist countries of Eastern Europe. In Yugoslavia, Milojic (1963) concluded that this system could not produce high yields of hay or improve soil fertility in the arid regions. Since shallow-rooted grasses did not fare well in arid conditions, the mixture of grasses and legumes was replaced with alfalfa or the ratio of grasses in the mixture was reduced. The introduction of the 'crop-grass system' hence reduced the acreages of other crops. Since yields were not increased, the total production of all crops, especially cereals, was reduced. This explains why the 'crop-grass system' was never adopted in Eastern Europe outside of the former USSR.

CROPPING SYSTEMS AFTER WORLD WAR II

The Cash, Industrial, and Continuous Cropping Systems

The 'cash cropping system' is alternatively called the 'market-oriented management system' because the available land is mostly used for production of high-value crops adapted to the local conditions. In the developed countries agricultural produce started to be treated as market commodities after World War I, and especially after World War II. In Eastern Europe, except for Poland and to some extent Yugoslavia, the change in the social system brought about 'collectivization' and the establishment of large state farms, cooperatives, 'kolkhozes' and 'sovkhozes.' Simultaneously, industrialization reduced the available agricultural workforce and led to the introduction of highly efficient agricultural equipment. Agriculture thus became increasingly specialized and crop production gradually disintegrated from livestock production. Specialized estates with no livestock were forced to reduce the number of crops because they had no reason to produce feed crops. The number of fields in the rotations was also reduced. Such estates tried to make their production simpler and to minimize costs. This approach encouraged the production of cereals with reduced manual labor and introduced herbicides. In the 1960s, these changes led towards industrial cropping and in extreme cases towards continuous cropping.

The characteristic feature of the 'cash cropping system' was the absence of a firm long-term crop rotation plan. Instead, cropping plans were made on an annual basis and crops expected to bring highest profits were preferred. Under such conditions, the proportions of planted crops were not stable but varied from year to year depending on the market demand. Heavy emphasis on the most profitable crops led to frequent growing of crops in the same field. Crops were grown in short rotations and the principles of crop rotation were ignored. As a consequence, pest, disease, and weed occurrences increased. This ultimately resulted in reduced crop yields and increased production risks. Alleviation of this situation necessitated restoration of suitable crop rotations. The 'cash cropping system' required increased knowledge of modern crop husbandry technology. Under unfavorable ecological conditions, the system was risky and frequently caused economic disasters.

The 'industrial cropping system' evolved from the 'cash cropping system' through specialization and concentration of production. The main characteristics of this system included repeated growing of a single crop (either corn, soybean, cereals, or potato) and intensive use of machines and agrochemicals. The 'industrial cropping system' was developed in the technically advanced midwest United States, i.e., the Corn Belt. High specialization, intensive cropping of corn and soybean, high mechanization, and complex protection practices against diseases, pests, and weeds were characteristics of this cropping system (Molnar, 1975). The basic premise of the system was complete mechanization of all operations in crop production, from tillage to harvest (Molnar, 1974). However, factors such as intensive chemical inputs and soil compaction, reduced organic matter levels and eventually caused a shift to alternate systems such as integrated cropping system (Poincelot, 1986; Altieri, 1987).

Cropping Systems in Some Specific East European Countries

Poland, Czech Republic, and Slovakia

Ecological conditions in these countries are fairly different. In Poland, where private land ownership had been preserved (the only socialist country that went without 'collectivization'), crop production was distributed according to the soil, climatic, and physiographic conditions. In the flat part of Silesia, sugar beet and wheat were grown on fertile soils, whereas potatoes and rye on soil with poor fertility. Examples of crop rotations (Dziezyc, 1961) in the flat parts of Silesia include: sugar beet-spring cereals-legumes-winter wheat-oil crops-spring or winter wheat, or potato-oat and spring cereals-legumes-rye-oil crops-rye.

In the mountainous part of Silesia, the proportions of sugar beet, winter and spring wheat and winter barley in rotations decreased with altitude, while the proportions of rye, oats, spring barley, clover and clover-grass mixtures increased (Table 1).

In the Czech Republic and Slovakia, 'crop rotation systems' have a long tradition. From the beginning of the 19th century, the three-field rotation was systematically replaced with more intensive systems. Various modifications of the 'Norfolk system' played important roles. The 'crop-long term fallow system' remained in use only in the mountainous areas (Shimon, 1961). State farms tested and introduced various rotations containing alfalfa and clover and the 'crop-grass system' was also experimented. In the potato-growing region, rotations included clover or clover-grass mixtures and six- and twelve-field rotations were frequently used. The row crop alfalfa/clover rotation maintained soil fertility.

From the early 1970s till the late 1980s, the 'industrial cropping system' was successfully used on many large state farms in the southern flat parts of Slovakia. These farms which used rotations of corn and cereals achieved yield levels close to those obtained in Western Europe. Southern Slovakia is a fairly

TABLE 1. Examples of crop rotations used in the mountainous part of Silesia (Adapted from Dziezyc, 1961).

Altitude (m)		
300-400	400-600	Above 600
Potatoes	Potatoes	Potatoes
Sugar beet	Barley, oats	Oats, barely
Spring cereals	Grain legumes	Clover-grass mixture
Grain legumes	Clover-grass mixture	Clover-grass mixture
Rye	Clover-grass mixture	Clover-grass mixture
Clover, alfalfa	Wheat, rye	Rye
Winter wheat	Rye	
Oats		

dry region (annual rainfall less than 600 mm, with high oscillations in some years) with frequent droughts. Hence, irrigation systems were established and crop rotations suitable for irrigation were intruded.

Hungary, Romania, Bulgaria

Hungary, Romania, and Bulgaria differ fairly in ecological conditions. However, the cropping systems in these countries were similar because of their social system and planned economy. Although, some characteristics were specific for each country. Hungary, Romania, and Bulgaria had favorable growing conditions for corn and cereals, and these crops occupied 60-70% of the rotation. Other row crops such as sunflower and sugar beet occupied 10-15% of the rotation (Nyiri et al., 1995; Dzumalieva and Vasilev, 1986).

In the 1960s, experiments were conducted in Hungary to assess possibilities of specializing and concentrating the cropping system to four-five crops (Belák, 1961). As a result of these experiments, 'industrial cropping systems' were introduced in the 1970s with modern farm machines, extensive use of mineral fertilizers and pesticides, and high yielding varieties and hybrids. In this period, Hungary had the highest yields of wheat and corn among the socialist countries. The average yields of winter wheat and corn were 4.90 Mg ha^{-1} and 5.96 Mg ha^{-1}, respectively. The total annual Hungarian wheat production was nearly 15 million metric tons. Sandy soils of Southern Hungary between the Danube and Drava rivers had a different cropping system. Rye intercropped with vetch or cowpea (*Vigna sinensis* L.)-potato-rye were common rotations.

Romania attempted to copy the Hungarian system of industrial cropping and several Romanian state farms showed that it was possible to do so. However, the lack of technical knowledge and organizational skills limited the large-scale adoption of this system. This explains why the Romanian winter wheat and corn yields have lagged far behind the yields of these crops obtained in other southeastern European countries (Starcevic et al., 1999). For example, during 1983-1985, the average yields of winter wheat and corn in Romania were 2.57 Mg ha^{-1} and 3.52 Mg ha^{-1}, respectively. Examples of crop rotation in Romania (Vasiliu, cit. Könnecke, 1967) include: forage crop-forage crop-winter wheat-corn-winter wheat-corn-barley-pea-winter wheat-corn; pea or cereal-winter wheat-corn-winter wheat-corn-winter wheat-sunflower-corn-barley-corn; and alfalfa/grass mixture-forage barley/rye-vetch/cereal mixture/Sudan grass-fodder beet-silo corn.

Bulgaria adopted the Soviet 'collectivization' and planned economy model soon after World War II (1946), went through a period of 'collectivization' after the Soviet model, and adopted the planned economy. The success of this system was similar to that of the other socialist countries. Bulgaria has a warmer climate than the other East European countries and half of its arable land is in the lowlands with favorable conditions for intensive production of crops like corn, rice (*Oryza sativa* L.), cotton (*Gossypium hirsutum* L.), tobacco (*Nicotiana tabacum* L.), and vegetables (Dzumalieva and Vasilev, 1986). In the mountainous regions, forages were included in the rotation to prevent erosion (Garbutshev, 1961). Among other rotations, bean-wheat-sugar beet-oat-spring barley was suitable. Bulgaria also had a long tradition of vegetable production, notably tomato (*Lycopersicon esculentum* Mill.), pepper (*Capsicum annuum* L.), onion (*Allium cepa* L.), cucumber (*Cucumis sativus* L.), carrot (*Daucus carota* subsp. *sativus* Hoffm.), melon (*Cucumis melo* L.), and watermelon (*Citrulus vulgaris* L.) were grown under irrigation.

Croatia and Yugoslavia

Croatia is one of the six republics constituting the former Yugoslavia. Its main agricultural region, Slavonia, is located in the eastern part of the country and it borders Vojvodina Province, the main agricultural region of the present Yugoslavia. After World War II, Yugoslavia underwent the Soviet-model of 'collectivization.' In 1952-1953, the cooperatives 'sovkhozes' and 'kolkhozes' were dissolved and most of the land was returned to farmers, although, about 15% of the arable land remained state-owned. In the late 1950s, agricultural production was intensified by replacing low-yielding domestic varieties with Italian wheat varieties, American corn hybrids, and Soviet sunflower hybrids. Use of mineral fertilizers and pesticides were intensified and modern agricultural equipment were purchased. By mid-1960s, Yugoslavia became self-sufficient in basic food items (wheat, sugar, meat) and started exporting agricultural commodities. A liberal economic system and a flexible agrarian policy stimulated rapid development of agricultural production.

Two-crop rotation remained in use on the private farms and the prevalent sequence was corn-winter wheat. Other row crops such as sunflower or sugar beet were grown as an option to corn. Small farmers most frequently grew corn in monoculture. In the Vojvodina Province, corn was grown in 43 to 48% of the total arable land. Specialized production started on state farms in the early 1970s and livestock production was separated from crop production. Consequently, the number of crops grown was reduced and rotations were narrowed down to four-five crops (winter wheat, corn, sunflower, sugar beet, and, since 1975, soybean). In the late 1970s, some state farms in the Vojvodina Province and Slavonia achieved the technical, professional, and organizational skills essential for the introduction of the 'industrial cropping system' used in Hungary. These farms obtained yields equivalent to those obtained in Western Europe, but at much higher production costs. Total production was considerably increased in both Yugoslavia and Croatia. During 1983-1985, the average winter wheat yield in Yugoslavia and Croatia was 3.76 Mg ha^{-1} and 4.09 Mg tha^{-1}, respectively, and average corn yield was 4.71 Mg ha^{-1} and 5.00 Mg ha^{-1}, respectively (Starcevic et al., 1999).

Countries of the Former USSR

Vast differences in climatic and soil conditions exist in the huge European sector of the former USSR. Liste (1964) distinguished the following zones in the European part of the former USSR.

Semidesert Zone

This zone is characterized by hot summers and cold winters, conditions too harsh for crop production. The annual precipitation is less than 200 mm and crop production is possible only with irrigation (Berg, 1958). Irrigation systems had been constructed for intensive production of cotton, alfalfa, sugar beet, corn, tobacco, forage, and vegetable crops. Alfalfa was grown in pure stands or intercropped with grass crops for three years, followed by four years of monoculture cotton. The role of alfalfa in this rotation was to increase soil fertility and to mitigate diseases caused by *Fusarium* spp. and *Verticillium* spp. Irrigated sugar beet production was concentrated in the Caucasian countries, often as a monoculture comprising 33% of the rotation. Despite this intensity no serious nematode problem occurred (Jerlepesov, 1963).

Steppe Zone

The main crops in this zone are wheat, corn, sunflower, sugar beet, and sorghum. The eastern part of this zone is drier than the western. Crops such as spring wheat, sunflower (*Helianthus annuus* L.), and sorghum (*Sorghum vulgare* L.) are grown in the east. Although sugar beet comprised 33% of the rotation, forages were also included in rotations. Fallowing was practiced to conserve moisture in the arid southern region of Ukraine.

Forest-Steppe Zone

This zone has a humid climate with deep, fertile chernozem soils rich in humus. The annual precipitation is lower and the growing season is shorter in the eastern than in the western part of the zone. Although the severe climate permits growing of spring wheat only, considerable areas are planted to corn (for forage and ensiling), sunflower, sugar beet, sorghum, pea, and faba bean (*Vicia faba* L.) (Table 2). The central part of the forest-steppe zone in Central Russia has a moderate continental climate. This is the main region for hemp which is often grown in monoculture (Vorobyev, 1970). Although, it is sometimes preferred in rotations because of its ability to suppress weeds. Erosion is a major problem in about 10 million hectares in Ukraine. Rotations were introduced to control erosion in the western part of the forest-steppe zone where the conditions for crop production are particularly favorable (Pastushenko, 1955).

TABLE 2. Examples of crop rotations used in different regions of the forest-steppe zone (Adapted from Pastushenko, 1955).

Easter European region	Central Russian region
Perennial forage crops	Hemp
Perennial forage crops	Sugar beet
Winter rye	Hemp
Pea, oats	Potato
Winter rye	Hemp
Barley	Faba bean
Winter rye (for forage)	
Winter rye	
Spring cereals	

Southern region	Western region
Black fallow	Sugar beet
Winter wheat intercropped with grass crops	Spring cereal with clover
Grasses	Clover
Grasses	Winter wheat
Spring wheat	Sugar beet
Sunflower	Grain legumes (pea, vetch)
Spring wheat	Winter wheat
Sainfoin	Corn
Winter rye	Pea
Spring cereals	Winter wheat

Mixed Forest Zone

This zone includes northern Ukraine, eastern Russia, Belarus, and Baltic countries in the non-chernozem zone of the former USSR. The growing season is very short (less than 135 days) in this region (Golcberg and Pokrovskaya, 1972) and the main crops are winter wheat, rye, potato, lupine, faba bean, flax (*Linum usitatissimum* L.), silo corn, and perennial forages. Some examples (Liste, 1964) of crop rotations of this region are: perennial forage-perennial forage-flax-green fallow-winter wheat-potato-oats; green fallow-winter wheat-winter wheat-oats-potato-spring wheat-green fallow-winter wheat; green fallow-winter rye, winter rye-potato-legumes-spring cereal; and potato-winter rye-lupine (in Belarus). The Baltic countries have a favorable maritime climate for cereals, faba bean, forage kale (*Brassica oleracea* L.), and silo corn.

Coniferous Forest Zone

This zone covers the northern part of Russia, with the 60th parallel (northern latitude) as its northern border. The main crops are spring cereals, potato, annual and perennial forages, and vegetables. Because of a short (between 90 and 120 days) growing season (Golcberg and Pokrovskaya, 1972), winter cereals can be grown only after fallow. Potato is frequently grown repeatedly in the same field because there is no risk of nematode. Flax production is a long tradition of the Jaroslaw region and is often rotated with a perennial forage, potato, spring cereal, or grain legume.

PRESENT CROPPING SYSTEMS

The dissolution of the USSR on 31 December 1991 led to far-reaching political, social, and economic turmoil in the Eastern European countries. This inevitably affected their agricultural production in general. The land that had been collectivized after World War II was returned to their rightful owners or descendants. However, the rate and extent of privatization differed from one country to another. Although the process is in its final stage in Hungary, the Czech Republic, Slovakia, Bulgaria, and Romania, privatization still has a long way to go in the other East European countries. Poland does not have this problem because it never adopted the Soviet 'collectivization' model. After privatization, large state farms were typically divided into smaller private fields, but the new land owners were not professional farmers and lacked the skills and financial resources to start crop production. It will hence take time to consolidate agricultural production and this is why privatization of farms in these countries resulted in drastic yield reductions of major field crops.

Starcevic et al. (1999) compared production trends of several southeastern European countries (Bulgaria, Croatia, Hungary, Romania, and Yugoslavia) with some European Union (EU) countries (France, Germany, Greece, and Italy). They reported that, in the mid-1960s, the average yields of the major field crops were low in both the EU and the southeastern European countries. However, twenty years later, wheat yields in EU countries increased by 103% whereas, they increased by only 78% in southeastern Europe. The increasing trend continued in the EU countries in the mid-1990s. In the southeastern European countries, however, the trend was reverse and the yields declined further. Similar trends occurred in corn, sugar beet, sunflower, and soybean yield. Hence, the present crop production in southeastern Europe is far from satisfactory. Reasons for this situation vary from one country to another, but there is one reason in common-limited financial inputs in crop production. Although the five countries analyzed make only a part of the region of East Europe, they may serve as a good example for the entire region. The average yields of major field crops in the period 1991-1995 in some Eastern European countries are shown in Table 3.

At present, various cropping systems are used in the East European countries. Fixed rotations have been abandoned everywhere, including the countries of the former USSR. Large efforts were made to reorganize 'kolkhozes' and 'sovkhozes' to ensure sufficient food supply and to meet the demands of the market economy. The success of these reorganizations depend on the polit-

TABLE 3. Average yields of major field crops in some East European countries during 1991-1995 (FAO Statistics).

Country	Wheat Mg ha^{-1}	Corn Mg ha^{-1}	Sugar beet Mg ha^{-1}	Soybean Mg ha^{-1}	Sunflower seed Mg ha^{-1}
Belarus	2.55	1.78	22.57	-	1.41
Bulgaria	3.09	3.25	15.96	1.04	1.26
Croatia	3.93	4.52	35.02	2.12	2.01
Czech Republic	4.47	4.18	38.60	1.14	2.03
Estonia	1.91	-	26.40	-	-
Georgia	1.38	2.42	11.54	0.16	0.36
Hungary	4.21	4.44	30.53	1.82	1.76
Kazakhstan	0.89	2.64	10.66	1.04	0.33
Lithuania	2.44	-	22.38	-	-
Poland	3.39	4.50	32.78	-	-
Romania	2.53	3.01	20.55	1.34	1.26
Russian Fed.	1.59	2.52	16.76	0.72	0.96
Slovakia	4.37	4.54	34.35	1.35	1.79
Ukraine	3.22	2.65	20.30	0.94	1.22
Yugoslavia	3.37	3.37	28.43	1.60	1.83

ical will of the governments of these countries to introduce democratic changes in the deep rooted negative attitudes toward the agricultural sector. In the countries that are in the final stages or that have completed privatization of agricultural land (Poland, the Czech Republic, Hungary, Bulgaria, Romania, and Slovakia), the process of farm augmentation is being rapidly conducted. Vacant land is either bought or leased by people with money and those who understand the meaning of private ownership. Farmers are making cooperatives to pool their finances in order to purchase expensive equipment and to use them more efficiently. Cooperatives are also being made in the sphere of marketing (Stefanovits, 1998; Kovács, 1997; Sipos, 1998) and 'cash cropping system' is most prevalent in these countries. Private farmers are growing crops without considering crop rotations. For example, sunflower is being frequently grown in a two-crop rotation, three-crop rotation, or even in monoculture (Nyiri et al., 1995). This practice has brought about significant yield reductions in a short period because of the occurrence of pests such as *Phomopsis* ssp., *Sclerotinia* ssp., and a new broomrape (*Orobanche cumana* Walbr. race E) ecotype (Molnar, 1999).

Corn has also been attacked by a new insect, corn rootworm (*Diabrotica virgifera virgifera* Le Conte), in Yugoslavia. This insect has spread to neighboring countries (Croatia, Hungary, Romania, and Bulgaria) and has already caused significant economic damage. The insect is expected to spread further to Ukraine, Moldavia, southern Russia, and the Caucasian countries. Chemical control of the insect is possible but very expensive. Rotating corn with other crops is the most efficient control practice but this may cause corn growers to reduce corn acreage and avoid monoculture corn (Camprag, Baca, and Sekulic, 1994). Some examples of rotations (Molnar, 1999) to avoid this insect are: corn-soybean-winter wheat-corn-sunflower-corn; corn-field pea-corn-soybean-winter wheat-corn; corn-sunflower-winter wheat-corn-soybean-corn-winter wheat-corn; corn-sunflower-winter wheat-corn-spring wheat-sugar beet-spring barley-corn.

In the northern Eastern European countries (Poland, Belarus, the Baltics, and Russia), ecological conditions favor cereal production, e.g., wheat, barley, rye, and oat. Mechanization has eased the production of cereals and farmers are now tempted to include a high proportion of cereals in their cropping system, and this may lead to problems. Yield reductions are large in cereal monocultures but are less pronounced on more fertile soils than on soils with poor fertility. Fungal diseases and pests can cause large yield reductions in rotations predominated by cereals. For example, *Pseudocercosporella* spp. and *Ophibolus* spp. are the most frequent pests followed by *Fusarium* spp., *Erysiphe* spp., *Typhula* spp., and *Puccinia* spp. (Glynne, 1963; Boyarczuk, 1970). Other important pests are nematodes (*Heterodera* spp. and *Pratylenchus* spp.), cereal leaf beetle (*Oulema melanopa* L.), and cereal ground beetle (*Zabrus tenebrioides* Goeze) (Fischer, 1971; Drews, 1972).

Inclusion of cereals too frequently in rotations encourage shifts in weed species that cannot be easily controlled with herbicides, for example, silky bentgrass (*Apera-spica venti* L.), wild oats (*Avena fatua* L.), goose grass (*Galium aparine* L.), and blackgrass (*Alopecurus myosuroides* Huds.) (Koch, 1964; Bachthaler, 1968). To increase the portion of cereals above 67% in the rotation, it is necessary to increase the number of crops (6 to 10), e.g., sugar beet, potato, rapeseed (*Brassica napus* L.), legumes, forage crops, vegetables, or green manure (Molnar, 1999). It can be concluded that drastic reductions in yields of the major field crops in the East European countries in recent decades were not only due to limited inputs but also because of the abandonment of ecologically sound cropping systems.

FUTURE CROPPING SYSTEMS

Analysis of the causes for low crop yields shows that crop production is in a precarious position in Eastern Europe. Romania and some countries of the former USSR are large importers of wheat. Nevertheless, it is estimated that these countries, when they overcome the current economic and political crisis, may become important exporters of wheat and other agricultural commodities (Haen and Lindland, 1996). These estimates seem to be realistic because these countries have favorable agroecological conditions for crop production. The southern East European countries have higher mean annual temperatures and longer growing season than the West European countries and offer possibilities of double cropping (Shashko, 1967). However, frequent droughts can cause large yield variations in the southern East European countries.

Irrigation water is available from large rivers (Danube, Drava, Sava, Tisza, Dnyester, Dnyeper, Don, and Volga) flowing through these countries. Yet, because of outdated and crippled irrigation systems, the irrigated acreage is negligible when compared with some Western European countries such as Italy and Greece where 39% of the arable land is irrigated (Starcevic et al., 1999). It would be advantageous to intensify irrigation because in countries such as Croatia, Hungary, Romania, and Yugoslavia, row crops are grown in 45% to 65% of the acreage while in Bulgaria vegetables are grown on a large scale. Crop yields could also be increased and stabilized by introducing intensive irrigated field crop-vegetable crop rotations, field crop-forage crop rotations, and multiple cropping systems (Francis, 1986), double cropping and intercropping systems (Vucic, 1981; Kahnt, 1982; Momirovic et al., 1998) and companion cropping (Oljaca, 1997). Unfortunately, the countries in transition cannot afford to construct new irrigation systems and have serious problems in maintaining the existing systems.

The high yield levels achieved after World War II in some East European countries (Hungary, the Czech Republic, Slovakia, Croatia, and northern Yugoslavia) resulted from the intensive use of mineral fertilizers, pesticides, heavy mechanization, and from adoption of cash cropping and industrial cropping systems. In this process, however, crop rotation with legumes and the use of organic fertilizers were neglected. As a result, in recent decades, the level of soil fertility and humus content have decreased. Similarly, soil compaction and reduced water retention capacity have reduced microbiological activity of the soil. Intensive pesticide applications have resulted in high levels of pesticide residues in the soil and groundwater and have reduced beneficial insects and fauna. Herbicide-resistant weeds have also been reported. People have become concerned of pesticide residues entering food chains and affecting human health. Short rotations, especially monocultures, have increased soil erosion and reduced biological diversity.

Negative effects of intensive crop production systems were first felt in Western countries. Consequently, alternative crop production systems are being developed in Western Europe, USA, and Canada (Diercks, 1983; Altieri, 1987; Francis, 1990). The political and economic upheavals in the late 1980s and the early 1990s delayed the development of alternative crop production systems in Eastern Europe. However, it appears that East European countries will also alter their crop productions systems after the model of the developed western countries. The feeling is that they will not be able to compete on the world market if they do not alter their production practices. This is the right time for the East European countries to develop their own strategies of agricultural production based on sound ecological and biological principles. Crop rotation will play an important role because it is the most effective and economic measure in the control of diseases, pests, and weeds. At present in the East European countries, integrated management systems such as integrated pest management (IPM) seem to be the most applicable. In the integrated cropping systems, the portions of certain crops or groups of related crops should not exceed limits imposed by the conditions of the location (Diercks and Heitefuss, 1990). Taking in consideration the prevailing sowing plans and the agroecological conditions, the following proportions of crops are recommended for the East European countries (Table 4).

According to an estimate, the International Federation of Organic Agriculture Movements (IFOAM) covers about 5% of the total agricultural production in the EU. The corresponding figures in the East European countries range from 0.1% to 1% (Molnar and Kastori, 1999). The IFOAM recommends systems of organic-biological cropping for developing countries in order to reduce chemical pollution. It seems logical for the East European countries to

TABLE 4. Maximum portions of field crops in the systems of integrated cropping in percentage.

| Crop | Conditions of the site | | Note |
	arid warm	humid cool	
	%	%	
Corn for grain	40	50	Without rotation with
Corn for grain	20	20	perennial grasses
Total small grains	75	67	
Winter wheat	50	40	
Winter barley	33	25	
Winter rye, triticale	33	50	
Spring wheat	33	40	
Spring barley	33	40	
Oats	25	25	
Beta beets	20	25	Varieties resistant to
Potato	25	33	nematodes
Seeds potato	20	25	
Rapeseed	20	33	
Beta beets and all	25	33	
Crucifers			
Sunflower	17	12	
Soybean	25	25	
Pea for grain	20	25	Without rotation with other legumes
Broad bean	20	33	
Alfalfa	20	17	Without rotation with other
Red clover	12	17	legumes
Total grain legumes	17	20	
Flax	14	12	
Hemp	33	33	

accept IFOAM's recommendation and try to find most suitable organic-biological cropping systems. However, this is an absurd notion because what may look like organic agriculture is merely a consequence of poverty, vis-a-vis a major lack of better alternatives. The current practices have nothing in common with biological cropping since most East European countries have neither the concept, conditions, nor the necessary knowledge for adopting biological cropping systems. Some countries such as Hungary, Yugoslavia, Croatia, Poland, the Czech Republic, and Slovakia, have somewhat unrealistically opted for sustainable agriculture. However, in these countries, sustainability remains as a concept and not a practice (Angyán, 1994). Although it seems reasonable to expect that after consolidation the acreages under biological cropping systems will increase in these countries, such systems may never become the basic and dominant systems of land use (Fischbeck, 1993).

In view of the theoretical background of the alternative cropping systems, a question may be raised about the future potential of these systems in Eastern Europe. Milojic (1989) considers integrated cropping systems as a more ambitious and less viable system than conventional systems. Intensive multidisciplinary studies are necessary for the development of new cropping systems based on biological principles and natural laws. To meet the current challenges, sustainable systems must have four objectives, i.e., economy, ecological harmony, political and social acceptability, and protection of environment and natural resources. A major difficulty in developing new cropping systems is the incompatibility among the objectives, i.e., the necessity of making sound compromises. It would be unrealistic to expect an ultimate system (Geng, Hess, and Auburn, 1990). Future systems and technologies must enforce harmony between man and environment, ensure sufficient production of quality food, and protect agroecosystems and the environment from pollution and further degradation.

In cooperation with Western manufacturers of seed drills, some East European countries have made initial steps towards introducing conservation tillage systems. Large private farmers in Bulgaria, the Czech Republic, Hungary, Poland, Romania, and Slovakia have been offered such drills, especially those for cereals. Since 1996, branches of ISTRO (International Soil and Tillage Research Organization) exist in these countries. Members of this organization study intensively the applicability of conservation tillage systems. Adoption of these new tillage systems is expected to intensify in proportion with the development of financial strength of the private farmers.

The East European countries share the concerns of EU with regard to the use of genetically modified crops. Only France has considerable acreage under transgenic corn hybrids. Experimentation with genetically modified crops was allowed in the other EU countries in February 2001, but not their commercial growing. With the exception of Hungary, experiments with genetically modified crops are not permitted in the East European countries. In the year 2000, a law on genetically modified crops was legislated in Hungary. This law permitted experiments with genetically modified crops, but under stringent control. Yugoslavia plans to legislate such a law in 2001 and similar laws are in preparation in other East European countries.

CONCLUSION

Historically, the development of cropping systems in Eastern Europe has been a long and gradual process often interlaced in time and space. The first cropping system to be used in Eastern Europe was the 'crop-long term fallow system.' Increases in population and food demand gave rise to the 'crop rotation system,' which became dominant in Eastern Europe in the 10th and 11th centuries. In this system, the cropped area occupied 50-60% (sometimes up to 80%) of the total arable area, and the area was divided into two or three tracts. This was the beginning of two- and three-field rotations. After the discovery of America, corn or potato was alternated in two-field rotations, while annual legumes were added to corn and potato in three-field rotations. The systems were further improved by the introduction of other row crops and cereals. Perennial legumes were grown outside the rotations.

Before World War II, a modified 'Norfolk rotation' was most important in Eastern Europe. Corn was the major row crop in the more arid southern parts of Eastern Europe, and potato was the major row crop in the northern parts. Vilyams (1935) developed the 'crop-grass system' to accelerate soil fertility restoration and improve soil structure. This system did not perform well in arid regions, and in spite of the support from the Soviet regime it never became widespread in the other socialist countries of Eastern Europe. 'Cash cropping and industrial cropping systems' were introduced after World Wars I and II, respectively. Despite significant increases in crop yields, these systems negatively affected the soil fertility, caused environmental pollution, and enhanced the risk of pesticide residues entering the food chains. A solution was sought in the form of alternative cropping systems in the 1970s.

After the dissolution of the USSR, large political, social, and economic changes occurred in all East European countries. In the countries which had resorted to 'collectivization,' land is being returned to the rightful owners. This process is at an advanced stage in Hungary, the Czech Republic, Slovakia, Bulgaria, and Romania, while the other countries have lagged behind. After privatization, large state farms were dissolved while new owners of small farms lacked input resources and knowledge to maintain the intensive crop production. The results of lower inputs and intensified attacks of diseases, insects, and weeds were drastic yield reductions, and this was experienced in all East European countries.

It is estimated that these countries, when they overcome the current economic and political crisis, may become important exporters of agricultural commodities by 2020. To achieve that, however, they must design new cropping systems based on the principles of market economy. The concept of integrated cropping system is most acceptable for the East European countries because neither organic agriculture nor biological cropping can become dominant production systems. These systems, however, should not be disregarded but should be intensively studied in order to help in reducing the application of expensive mineral fertilizers and toxic pesticides. The difficulty in developing new, improved cropping systems lies in the incompatibility among their objectives. However, new systems and technologies based on biological principles can establish harmony between man and environment, ensure sufficient production of quality food, and protect the soil from further pollution and degradation in Eastern Europe. Experiments with conservation soil tillage systems have been started in several East European countries (Bulgaria, the Czech Republic, Hungary, Poland, Romania, and Slovakia). Increase in the economic power of private farmers is expected to intensify the implementation of new soil tillage systems. The growing of genetically modified crops is prohibited in the East European countries. In the year 2000, Hungary legislated a law which permits experiments with genetically modified crops, but under rigorous control.

REFERENCES

Altieri, M.A. (1987). *Agroecology: The Scientific Basis of Alternative Agriculture.* Boulder, CO: Westview Press.

Ángyán, J. (1994). Környezetbarát gazdálkodási rendszer- és sturkturváltás a szántóföldi Növénytermesztésben (in Hungarian). *Agrártudományi Egyetem,* Gödöll.

Bachthaler, G. (1968). Entwicklung der Aackerunkrautflora in Abhängigkeit von veränderten Feldbaumethoden (I. und II. Teil). *Zeitschrift Acker und Pflanzenbau* (in German)127:149-170, 326-353.

Belák, A. (1961). Die Pflanzenfolge und die Vorfruchtwirkung bei intensiven Ackerfruchtfolgen. *Wissenschaftlicher Zeitschrift Universität Halle* (in German) 2/3: 281-283.

Berg, L.S. (1958). *Die gerographischen Zonen der Sowjetunion* (in German). Bd. I. Leipzig.

Boyarczuk, J. (1970). Badania nad aspornoscia pszenicv ozimei na lamliwoosc zdzbla. (*Cercosporella herpotrichoides* Fron.). Czési I. Znaczenie gospodarcze choroby. Biologija-patogena. Warunki infekciji. *Hodowla Roslin, Aklimatiyzacja i Nasiennictnjo* (in Polish) 14:327-341.

Camprag, D. (2000). *Integralna zastita ratarskih kultura od stetocina* (in Serbian). Poljoprivredni fakultet, Novi Sad.

Camprag, D., F. Baca, and R. Sekulic (1994). *Diabrotica virgifera virgifera* Le Conte nova stetocina u Jugoslaviji. *Pesticidi* (in Serbian) 9:45-50.

Diercks, R. (1983). *Alternativen im Landbau* (in German). Stuttgart: Verlag Eugen Ulmer.

Diercks, R. and R. Heitefuss (1990). *Integrierter Landbau* (in German). München: BLV Verlagsgesellschaft GmbH.

Drews, F.W. (1972). *Untersuchungen über wandernde Wurzelnematoden der Gattung Pratylenchus Filipjew 1934 (Nematoda) an Getreide* (in German). Halle: Martin-Luther-Univ., Sektion Pflanzenproduktion.

Dziezyc, J. (1961). Das Problem der geeigneten (typischen) regionalen Fruchtfolgen und deren Untersuchung. *Wissenschaftlicher Zeitschrift Universit@t Halle* (in German) 2/3:465-471.

Dzumalieva, D. and A. Vasilev (1986). *Seitboobrasenija pri intenzivnoto zemledelie* (in Bulgarian). Sofia: Zemizdat.

Fischbeck, G. (1993). Pflanzenbauliche Konzepte des ökologischen Landbaues. *Rundgespräche der Kommision für Ökologie* (in German) 7:205-220.

Fischer, R. (1971). Untersuchungen über die Vermehrung pflanzenparasitärer Nematoden und den Ertragsverlauf bei fortgesetztem Getreidebau. II. Untersuchungen über die Vorfruchtbeziehungen und Ertagsverlauf bei fortgesetztem Getreidebau. *Archiv Acker und Pflanzenbau und Bodenkunde* (in German) 15:479-512.

Francis, C.A. (1986). *Multiple-Cropping Systems*. New York, NY: Macmillan Publishing Company.

Francis, C.A. (1990). Sustainable agriculture: Myths and realities. *Journal of Sustainable Agriculture* 1:97-106.

Geng, S., C.E. Hess, and J. Auburn (1990). Sustainable agricultural systems: Concepts and definitions. *Zeitschrift für Acker und Pflanzenbau* (in German) 165:73-85.

Golcberg J.A. and T.V. Pokrovskaya (1972). *Agroklimaticheskiy atlas mira* (in Russian). Moscva-Leningrad: Gidrometeoizdat.

Glynne, M. (1963). Eyespot (*Cercosporella herpotrichoides*) and other factors influencing yield of wheat in the sixcourse rotation experiment at Rothamsted (1930-1960). *Annals of Applied Biology* 51:198-224.

Grigg, D.B. (1974). *Agricultural Systems of the World: An Evolutionary Approach*. London, UK: Cambridge University Press.

Haen, A. and J. Lindland (1996). Word cereal utilization, production and trade in year 2020. *Entwicklung landlicher Raum* 30:25-34.

Jerlepesov, M.N. (1963). Ergebnisse der Agrarwissenschaft in Südost-Kasachstan. *Nachrichten Landwirtschaftliche-Wissenschaft* (in German) 1:97-104.

Kahnt, G. (1982). Biologischer Landbau. *Der Biologie Unterricht* (in German) 18:6-17.

Koch, W. (1964). Einige Beobachtungen zur Verukrautung während mehrjährigen Getreidebaus und vershiedenartiger Unkrautbekämpfung. *Weed Research* (in German) 4:351-356.

Könnecke, G. (1967). *Fruchtfolgen* (in German). Berlin: WEB Deutscher Landwirtschaftsverlag.

Kovács, F. (1997). Agrárstratégiát–tudományos háttérrel. In *A magyar agrárgazdaság jelene és kilátásai* (in Hungarian), ed. F. Glatz, Budapest: Magyar Tudom<nyos AkadJmia., pp. 178-179.

Liste, H.J. (1964). *Fruchtfolgeforschung und Fruchtfolgen in der UdSSR* (in German). Halle/Saale: Halle/Saale Universit@t.

Lörinc, J., G. Sipos, and S. Sipos (1978). *Földmhveléstan* (in Hungarian). Budapest: Mezögazdasági kiadó.

Mándy, Gy. (1972). *Hogyan jöttek létre kultúrnövényeink?* (in Hungarian) Budapest: Mezögazdasági kiadó.

Martin, J.H., W.H. Leonard, and D.L. Stamp (1976). *Principles of Field Crop Production*. New York, NY: Macmillan Publishing Company.

Mihalic, V. (1970). Plodored. *Poljoprivredna enciklopedija 2* (in Croatian), Zagreb: Jugoslovenski leksikografski zavod. pp. 522-528.

Mihalic, V. (1976). *Opca proizvodnja bilja* (in Croatian). Zagreb: Skolska knjiga.

Milojic, B. (1963). Prilog proucavanju polja u plodoredima. *Zbornik radova Poljoprivrednog fakulteta*, Beograd-Zemun: Poljoprivredni fakultet (in Serbian) 368:1-9.

Milojic, B. (1989). *Aktuelni problemi biljne proizvodnje* (in Serbian). Beograd: Naucna knjiga.

Milojic, B. (1990). *Sistem bioloskog ratarenja* (in Serbian). Beograd: Knjizevne novine.

Molnar, I. (1974). Industrijska traka na njivi. *Poljoprivrednik* (in Serbian) XII:987, 988, 989, 990.

Molnar, I. (1975). Industrijskim ratarenjem protiv stagnacije prinosa. *Poljoprivrednik* (in Serbian) XIX:1052.

Molnar, I. (1995). *Opste ratarstvo* (in Serbian). Novi Sad: Feljton.

Molnar, I. (1999). *Plodoredi u ratarstvu* (in Serbian). Novi Sad: Naucni Institut za ratarstvo i povrtarstvo.

Molnar, I. and R. Kastori (1999). Pravci razvoja biljne proizvodnje-Biljna proizvodnja na raskrscu. *Strategijski menadzment* (in Serbian) 2-3:4-10.

Momirovic, N., R. Cvetkovic, Z. Radosevic, and S. Oljaca (1996). Double cropping–a field production method toward agricultural intensification and agroecosystem protection. *Ekologia* 33:55-62.

Nyiri, L., M. Birkás, T. Kismányoky, I. Lányski, and J. Nagy (1995). *Földmáveléstan* (in Hungarian). Budapest: Mezögazdasági kiadó.

Oljaca, S. (1997). *Produktivnost kukuruza i pasulja u zdruzenom usevu u uslovima prirodnog i irigacionog vodnog rezima* (in Serbian). Beograd-Zemun: Poljoprivredni fakultet.

Pastushenko, W.O. (1955). Bodenschützende Fruchtfolgen im Karpatengebirgskolchos "Stalin." *Ackerbau* (in German). 2:57-61.

Poincelot, R. (1990). Agriculture in transition. *Journal of Sustainable Agriculture* 1:9-40.

Poincelot, R.P. (1986). *Toward a More Sustainable Agriculture*. Westport, CT: AVI Publishing Company Inc.

Pryanisnikov, D. (1945). *Azot v zhizhnjij rastenija i v zemledelii* (in Russian). Moskva: Doklad Akademii Nauk. SSSR, I-III.

Shashko, I.D. (1967). *Agroklimaticheskoye rajonirovanye SSSR* (in Russian). Moskva-Leningrad: Gidrometeoizdat.

Shimon, J. (1961). Einige Erfahrungen und Erkenntnnisse von Fruchtfolgeversuchen *Wissenschaftlicher Zeitschrift Universit@t Halle* (in German) 2/3:433-436.

Sipos, A. (1998). Az agrártermelés alapozásának közgazdasági tényezö. A birtokviszonyok korszerásitése. In *Az agrártermelés tudományos alapozása* (in Hungarian), ed. F. Glatz, Budapest: Magyar Tudom<nyos AkadJmia. pp. 49-62.

Sipos, G. (1964). *Földmáveléstan* (in Hungarian). Budapest: Mezögazdasági kiadó.

Sipos, G. (1972). *Földmáveléstan* (in Hungarian). Budapest: Mezögazdasági kiadó.

Starcevic, Lj., J. Crnobarac, M. Malesevic, and B. Marinkovic (1999). Stanje i perspektiva ratarske proizvodnje u zemljama Jugoistocne Evrope (in Serbian). *Proceedings of the 2nd International Scientific Conference "Proizvodnja njivskih biljak na pragu XXI veka,"* Novi Sad: Poljoprivredni fakultet, pp. 7-21.

Stefanovits, P. (1998). A termöföld szerepe az agrárgazdaságban. In *Az agrártermelés tudományos alapozása* (in Hungarian), ed. F. Glatz, Budapest: Magyar Tudományos AkadJmia. pp. 63-70.

Stojkovic, L. (1951). Plodored u zitorodnim krajevima. *Socijalisticka poljoprivreda* (in Serbian) 4:13-32.

Stojkovic, L., B. Belic, I. Molnar, K. Smiljanski, and S. Dzilitov (1975). Kukuruz u sistemu iskoriscavanja zemljista na cernozemu u Vojvodini. *Zemljiste i biljka poljoprivreda* (in Serbian) 1-2:57-67.

Tulaykov, N.M. (1963). *Kritika travopoljnog sistemi zemledelije* (in Russian). Moskva: Izdateljstvo seljkohozjastvennog literaturi, zhurnal i plakov.

Vilyams, V.R. (1935). *Travopoljnie sevooboroti* (in Russian), Moskva: Doklad Akademii Nauk SSSR.

Vorobyev, S.A. (1970). O teorii sevoobarotov v intensivnom zemledelii. *Vestnik selkohozjajstvennoj nauki* (in Russian) 12:1-12.

Vucic, N. (1981). *Navodnjavanje i dve zetve godisnje* (in Serbian). Novi Sad: NISRO Dnevnik-OOUR *Poljoprivrednik*.

Socioeconomic and Agricultural Factors Associated with Mixed Cropping Systems in Small Farms of Southwestern Guatemala

Francisco J. Morales
Edin Palma
Carlos Paiz
Edgardo Carrillo
Iván Esquivel
Vianey Guillespiez
Abelardo Viana

SUMMARY. Many small farming communities in Latin America have modified their traditional cropping systems to incorporate non-tradi-

---

Francisco J. Morales is Senior Scientist, Centro Internacional de Agricultura Tropical (CIAT), Cali, Colombia.

Edin Palma is a student, Universidad Rafael Landivar.

Carlos Paiz, Edgardo Carrillo, Iván Esquivel, and Vianey Guillespie are Agronomists, Instituto de Ciencia y Tecnología Agropecuaria (ICTA), Guatemala.

Abelardo Viana is Economist, Profrijol (COSUDE), Guatemala.

Address correspondence to: Francisco J. Morales, Senior Scientist, Centro Internacional de Agricultura Tropical (CIAT), AA 6713, Cali, Colombia (E-mail: F.MORALES@cgiar.org).

The authors acknowledge the financial support of the Danish International Development Assistance (DANIDA) during the planning and execution of this case-study.

[Haworth co-indexing entry note]: "Socioeconomic and Agricultural Factors Associated with Mixed Cropping Systems in Small Farms of Southwestern Guatemala." Morales, Francisco J. et al. Co-published simultaneously in *Journal of Crop Production* (Food Products Press, an imprint of The Haworth Press, Inc.) Vol. 9, No. 1/2 (#17/18), 2003, pp. 649-659; and: *Cropping Systems: Trends and Advances* (ed: Anil Shrestha) Food Products Press, an imprint of The Haworth Press, Inc., 2003, pp. 649-659. Single or multiple copies of this article are available for a fee from The Haworth Document Delivery Service [1-800-HAWORTH, 9:00 a.m. - 5:00 p.m. (EST). E-mail address: docdelivery@haworthpress.com].

tional export crops (NTEC). The shift from subsistence to commercial agriculture is perceived by development agencies as an opportunity to alleviate poverty in rural areas. However, most small-scale farmers are not familiar with the production problems of NTEC, such as *Bemisia tabaci* and various geminiviruses transmitted by this whitefly species. In the absence of adequate technical assistance, due to drastic budgetary reductions in national agricultural research programs, farmers have relied on agrochemicals to protect their NTEC. This situation has led to considerable pesticide abuse and rejection of contaminated produce in international markets. This study analyzes some of the factors determining the adoption of NTEC and displacement of traditional food crops in southwestern Guatemala, and suggests possible measures to allow small farming communities to benefit from broad-based cropping systems that include both traditional and non-traditional food and cash crops. *[Article copies available for a fee from The Haworth Document Delivery Service: 1-800-HAWORTH. E-mail address: <docdelivery@haworthpress.com> Website: <http://www.HaworthPress.com> © 2003 by The Haworth Press, Inc. All rights reserved.]*

KEYWORDS. Pesticide abuse, whitefly, *Bemisia tabaci*, geminivirus, Bean golden yellow mosaic, export crops, globalization, sustainability

INTRODUCTION

The "lost decade" or economic crisis of the 1980s caused drastic changes in traditional cropping systems throughout Latin America. In 1982, the total external debt of Latin America was US $333.5 billion, up from a total of US $2.3 billion in 1950 (BID, 1994). The impact of this crisis was particularly felt by the poorer countries in the region, which were still dependent on traditional agricultural products, such as coffee (*Coffea arabica* L.), cotton (*Gossypium hirsutum* L.), tobacco (*Nicotiana tabacum* L.), and bananas (*Musa* spp.), to counteract their growing trade deficits. Unfortunately, the value of most traditional export crops has been steadily decreasing due to economic factors beyond the control of developing countries. For Central America, one of the most economically-affected regions in Latin America, the growing demand for fresh vegetables and fruits in North America presented a timely opportunity for crop diversification (Stanley, 1999). This region increased the production of non-traditional export crops (NTEC), such as tomato (*Lycopersicon esculentum* Mill.), chili peppers (*Capsicum* spp.), melon (*Cucumis melo* L.), broccoli (*Brassica oleracea* L.), eggplant (*Solanum melongena* L.), and okra (*Abelmoschus esculentus* L. Moench), over 75% between 1984 and 1989 (Thrupp, Bergeron, and Waters, 1995). Guatemala, one of the most representative Central American countries, had approximately 250 organized exporters and over 100,000 farmers producing NTEC by 1996, which created an additional 83,000 jobs associated with these mixed cropping systems (GEXPRONT, 1996).

Unfortunately, the boom of NTEC in Latin America came at a time when most national agricultural research institutions were being drastically downsized, as a result of the economic recession and shift in research priorities from food production to natural resource management. The globalization of the economy and dismantling of trade barriers further downplayed the issue of food security as a regional priority (Robinson, 1998). A negative consequence of these new policies was the disruption of effective technical support for producers of NTEC. Thus, producers had to rely on pesticides to meet the strict quality standards for fresh vegetables in international markets. As a result, many shipments of NTEC were rejected due to high levels of pesticide residues, with consequent losses to both exporters and producers (Thrupp, Bergeron, and Waters, 1995). In the mean time, traditional food crops, such as corn (*Zea mays* L.) and beans (*Phaseolus vulgaris* L.), were displaced to marginal lands in most countries in Latin America, with a subsequent drop in their productivity and capacity to satisfy the internal food demand. Even the main common bean- and corn-producing countries in Latin America found themselves importing these food staples from Asia and some industrialized countries. Ultimately, these unsustainable mixed cropping systems started to collapse due to severe outbreaks of pests and diseases, which caused significant and often total production losses in both traditional and non-traditional food and industrial crops.

The objective of this paper is to point out some of the biological and socioeconomic factors associated with the mixed cropping systems currently found in southwestern Guatemala, and their impact on the environment and wellbeing of small-scale farming communities.

MATERIALS AND METHODS

Southwestern Guatemala was selected as the case-study area because of the existence of traditional and non-traditional crops, and the presence of whitefly (*Bemisia tabaci*). This whitefly species is currently considered the most important pest and vector of plant viruses affecting both traditional and non-traditional crops in Middle America (Morales and Anderson, 2001). Moreover, *B. tabaci* has been largely responsible for the excessive application of pesticides in broad-based cropping systems; serious environmental and food contamination; and high crop production costs. Finally, the agricultural communities found in this region have undergone a gradual process of cultural change, from an indigenous to "ladino" (defined as people who have never been or are not anymore part of an indigenous community) status, which has paralleled the shift from subsistence to commercial agriculture.

The municipality of Monjas has a population of approximately 28,000, of whom 55% live in the rural area. Additionally, three neighboring villages in the municipality of Santa Catarina Mita, and one village in the municipality of El Progreso, were also included because these villages were socioeconomically integrated to the community of the Valley of Monjas. Considering the need to conduct a limited but detailed examination of a relatively small number of people in each village, a "case study" methodology was chosen. Preliminary interviews with representative growers selected by local extension specialists of the national agricultural research program (ICTA) were conducted in each village to guide the survey and test the questionnaire. The "geographical area of coverage" was each of the 16 villages in the three municipalities selected for this study (Table 1). At the village level, the number of respondents was usually less than 100, and the subjects of the study were individual farmers selected at random. Sample size (n) was determined according to the formula: $n = Z^2 \times p \times q/S_E$, where Z = confidence interval (95%); p = proportion of growers reporting whitefly/geminivirus problems with their crops; q = proportion of growers that do not report whitefly/geminivirus problems with their crops; and S_E = Standard error (Palma, 1999). The frequency of enumeration was a single visit to each respondent, and data was collected through individ-

TABLE 1. Geographic location of villages surveyed and number of farmers interviewed

Department	Municipality	Village	No. Farmers
Jalapa	Monjas	Achiotes	8
		Achiotillos	4
		Garay Viejo	6
		La Campana	13
		Llano Grande	23
		Mojarritas	6
		Morazán	9
		Piedras Blancas	5
		Plán de la Cruz	4
		ZSan Antonio	18
		San Juancito	9
		Terrones	14
Jutiapa	Santa Catarina Mita	Jocote Dulce	7
		Magueyes	4
		Uluma	1
	El Progreso	El Ovejero	7
Total			138

ual interviews (Casley and Lury, 1989). The questionnaire was designed to collect basic information on: respondents' basic data, geographic location of farms; cropping systems; whitefly problem; climatic conditions; pest/disease management practices; production costs; and farmers' decision-making process. A total of 56 variables were coded for statistical analysis. Some of the variables were further subdivided according to the various crops mentioned by farmers. The questionnaire had 43 questions, and was designed to be completed in approximately 30 minutes. The final survey was completed in two weeks.

RESULTS AND DISCUSSION

The total number of respondents was 138, with the majority (119) from the municipality of Monjas, department of Jalapa. This municipality spans an area of 256 km², and includes the main agricultural ecosystem of the region, the Valley of Monjas (14° 29′ 34″ and 89° 52′ 32″ W), located at 960 m above sea level. The annual mean temperature is 23.7°C, and the mean annual rainfall is 900 mm, distributed between May and October. The soils are fertile and there is an irrigation district in the valley. The analysis of the individual respondent variables, revealed that all of the 138 farmers interviewed were males. Of these farmers, 42.7% owned the land, 18.8% were tenants (cost of renting land ranged between US $50 and 220/ha), and 37.6% were 'medianeros' (i.e., farmers who contribute all or most of the inputs and shared the profits with the owner of the land). The majority (47.4%) of the farmers had been working the same land for over 10 years, and only 10% of the respondents had been working the fields less than two years. The 'medianeros' constitute a relatively new group of farmers (the mode for this group was three years working the land in the Monjas area). A personal variable added later showed that only 3.6% of the farmers interviewed had any technical training, and that 54.7% of the respondents had only a primary school education. About 41% of the farmers had no schooling. The average age of the farmers was 45 years and the size of their families was six members. There was a negative correlation (−0.4) between the age of the farmers and the cultivation of high value crops, such as tomato and chili pepper.

In the early 1970s, the major crop in this area used to be tobacco (approximately 1,400 ha), followed by common bean (720 ha), tomato (650 ha), corn (498 ha), broccoli (280 ha), and chili peppers (5 ha). At present, tobacco and tomato production were significantly reduced due to lower market prices and whitefly/geminivirus problems, respectively. Nevertheless, tobacco was still the predominant crop in the region because it is grown under contract for private companies, which guarantees its commercialization. Tomato and chili pepper were produced by the more advanced and solvent farmers in the region. The production of broccoli, on the contrary, increased relative to that of tobacco and tomato. However, the majority of farmers growing non-traditional crops had changed the composition of their cropping systems and corn was their main food crop (43.8% of the respondents). Common bean, on the contrary, was only cited by 5.1% of the farmers interviewed as their main food or cash crop. As for commercial crops, tobacco was mentioned as the main commodity by 32.8% of the growers, followed by tomato, broccoli, chili peppers, and cucurbits, according to 7.3, 5.8, 3.6, and 0.7% of the respondents, respectively. Average areas for these crops were: tobacco (4.1 ha), tomato (1.7 ha), common bean (0.9 ha), broccoli (2.3 ha), chili peppers (1 ha), corn (2.8 ha), and watermelon and cucumber (1.4 ha). Approximately 88% of the farmers interviewed responded that they practiced crop rotation (using maize, common bean, tomato, broccoli, cucurbits, etc.) to maintain soil fertility and reduce the incidence of pests and diseases. The average number of crops grown per farm was 2.9. Table 2 shows the distribution of crops, production systems, proportion of producers, area and presence or absence of whitefly/geminivirus problems for the four major crops grown in the valley of Monjas. It is apparent from these data that irrigation was primarily used for tobacco production, whereas food crops were grown mainly under rainfed conditions. Regarding farmers' perception about profitability, the order of crops closely reflected the farmers' choice for their preferred cash crop: tobacco, tomato, broccoli, common bean, chili peppers, and corn. In the case of food crops, corn occupied the last place of all crops in terms of profitability, whereas common bean was in third place.

In reference to the *B. tabaci* problem, nearly all farmers (99.3%) recognized the whitefly as an insect, and 96.4% considered it as an important pest. Approximately 65% of the farmers cited tomato, common bean, chili peppers, and tobacco as the crops most affected by *B. tabaci*. Most of the growers believed that the whitefly and/or geminiviruses this vector transmits, could cause between 50 and 75% yield losses in tobacco, tomato, common bean, broccoli,

TABLE 2. Cropping systems, producers and whitefly/geminivirus problems in the valley of Monjas, Jalapa, Guatemala.

Crops	% Growers		Area (ha)		Wf/Gv* problem
	Irrigated	Rainfed	Irrigated	Rainfed	
Tobacco	29.7	31.4	145	206	Yes
Maize	2.9	47.0	15	287	No
Tomato	8	10	19	71	Yes
Common bean	6.5	21	8	65	Yes

* Wf = whitefly; Gv = geminivirus

and chili. Hence, 90% of the tobacco and tomato growers used insecticides to control *B. tabaci*, whereas only 10% of the farmers interviewed utilized geminivirus- or whitefly-resistant cultivars. This observation is explained by the fact that only common beans have been bred for resistance to whitefly-transmitted geminiviruses in this region of the world. Most bean growers (84%) controlled whitefly-transmitted viruses with insecticides against the insect vector, and only 16% used resistant varieties. In the case of chili peppers, all of the farmers used pesticides to control the whitefly/geminivirus problems. Interestingly, only biological insecticides were used on broccoli due to a specific market demand.

The problem of pesticide abuse was notorious in these mixed cropping systems. Approximately 55% of the farmers claimed that agricultural 'technicians' helped them select agrochemicals. However, there was no common criterion among farmers for pesticide application. Approximately 23% of the farmers applied pesticides as a preventive measure; 12-15% applied pesticides on a calendar basis; and 1.5 to 8% applied when farmers noticed insect damage. The remaining farmers applied pesticides in an irregular pattern. Pesticide abuse was particularly apparent in the number of applications per crop. The number of pesticide applications in tobacco, tomato, common bean, broccoli, and chili ranged from 1 to 50 per crop cycle. According to the majority of the farmers interviewed, tobacco was usually sprayed 1 to 6 times, whereas tomato was treated from 3 to 30 times. Common bean was sprayed between two and four times, and chili peppers between four and 15 times. The more expensive systemic pesticides (e.g., carbamates and imidacloprid) were predominantly used in crops such as, tomato, tobacco, and watermelon. The remaining insecticides were mostly pyrethroid, organo-phosphate or organo-chlorinated compounds. Broccoli is an interesting case because growers are required to use only biological insecticides on this crop. For the remaining crops, farmers often use 'cocktails' of insecticides, fungicides and bactericides. The number of different pesticide products applied to each crop were: 11 commercial products in tomato, nine in tobacco, nine in cucurbits, eight in common bean, four in chili pepper, one in corn, and three biological products in broccoli. Pesticide abuse also resulted in high production costs (Table 3).

The apparent gender bias observed in this survey can be explained by the fact that males are in charge of most agricultural activities in these rural communities, except in home gardens (Noval, 1992). The farmers interviewed can be classified as traditional, small-scale farmers, according to the number of years they had been working in this region, and the average area (0.9-4.1 ha) occupied by the different crops analyzed. Corn and common bean are traditional food crops grown since pre-Columbian times, whereas cash crops, such as tobacco, were introduced in the region soon after the creation of an irrigation district in the early 1970s. Thus, tobacco could still be considered as a

TABLE 3. Average crop production costs,* including crop and whitefly protection costs (USD/ha), in the case-study region selected in southwestern Guatemala.

Crop	TCPC ($)	CPC ($)	CPC%	WFCC ($)	WFCC%
Tobacco	2,259	1,950	86.3	1,374	60.8
Tomato	2,018	1,968	97.5	502	24.8
Common bean	414	219	52.9	52	12.5
Broccoli	1,363	-	-	191	14.0
Chili	2,653	1,559	58.8	505	19.0
Cucurbits	918	437	47.6	160	17.4
Corn	850	299	35.2	-	-

* TCPC: Total Crop Production Cost; CPC: Crop Protection Cost; WFCC: Whitefly Control Cost; % of TCPC; - = Data not available.

NTEC for this area, although it is classified as a traditional export crop in national terms. The expansion of tobacco plantings in the 1970s was associated with the emergence of *B. tabaci* as a new pest in this region. The whitefly outbreaks brought about a new problem to common bean production, Bean golden yellow mosaic virus (BGYMV). This virus attacked the local common bean landraces causing significant yield losses. Fortunately, BGYMV-resistant common bean varieties were released in Guatemala in the early 1980s, and were rapidly adopted by most farmers in southwestern Guatemala (Viana, 1997).

With the collapse of tobacco prices in the mid-1970s, there was a significant reduction in the area planted to this crop. As an alternative, tomato production was intensified until the late 1980s, as one of the first NTEC grown in this region. Unfortunately, a geminivirus disease known as "acolochamiento" reduced the tomato area from 650 ha to about 70 ha, and tomato was no longer considered a viable NTEC (GEXPRONT, 1996). At that time, broccoli became the predominant NTEC in this region, occupying approximately 650 ha in 1998. Although broccoli is not significantly affected by *B. tabaci* or geminiviruses, the whitefly reproduces abundantly on the stubble of this crop after harvest (February-March), and generates *B. tabaci* outbreaks in April. Corn is not affected by either *B. tabaci* or geminiviruses and it remains the predominant food crop in this area, and most of Latin America. Nevertheless, both corn and common bean, have been gradually displaced from the irrigated and more fertile soils to marginal areas in this region of the world.

The composition of crops in southwestern Guatemala reflects the trend from subsistence to commercial agriculture in small-scale farming communities. This trend has been facilitated by the availability of additional sources of income from private investors or migrant labor (i.e., relatives working in the United States). This has helped to defray the high costs of producing non-traditional crops. For example, the cost of planting a hectare of common bean ranges between US $300-400, whereas tomato costs between US $2,000-4,000/ha. However, the net profit expected for common bean ranges between US $100-200/ha, whereas the net profit for tomato often exceeds US $2,000/ha (Morales et al., 2000; Thrupp, Bergeron, and Waters, 1995).

Considering the high investment required to grow non-traditional crops, it is easy to understand the emphasis on preventive chemical control. In fact, up to 80% of the total cost of producing some NTEC corresponds to pesticides. Whitefly control accounts for 35% of the average chemical protection costs associated with the production of NTEC. The relatively low use of pesticides and crop protection costs observed for corn, are probably related to the crop's resistance to *B. tabaci* and the geminiviruses that this vector transmits. The cost for pest and disease protection in common bean was lower than for the other crops, which suggests that farmers are using improved bean cultivars (although only 16% of the bean growers interviewed acknowledged their use, probably due to seed exchange among farmers without proper identification of the materials). A preliminary field survey conducted prior to the implementation of this case-study, showed that most farmers were planting improved common bean cultivars (e.g., upright architecture). The lack of proper technical assistance is blamed by 43.5% of the farmers for the abandonment of various crops affected by *B. tabaci* and/or geminiviruses transmitted by this whitefly species. Only 24% of these farmers have attempted to grow whitefly/geminivirus-susceptible crops again, using new pesticides currently available to control *B. tabaci* (e.g., imidacloprid).

CONCLUSION

It is apparent that the traditional cropping systems of many small farming communities in Guatemala incorporated non-traditional, high value crops, as observed in previous studies (Carletto, Janvry, and Sadoulet, 1996; Carter, Barham, and Mesbah, 1996; Morales et al., 2000). The trend from subsistence to commercial agriculture is interpreted by some development agencies as an indication of agricultural development. According to these agencies, NTEC help repay foreign loans, reduce dependence on traditional export crops, create jobs, and stabilize the economy (Thrupp, Bergeron, and Waters, 1995; Stanley, 1999). However, the shift to commercial agriculture has often taken place at the expense of traditional food crops, such as common bean and corn. Fortunately, the price fluctuations and high risk associated with the production of

NTEC have convinced most small-scale farmers of the need to reserve a significant portion of their land for food crops (Carter, Barham, and Mesbah, 1996; Morales et al., 2000). In this and a previous study conducted in the department of Baja Verapaz, Guatemala (Morales et al., 2000), it became evident from the proportion of the available land planted to corn, that this cereal is the basis of food security in these communities. Common bean was regarded both as a food staple and alternative cash crop by over 60% of the farmers interviewed in the case-study conducted in Baja Verapaz.

Small-scale farmers were initially able to participate in the NTEC boom because the use of family labor reduced production costs for commercial exporters, who contracted them for production of NTEC. However, the tendency of small-scale farmers to abuse pesticides forced commercial exporters to re-introduce direct production of NTEC or contract their production only with large-scale farmers (Carter, Barham, and Mesbah, 1996). The pesticide abuse problem is linked to the lack of technical assistance for most small-scale farmers. In this case study, only 23.4% of the farmers interviewed had received technical assistance. These problems have also prevented small-scale farmers involved in NTEC-production from acquiring more land (Carletto, Janvry, and Sadoulet, 1996). Thus, policy makers will have to provide small-scale farmers with adequate support (e.g., credit, crop insurance, and technical assistance), if they are to improve their socioeconomic situation through a "broad-based sustainable development." This concept is defined as "equitable opportunities for poor farmers; guaranteeing food security, and developing agricultural practices that are economically viable and environmentally sound" (Thrupp, Bergeron, and Waters, 1995). It is evident from various economic studies conducted on this theme that proper technical assistance in pest and disease control practices is critical for the production of both traditional and non-traditional crops by small-scale farmers (Thrupp, Bergeron, and Waters, 1995; Carletto, Janvry, and Sadoulet, 1996; Carter, Barham, and Mesbah, 1996).

REFERENCES

BID (1994). *Progreso Económico y Social en América Latina.* Informe 1994. BID, Washington.

Carletto, C., A. Janvry, and E. Sadoulet. (1996). *Knowledge, toxicity, and external shocks: the determinants of adoption and abandonment of non-traditional export crops by small-holders in Guatemala.* Department of Agricultural and Resource Economics, University of California, Berkeley, Working Paper No. 791. 32 p.

Carter, M.R., B.L. Barham, and D. Mesbah. (1996). Agricultural export booms and the rural poor in Chile, Guatemala and Paraguay. *Latin American Research Review* 31: 33-63.

Casley, D.J. and D.A. Lury. (1989). *Data collection in developing countries.* Second edition. Oxford, UK: Clarendon Press.

GEXPRONT. (1996). *Exportaciones Agícolas no Tradicionales: Situación Actual y Estrategia Futura.* Gremial de Exportadores de Productos No Tradicionales. Guatemala.

Morales, F.J., J.A. Sierra, R. Ruano, M. Osorio, M. Landaverri, J.L. Ordoñez, and Viana, A. (2000). *The socioeconomic and environmental impact of non-traditional cropping systems on small farming communities in the department of Baja Verapaz, Guatemala.* CIAT, Cali, Colombia.

Morales, F.J. and P.K. Anderson. (2001). The emergence of whitefly-transmitted geminiviruses in Latin America. *Archives of Virology,* 146: 415-441.

Noval, J. (1992). *Resumen etnográfico de Guatemala.* Editorial Piedra Santa, Guatemala.

Palma, E.A. (2000). *Implicaciones socioeconómicas generadas por el ataque de mosca blanca (Bemisia spp.), Monjas, Jalapa.* Tesis, Universidad Rafael Landivar, Guatemala.

Robinson, W.I. (1998). (Mal)Development in Central America: Globalization and social change. *Development and Change* 29:467-497.

Stanley, D.L. (1999). Export diversification as a stabilization strategy: The Central American case revisited. *The Journal of Developing Areas* 33: 541-548.

Thrupp, L.A., G. Bergeron, and W.F. Waters. (1995). *Bittersweet harvests for global markets: Challenges in Latin America's export boom.* World Resources Institute. Washington, DC.

Viana, A. (1997). *Estudios socioeconomicos.* pp. 51-57 In: Informe Técnico por Resultados, POA 1996-1997, Profrijol. Guatemala.

The Future of Cereal Yields and Prices: Implications for Research and Policy

Mark W. Rosegrant
Michael S. Paisner
Siet Meijer

SUMMARY. Crop yields experienced significant increases in much of the world over the last 30 years, in large part due to investment in agricultural research, irrigation, and rural infrastructure. However, recent declines in yield growth are major causes of concern for future food security. Especially in Asia and sub-Saharan Africa the future prospects for food production are problematic. Both regions must meet strong growth in food demand–driven mainly by income growth in Asia and by population growth in sub-Saharan Africa. The production challenges in the two regions are fundamentally different: much of Asia is facing post-Green Revolution challenges of how to sustain crop yield growth following widespread adoption of modern varieties while dealing with environmental degradation in some high intensity crop production systems. Sub-Saharan Africa has yet to experience a Green Revolution, and must boost crop yield growth in the face of less favorable agroclimatic environments than in Green Revolution breadbaskets of Asia. In this paper, we review the past and possible future developments in yield growth

Mark W. Rosegrant (E-mail: m.rosegrant@cgiar.org) is Senior Research Fellow, Michael S. Paisner is former Senior Research Assistant, and Siet Meijer is Research Analyst, International Food Policy Research Institute (IFPRI), 2033 K Street, NW, Washington, DC 20006-1002.

[Haworth co-indexing entry note]: "The Future of Cereal Yields and Prices: Implications for Research and Policy." Rosegrant, Mark W., Michael S. Paisner, and Siet Meijer. Co-published simultaneously in *Journal of Crop Production* (Food Products Press, an imprint of The Haworth Press, Inc.) Vol. 9, No. 1/2 (#17/18), 2003, pp. 661-690; and: *Cropping Systems: Trends and Advances* (ed: Anil Shrestha) Food Products Press, an imprint of The Haworth Press, Inc., 2003, pp. 661-690. Single or multiple copies of this article are available for a fee from The Haworth Document Delivery Service [1-800-HAWORTH, 9:00 a.m. - 5:00 p.m. (EST). E-mail address: docdelivery@haworthpress.com].

in Asia and sub-Saharan Africa, the possible causes of the current yield growth problems and the management challenges both regions face to overcome them. *[Article copies available for a fee from The Haworth Document Delivery Service: 1-800-HAWORTH. E-mail address: <docdelivery@ haworthpress.com> Website: <http://www.HaworthPress.com> © 2003 by The Haworth Press, Inc. All rights reserved.]*

KEYWORDS. Food security, Asia, sub-Saharan Africa, cropping systems, IMPACT model

INTRODUCTION

The last 30 years have shown the remarkable resiliency and growth potential of world agriculture. Between 1967 and 1997, global cereal production increased 84%, significantly surpassing the increase in population of 67%. As a result, global per capita cereal production rose from 294.4 kg capita^{-1} in 1967 to 325.2 kg capita^{-1} in 1997. Meanwhile, between 1982 and 1997, real world wheat (*Triticum aestivum* L.) prices declined by 28%, rice (*Oryza sativa* L.) prices by 29%, and corn (*Zea mays* L.) prices by 30%. Malnutrition among children under five in developing countries realized impressive declines from an aggregate rate of over 45% in 1970 to 31% in 1995 (FAO, 2000; Smith and Haddad, 2000). Agricultural research, particularly the development of improved wheat and rice varieties during the Green Revolution and the continued improvement of hybrid and improved corn varieties, played a fundamental role in bringing about these tremendous improvements in global per capita cereal availability and food security.

Serious questions remain, however, about the ability of world agriculture to continue to realize significant increases in cereal availability, particularly in developing countries, over the next decades. Suitable arable area throughout much of the developing world, particularly Asia, is already under cereal production, and the increasing use of externally derived inputs on most cropping systems has had a negative impact on the health of the natural resource base. At the other end of the spectrum, some cropping systems, particularly in sub-Saharan Africa (SSA), remain enmeshed in a vicious cycle of low input application, low productivity, and deteriorating soil nutrient levels (Cleaver and Schreiber 1994). These concerns raise the question whether it will be possible to keep producing enough food for the growing population while keeping prices low. The first section of this paper will discuss yield trends and threats to agricultural production systems in Asia and SSA, touching on the role that crop management research can play in alleviating some worrisome trends in these regions. The next section will describe the International Model for Policy

Analysis of Agricultural Commodities and Trade (IMPACT) and present highly condensed results from the IMPACT baseline (see Rosegrant et al. 2001a, for a more comprehensive treatment) followed by a series of alternative IMPACT scenarios estimating the effects of dramatic shifts from the baseline for yield growth rates in both the developed and the developing worlds on cereal prices and food security. We will conclude the paper with a comprehensive conclusion based on the issues raised in the regional discussions as well as the IMPACT projection results.

DECLINING CROP YIELD GROWTH:
THREATS TO FUTURE FOOD SUPPLY AND PRICES

Concern with declining yield growth rates over the last two decades has dominated much recent thinking about the future of agricultural production, but dire predictions regarding the imminent collapse of agricultural yield growth should receive somewhat skeptical treatment for two main reasons. First, declining yield growth rates can be consistent with long-term linear increases in cereal yields, and may simply reflect constant unitary increases over an increasing base level. Second, as the share of lower yield regions in world agricultural production grows, the aggregate worldwide yield growth rate may decline even as yield growth rates in each individual region do not (Dyson, 1996).[1] Despite these important structural elements behind declining yield growth rates, however, several worrisome trends are evident, particularly the fact that yield growth has clearly slowed somewhat at the regional level. It appears that many yield gains in recent decades were attributable to advances that may not be replicable, including higher crop planting density through changes in plant architecture, higher usable food product weight as a fraction of total plant weight, multiple harvesting, introduction of strains with greater fertilizer responsiveness, and better management practices. Crop yields may be approaching their physical limitations in some high yield systems, primarily in developed countries, and the maximum yield potentials of rice and corn have changed little over the past three decades (Fedoroff and Cohen, 1999). These real causes for concern lead to the almost unavoidable conclusion that crop yield growth will be slower over the projections period than it has been in the past.

A large number of complex factors will determine yield growth to the year 2020, including: (a) rates of fertilizer and pesticide application, (b) the pace of research investment and advances in biotechnology, (c) physical and human capital development, (d) the degradation and misuse of the natural resource base, with water shortages particularly relevant, and (e) long-run trends in cereal prices. While technological developments and large increases in external inputs will play a significant role in determining future yields, enhanced crop management techniques may hold the key to maintaining strong yield growth by reducing the need for higher input applications, permitting the widespread adoption of enhanced hybrids, and slowing the degradation of the natural resource base. The following two sections will explore past yield trends and future prospects in two particularly fragile developing regions: Asia and SSA.

Management Challenges of the Post-Green Revolution Period in Asia

Impacts of the Green Revolution versus Recent Developments

During the Green Revolution, growth in cereal crop productivity resulted from an increase in land productivity; it occurred through strong policy support and good market infrastructure in areas of growing land scarcity and/or areas with high land values. High levels of research and infrastructure investment, especially in irrigation infrastructure, led to rapid intensification of the lowlands, with the result that both irrigated and high-rainfall lowland environments became the primary source of food supply for Asia's escalating population (Rosegrant and Pingali, 1994; Pingali, Hossain, and Gerpacio, 1997).

However, rapid factor accumulation is often downplayed in discussions of the new seed technologies that represented the most visible catalyzing agent for the Green Revolution. In fact, Byerlee (1998) points out that two non-technological paradigms dominated efforts to raise rice and wheat yields throughout the Green Revolution period: the first emphasized heavy application of external inputs, particularly fertilizer and pesticides in combination with high yielding varieties, while the second emphasized the "package approach" which promoted crop-specific input packages to be applied across wide geographic areas. Government policy throughout Asia has historically encouraged high rates of input application through subsidies, price controls and over-valued exchange rates (Rosegrant and Hazell, 2000). As Murgai (1999) points out for the Punjab region of India, fertilizer and capital input accumulation, rather than total factor productivity (TFP) growth was responsible for the preponderance of yield growth during the initial period of high yielding varieties adoption between 1965 and 1973, and TFP growth only accelerated rapidly between 1974 and 1984. While Japan and Taiwan accounted for almost all Asian fertilizer use before the Green Revolution, fertilizer use is now ubiquitous across the region, and accounts for approximately one-half of total world use at 125 kg of nutrients ha^{-1} (Byerlee, 1998).

The Green Revolution paradigms were remarkably successful, and had a dramatic effect on food security in Asia, permitting the two most populous countries in the region, China and India, to escape rising import dependence and periodic food shortages. However, recent signs indicate that phenomenal Green Revolution growth in wheat and rice productivity has slowed, especially in the intensively cultivated lowlands. Between 1967 and 1997, cereal yield growth rates declined steadily throughout almost all of Asia (except India). Though rice yields took off later in India than in China, the precipitous drop from 3.4% annual yield growth in the 1982 to 1990 period, to 1.3% growth between 1990 and 1997 is a cause of concern for a region that still has a massive amount of food insecurity despite overall cereal self-sufficiency (see Table 1 for yield data on cereals in other Asian regions).

The threat to South and Southeast Asian rice and wheat yields seems particularly acute, since despite the slowing of Chinese rice yield growth rates in the 1990s, a process of convergence between South and Southeast Asian rice yields and those in China does not seem to have occurred over the past 30 years. While only 1.1 Mg ha^{-1} separated Chinese rice yields from Indian rice yields in 1967, the introduction of hybrid rice cultivars specifically suited to irrigated temperate climates spurred a process whereby Chinese rice yields were 2.3 Mg ha^{-1} higher than Indian yields and a full 2.0 Mg ha^{-1} higher than Southeast Asian yields in 1997 (Table 1).

TABLE 1. Growth of rice, wheat, and corn production and yields in Asia, 1967-97.

	Crop yield growth			Crop yield			
	1967-82	1982-90	1990-97	1967	1982	1990	1997
	% year^{-1}			Mg ha^{-1}			
Rice							
China	2.8	2.1	1.6	2.1	3.2	3.7	4.2
India	2.0	3.4	1.3	1.0	1.3	1.7	1.9
East Asia	2.8	2.0	1.5	2.1	3.3	3.8	4.2
South Asia	1.9	3.1	1.3	1.1	1.3	1.7	1.9
Southeast Asia	3.1	1.8	1.2	1.1	1.8	2.0	2.2
All Asia	2.5	2.1	1.1	1.4	2.0	2.4	2.6
Wheat							
China	5.5	3.0	3.0	1.1	2.5	3.1	3.8
India	4.1	3.3	2.0	0.9	1.7	2.2	2.6
East Asia	5.5	3.0	3.1	1.1	2.4	3.1	3.8
South Asia	4.1	2.8	2.0	0.9	1.7	2.0	2.4
Southeast Asia	5.3	−2.9	−0.7	0.6	1.2	1.0	0.9
All Asia	4.7	2.9	2.5	1.0	2.0	2.6	3.1
Maize							
China	4.1	3.4	1.9	1.8	3.3	4.3	5.0
India	1.1	2.7	1.6	1.0	1.2	1.5	1.7
East Asia	4.1	3.4	1.6	1.8	3.3	4.4	4.9
South Asia	0.9	2.4	1.4	1.1	1.2	1.5	1.7
Southeast Asia	2.6	2.7	4.0	1.0	1.5	1.8	2.4
All Asia	3.4	3.3	2.3	1.5	2.5	3.2	3.8

Source: Computed from FAOSTAT data, FAO (2000).

One reason behind India's relatively slow crop yield growth *vis a vis* China has been the fact that adoption of Green Revolution technologies in India has until recently been limited to a few geographical regions in the northwestern breadbasket, namely Punjab, Haryana, and parts of Uttar Pradesh, and to some southern rice areas such as Tamil Nadu. The northwestern breadbasket contributes 70% of national wheat output on just 14% of total cropped area, thus accounting for the relatively better performance of Indian wheat yields, which only lagged those in China by 1.3 Mg ha^{-1} in 1997 (Hopper, 1999).[2]

Main Causes of Slowing Yield Growth

Declining world cereal prices and factors related to the increasing intensification of cereal production have caused the slowdown in cereal yield growth in developing Asian countries since the early 1980s. Declining cereal prices caused a direct shift of land out of cereals into more profitable cropping alternatives and slowed growth in input use, thus hurting yields. More importantly over the long run, declining world prices have also slowed investment in crop research and irrigation infrastructure, with consequent effects on yield growth (Rosegrant and Pingali, 1994; Rosegrant and Svendsen, 1993).

Historically, agricultural investments in Asia and most developing countries have focused on irrigated and high-potential rainfed areas in order to increase food production. This strategy has been widely used with the idea that investments in these areas will trickle down to help reduce poverty in the less favored areas (LFA). Increased food production in the irrigated and high-potential rainfed areas is expected to lead to a reduction in food prices, thus helping to alleviate poverty in LFAs as well. Under the assumption that the potential for agricultural development is limited in many LFAs, policies that emphasize the development of non-farm sectors of the economy and migration out of these areas are often suggested as long-term solutions in LFAs. Although these strategies have worked in some areas, many LFAs have fallen even further behind due to the poor growing conditions, inadequate rainfall and lack of investment (Rosegrant and Hazell, 2000). Despite some out-migration to more rapidly growing areas, population size continues to grow in many LFAs and this growth has not been matched by increased yields. The result is often worsening poverty and food-insecurity problems, as well as the widespread degradation of natural resources. The use of high levels of inputs and achievement of relatively high wheat and rice yields in parts of Asia have made it more difficult to sustain the same rates of yield gains, as yields in these regions approach the economic optimum levels.

By 1990, modern varieties of rice occupied 74% of rice area in Asia, accounting for all irrigated area plus about one-third of the rainfed lowlands, while three-quarters of the more recent adoption took place on rainfed land,

and adoption rates for improved varieties of corn and wheat in rainfed environments are approaching those in irrigated areas (Byerlee, 1996). Opportunities for further expansion of modern variety use are essentially exhausted in existing irrigated areas but have potential in rainfed areas, though drought and submergence risk severely constrain dissemination in these environments (Pingali, Hossain, and Gerpacio, 1997). Overall, the decline in yield growth potential has been particularly evident in India, representing a combination of both the full diffusion of modern technologies over the more advanced northwest and the aforementioned continued stagnation of agricultural productivity in most of the rest of the country (Hopper, 1999).

Environmental and resource constraints have also contributed significantly to the slowdown in yield growth evident over the last two decades. Increased land-use intensity has led to the necessity of higher input requirements to sustain current yield gains. Moreover, Pingali, Hossain, and Gerpacio (1997) argue that the practice of intensive rice monoculture itself contributes to the degradation of the paddy resource base and hence declining productivities. Declining yield growth trends can be directly associated with the ecological consequences of intensive rice monoculture systems, including buildup of salinity and water-logging, use of poor quality groundwater, nutrient depletion and mining, increased soil toxicities, and increased pest buildup, especially soil pests. Salinization affects an estimated 4.5 million ha in India, and water-logging affects a further 6 million ha (Abrol, 1987; Chambers, 1988; and Dogra, 1986 as cited in Pingali, Hossain, and Gerpacio, 1997). Cassman and Pingali (1995) estimate a decline in yields due to nitrogen (N) depletion at all N levels of 30% over a 20-year period. Many of these degradation problems are also prevalent in the irrigated lowlands, where farmers grow wheat in rotation after rice (Hobbs and Morris, 1996). Experimental evidence from India shows that constant application of a low level of inputs over an extended period has led to declining yields in rice-wheat systems (Paroda, 1998).[3]

However, intensification *per se* is not the root cause of lowland resource base degradation. A policy environment that encourages monoculture systems and excessive or unbalanced input use is partially to blame. Trade policies, output price policies, and input subsidies, particularly for water and fertilizer, have all contributed to the unsustainable use of the land base. The dual goals of food self-sufficiency and sustainable resource management are often mutually incompatible. Policies designed for achieving food self-sufficiency tend to undervalue goods not traded internationally, especially land, water, and labor resources. As a result, the successful drive for food self-sufficiency in Asian countries with an exhausted land frontier came at a high ecological and environmental cost. Appropriate policy reform, both at the macro as well as the sector level, will go a long way towards arresting current degradation trends, but the degree of degradation in many regions will pose severe policy challenges (Pingali and Rosegrant, 1998). Also input markets have been inefficient and property rights weak throughout the region since independence, thus limiting long-term entrepreneurial planning and undermining both the will and ability of farmers to invest in the profitability of their land (World Bank, 2000). However, even if environmental degradation in intensive Asian cropping systems is stabilized, it is unlikely that previous crop yield growth rates will be restored, given continued declines in research and infrastructure investments.

Integrated Cropping Management

As cropping systems management research will play an essential role in halting and perhaps reversing degradation trends throughout Asia, enhancement of the knowledge base of individual farmers is the key to practically implementing theoretical concepts (Price and Balasubramanian, 1998). Several broad approaches will enable more efficient input use, including the adjustment of input levels to site and season-specific conditions, emphasis on the timing and methods of input application, and integration of a range of practices to enhance nutrient availability without increasing levels of input application (Byerlee, 1998). Diversity in cropping systems can increase production through biochemical and ecological complementation, and the introduction of additional species can also enhance overall system stability (Trenbath, 1999). For example, Trenbath (1999) reports that use of a two component intercrop can reduce the land area required to produce a certain yield at a fixed level of risk from what would be required if the crops were grown separately. Specific applications of cropping system management also include integrated pest management (IPM) strategies, involving the use of host-plant resistance, cultural practices, and biological control to maintain low pest populations; and integrated weed management, employing minimum tillage, minimum effective rates of herbicide use, and integrated agronomic practices to maintain the competitiveness of crops (Kon, 1993; Schoenly, Mew, and Reichardt, 1998).

So far, however, integrated management practices have not disseminated widely across Asia. The nature of these practices almost by definition limits them to the small scale, although the benefits can be substantial. For instance, Garrity (1999) reports for several thousand Philippine households in Mindanao, that use of vegetative buffer strips to conserve soil and sustain yields on steeply sloping cropland increased yields by about 0.5 Mg ha^{-1}. One recent wide-scale success has been the implementation of IPM in Indonesia, which achieved reduced pesticide use with no reported decline in rice productivity (Bhuiyan, Tuong, and Wade, 1998). IPM techniques have also spread to 42,500 ha in the Yunnan Province in Southern China and to 92% of farmers on the Mekong Delta, with the number of pesticide applications in the Mekong falling from 3.4 per farmer per season to one per farmer per season (Mew, 2000 and IRRI, 2000, as cited in Wood, Sebastian, and Scherr, 2001).

Integrated management strategies are highly knowledge intensive, and the ability of farmers to make use of available techniques will depend on a number of crucial variables, including the quality and reach of national extension services, farmer education, availability of alternative sources of information to those provided by the government, and development of communications infrastructure (Byerlee, 1998). Management techniques can have negative consequences on crop productivity if improperly implemented, and multi-species systems under the intensification stress that characterizes much of Asian rice production can be at risk of collapse in the absence of good management (Trenbath, 1999). In the case of IPM, neither the introduction of perennial plants nor higher plant diversity necessarily achieves reduced pest and disease risk in the targeted crop. For instance, the introduction into the cropping system of a plant species that harbors pests or diseases of other species may increase the risk of wider outbreak (Schroth, Krauss, and Gasparott, 2000). In order to gain the benefits from and reduce the risk from adoption of more complex crop systems extensive reform of national institutions, support services and markets will be required to better enable farmers to access and implement information specific to their circumstances (Byerlee, 1998). The strength of local research is especially relevant to the pace at which local farmers' adopt available technologies.

Boosting Yields in the Less Favored Areas of Asia

As noted above, the less favored areas of Asia have received relatively little attention as a source of crop production growth. The relative neglect of these areas appears increasingly untenable, given the failure of past patterns of agricultural growth to resolve growing poverty, food insecurity and environmental problems in many LFAs, and emerging evidence that the right kinds of investments can increase agricultural productivity to much higher levels than previously thought in many less-favored lands. Increased public investment in many LFAs may have the potential to generate competitive if not greater agricultural growth on the margin than comparable investments in many high-potential areas, and could have a greater impact on the poverty and environmental problems of the LFAs in which they are targeted (Hazell, Jagger, and Knox, 2000). If so, then additional investments in LFAs may actually give higher aggregate social returns to a nation than additional investments in high-potential areas (Fan, Hazell, and Haque, 2000; Fan, Zhang, and Zhang, 2002).

Rosegrant and Hazell (2000) note that the general development strategies that worked for the more favorable areas of Asia are also appropriate for LFAs, including promotion of broad-based agricultural development; improvement of technology and farming systems; ensuring equitable and secure access to natural resources; promotion of effective risk management; investment in rural populations and infrastructure; providing the appropriate policy environment; and reinforcing public institutions.

Sub-Saharan Africa: New Approaches to Reverse Yield Stagnation

Causes of Yield Stagnation and Potential Remedies

Cereal yield growth rates in SSA have consistently lagged well behind those in every other developing region, and yields only rose at an annual rate of 0.3% between 1990 and 1997. In the 1960s, 1970s, and early-1980s, most African governments attempted to accelerate the process of industrial development and ensure cheap urban food prices by taxing agriculture through measures that included overvalued exchange rates and price-depressing marketing board interventions in food markets (World Bank, 1994). Market reforms were expected to boost productivity by increasing the availability and encouraging the use of modern inputs. However, while strong production growth in both Northern, and Central and Western SSA during the 1990s may be an indication that market reforms are finally having a positive effect, results in the food crop sector have been disappointing (Kherallah et al., 2000). Structural and institutional constraints are continuing to exert a severe drag on agricultural performance in the region despite some degree of market liberalization. Two key problematic areas requiring immediate direct attention are ongoing soil fertility depletion and the need for intensification and diversification of smallholder farming (Sanchez et al., 2001).

The ongoing nutrient depletion of soils throughout SSA lies at the root of the failure of successive interventions to raise regional cereal yields. In particular, research into improved crop germplasm will not be effective if new varieties are planted in nutrient-depleted soil (Sanchez et al., 2001). Severe soil fertility depletion and erosion affect both marginal and high-quality rainfed lands, which are both frequently farmed under conditions characterized by inappropriate nutrient replacement or conservationist practices, pest problems, over-dependence on corn monoculture, and high water variability. Studies indicate that productivity losses from soil degradation since WWII have amounted to 25% across SSA (Oldeman, 1998 and Sanchez et al., 1997 as cited in Scherr, 1999). Bojo (1996) estimates economic loss from soil degradation as ranging from under 1% of agricultural GDP in Madagascar, Mali, and South Africa to between 2 and 5% in Ethiopia and Ghana to over 8% of agricultural GDP in Zimbabwe (Scherr, 1999). On originally fertile land, farm-survey data for the region show declines in corn yields ranging up to 4 Mg ha^{-1} under continuous corn-bean (*Phaseolus* sp.) rotation in the absence of fertilizer inputs. Scherr and Yadav (1996) estimate cumulative crop yield reductions from erosion at 6.2% across all SSA countries. Soil degradation will undoubtedly remain a serious problem in the next 20 years, particularly on marginal lands with rapidly growing populations in parts of the Sahel, mountainous East Africa, and the dry belt stretching from the coast of Angola to southern Mozambique (Cleaver and Schreiber, 1994). For instance, Lal (1995) predicts that water erosion alone will reduce crop productivity in SSA by 14.5% between 1997 and 2020.

Theories of induced technological innovation predict that growing population pressure will lead to higher yield growth rates as low-input agriculture becomes increasingly non-viable (Boserup, 1981). Population pressures throughout SSA are clearly intensifying, as shown by the decline in total per capita cropped area from 0.40 ha per capita in 1967 to 0.27 ha per capita in 1997.[4] However, population-induced innovation requires land-constrained populations to possess ready access to technologies and inputs. Cleaver and Schreiber (1994) point out that even as populations throughout SSA are losing their ability to practice shifting cultivation due to high population densities, they also continue to practice other elements of extensive cultivation, including low levels of technological and capital inputs, traditional land tenure and land husbandry practices, and traditional methods of resource acquisition. Exchange rate, tax, trade, and pricing policies in the context of stifling government control have squeezed out the private sector and stifled farmer investment in agricultural land, hindering uptake of the methods necessary for agricultural intensification and improved soil management (Cleaver, 1993).

Throughout SSA, low-levels of fertilizer use, only 8 kg ha^{-1} of arable and permanent cropland in 1997, represent a primary factor behind declining soil fertility. The past trends of fertilizer use and cereal yields in SSA compared to those in India are presented in Figure 1. While the removal of subsidies during the late-1980s and early-1990s was necessary to stimulate private sector participation in the market, the benefits of liberalization have not yet been realized due to a variety of factors, including trade barriers, political indifference, foreign exchange shortages, low crop prices, and a lack of institutional and physical infrastructure (World Bank, 2000). National governments must encourage fertilizer use in high-potential areas, put in place proper measures to ensure environmental sustainability, and address the high cost of fertilizers by lowering transport costs and raising scale-economies of international purchasing and shipment (Byerlee and Heisey, 1996; Bumb and Baanante, 1996). Given the fact that the region's total supply potential is only 8.4 million Mg (with demand of 3.5 million Mg in 1994/1995), SSA will have to import large quantities of fertilizer over the foreseeable future, thus necessitating stable and timely supplies of foreign exchange (Bumb and Baanante, 1996).

A number of researchers (Versteeg, Adegbola, and Koudokpon, 1993; Janssen 1993) have pointed to the suitability of labor-intensive techniques

FIGURE 1. Fertilizer consumption and cereal yield indices for sub-Saharan Africa and India, 1966-1998.

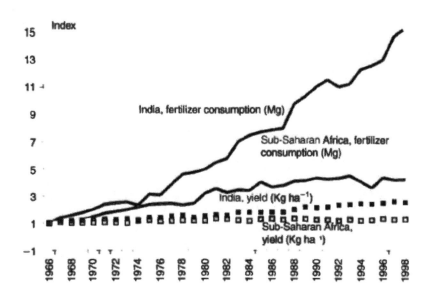

such as legume rotations, animal manures, and alley cropping as potential short-term fertilizer substitutes in cases of labor surplus (Byerlee and Heisey, 1996). Such organic solutions are inadequate on their own due to the high-level of crop nutrients required by much of SSA's degraded soil base, but organic and synthetic solutions have additive properties that could enhance overall nutrient replacement and provide tradeoffs between capital and labor (Vanlauwe, Aihou, and Houngnandan, 2001). Some potential strategies include the use of leguminous short term and tree fallows to increase N concentrations and maximize potassium recycling (Sanchez et al., 2001). Becker and Johnson (1998) highlight the effectiveness of site-specific, multi-purpose cover legumes as short-duration fallows capable of sustaining rice yields under intensified cropping, with use of these fallows in one study area increasing rice yields by 29% above the control.

As was pointed out in the Asian context, these alternative nutrient-replenishing practices are generally site-specific and highly knowledge-intensive, requirements that are problematic for smallholders lacking significant management capacity. Further research into how best to empower farmers to carry out proper cropping systems management is clearly required, and much of this research will have to be publicly funded and performed by researchers at the national level, since private sector research capacity in SSA is minimal. Areas in need of attention include the development of nutrient management systems for specific soils, low-cost soil rehabilitation techniques, methods for incorporating perennial crops in farming landscapes, and innovative incentive-structures to encourage long-term conservation of forest and grazing land (Scherr, 1999).

Alston et al. (2000) calculated the mean rate of return to local research in SSA at a high 34.3%, despite the lack of staff continuity and breeding strategies in many national programs. This indicates that local research and extension services have the capability to perform the tasks required of them, but need to be strengthened and better funded if they are to provide consistent and expanded services. Dynamic crop management programs involving extensive on-farm research have not traditionally been a national priority in SSA, with most governments preferring to place resources behind relatively less complex and costly research into improved germplasm, particularly high-yielding corn varieties (Byerlee and Heisey, 1996; Dowswell, Paliwal, and Cantrell, 1996).

Agricultural Research and Policies for Growth in Sub-Saharan Africa

As in Asia, rainfed and LFAs in SSA have not benefited much from the Green Revolution, although in recent years modern varieties also started to spread in the region. Both conventional and non-conventional breeding techniques are used to increase rainfed cereal yields. Three major breeding strategies include research to increase the harvest index, to increase plant biomass, and to increase stress tolerance (particularly drought resistance). The first of these may have only limited potential for generating further yield growth due to physical limitations, but the latter two exhibit considerable potential (Cassman, 1999; Evans, 1998; Rosegrant et al., 2001b). For example, the "New Rice for Africa," a hybrid between Asian and African species, was bred to fit the rainfed upland rice environment in West Africa. It produces over 50% more grain than current varieties when cultivated in traditional rainfed systems without fertilizer. In addition to higher yields, these varieties mature 30 to 50 days earlier than current varieties and are far more disease and drought tolerant than previous ones (WARDA, 2001).

If agricultural research investments can be sustained, the continued application of conventional breeding and the recent developments in non-conventional breeding offer considerable potential for improving cereal yield growth in rainfed environments in Africa. Cereal yield growth in farmers' fields will come both from incremental increases in the yield potential in rainfed and irrigated areas and from improved stress resistance in diverse environments, including improved drought tolerance. The rate of growth in yields will be enhanced by extending research both downstream to farmers and upstream to the use of tools derived from biotechnology to assist conventional breeding, and, if concerns over risks can be solved, from the use of transgenic breeding.

Extending research downstream through participatory plant breeding could play a key role for successful yield increases through genetic improvement in rainfed environments (particularly in dry and remote areas). Farmer participation in the very early stages of selection helps to fit the crop to a multitude of target environments and user preferences (Ceccarelli et al., 1996). Participatory plant breeding may be the only possible type of breeding for crops grown in remote regions; a high level of diversity is required within the same farm, or for minor crops that are neglected by formal breeding.

In order to assure effective breeding for high stress environments, it is also essential to move upstream to ensure deployment of diverse genes in new plant varieties. It is, therefore, crucial that the tools of biotechnology, such as marker-assisted selection and cell and tissue culture techniques, be employed, even if these countries stop short of true transgenic breeding. To date, however, application of molecular biotechnology has been limited to a small number of traits of interest to commercial farmers, mainly developed by a few life science companies operating at a global level (Byerlee and Fisher, 2000). Very few applications with direct benefits to poor consumers or to resource-poor farmers in developing countries have been introduced, although, the New Rice for Africa described above may show the way for the future in using biotechnology tools to aid breeding for breakthroughs beneficial to production in developing countries. Much of the science and many tools and intermediate products of biotechnology are transferable to solve high priority problems in the tropics and subtropics, but it is generally agreed that the private sector will not invest sufficiently to make the needed adaptations in these regions. Consequently, national and international public sectors in the developing world will have to play a key role, much of it by accessing proprietary tools and products from the private sector.

By the late 1990s, improved corn varieties were being grown on approximately 40% of the corn area in SSA, but the lagging development of sustainable management practices has kept yields well below the developing world average (Byerlee and Eicher, 1997). Accumulated farming experience simply cannot provide farmers with the knowledge-base necessary to effectively apply fertilizer, plant high-yield varieties with appropriate density, and weed early once soil fertility is restored. As corn production systems becoming increasingly science-based, it will be up to a variety of publicly and internationally funded information disseminators to enhance farmer knowledge, technical skills, and managerial capacity (Dowswell, Paliwal, and Cantrell, 1996).

In addition to greater attention to cropping systems and breeding techniques, future research efforts must diversify from a focus on corn to explore opportunities for alternative crops such as cassava (*Manihot esculenta* Crantz) and rice that have particular problems associated with African agro-ecological conditions. Technological diffusion has proven to be a major problem in SSA. Goldman and Block (1993) identified a number of commodities for which there exist under-utilized high-yielding varieties, including cassava (with potential yield increases of 50% on half of currently planted area), sweet potato (*Ipomoea batatas* L. Lam.), and rice (for both irrigated and mangrove environments) (cited in Spencer, 1994). Nevertheless, the experience with corn in SSA shows that small farmers will make use of improved seeds and complementary inputs provided the technology, infrastructure, and appropriate macroeconomic environment. In order to address the conditions particular to the region, improved technology packages need to place a premium on efficient input use and maximizing returns to labor and cash during early adoption. Above all, effective research must be embedded within an overall framework for agricultural development that emphasizes smallholder commercialization, private sector initiative at all levels, decentralized public participation, trade, and poverty alleviation (World Bank, 2000).

As explained above, crop research is essential to increasing crop yields. However, cereal yields can also be increased through improved policies and increased investment in areas with exploitable yield gaps (the difference between the genetic yield potential and actual farm yields). Such exploitable gaps may be relatively small in high intensity production areas such as most irrigated areas, where production equal to 70% or more of the yield gap is achieved (Cassman, 1999). However, with yield potential growing significantly in rainfed environments exploitable yield gaps are still considerably higher in rainfed areas, because remoteness, poor policies and a lack of investments have often isolated these regions from access to output and input markets, so farmers face depressed prices for their crops and high prices or lack of availability of inputs. Significant increases in rural roads and communications will therefore be necessary to complement agricultural research in boosting crop productivity in SSA (Rosegrant et al., 2001a). Finally, it must be noted that successful crop production growth based on agricultural research, policy reform, and infrastructure investment in Africa will be dependent on a reversal of the endemic cycles of violence and weak governance inflicting many countries in the region.

THE IMPACT MODEL:
BASELINE RESULTS AND ALTERNATIVE SCENARIOS

Model Description

In this section the baseline projections of the IMPACT model developed by IFPRI will be discussed as well as high and low yield scenario projections.

This will give some insight in what the effect of different yields can have on future food security and prices.

The IMPACT model covers 36 countries and regions and 16 commodities, including all cereals, meats and milk, roots and tubers, soybean (*Glycine max* L. Merr.), meals and oils. IMPACT is a non-spatial partial equilibrium model focusing on the agricultural sector. The model represents a competitive agricultural market for crops and livestock. It is specified as a set of country or regional sub-models, within each of which supply, demand and prices for agricultural commodities are determined. The country and regional agricultural sub-models are linked through trade, a specification that highlights the inter-dependence of countries and commodities in global agricultural markets. The model uses a system of supply and demand elasticities, incorporated into a series of linear and nonlinear equations, to approximate the underlying production and demand functions. World agricultural commodity prices are determined annually at levels that clear international markets. Demand is a function of prices, income and population growth. Growth in crop production in each country is determined by crop prices and the rate of productivity growth. The rate of productivity growth is primarily determined by the projected investments in agricultural research, including crop breeding and crop management research. Other sources of growth considered include agricultural extension and education, roads, and irrigation. In order to explore food security effects, IMPACT projects the percentage and number of malnourished preschool children (0 to 5 years old) in developing countries as a function of average per capita calorie availability, the share of females with secondary schooling out of the corresponding female school-age population, the status of females relative to men as captured by the ratio of female to male life expectancy at birth, and the percentage of the population with access to clean water (see also Smith and Haddad, 2000).

Baseline Projections: Summarized Results

According to new baseline results from the model, which presents our best estimate of projections to 2020, global cereal production is projected to increase at an annual rate of 0.9% between 1997 and 2020, rising by a total of 33% from 1,871 million Mg to 2,497 million Mg. This growth will represent a sharp decline from 2.1% annual growth achieved between 1967 and 1997. Global meat production will grow by 56%, i.e., from 210 million Mg in 1997 to 327 million Mg in 2020. Developing countries will account for an increasing share in global cereal demand in 2020: 59%, up from 54% in 1997, and the role of developing countries in global meat demand will also increase rapidly from 52% in 1997 to 63% by 2020.

Substantial imports from developed countries will be required to meet a growing share of cereal and meat demand in developing countries. Under the IMPACT baseline, the developing world will import a projected 202 million Mg of cereals in 2020, an increase of 94% (98 million Mg) above cereal imports in 1997. Major net cereal importing regions will include West Asia and North Africa at 73.1 million Mg, East Asia at 67.1 million Mg, and SSA at 27.3 million Mg.

Yield growth will be responsible for driving cereal production growth over the projection period, as cereal area expands little with the exception of Latin America and SSA.[5] Cereal yields are projected to increase at an annual rate of 1.1% in the developing world and 0.7% in the developed world. This yield performance is sharply below growth achieved between 1967 and 1997, when cereal yields rose at an annual rate of 2.4% in the developing world and 1.7% in the developed world. As a result of these trends, real world cereal prices are projected to decline only slowly over the projections period (Table 2). Wheat prices are projected to decline by 8% between 1997 and 2020, rice prices by 12%, and other coarse grain prices by 11%. Corn prices are only projected to decline by 1% between 1997 and 2020. These cereal price developments show a significant break from the past trends for all cereals. Figure 2, showing past trends and projections for wheat, illustrates the dramatic change. Similar price developments hold for rice and corn.

Ultimately, despite falling real food prices and expanding world trade, the poor in the developing world will realize only modest improvement in food security. While most regions will have varying declines in the number of malnourished children under the age of five, IMPACT projects that SSA will actually have a slight increase in the number of malnourished children, from 32.7 million in 1997 to 39.3 million in 2020.

Alternative Scenarios for Crop Yield Growth: Assumptions and Projections

With real causes for concern about the future of cereal yield growth rates over the next two decades, it is imperative to determine the relative importance

TABLE 2. Global Cereal prices, 1997 and projected 2020

Cereal	1997	2020
	US$ Mg^{-1}	
Wheat	133	123
Maize	103	102
Rice	285	250
Other coarse grains	97	86

Source: IMPACT projections, June 2001.

FIGURE 2. Global wheat price, 1961-2020

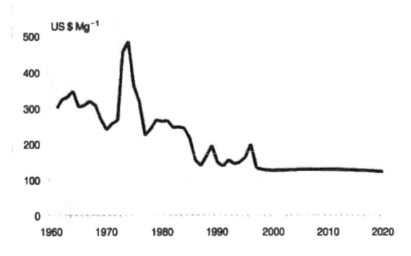

Source: 1961-1997: World Bank, Commodity Price Outlook, Various Years; 1998-2020: IMPACT Projections, June 2001.

of cereal yields to international prices and food security. In order to assess the sensitivity of cereal prices to yield growth rates, we modeled two alternative yield scenarios and charted the projected price trends for the four cereals over the 23-year period between 1997 and 2020. One possible general global trend over the next 23 years involves significant declines in the resources available for agricultural research and irrigation development, with continued environmental degradation taking a toll on cereal production systems. We assess this trend in the IMPACT model for the so-called low-yield scenario, of which assume no growth in irrigated area during the projections period. Additionally, the low yield scenario assumes a decline in specified yield growth rates for meats, milk, and all crops in the developed world of 50% from the baseline level, and a decline in specified yield growth rates for meats, milk, and all crops of 40% from the baseline in all developing regions.

It is possible, on the other hand, that potential threats to world agricultural production will galvanize governments, multilateral institutions, and private firms into increasing their investments into genetic research, crop management and irrigation development, thus leading to significant yield increases and expansion of irrigation infrastructure. The IMPACT high-yield scenario assumes an increase in the expansion of irrigated area of 1% per year greater than baseline growth rates. Additionally, the high yield scenario assumes an increase in specified yield growth rates for all livestock, milk, and crops of

20% above baseline assumptions in the developed world and 40% above baseline assumptions in the developing world. Tables 3 and 4 summarize projected crop yields and crop yield growth rates under the alternative scenarios. It should be noted that IMPACT simulations lead to changes in realized yields different from those that would result from straight-line calculations made from the initially specified yield growth rates, because the model captures the feedback effects between changes in yields and output prices.[6] For example, if cereal yield growth assumptions are lowered relative to the baseline, cereal prices increase relative to baseline prices, which subsequently leads to partially countervailing increases in cereal yields (and area) in response to higher price incentives.

Tables 5 and 6 summarize price trends in absolute numbers and percentages, respectively, under the various yield scenarios. The scenarios show that

TABLE 3. Cereal yields for the 2020 baseline and various scenarios.

Region	Baseline	Low-yield	High-yield
		Mg ha^{-1}	
Latin America	3.7	3.4	4.0
Sub-Saharan Africa	1.4	1.3	1.5
West Asia/North Africa	2.6	2.4	2.7
China	5.5	5.0	6.2
India	2.4	2.2	2.8
Asia	3.6	3.3	4.0
Developing Countries	3.1	2.8	3.9
Developed Countries	4.0	3.8	3.4

Source: IMPACT projections, June 2001.

TABLE 4. Projected regional cereal yield growth rates under the baseline and alternative yield scenarios, 1997-2020.

Region	Baseline	Low-yield	High-yield
		% year^{-1}	
Latin America	1.6	1.2	2.0
Sub-Saharan Africa	1.8	1.3	2.0
West Asia/North Africa	1.1	0.9	1.4
China	1.2	0.8	1.7
India	1.3	0.8	1.9
Asia	1.2	0.8	1.7
Developing Countries	1.2	0.8	1.7
Developed Countries	0.7	0.6	0.7

Source: IMPACT projections, June 2001.

TABLE 5. Cereal prices in 1997 and 2020 under various yield scenarios.

	1997	Baseline	Low-yield	High-yield
	US$ Mg^{-1}			
Wheat	133	123	164	92
Maize	103	102	140	75
Other coarse grains	97	86	122	62
Rice	285	250	392	156

Source: IMPACT projections, June 2001.

TABLE 6. Difference in cereal prices from baseline levels in 2020 under alternative yield scenarios.

	Low-yield	High-yield
	% difference	
Wheat	33	−29
Maize	37	−31
Other coarse grains	42	−36
Rice	46	−40

Source: IMPACT projections, June 2001.

cereal prices are highly sensitive to the projected growth rates in crop yields. As would be expected, prices for all crops decline under the high yield scenarios, although the individual cereals have significantly varying sensitivity to yield growth rate changes.

The yield shifts and resulting price changes have significant impacts on cereal production, demand and trade in both the developing and developed worlds. Cereal production in the developing world in 2020 under the low yield scenario is projected to decline 10% from the baseline, from 1,473 million Mg to 1,343 million Mg, with a corresponding increase of 11% under the high yield scenario to 1,638 million Mg (Table 7). Cereal production in both South Asia and China displays a particularly high sensitivity to yield shocks under the alternative scenarios. Cereal production in the developed world in 2020 under the low yield scenario is projected to decline only 2% from the baseline level, from 1,024 million Mg to 1,005 million Mg, and is also projected to decline by 1% under the high yield scenario.

Total per capita cereal demand in the developing world will also be significantly affected by the alternative yield scenarios, falling 7% from the baseline in 2020 to 255 kg capita^{-1} under the low yield scenario and rising 8% above the baseline in 2020 to 297 kg capita^{-1} under the high yield scenario (Table 8). Total daily per capita kilocalorie (Kcal) consumption in the developing world

TABLE 7. Cereal production in 1997 and under baseline and alternative yield scenarios in 2020.

Regions	1997	Baseline	Low-yield	High-yield
		million Mg		
Latin America	125	207	192	224
Sub-Saharan Africa	69	129	119	139
West Asia/North Africa	86	123	117	126
China	383	518	469	587
India	182	254	217	299
Asia	738	1,014	915	1,149
Developing countries	1,017	1,473	1,343	1,638
Developed countries	854	1,024	1,005	1,010

Source: IMPACT projections, June 2001.

TABLE 8. Per capita demand in 1997 and under baseline and alternative yield scenarios in 2020.

Regions	1997	Baseline	Low-yield	High-yield
		Kg capita ^{1}		
Latin America	284	323	297	354
Sub-Saharan Africa	147	163	146	180
West Asia/North Africa	381	387	367	410
China	315	387	361	418
India	189	205	191	221
Asia	248	280	260	302
Developing countries	249	275	255	297
Developed countries	559	604	583	615

Source: IMPACT projections, June 2001.

undergoes similar shifts, falling 4% from the baseline level of 3,274 Kcal capita^{-1} day^{-1} in 2020 to 3,137 Kcal capita^{-1} day^{-1} under the low yield scenario and rising 5% to 3,429 Kcal capita^{-1} day^{-1} in 2020 under the high yield scenario. Sub-Saharan Africa has the largest shift in per capita demand and daily per capita Kcal consumption under the alternative yield scenarios due to low initial cereal demand and high regional demand elasticities. Meanwhile, per capita cereal demand in the developed world falls 3% from the baseline in 2020 to 583 kg capita^{-1} under the low yield scenario and rises 2% above the baseline in 2020 to 615 kg capita^{-1} under the high yield scenario (Table 9).

Cereal trade is influenced by the yield growth shocks, but the overall magnitude of trade shifts is not as large as might be expected, because the price effects of yield shocks move demand in the same direction as supply. The relative supply and demand elasticities determine the net change for different regions. Net cereal imports into the developing world from the developed world in 2020 increase from 202 million Mg under the baseline to 212 million Mg under the low yield scenario, while net cereal imports under the high yield scenario decline by substantially more to 173 million Mg (Table 10). South Asia, Southeast Asia, and China all have higher net imports under the low yield scenarios and lower net imports under the high yield scenarios, indicating that the combined yield and price effects of the various scenarios on domestic production outweigh the price effects on domestic demand. Sub-Saharan Africa, Latin America, and West Asia/North Africa (WANA), on the other hand, all have lower net imports under the low yield scenarios and higher net

TABLE 9. Kilocalorie (Kcal) availability in 1997 and under the baseline and alternative yield scenarios in 2020.

Regions	1997	Baseline	Low-yield	High-yield
		Kcal capita^{-1} day^{-1}		
Latin America	2,931	3,274	3,137	3,429
Sub-Saharan Africa	2,232	2,442	2,314	2,582
West Asia/North Africa	3,052	3,208	3,106	3,315
China	2,927	3,536	3,414	3,669
India	2,477	2,868	2,706	3,042
Asia	2,665	3,091	2,957	3,236
Developing countries	2,667	3,015	2,885	3,158
Developed countries	2,931	3,274	3,137	3,429

Source: IMPACT projections, June 2001.

TABLE 10. Net cereal imports in 1997 and under the baseline and alternative yield scenarios in 2020.

Regions	1997	Baseline	Low-yield	High-yield
		million Mg		
Latin America	−15	−4	−1	−7
Sub-Saharan Africa	−12	−27	−21	−34
West Asia/North Africa	−45	−73	−68	−81
China	−8	−48	−58	−23
India	2	−7	−24	19
Asia	−31	−97	−120	−50
Developing countries	−104	−202	−212	−173
Developed countries	−15	−4	−1	−7

Source: IMPACT projections, June 2001.

imports under the high yield scenarios, indicating that the price effects on domestic demand under the various scenarios in these countries outweigh the combined yield and price effects on domestic production. While WANA has very low production and demand responses to the yield scenarios, both Latin America and SSA have very high demand responses relative to the Asian countries. Thus, Latin America and SSA have particularly high stakes in sustaining strong yield growth, since weak yield growth hurts them both directly through the production effects and indirectly through relatively large demand declines due to higher world prices.

Projections for the number of malnourished children under the age of five in the developing world under the alternative yield scenarios reveal that the developing world has a very high stake indeed in strong yield growth (see Table 11). The baseline scenario projects the percent of malnourished children in the developing world to decline from 31.4% in 1997 to 24.6% in 2020, with the number of malnourished children in South Asia declining from 85.0 million children (50.1%) to 63.3 million children (40.7%), representing 62% of the total. Under the low yield scenario, the percent of malnourished children under the age of five in the developing world is only projected to decline to 25.6%. South Asia and SSA will account for the majority of this increase, with the percent of malnourished children in South Asia in 2020 increasing from 40.7% under the baseline to 42.2% under the low yield scenario, and the percent of malnourished children in SSA increasing from 29.2% to 30.6% million children. Conversely, the high yield scenario leads to a decline in the number of malnourished children under the age of five in the developing world to 124.8 million children compared to 166.3 million in 1997 and 131.5 million under the baseline. South Asia and SSA again have the largest declines, with the number of malnourished children in South Asia falling to 60.9 million children

TABLE 11. Number of malnourished children in 1997 and projected under the baseline and alternative yield scenarios in 2020.

Regions	1997	Baseline	Low-yield	High-yield
		Million		
Latin America	5.1	2.5	3.0	1.8
Sub-Saharan Africa	32.7	39.3	41.2	37.4
West Asia/North Africa	5.9	4.0	4.4	3.7
China	18.4	8.5	9.3	7.6
India	64.1	44.2	45.7	42.6
Asia	122.6	85.7	89.3	81.9
Developing countries	166.3	131.5	137.9	124.8
Developed countries	5.1	2.5	3.0	1.8

Source: IMPACT projections, June 2001.

and the number of malnourished children in SSA falling to 37.4 million children (see Table 11).

As these results show, the small declines in cereal prices projected in the IMPACT baseline are highly sensitive to changes in the rates of yield growth achieved over the next two decades. Rice prices are particularly sensitive to the yield growth scenarios because of the relative thinness of international rice markets as a share of total production. Cereal production shifts and the resulting changes in prices in turn have a significant impact on projected levels of childhood malnutrition and patterns of international cereal trade. Since price shifts are largely a matter of concern to those for whom food expenditures represent a large share of income, rates of crop yield growth achieved over the next few decades. Therefore, rates of investment growth for agricultural research and infrastructure development will determine in large measure the price of food for the poor.

CONCLUSIONS

IMPACT results show that achievement of the levels of yield growth are necessary to at the very least maintain international cereal prices at constant levels is by no means preordained, and that complacency could result in sharp increases in long-term prices. Developing countries cannot rely on the developed world to supply the preponderance of their future food needs. Trade will be an important component of food security in the developing world into the next century, but trade cannot substitute for well-targeted and funded investments for domestic food production as well as proper policies that provide incentives to local farmers. Production shortfalls in the developed world due to shifting market conditions and farm-support policies could have significant price implications and a devastating impact on the poor if the degree of food dependency in the developing world is too high. The corollary of the projected price increases under the low yield scenario is that the potential upside of investments in agricultural production are very high, with significant empowerment of local research capacity necessary to ensure the widespread diffusion of technologies from the developed to developing worlds.

In order to promote crop yield growth, policy makers will need to pursue appropriate and sustainable methods of agricultural intensification for both favored and less-favored areas. This dual strategy will be particularly challenging if government budgets for investment in agriculture and rural areas continue to remain tight; striking the right investment balance between irrigated and rainfed regions and between high- and low-potential rainfed areas will be particularly important. A significant part of the necessary funds could be met in some countries by reducing wasteful public expenditure in rural areas, particularly on subsidies for fertilizers, pesticides, electricity, and irrigation water. These subsidies may have played an important role in launching the Green Revolution, but today they are rarely needed and can be counter-productive because they create incentives for the overuse of water and farm chemicals, leading to environmental degradation. Investments in irrigated and high-potential rainfed areas cannot be neglected, because these areas still provide much of the food needed to keep prices low and to feed growing urban populations and livestock. As is the case in India, only the more advanced northwest uses more varieties and technologies while the rest of the country experiences continued stagnation of agricultural productivity. On the other hand, rainfed areas will also contribute a significant share of future food production, as well as income and employment. Successful development of rainfed areas is likely to be more complex than in high-potential irrigated areas because of their relative lack of access to infrastructure and markets, and their more difficult and variable agroclimatic environments. Crop research targeted to rainfed areas should be accompanied by increased investment in crop management, rural infrastructure and policies to close the gap between economic potential yields in rainfed areas and the actual yields achieved by farmers. Finally, it is important that public agencies are able to deal with the unique problems of rainfed areas. Many extension and agricultural research agencies have traditionally focused on irrigated and high-potential areas. The top-down approach to management that is used in many of these agencies also generally does not work well in the less-favored rainfed areas. A more participatory approach by these agencies with more accountability to the farmers may help in the development of less-favored rainfed areas.

Effective cropping system research to shrink the gap between potential and actual cereal yields will be an increasingly important factor affecting future cereal production increases and food security over the next two decades. As farmers in many major producing areas reach the limits of further intensification, and farmers in other areas struggle to maintain their marginal livelihoods, continuing yield growth will depend on the ability of cropping systems to produce higher yields with the same levels of input use, and in some cases the same yields at lower levels of input use.

NOTES

1. However, the sword cuts both ways: the base level for demand growth has also increased, and growth in total demand requirements is projected to be as high over the next 25 years as over the previous 25 years.

2. A possible problem with Chinese agricultural statistics should be noted; Chinese yields may be significantly overstated, by perhaps 10% for rice and up to 40% in some provinces for maize (Smil, 2000).

3. For instance, Cassman and Pingali (1996) estimate a decline in yields due to N depletion at all N levels of 30% over a 20-year period.

4. Although sub-Saharan Africa has the highest per capita cereal area harvested of any IMPACT region at 0.13 ha per capita, and this figure has actually increased from 0.12 ha per capita in 1982.

5. Technological change, investments, and output and input prices determine crop yield trends in the IMPACT model. The non-price component of yield projections are broken down into public research components, including management research, conventional plant breeding, wide-crossing/hybridization breeding, and biotechnology (transgenic) breeding; private-sector agricultural research and development; agricultural extension; markets; infrastructure; and irrigation. To generate the projected time path of yield growth, use has been made of 'before-the-fact' and 'after-the-fact' studies of agricultural research, priority setting, sources of agricultural productivity growth, and 'expert opinion.' Projections are also made for harvested crop area. Estimation of non-price area growth rates depends on the availability of cultivable land, irrigation, infrastructure investment, and productivity gains, as well as on prices (for additional details see Rosegrant et al., 2001a).

6. Please also note that the 1997 base yield remains constant in all scenarios, so that the shock to yield growth rates only impacts projected yield increases between 1997 and 2020.

REFERENCES

Abrol, I.P. (1987). Salinity and food production in the Indian Sub-Continent. In *Water and Water Policy in World Food Supplies*, ed., W.R. Jordan, College Station, TX: Texas A & M University Press.

Alston, J.M., C. Chan-Kang, M.C. Marra, P.G. Pardey, and T.J. Wyatt. (2000). A meta-analysis of rates of return to agricultural R& D: Ex Pede herculem? IFPRI Research Report 113, Washington, DC: International Food Policy Research Institute (IFPRI).

Becker, M. and D.E. Johnson. (1998). The role of legume fallows in intensified upland rice-based systems of West Africa. *Nutrient Cycling in Agroecosystems* 53:11-82.

Bhuiyan, S.I., T.P. Tuong, and L.J. Wade. (1998). Management of water as a scarce resource: Issues and options in rice culture. In *Sustainability of Rice in the Global Food System*, ed., N.G. Dowling, S.M. Greenfield, and K.S. Fisher. Los Baños, Philippines: Pacific Basin Study Center and International Rice Research Institute.

Bojo, J. (1996). The costs of land degradation in Sub-Saharan Africa. *Ecological Economics* 16:161-173.

Boserup, E. (1981). *Population and Technological Change: A Study of Long-Term Trends*. Chicago, IL: University of Chicago Press.

Bumb, B.L. and Baanante, C.A. (1996). *The Role of Fertilizer in Sustaining Food Security and Protecting the Environment to 2020*. 2020 Food, Agriculture, and the Environment Discussion Paper 17. Washington, DC: International Food Policy Research Institute.

Byerlee, D. (1998). Knowledge-intensive crop management technologies: Concepts, impacts and prospects in Asian agriculture. In *Impact of Rice Research*, ed., P. Pingali and M. Hossain, Bangkok, Thailand: Thailand Development Research Institute; Los Baños, Philippines: International Rice Research Institute.

Byerlee, D. (1996). Modern varieties, productivity and sustainability. *World Development* 24(4): 697-718.

Byerlee, D. and K. Fischer. (2000). *Accessing Modern Science: Policy and Institutional Options for Agricultural Biotechnology in Developing Countries*. AKIS (Agricultural Knowledge and Information Systems) Discussion Paper. Washington, DC: The World Bank.

Byerlee, D. and C.K. Eicher. (1997). *Africa's Emerging Maize Revolution*. Boulder, CO: Lynne Reinner Publishers.

Byerlee, D. and P.W. Heisey. (1996). Past and potential impacts of maize research in Sub-Saharan Africa: A critical assessment. *Food Policy* 21:255-277.

Cassman, K.G. (1999). Ecological intensification of cereal production systems: Yield potential, soil quality, and precision agriculture. *Proceedings of the National Academy of Sciences of the United States of America* 96:5952-5959.

Cassman, K. and P.L. Pingali. (1995). Intensification of irrigated rice systems: Learning from the past to meet future challenges. *GeoJournal* 35:299-305.

Ceccarelli, S., S. Grabdo, R. Tutwiler, J. Baha, A.M. Martini, H. Salahieh, A. Goodchild, and M. Michael. (2000). Methodological study on participatory barley breeding: I. Selection phase. *Euphytica* 111:91-104.

Chambers, R. (1988). *Managing Canal Irrigation: Practical Analysis from South Asia*. Cambridge, UK: Press Syndicate of the University of Cambridge.

Cleaver, K.M. (1993). *A Strategy to Develop Agriculture in Sub-Saharan Africa and a Focus for the World Bank*. World Bank Technical Paper No. 203. Washington, DC: Africa Technical Department Series, World Bank.

Cleaver, K.M. and G.A. Schreiber. (1994). *Reversing the Spiral: The Population, Agriculture, and Environment Nexus in Sub-Saharan Africa*. Directions in Development Series. Washington, DC: World Bank.

Dogra, B. (1986). The Indian experience with large dams. In *The Social and Environmental Effects of Large Dams Volume 2*, ed., E. Goldsmith and N. Hildyard. London, UK: Wadebridge Ecological Center.

Dowswell, C.R., R.L. Paliwal, and R.P. Cantrell. (1996). *Maize: The Third World*. Winrock Development-Oriented Literature Series, Boulder, CO: Westview Press.

Dyson, T. (1996). *Population and Food: Global Trends and Future Prospects*. London and New York: Routledge.

Evans, L.T. (1998). *Feeding the Ten Billion: Plants and Population Growth*. Cambridge, UK: Cambridge University Press.

Fan, S., L. Zhang, and X. Zhang. (2002). *Growth, Inequality, and Poverty in Rural China: The Role of Public Investments*. Washington, DC: International Food Policy Research Institute, Research Report 125.

Fan, S., P. Hazell, and T. Haque. (2000). Targeting public investments by agroecological zone to achieve growth and poverty alleviation goals in rural India. *Food Policy* 25: 411-428.

FAO (Food and Agriculture Organization of the United Nations). (2000). FAOSTAT database. <*http://ffaostat.fao.org*>.

Fedoroff, N.V. and J.E. Cohen. (1999). Papers from a National Academy of Sciences Colloquium on plants and population: Is there time? *Proceedings of the National Academy of Sciences of the United States of America* 96(11): 5903-5908.

Garrity, D.P. (1999). Contour farming based on natural vegetative strips: Expanding the scope for increased food crop production on sloping lands in Asia. *Environment, Development and Sustainability* 1:323-336.

Goldman, R.H. and S. Block. (1993). *Proceedings from the Symposium on Agricultural Transformation in Africa*, APAP Phase II, Technical Report No. 137,Washington, DC: USAID Bureau for Research and Development.

Hazell, P.B.R., P. Jagger, and A. Knox. (2000). *Technology, Natural Resource Management and the Rural Poor*. IFAD (International Fund for Agricultural Development) Working Paper. Rome, Italy: IFAD.

Hobbs, P.R. and M. Morris. (1996). *Meeting South Asia's Future Food Requirements from Rice Wheat Cropping Systems: Priority Issue Facing Researchers in the Post-Green Revolution Era*. Natural Resources Group Working Paper 96-01. Mexico, D. F.: International Maize and Wheat Improvement Center (CIMMYT).

Hopper, G.R. (1999). Changing food production and quality of diet in India, 1947-1998. *Population and Development Review* 25:443-477.

IRRI (International Rice Research Institute). (2000). *Something to Laugh About*. Los Baños, Philippines: International Rice Research Institute.

Janssen, B.H. (1993). Integrated nutrient management: The use of organic and mineral fertilizers. In *The Role of Plant Nutrients for Sustainable Food Crop Production in Sub-Saharan Africa*, ed., H. van Reuler and W.H. Prins, Leidschendam, The Netherlands: Dutch Association of Fertilizer Producers (VKP).

Kherallah, M., C. Delgado, E. Gabre-Madhin, N. Minot, and M. Johnson. (2000). *Agricultural Market Reforms in Sub-Saharan Africa: A Synthesis of Research Findings*. Washington, DC: International Food Policy Research Institute.

Kon, K.F. (1993). Weed management: Towards tomorrow (with emphasis on the Asia-Pacific Region). *Proceedings of the 14th Asia-Pacific Weed Science Society Conference*. Brisbane, Australia: Asia Pacific Weed Society.

Lal, R. (1995). Erosion-crop productivity relationships for soil of Africa. *Soil Science Society of America Journal* 59:661-667.

Mew, T. (2000). Research initiatives in cross ecosystems: Exploiting bio-diversity for pest management. Los Baños, Philippines: International Rice Research Institute.

Murgai, R. (1999). The green revolution and the productivity paradox: Evidence from the Indian Punjab. Policy Research Working Paper 2232. Washington, DC: World Bank.

Oldeman, L.R. (1998). Soil degradation: A threat to food security? Report 98/01. Wageningen, The Netherlands: International Soil Reference and Information Centre.

Paroda, R.S. (1998). Priorities and opportunities of rice production and consumption in India for self-sufficiency. In *Sustainability of Rice in the Global Food System*, ed., N.G. Dowling, S.M. Greenfield, and K.S. Fischer. Los Baños, Philippines: Pacific Basin Study Center and International Rice Research Institute.

Pingali, P.L., M. Hossain, and R.V. Gerpacio. (1997). *Asian Rice Bowls: The Returning Crisis?* Wallingford, UK: CAB International.

Pingali, P.L. and M.W. Rosegrant. (1998). Supplying wheat for Asia's increasing westernized diets. *American Journal of Agricultural Economics* 80:954-959.

Price, L.M.L and V. Balasubramanian. (1998). Securing the future of intensive rice systems: A knowledge-intensive resource management and technology approach. In *Sustainability of Rice in the Global Food System*, ed., N.G. Dowling, S.M. Greenfield, and K.S. Fischer. Los Baños, Philippines: Pacific Basin Study Center and International Rice Research Institute.

Rosegrant, M.W. and P.B.R. Hazell. (2000). *Transforming the Rural Asian Economy: The Unfinished Revolution*. ABD: *A Study of Rural Asia*, Volume 1. Oxford, UK: Oxford University Press.

Rosegrant, M.W., M.S. Paisner, S. Meijer, and J. Witcover. (2001a). *Global Food Projections to 2020: Emerging Trends and Alternative Futures*. Washington, DC: International Food Policy Research Institute.

Rosegrant, M.W., X. Cai, S. Cline, and N. Nakagawa. (2001b). *The Role of Rainfed Agriculture in the Future of Global Food Production*. Invited background research paper for The International Freshwater Conference, Bonn, Germany, December, 2001.

Rosegrant, M.W. and P.L. Pingali. (1994). Policy and technology for rice productivity growth in Asia. *Journal of International Development* 6:665-688.

Rosegrant, M.W. and M. Svendsen. (1993). Asian food production in the 1990's: Irrigation investment and management policy. *Food Policy* 18:13-32.

Sanchez, P.A., K.D. Shepherd, M.J. Soule, F.M. Place, A.U. Mokwunye, R.J. Buresh, F.R. Kwesiga, A.N. Izac, C.G. Ndiritu, and P.L. Woomer. (1997). Soil fertility replenishment in Africa: An investment in natural resource capital. In *Replenishing Soil Fertility in Africa*, ed., R.J. Buresh and P.A. Sanchez. Special Publication No. 51, Madison, WI: Soil Science Society of America and American Society of Agronomy.

Sanchez, P.A., B. Jama, A.I. Niang, and C.A. Palm. (2001). Soil fertility, small-farm intensification and the environment in Africa. In *Tradeoffs or Synergies: Agricultural Intensification, Economic Development and the Environment*, ed., D.R. Lee and C.B. Barrett. New York, NY: CABI Publishing.

Scherr, S.J. (1999). *Soil Degradation: A Threat to Developing-Country Food Security by 2020*. 2020 Vision for Food, Agriculture, and the Environment Discussion Paper No. 27, Washington, DC: International Food Policy Research Institute.

Scherr, S. and S. Yadav. (1996). *Land Degradation in the Developing World: Implications for Food, Agriculture, and the Environment to 2020*. 2020 Vision for Food, Agriculture, and the Environment Discussion Paper No. 14, Washington, DC: International Food Policy Research Institute.

Schoenly, K., T.W. Mew, and W. Reichardt. (1998). Biological diversity of rice landscapes. In *Sustainability of Rice in the Global Food System*, ed., N.G. Dowling, S.M. Greenfield, and K.S. Fisher, Los Baños, Philippines: Pacific Basin Study Center and International Rice Research Institute.

Schroth, G., U. Krauss, and L. Gasparott. (2000). Pests and diseases in agroforestry systems of the humid tropics. *Agroforestry Systems* 50(3):199-241.

Smil, V. (2000). *Feeding the World: A Challenge for the Twenty-First Century.* Cambridge, MA: MIT Press.

Spencer, D.S.C. (1994). *Infrastructure and Technology Constraints to Agricultural Development in the Humid and Sub-Humid Tropics of Africa.* Environment and Production Technology Discussion Paper 3, Washington, DC: International Food Policy Research Institute.

Smith, L. and L. Haddad. (2000). *Explaining Child Malnutrition in Developing Countries: A Cross-Country Analysis.* IFPRI Research Report No. 111: Washington, DC: International Food Policy Research Institute.

Trenbath, B.R. (1999). Multispecies cropping systems in India: Predictions of their productivity, stability, resilience and ecological sustainability. *Agroforestry Systems* 45:81-107.

Vanlauwe, B., K. Aihou, and P. Houngnandan. (2001). Nitrogen management in 'adequate' input maize-based agriculture in the derived savanna benchmark zone of Benin Republic. *Plant and Soil* 228:61-71.

Versteeg, M.N., P. Adegbola, and V. Koudokpon. (1993). Investigación participatoria en la República de Benín: el método investigación-desarrollo (con un ejemplo de caso sobre mucuna) In *Gorras y Sombreros: Caminos Hacia la Colaboración Entre Técnicos y Campesinos,* ed., D. Buckles. Memoria del Taller Sobre Mos Métodos Participativos de Investigación y Extensión Aplicados a las Tecnologías Basadas en Abonos Verdes, Mexico, D.F.: International Maize and Wheat Improvement Center (CIMMYT) (in Spanish).

WARDA (West Africa Rice Development Association) (2001). *NERICA: Rice for Life.* M'bé, Côte d'Ivoire: WARDA.

Wood, S., K. Sebastian, and S. Scherr. (2001). *Pilot Analysis of Global Ecosystems: Agroecosystems.* Washington, DC: World Resources Institute and International Food Policy Research Institute.

World Bank. (1994). *Adjustment in Africa: Reforms, Results and the Road Ahead.* Washington, DC: World Bank.

World Bank. (2000). *Can Africa Claim the 21st Century?* Washington, DC: World Bank.

World Bank. (Various Years). *Commodity Price Outlook.* Washington, DC: World Bank.

Index

Abdin, O.A., 463
Aber, J.D., 131
Acid detergent fiber (ADF), 478
ACT Network. *See* African
    Conservation Tillage (ACT)
    Network
ADF. *See* Acid detergent fiber (ADF)
Aflatoxin, in corn, 292
Africa, yam-based cropping systems
    in, 531-558. *See also*
    Yam-based cropping
    systems, in Africa
African Conservation Tillage (ACT)
    Network, 226
African Yam Network, 554
Agricultural research, and policies for
    growth, in sub-Saharan
    Africa, 673-675
Agricultural systems
    future, 28-30
    sustainability as concept in,
        308-309
Agriculture
    Brazilian, 593-621. *See also*
        Brazilian agriculture
    precision, in cropping systems,
        361-381. *See also* Precision
        agriculture, in cropping
        systems
    smallholder, in Brazil, 612-615,614f
    sustainable, evaluation of,
        conceptual framework for,
        433-454. *See also* Sustainable
        agriculture, evaluation of,
        conceptual framework for

U.S., transgenic crops in, 501-530.
    See also Transgenic crops, in
    U.S. agriculture
zero tillage, 602
Agriculture and Agri-Food Canada, 77
Agroecology, defined, 23
Agroecosystem(s)
    cereal-based, NUE in, assessment
        of, 158-185. *See also*
        Nitrogen use efficiency
        (NUE), in cereal-based
        agroecosystems
    forage legumes use in, 189-191,190t
        benefits of, 191-197,192t,193t.
        See also Forage legumes, for
        sustainable cropping systems,
        benefits of
Agroecosystem productivity, elevated
    CO_2-induced changes in,
    217-244. *See also* CO_2,
    increasing atmospheric,
    agroecosystem productivity
    effects of
Agronomic Institut of Parana
    (IAPAR), 602
Agronomy, chemical ecology to,
    455-499. *See also* Northeast,
    cropping systems in
Agronomy Journal, 18
Aguilar, I., 196
Ahuja, L.R., 386
Air quality, interdependence between
    soil quality and water quality,
    37f
Akobunudu, I.O., 246

Alfalfa, in agroecosystems, 189,190t
Allen, R.R., 94
Allmaras, R.R., 75,78,82,86,91,94,112
Alston, J.M., 673
Altieri, M.A., 7
Alves, B.J.R., 593
Amemiya, M., 91
American Journal of Alternative Agriculture, 19
American Society of Agricultural Engineering (ASAE) Standards, 413
American Society of Agricultural Engineers (ASAE) Standards, 79
"Amigos da Terra," 605
Angers, D.A., 101,104,105,112
Angler, D.A., 195
Animal and Plant Inspection Service (APHIS), 279
Anthropod(s), 62
Aphid(s), Russian wheat, 289
APHIS. *See* Animal and Plant Inspection Service (APHIS)
Aquatic guidelines, for water quality, 339
ArcView, 366
Area Time Equivalent Ratio (ATER), 47-48
Aref, S., 93
Arid West of U.S., tillage systems in, overview of, 88f,99-100
ASAE Standards, 79
Asia
  in less favored areas of, growth yields in, boosting of, 669-670
  post-Green Revolution period in, management challenges of, 664-670,665t
Asiedu, R., 531
ATER. *See* Area Time Equivalent Ratio (ATER)
Atkinson, D., 130,143
Auernhammer, H., 364

Axinn, G.H., 4
Axinn, N.W., 4
Ayoub, M., 473

Bacillus thuringiensis (Bt), insecticidal crops derived from
  in development, 507-509
  in U.S. agriculture, 503-509,504t, 505t
  with regulatory approval, 503-507,504t,505t
Bacteria, in water quality control, 348-349
Bahia grass, 315
Baidu-Forson, J., 449
Bailey, M.J., 235
Baker, F.H., 25
Balancing-transfer method, in root production and mortality calculations, 131
Ballantyne, P., 436
Bar gene, 511
Barbosa, R.I., 610
Barker, K.R., 271
Barnes, D.K., 192,199,205
Bauer, P.J., 77,91
Bawden, R.J., 4
Bazaaz, F.A., 138
Bean golden yellow mosaic virus (BGYMV), 656
Bean leaf beetle, 290
Beet(s), sugar, in U.S. agriculture, 512t,515
Beetle(s)
  bean leaf, 290
  carabid, 62
  cereal leaf, 289
  Colorado potato, 162,295,506,508
Begna, S.H., 469
Belford, R.K., 146
Beneficial management principles (BMPs), in water quality control, 345
Bennett, C.F., 440

Bennett, O.L., 87
Bennett's Hierarchy model, 434,439,
    440-441,440f
Berg, R.D., 98
Berntson, G.M., 138
BGYMV. *See* Bean golden yellow
    mosaic virus (BGYMV)
Bidart, M.G., 93
Bihler, D.D., 251
Biodiversity levels, in sustainable
    agriculture evaluation, 446
Biological nitrogen fixation (BNF),
    legumes in agroecosystems
    and, 191-193,192t,193t
Bioremediation, legumes in
    agroecosystems and, 195-196
Biotechnology, 8
Birdsfoot trefoil, in agroecosystems,
    189,190t
Black, A.L., 91,94
Black, K.E., 131
Black, P.E., 331
Black root rot, 61
Blackshaw, R.E., 60,199,254
Blade plow, in tillage, 81
Blair, J.M., 29
Blevins, R.L., 87
Block, S., 675
Blum, U., 60,197
BMPs. *See* Beneficial management
    principles (BMPs)
BNF. *See* Biological nitrogen fixation
    (BNF)
Bock, B.R., 176,474
Boddey, R.M., 593,597
Bojo, J., 670
Boll Weevil Eradication Loan
    Deficiency Payments, 311-312
Boll Weevil Eradication program, 311
Borer(s), corn, European, 291
Bowman, G., 286
Bowman, R.A., 96
Brachiaria grasses, in Brazil, 609-610
Brachiaria pastures, in Brazil,
    610-611,612

Bradley, E.B., 291
Bradley, J.B., 448
Brandt, J., 27
Brassicas grasses, 58, 60
Brazil, described, 593-595
Brazilian agriculture, 593-621
    coffee, 600-601,601t
    history of, 595-596
    introduction to, 594-595
    planted pastures, 608-612
    smallholder, 612-615,614f
    soybean, 602-608,603f,604t,605t,
        606f,607f
    sugarcane, 596-600,599f
Broccoli, in U.S. agriculture, 509
Bromoxynil, 511
Bronson, J.A., 207,208
Brouwer, R., 144
Brust, G.E., 252
Bryant, J.P., 136
BT. *See* Bacillus thuringiensis (Bt)
Buffer zones, in water quality control,
    351-352
Bug(s), chinch, 289
Buhler, D.D., 205,245
Bulgaria, cropping systems in, 632-633
Bulman, P., 473
Bunce, J.A., 235
Busscher, W.J., 91
Butler, G.W., 144
Butts, C.L., 321
Byerlee, D., 664

C factor, of Universal Soil Loss
    Equation, 195
C_4 grasses, in Northeast, 476-479,477t
Cabral, P.A., 595
Caldwell, C.D., 473
Caldwell, R.M., 383
Camberato, J.J., 474
Campbell, C.A., 107
Campbell, R.B., 90
Campbell, R.H., 364
CAN. *See* Controller Area Network
    (CAN)

Canadian Drinking Water Guidelines, 337

Canadian Water Quality Guidelines, 337

Canola, in U.S. agriculture, 512t,514

Carabid beetles, 62

Carbon, in roots, 129

Cardina, J., 252,259

Cardoso, F.H., 613

Carlson, H.L., 256

Carpenter, J.E., 501,504,504t,505t

Carrillo, E., 649

Carruthers, K., 465

Carson, R., 272,273

Carter, D.L., 86,98

Carter, M.R., 100,101,102,103

Cash cropping system, in Eastern Europe, 630

Cation-exchange capacity (CEC), 602

Cato, in forage legume utilization, 198

Cavigelli, M.A., 5

CEC. *See* Cation-exchange capacity (CEC)

Central Canadian provinces, tillage systems in, overview of, 88f,103-105

Central Tuber Crops Research Institute (CTCRI), 552

Central U.S., tillage systems in, overview of, 88f,91-93

Cereal leaf beetle, 289

Cereal rye, in North American cropping systems, 55

Cereal yields, future of, 661-690
  impacts of Green Revolution vs. recent developments, 664-666,665t
  introduction to, 662-663

Cerealbased agroecosystems, NUE in, assessment of, 158-185. *See also* Nitrogen use efficiency (NUE), in cereal-based agroecosystems

CERES model, 420

CERES-Maize, 391

Certified Crop Advisors, 388

Children's Health and Environment, 334

Chinch bug, 289

Chisel plow, in tillage, 81

Christensen, N.B., 99

Christensen, S., 370

Chronic injection system development, in Northeast, 479-481

CHTS. *See* Crop Harvest Tracking System (CHTS)

CHU system. *See* Corn heat unit (CHU) system

Cicer milk vetch, in agroecosystems, 189,190t

CIPP Evaluation model, 438

CISG program, 322,323

Claveran, A.R., 86,107

Clay, S.A., 196

Cleaver, K.M., 671

Clover(s)
  crimson, 62
  Kura, in agroecosystems, 189,190t
  red, in agroecosystems, 189,190t
  white, in agroecosystems, 189,190t

"Clube da Minhoca," 605

CMV. *See* Cucumber mosaic virus (CMV)

CO_2, increasing atmospheric, agroecosystem productivity effects of, 217-244
  introduction to, 218-220
  study of
    discussion of, 223-237
    materials and methods in, 221-223
    results of, 223-237,224t,225t, 227f,228f,230f,234f

Coastal Plain, sustainable cotton production system for, 312-317. *See also* Cotton system, sustainable, for Coastal Plain

Coccinellid(s), 62

Coffee, in Brazil, 600-601,601t

Cole, D.W., 136
Colorado potato beetle (CPB), 162,
    295,506,508
Colorado State University, 375
Columbia Lance nematode, 294
Columnella
    in forage legume utilization, 198
    in soil classification, 34
Colvin, T.S., 7
Comair Root Length Scanner, 222
Common vetch, in agroecosystems,
    189,190t
Communities in Schools of Georgia
    (CISG) program, 308,322,
    323
Community(ies), rural, sustainability
    in, 321-322
Community support, in sustainable
    cropping systems principles
    and practices, 309-310
Compartment flow method, in root
    production and mortality
    calculations, 131
Conklin, A.E., 60,258
Conservation, in sustainable
    agriculture evaluation,
    445-446
Conservation Reserve Program (CRP),
    5,44,161
Conservation Technology Information
    Center (CTIC), 79
Conservation tillage, 45,46t
    demerits of, 77-78
    described, 77
    merits of, 77-78
Context evaluation, defined, 438
Controller Area Network (CAN), 387
Conventional tillage (CT), 604
    defined, 79
    described, 76-77
Conway, G.R., 5
Cook, R.J., 287,289
Cooke, G.W., 3
Cool season grasses, 58
Cordgrass, 476

Corn
    aflatoxin in, 292
    field, in U.S. agriculture, 504,504t,
        505t,512t,515
    leafy reduced-stature, in Northeast,
        465-472,467f,468t,470f,471t
        development of, 466-467,468t
        grain yield in, 468t,469
        harvest index in, 468t,469
        leaf area index in, 467-469,468t
        number of leaves in, 467,467f,
            468t
        root morphology and fractal
            dimension in, 469-472,470f,
            471t
    pest management in, 291-293
    soybean and, intercropping of
        legumes with, 202-205,203t,
        204t
    sweet, in U.S. agriculture, 507
Corn Belt, 393,402,631
Corn borer, European, 291
Corn heat unit (CHU) system, 458
Corn rootworm complex, 291-292
Costa, C., 455,469,472
Coté, D., 104
Cotton
    historic perspective on, 311-312
    pest management in, 293-294
    in U.S. agriculture, 505,511-513,
        512t
Cotton system, 307-327
    for community support of
        sustainable principles and
        practices, 309-310
    on farm research to promote
        sustainable practices, 310-311
    introduction to, 308-312
    sustainable
        for Coastal Plain, 312-317
            economics of, 316-317
            environmental impact of,
                315-316
            habitat management in,
                314-315

reduced tillage and cover
crops in, 312-314
concept of, 308-309
on-farm research development
of, 317-322
economics of, 321
environmental impact of, 320
Internet-based technology
transfer in,
318-320,319f
in rural communities,
321-322
Coulter carts, in tillage, 82,84
Coutts, M.P., 139
Cover crops, 47-48
defined, 47,54
in North American cropping
systems
benefits of, 53
cereal rye, 55
constraints in use of, 65-67
in erosion control, 54-56
establishment of, 63-64,63f
historical background of, 54
in insect and pathogen
management, 60-62
introduction to, 54
living mulches, 65
management options for, 63f
in nutrient management, 57-58
role of, 53-74
selection of, 65
in soil quality enhancement,
56-57
in weed management, 58-60
sustainable cotton system for
Coastal Plain, 312-314
uses of, 54
in water quality control, 347-348
in weed management, 250-252
Cox, W.J., 89,199,202
CPB. *See* Colorado potato beetle (CPB)
Cramer, C., 286
Crews, T.E., 448
Crimson clover, 62

Croatia, cropping systems in, 633-634
Cromar, H.E., 252
Crop(s)
herbicide-resistant, in U.S.
agriculture, 509-516,512t.
See also Herbicide-resistant
crops, in U.S. agriculture
transgenic, in U.S. agriculture,
501-530. *See also* Transgenic
crops, in U.S. agriculture
Crop competitiveness, in weed
management, 253-255
Crop cultivar decisions, 19
Crop Harvest Tracking System
(CHTS), 365
Crop physiology investigations, in
Northeast, 479-481
Crop production, efficiency of,
improved, 18-23
Crop production research in
southwestern Quebec,
456-462,457f,459f
Crop rotation
in water quality control, 346
in weed management, 248-249
Crop rotation system, in Eastern
Europe, 627-628
Crop rotations, 24-25
Crop yield
growth of
alternative scenarios for,
677-684,679t-683t
decline in, threats to future food
supply and prices and,
663-675,665t,672f
maximizing of, cropping systems
for, 3
Cropgrass management system, in
Eastern Europe, 628-630
Croplong term fallow system, in
Eastern Europe, 625
Cropping
relay, 24
strip, 24

Cropping practices, human health and, 335-337

Cropping systems
  analysis of, DSSs in, 383-407. *See also* Decision support systems (DSSs), for cropping systems analysis
  defined, 3,33
  in Eastern Europe, 623-647. *See also* Eastern Europe, cropping systems in
  economic analysis of, 409-432
    agronomic framework for, 418-422,421t,422t
    case example, 422-426,424t, 426f
    crop yield in, 420-421,421t,422t
    economic framework for, 413-418
    historical background of, 411-412
    introduction to, 410-411
    profitability in, 413-416
    resource allocation in, 418
    risk analysis in, 416-418
    suitable field day results in, 421-422
    underlying production environment in, 419-420
  IPM and, 271-305. *See also* Integrated pest management (IPM), cropping systems and
  for maximizing crop yields, 3
  mixed, in small farms of southwestern Guatemala, socioeconomic and agricultural factors associated with, 649-659
  mixed, in small farms of southwestern Guatemala, socioeconomic and agricultural factors associated with. *See also* Southwestern Guatemala, small farms of, mixed cropping systems in

in North America, cover crops in, role of, 53-74. *See also* Cover crops, in North American cropping systems
in Northeast, 455-499. *See also* Northeast, cropping systems in
precision agriculture in, 361-381. *See also* Precision agriculture, in cropping systems
research on
  emerging trends in, 1-13. *See also* Research, cropping systems-related, emerging trends in
  root dynamics in, 127-155. *See also* Root(s), dynamics of
resource-efficient, design of, advances in, 15-32. *See also* Resource-efficient cropping systems, design of, advances in
soil quality and, 44-48,44f,45f,46t. *See also* Soil quality, cropping systems and
sustainable
  forage legumes for, 187-216. *See also* Forage legumes, for sustainable cropping systems
  root dynamics and, 145-146
  in Southeast, conceptual model for, 307-327
water quality concerns related to, 329-359. *See also* Water quality, concerns related to, cropping strategies and
weeds within, 245-270. *See also* Weed(s)
yam-based, in Africa, 531-558. *See also* Yam-based cropping systems, in Africa
CropSys, 389-390
Crown vetch, in agroecosystems, 189,190t
CRP. *See* Conservation Reserve Program (CRP)

Cruse, R.M., 463
CT. *See* Conventional tillage (CT)
CTCRI. *See* Central Tuber Crops
    Research Institute (CTCRI)
CTIC. *See* Conservation Technology
    Information Center (CTIC)
Cucumber mosaic virus (CMV), in
    U.S. agriculture, 518
Cultivar selection, in ICM, 473
Cultivation, in weed management,
    252-253
Cultivator(s)
    field, in tillage, 81
    uses of, 20-21
Czech Republic, cropping systems in,
    631-632

da Silva, J.E., 602
Dabney, S.M., 60
Damm, E., 140
Dao, T.H., 96
DAVCO Farming, 366
Davidson, J.I., 321
DBM. *See* Diamond-back moth (DBM)
de Bos, R., 30
De Gaetano, A.T., 89-90
de Ruiter, H.E., 232
Deacon, J.W., 137
Dean, W., 600
Decision Support System for
    Agrotechnology Transfer
    (DSSAT), 384,388,402-403
Decision support systems (DSSs), 8
    for cropping systems analysis,
        383-407
    decision support in, 399-402,
        401f
    introduction to, 383-388
    modeling competition among
        individual plants, 388-391,
        391ft,392f
    modeling historical trends in
        county-level yield, 393-395,
        394t

modeling yield maps, 395-399,
    396f,397f,399f
    in nitrogen management, 159,160f
Decker, A.M., 205,206
DeFrank, J., 60
DeHaan, K.R., 364
DeHaan, R.L., 205,251
DEM. *See* Digital Elevation Model
    (DEM)
den Biggelaar, C., 433
Deng, 469
Department of Commerce, 83,83f
Derksen, D.A., 253
Detergent fiber
    acid, 478
    neutral, 478
Development Agency Monitoring and
    Evaluation model, 438
Dewey, J., 384
Diamond-back moth (DBM), 509
Dieleman, J.A., 255
Differential GPS (DGPS) receiver, 362
Digital Elevation Model (DEM), 389
Digital Orthophoto Quarter
    Quadrangle, 395,397f
Dillon, C.R., 409,419,421,423
Dinitrogen fixation, legumes in
    agroecosystems and, 191
Doty, C.W., 90
Driving Force-State-Response (DSR)
    model, 434,439,441,442f
DSSAT. *See* Decision Support System
    for Agrotechnology Transfer
    (DSSAT)
DSSs. *See* Decision support systems
    (DSSs)
Duffy, M., 291
Duffy, P.A., 410,411
Dumanski, J., 445
Dury, C.F., 200
Dyck, E., 197

Eagle, A.J., 100
EARO, 573

Earthworm Club, 605
Eastern Canadian provinces, tillage
systems in, overview of,
88f,100-103
Eastern Europe
coniferous forest zone of, cropping
systems in, 636
cropping systems in, 623-647
after World War II, 630-636,
632t,635t
cash cropping system, 630
industrial cropping system,
630-631
current, 636-639,637t
future, 639-642,641t
introduction to, 624
IPM in, 640,641t
before World War II, 625-630
crop rotation system, 627-628
crop-grass management system,
628-630
croplong term fallow system,
625
field rotation system, 626-627
forest-steppe zone of, cropping
systems in, 635,635t
mixed forest zone of, cropping
systems in, 636
semidesert zone of, cropping
systems in, 634
steppe zone of, cropping systems in,
634-635
Eberlein, C.V., 207-208
EBPM. *See* "Ecologically Based Pest
Management" (EBPM)
"Ecological Management of Weeds,"
272
"Ecologically Based Pest
Management" (EBPM),
272,300
Economic analysis, of cropping
systems, 409-432. *See also*
Cropping systems, economic
analysis of
Economic Evaluation model, 438

Economic goals, of nitrogen
management, 176-180
Economic issues
as factor in on-farm developing
sustainable cotton production
system, 321
sustainable cotton system for
Coastal Plain, 316-317
Economic Research Service, 80
Ecosystem(s), natural, diversity
among, 23-24
Edaphic factors, in weed management,
255-256
Edwards, C.A., 4
EEC thresholds, 337-338
Eggplant, in U.S. agriculture, 508
EIQ. *See* Environmental Index
Quotient (EIQ)
Eissenstat, D.M., 131,134,143
Ekanayake, I.J., 531,533
ELISA (enzyme-linked
immunosorbent assay),
279-280
Ellert, B.H., 112
Elliot, J.A., 347
Ellis, R.J., 235
Elmer, W.H., 61-62
Embrapa Beef Cattle center, 609
Energy feedback, legumes in
agroecosystems and, 194
Enting, I.G., 232
Entz, M.H., 196,199
Environment
as factor in on-farm research
developing sustainable cotton
production system, 320
as factor in tillage, 111-113
Environmental
as factor in nitrogen management,
176-180
impact on sustainable cotton system
for Coastal Plain, 315-316
Environmental Index Quotient (EIQ),
320
Environmental McCarthyism, 335

Enzyme-5-enolpyruvyl-shikimate-3-
    phosphate synthase (EPSPS),
    510
Enzyme-linked immunosorbent assay
    (ELISA), 279-280
EPSPS. *See* Enzyme-5-enolpyruvyl-
    shikimate-3-phosphate
    synthase (EPSPS)
Erbach, D.C., 7
Erenstein, O., 109
Erich, M.S., 197
Erickson, A.E., 207,208
Erosion control, cover crops in, 54-56
Esquivel, I., 649
Euclides, V.P.B., 611
Europe, Eastern, cropping systems in,
    623-647. *See also* Eastern
    Europe, cropping systems in
European Commission's 1980
    Drinking Water Directive,
    337-338
European corn borer, 291
European Economic Community
    (EEC) thresholds, 337-338
Evaluation
    concepts of, 436-437
    context, defined, 438
    defined, 436
    input, defined, 438
    process, defined, 438
    product, defined, 438
Evans, S.D., 93
Exner, D.N., 463
eXtensible Markup Language (XML),
    400
Extension Educators, 388

FACE. *See* Free-air CO_2 enrichment
    (FACE)
Fageria, N.K., 4
Fahey, R., 139
Farm research, in promotion of
    sustainable practices, 310-311

Farm-A-Syst checklist, 318
Farming, mixed, 25
Farming in Nature's Image: An
    Ecological Approach to
    Agriculture, 27
Fausey, N.R., 92
Fava beans, 61-62
FD. *See* Fractal dimension (FD)
Fearnside, P.M., 610
Federal Agriculture Improvement and
    Reform Act of 1996, 435
Feeding value, legumes in
    agroecosystems and, 193-194
Fenster, C.R., 94
Fernandex-Cornejo, J., 275
Fertility
    management of, in weed
        management, 256-257
    nitrogen, in ICM, 473-474
    soil, in Semiarid West Africa,
        technologies for, 569-572,
        570t,571f
Fertilizer(s), 21
    inorganic, alternatives to, 579-580
    N, 66
Fiber(s), detergent
    acid, 478
    neutral, 478
Field corn, in U.S. agriculture,
    504,504t,505t,512t,515
Field cultivator, in tillage, 81
Field rotation system, in Eastern
    Europe, 626-627
Finkel, A.M., 332
Fitter, A.H., 139
Fleischer Manufacturing, 20
Flores, S., 236
Fly(ies), Hessian, 289
Fogel, R., 139
Food, Agriculture, Conservation and
    Trade Act of 1990, 435
Food Quality Protection Act (FQPA),
    277,299
Forage legumes
    in agroecosystems, 189-191,190t

annual and green manures,
199-202,200t
for sustainable cropping systems,
187-216
alfalfa, 189,190t
benefits of, 191-197,192t,193t
bioremediation, 195-196
dinitrogen fixation, 191-193,
192t,193t
energy feedback, 194
feeding value, 193-194
soil and water conservation,
194-195
weed control, 196-197
birdsfoot trefoil, 189,190t
cicer milk vetch, 189,190t
common vetch, 189,190t
commonly used, characteristics
of, 189-191,190t
crown vetch, 189,190t
hairy vetch, 189,190t
historical background of,
197-198
intercropping with corn and
soybean, 202-205,203t,204t
introduction to, 188-189
Kura clover, 189,190t
monocropping, 199-202,200t
perennial living mulch systems,
206-208,207t
red clover, 189,190t
small grain intercrops with,
199-202,200t
sweetclover, 189,190t
traditional rotations in, 198-199
white clover, 189,190t
winter annual cover crops,
205-206
Foster, M.S., 205,251
Foster, W.G., 333
Foulkes, M.J., 146
FQPA. *See* Food Quality Protection
Act (FQPA)
Fractal dimension (FD), 472
Francis, C.A., 6,7,15

Fraser, D.G., 258
Frederick, J.R., 91
Free-air CO_2 enrichment (FACE), 219
Freeman, H.E., 436
Freyenberger, S., 318
Friends of the Land, 605
Frye, W.W., 58,206
Fungus(i), mycorrhizal, 464

Gallandt, E., 53
GAPs. *See* "Good Agricultural
Practices" (GAPs)
Garrity, D.P., 668
Gassen, D.N., 608
Gassen, F.R., 608
GCTA. *See* Georgia Conservation
Tillage Alliance (GCTA)
Gene(s)
bar, 511
"leafy," discovery of, 466
Genetically modified organism (GMO)
cultivars, 20
Genetically modified organisms
(GMOs), 85,298
GeoFocus, 365
Geographic information systems
(GISs), 21,277,362,383,386,
387,402,549
Georgia Conservation Tillage Alliance
(GCTA), 308,312
Gerpacio, R.V., 667
Ghaffarzadeh, M., 201
Gianessi, L., 504,504t,505t
Gianessi, L.P., 501
Gibbs, R.J., 136
GISs. *See* Geographic Information
Systems (GISs)
Gliessman, S.R., 7
Global hunger, soil quality and,
42-43,43t
Global positioning systems (GPSs),
8,18,19,277,361,362,383,
386,387,402
Glufosinate, 510-511

Glyphosate, 510
Glyphosate-resistant lettuce, in U.S. agriculture, 516
GMOs. See Genetically modified organisms (GMOs)
Goins, D.G., 140,144
Goldman, R.H., 675
"Good Agricultural Practices" (GAPs), 273
Gordon, R., 207
Goss, M.J., 127,140
Goudriaan, J., 232
GPSs. See Global positioning systems (GPSs)
Graham, J.H., 143
Grain N accumulation efficiency, 173-175,173t,174t
Grain rye, in North American cropping systems, 55
Grain yield, 468t,469
Grass(es)
  Bahia, 315
  Brachiaria, in Brazil, 609-610
  Brassicas, 58,60
  cool season, 58
  Sorghum-Sudan, 61
  Sudan, 61
Grazing, intensive rotational, 25
Great Plains of U.S., tillage systems in, overview of, 88f,94-97
Greaves, M.P., 258
Green manures, 199-202,200t
Green Revolution, vs. recent developments, impacts of, 664-666,665t
Greenwood, R.M., 144
Greyson, P.R., 207
Grier, C.C., 136
Griffith, D.R., 91,92
Grigg, D.B., 624
Grove, J.H., 58
Growth yield
  in less favored areas of Asia, boosting of, 669-670
  slowing, causes of, 666-668

Guatemala, southwestern. See Southwestern Guatemala
Guertal, E.A., 410,411
Guillespiez, V., 649
Gupta, U.C., 102
Guy, S.O., 475
Guyer, D.E., 370

Habitat management, sustainable cotton system for Coastal Plain, 314-315
Hairy vetch, in agroecosystems, 189,190t
"Half-life," in root dynamics, 131
Halvorson, A.D., 96
Hamel, C.V., 464
Han, S., 364
Hansen, J.W., 412
Hao, X.Y., 106
Hard red spring wheat (HRSW) study, 160-161,164,165t,166,169, 171,177,178
Hardy, W.F., 232
Harker, D.B., 329
Hartman, G.P., 96
Harvest index, 468t,469
HarvestMaster, 365
Hatfield, J.L., 86,93
Hazell, P.B.R., 418,669
Heagle, A.S., 221
Health Canada, 338
Heck, W.W., 221
Heichel, G.H., 192,199
Heisel, T., 370
Hendrick, R.L., 130,136,139,142
Herbicide(s)
  mechanisms of resistance of, 510-511
  modes of action of, 510-511
Herbicide-resistant crops, in U.S. agriculture, 509-516,512t
  canola, 512t,514
  cotton, 511-513,512t
  in development, 516

field corn, 512t,515
with regulatory approval,
511-515,512t
soybean, 512t,513-514
sugar beet, 512t,515
Herdt, R.W., 449
Hessian fly, 289
Hesterman, O.B., 199,205
Higgins, D.R., 161,163
High Plains Climate Center, 393
Higley, L.G., 448
Hill, J.E., 256
Hively, W.D., 202
Hodge, A., 139
Hofman, A.R., 364
Homestead Act of 1862, 16
Hooker, J.E., 130,143
Horton, D., 436
Hossain, M., 667
House, G.J., 252
Howell, T.A., 386
Hoy, M.A., 287
HRSW study, 160-161,164,165t,166,
169,171,177,178
Huggins, D.R., 157,196
Hughes, J.W., 139
Huisman, O.C., 145
Hungaria, M., 484
Hungary, cropping systems in, 632-633
Hunger, global, soil quality and, 42-43,43t
Huntington, T.G., 58
Hurdey, S.E., 332

IAPAR. *See* Agronomic Institut of
Parana (IAPAR)
Ibn-Al-Awan, in soil classification, 35
IBSNAT. *See* International Benchmark
Sites Network for
Agrotechnology Transfer
(IBSNAT)
ICRISAT, 561. *See* International Crops
Research Institute for the
Semi-Arid Tropics
(ICRISAT)

Idaho National Engineering and
Environmental Laboratory,
370
IDW. *See* Inverse Distance Weighing
(IDW)
IFDC (International Fertilizer
Development Center), 571
IFOAMSs. *See* International
Federation of Organic
Agriculture Movements
(IFOAMs)
IFSs. *See* Integrated Farming Systems
(IFSs)
IITA. *See* International Institute of
Tropical Agriculture (IITA)
IKONOS satellite, 386
Imantas Spader, 60
IMPACT. *See* International Model for
Policy Analysis of
Agricultural Commodities
and Trade (IMPACT)
IMPACT model,
675-684,677t,678t-683t
baseline projections for,
676-677,677t,678f
crop yield growth in, alternative
scenarios for,
677-684,679t-683t
described, 675-676
Indicator(s)
defined, 437
in evaluation, 437-438
in sustainable agriculture
evaluation, 442-447,443f
linkage between, 447
Industrial cropping system, in Eastern
Europe, 630-631
Industrial Revolution, 236
Inorganic fertilizers, alternatives to,
579-580
Input evaluation, defined, 438
Insect management, cover crops in,
60-62
Insecticidal crops, in U.S. agriculture,
503-509,504t,505t

Bt crops, with regulatory approval,
    503-507,504t,505t
Institut des Recherches d'Agriculture
    Tropicale/Institut National
    d'Investigation Agricole
    (IRAT/INIA), 579
"Integrated Crop Management" (ICM),
    273
Integrated cropping management,
    668-669
Integrated Farming Systems (IFSs),
    273
Integrated pest management (IPM),
    260
  assessments of, 277
  cropping systems and, 271-305
    biological control in, 287
    challenges facing, 298-300
    crop resistance in, 286-287
    decision-making aids in,
      281-282
    emerging opportunities in,
      298-300
    habitat management in,
      282-286,285t,286f
    host-plant resistance/tolerance
      in, 286-287
    introduction to,
      272-277,275t,276t
    nematode diagnostics, 280
    pesticides in, 287-288
    rotation/cultural practices in,
      282-286,285t,286f
    tools in, 282-288,283f,285t,286f
  in crops of U.S., 288-298
    corn, 291-293
    cotton, 293-294
    potato, 294-296
    soybean, 290-291
    strawberries, 296-298
    tillage in, 289
    wheat, 288-290
  defined, 272
  diagnostics in, 277
  in Eastern Europe, 640,641t

history of, 272
infestations of, 277-279,278f
pathogen detection and diagnoses
    in, 279-280
tools in, 277-282,278f
in yam-based cropping systems in
    Africa, 552
Intensive cereal management, in
    Northeast, 472-476
  cultivar selection in, 473
  nitrogen effect in, 473-474
  as package, 472-473
  pesticides in, 475-476
  plant growth regulators in, 474-475
Intensive mixed intercropping, 24
Intensive rotational grazing, 25
Intercropping
  intensive mixed, 24
  of legumes, with corn and soybean,
    202-205,203t,204t
  in Northeast, 462-465
  in weed management, 250
International Benchmark Sites
    Network for Agrotechnology
    Transfer (IBSNAT), 384-385
International Crops Research Institute
    for the Semi-Arid Tropics
    (ICRISAT), 579
International Federation of Organic
    Agriculture Movements
    (IFOAMs), 640-641
International Fertilizer Development
    Center (IFDC), 571
International Institute of Tropical
    Agriculture (IITA), 533
International Model for Policy
    Analysis of Agricultural
    Commodities and Trade
    (IMPACT), 662-663
International Organization for
    Standardization (ISO), 387
International Soil and Tillage Research
    Organization (ISTRO), 642
Internet, 398

Internet-based technology transfer, in
    on-farm research
    development of sustainable
    cotton production system,
    318-320,319f
Invasive Species Council, 299
Inverse Distance Weighing (IDW),
    372
Iowa State University's Department of
    Agronomy and Statistics, 368
IPM. *See* Integrated pest management
    (IPM)
Iqbal, M., 92
IRAT/INIA. *See* Institut des
    Recherches d'Agriculture
    Tropicale/Institut National
    d'Investigation Agricole
    (IRAT/INIA)
Islas Guitiérrez, J., 109
ISO. *See* International Organization for
    Standardization (ISO)
ISTRO (International Soil and Tillage
    Research Organization), 642

Jackson, W.A., 161
Jacobson, E.T., 476
Janke, R.R., 64,318
Janovicek, K., 100,101,102,103
Jans, S., 275
JanuSys, 389,390,393,394,394t,395,
    397f,399
Janzen, H.H., 105,106,112
Jay, M., 307
Jefferson, T., in forage legume
    utilization, 198
Jenny, H., 35,87
Jeranyama, P., 205
Johnson, G.A., 259
Johnson, G.V., 172
Johnson, M.G., 131
Johnson, W.E., 105
Jones, J.W., 412
Jones, R.J., 25

Journal of Soil and Water
    Conservation, 18
Journal of Sustainable Agriculture, 19

Kachanoski, R.G., 110
Kaiser, H.M., 395
Kamprath, E.J., 161
Karisto, L.A.M., 482
Karlen, D.L., 93
Karunatilake, U., 89
Keeney, D.R., 172
Kelner, D.J., 199
Kemper, W.D., 7
Kennedy, A.C., 258
Kentel, N., 196
Kentucky Agricultural Statistics, 420
Kentucky Cooperative Extension
    Service specialists, 419
Kentucky Farm Business Management
    Association, 420
Ketcheson, J., 100,103
Khosla, R., 361
Kimmons, J.P., 131
Kinsella, J., 27
Kirkwood, V., 445
"Kitab al-Felha," 35
Koch, B., 361
Kohler, K.A., 205,251
Kono, Y., 472
Koo, S.J.R., 7
Kosola, K.R., 143
Kovach, J., 320
Krauss, H.A., 110
Krauss, U., 137
Kremer, R.J., 258
Krewski, D., 332
Kriging, 372
Kühn, 272
Kunelius, H.T., 103
Kura clover, in agroecosystems,
    189,190t
Kurz, W.A., 131
Kurzweil, R., 388,399

LAI. See Leaf area index (LAI)
Lal, R., 33,35,86,91,92,671
Lamb, M.C., 321
Lamborg, M.R., 232
LaMondia, J.A., 61-62
Lana woolypod vetch, 61
Land Equivalent Ratio (LER), 47
Land management practices, in water
        quality control, 345-346
Land use, in sustainable agriculture
        evaluation, 445-446
Land use factor (L), defined, 47
Landscape(s), rural
    functions of, 27
    multifunctional, future recognition
        of, 27-28
Langdale, G.W., 90
Larney, F.J., 105,106
Lascano, C.E., 611
LCB. See Lesser cornstalk borer
        (LCB)
LCOs. See Lipo-chitooligosacchairdes
        (LCOs)
Leaf area index (LAI), 467-469,468t
"Leafy" gene, discovery of, 466
Ledgard, S.F., 193
Légère, A., 104, 105
Legume(s)
    described, 188
    forage, for sustainable cropping
        systems, 187-216. See also
        Forage legumes, for
        sustainable cropping systems
    historical background of, 188
Lehman, M.E., 60
Lehmann, J., 131
Leighty, C.E., 249
Lemainski, J., 602
Lemunyon, J.L., 55
LER. See Land Equivalent Ratio
        (LER)
Lesser cornstalk borer (LCB), 508-509
Lettuce, glyphosate-resistant, in U.S.
        agriculture, 516
Lewis, J., 307

Liang, B.C., 169-170
Liebman, M., 197,249
Lightfoot, C., 412
Lindstrom, K., 482
Lindstrom, M.J., 110
Lipo-chitooligosacchairdes (LCOs),
        for crop production in
        Northeast, 483-485
LISA. See Low-Input Sustainable
        Agriculture (LISA)
Liste, H.J., 634
Living mulches, 65
Lobb, D.A., 110
Logical Framework model, 438-439
Low-Input Sustainable Agriculture
        (LISA), 435
Lynam, J.K., 449
Lynch, V., 146

Ma, B., 455
Ma, L., 386
Macedo, I., 597
MacKenzie, A.F., 169-170
Mackie-Dawson, L.A., 130
Madakadze, I.C., 455,476,477,478,479
Magdoff, F., 86,286
Majdi, H., 140
Malik, V.S., 253
Manering, J.V., 91,92
Manure(s), green, 199-202,200t
Map Unit Interpretation Record,
        393,394
Martin, A.R., 259
Martin, R.C., 207,462,464
Max-Min method, in root production
        and mortality calculations,
        131
MBC contents, 37
McCarl, B.A., 421
McConkey, B., 329
McCool, R.I., 110
McCoy, E.L., 259
McCracken, D.V., 58
McDuffie, H.H., 329

McGill University, 482
McKay, H., 139
McLeod, J.A., 101
Meier, C.E., 136
Meijer, S., 661
Mellish, D.R., 473
Merckx, R., 136
Mexico
    "The Farmer Researcher" program
        in, 108
    tillage systems in, 107-109
Michaels, T.E., 253
Microbial biomass carbon (MBC)
        contents, 37
Mielke, L.N., 91
Miller, M.H., 110
Milojic, B., 629,642
Minimization of total absolute
        deviation (MOTAD), 418
Minnesota Land Stewardship Project,
        318
Miralles, D.J., 146
Mite(s), 62,297
Mixed farming, 25
Mjelde, J.W., 421
Moen, T.N., 395
Mohler, C.L., 448
Moldboard plow, in tillage, 80
    in U.S.(1977-1991), 83,83f
Moldenhauer, W.C., 87,90,91,92,94
Moll, R.H., 161
Mollison, B., 26
Mondardo, A., 604
Monitoring
    concepts of, 436-437
    defined, 436
Morales, F.J., 649
Morford, S., 433
Morrison, J.E., Jr., 86
Mortensen, D.A., 259
Mortensen, L.F., 441
Moseley, G., 194
MOTAD (minimization of total
        absolute deviation), 418
Moth, diamond-back, 509

"Movimento sem Terra" (MST), 613
MST. *See* "Movimento sem Terra"
        (MST)
Mtengeti, E.J., 194
Mulch(es), living, 65
Mulch tillage, defined, 79-80
Muller-Scharer, H., 258
Muntifering, R.B., 410,411
Murgai, R., 664
Murphy, S.D., 252
Mycorrhizal fungi, 464

N fertilizer, 66
N fertilizer replacement values
        (N-FRV), 65
N residues, reduced, in Northeast
        cropping systems, 465
Nambiar, E.K.S., 136
National Academy of Sciences, 333
National Center for Engineering in
        Agriculture, 369
National Coalition on Integrated Pest
        Management, 272
National Root Crops Research Institute
        (NRCRI), 550
Natural ecosystems, diversity among,
        23-24
Natural Systems Agriculture, 1,15,18,26
"Natural Systems Agriculture," 7
N'dayegamiye, A., 104
Neave, P., 445
Nebraska Soil Fertility Network, 395
Nebraska Soybean and Feed Grains
        Profitability Project
        (NSFGPP), 395
Neill, C., 608
Nelson, W.W., 110
Nematode(s), 315
    Columbia Lance, 294
Nematode diagnostics, 280
Neutral detergent fiber, 478
New Roots for Agriculture, 26
N-FRV. *See* N fertilizer replacement
        values (N-FRV)

Nitrate guideline, for water quality, 340

Nitrogen, sources of, evaluation of, 161-175,162f,163t,165t,167t, 168t,173t,174t. *See also* Nitrogen use efficiency (NUE), in cereal-based agroecosystems, assessment of

Nitrogen balance index, 179

Nitrogen fertility, in ICM, 473-474

Nitrogen fertilizer utilization efficiency, 177-178

Nitrogen harvest efficiency, 173-175,173t,174t

Nitrogen harvest index, 171-172

Nitrogen loss index, 179-180

Nitrogen reliance index, 178-179

Nitrogen retention efficiency, 164-168,165t,167t,168t

Nitrogen sinks, evaluation of, 161-175, 162f,163t,165t,167t,168t,173t, 174t. *See also* Nitrogen use efficiency (NUE), in cereal-based agroecosystems, assessment of

Nitrogen uptake efficiency, available, 169-170

Nitrogen use efficiency (NUE)
  in cereal-based agroecosystems
    assessment of, 158-185
      available uptake efficiency in, 169-170
      components in, 163-164, 168-169
      economic and environmental quality factors in, 176-180
      grain N accumulation efficiency in, 173-175,173t,174t
      indicators for, application of, 160-161
      nitrogen balance index in, 179

nitrogen fertilizer utilization efficiency in, 177-178
      nitrogen harvest index, 171-172
      nitrogen loss index in, 179-180
      nitrogen reliance index in, 178-179
      nitrogen retention efficiency in, 164-168,165t, 167t,168t
    introduction to, 158-161,160f
    components of, 163-164,168-169
      derived from plant physiologic processes, 168-169
      to soil processes–related, 163-164
    defined, 176
    described, 170-171,172-173
    DSS in, 159,160f
    terminology related to, 163t

Non-traditional export crops (NTEC), 649-651,656,657,658

Norman, J.M., 389,390

Norman, J.R., 143

Norrie, J., 473

North America, tillage systems in, regional overview of, 86-109. *See also* Tillage systems, in North America

North American cropping systems
  cover crops in, role of, 53-74. *See also* Cover crops, in North American cropping systems
  tillage research in, advances in, 75-125. *See also* Tillage; Tillage research, in North American cropping systems, advances in

Northeast
  crop production in, LCOS for, 483-485
  cropping systems in, 455-499
    C_4 grasses, 476-479,477t

development of, 476-479,
477t
nutritive value of, 476-479,
477t
paper production, 479
persistence in, 476-479,477t
seed germination, 478-479
capture of resources and
resource utilization in,
462-463
chronic injection system
development, 479-481
crop physiology investigations
in, 479-481
effect of suboptimal conditions
on symbiotic nodulation,
481-482
future research in, 485-487
generate and convert resources
in, 463-465
ICM in, 472-476. *See also*
Intensive cereal management
(ICM), in Northeast
improvement of nitrogen
fixation by manipulation
exchange of symbiotic
signals, 482-483
improving legume nodulation
under nodulation inhibitory
conditions, 481-483
intercropping systems, 462-465
introduction to,
456-462,457f,459f
leafy reduced-stature corn,
465-472,467f,468t,470f,471t.
See also Corn, leafy
reduced-stature, in
Northeast
PGPR, 483
PGPR associated with legume
nodulation and nitrogen
fixation, 483
reduced soil N residues with, 465
rhizobia-to-legume signals, 483-485
weed control in, 465

Northeastern U.S., tillage systems in,
overview of, 87-90,88f
No-till Buffalo planters and
cultivators, 20
No-tillage (NT)/zero-tillage, defined,
80
NRCRI. *See* National Root Crops
Research Institute (NRCRI)
NSFGPP. *See* Nebraska Soybean and
Feed Grains Profitability
Project (NSFGPP)
NTEC. *See* Non-traditional export
crops (NTEC)
NUE. *See* Nitrogen use efficiency
(NUE)
Null Hypothesis, 334,335
Nutrient(s)
cycling of, root "turnover" effects
on, 136
in water quality control, 348-349
Nutrient management, cover crops in,
57-58
Nyakatawa, E.Z., 55
Nylund, J.-E., 140

Oat(s), "Saia," 62
Oberle, S.L., 172
Occupational Health and Safety
Administration (OHSA), 332
OECD. *See* Organization for
Economic Cooperation and
Development (OECD)
Ohio State University, 372
Ohno, T., 249
Okigbo, B.N., 3
Oliveira, V.F., 207-208
Olson, D., 307
Ominski, P.E., 196
On-farm research, in development of
sustainable cotton production
system, 317-322. *See also*
Cotton system, sustainable,
on-farm research development
of

Ontario Environment Farm Plan, 318
Open top chambers (OTC), 219
Oplinger, E.S., 475
Opoku, G., 104
OrbView-3 satellite, 386
Organic Gardening, 19
Organic matter amendments, in weed
    management, 257-258
Organization for Economic
    Cooperation and
    Development (OECD), 441
Orkwor, G.C., 533
OSHA. *See* Occupational Health and
    Safety Administration (OHSA)
OTC. *See* Open top chambers (OTC)

Paau, A.S., 482
Pacific Northwest U.S., tillage systems
    in, overview of, 88f,97-99
Paisner, M.S., 661
Paiz, C., 649
Palma, E., 649
PAMS. *See* Prevention, avoidance,
    monitoring, and suppression
    (PAMS)
Pan, B., 482
Pan, W.L., 157,161,163
Papaya, in U.S. agriculture, 517-518
Papaya ringspot virus (PRSV), in U.S.
    agriculture, 517-518
Papendick, D.K., 110
Parajulee, M.N., 62
Parallel swath navigation, 371-372
Parr, J.F., 446
Pasture(s)
    Brachiaria, in Brazil, 610-611,612
    planted, in Brazil, 608-612
PAT. *See* Phosphinothricin acetyl
    transferase (PAT)
Pathogen(s), elimination of, 280-281
Pathogen management, cover crops in,
    60-62
Pathogen-resistant crops, in U.S.
    agriculture, 516-521

in development, 519-521
papaya, 517-518
with regulatory approval, 517-518
summer squash, 518
Paul, E.A., 86,232
Peanut, in U.S. agriculture, 508-509
Pearman, G.I., 232
Pelletier, G., 364
Perennial living mulch systems,
    206-208,207t
Permaculture, 26
Pesic-van Esbroeck, 281
Pest management, in sustainable
    agriculture evaluation, 445
Pesticide(s)
    in cropping systems, 287-288
    in ICM, 475-476
    in water quality control, 349-350
Peterson, G.A., 96
Peterson, T.A., 191
Peterson, W., 436
PFRA (Prairie Farm Rehabilitation
    Administration), 341-342
PGPR. *See* Plant growth promoting
    rhizobacteria (PGPR)
PGRs. *See* Plant growth regulators
    (PGRs)
Pharak, S., 307
Phillips, D.L., 131
Phosphinothricin acetyl transferase
    (PAT), 511
Phosphorus levels, for water quality,
    340-341
Pingali, P.L., 667
PLANETOR, 318
Plant growth promoting rhizobacteria
    (PGPR), 483
Plant growth regulators (PGRs), in
    ICM. 474-475
Pliny, in forage legume utilization, 198
Plow(s)
    blade, in tillage, 81
    chisel, in tillage, 81
    moldboard, in tillage, 80
        in U.S.(1977-1991), 83,83f

Rome-Pegasus, 99
sweep, in tillage, 81
Poland, cropping systems in, 631-632
Posner, J.L., 201
Potato(es)
pest management in, 294-296
in U.S. agriculture, 506-507
Potato beetle, Colorado,
162,295,506,508
Potter, K.N., 96
Power, A.G., 448
Prairie Farm Rehabilitation
Administration (PFRA),
341-342
Prayitno, J., 484
Precautionary Principle, in water
quality effects, 335
Precision agriculture, in cropping
systems, 361-381
grid sampling and management
zones in, 372-376,373f-375f
introduction to, 362-363
variable rate technology in,
367-372. *See also* Variable
rate technology (VRT), in
precision agriculture
yield monitoring in, 363-367,366t
Pregitzer, K.S., 139,142
Prevention, avoidance, monitoring,
and suppression (PAMS),
275,277
Primary tillage, 79
Primavesi, A.C.P., 610
Primavesi, O., 610
Prior, S.A., 217
Prithiviraj, B., 455
Proactive agriculture, in water quality
control, 344-345
"ProAlcool" program, 597,598
Process evaluation, defined, 438
Product evaluation, defined, 438
Progra Evaluation model, 434
Program Evaluation model,
439,440-441,440f
Promeristem, 128

PRSV. *See* Papaya ringspot virus
(PRSV)
Publicover, D.A., 131
Purple vetch, 61
Putnam, A.R., 60

Quality of life, sustainable agriculture
effects on, 447
Quebec, southwestern, crop production
research in, 456-462,457f,
459f
Quick Bird 1 satellite, 386

Ragsdale, N.N., 299
Raimbault, B.A., 102
Rambo, A.T., 4
Randall, G.W., 196
Raun, W.R., 95,172,387,389
Rawlins, S.L., 364
Real-time-kinematic (RTK) DGPS,
362-363
Reardon, T., 564-565
Red clover, in agroecosystems,
189,190t
Reddy, K.C., 55
Reduced tillage/minimum tillage,
defined, 79
Reeder, J.D., 86,87
Reeder, R., 91,96
Reicosky, D.C., 75,90,112
Reid, J.B., 136,140,141f
Relay cropping, 24
"Report and Recommendations on
Organic Farming," 434
Resck, D.V.S., 602
Research, cropping systems-related
emerging trends in, 1-13
introduction to, 2-3
holistic model, 5-7,6f
interdisciplinary approach to,
development of, 4-7,6f
learning cycle in, 8-10,9f
systems approach to, 4-5

trends in, 7-8
Research Project Management model,
     439
Resource-efficient cropping systems
     design of
          advances in, 15-32
          introduction to, 16-18,17t
          using agroecological principles,
               23-27
          improving of, 18-21
          innovations in, 21-23
Reverse yield, stagnation in,
          approaches to, in sub-Saharan
               Africa, 670-675,672f
Revised Universal Soil Loss Equation
     (RUSLE), 55
Rhizobacteria, plant growth
     promoting, 483
Rhizobia, 191
Rhizobia-to-legume signals, in
          Northeast, 483-485
Rice, in U.S. agriculture, 516
Richardson, R.H., 448
Richter, J., 35
Ridge tillage, defined, 79
Riha, S.J., 395
Risk assessment, defined, 332
Roberts, W.S., 411,415-416
Robinson, A.D., 139
Rochette, P., 112
Rockwell, S.K., 440
Rodale Institute, 56
Rodale Institute Farming System Trial,
     56-57
Rogers, H.H., 217,221
Romania, cropping systems in, 632-633
Rome-Pegasus plow, 99
Root(s)
     architecture of, root dynamics and,
          133-134,133f
     carbon in, 129
     cells of, 128-129
     described, 128
     dynamics of
          assessment of, 130-131,132t

in cropping systems research,
     127-155
factors affecting,
     137-145,137f,141f
     aboveground, 137f,143-145
     inherent, 137-138,137f
     soil, 137f,138-143,141f
"half-life" in, 131
introduction to, 128-129
root architecture and, 133-134,
     133f
sustainable cropping systems
     and, 145-146
growth of, stage of development in,
     133,133f
longevity of, in different crop
     species, 134,135t
tomato
     density of, 141f
     longevity of, 141f
Root "turnover"
     definitions of, 132t
     in different plant species,
          significance of, 134-136,135t
     effect on nutrient cycling, 136
Rosegrant, M.W., 661,669
Rossi, P.H., 436
Rotation, crop, in weed management,
     248-249
Rottmeirer, J., 364
Ruess, R.W., 136
Runion, G.B., 217
Rural communities, sustainability in,
     321-322
Rural landscapes
     functions of, 27
     multifunctional, future recognition
          of, 27-28
RUSLE. *See* Revised Universal Soil
     Loss Equation (RUSLE)
Russelle, M.P., 140,144,191,196
Russian wheat aphid, 289
Rye
     cereal, in North American cropping
          systems, 55

grain, in North American cropping
systems, 55
winter, in North American cropping
systems, 55
Rytter, L., 130,143
Rytter, R.-M., 130,143

Sahelo-Sudanian zone, new technology
for, 576-579,576t
"Saia" oats, 62
Sajise, P.E., 4
Salinas-Garcia, J.R., 95,108,109
Samson, N., 105
Sanders, B.J., 98
Sanders, J.H., 559
Sanderson, J.B., 101
Sandman, P.M., 332
Sankula, S., 501
SARE. *See* Sustainable Agriculture
Research and Education
Program (SARE)
Sarrantonia, M., 53
Saxena, D., 236
Scheepens, P.C., 258
Scherer, C.W., 332
Scherr, S., 670-671
Schindlebeck, R.R., 89
Schlesinger, W.H., 232
Schmitt, M.A., 201
Schomberg, H.H., 307
Schreiber, G.A., 671
Schreiber, M.M., 249
Schueller, J.K., 365
Schultz, M.A., 207,208
Schumacher, T.E., 110
Schuman, G.E., 96
Scott, R.K., 146
Scott, T.W., 202,206
Secondary tillage, 79
Seeding equipment, in tillage, 583
Semiarid West Africa
crop technology introduction in,
559-592
agricultural technology, 584-585

described, 561-564,562f,564f
fertilizer policy, 581
increasing demand for
traditional cereals, 583-584
input markets, 584-585
introduction to, 560-561
new technologies, 585-587
organic fertilizer, after structural
adjustment, profitability of,
581-583
strategy in, preliminary
evaluation of, 566-569,568t
welfare of women and, 585-587
crops in, 561
described, 561,562f
intensive agricultural technology
development for, strategy for,
564-566
water and soil fertility technologies
for, 569-572,570t,571f
Sen, D.N., 4
Sequin, P., 187
Shapiro, B.I., 559
Sharpley, A.N., 344-345
Sheaffer, C.C., 187,201,205,207-208
Shelterbelts, in water quality control,
347-348
Shirley, C., 286
Shrestha, A., 201
Siddique, K.H.M., 146
Sijtsma, C.H., 103
Silent Spring, 272
Silesia, cropping systems in, 631,632t
Silvers, C.S., 501
Simmons, S.R., 201
Simon, H., 384,401
Singer, J.W., 199
Sipos, A., 625
Skidmore, E.L., 77,86,94
Slafer, G.A., 146
Slosser, J.E., 162
Slovakia, cropping systems in, 631-632
Small grain intercrops, with forage
legumes, 199-202,200t

Smallholder agriculture, in Brazil,
    612-615,614f
Smith, D.L., 455,464,472,473,479,481,
    482
Smith, J.A., 96
Smith, R.F., 272
Smother crops, in weed management,
    250-252
SOC. *See* Soil organic carbon (SOC)
Soil, in root dynamics, 137f,138-143,
    141f
Soil and water conservation, legumes
    in agroecosystems and,
    194-195
Soil degradation, 41-42,41f,42t
    defined, 41
Soil erosion, tillage and, 109-111
Soil fauna and flora, species diversity
    of, 37
Soil fertility, in Semiarid West Africa,
    technologies for,
    569-572,570t,571f
Soil fertility enhancement,
    technological options for,
    44,44f
Soil fertility management, 45-47,46t
Soil N residues, reduced, in Northeast
    cropping systems, 465
Soil organic carbon (SOC), 96
Soil organic carbon (SOC) contents,
    37-38
Soil organic matter (SOM), in cover
    crops, 55
Soil quality, 33-52
    attributes of, 37-38,38f
    attributes related to specific soil
        functions, 36t
    basic concepts of, 34-38,36t,37f,38f
    classification of, 34-35
    components of, 38f
    cropping systems and,
        44-48,44f,45f,46t
    conservation tillage, 45,46t
    cover crops, 47-48

soil fertility management,
    45-47,46t
decline in, effects of, 33
definitions of, 33,35,36t
described, 56
determinants of, 33,38-40,40t
global hunger and, 42-43,43t
impacts on changes in crop yield,
    40,40t
inter-dependence between water
    quality and air quality, 37f
introduction to, 34
in sustainable agriculture
    evaluation, 446
Soil Survey Geographic (SSURGO)
    database map, 393
Soils and Crops Research Centre, 368
Soil-water interactions, water quality
    effects of, 343-344
SOM. *See* Soil organic matter (SOM)
Soper, K., 144
Sorenson, C., 271
Sorghum-Sudan grass, 61
Southeastern U.S.
    sustainable cropping systems in,
        conceptual model for,
        307-327
    tillage systems in, overview of,
        88f,90-91
Southwestern Guatemala, small farms
    of, mixed cropping systems
    in, socioeconomic and
    agricultural factors associated
    with, 649-659
    introduction to, 650-651
    materials and methods in study of,
        651-653,652t
    results and discussion of,
        653-657,654t,656t
Soybean
    in Brazil, 602-608,603f,604t,605t,
        606f,607f
    corn and, intercropping of legumes
        with, 202-205,203t,204t
    pest management in, 290-291

in U.S. agriculture, 508,512t, 513-514

SOYGRO, 420

Sparrow, D.H., 259

Spedding, C.R.W., 5

Squash, summer, in U.S. agriculture, 518

SSA. *See* Sub-Saharan Africa (SSA)

SSURGO database map, 393

Stacey, G., 484

Starcevic, Lj., 637

Starratt, C.E., 473

Statistics Canada, 350

Stern, V.M., 272

Stewart, B.A., 86,94

Stewart, K., 479

Stinner, B.R., 29

Stojkovic, L., 626

Stoller, E.W., 253

Stotzky, G., 236

Strawberries
  pest management in, 296-298
  in U.S. agriculture, 516

Strip cropping, 24

Strip tillage/zone tillage, defined, 80

Stute, J.K., 201

Sub-Saharan Africa (SSA), 559-592.
      See also Semiarid West
      Africa
  agricultural research and policies
      for growth in, 673-675
  new approaches to reverse yield
      stagnation in, 670-675,672f
  yam-based cropping systems in,
      531-558. *See also* Yam-based
      cropping systems, in Africa

Successful Farming, 19

Sudan grass, 61

Sudanian zone, new technology for,
      572-576,572t,574f

Sugar beet, in U.S. agriculture,
      512t,515

Sugarcane, in Brazil, 596-600,599f

Summer squash, in U.S. agriculture,
      518

Sunshine Farm, at The Land Institute,
      26

Sustainability, in rural communities,
      321-322

Sustainable agriculture, 273
  defined, 435
  evaluation of
    biodiversity levels in, 446
    CIPP Evaluation model in, 438
    concepts of, 436-437
    conceptual framework for,
        433-454
      introduction to, 434-436
    Development Agency
        Monitoring and Evaluation
        model in, 438
    The Driving Force-State-
        Response model in, 439
    Economic Evaluation model in,
        438
    evaluation framework for,
        439-442,440f,442f
    farm financial resources and
        economic performance in,
        446-447
    frameworks for, 438-439
    indicators in, 437-438,442-447,
        443f
    indicators of, linkage between,
        447
    land use and conservation in,
        445-446
    limitations of, 447-450
    Logical Framework model in,
        438-439
    measurements in, 442-447,443f
    nutrient use and management in,
        445
    pest management in, 445
    Program Evaluation model in,
        439
    quality of life, 447
    Research Project Management
        model in, 439

soil quality and management in, 446
Sustainable Agriculture Research and Education Program (SARE), 435
Suvedi, M., 433
Swanton, C.J., 104,252,253
Swath bars, parallel, 371-372
Sweep plow, in tillage, 81
Sweet corn, in U.S. agriculture, 507
Sweetclover, in agroecosystems, 189,190t
Swinnen, J., 136
Swinton, S., 411,415-416
Switchgrasses, 476
Sylvester-Bradley, R., 146

Tandem disk harrow, in tillage, 81-82
Tapia-Vargas, M., 86
Tatsumi, J., 472
Teasdale, J.R., 59,235,251,253
Tennant, D., 146
TFP growth. *See* Total factor productivity (TFP) growth
Thaer, A.D., 628
The Driving Force-State-Response model, 439
"The Farmer Researcher" program, in Mexico, 108
The Land Institute, 26
   Sunshine Farm at, 26
Theophrastus, in soil classification, 34-35
Thomas, S.M., 144
Thompson, I.P., 235
Tillage, 75-125
   blade plow in, 81
   chisel plow in, 81
   conservation, 45,46t
      demerits of, 77-78
      described, 77
      merits of, 77-78
   conventional, 604
      defined, 79

described, 76-77
coulter carts in, 82,84
defined, 76
equipment in, 80-82
   need for larger, 84
   seeding, 583
   trends in, 82-86,83f
field cultivator in, 81
intensive, effects of, 76
management of, 111-113
moldboard plow in, 80
   in U.S. (1977-1991), 83,83f
mulch, defined, 79-80
in North American cropping systems, 76-86,83f
no-tillage (NT)/zero-tillage, defined, 80
in pest management, 289
primary, 79
reduced
   sustainable cotton system for Coastal Plain, 312-314
   in water quality control, 346-347
reduced tillage/minimum tillage, 79
ridge, defined, 79
secondary, 79
soil erosion and, 109-111
strip tillage/zone tillage, defined, 80
sweep plow in, 81
tandem disk harrow in, 81-82
terminology related to, conventional vs. conservation, 76-77
types of, 79-80
Tillage research, 75-125
   environmental concerns related to, 111-113
   in North American cropping systems, advances in, 75-125. *See also* Tillage
   introduction to, 76-86,83f
Tillage rotation, 85
Tillage systems
   defined, 78-80
   in Mexico, 107-109

in North America, regional
  overview of, 86-109,88f
  arid West, 88f,99-100
  central Canadian provinces,
    88f,103-105
  central U.S., 88f,91-93
  described, 86-87,88f
  Eastern Canadian provinces,
    88f,100-103
  Great Plains of U.S., 88f,94-97
  northeastern U.S., 87-90,88f
  Pacific Northwest U.S.,
    88f,97-99
  southeastern U.S., 88f,90-91
  western Canadian provinces,
    88f,105-107
in weed management, 252-253
Tillman, G., 307
Timper, P., 307
Tingey, D.T., 131
Tiscareno-López, M., 107,108
Tollenaar, M., 102
Tomato(es)
  root density in, 141f
  root longevity in, 141f
Torbert, H.A., 217,233
Total factor productivity (TFP)
    growth, 664
Transgenic crops, in U.S. agriculture,
    501-530
  broccoli, 509
  canola, 512t,514
  cotton, 505,511-513,512t
  eggplant, 508
  field corn, 504,504t,505t,512t,515
  herbicide-resistant crops,
    509-516,512t. *See also*
    Herbicide-resistant crops, in
    U.S. agriculture
  insecticidal crops, 503-509,504t,
    505t. *See also* Insecticidal
    crops, in U.S. agriculture
  introduction to, 502-503
  pathogen-resistant crops, 516-521

peanut, 508-509
potato, 506-507
soybean, 508,512t,513-514
sugar beet, 512t,515
sweet corn, 507
Trenbath, B.R., 668
"Trend is not destiny!", 30
Tress, B., 27
Tress, G., 27
Trimble, S.W., 311
Triplett, G.B., 202
Tull, J., in cover crops, 54
"Turnover," root. *See* Root "turnover"
Tyler, D.D., 90

Unger, P.W., 77,86,94
United States Environmental
    Protection Agency (USEPA),
    332
Universal Soil Loss Equation, C factor
    of, 195
University of Georgia, 508
University of Idaho, 370
University of Minnesota Southwest
    Research and Outreach
    Center, 161
University of Nebraska, 20
  Lincoln East Campus, 390
University of Southern Queensland,
    366
Upadhyaya, S.K., 364
Urquiaga, S., 593
U.S. Coast Guard, 362
U.S. Congress, 435
U.S. Corn Belt, 22
U.S. Department of Agriculture
    (USDA), 434,505,506,507
USDA. *See* U.S. Department of
    Agriculture (USDA)
USDA Extension Service, 440
USDA-ARS, 375
USDA-ARS National Soil Dynamics
    Laboratory, 221
USDA-CSREES, 435

USEPA. *See* United States
    Environmental Protection
    Agency (USEPA)
USSR, former, countries of, cropping
    systems in, 634

Valdivia, R., 108
Valdrighi, M.M., 257-258
Value(s), feeding, legumes in
    agroecosystems and, 193-194
van den Bosch, R., 272
Van der Vlugt, J.L.F., 138
Van Es, H.M., 86,89-90,286
Van Noordwijk, M., 142
Van Veen, J.A., 136
Variable rate technology (VRT), in
    precision agriculture,
    367-372
    parallel swath navigation in,
    371-372
    site-specific nutrient management
    in, 367-369
    site-specific planting in, 370-371
    site-specific weed management in,
    369-370
Veseth, R.J., 289
Vessey, J.K., 199
Vetch
    cicer milk, in agroecosystems,
    189,190t
    common, in agroecosystems,
    189,190t
    crown, in agroecosystems, 189,190t
    hairy, in agroecosystems, 189,190t
    Lana wollywood, 61
    purple, 61
Viana, A., 649
Villarreal, E., 108
Vilyams, V.R., 628,629,643
Virgil
    in cover crops, 54
    in forage legume utilization, 198
    in soil classification, 35
Virginia Polytechnic Institute, 56

Vogt, K.A., 131,136
Voldeng, H.D., 464
VRT. *See* Variable rate technology
    (VRT)
Vyn, T.J., 100,101,102,103,104,200

Walter, A.M., 370
Wander, M.M., 93
Wang, X., 469,472
Warm-season grasses, 476-477,477t
Washington State University, 365
Water, in Semiarid West Africa,
    technologies for,
    569-572,570t,571f
Water and soil conservation, legumes
    in agroecosystems and,
    194-195
Water quality
    acceptable risk in, 331-332
    agricultural contributions to,
    quantification of, 342-344
    *Canadian Drinking Water
    Guidelines*, 337
    *Canadian Water Quality
    Guidelines*, 337
    concerns related to
    clarification of, 342-343
    compounding effects of,
    343-344
    cropping strategies and, 344-352
    bacteria, 348-349
    beneficial management
    principles, 345
    buffer zones, 351-352
    cover crops, 347-348
    crop rotations, 346
    land management practices,
    345-346
    managing inputs, 348
    nutrients, 348-349
    pesticides, 349-350
    proactive agriculture, 344-345
    reduced tillage, 346-347
    shelterbelts, 347-348

cropping systems and, 329-359
   cropping practices, 335-337
      human health, 335-337
      of farmers, 344
      protocols for, 343-344
      soil-water interactions, 343-344
   effects of, weighing of, 334-337
      classical science approach to, 334
      precautional principle in, 335
   guidelines and standards for
      application of, 339-342
      aquatic guidelines, 339
      impact of different standards on
         interpretation, 341-342
      nitrate guideline, 340
      phosphorus levels, 340-341
   indicators of, 337-339
      less stringent, intermediate
         approach, 338
   interdependence between soil
      quality and air quality, 37f
   known hazards vs. unknown in,
      333-334
   risk assessment related to, 332-333
   state of, 330-331
   uncertainty and, 330-334
Watson, C.A., 127,142
Wauchope, D., 307
Weaver, J.E., 137
Weed(s), 279
   cropping practices and, 248-259
   within cropping systems, 245-270
      behavior of, 247
      introduction to, 246-248
      management of
         approaches to, introduction
            to, 246-248
         cover crops, 58-60,250-252
         crop competitiveness in,
            253-255
         crop rotation in, 248-249
         cultivation in, 252-253
         edaphic factors in, 255-256
         fertility management in,
            256-257

         intercropping in, 250
         organic matter amendments
            in, 257-258
         site-specific, 258-259
         smother crops, 250-252
         systems for, development of,
            259-261
         tillage systems in, 252-253
         weed suppressive soils in,
            257-258
      problems associated with,
         introduction to, 246-248
Weed, control of
   legumes in agroecosystems and,
      196-197
   in Northeast cropping systems, 465
Weed suppressive soils, 257-258
Welles, J.M., 389,390
Wells, C.E., 131,134
West Africa
   farming in, indigenous knowledge
      of, 534-536
   semiarid, crop technology introduction
      in, 559-592. *See also* Semiarid
      West Africa, crop technology
      introduction in
Westerman, A.L., 172
Western Canadian provinces, tillage
      systems in, overview of,
      88f,105-107
Wheat, pest management in, 288-290
White clover, in agroecosystems,
      189,190t
White, D.C., 448
White, R.H., 197
Wiese, A.F., 94
Wild, K., 364
Wilkins, D.W., 86
Wilks, D.S., 89-90
Wilman, D., 194
Winrock, 25
Winter annual cover crops, 205-206
Winter rye, in North American
      cropping systems, 55

Wireworm(s), 162
Within-field diversity, 24
World Bank, 561
World War II
    cropping systems after, in Eastern
        Europe, 630-636,632t,635t.
        See also Eastern Europe,
        cropping systems in, after
        World War II
    cropping systems before, in Eastern
        Europe, 625-630
Worsham, A.D., 197

Xavier, D.F., 593
XML. *See* eXtensible Markup
        Language (XML)

Yadav, S., 670-671
Yam(s), agroecological distribution of,
        537-538
Yamauchi, A., 472
Yam based cropping systems, in
        Africa, 531-558
    agronomic perspectives on,
        551-552
    challenges to, 538-549,542f-548f
        harvest- and storage-related,
            546-548,547f-548f
        labor-related, 539
        land preparation-related,
            539-540
        mulching-related, 543-544
        pest- and disease-related,
            548-549
        planting material-related,
            538-539

soil amendments-related,
        545-546
    staking-related, 541-543,
        542f-546f
    weeding-related, 540-541
consumption of, 538
institutional issues in, 549
integrated pest management in, 552
introduction to, 532-533
physiological considerations in,
        550-551
plant breeding in, 552-553
policy constraints in, 549
socioeconomic issues in, 549
technological options for, 549-553
Yield monitoring
    errors in, 366-367
    in precision agriculture,
        363-367,366t
Young, A., 627
Yugoslavia, cropping systems in,
        633-634

Zadoks, J.C., 474
Zahran, H.H., 482
Zarkadas, C.G., 473
Zech, W., 131
        Zero tillage (ZT) agriculture,
        602
Zhang, F., 455,481,482,483
Zhou, X.M., 455,463,480
Zhu, Y., 201
Ziska, L.H., 235
ZMV. *See* Zucchini mosaic virus
        (ZMV)
Zucchini mosaic virus (ZMV), in U.S.
        agriculture, 518

Printed and bound by CPI Group (UK) Ltd, Croydon, CR0 4YY

23/10/2024

01777672-0002